Macrophages and Dendritic Cells

Macrophages and Dendritic Cells

Edited by David Johnson

STATES
ACADEMIC PRESS
www.statesacademicpress.com

States Academic Press,
109 South 5th Street,
Brooklyn, NY 11249, USA

Visit us on the World Wide Web at:
www.statesacademicpress.com

This book contains information obtained from authentic and highly regarded sources. Copyright for all individual chapters remain with the respective authors as indicated. All chapters are published with permission under the Creative Commons Attribution License or equivalent. A wide variety of references are listed. Permission and sources are indicated; for detailed attributions, please refer to the permissions page and list of contributors. Reasonable efforts have been made to publish reliable data and information, but the authors, editors and publisher cannot assume any responsibility for the validity of all materials or the consequences of their use.

ISBN: 978-1-63989-337-9

Trademark Notice: Registered trademark of products or corporate names are used only for explanation and identification without intent to infringe.

Cataloging-in-Publication Data

Macrophages and dendritic cells / edited by David Johnson.
p. cm.
Includes bibliographical references and index.
ISBN 978-1-63989-337-9
1. Macrophages. 2. Dendritic cells. 3. Antigen presenting cells. 4. Cytology. I. Johnson, David.
QR185.8.M3 M33 2022
616.079--dc23

Table of Contents

Preface

This book aims to highlight the current researches and provides a platform to further the scope of innovations in this area. This book is a product of the combined efforts of many researchers and scientists, after going through thorough studies and analysis from different parts of the world. The objective of this book is to provide the readers with the latest information of the field.

The white blood cells which are involved in the detection, phagocytosis and destruction of harmful organisms such as bacteria are termed as macrophages. Anything within the body which does not have the proteins on its surface which are found in healthy human cells are digested by these cells. Some of the harmful organisms and cells digested by macrophages are cancer cells, microbes and foreign substances. The cells which are responsible for processing antigen material and presenting it to the T cells of the immune system on the surface of the cell are termed as dendritic cells. Macrophages, dendritic cells and monocytes together make up the mononuclear phagocyte system. This book contains some path-breaking studies on macrophages and dendritic cells. It consists of contributions made by international experts. This book will serve as a valuable source of reference for graduate and post graduate students.

I would like to express my sincere thanks to the authors for their dedicated efforts in the completion of this book. I acknowledge the efforts of the publisher for providing constant support. Lastly, I would like to thank my family for their support in all academic endeavors.

Editor

Dendritic cells: The double agent in the war against HIV-1

Alba Martín-Moreno [1,2] and Mª Angeles Muñoz-Fernández [1,2,3,4†]*

[1] *Sección de Inmunología, Laboratorio InmunoBiología Molecular, Hospital General Universitario Gregorio Marañón (HGUGM), Madrid, Spain,* [2] *Instituto Investigación Sanitaria Gregorio Marañón (IiSGM), Madrid, Spain,* [3] *Spanish HIV-HGM BioBank, Madrid, Spain,* [4] *Networking Research Center on Bioengineering, Biomaterials and Nanomedicine (CIBER BBN), Madrid, Spain*

Edited by:
Constantinos Petrovas,
Vaccine Research Center (NIAID),
United States

Reviewed by:
Tracy Ruckwardt,
National Institutes of Health (NIH),
United States
Catarina E. Hioe,
Icahn School of Medicine at Mount
Sinai, United States

*Correspondence:
Mª Angeles Muñoz-Fernández
mmunoz.hgugm@gmail.com

†ORCID:
Mª Angeles Muñoz-Fernández
orcid.org/0000-0002-0813-4500

Human Immunodeficiency Virus (HIV) infects cells from the immune system and has thus developed tools to circumvent the host immunity and use it in its advance. Dendritic cells (DCs) are the first immune cells to encounter the HIV, and being the main antigen (Ag) presenting cells, they link the innate and the adaptive immune responses. While DCs work to promote an efficient immune response and halt the infection, HIV-1 has ways to take advantage of their role and uses DCs to gain faster and more efficient access to CD4$^+$ T cells. Due to their ability to activate a specific immune response, DCs are promising candidates to achieve the functional cure of HIV-1 infection, but knowing the molecular partakers that determine the relationship between virus and cell is the key for the rational and successful design of a DC-based therapy. In this review, we summarize the current state of knowledge on how both DC subsets (myeloid and plasmacytoid DCs) act in presence of HIV-1, and focus on different pathways that the virus can take after binding to DC. First, we explore the consequences of HIV-1 recognition by each receptor on DCs, including CD4 and DC-SIGN. Second, we look at cellular mechanisms that prevent productive infection and weapons that turn cellular defense into a Trojan horse that hides the virus all the way to T cell. Finally, we discuss the possible outcomes of DC-T cell contact.

Keywords: dendritic cells, HIV-1, endocytosis, trans-infection, cis-infection, immune response

INTRODUCTION

Human Immunodeficiency Virus (HIV) infects cells from the immune system, its main target being CD4$^+$ T cells. For this purpose, the virus has developed tools to circumvent the host immunity and even use it in its advance. The disease progression can differ dramatically depending on the first interaction between the virus and the immune cells, and the early host response, which can produce neutralizing antibodies and cytotoxic responses. HIV-1 hijacks the host's immune mechanisms and uses them to spread faster, resulting in a two-edged fight by the immune system, stuck between not responding or responding and facilitating the viral spread.

Nowadays, and only for the lucky patients who have access to treatment, HIV-1 infection is a chronic disease instead of a death sentence. However, less than 60% of the 37 million people infected with HIV-1 are on antiretroviral therapy according to UNAIDS data. Adherence to uninterrupted treatment is vital, as viral load and lymphocyte death rapidly increase upon treatment interruption. Economic issues and a wide range of social and psychological side effects result in scarcity of treatment available or lack of adherence to treatment. For these reasons, finding a more accessible and long-lasting solution against HIV-1 is one of the biggest current challenges of the biomedical community.

Viral latency, high mutability and variability highly hinder the finding of a cure or an efficient prophylactic vaccine (1–3), making the possibility of either option controversial among researchers (4, 5). Therefore, finding a functional cure appears as a very appealing option. Functional cure means that HIV-1 is still present in the host organism and remains latent in the genome of many cells, but the host immune system keeps the infection under control in the absence of treatment, which requires a specific and efficient immune response.

The idea of modulating the immune system as a way to fight disease was first formulated in the late 19th century by William B. Coley, and it is now known as immunotherapy or biological therapy. It is often successfully used in cancer therapies, which use immune modulators to halt the growth of cancer cells, and to enhance the cytotoxic immunity and cancer recognition to fight the tumor. The protocols for *ex-vivo* (in a laboratory) or *in-vivo* modulation of the patient's immune cells are rapidly increasing in the era of personalized medicine.

Due to their role as antigen presenting cells (APCs), dendritic cells (DCs) are promising candidates to achieve the functional cure of HIV-1 infection. DCs are innate immune cells that patrol tissues, recognize Ag, participate in early immune response, and, upon Ag uptake and processing, present Ag and activate T cells, serving as a link between general innate immunity and specific adaptive immune cells. DCs are localized in all tissues in the body, and undergo maturation and migrate to the lymph nodes upon encountering an Ag (6, 7). Once in the lymph nodes, they connect with naïve T cells through what is known as immune synapse, which serves to both present Ag and activate the lymphocyte. If this process is successful, it triggers a specific immune response (8). However, HIV-1 also exploits DCs as a means of transportation from the site of infection to the lymph nodes, where the high density of CD4$^+$ T cells and direct cell-to-cell contact through immune synapses ease the spread of the virus and fast infection of a high number of cells.

In order to successfully design a DC-based immunotherapy, it is essential to understand all the diverse interactions between DCs and HIV-1, and the factors that determine the outcomes of those interactions. In this review, we summarize the current state of knowledge on DCs and their role and behavior during HIV-1 infection.

DENDRITIC CELLS

Dendritic cells represent 0.5–2% of peripheral blood mononuclear cells (PBMCs) (9). DCs are less susceptible to HIV-1 infection than CD4$^+$ T cells, as only around 1% of DCs are infected (10), and the HIV-1 infection is less productive than in CD4$^+$ T cells. Nonetheless, DCs are of utmost importance for the immune response to HIV-1 as they are among the first cells to encounter the virus after the infection through the mucosa and play a pivotal role in the establishment of HIV-1 infection, and progression of the disease (11).

Immature DCs (iDCs) are located in the mucosa and peripheral tissues, where they capture and process antigens. The encounter of an iDC with the stimulus of an Ag causes the maturation and the subsequent migration of the now mature DCs (mDCs) to the secondary lymphoid tissues, where they present the Ag to lymphocytes and prime naïve T cells (12, 13).

As key immune cells, DCs secrete a diverse group of interleukins, aimed to orchestrate the immune response. Most of these cytokines, including IL-2, IL-7, IL-12, IL-15, IL-18, IL-23, and IL-27, induce or enhance maturation, activation and proliferation of Th1 cells, and cytotoxic responses. DCs also secrete the immunosuppressive IL-10 (14).

Classically, DCs were described as HLA-DR$^+$ lineage$^-$ cells, due to the high expression of major histocompatibility complex (MHC) class II (HLA-DR) and the lack of typical lineage markers, such as CD3 (T cells), CD19/20 (B cells) and CD56 (Natural Killer (NK) cells). However, more recently different subtypes of DCs were identified, and a number of DCs lineage markers were recognized (15). Nowadays, there is some consensus on this topic, and, as it has been recently reviewed by Rhodes et al. (16) and Collin and Bigley (17), DCs are divided in three well-differentiated subsets with specific functions and characteristic markers. This classification recognizes plasmacytoid DCs (pDCs) and two types of "classical" or "conventional" DCs (cDCs), previously known as myeloid DCs (15, 18, 19), known as cDC1 and cDC2 (**Table 1**).

cDCs express the myeloid antigen CD1a, b, c, and d, together with CD14, CD209 (Dendritic Cell-Specific Intercellular adhesion molecule-3-Grabbing Non-integrin (DC-SIGN)), and Factor XIIIA, at expression levels similar to those of monocytes. A small subset of cDCs, the cross-presenting DCs, also expresses CD141, CLEC9A, and XCR1. However, pDCs are characterized by the expression of CD303 (CLEC4C), CD304 (neuropilin), and CD123 (IL-3 receptor) (20).

Both pDCs and cDCs display characteristic surface markers on their membrane whose expression levels correlate with their maturation/activation state: CD83 is a maturation marker; CD80 and CD86 are activation markers involved in Ag presentation and activation of T cells; and CCR7 is a chemokine receptor up-regulated in mDCs implicated in migration to secondary lymphoid organs (21). Furthermore, all DCs express the receptor that participates in binding and internalization of HIV-1, CD4, as well as the co-receptors CCR5 and CXCR4 (22).

Plasmacytoid Dendritic Cells

pDCs present a lack of lineage markers, CD19-, CD3-, CD11c-, CD14-, but express MHC-II (HLA-DR$^+$). pDCs can be recognized among DCs because they express selective markers including BDCA-4, BDCA-2, LILRA4, and CD123 (20).

Immature pDCs express the chemokine receptors CCR1, CCR2, CCR5, CCR6, and CXCR1. Thus, pDCs are recruited to areas of infection by chemokines such as MIP-3 alpha/CCL20 (23). After virus uptake via endocytosis, pDCs recognize RNA viruses (like HIV-1) via TLR7, activating a signaling cascade that results in pDCs maturation, IFN-α, IFN-β, IL-6, and TNF-α production, and changes in expression of membrane proteins, including an increase in expression of chemokine receptor CCR7 (24) and CD40, CD80, and CD86 co-stimulatory molecules. CCR7 has a key role in driving the maturing DCs to the lymph nodes. Expression of MHC-II and the co-stimulatory molecules

TABLE 1 | Comparison between plasmacytoid and conventional DCs.

	Plasmacytoid DCs	Myeliod/conventional DCs	
		cDC1	cDC2
Origin	Plasmacytoid cells	Myeloid precursor cells or Axl+ cells	
Markers	CD303 (CLEC4C) CD304 (neuropilin) CD123 (IL-3R) BDCA-2 BDCA-4 LILRA4	CD13, CD33 CD11b,c (low) SIRPa (low) CD1a,b,c, and d CD209 (DC-SIGN) BDCA-3 (Clec)9A CADM1 FactorXIIIA	CD13, CD33 CD11b,c SIRPa CD1c, CD2 FceR1 Clec4A,10A,12A,13B VEGFA CD32a CD14
TLR expression	TLR7, TLR9	TLR3, TLR9, TLR10	TLR2, TLR4, TLR5, TLR6, TLR8, TLR9
Chemokine receptors	CCR1, CCR2, CCR5, CCR9 CCR7 (mature cells) CXCR1, CXCR3	CCR1, CCR2, CCR4, CCR5, CCR6, CCR9 CCR7 (mature cells) CX3CR1	
Cytokine production	IFN-α, IFN-β IL-6, IL-8, IL-12, IL-18 TNF-α CCL3, 4, 5 CXCL10,11 CXCL11	IFN (type I) IL-12	IL-1, IL-6, IL-8, IL-10, IL-12, IL-15, IL-18, IL-23 TNF-α CCL3, 4 CXCL8
Main function	IFN production (activation of antiviral immune response)	Antigen presentation via MHC class I (priming and activation of CD8+ T cells)	Antigen presentation (lymphocyte priming and activation)
Anti-HIV-1 functions	Inhibition of viral replication (type I IFN)	Generation of specific adaptive immune response (mainly cytotoxic)	Generation of specific adaptive immune response (humoral and cytotoxic)
Pro-HIV-1 functions	T cell recruitment to infection site T cell activation (increasing susceptibility to HIV-1 infection) Long-term immune suppression (IDO production)	HIV-1 transport to lymph nodes Cell-to-cell transfer to T cells	

allows pDCs to present Ags to CD4+ T cells, although not as efficiently as cDCs (25, 26). However, the most characteristic function of pDCs is type I IFN production, which stimulates a strong anti-viral response, but may also contribute to the chronicity of HIV-1 infection (27). Other soluble factors released by pDCs upon activation are CCL3, CCL4, CCL5, IL-8, CXCL10, and CXCL11 chemokines (28, 29). The multiple functions attributed to pDCs including IFN production, NK cell activation via IL-12 and IL- 18 secretion and Ag presentation have been reviewed by Swiecki and Colonna (30).

pDCs are among the first cells to encounter HIV-1 after infection (27). pDCs detect HIV-1 through TLR 7, which leads to activation, but can also be infected by HIV-1, due to the fact that they express CD4, CXCR4, and CCR5 (11). The role of pDCs in HIV-1 infection and disease development is both beneficial and detrimental because HIV-1 manipulates the anti-viral mechanisms of the cell to its own benefit. The functions of pDCs in the HIV-1 infection can be simplified in four steps (27): first, pDCs produce high levels of type I IFN, inhibiting viral replication and inducing bystander T cell activation which means to be a defensive

response but also leaves the lymphocytes susceptible to HIV-1 infection. Second, pDCs release chemokines such as CCL5 that recruit CCR5+ CD4+ T cells to the site of infection and thus facilitate the spread. However, it is also noteworthy that CCL5, similarly to Maraviroc and other CCR5 ligands, competes with HIV-1 protein gp120 for binding to CCR5 co-receptor, thus having a inhibitory effect on HIV-1 entry and infection (31). Third, during chronic HIV-1 infection, the production of IFN persists, causing apoptosis of T cells and contributing to decrease in CD4+ cell count. And fourth, pDCs produce the enzyme indoleamine-pyrrole 2,3-dioxygenase (IDO) that skews Treg/Th17 homeostasis toward the increase of immunosuppressive Treg cells (32).

Although the number of pDCs in peripheral blood drops during HIV-1 infection, in parallel with the drop of CD4+ T cells, the remaining pDCs maintain their phenotype and are, in fact, hyper activated. The decrease of pDC number in blood is believed to be caused partially because of high cell death, but also because they migrate to the lymph nodes and accumulate there, where they keep a high activation state that eventually leads to their own apoptosis (33, 34).

It has been shown that DCs from blood from HIV-1-infected individuals not only present elevated expression levels of TLR and their downstream components, but also they are hyper-responsive to TLR7 agonists and produce high levels of cytokines after stimulation (24). In the same study, they showed that pDCs from uninfected individuals up-regulate CCR7, CD40, and CD86 in response to TLR7/8 agonist R848 and aldrithiol-2 (AT-2)-inactivated HIV-1, and pDCs from HIV-1-infected individuals retain their capacity to up-regulate these receptors (24). Moreover, it has been shown that when comparing HIV-1-infected individuals and healthy controls, there is no difference in the expression of CCR7, CD83, CD80, and CD86 after TLR7 stimulation, and likewise, there is no significant difference in IFN-α production per cell (35).

Due to their role in IFN production, a dysregulation in pDC function may result in uncontrolled viral infections (36). For most viruses, a lower IFN production would result in a lower antiviral response and a stronger infection, but in the opposite case, when the IFN release is maintained in time, it would result in T cell apoptosis, and a decrease in the host defenses. More strongly, in the case of HIV-1 infection, IFN release is also responsible for T cell activation, rendering them susceptible to HIV-1 infection. Not only the IFN production, but also the multiple roles of pDCs make a pivotal difference in the fate of HIV-1 infection and viral spread or immune response.

Conventional Dendritic Cells

Unlike pDCs, which mainly respond to pathogens by secreting large amounts of IFN-α, cDCs are specialized APCs that serve as an important link between the innate immune system and the adaptive immune response. Upon Ag recognition, cDCs produce different inflammatory cytokines, including IL-12, IL-15, IL-23, IL-6, TNF, and IL-1b (14). Similarly to pDCs, cDCs express the chemokine receptors CCR1, CCR2, CCR5, CCR7, and CCR9, and they also express CCR4, CCR6, and CX3CR1 (37). cDCs are rarely infected by HIV-1, but they capture and internalize virus. As we will explain in the next sections, internalized virus can be processed and presented in the cell surface for T cell priming, or can on the contrary be transferred to CD4$^+$ T cells, promoting this way the HIV-1 infection and spread. Similar to pDCs, cDCs play a dual role, simultaneously restraining and potentiating HIV-1 infection (38).

During HIV-1 infection, cDCs are mostly depleted from the blood, partially because of apoptosis and partially because they migrate and accumulate in lymph nodes (39, 40). Moreover, the differentiation of cDCs from monocytes, which usually happens upon inflammation, may be impeded due to aberrant IFN-α production (41). The functionality of the remaining circulating cDCs is controversial. Some studies report that maturation of cDCs is impaired (42–44), whereas other studies claim that cDCs are not functionally defective, respond to TLR7 stimulation by up-regulating CCR7, CD40, CD80, CD83, and CD86 (21, 24), and may even be hyper-responsive to TLR stimulation (10). Binding and internalization of HIV-1 is mediated by a variety of DCs-expressed surface receptors. While activation of pDCs by HIV-1 is mainly thought to involve intracellular members of the TLR family, it has been shown that HIV-1 does not induce

maturation of cDCs by activating TLRs (45, 46). HIV-1 also prevents the activating effect of cDCs on other cell types. In normal conditions, cDCs mediate NK cell activation and fuel its lytic activity, but HIV-1 disrupts cDC-NK cell cross talk, weakening this innate line of defense (47).

The few cDCs that are infected by HIV-1 sustain a substantial viral burden and have a relatively long half-life, which suggests that they may have a role in latency. HIV-1 can signal through TLR-8, which initiates transcription from integrated provirus via MAPK pathway and nuclear factor kB (NF-kB) (48).

It has been observed that early HIV-1 infection induces a strong and simultaneous increase of LILRB2 and MHC-I expression on the surface of blood cDCs. Analysis of factors involved indicates that HIV-1 replication, TLR7/8 triggering, and treatment by IL-10 or type I IFNs increase LILRB2 expression (49). *In vitro* experiments suggest that strong LILRB2-HLA binding negatively affects Ag-presenting properties of DCs (50).

Although all cDCs share the same precursor cells and many characteristics, they can be divided in two types: cDC1 and cDC2.

Conventional DC1 (cDC1)

Peripheral blood myeloid cDC1 are a rare population of DCs [0.05% of PBMCs (51) or one tenth the amount of cDC2 (52, 53)]. They are identified by the expression of high levels of CD141 (BDCA-3) and XCR1 (54), together with expression of C-type lectin-like receptor (Clec)9A and cell adhesion molecule 1 (CADM1) (28, 29). cDC1 express CD13 and CD33 in common with cDC2, but lower CD11c, CD11b, and SIRPa (CD172). They detect intracellular dsRNA and DNA thanks to their high expression levels of TLR3, TLR9, and TLR10 (28, 55, 56), leading to IRF3-dependent production of type I IFNs and IL-12 (57, 58). Secretion of IL-12 promotes Th1 and NK cell responses. However, cDC1 are most characterized by their superior ability to cross-present and efficiently prime CD8$^+$ T cells via MHC class I (51, 59). Interestingly, it has been shown that cDC1 are constitutively resistant to productive viral infection (60).

Conventional DC2 (cDC2)

cDC2 cells are the major subset of myeloid DCs in blood, and are characterized for being potent stimulators of naïve T cells. They are identified by the expression of specific markers such as CD1c, CD2, FceR1, SIRPa, and myeloid antigens CD11b, CD11c, CD13, and CD33. Also, transcriptional profiling studies have recently identified CLEC10A (CD301a), VEGFA, and FCGR2A (CD32A) as cDC2 markers (60). cDC2 resemble monocytes in the expression of a wide range of lectins, TLRs, and NOD-like receptors (28). The lectins expressed by cDC2 include Clec4A, Clec10A, Clec12A, and DEC205 (Clec13B) (61, 62). cDC2 express TLR2, 4–6, 8, and 9 and, upon stimulation, they produce several soluble factors such as TNF-a, IL-1, IL-6, IL-8, IL-10, IL-12, IL-18, and IL-23 (63, 64) and chemokines including CCL3, CCL4, and CXCL8 (28, 65). Due to the wide range of factors produced and strong cross-presenting abilities, cDC2 promote a potent activation of Th1, Th2, Th17, and CD8$^+$ T cell responses (66).

Interestingly, a population of CD11c$^+$ DCs has been found in the anogenital tissue. These cells are transcriptionally very similar to cDC2, but present a higher expression of CD4 and CCR5, which could explain the higher susceptibility to HIV-1 productive infection and consequent transmission to T cells (67). Their presence in the anogenital tissues (where HIV-1 infection occurs) may be key in the initial spread of the virus and establishment of the infection during sexual transmission.

DCs can be obtained *in vitro* from CD14$^+$ monocytes after stimulating with granulocyte/macrophage colony-stimulating factor (GM-CSF) and IL-4 (68). These cells express the same maturation and function markers (69), and resemble the behavior of cDC2.

ROLE OF DENDRITIC CELLS IN HIV-1 INFECTION

As it was mentioned before, DCs are the APCs par excellence. Their most recognized role is to process Ags and present them on MHC on the cell surface, together with co-stimulatory molecules, aiming to present them to T cells and activate a specific and efficient immune response. In order to achieve their objective, once they find an Ag and become mature, they travel to secondary lymphoid tissues, where most resting or naïve T cells are accumulated.

The APC role of DCs that have been in contact with HIV-1 is a two-faced job. On one side, DCs can activate an immune response against HIV-1 in very early stages of infection, by presenting HIV-1 peptides to lymphocytes and thus promoting differentiation, activation and proliferation of HIV-1 recognizing T and B cells. However, HIV-1 also uses DCs to travel to T cell clusters and boost infection of T cells, through a process known as trans-infection which was first described in 1992 (70).

The pathway leading to DC-T cell contact can be divided in three steps: HIV-1 capture by DCs (**Figure 1**), intracellular processing and trafficking of captured virus (**Figure 2**), and connection with T cells (**Figure 3**).

HIV-1 Capture by Dendritic Cells

The recognition and capture of HIV-1 by DCs is a key step to determine the interaction between virus and immune system. The cell receptor that first recognizes HIV-1 and captures virus will determine the fate of cell and virus (71) (**Figure 1**). The main players are CD4 and DC-SIGN, leading the first one to conventional infection and the latter to either trans-infection or Ag presentation.

CD4 is the HIV-1 receptor par excellence. Binding of HIV-1 to this receptor, and a CCR5 or CXCR4 co-receptor will cause the virus to enter the cell and follow its conventional viral cycle, including retrotranscription of its genetic material, integration in the host genome and eventually, viral production. Productive infection of DCs is a rather rare event, 10- to 100-fold less frequent *in vivo* than CD4$^+$ T cell HIV-1 infection (72). In fact,

an *ex-vivo* study using DCs from healthy donors found that on average only 1–3% of both cDCs and pDCs are susceptible to HIV-1 infection, as observed by intracellular staining of HIV-1 protein p24 (10). However, even DC infection levels of <1% are more than sufficient to cause an explosive viral infection in CD4 lymphocytes (73).

An alternative DCs-specific HIV-1 binding protein was first described in 2000 (74). The C-type lectin receptor (CLR) DC-SIGN, which is highly expressed on cDCs located in mucosal tissues, recognizes mannose and fucose on a wide range of pathogens, including the high-mannose oligosaccharides of HIV-1 gp120. DC-SIGN is the key to understanding the dual and ambiguous role of DCs in HIV-1 infection. It acts both as HIV-1 receptor, enhancing trans-infection, and as pattern recognition receptor (PRR), enabling the development of a specific adaptive immune response (75). The fate between these two possibilities has been shown to depend on the N-glycan composition of HIV-1 (76). This was first noted after comparing the fate of normally glycosylated HIV-1 (containing a heterogeneous glycan composition) with that of oligomannose-enriched virus, and observing that the modified virus leads to enhanced viral capture by DCs, enhanced viral degradation and more efficient presentation to CD4$^+$ T cells, instead of leading to HIV-1 transmission. Different viral strains differ in their glycan composition, and more recent studies have proven its relevance, by showing that viral strains containing high-mannose glycans were more efficiently bound to DC-SIGN, degraded and presented as Ags (77, 78). Interestingly, the glycan composition of the HIV-1 envelope, and consequent sensitivity to lectins, also correlates with their sensitivity to neutralizing antibodies (Abs) (78, 79).

Binding of HIV-1 to DC-SIGN has been linked to trans-infection rather than to productive HIV-1 infection of DCs. The main mechanism for trans-infection is the transmission of viral particles through the infectious synapse between cDCs and T cells, which will be explained in the next sections. Leukemia-associated Rho guanine nucleotide exchange factor (LARG) is essential for the formation of synapse and has been shown to be activated by DC-SIGN triggered by HIV-1. LARG activation then leads to recruitment and activation of the small GTPase Rho A, but the consequences of this signaling have not been fully understood yet (80). Accordingly, a recent study (81) showed that blockage of DC-SIGN by binding to other proteins such as surfactant protein D, which is secreted by epithelial cells to mucosal surfaces, suppresses binding to gp120 and thus decreases the transmission to lymphocytes.

On the other hand, DC-SIGN bound HIV-1 can be internalized and processed for Ag presentation both to CD4$^+$ T cells via MHC-II (82) and to cytotoxic T cells via MHC-I (83). The signaling cascades initiated after HIV-1 binding to DC-SIGN lead to an increase in IL-10 production, which, due to its role in differentiation of naïve T cells into helper cells, is in concordance with the APC role of cDCs. DC-SIGN triggering activates RAF1, a serine/threonine protein kinase, which induces the phosphorylation of NF-κB subunit p65 at Ser276. This phosphorylation enables the binding of histone acetyl transferases CREB-binding protein (CBP) and

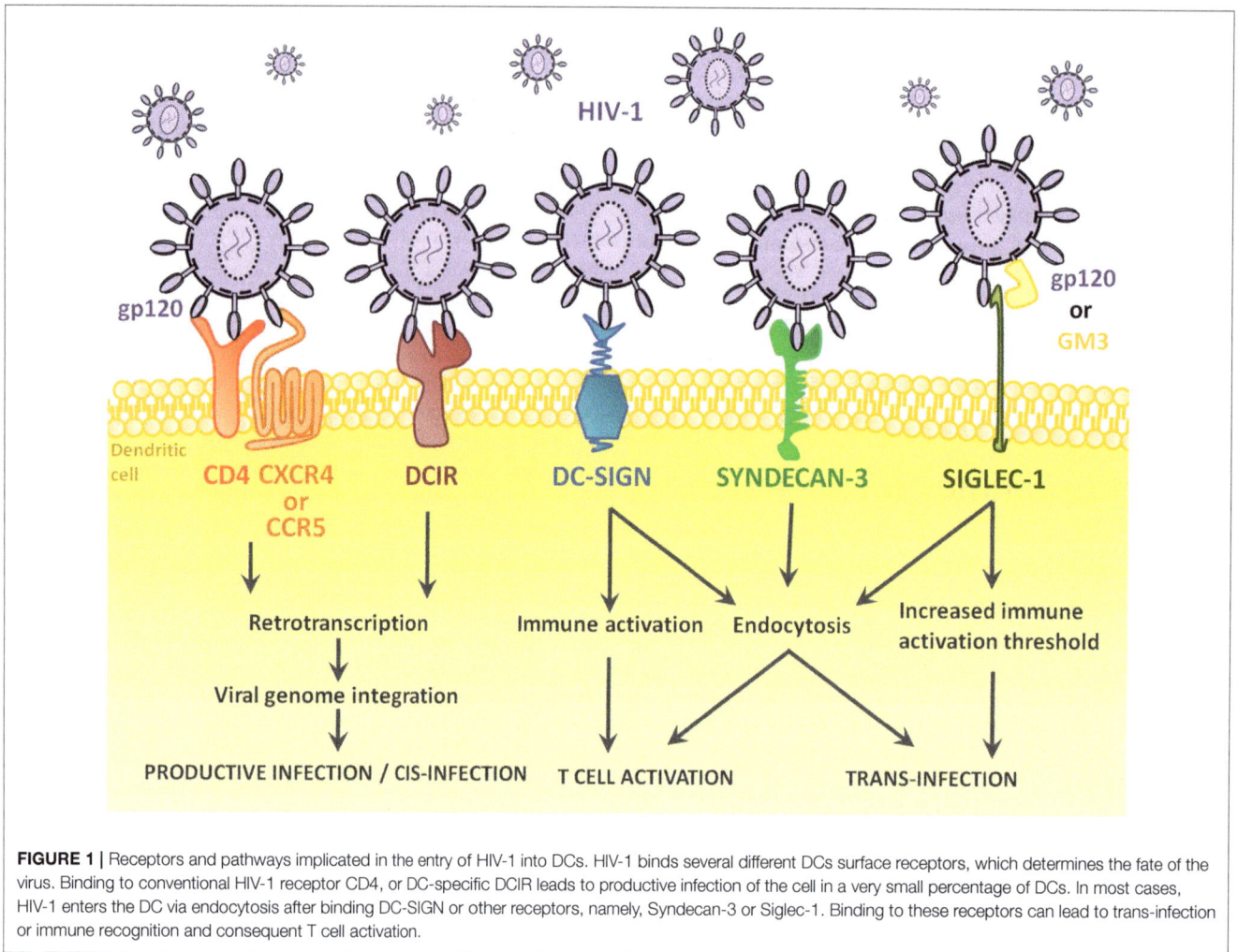

FIGURE 1 | Receptors and pathways implicated in the entry of HIV-1 into DCs. HIV-1 binds several different DCs surface receptors, which determines the fate of the virus. Binding to conventional HIV-1 receptor CD4, or DC-specific DCIR leads to productive infection of the cell in a very small percentage of DCs. In most cases, HIV-1 enters the DC via endocytosis after binding DC-SIGN or other receptors, namely, Syndecan-3 or Siglec-1. Binding to these receptors can lead to trans-infection or immune recognition and consequent T cell activation.

p300 to p65 and, thus, its acetylation, which results in an increased activity of the NF-κB transcription factor, this way increasing the transcription and expression of IL-8 and IL-10 (84). Surprisingly, phosphorylated p65 recruits pTEF-b, containing cyclin T1 and CDK9, to HIV-1 long terminal repeats, promoting transcriptional elongation, and thus, favoring productive replication if the viral genome has already been integrated in the DC genome (85). Even though TLR signaling is not required for RAF1 activation by DC-SIGN, NF-κB translocates to the nucleus as a consequence of TLR-4 and TLR-3 signaling pathways, suggesting a crosstalk between DC-SIGN and TLR signaling (84). The combined effects of Ag sensing by TLR-3, TLR-4, NOD2, and DC-SIGN have been dissected in a recent study (86) strongly proving that the ligand-driven triggering of these PRRs reduces viral replication and increases the ability of cDCs to stimulate HIV-specific cytotoxic T cells. Although, RAF1 is a well-known component of the RAF1-MEK-ERK signaling cascade (87) and HIV-1 has been linked to ERK activation (88), binding of mannose-containing pathogens by DC-SIGN does not trigger ERK activation, and to the extent of our knowledge, the involvement of DC-SIGN in ERK activation has not been proven, so it is unlikely that DC-SIGN is the initiator of the cascade leading to ERK activation.

Additionally, other receptors could be implicated in the binding of HIV-1 to DCs. For example, cell-surface heparan sulfate proteoglycan, syndecan-3, has been shown to bind to HIV-1 gp120 protein and to be involved in the capture of the virus (89). Another recently discovered C-type lectin receptor, named DCIR (for DCs immunoreceptor) has been demonstrated to bind to HIV-1, to participate in its capture and to be involved in processes leading to the productive infection of monocyte-derived DCs, and thus, to cis-infection of T cells (90). Finally, the sialic acid-binding immunoglobulin-like lectin Siglec-1 was shown to bind the host-derived ganglioside GM3 that is incorporated into the viral membrane and is a key interaction involved in HIV-1 transfer to CD4$^+$ T cells via trans-infection (91–93). Siglecs are immune modulatory transmembrane proteins that bind sialylated carbohydrates and set the activation thresholds for immune response. HIV-1, similar to many other pathogens, has evolved to incorporate sialic residues on gp120, and can not only bind to Siglec-1 on DCs and monocytes/macrophages through gangliosides on the viral lipid envelope, but also to Siglec-7 on NK cells and monocytes/macrophages (94), as has been reviewed by Mikulak et al. (95). The binding of HIV-1 to Siglec-1 also increases the entry to DCs, contributing to HIV-1 dissemination.

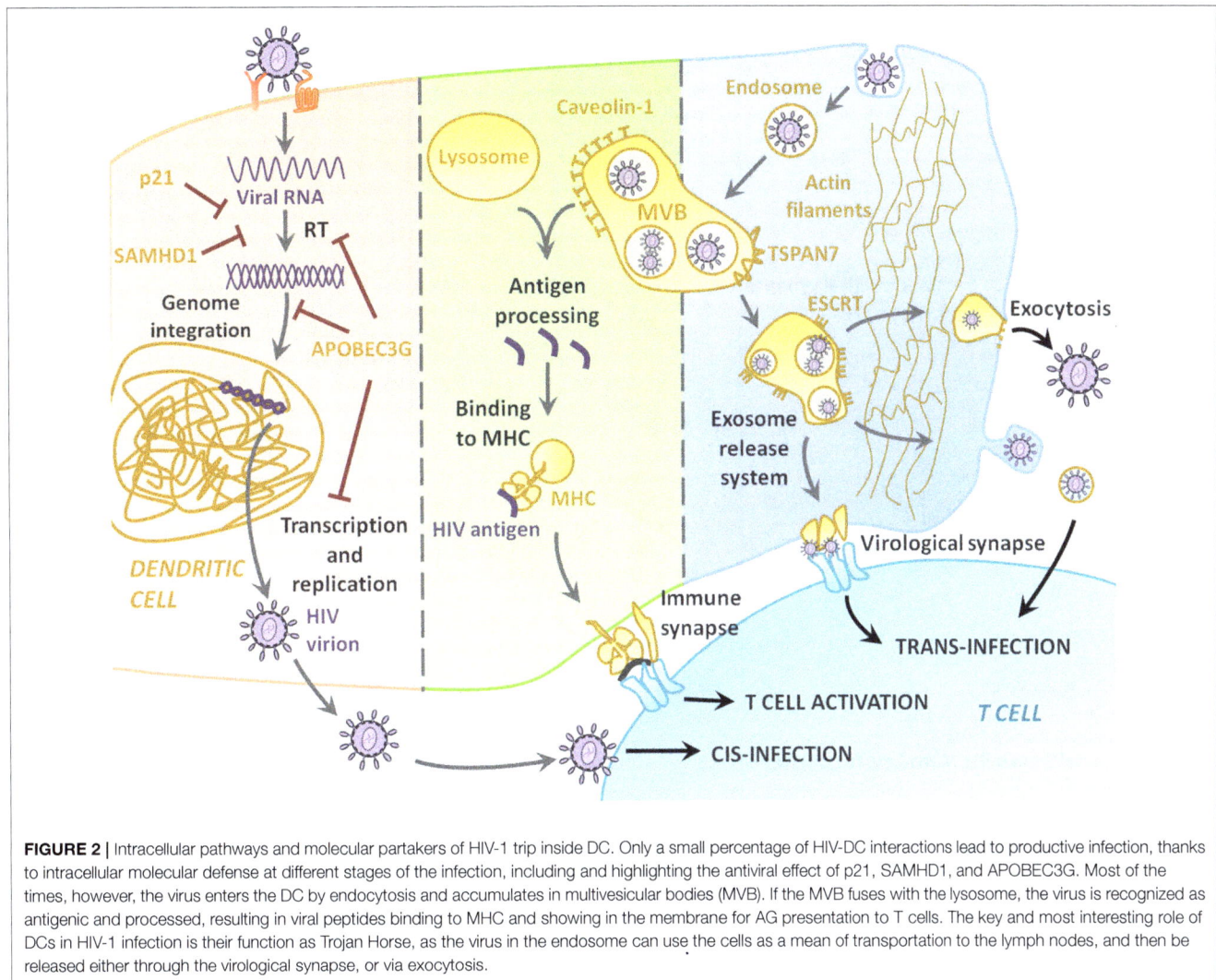

FIGURE 2 | Intracellular pathways and molecular partakers of HIV-1 trip inside DC. Only a small percentage of HIV-DC interactions lead to productive infection, thanks to intracellular molecular defense at different stages of the infection, including and highlighting the antiviral effect of p21, SAMHD1, and APOBEC3G. Most of the times, however, the virus enters the DC by endocytosis and accumulates in multivesicular bodies (MVB). If the MVB fuses with the lysosome, the virus is recognized as antigenic and processed, resulting in viral peptides binding to MHC and showing in the membrane for AG presentation to T cells. The key and most interesting role of DCs in HIV-1 infection is their function as Trojan Horse, as the virus in the endosome can use the cells as a mean of transportation to the lymph nodes, and then be released either through the virological synapse, or via exocytosis.

The entry way of HIV-1 into cells has been historically believed to be via membrane fusion after receptor and co-receptor binding. However, recent studies have proven that HIV-1 can enter the cell via dynamin-dependent endocytosis, and that membrane fusion can occur inside the cell, starting controversy (96). These different entry pathways could depend on cell type, and in the case of DCs, there is consensus in that the virus can either fuse at plasma membrane or be endocytosed, leading to productive infection or internalization and trans-infection of T cells, respectively (97). The endocytosed virus has been demonstrated to cause productive infection of CD4$^+$ T cells (98). HIV-1 entry to DC by endocytosis is also the path leading to lysosomal fusion, processing and Ag presentation.

Pritschet et al. (99) highlighted the importance of the endocytic pathway in pDCs when they demonstrated that endocytosis, but not membrane fusion, increased the IFN-α production. They treated pDCs with dynasore, a GTPase inhibitor that rapidly inhibits dynamin and thus, endocytosis, or

with fusion inhibitor T-20, and observed that only the former decreased IFN-α production.

Interestingly, binding of HIV-1 particle to C-type lectin receptors, mainly the aforementioned DC-SIGN, but also the Langerhans cell specific langerin, preferentially route the HIV-1 toward the endocytic pathway. Langerhans cells are DCs that reside in the epithelium characterized by their expression of langerin and the presence of endosome-related Birbeck granules in their cytoplasm. Binding of HIV-1 to langerin leads to internalization via caveolin-1 positive vesicles, linked to lysosomal degradation pathway, thus representing a restriction mechanism for HIV-1 infection (100).

The most recent studies in DCs, suggest that HIV-1 fusion occurs mostly at cell surface, instead of endosomic vesicles, and depends on the amount of CD4 expressed (101). Low levels of CD4 expression limit the fusion and divert DCs-HIV-1 interaction toward the competing binding to DC-SIGN and the consequent not productive pathway. The interaction of DCs with HIV-1 is strongly determined by the maturation state of the cells.

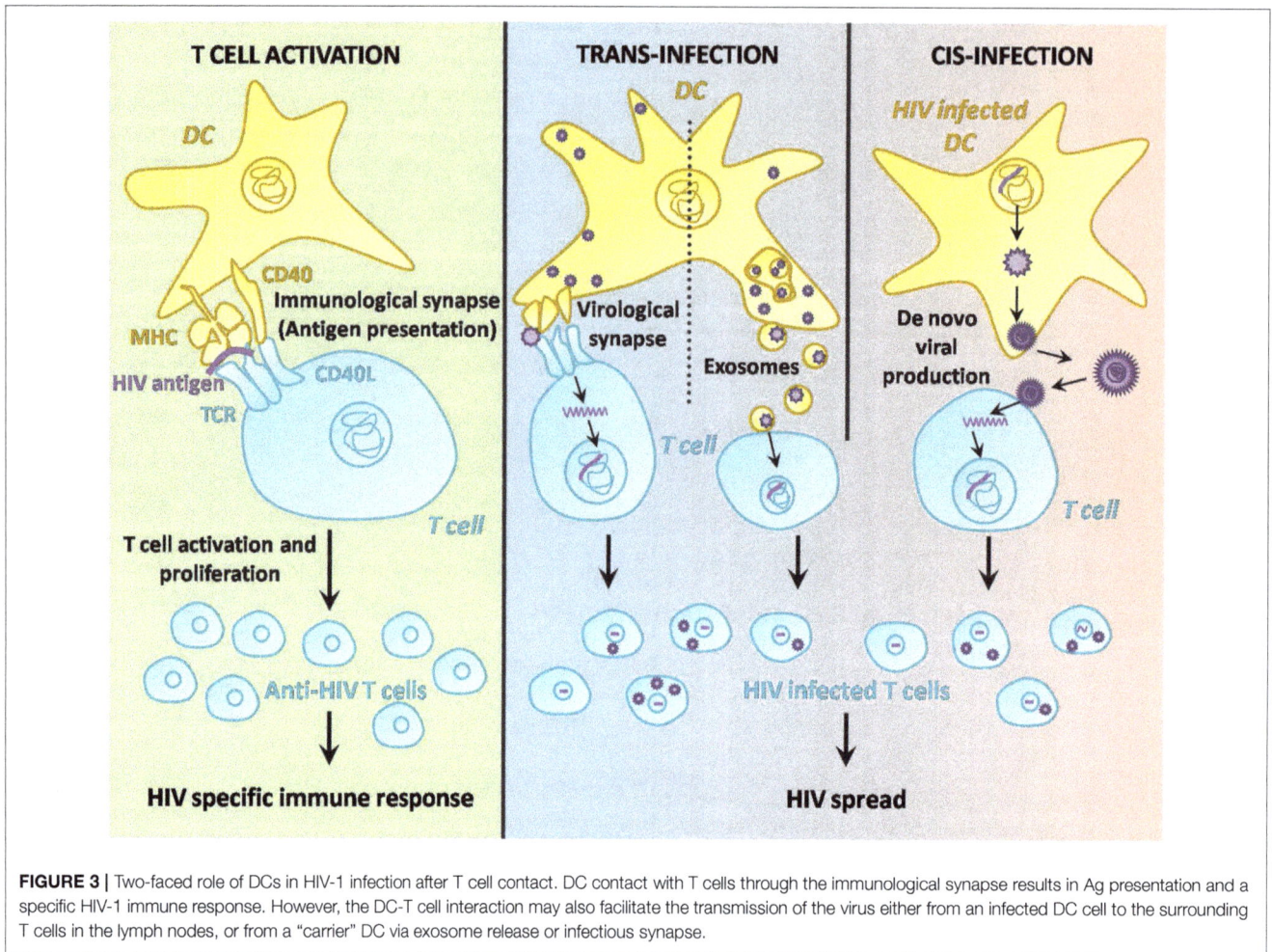

FIGURE 3 | Two-faced role of DCs in HIV-1 infection after T cell contact. DC contact with T cells through the immunological synapse results in Ag presentation and a specific HIV-1 immune response. However, the DC-T cell interaction may also facilitate the transmission of the virus either from an infected DC cell to the surrounding T cells in the lymph nodes, or from a "carrier" DC via exosome release or infectious synapse.

Viral fusion is even lower and shows slower kinetics in mDCs when comparing with iDCs (102).

Intracellular Trip
Dendritic Cell Infection or Lack Thereof

As it has been mentioned before, the interaction between HIV-1 and host cell can lead to productive infection, but more often than not, HIV-1 follows other alternative pathways. Some intracellular factors that play a key role in determining the fate of such interaction have been discovered (**Figure 2**). Intracellular antiviral responses represent a first line of natural defense in preventing infection as proved by the fact that HIV-1 has developed efficient counteracting measures.

Similar to viral fusion, viral replication is less effective in DCs than in CD4$^+$ T cells, but can still occur in DCs. iDCs can be productively infected, although only by R5 virus, while mDC can internalize both R5 and X4 HIV-1 strains, but the replication is blocked (103). It is controversial whether the transcription in mDC is blocked at the step of reverse transcription (104) or inhibited post-integration (105). A study from 1999 showed that the transcription blockage in mDC is surpassed after binding to T cells, via binding to CD40L, allowing and promoting cis-infection (106).

Despite binding and internalization by DCs, HIV-1 fails to induce a strong innate immune response or inflammation. The mechanisms that HIV-1 uses to evade detection inside the DCs are still to be fully understood. The inflammatory signaling pathways in DCs are stimulated by HIV-1 after binding to TLR-8. Using a phosphoproteomics approach followed by a siRNA screen, a recent study found that HIV-1 exploits the dynein motor protein Snapin, a natural cellular inhibitor of TLR-8 signaling and a general regulator of endosome maturation, by enhancing dynein expression. The increase in Snapin expression facilitates the accumulation of HIV-1 particles in endosomes containing low TLR-8, hence evading the sensing and inflammatory response (107).

The low HIV-1 infection is presumably due to low expression of CD4 receptor and co-receptors on the surface of the DCs, fast and efficient degradation of internalized virus, limited availability of dNTPs on non-cycling cells, and the expression of host factors that could block HIV-1 infection and/or replication (108).

The fact that HIV-1 replicates inefficiently in non-dividing cells, and the discovery that the infection can be enhanced by Vpx [a protein encoded by HIV-2 and several lineages of simian immunodeficiency virus (SIV)], pointed at the existence of a cellular HIV-1 restriction factor (109). Sterile alpha motif and HD-domain containing protein 1 (SAMHD1) was first described in 2011 as the restriction factor that gives non-dividing myeloid cells certain resistance against HIV-1 infection (110, 111). It was identified by mass spectrometry studies of proteins immunoprecipitated from cells expressing or not Vpx. Vpx was later shown to interact with the C-terminal domain of SAMHD1, driving it to E3 ubiquitin ligase complex, leading to proteosomal degradation (112). SAMHD1 is expressed in the majority of the nucleated hematopoietic cells, and it is especially abundant in the HIV-1 target cells residing in the anogenital mucosa (113). Keeping in mind that HIV-1 does not contain Vpx, all this information brings up the questions of how HIV-1 avoids the SAMHD1 barrier, what is the role of SAMHD1 in HIV-1 infection, and whether it could contribute to the inefficient immune response against HIV-1 (114).

SAMHD1 is a phosphohydrolase (115) that dephosphorylates, and thus depletes, the pool of deoxynucleotide triphosphates (dNTP) (115, 116) to a level below that required for HIV-1 reverse transcription (117). Structural studies have shown that SAMHD1 works as a highly efficient catalytic tetramer, which forms after binding to GTP and dGTP (118–121). This allosteric regulation by GTP warrants a close communication and coordinated regulation with the ribonuclease reductase (the enzyme responsible for *de novo* dNTP synthesis) in order to maintain balanced intracellular dNTP levels. As can be intuited, SAMHD1 regulation is cell-cycle dependent, and it has several means of control. Mainly, it is inhibited by phosphorylation in Threonine 592 by the complex of cyclin A with cyclin-dependent kinase 1 (CDK1) and CDK2 (122–124). Binding of single-stranded nucleic acids to dimerizing interface inhibit formation of the catalytic tetramer (125), and also oxidation caused by proliferation-induced hydrogen peroxide forms disulfide bonds and blocks catalytic activity (126).

These regulatory mechanisms keep SAMHD1 inactive during the G1 and S phase of the cell cycle in proliferating cells and explain why HIV-1 infects dividing cells more efficiently. The same study that showed that SAMHD1 is expressed in most tissues and hematopoietic cells, also pointed out that, predictably, activated CD4$^+$ T cells contain the phosphorylated form pSAMHD1 (T592), whereas cells, which are not HIV-susceptible such as resting T cells and macrophages, carried the HIV-restrictive unphosphorylated protein (113). In macrophages, cyclin L2 has been found to bind to SAMHD1, leading to its degradation and thus controlling its abundance (127). Knockdown of cyclin L2 highly decreased HIV-1 replication in macrophages, but not in cyclin cells, which suggest the possibility that it might also be found in DCs.

Several functions and roles related to immune response have been attributed to SAMHD1, including down-regulating type I interferon response, which depends on the residue G209 of SAMHD1 (128), and blocking the replication of virus that uses retrotranscription in their replication cycle (retroviruses and hepatitis B virus) (129, 130). Although SAMHD1 had been reported as a type I interferon (IFN)-inducible protein, its expression or phosphorylation levels have been demonstrated

to show no variation after IFN stimulation (131). However, its relationship with IFN is still controversial. Other studies claim that the constitutive signaling through the type-I IFN receptor in pDC blocks the degradation of SAMHD1 and counteracts the effect of Vpx (132).

SAMHD1 has a dual enzymatic function, degrading not only dNTPs (dNTPase activity) but also viral nucleic acids (RNase activity). The nuclease activity is also implicated in HIV-1 restriction (133), although there is no consensus about whether it is required (134) or not (135) for HIV-1 restriction. Recent studies on this topic showed that an increased expression of cyclin-dependent kinase inhibitor p21$^{Wafl/Cip1}$ (p21) also decreases the size of the intracellular dNTP pool by suppressing several enzymes involved in dNTP synthesis. Interestingly, p21 blocks SAMHD1 phosphorylation and promotes its dNTPase-independent antiviral activity without affecting its dNTPase activity (136). The expression of p21 is low in responding iDC, and increases gradually during differentiation (137). This correlates with the fact that some iDCs can be productively infected, while the viral replication is blocked in the mDCs that internalized the virus (103). Altogether, induction of p21 may result in an effective inhibition of HIV-1 replication in DCs, appearing to be a key regulator of HIV-1 infection and a potential target for drug design.

Another innate antiviral factor was first isolated in 2002, and identified as APOBEC3G (apolipoprotein B mRNA editing enzyme catalytic polypeptide-like 3G) (138). It was then described to act at the late stages of viral production and to be counteracted by the HIV-1 protein Vif (139). Two other components of APOBEC family, APOBEC3F, and APOBEC3B, exhibit anti-HIV-1 activities (140). Interestingly, APOBEC3G, together with APOBEC3F, can be encapsulated into HIV-1 virions, in absence of Vif, decreasing the infectivity of the virion (141), and it has been long studied as a candidate target for drug discovery (142). Both APOBEC3G and APOBEC3F are expressed in cell populations susceptible to HIV-1 infection, whereas APOBEC3B is not normally expressed in lymphoid cells and has been shown to be resistant to HIV-1 Vif. Therefore, activation of endogenous APOBEC3B gene in primary human lymphoid cells has been proposed as a novel and effective strategy for inhibition of HIV-1 replication *in vivo* (143).

APOBEC3G is a cytidinedeaminase of ssDNA, which means that it binds to ssDNA and catalyzes the deamination of cytidine to uridine, and it exerts its antiviral activity by interfering with the proper replication of viruses. However, in the case of HIV-1, it has been shown that APOBEC3G inhibits infection in two previous steps of the viral cycle. First, it interferes with reverse transcription, in a deaminase independent manner, by blocking tRNA$_3^{Lys}$ priming, thereby inhibiting the production of early minus-sense ssDNA (144). This implies that it cannot just bind to DNA, but also to RNA, through at least two different described RNA binding sites (145). Nonetheless, five to seven-fold reduction in viral DNA synthesis in the presence of APOBEC3G and absence of Vif only partially accounts for the 35-fold decrease in HIV-1 infectivity, which led to the discovery that there is an additional role of APOBEC3G that results in a five-fold decrease in the amount of integrated DNA. APOBEC3G modifies the processing and removal of tRNA by reducing its efficiency and specificity, resulting in aberrant viral DNA ends, inefficient for integration (146). Vif triggers the poly-ubiquitination of APOBEC3G and targets it for proteasomal degradation (147).

Despite being present in most cell types, APOBEC3G has an especially important role in DCs. Pion et al. (148) proved in 2006 that APOBEC3G levels correlate with HIV-1 infection in DCs. They showed that the small subset of iDC susceptible to HIV-1 infection were deficient in APOBEC3G, and they also noted that APOBEC3G levels increased during maturation, which further reduces susceptibility to infection. Likewise, APOBEC3G expression is up-regulated by DCs stimulation with CD40 and CCR5 (149), and poly (I:C) and TNF-α (150), which are known to induce DCs maturation.

Due to the importance of IFN-α in the innate immune response against viral infections, including HIV-1, its effect on APOBEC3G expression appears particularly interesting. A study published in 2013, found that low amounts of IFN-α significantly induced APOBEC3A, F and G expression in monocyte-derived iDC, without causing maturation (151). In the same study, they showed a significantly reduced transmission of HIV-1 from DCs to autologous T cells in the presence of IFN-α, presumably caused by the increased APOBEC expression levels. Similar results were found in pDCs, in which IFN-α increases APOBEC3G, thus decreasing HIV-1 infectivity (152).

APOBEC3G is only enzymatically active when in low molecular mass (LMM) configuration, but its recruitment into high molecular mass (HMM) RNA-protein complexes results in a block in its deaminase function. Activation of CD4$^+$ T cells enhances APOBEC3G expression and recruitment into HMM, rendering the cells permissive to HIV-1 infection. On the contrary, maturation of DCs leads to an increased expression and accumulation of LMM APOBEC3G, which then contributes to halting HIV-1 infection (150).

Another characteristic of APOBEC3G which differs in DCs with regards to other cell types is the subcellular location. In CD4$^+$ T cells, the expression APOBEC3G increases with activation, but it translocated to the nucleus, which is compatible with the increase in HIV-1 infection susceptibility. In DCs, APOBEC3G is localized in the cytoplasm, where it performs its functions, regardless of the maturation state of cells (153).

Although less efficient than SAMHD1 or APOBEC3G, TRIM5α is another known HIV-1 restriction factor that was first described as a rhesus macaque protein responsible for blocking infection by HIV-1 (154). TRIM5α restricts HIV-1 infection by disrupting the viral capsid and inducing its proteasome-dependent degradation (155–157). Its function in DCs is regulated by SUMOylation and nuclear sequestration (158). Recent studies showed that IFN-α-mediated stimulation of immunoproteasome facilitates the degradation of the viral capsid and inhibition of HIV-1 DNA synthesis by TRIM5α (159), thus suggesting a possible mechanism of action of this restriction factor. However, TRIM5α does not have a strong role in DCs, as it has been shown to only restrict HIV-1 infection upon C-type-lectin-receptor-dependent uptake of HIV-1 (160). DC-SIGN dependent entry of the virus leads to dissociation of TRIM5α from DC-SIGN, and consequent abrogation of restriction.

Interestingly, the effect of HIV-1 productive or restrictive infection on gene expression pattern depends on the maturation state of DCs. A study based on whole-genome microarrays recently found that expression of interferon-stimulated genes involved in control of viral replication, was induced after productive HIV-1 infection of iDCs with Vpx-loaded HIV-1 particles, thus inducing an antiviral state in surrounding cells. On the contrary, in the case of mDCs, productive HIV-1 infection decreased expression of interferon-stimulated genes CXCR3-binding chemokines, suggesting a diminished trans-infection of CD4 lymphocytes. Paradoxically, restrictive HIV-1 infection had the opposite effect on mDCs, increasing the aforementioned gene expression, which would result in lymphocyte attraction and enhanced trans-infection (161).

Summing up, the currently known restriction factors and their mechanisms of action may represent only the tip of the iceberg of mechanisms evolved to protect eukaryotic cells from HIV-1 infections, and to preserve the integrity of host cells and of its genome. Equivalently, SAMHD1, APOBEC3G and TRIM5α are well-known cellular factors that restrain HIV-1 from productively infecting and replicating in DCs, but further research could shed light on other mechanisms or ways to take advantage of these ones with a therapeutic or prophylactic aim.

Dendritic Cells as HIV-1 Carriers

Although DC infection is rare, HIV-1 frequently contacts DCs and enters to the cell, harnessing cell migration to reach more infection-prone cells. This pathway is the most favored after HIV-1 binding to DC-SIGN or other C-type lectin receptors, or when the virus enters via endocytosis. In this case, the virus uses the DC as a mere mean of transportation. The endosomes develop into multivesicular bodies (MVBs), and can then follow two paths: lysosomal fusion and degradation, probably leading to antigen presentation, or re-fusion with plasma membrane and release of exosomes or viral particles, spreading the infection.

In the specific case of langerin mediated HIV-1 capture, the viral particles have been shown to accumulate in vesicles in co-localization with langerin and caveolin-1, which was related to the lysosomal degradation pathway (100). A similar pathway was described in pDCs, after dynamin-dependent endosomal entry of HIV-1. The viral particles were found to localize in vesicles containing caveolin; early endosomal Ag 1; RabGTPases 5, 7, and 9, and lysosomal-associated membrane protein 1 (99). These data suggest that HIV-1 containing endosomes also follow the lysosomal pathway, thus triggering Ag presentation and an adaptive immune response. Lysosome fusion and Ag degradation is a key part of antigenic peptide binding to MHC-I or MHC-II and consequent Ag cross-presentation to CD8$^+$ or CD4$^+$ T cells, respectively (162). Further research on this topic is needed, as promotion of this pathway offers a potential road to anti-HIV drug design.

Binding of HIV-1 peptides on MHC-I molecules is remarkably interesting for the design of a therapeutic vaccine (163). The cytotoxic responses triggered by MHC-I presentation to CD8$^+$ T cells have a key role in the specific immune response against HIV-1, and cDCs have been shown to process antigen, present it on MHC-I and generate a specific anti-HIV-1 cytotoxic immune response (164). Intracellular processing is required for optimal presentation, and the ability of DCs to process and present antigen on MHC-I remains intact in DCs derived from HIV-1 infected patients on cART (165). This processing pathway could be exploited for the design of DC-based therapies against HIV-1.

However, HIV-1 avoids lysosomal degradation and exploits the endosomal pathway in DCs, turning it into an efficient and dangerous way to spread to lymph nodes, where it transfers to T cells, its favored target. HIV-1 transfer from DCs to T cells is regulated by actin dynamics (166). Recent studies have revealed the role of TSPAN7, a member of the tetraspanin family which promotes actin nucleation and stabilization via the ARP2/3 complex, in halting HIV-1 internalization in the endosome, keeping it close to plasma membrane, in actin-rich dendrites, favoring an efficient transfer to T cells (167).

TSPAN7 increases the list of tetraspanins related with the HIV-1 infection cycle (168). The HIV-1 budding sites, virological synapses, and virions have long been known to be enriched in the tetraspanin family members CD9, CD63, and CD81 (169). HIV-1 uses these host molecules for its own benefit. CD9 regulates trafficking of MHC-II and its surface expression levels, thus affecting Ag presentation (170). CD81 is associated with SAMHD1, decreasing its activity, increasing the availability of dNTP, and thus enhancing HIV-1 infection (171). CD63 has a dual activity and its function is still not clear. It regulates HIV-1 replication (172) and it is required for reverse transcription (173).

Finally, there is a group of proteins that is booming in the last years, but whose role in the context of DCs and HIV-1 has been overlooked for now. The endosomal sorting complexes required for transport (ESCRT) machinery consists on a group of cytosolic protein complexes that enable membrane remodeling and budding away from the cytoplasm, and thus orchestrate cellular processes such as MVB and exosome biogenesis (174), cytokinesis and viral budding (175). HIV-1 is not an exception, and takes advantage of this tool to be released from infected cells (176, 177). Briefly, HIV-1 gag has been shown to engage tsg101 (178, 179) or ALIX (180) in order to recruit the rest of the ESCRT machinery to the assembly sites where they mediate budding. High-throughput screening is a useful tool for finding inhibitors for these interactions (181). When looking at DCs, the ESCRT machinery has an extra function, as it regulates Ag presentation by MHC class-II (182). The ESCRT proteins drive MHC class-II complex to either lysosomal degradation in non-activated DCs, or transfer to cell surface for Ag presentation in activated DCs, thus standing out as key figures for DC function. It would be interesting to study the specific role of the ESCRT machinery in DCs in presence of HIV-1, as a potential start point for the design of anti-HIV-1 drugs or therapeutic vaccines.

Although the main questions that come to mind when in the rational search of an HIV-1 therapeutic or prophylactic are usually focused on virus-cell interactions and viral replication, the multiple and diverse intracellular molecular players that determine the fate of the virus and the outcome of infection also form a rich and vastly unexplored field with a big potential for drug design. Altering the HIV-1 intracellular pathway may serve to modulate the immune response and control the consequences

of the infection, pointing out the importance of a detailed study of all the implicated molecules.

Dendritic Cell-T Cell Contact

The ultimate function of DCs is to present Ag to the T cells, for which they establish contact either in the mucosa or after migration to the lymph nodes. HIV-1 subverts the Ag presentation process to increase transmission and infection of T cells (**Figure 3**). Although some DCs can retain viruses up to 6 days, most viruses are degraded within 24 h after exposure, which does not match the timeline of T cell infection in co-culture. A 2-phase transmission pathway has been suggested, initiated by iCD and mCD uptake and transmission of HIV-1 to T cells via trans-infection that decays after 24 h, and followed by a long-term second round of cis-infection after 48 h, when the infection has resulted productive in iDCs, and a new generation of virus is transmitted to T cells (183).

Dendritic Cell-T Cell Activation

As we have mentioned before, cDCs are mostly known for their role as APCs. Although the molecular mechanisms remain controversial, the accepted dogma is that cDCs capture Ag by endocytosis, degrade it into peptides after fusion of the endosome with the lysosome, and present it to T cells bound to MHC complex. Recognition of MHC bound Ag, together with binding to co-stimulatory molecules on the cDC surface, in a molecular cluster known as the "immunological synapse," induces specific T cell responses, and thus, the adaptive immune response. The state of knowledge about this process has been reviewed in other recent publications (162, 184–186).

In spite of the extensive research and accumulated knowledge in this field, there are still a lot of questions that remain to be answered and are of great interest for therapy design, among other things. Hence, more detailed molecular pathways and alternative circumstances are being explained. A study from 2018, for example, described a new pathway leading to initiation of adaptive immune responses *in vivo*, which occurs after the infective agent has bypassed capture by the innate immune cells at the infection site (187), providing evidence of the existence of an alternative lymphocyte activating pathway that may function in parallel or as a backup to the conventional pathway. They found that the Ag can travel to lymph nodes, where it is phagocytosed by macrophages, causing their death, and the debris released activates Ag presentation by monocytes in the blood. Further studies about the factors contained in the debris that enhance Ag presentation could provide clues in the search of an HIV-1 therapeutic vaccine.

In the case of HIV-1, cDCs from infected patients are unable to generate lymphocyte activation and proliferation *in vitro*, even when they show all the signs of activation (42). However, cDCs from elite controllers of HIV-1 infection have been found to effectively prime T cell responses (188). These recent results corroborate their previous findings that DCs from elite controllers produced type I IFN in a rapid and sustained way, expressed less SAMHD1, and accumulated viral reverse transcripts after exposure to HIV-1. Those signs of improved cell-intrinsic immune recognition of HIV-1 translated into a strong stimulation of a specific CD8$^+$ T cell response, suggesting that improved DC-T cell Ag presentation contributes to control of HIV-1 infection in elite controllers (189).

Recent studies are shedding some light on the mechanisms of induction of HIV-1 specific T cell responses by DCs. The programmed cell death ligand 1 (PD-L1) is highly expressed in highly stimulatory cDCs, and it has been shown to have an enhancing role in priming naïve T cells and activation of CD8$^+$ T cells into effector T cells, while it has a negative role in later phases, as it decreases the magnitude of memory HIV-specific cytotoxic T cell responses (190). These findings contradict the previous suggestion that PD-L1 expressing DCs in the lymph nodes may hinder the generation of HIV-specific T cell response (191).

Manipulating cross-presentation is a promising tool for effective vaccine design against cancer and infectious diseases, including HIV-1. For this reason, further exploring and describing the mechanism and main molecular partakers of DC-T cell crosstalk is indispensable in the search of DC based vaccines.

Dendritic Cell-T Cell HIV-1 Transmission

Trans-infection via infectious synapse

The first studies about DC-T cell HIV-1 transmission claimed that cell-to-cell contact was necessary for the T cell infection to occur (192). This fact was corroborated by several later studies, and the contact phase was named "infectious" or "virological" synapse (193). Although the exact molecule composition and structure of the infectious synapse has not been fully described yet, it is believed to be similar to that of the immunological synapse, through which the APCs present Ag and activate T cells (194, 195). Interestingly, DC-SIGN is detected in and essential for the formation of the infectious synapse, as demonstrated by Arrighi et al. (196, 197) in several experiments suppressing the expression of DC-SIGN and showing the lack of synapses and inefficient HIV-1 transmission. However, it has been reported that in this DC-SIGN mediated HIV-1 transmission, the expression of MHC class II molecules on virus donor cells is not required, setting considerable differences with the immunological synapses (198).

Besides being essential for the formation of the synapse, DC-SIGN binding to HIV-1 particle in the first place also sets a signaling cascade that favors the trans-infection pathway over productive infection or viral degradation. DC-SIGN is known to associate with LARG, whose activation promotes activation of RhoA and the focal adhesion molecules FAK, Pyk2, and paxillin, all of them structural components of the infectious synapse (199).

After binding to DC-SIGN, among other receptors, HIV-1 enters the cell via endocytosis, and remains in MVBs inside the cell. Contact of cDCs with CD4$^+$ T cells causes the internalized vesicles containing the HIV-1 particles to migrate to cell-cell junction, facilitating the transfer of virions across the synapse (193). This appears to be more efficient with mature than immature monocyte-derived DCs (183).

Studies comparing HIV-1 non-progressors with HIV-1-infected progressors and HIV-1-seronegative donors showed that cDCs from HIV-1 non-progressors failed to mediate trans-infection, highlighting the importance of this process for HIV-1 infection and dissemination. They also linked the lack of trans-infection to lower cholesterol levels and increased expression of the reverse cholesterol transporter ABCA1 (ATP-binding cassette transporter A1) in the DCs, but not in T cells, thus opening the door to new therapeutic approaches involving lipid metabolism enhancement (200).

Trans-infection via the virological synapse is more efficient for R5-tropic HIV-1 than X4-tropic HIV-1 strains (201). Independently of the viral entry to DCs or productivity of the infection, co-culture of CD4$^+$ T cells with R5-infected cDCs resulted in a higher viral expansion than co-culture with X4-infected cDCs. These results were also found to depend on T cell activation state. The maturation state of DCs also dictates the efficiency and selectivity of the viral transmission. mDCs are more efficient than iDCs in transferring HIV-1 to T cells, partially due to the concentration of viral particles at the virological synapse that is more frequently observed in mDCs (202). Studies on DCs activated by different maturation factors showed that iDCs and CD40L-induced mDCs were more susceptible to productive infection and lead to cis-infection, while LPS and TNF-α-induced mDCs mediated efficient trans-infection of CD4$^+$ T cells (203). One very important reason to consider halting trans-infection in the search of new effective anti-HIV-1 therapies is the fact that this mechanism of T cell infection is insensitive to the antiretroviral drugs (204). So even under cART or pre-exposure prophylaxis (PreP), HIV-1 could spread to T cells and increase the reservoir, thus hindering the so desired cure.

Trans-infection via infectious exosomes

Wiley and Gummuluru discovered an HIV-1 trans-infection pathway mediated not by cell-to-cell contact but by DC-derived exosomes (205). They reported that HIV-1 particles that were captured by iDCs and rapidly endocytosed to tetraspanin-rich compartments could be constitutively released to extracellular milieu in association with exosomes (HLA-DR1$^+$, CD1b$^+$, CD9$^+$, and CD63$^+$ vesicles). They also discovered that the HIV-1 particles associated with exosomes from DCs could fuse with target-cell membranes, and were 10-fold more infectious than cell-free virus particles.

Exosomes are small vesicles (50–200 nm) containing genomic, proteomic or lipid cargos, which are released to the extracellular milieu by a broad range of cell types and facilitate cell-cell communication (206). In particular, DC-derived exosomes have been shown to be remarkably efficient at HIV-1 trans-infection. This could be explained by the fact that the exosomes carrying the virus also carry infection enhancing factors, including T cell activators and binding molecules. MHC class II molecules are detected in exosomes, which could be sufficient to activate T cells, increasing their susceptibility to HIV-1 infection (207). A comparative study demonstrated that the presence of fibronectin and galectin-3 in exosomes from DCs was key for HIV-1 trans-infection of T cells, as blockage of both receptors significantly inhibited this process (208).

Recent studies suggest that exosomes from HIV-1 infected DCs can not only activate and infect resting CD4$^+$ T lymphocytes via trans-infection, but also reach and reactivate the HIV-1 reservoir, further boosting the progression of infection (209). Those studies further corroborate an almost contemporary publication that describes how exosomes released by uninfected cells already activate transcription of latent reservoirs (210).

Years before the exosome-bound HIV-1 trans-infection pathway was described, exosomes were already in the spotlight of retroviral infection through the widely accepted "Trojan exosome hypothesis" (211). This hypothesis postulates that retroviruses, including HIV-1, exploit the cell-encoded machinery of vesicle traffic and exosome exchange for both the biogenesis of viral particles and mode of infection and spread. Strong evidence supporting this hypothesis is the huge similarity between exosomes and HIV-1 particles regarding their host cell lipid and protein composition and biogenesis. Both particles contain higher levels of cholesterol and glycosphingolipids than the plasma membrane (206, 212). Both exosomes and HIV-1 particles are enriched in many protein components in comparison with the plasma membrane, including tetraspannins (213), GPI proteins and Lamps (214, 215). HIV-1 also exploits the mechanisms that target proteins to MVBs, previous to formation of exosomes. HIV-1 protein Gag not only binds to exosome biogenesis factors, such as cyclophilin and tsg101, but it also forms aggregates. It is N-terminally myristoylated and monoubiquitylated, which are known MVB targeting mechanisms (216). Even more relevant evidence supporting the Trojan exosome hypothesis is the observation of infection independent of *Env* proteins and cell receptors (217, 218). Although depletion of HIV-1 *Env* decreases its infectivity to <1%, *Env* depleted HIV-1 particles still infect CD4$^+$ and CD4$^-$ cells with the same efficiency. This could be explained by the exchange of exosomes, especially common among immune cells as a communication system for immune surveillance (219, 220), which after being released into the extracellular milieu can fuse with membranes of neighboring cells by a clathrin-mediated mechanism.

In spite of the many similarities regarding biogenesis and molecular composition between HIV particles and exosomes, the presence on exosomes of T cell activators and binding molecules, such as HLA-DR, explains the higher infectivity of HIV carrying exosomes in comparison with free HIV particles (207). As HIV-1 can only infect dividing cells, T cell activators present on exosomes allow the HIV-1 virus to retrotranscribe and infect the T cell after exosome binding. On the other hand, binding molecules increase the affinity of the particle for the host cell, and also allow the HIV-1 virus to enter the T cell, sometimes without even binding to its receptors.

The study of exosomes has increased exponentially recently, and there is only new evidence of the multiple and ambiguous role they play during HIV-1 infection. Some studies, for example, suggest that exosomes carrying CD4 or DC-SIGN receptors compete for binding to the virus, thus pointing to a protective role of exosomes against HIV-1 spreading (221, 222). However, most of the new findings are related to ways by which HIV-1 hijacks the exosomal system and increases infection (223).

Surprisingly, HIV-1 binding to DCIR in DCs, which leads to productive infection as explained before, also triggers the release of exosomes containing the pro-apoptotic protein DAP-3, which increased spontaneous apoptosis in uninfected CD4$^+$ T cells (224).

Although the relative importance of this pathway remains to be determined, it definitely also contributes to infectious synapse mediated trans-infection by providing a pathway by which the intracellular virus reaches the synapse, avoiding its degradation in the lysosomal pathway.

Cis-infection after viral replication in dendritic cells

Trans-infection can persist up to 24 h after HIV-1 binding to the surface of DCs, with a peak around 6 to 12 h, but the majority of the virus is degraded after that time by fusion with the lysosomal compartment in DC. As DCs require around 12–24 h to migrate to the lymph nodes, some scientists claim that it is unlikely that trans-infection is an effective "Trojan horse" for the virus, and that it does not explain the peak of CD-T cell transmission observed after 24 h in co-culture. The second phase of HIV-1 transmission is believed to imply DC productive infection, which, although less efficient than in CD4$^+$ T cells, has a key role in long-term HIV-1 transmission (183). *In vitro*, cis-infection is more efficient than trans-infection. This is probably due to a combination of factors including a higher concentration of viral particles in each donor cell, longer duration of HIV-1 transfer (with a peak around 72–96 h), long duration of viral production and cell survival up to 40 days (225), and the need for only less than 0.1% HIV-1 infected DCs.

HIV-1 protein Nef plays an important role at enhancing cis-infection. By inhibiting DC-SIGN endocytosis and thus upregulating surface levels of DC-SIGN in HIV-1-infected DCs, Nef promotes lymphocyte clustering and viral transmission (226). Not only does Nef facilitate DC-T cell contact, but it also promotes T cell activation, which is necessary for HIV-1 infection. A study comparing the effect of WT HIV-1 vs. Nef-mutated HIV-1 on the activation of resting CD4$^+$ T cells found that Nef was required for T cell activation and productive HIV-1 infection (227). Contradictorily, Nef expression in HIV-1 infected DCs induces the expression of the interferon-induced protein tetherin (228), which restricts virion release (229–231) and cell-to-cell infection (232).

One of the biggest challenges in studying *de novo* HIV-1 production is the lack of tools to distinguish between cytosolic immature and endocytic mature virus. Turville et al. (233) developed an assay that allows differentiating between the newly synthesized viral particles and the mature virus traveling through the endocytic path. They developed a modified infectious HIV-1 construct that could be stained with biarsenical dyes and allowed the restricted observation of newly synthesized uncleaved gag protein. Combination of this system with HIV-1 gag antibodies, which stain mature virions, can be used to track viruses at different maturation stages.

As it has been mentioned before, mDCs are more permissive of HIV-1 replication than iDCs, and also display an increased surface expression of CXCR4, favoring the infection by X4 tropic viruses. Thus, it seems logical that the maturation stage of DC plays a crucial role in HIV-1 infection, spreading of viral strains and disease progression. This was shown by a study promoting DC maturation, by treatment of DCs with dectin-1/TLR2 and NOD2 ligands, which resulted in an increase of cis-infection of autologous CD4$^+$ T cells by X4-tropic HIV-1 (234).

IMPLICATIONS AND CONCLUSIONS

The modification of DCs and their use as immunotherapy is increasing in the last years, especially in the field of cancer therapies (235). DC-based therapy has been shown to be safe and feasible in several phase I/II clinical trials (236–238). Due to their many implications and key role in adaptive immune system activation and also in HIV-1 infection, DCs are also promising immunotherapy candidates in the search of an HIV-1 functional cure or vaccine.

Despite the growing knowledge in DC biology and HIV-1 infection, the exact mechanisms that determine the type of response generated are still poorly understood and further research is required. Understanding these immune behaviors would empower rational frameworks for the design of immune modulators working on a systems level that serve a prophylactic or therapeutic purpose.

Future therapies could work at any stage and take advantage of the peculiarities of DCs for a highly specific and targeted treatment. For example, specifically targeting DC-SIGN has been achieved thanks to advances in drug design (239). This could be useful for *in vivo* targeting of drugs to DCs. According to what has been summarized in this review, harnessing certain entrance mechanisms could determine the fate of the HIV-1 or Ag, and could promote Ag presentation instead of viral spread (162). Natural DC responses can be manipulated in order to create a specific T cell response, both cytotoxic and humoral. In fact, DC membrane vesicles are already being used as a delivery platform for CD8$^+$ activation (240).

All these steps of DC function should be exploited but the design of controlled, safe and efficient DC-based therapies require more molecular, and systems knowledge. Further studies in the field of DC modulation, Ag presentation and DC-HIV-1 interaction are essential to set the bases for the so desired HIV-1 functional cure.

AUTHOR CONTRIBUTIONS

Both authors conceived the review and have participated in writing, graphics, review, and discussion of the article. Both authors read and approved the final version or article.

FUNDING

This work has been (partially) funded by the RD16/0025/0019, projects as part of Acción Estratégica en Salud, Plan Nacional de Investigación Científica, Desarrollo e Innovación Tecnológica (2013–2016) and cofinanced by Instituto de Salud Carlos III (Subdirección General de Evaluación) and Fondo Europeo de Desarrollo Regional (FEDER), RETIC PT17/0015/0042, Fondo de Investigacion Sanitaria (FIS) (grant number PI16/01863) and EPIICAL project. CIBER BBN is an initiative funded by the VI National R&D&i Plan 2008–2011, IniciativaIngenio 2010, the Consolider Program, and CIBER Actions and financed by the Instituto de Salud Carlos III with assistance from the European Regional Development Fund. This work has been supported partially by a EUROPARTNER: Strengthening and spreading international partnership activitiesof the Faculty of Biology and Environmental Protection for interdisciplinary research andinnovation of the University of Lodz Programme: NAWA International Academic Partnership Programme. This article/publication is based upon work from COST Action CA 17140 Cancer Nanomedicine from the Bench to the Bedside supported by COST (European Cooperation in Science and Technology).

REFERENCES

1. Hoth DF. Issues in the development of a prophylactic HIV vaccine. *Ann N Y Acad Sci.* (1993) 685:777–83. doi: 10.1111/j.1749-6632.1993.tb35943.x

2. Kippax S. HIV and technology: the issue of prophylactic vaccines. *Dev Bull.* (2000) 52:24–5.

3. Xu W, Li H, Wang Q, Hua C, Zhang H, Li W, et al. Advancements in developing strategies for sterilizing and functional HIV cures. *Biomed Res Int.* (2017) 2017:6096134. doi: 10.1155/2017/6096134

4. Liu C, Ma X, Liu B, Chen C, Zhang H. HIV-1 functional cure: will the dream come true? *BMC Med.* (2015) 13:284. doi: 10.1186/s12916-015-0517-y

5. Davenport MP, Khoury DS, Cromer D, Lewin SR, Kelleher AD, Kent SJ. Functional cure of HIV: the scale of the challenge. *Nat Rev Immunol.* (2019) 19:45–54. doi: 10.1038/s41577-018-0085-4

6. Mildner A, Jung S. Development and function of dendritic cell subsets. *Immunity.* (2014) 40:642–56. doi: 10.1016/j.immuni.2014.04.016

7. Worbs T, Hammerschmidt SI, Forster R. Dendritic cell migration in health and disease. *Nat Rev Immunol.* (2017) 17:30–48. doi: 10.1038/nri.2016.116

8. Ugur M, Mueller SN. T cell and dendritic cell interactions in lymphoid organs: More than just being in the right place at the right time. *Immunol Rev.* (2019) 289:115–28. doi: 10.1111/imr.12753

9. Kleiveland CR. Peripheral blood mononuclear cells. *Impact Food Bioactives Health.* (2015) 2015:161–7. doi: 10.1007/978-3-319-16104-4_15

10. Smed-Sorensen A, Lore K, Vasudevan J, Louder MK, Andersson J, Mascola JR, et al. Differential susceptibility to human immunodeficiency virus type 1 infection of myeloid and plasmacytoid dendritic cells. *J Virol.* (2005) 79:8861–9. doi: 10.1128/JVI.79.14.8861-8869.2005

11. Manches O, Frleta D, Bhardwaj N. Dendritic cells in progression and pathology of HIV infection. *Trends Immunol.* (2014) 35:114–22. doi: 10.1016/j.it.2013.10.003

12. Banchereau J, Steinman RM. Dendritic cells and the control of immunity. *Nature.* (1998) 392:245–52. doi: 10.1038/32588

13. Banchereau J, Briere F, Caux C, Davoust J, Lebecque S, Liu YJ, et al. Immunobiology of dendritic cells. *Annu Rev Immunol.* (2000) 18:767–811. doi: 10.1146/annurev.immunol.18.1.767

14. Akdis M, Aab A, Altunbulakli C, Azkur K, Costa RA, Crameri R, et al. Interleukins (from IL-1 to IL-38), interferons, transforming growth factor beta, and TNF-alpha: Receptors, functions, and roles in diseases. *J Allergy Clin Immunol.* (2016) 138:984–1010. doi: 10.1016/j.jaci.2016.06.033

15. Ziegler-Heitbrock L, Ancuta P, Crowe S, Dalod M, Grau V, Hart DN, et al. Nomenclature of monocytes and dendritic cells in blood. *Blood.* (2010) 116:e74–80. doi: 10.1182/blood-2010-02-258558

16. Rhodes JW, Tong O, Harman AN, Turville SG. Human dendritic cell subsets, ontogeny, and impact on HIV infection. *Front Immunol.* (2019) 10:1088. doi: 10.3389/fimmu.2019.01088

17. Collin M, Bigley V. Human dendritic cell subsets: an update. *Immunology.* (2018) 154:3–20. doi: 10.1111/imm.12888

18. Dzionek A, Fuchs A, Schmidt P, Cremer S, Zysk M, Miltenyi S, et al. BDCA-2, BDCA-3, and BDCA-4: three markers for distinct subsets of dendritic cells in human peripheral blood. *J Immunol.* (2000) 165:6037–46. doi: 10.4049/jimmunol.165.11.6037

19. MacDonald KP, Munster DJ, Clark GJ, Dzionek A, Schmitz J, Hart DN. Characterization of human blood dendritic cell subsets. *Blood.* (2002) 100:4512–20. doi: 10.1182/blood-2001-11-0097

20. Collin M, McGovern N, Haniffa M. Human dendritic cell subsets. *Immunology.* (2013) 140:22–30. doi: 10.1111/imm.12117

21. Ebner S, Lenz A, Reider D, Fritsch P, Schuler G, Romani N. Expression of maturation-/migration-related molecules on human dendritic cells from blood and skin. *Immunobiology.* (1998) 198:568–87. doi: 10.1016/S0171-2985(98)80079-X

22. van Montfort T, Thomas AA, Pollakis G, Paxton WA. Dendritic cells preferentially transfer CXCR4-using human immunodeficiency virus type 1 variants to CD4$^+$ T lymphocytes in trans. *J Virol.* (2008) 82:7886–96. doi: 10.1128/JVI.00245-08

23. Dieu MC, Vanbervliet B, Vicari A, Bridon JM, Oldham E, Ait-Yahia S, et al. Selective recruitment of immature and mature dendritic cells by distinct chemokines expressed in different anatomic sites. *J Exp Med.* (1998) 188:373–86. doi: 10.1084/jem.188.2.373

24. Sabado RL, O'Brien M, Subedi A, Qin L, Hu N, Taylor E, et al. Evidence of dysregulation of dendritic cells in primary HIV infection. *Blood.* (2010) 116:3839–52. doi: 10.1182/blood-2010-03-273763

25. Villadangos JA, Young L. Antigen-presentation properties of plasmacytoid dendritic cells. *Immunity.* (2008) 29:352–61. doi: 10.1016/j.immuni.2008.09.002

26. Reizis B, Bunin A, Ghosh HS, Lewis KL, Sisirak V. Plasmacytoid dendritic cells: recent progress and open questions. *Annu Rev Immunol.* (2011) 29:163–83. doi: 10.1146/annurev-immunol-031210-101345

27. O'Brien M, Manches O, Bhardwaj N. Plasmacytoid dendritic cells in HIV infection. *Adv Exp Med Biol.* (2013) 762:71–107. doi: 10.1007/978-1-4614-4433-6_3

28. Villani AC, Satija R, Reynolds G, Sarkizova S, Shekhar K, Fletcher J, et al. Single-cell RNA-seq reveals new types of human blood dendritic cells, monocytes, and progenitors. *Science.* (2017) 356:6335. doi: 10.1126/science.aah4573

29. Zoccali C, Moissl U, Chazot C, Mallamaci F, Tripepi G, Arkossy O, et al. Chronic fluid overload and mortality in ESRD. *J Am Soc Nephrol.* (2017) 28:2491–7. doi: 10.1681/ASN.2016121341

30. Swiecki M, Colonna M. The multifaceted biology of plasmacytoid dendritic cells. *Nat Rev Immunol.* (2015) 15:471–85. doi: 10.1038/nri3865

31. Secchi M, Vassena L, Morin S, Schols D, Vangelista L. Combination of the CCL5-derived peptide R4.0 with different HIV-1 blockers reveals wide target compatibility and synergic cobinding to CCR5. *Antimicrob Agents Chemother.* (2014) 58:6215–23. doi: 10.1128/AAC.03559-14

32. Manches O, Munn D, Fallahi A, Lifson J, Chaperot L, Plumas J, et al. HIV-activated human plasmacytoid DCs induce Tregs through an indoleamine 2,3-dioxygenase-dependent mechanism. *J Clin Invest.* (2008) 118:3431–9. doi: 10.1172/JCI34823

33. Fitzgerald-Bocarsly P, Jacobs ES. Plasmacytoid dendritic cells in HIV infection: striking a delicate balance. *J Leukoc Biol.* (2010) 87:609–20. doi: 10.1189/jlb.0909635

34. Reeves RK, Evans TI, Gillis J, Wong FE, Kang G, Li Q, et al. SIV infection induces accumulation of plasmacytoid dendritic cells in the gut mucosa. *J Infect Dis.* (2012) 206:1462–8. doi: 10.1093/infdis/jis408

35. Kaushik S, Teque F, Patel M, Fujimura SH, Schmidt B, Levy JA. Plasmacytoid dendritic cell number and responses to Toll-like receptor 7 and 9 agonists vary in HIV Type 1-infected individuals in relation to clinical state. *AIDS Res Hum Retroviruses.* (2013) 29:501–10. doi: 10.1089/aid.2012.0200

36. Miller E, Bhardwaj N. Dendritic cell dysregulation during HIV-1 infection. *Immunol Rev.* (2013) 254:170–89. doi: 10.1111/imr.12082

37. Griffith JW, Sokol CL, Luster AD. Chemokines and chemokine receptors: positioning cells for host defense and immunity. *Annu Rev Immunol.* (2014) 32:659–702. doi: 10.1146/annurev-immunol-032713-120145

38. Derby N, Martinelli E, Robbiani M. Myeloid dendritic cells in HIV-1 infection. *Curr Opin HIV AIDS.* (2011) 6:379–84. doi: 10.1097/COH.0b013e3283499d63

39. Macatonia SE, Lau R, Patterson S, Pinching AJ, Knight SC. Dendritic cell infection, depletion and dysfunction in HIV-infected individuals. *Immunology.* (1990) 71:38–45.

40. Donaghy H, Pozniak A, Gazzard B, Qazi N, Gilmour J, Gotch F, et al. Loss of blood CD11c$^+$ myeloid and CD11c$^-$ plasmacytoid dendritic cells in patients with HIV-1 infection correlates with HIV-1 RNA virus load. *Blood.* (2001) 98:2574–6. doi: 10.1182/blood.V98.8.2574

41. Kodama A, Tanaka R, Zhang LF, Adachi T, Saito M, Ansari AA, et al. Impairment of *in vitro* generation of monocyte-derived human dendritic cells by inactivated human immunodeficiency virus-1: Involvement of type I interferon produced from plasmacytoid dendritc cells. *Hum Immunol.* (2010) 71:541–50. doi: 10.1016/j.humimm.2010.02.020

42. Donaghy H, Gazzard B, Gotch F, Patterson S. Dysfunction and infection of freshly isolated blood myeloid and plasmacytoid dendritic cells in patients infected with HIV-1. *Blood.* (2003) 101:4505–11. doi: 10.1182/blood-2002-10-3189

43. Granelli-Piperno A, Golebiowska A, Trumpfheller C, Siegal FP, Steinman RM. HIV-1-infected monocyte-derived dendritic cells do not undergo maturation but can elicit IL-10 production and T cell regulation. *Proc Natl Acad Sci USA.* (2004) 101:7669–74. doi: 10.1073/pnas.0402431101

44. Martinson JA, Roman-Gonzalez A, Tenorio AR, Montoya CJ, Gichinga CN, Rugeles MT, et al. Dendritic cells from HIV-1 infected individuals are less responsive to toll-like receptor (TLR) ligands. *Cell Immunol.* (2007) 250:75–84. doi: 10.1016/j.cellimm.2008.01.007

45. Beignon AS, McKenna K, Skoberne M, Manches O, DaSilva I, Kavanagh DG, et al. Endocytosis of HIV-1 activates plasmacytoid dendritic cells via Toll-like receptor-viral RNA interactions. *J Clin Invest.* (2005) 115:3265–75. doi: 10.1172/JCI26032

46. Hertoghs N, van der Aar AM, Setiawan LC, Kootstra NA, Gringhuis SI, Geijtenbeek TB. SAMHD1 degradation enhances active suppression of dendritic cell maturation by HIV-1. *J Immunol.* (2015) 194:4431–7. doi: 10.4049/jimmunol.1403016

47. Altfeld M, Fadda L, Frleta D, Bhardwaj N. DCs and NK cells: critical effectors in the immune response to HIV-1. *Nat Rev Immunol.* (2011) 11:176–86. doi: 10.1038/nri2935

48. Schlaepfer E, Speck RF. TLR8 activates HIV from latently infected cells of myeloid-monocytic origin directly via the MAPK pathway and from latently infected CD4$^+$ T cells indirectly via TNF-alpha. *J Immunol.* (2011) 186:4314–24. doi: 10.4049/jimmunol.1003174

49. Alaoui L, Palomino G, Zurawski S, Zurawski G, Coindre S, Dereuddre-Bosquet N, et al. Early SIV and HIV infection promotes the LILRB2/MHC-I inhibitory axis in cDCs. *Cell Mol Life Sci.* (2018) 75:1871–87. doi: 10.1007/s00018-017-2712-9

50. Bashirova AA, Martin-Gayo E, Jones DC, Qi Y, Apps R, Gao X, et al. LILRB2 interaction with HLA class I correlates with control of HIV-1 infection. *PLoS Genet.* (2014) 10:e1004196. doi: 10.1371/journal.pgen.1004196

51. Jongbloed SL, Kassianos AJ, McDonald KJ, Clark GJ, Ju X, Angel CE, et al. Human CD141$^+$ (BDCA-3)$^+$ dendritic cells (DCs) represent a unique myeloid DC subset that cross-presents necrotic cell antigens. *J Exp Med.* (2010) 207:1247–60. doi: 10.1084/jem.20092140

52. Guilliams M, Dutertre CA, Scott CL, McGovern N, Sichien D, Chakarov S, et al. Unsupervised high-dimensional analysis aligns dendritic cells across tissues and species. *Immunity.* (2016) 45:669–84. doi: 10.1016/j.immuni.2016.08.015

53. Granot T, Senda T, Carpenter DJ, Matsuoka N, Weiner J, Gordon CL, et al. Dendritic cells display subset and tissue-specific maturation dynamics over human life. *Immunity.* (2017) 46:504–15. doi: 10.1016/j.immuni.2017.02.019

54. Poulin LF, Salio M, Griessinger E, Anjos-Afonso F, Craciun L, Chen JL, et al. Characterization of human DNGR-1$^+$ BDCA3$^+$ leukocytes as putative equivalents of mouse CD8alpha$^+$ dendritic cells. *J Exp Med.* (2010) 207:1261–71. doi: 10.1084/jem.20092618

55. Hemont C, Neel A, Heslan M, Braudeau C, Josien R. Human blood mDC subsets exhibit distinct TLR repertoire and responsiveness. *J Leukoc Biol.* (2013) 93:599–609. doi: 10.1189/jlb.0912452

56. Colletti NJ, Liu H, Gower AC, Alekseyev YO, Arendt CW, Shaw MH. TLR3 signaling promotes the induction of unique human BDCA-3 dendritic cell populations. *Front Immunol.* (2016) 7:88. doi: 10.3389/fimmu.2016.00088

57. Nizzoli G, Krietsch J, Weick A, Steinfelder S, Facciotti F, Gruarin P, et al. Human CD1c$^+$ dendritic cells secrete high levels of IL-12 and potently prime cytotoxic T-cell responses. *Blood.* (2013) 122:932–42. doi: 10.1182/blood-2013-04-495424

58. Liu S, Cai X, Wu J, Cong Q, Chen X, Li T, et al. Phosphorylation of innate immune adaptor proteins MAVS, STING, and TRIF induces IRF3 activation. *Science.* (2015) 347:aaa2630. doi: 10.1126/science.aaa2630

59. Bachem A, Guttler S, Hartung E, Ebstein F, Schaefer M, Tannert A, et al. Superior antigen cross-presentation and XCR1 expression define human CD11c$^+$CD141$^+$ cells as homologues of mouse CD8$^+$ dendritic cells. *J Exp Med.* (2010) 207:1273–81. doi: 10.1084/jem.20100348

60. Silvin A, Yu CI, Lahaye X, Imperatore F, Brault JB, Cardinaud S, et al. Constitutive resistance to viral infection in human CD141$^+$ dendritic cells. *Sci Immunol.* (2017) 2:eaai8071. doi: 10.1126/sciimmunol.aai8071

61. Harman AN, Bye CR, Nasr N, Sandgren KJ, Kim M, Mercier SK, et al. Identification of lineage relationships and novel markers of blood and skin human dendritic cells. *J Immunol.* (2013) 190:66–79. doi: 10.4049/jimmunol.1200779

62. Vu Manh TP, Elhmouzi-Younes J, Urien C, Ruscanu S, Jouneau L, Bourge M, et al. Defining mononuclear phagocyte subset homology across several distant warm-blooded vertebrates through comparative transcriptomics. *Front Immunol.* (2015) 6:299. doi: 10.3389/fimmu.2015.00299

63. Nizzoli G, Larghi P, Paroni M, Crosti MC, Moro M, Neddermann P, et al. IL-10 promotes homeostatic proliferation of human CD8$^+$ memory T cells and, when produced by CD1c$^+$ DCs, shapes naive CD8$^+$ T-cell priming. *Eur J Immunol.* (2016) 46:1622–32. doi: 10.1002/eji.201546136

64. Sittig SP, Bakdash G, Weiden J, Skold AE, Tel J, Figdor CG, et al. A comparative study of the T cell stimulatory and polarizing capacity of human primary blood dendritic cell subsets. *Mediators Inflamm.* (2016) 2016:3605643. doi: 10.1155/2016/3605643

65. Piccioli D, Tavarini S, Borgogni E, Steri V, Nuti S, Sammicheli C, et al. Functional specialization of human circulating CD16 and CD1c myeloid dendritic-cell subsets. *Blood.* (2007) 109:5371–9. doi: 10.1182/blood-2006-08-038422

66. Di Blasio S, Wortel IM, van Bladel DA, de Vries LE, Duiveman-de Boer T, Worah K, et al. Human CD1c$^+$ DCs are critical cellular mediators of immune responses induced by immunogenic cell death. *Oncoimmunology.* (2016) 5:e1192739. doi: 10.1080/2162402X.2016.1192739

67. Bertram KM, Botting RA, Baharlou H, Rhodes JW, Rana H, Graham JD, et al. Identification of HIV transmitting CD11c$^+$ human epidermal dendritic cells. *Nat Commun.* (2019) 10:2759. doi: 10.1038/s41467-019-10697-w

68. Sallusto F, Lanzavecchia A. Efficient presentation of soluble antigen by cultured human dendritic cells is maintained by granulocyte/macrophage colony-stimulating factor plus interleukin 4 and downregulated by tumor necrosis factor alpha. *J Exp Med.* (1994) 179:1109–18. doi: 10.1084/jem.179.4.1109

69. Zhou LJ, Tedder TF. CD14$^+$ blood monocytes can differentiate into functionally mature CD83$^+$ dendritic cells. *Proc Natl Acad Sci USA.* (1996) 93:2588–92. doi: 10.1073/pnas.93.6.2588

70. Cameron PU, Freudenthal PS, Barker JM, Gezelter S, Inaba K, Steinman RM. Dendritic cells exposed to human immunodeficiency virus type-1 transmit a vigorous cytopathic infection to CD4$^+$ T cells. *Science.* (1992) 257:383–7. doi: 10.1126/science.1352913

71. Jan M, Arora SK. Innate sensing of HIV-1 by dendritic cell-specific ICAM-3 grabbing nonintegrin on dendritic cells: degradation and presentation versus transmission of virus to T cells is determined by glycan composition of viral envelope. *AIDS Res Hum Retroviruses.* (2017) 33:765–7. doi: 10.1089/aid.2016.0290

72. McIlroy D, Autran B, Cheynier R, Wain-Hobson S, Clauvel JP, Oksenhendler E, et al. Infection frequency of dendritic cells and CD4$^+$ T lymphocytes in spleens of human immunodeficiency virus-positive patients. *J Virol.* (1995) 69:4737–45.

73. Kawamura T, Gatanaga H, Borris DL, Connors M, Mitsuya H, Blauvelt A. Decreased stimulation of CD4$^+$ T cell proliferation and IL-2 production by highly enriched populations of HIV-infected dendritic cells. *J Immunol.* (2003) 170:4260–6. doi: 10.4049/jimmunol.170.8.4260

74. Geijtenbeek TB, Kwon DS, Torensma R, van Vliet SJ, van Duijnhoven GC, Middel J, et al. DC-SIGN, a dendritic cell-specific HIV-1-binding protein that enhances trans-infection of T cells. *Cell.* (2000) 100:587–97. doi: 10.1016/S0092-8674(00)80694-7

75. Geijtenbeek TB, Gringhuis SI. Signalling through C-type lectin receptors: shaping immune responses. *Nat Rev Immunol.* (2009) 9:465–79. doi: 10.1038/nri2569

76. van Montfort T, Eggink D, Boot M, Tuen M, Hioe CE, Berkhout B, et al. HIV-1 N-glycan composition governs a balance between dendritic cell-mediated viral transmission and antigen presentation. *J Immunol.* (2011) 187:4676–85. doi: 10.4049/jimmunol.1101876

77. Jan M, Upadhyay C, Alcami Pertejo J, Hioe CE, Arora SK. Heterogeneity in glycan composition on the surface of HIV-1 envelope determines virus sensitivity to lectins. *PLoS ONE.* (2018) 13:e0194498. doi: 10.1371/journal.pone.0194498

78. Hioe CE, Jan M, Feyznezhad R, Itri V, Liu X, Upadhyay C. HIV-1 envelope glycan composition influences virus-host interaction. *J Immunol.* (2019) 202 (Suppl. 1):117–97.

79. Wagh K, Kreider EF, Li Y, Barbian HJ, Learn GH, Giorgi E, et al. Completeness of HIV-1 envelope glycan shield at transmission determines neutralization breadth. *Cell Rep.* (2018) 25:893–908.e897. doi: 10.1016/j.celrep.2018.09.087

80. Hodges A, Sharrocks K, Edelmann M, Baban D, Moris A, Schwartz O, et al. Activation of the lectin DC-SIGN induces an immature dendritic cell phenotype triggering Rho-GTPase activity required for HIV-1 replication. *Nat Immunol.* (2007) 8:569–77. doi: 10.1038/ni1470

81. Dodagatta-Marri E, Mitchell DA, Pandit H, Sonawani A, Murugaiah V, Idicula-Thomas S, et al. Protein-protein interaction between surfactant protein D and DC-SIGN via C-type lectin domain can suppress HIV-1 Transfer. *Front Immunol.* (2017) 8:834. doi: 10.3389/fimmu.2017.00834

82. Engering A, Geijtenbeek TB, van Vliet SJ, Wijers M, van Liempt E, Demaurex N, et al. The dendritic cell-specific adhesion receptor DC-SIGN internalizes antigen for presentation to T cells. *J Immunol.* (2002) 168:2118–26. doi: 10.4049/jimmunol.168.5.2118

83. Moris A, Nobile C, Buseyne F, Porrot F, Abastado JP, Schwartz O. DC-SIGN promotes exogenous MHC-I-restricted HIV-1 antigen presentation. *Blood.* (2004) 103:2648–54. doi: 10.1182/blood-2003-07-2532

84. Gringhuis SI, den Dunnen J, Litjens M, van Het Hof B, van Kooyk Y, Geijtenbeek TB. C-type lectin DC-SIGN modulates Toll-like receptor signaling via Raf-1 kinase-dependent acetylation of transcription factor NF-kappaB. *Immunity.* (2007) 26:605–16. doi: 10.1016/j.immuni.2007.03.012

85. Gringhuis SI, van der Vlist M, van den Berg LM, den Dunnen J, Litjens M, Geijtenbeek TB. HIV-1 exploits innate signaling by TLR8 and DC-SIGN for productive infection of dendritic cells. *Nat Immunol.* (2010) 11:419–26. doi: 10.1038/ni.1858

86. Cardinaud S, Urrutia A, Rouers A, Coulon PG, Kervevan J, Richetta C, et al. Triggering of TLR-3,−4, NOD2, and DC-SIGN reduces viral replication and increases T-cell activation capacity of HIV-infected human dendritic cells. *Eur J Immunol.* (2017) 47:818–29. doi: 10.1002/eji.201646603

87. Wellbrock C, Karasarides M, Marais R. The RAF proteins take centre stage. *Nat Rev Mol Cell Biol.* (2004) 5:875–85. doi: 10.1038/nrm1498

88. Shan M, Klasse PJ, Banerjee K, Dey AK, Iyer SP, Dionisio R, et al. HIV-1 gp120 mannoses induce immunosuppressive responses from dendritic cells. *PLoS Pathog.* (2007) 3:e169. doi: 10.1371/journal.ppat.0030169

89. de Witte L, Bobardt M, Chatterji U, Degeest G, David G, Geijtenbeek TB, et al. Syndecan-3 is a dendritic cell-specific attachment receptor for HIV-1. *Proc Natl Acad Sci USA.* (2007) 104:19464–9. doi: 10.1073/pnas.0703747104

90. Lambert AA, Gilbert C, Richard M, Beaulieu AD, Tremblay MJ. The C-type lectin surface receptor DCIR acts as a new attachment factor for HIV-1 in dendritic cells and contributes to trans- and cis-infection pathways. *Blood.* (2008) 112:1299–307. doi: 10.1182/blood-2008-01-136473

91. Izquierdo-Useros N, Lorizate M, Puertas MC, Rodriguez-Plata MT, Zangger N, Erikson E, et al. Siglec-1 is a novel dendritic cell receptor that mediates

HIV-1 trans-infection through recognition of viral membrane gangliosides. *PLoS Biol.* (2012) 10:e1001448. doi: 10.1371/journal.pbio.1001448

92. Pino M, Erkizia I, Benet S, Erikson E, Fernandez-Figueras MT, Guerrero D, et al. HIV-1 immune activation induces Siglec-1 expression and enhances viral trans-infection in blood and tissue myeloid cells. *Retrovirology.* (2015) 12:37. doi: 10.1186/s12977-015-0160-x

93. Hammonds JE, Beeman N, Ding L, Takushi S, Francis AC, Wang JJ, et al. Siglec-1 initiates formation of the virus-containing compartment and enhances macrophage-to-T cell transmission of HIV-1. *PLoS Pathog.* (2017) 13:e1006181. doi: 10.1371/journal.ppat.1006181

94. Varchetta S, Lusso P, Hudspeth K, Mikulak J, Mele D, Paolucci S, et al. Sialic acid-binding Ig-like lectin-7 interacts with HIV-1 gp120 and facilitates infection of CD4pos T cells and macrophages. *Retrovirology.* (2013) 10:154. doi: 10.1186/1742-4690-10-154

95. Mikulak J, Di Vito C, Zaghi E, Mavilio D. Host immune responses in HIV-1 infection: the emerging pathogenic role of siglecs and their clinical correlates. *Front Immunol.* (2017) 8:314. doi: 10.3389/fimmu.2017.00314

96. Miyauchi K, Kim Y, Latinovic O, Morozov V, Melikyan GB. HIV enters cells via endocytosis and dynamin-dependent fusion with endosomes. *Cell.* (2009) 137:433–44. doi: 10.1016/j.cell.2009.02.046

97. Janas AM, Dong C, Wang JH, Wu L. Productive infection of human immunodeficiency virus type 1 in dendritic cells requires fusion-mediated viral entry. *Virology.* (2008) 375:442–51. doi: 10.1016/j.virol.2008.01.044

98. Clotet-Codina I, Bosch B, Senserrich J, Fernandez-Figueras MT, Pena R, Ballana E, et al. HIV endocytosis after dendritic cell to T cell viral transfer leads to productive virus infection. *Antiviral Res.* (2009) 83:94–8. doi: 10.1016/j.antiviral.2009.03.009

99. Pritschet K, Donhauser N, Schuster P, Ries M, Haupt S, Kittan NA, et al. CD4- and dynamin-dependent endocytosis of HIV-1 into plasmacytoid dendritic cells. *Virology.* (2012) 423:152–64. doi: 10.1016/j.virol.2011.11.026

100. van den Berg LM, Ribeiro CM, Zijlstra-Willems EM, de Witte L, Fluitsma D, Tigchelaar W, et al. Caveolin-1 mediated uptake via langerin restricts HIV-1 infection in human Langerhans cells. *Retrovirology.* (2014) 11:123. doi: 10.1186/s12977-014-0123-7

101. Chauveau L, Donahue DA, Monel B, Porrot F, Bruel T, Richard L, et al. HIV fusion in dendritic cells occurs mainly at the surface and is limited by low CD4 levels. *J Virol.* (2017) 91:e01248–e01217. doi: 10.1128/JVI.01248-17

102. Cavrois M, Neidleman J, Kreisberg JF, Fenard D, Callebaut C, Greene WC. Human immunodeficiency virus fusion to dendritic cells declines as cells mature. *J Virol.* (2006) 80:1992–9. doi: 10.1128/JVI.80.4.1992-1999.2006

103. Granelli-Piperno A, Delgado E, Finkel V, Paxton W, Steinman RM. Immature dendritic cells selectively replicate macrophagetropic (M-tropic) human immunodeficiency virus type 1, while mature cells efficiently transmit both M- and T-tropic virus to T cells. *J Virol.* (1998) 72:2733–7.

104. Canque B, Bakri Y, Camus S, Yagello M, Benjouad A, Gluckman JC. The susceptibility to X4 and R5 human immunodeficiency virus-1 strains of dendritic cells derived *in vitro* from CD34$^+$ hematopoietic progenitor cells is primarily determined by their maturation stage. *Blood.* (1999) 93:3866–75.

105. Bakri Y, Schiffer C, Zennou V, Charneau P, Kahn E, Benjouad A, et al. The maturation of dendritic cells results in postintegration inhibition of HIV-1 replication. *J Immunol.* (2001) 166:3780–8. doi: 10.4049/jimmunol.166.6.3780

106. Granelli-Piperno A, Finkel V, Delgado E, Steinman RM. Virus replication begins in dendritic cells during the transmission of HIV-1 from mature dendritic cells to T cells. *Curr Biol.* (1999) 9:21–9. doi: 10.1016/S0960-9822(99)80043-8

107. Khatamzas E, Hipp MM, Gaughan D, Pichulik T, Leslie A, Fernandes RA, et al. Snapin promotes HIV-1 transmission from dendritic cells by dampening TLR8 signaling. *EMBO J.* (2017) 36:2998–3011. doi: 10.15252/embj.201695364

108. Wu L, KewalRamani VN. Dendritic-cell interactions with HIV: infection and viral dissemination. *Nat Rev Immunol.* (2006) 6:859–68. doi: 10.1038/nri1960

109. Goujon C, Arfi V, Pertel T, Luban J, Lienard J, Rigal D, et al. Characterization of simian immunodeficiency virus SIVSM/human immunodeficiency virus type 2 Vpx function in human myeloid cells. *J Virol.* (2008) 82:12335–45. doi: 10.1128/JVI.01181-08

110. Hrecka K, Hao C, Gierszewska M, Swanson SK, Kesik-Brodacka M, Srivastava S, et al. Vpx relieves inhibition of HIV-1 infection of macrophages mediated by the SAMHD1 protein. *Nature.* (2011) 474:658–61. doi: 10.1038/nature10195

111. Laguette N, Sobhian B, Casartelli N, Ringeard M, Chable-Bessia C, Segeral E, et al. SAMHD1 is the dendritic- and myeloid-cell-specific HIV-1 restriction factor counteracted by Vpx. *Nature.* (2011) 474:654–7. doi: 10.1038/nature10117

112. Ahn J, Hao C, Yan J, DeLucia M, Mehrens J, Wang C, et al. HIV/simian immunodeficiency virus (SIV) accessory virulence factor Vpx loads the host cell restriction factor SAMHD1 onto the E3 ubiquitin ligase complex CRL4DCAF1. *J Biol Chem.* (2012) 287:12550–8. doi: 10.1074/jbc.M112.340711

113. Schmidt S, Schenkova K, Adam T, Erikson E, Lehmann-Koch J, Sertel S, et al. SAMHD1's protein expression profile in humans. *J Leukoc Biol.* (2015) 98:5–14. doi: 10.1189/jlb.4HI0714-338RR

114. Antonucci JM, St. Gelais C, Wu L. The dynamic interplay between HIV-1, SAMHD1, and the innate antiviral response. *Front Immunol.* (2017) 8:1541. doi: 10.3389/fimmu.2017.01541

115. Goldstone DC, Ennis-Adeniran V, Hedden JJ, Groom HC, Rice GI, Christodoulou E, et al. HIV-1 restriction factor SAMHD1 is a deoxynucleoside triphosphate triphosphohydrolase. *Nature.* (2011) 480:379–82. doi: 10.1038/nature10623

116. Powell RD, Holland PJ, Hollis T, Perrino FW. Aicardi-Goutieres syndrome gene and HIV-1 restriction factor SAMHD1 is a dGTP-regulated deoxynucleotide triphosphohydrolase. *J Biol Chem.* (2011) 286:43596–600. doi: 10.1074/jbc.C111.317628

117. Lahouassa H, Daddacha W, Hofmann H, Ayinde D, Logue EC, Dragin L, et al. SAMHD1 restricts the replication of human immunodeficiency virus type 1 by depleting the intracellular pool of deoxynucleoside triphosphates. *Nat Immunol.* (2012) 13:223–8. doi: 10.1038/ni.2236

118. Ji X, Tang C, Zhao Q, Wang W, Xiong Y. Structural basis of cellular dNTP regulation by SAMHD1. *Proc Natl Acad Sci USA.* (2014) 111:E4305–4314. doi: 10.1073/pnas.1412289111

119. Koharudin LM, Wu Y, DeLucia M, Mehrens J, Gronenborn AM, Ahn J. Structural basis of allosteric activation of sterile alpha motif and histidine-aspartate domain-containing protein 1 (SAMHD1) by nucleoside triphosphates. *J Biol Chem.* (2014) 289:32617–27. doi: 10.1074/jbc.M114.591958

120. Miazzi C, Ferraro P, Pontarin G, Rampazzo C, Reichard P, Bianchi V. Allosteric regulation of the human and mouse deoxyribonucleotide triphosphohydrolase sterile alpha-motif/histidine-aspartate domain-containing protein 1 (SAMHD1). *J Biol Chem.* (2014) 289:18339–46. doi: 10.1074/jbc.M114.571091

121. Li Y, Kong J, Peng X, Hou W, Qin X, Yu XF. Structural insights into the high-efficiency catalytic mechanism of the sterile alpha-motif/histidine-aspartate domain-containing protein. *J Biol Chem.* (2015) 290:29428–37. doi: 10.1074/jbc.M115.663658

122. Cribier A, Descours B, Valadao AL, Laguette N, Benkirane M. Phosphorylation of SAMHD1 by cyclin A2/CDK1 regulates its restriction activity toward HIV-1. *Cell Rep.* (2013) 3:1036–43. doi: 10.1016/j.celrep.2013.03.017

123. Pauls E, Ruiz A, Badia R, Permanyer M, Gubern A, Riveira-Munoz E, et al. Cell cycle control and HIV-1 susceptibility are linked by CDK6-dependent CDK2 phosphorylation of SAMHD1 in myeloid and lymphoid cells. *J Immunol.* (2014) 193:1988–97. doi: 10.4049/jimmunol.1400873

124. St. Gelais C, de Silva S, Hach JC, White TE, Diaz-Griffero F, Yount JS, et al. Identification of cellular proteins interacting with the retroviral restriction factor SAMHD1. *J Virol.* (2014) 88:5834–44. doi: 10.1128/JVI.00155-14

125. Seamon KJ, Bumpus NN, Stivers JT. Single-stranded nucleic acids bind to the tetramer interface of SAMHD1 and prevent formation of the catalytic homotetramer. *Biochemistry.* (2016) 55:6087–99. doi: 10.1021/acs.biochem.6b00986

126. Mauney CH, Rogers LC, Harris RS, Daniel LW, Devarie-Baez NO, Wu H, et al. The SAMHD1 dNTP triphosphohydrolase is controlled by a redox switch. *Antioxid Redox Signal.* (2017) 27:1317–31. doi: 10.1089/ars.2016.6888

127. Kyei GB, Cheng X, Ramani R, Ratner L. Cyclin L2 is a critical HIV dependency factor in macrophages that controls SAMHD1 abundance. *Cell Host Microbe.* (2015) 17:98–106. doi: 10.1016/j.chom.2014.11.009

128. White TE, Brandariz-Nunez A, Martinez-Lopez A, Knowlton C, Lenzi G, Kim B, et al. A SAMHD1 mutation associated with Aicardi-Goutieres syndrome uncouples the ability of SAMHD1 to restrict HIV-1 from its ability to downmodulate type I interferon in humans. *Hum Mutat.* (2017) 38:658–68. doi: 10.1002/humu.23201

129. Gramberg T, Kahle T, Bloch N, Wittmann S, Mullers E, Daddacha W, et al. Restriction of diverse retroviruses by SAMHD1. *Retrovirology.* (2013) 10:26. doi: 10.1186/1742-4690-10-26

130. Chen Z, Zhu M, Pan X, Zhu Y, Yan H, Jiang T, et al. Inhibition of Hepatitis B virus replication by SAMHD1. *Biochem Biophys Res Commun.* (2014) 450:1462–8. doi: 10.1016/j.bbrc.2014.07.023

131. St. Gelais C, de Silva S, Amie SM, Coleman CM, Hoy H, Hollenbaugh JA, et al. SAMHD1 restricts HIV-1 infection in dendritic cells (DCs) by dNTP depletion, but its expression in DCs and primary CD4$^+$ T-lymphocytes cannot be upregulated by interferons. *Retrovirology.* (2012) 9:105. doi: 10.1186/1742-4690-9-105

132. Bloch N, O'Brien M, Norton TD, Polsky SB, Bhardwaj N, Landau NR. HIV type 1 infection of plasmacytoid and myeloid dendritic cells is restricted by high levels of SAMHD1 and cannot be counteracted by Vpx. *AIDS Res Hum Retroviruses.* (2014) 30:195–203. doi: 10.1089/aid.2013.0119

133. Beloglazova N, Flick R, Tchigvintsev A, Brown G, Popovic A, Nocek B, et al. Nuclease activity of the human SAMHD1 protein implicated in the Aicardi-Goutieres syndrome and HIV-1 restriction. *J Biol Chem.* (2013) 288:8101–10. doi: 10.1074/jbc.M112.431148

134. Ryoo J, Choi J, Oh C, Kim S, Seo M, Kim SY, et al. The ribonuclease activity of SAMHD1 is required for HIV-1 restriction. *Nat Med.* (2014) 20:936–41. doi: 10.1038/nm.3626

135. Antonucci JM, St. Gelais C, de Silva S, Yount JS, Tang C, Ji X, et al. SAMHD1-mediated HIV-1 restriction in cells does not involve ribonuclease activity. *Nat Med.* (2016) 22:1072. doi: 10.1038/nm.4163

136. Valle-Casuso JC, Allouch A, David A, Lenzi GM, Studdard L, Barre-Sinoussi F, et al. p21 restricts HIV-1 in monocyte-derived dendritic cells through the reduction of deoxynucleoside triphosphate biosynthesis and regulation of SAMHD1 antiviral activity. *J Virol.* (2017) 91:e01324–17. doi: 10.1128/jvi.01324-17

137. Liu WM, Scott KA, Thompson M, Dalgleish AG. Dendritic cell phenotype can be improved by certain chemotherapies and is associated with alterations to p21(waf1/cip1.). *Cancer Immunol Immunother.* (2013) 62:1553–61. doi: 10.1007/s00262-013-1456-0

138. Sheehy AM, Gaddis NC, Choi JD, Malim MH. Isolation of a human gene that inhibits HIV-1 infection and is suppressed by the viral Vif protein. *Nature.* (2002) 418:646–50. doi: 10.1038/nature00939

139. Richards CM, Li M, Perkins AL, Rathore A, Harki DA, Harris RS. Reassessing APOBEC3G inhibition by HIV-1 Vif-derived peptides. *J Mol Biol.* (2017) 429:88–96. doi: 10.1016/j.jmb.2016.11.012

140. Zheng Y-H, Irwin D, Kurosu T, Tokunaga K, Sata T, Matija Peterlin B. Human APOBEC3F is another host factor that blocks human immunodeficiency virus type 1 replication. *J Virol.* (2004) 78:6073–76. doi: 10.1128/JVI.78.11.6073-6076.2004

141. Ara A, Love RP, Follack TB, Ahmed KA, Adolph MB, Chelico L. Mechanism of enhanced HIV restriction by virion coencapsidated cytidine deaminases APOBEC3F and APOBEC3G. *J Virol.* (2017) 91:e02230–e02216. doi: 10.1128/JVI.02230-16

142. Greene WC, Debyser Z, Ikeda Y, Freed EO, Stephens E, Yonemoto W, et al. Novel targets for HIV therapy. *Antiviral Res.* (2008) 80:251–65. doi: 10.1016/j.antiviral.2008.08.003

143. Doehle BP, Schafer A, Cullen BR. Human APOBEC3B is a potent inhibitor of HIV-1 infectivity and is resistant to HIV-1 Vif. *Virology.* (2005) 339:281–8. doi: 10.1016/j.virol.2005.06.005

144. Guo F, Cen S, Niu M, Saadatmand J, Kleiman L. Inhibition of tRNA(3)(Lys)-primed reverse transcription by human APOBEC3G during human immunodeficiency virus type 1 replication. *J Virol.* (2006) 80:11710–22. doi: 10.1128/JVI.01038-06

145. Pan Y, Sun Z, Maiti A, Kanai T, Matsuo H, Li M, et al. Nanoscale Characterization of Interaction of APOBEC3G with RNA. *Biochemistry.* (2017) 56:1473–81. doi: 10.1021/acs.biochem.6b01189

146. Mbisa JL, Barr R, Thomas JA, Vandegraaff N, Dorweiler IJ, Svarovskaia ES, et al. Human immunodeficiency virus type 1 cDNAs produced in the presence of APOBEC3G exhibit defects in plus-strand DNA transfer and integration. *J Virol.* (2007) 81:7099–110. doi: 10.1128/JVI.00272-07

147. Donahue JP, Vetter ML, Mukhtar NA, D'Aquila RT. The HIV-1 Vif PPLP motif is necessary for human APOBEC3G binding and degradation. *Virology.* (2008) 377:49–53. doi: 10.1016/j.virol.2008.04.017

148. Pion M, Granelli-Piperno A, Mangeat B, Stalder R, Correa R, Steinman RM, et al. APOBEC3G/3F mediates intrinsic resistance of monocyte-derived dendritic cells to HIV-1 infection. *J Exp Med.* (2006) 203:2887–93. doi: 10.1084/jem.20061519

149. Pido-Lopez J, Whittall T, Wang Y, Bergmeier LA, Babaahmady K, Singh M, et al. Stimulation of cell surface CCR5 and CD40 molecules by their ligands or by HSP70 up-regulates APOBEC3G expression in CD4$^+$ T cells and dendritic cells. *J Immunol.* (2007) 178:1671–9. doi: 10.4049/jimmunol.178.3.1671

150. Stopak KS, Chiu YL, Kropp J, Grant RM, Greene WC. Distinct patterns of cytokine regulation of APOBEC3G expression and activity in primary lymphocytes, macrophages, and dendritic cells. *J Biol Chem.* (2007) 282:3539–46. doi: 10.1074/jbc.M610138200

151. Mohanram V, Skold AE, Bachle SM, Pathak SK, Spetz AL. IFN-alpha induces APOBEC3G, F, and A in immature dendritic cells and limits HIV-1 spread to CD4$^+$ T cells. *J Immunol.* (2013) 190:3346–53. doi: 10.4049/jimmunol.1201184

152. Wang FX, Huang J, Zhang H, Ma X. APOBEC3G upregulation by alpha interferon restricts human immunodeficiency virus type 1 infection in human peripheral plasmacytoid dendritic cells. *J Gen Virol.* (2008) 89 (Pt 3):722–30. doi: 10.1099/vir.0.83530-0

153. Oliva H, Pacheco R, Martinez-Navio JM, Rodriguez-Garcia M, Naranjo-Gomez M, Climent N, et al. Increased expression with differential subcellular location of cytidine deaminase APOBEC3G in human CD4$^+$ T-cell activation and dendritic cell maturation. *Immunol Cell Biol.* (2016) 94:689–700. doi: 10.1038/icb.2016.28

154. Stremlau M, Owens CM, Perron MJ, Kiessling M, Autissier P, Sodroski J. The cytoplasmic body component TRIM5alpha restricts HIV-1 infection in Old World monkeys. *Nature.* (2004) 427:848–53. doi: 10.1038/nature02343

155. Yap MW, Nisole S, Lynch C, Stoye JP. Trim5alpha protein restricts both HIV-1 and murine leukemia virus. *Proc Natl Acad Sci USA.* (2004) 101:10786–91. doi: 10.1073/pnas.0402876101

156. Wu X, Anderson JL, Campbell EM, Joseph AM, Hope TJ. Proteasome inhibitors uncouple rhesus TRIM5alpha restriction of HIV-1 reverse transcription and infection. *Proc Natl Acad Sci USA.* (2006) 103:7465–70. doi: 10.1073/pnas.0510483103

157. Black LR, Aiken C. TRIM5alpha disrupts the structure of assembled HIV-1 capsid complexes *in vitro. J Virol.* (2010) 84:6564–9. doi: 10.1128/JVI.00210-10

158. Portilho DM, Fernandez J, Ringeard M, Machado AK, Boulay A, Mayer M, et al. Endogenous TRIM5alpha function is regulated by SUMOylation and nuclear sequestration for efficient innate sensing in dendritic cells. *Cell Rep.* (2016) 14:355–69. doi: 10.1016/j.celrep.2015.12.039

159. Jimenez-Guardeno JM, Apolonia L, Betancor G, Malim MH. Immunoproteasome activation enables human TRIM5alpha restriction of HIV-1. *Nat Microbiol.* (2019) 4:933–40. doi: 10.1038/s41564-019-0402-0

160. Ribeiro CM, Sarrami-Forooshani R, Setiawan LC, Zijlstra-Willems EM, van Hamme JL, Tigchelaar W, et al. Receptor usage dictates HIV-1 restriction by human TRIM5alpha in dendritic cell subsets. *Nature.* (2016) 540:448–52. doi: 10.1038/nature20567

161. Calonge E, Bermejo M, Diez-Fuertes F, Mangeot I, Gonzalez N, Coiras M, et al. Different expression of interferon-stimulated genes in response to HIV-1 infection in dendritic cells based on their maturation state. *J Virol.* (2017) 91:e01379–e01316. doi: 10.1128/JVI.01379-16

162. Embgenbroich M, Burgdorf S. Current concepts of antigen cross-presentation. *Front Immunol.* (2018) 9:1643. doi: 10.3389/fimmu.2018.01643

163. van Montfoort N, van der Aa E, Woltman AM. Understanding MHC class I presentation of viral antigens by human dendritic cells as a basis for rational design of therapeutic vaccines. *Front Immunol.* (2014) 5:182. doi: 10.3389/fimmu.2014.00182

164. Buseyne F, Gall SL, Boccaccio C, Abastado J-P, Lifson JD, Arthur LO, et al. MHC-I–restricted presentation of HIV-1 virion antigens without viral replication. *Nat Med.* (2001) 7:344–9. doi: 10.1038/85493

165. Huang XL, Fan Z, Colleton BA, Buchli R, Li H, Hildebrand WH, et al. Processing and presentation of exogenous HLA class I peptides by dendritic cells from human immunodeficiency virus type 1-infected persons. *J Virol.* (2005) 79:3052–62. doi: 10.1128/JVI.79.5.3052-3062.2005

166. Menager MM, Littman DR. Actin dynamics regulates dendritic cell-mediated transfer of HIV-1 to T cells. *Cell.* (2016) 164:695–709. doi: 10.1016/j.cell.2015.12.036

167. Menager MM. TSPAN7, effector of actin nucleation required for dendritic cell-mediated transfer of HIV-1 to T cells. *Biochem Soc Trans.* (2017) 45:703–8. doi: 10.1042/BST20160439

168. Thali M. Tetraspanin functions during HIV-1 and influenza virus replication. *Biochem Soc Trans.* (2011) 39:529–31. doi: 10.1042/BST0390529

169. Krementsov DN, Weng J, Lambele M, Roy NH, Thali M. Tetraspanins regulate cell-to-cell transmission of HIV-1. *Retrovirology.* (2009) 6:64. doi: 10.1186/1742-4690-6-64

170. Rocha-Perugini V, Martinez Del Hoyo G, Gonzalez-Granado JM, Ramirez-Huesca M, Zorita V, Rubinstein E, et al. CD9 regulates major histocompatibility complex class II trafficking in monocyte-derived dendritic cells. *Mol Cell Biol.* (2017) 37:e00202–e00217. doi: 10.1128/MCB.00202-17

171. Rocha-Perugini V, Suarez H, Alvarez S, Lopez-Martin S, Lenzi GM, Vences-Catalan F, et al. CD81 association with SAMHD1 enhances HIV-1 reverse transcription by increasing dNTP levels. *Nat Microbiol.* (2017) 2:1513–22. doi: 10.1038/s41564-017-0019-0

172. Fu E, Pan L, Xie Y, Mu D, Liu W, Jin F, et al. Tetraspanin CD63 is a regulator of HIV-1 replication. *Int J Clin Exp Pathol.* (2015) 8:1184–98.

173. Li G, Dziuba N, Friedrich B, Murray JL, Ferguson MR. A post-entry role for CD63 in early HIV-1 replication. *Virology.* (2011) 412:315–24. doi: 10.1016/j.virol.2011.01.017

174. Colombo M, Moita C, van Niel G, Kowal J, Vigneron J, Benaroch P, et al. Analysis of ESCRT functions in exosome biogenesis, composition and secretion highlights the heterogeneity of extracellular vesicles. *J Cell Sci.* (2013) 126 (Pt 24):5553–65. doi: 10.1242/jcs.128868

175. McDonald B, Martin-Serrano J. No strings attached: the ESCRT machinery in viral budding and cytokinesis. *J Cell Sci.* (2009) 122 (Pt 13):2167–77. doi: 10.1242/jcs.028308

176. Lippincott-Schwartz J, Freed EO, van Engelenburg SB. A consensus view of ESCRT-mediated human immunodeficiency virus type 1 abscission. *Annu Rev Virol.* (2017) 4:309–25. doi: 10.1146/annurev-virology-101416-041840

177. Hurley JH, Cada AK. Inside job: how the ESCRTs release HIV-1 from infected cells. *Biochem Soc Trans.* (2018) 46:1029–36. doi: 10.1042/BST20180019

178. Choudhuri K, Llodra J, Roth EW, Tsai J, Gordo S, Wucherpfennig KW, et al. Polarized release of T-cell-receptor-enriched microvesicles at the immunological synapse. *Nature.* (2014) 507:118–23. doi: 10.1038/nature12951

179. Strickland M, Ehrlich LS, Watanabe S, Khan M, Strub MP, Luan CH, et al. Tsg101 chaperone function revealed by HIV-1 assembly inhibitors. *Nat Commun.* (2017) 8:1391. doi: 10.1038/s41467-017-01426-2

180. Chaturbhuj D, Patil A, Gangakhedkar R. PYRE insertion within HIV-1 subtype C p6-Gag functions as an ALIX-dependent late domain. *Sci Rep.* (2018) 8:8917. doi: 10.1038/s41598-018-27162-1

181. Siarot L, Chutiwitoonchai N, Sato H, Chang H, Fujino M, Murakami T, et al. Identification of human immunodeficiency virus type-1 Gag-TSG101 interaction inhibitors by high-throughput screening. *Biochem Biophys Res Commun.* (2018) 503:2970–6. doi: 10.1016/j.bbrc.2018.08.079

182. ten Broeke T, Wubbolts R, Stoorvogel W. MHC class II antigen presentation by dendritic cells regulated through endosomal sorting. *Cold Spring Harb Perspect Biol.* (2013) 5:a016873. doi: 10.1101/cshperspect.a016873

183. Turville SG, Santos JJ, Frank I, Cameron PU, Wilkinson J, Miranda-Saksena M, et al. Immunodeficiency virus uptake, turnover, and 2-phase transfer in human dendritic cells. *Blood.* (2004) 103:2170–9. doi: 10.1182/blood-2003-09-3129

184. Soares H, Lasserre R, Alcover A. Orchestrating cytoskeleton and intracellular vesicle traffic to build functional immunological synapses. *Immunol Rev.* (2013) 256:118–32. doi: 10.1111/imr.12110

185. Pettmann J, Santos AM, Dushek O, Davis SJ. Membrane ultrastructure and T cell activation. *Front Immunol.* (2018) 9:2152. doi: 10.3389/fimmu.2018.02152

186. Saiz ML, Rocha-Perugini V, Sanchez-Madrid F. Tetraspanins as organizers of antigen-presenting cell function. *Front Immunol.* (2018) 9:1074. doi: 10.3389/fimmu.2018.01074

187. Scales HE, Meehan GR, Hayes AJ, Benson RA, Watson E, Walters A, et al. A novel cellular pathway of antigen presentation and CD4 T cell activation *in vivo. Front Immunol.* (2018) 9:2684. doi: 10.3389/fimmu.2018.02684

188. Martin-Gayo E, Cole MB, Kolb KE, Ouyang Z, Cronin J, Kazer SW, et al. A reproducibility-based computational framework identifies an inducible, enhanced antiviral state in dendritic cells from HIV-1 elite controllers. *Genome Biol.* (2018) 19:10. doi: 10.1186/s13059-017-1385-x

189. Martin-Gayo E, Buzon MJ, Ouyang Z, Hickman T, Cronin J, Pimenova D, et al. Potent cell-intrinsic immune responses in dendritic cells facilitate HIV-1-specific t cell immunity in HIV-1 elite controllers. *PLoS Pathog.* (2015) 11:e1004930. doi: 10.1371/journal.ppat.1004930

190. Garcia-Bates TM, Palma ML, Shen C, Gambotto A, Macatangay BJC, Ferris RL, et al. Contrasting roles of the PD-1 signaling pathway in dendritic cell-mediated induction and regulation of HIV-1-specific effector T cell functions. *J Virol.* (2019) 93:e02035–18. doi: 10.1128/JVI.02035-18

191. Carranza P, Del Rio Estrada PM, Diaz Rivera D, Ablanedo-Terrazas Y, Reyes-Teran G. Lymph nodes from HIV-infected individuals harbor mature dendritic cells and increased numbers of PD-L1+ conventional dendritic cells. *Hum Immunol.* (2016) 77:584–93. doi: 10.1016/j.humimm.2016.05.019

192. Tsunetsugu-Yokota Y, Yasuda S, Sugimoto A, Yagi T, Azuma M, Yagita H, et al. Efficient virus transmission from dendritic cells to CD4+ T cells in response to antigen depends on close contact through adhesion molecules. *Virology.* (1997) 239:259–68. doi: 10.1006/viro.1997.8895

193. McDonald D, Wu L, Bohks SM, KewalRamani VN, Unutmaz D, Hope TJ. Recruitment of HIV and its receptors to dendritic cell-T cell junctions. *Science.* (2003) 300:1295–7. doi: 10.1126/science.1084238

194. Bromley SK, Burack WR, Johnson KG, Somersalo K, Sims TN, Sumen C, et al. The immunological synapse. *Annu Rev Immunol.* (2001) 19:375–96. doi: 10.1146/annurev.immunol.19.1.375

195. Garcia E, Pion M, Pelchen-Matthews A, Collinson L, Arrighi JF, Blot G, et al. HIV-1 trafficking to the dendritic cell-T-cell infectious synapse uses a pathway of tetraspanin sorting to the immunological synapse. *Traffic.* (2005) 6:488–501. doi: 10.1111/j.1600-0854.2005.00293.x

196. Arrighi JF, Pion M, Garcia E, Escola JM, van Kooyk Y, Geijtenbeek TB, et al. DC-SIGN-mediated infectious synapse formation enhances X4 HIV-1 transmission from dendritic cells to T cells. *J Exp Med.* (2004) 200:1279–88. doi: 10.1084/jem.20041356

197. Arrighi JF, Pion M, Wiznerowicz M, Geijtenbeek TB, Garcia E, Abraham S, et al. Lentivirus-mediated RNA interference of DC-SIGN expression inhibits human immunodeficiency virus transmission from dendritic cells to T cells. *J Virol.* (2004) 78:10848–55. doi: 10.1128/JVI.78.20.10848-10855.2004

198. Wu L, Martin TD, Han YC, Breun SK, KewalRamani VN. Trans-dominant cellular inhibition of DC-SIGN-mediated HIV-1 transmission. *Retrovirology.* (2004) 1:14. doi: 10.1186/1742-4690-1-14

199. Prasad A, Kulkarni R, Jiang S, Groopman JE. Cocaine enhances DC to T-cell HIV-1 transmission by activating DC-SIGN/LARG/LSP1 complex and facilitating infectious synapse formation. *Sci Rep.* (2017) 7:40648. doi: 10.1038/srep40648

200. Rappocciolo G, Jais M, Piazza P, Reinhart TA, Berendam SJ, Garcia-Exposito L, et al. Alterations in cholesterol metabolism restrict HIV-1 trans infection in nonprogressors. *mBio.* (2014) 5:e01031–e01013. doi: 10.1128/mBio.01031-13

201. Yamamoto T, Tsunetsugu-Yokota Y, Mitsuki YY, Mizukoshi F, Tsuchiya T, Terahara K, et al. Selective transmission of R5 HIV-1 over X4 HIV-1 at the dendritic cell-T cell infectious synapse is determined by the T cell activation state. *PLoS Pathog.* (2009) 5:e1000279. doi: 10.1371/journal.ppat.1000279

202. Wang JH, Janas AM, Olson WJ, Wu L. Functionally distinct transmission of human immunodeficiency virus type 1 mediated by immature and mature dendritic cells. *J Virol.* (2007) 81:8933–43. doi: 10.1128/JVI.00878-07

203. Dong C, Janas AM, Wang JH, Olson WJ, Wu L. Characterization of human immunodeficiency virus type 1 replication in immature and mature dendritic cells reveals dissociable cis- and trans-infection. *J Virol.* (2007) 81:11352–62. doi: 10.1128/JVI.01081-07

204. Kim JT, Chang E, Sigal A, Baltimore D. Dendritic cells efficiently transmit HIV to T Cells in a tenofovir and raltegravir insensitive manner. *PLoS ONE.* (2018) 13:e0189945. doi: 10.1371/journal.pone.0189945

205. Wiley RD, Gummuluru S. Immature dendritic cell-derived exosomes can mediate HIV-1 trans infection. *Proc Natl Acad Sci USA.* (2006) 103:738–43. doi: 10.1073/pnas.0507995103

206. Denzer K, Kleijmeer MJ, Heijnen HF, Stoorvogel W, Geuze HJ. Exosome: from internal vesicle of the multivesicular body to intercellular signaling device. *J Cell Sci.* (2000) 113 (Pt 19):3365–74.

207. Gauvreau ME, Cote MH, Bourgeois-Daigneault MC, Rivard LD, Xiu F, Brunet A, et al. Sorting of MHC class II molecules into exosomes through a ubiquitin-independent pathway. *Traffic.* (2009) 10:1518–27. doi: 10.1111/j.1600-0854.2009.00948.x

208. Kulkarni R, Prasad A. Exosomes derived from HIV-1 infected DCs mediate viral trans-infection via fibronectin and galectin-3. *Sci Rep.* (2017) 7:14787. doi: 10.1038/s41598-017-14817-8

209. Chiozzini C, Arenaccio C, Olivetta E, Anticoli S, Manfredi F, Ferrantelli F, et al. Trans-dissemination of exosomes from HIV-1-infected cells fosters both HIV-1 trans-infection in resting CD4$^+$ T lymphocytes and reactivation of the HIV-1 reservoir. *Arch Virol.* (2017) 162:2565–77. doi: 10.1007/s00705-017-3391-4

210. Barclay RA, Schwab A, DeMarino C, Akpamagbo Y, Lepene B, Kassaye S, et al. Exosomes from uninfected cells activate transcription of latent HIV-1. *J Biol Chem.* (2017) 292:11682–701. doi: 10.1074/jbc.M117.793521

211. Gould SJ, Booth AM, Hildreth JE. The Trojan exosome hypothesis. *Proc Natl Acad Sci USA.* (2003) 100:10592–7. doi: 10.1073/pnas.1831413100

212. Aloia RC, Tian H, Jensen FC. Lipid composition and fluidity of the human immunodeficiency virus envelope and host cell plasma membranes. *Proc Natl Acad Sci USA.* (1993) 90:5181–5. doi: 10.1073/pnas.90.11.5181

213. Garcia E, Nikolic DS, Piguet V. HIV-1 replication in dendritic cells occurs through a tetraspanin-containing compartment enriched in AP-3. *Traffic.* (2008) 9:200–14. doi: 10.1111/j.1600-0854.2007.00678.x

214. Arthur LO, Bess JW Jr, Sowder RC II, Benveniste RE, Mann DL, Chermann JC, et al. Cellular proteins bound to immunodeficiency viruses: implications for pathogenesis and vaccines. *Science.* (1992) 258:1935–8. doi: 10.1126/science.1470916

215. Thery C, Boussac M, Veron P, Ricciardi-Castagnoli P, Raposo G, Garin J, et al. Proteomic analysis of dendritic cell-derived exosomes: a secreted subcellular compartment distinct from apoptotic vesicles. *J Immunol.* (2001) 166:7309–18. doi: 10.4049/jimmunol.166.12.7309

216. Garrus JE, von Schwedler UK, Pornillos OW, Morham SG, Zavitz KH, Wang HE, et al. Tsg101 and the vacuolar protein sorting pathway are essential for HIV-1 budding. *Cell.* (2001) 107:55–65. doi: 10.1016/S0092-8674(01)00506-2

217. Pang S, Yu D, An DS, Baldwin GC, Xie Y, Poon B, et al. Human immunodeficiency virus Env-independent infection of human CD4$^−$ cells. *J Virol.* (2000) 74:10994–1000. doi: 10.1128/JVI.74.23.10994-11000.2000

218. Chow YH, Yu D, Zhang JY, Xie Y, Wei OL, Chiu C, et al. gp120-Independent infection of CD4$^−$ epithelial cells and CD4$^+$ T-cells by HIV-1. *J Acquir Immune Defic Syndr.* (2002) 30:1–8. doi: 10.1097/00042560-200205010-00001

219. Stoorvogel W, Kleijmeer MJ, Geuze HJ, Raposo G. The biogenesis and functions of exosomes. *Traffic.* (2002) 3:321–30. doi: 10.1034/j.1600-0854.2002.30502.x

220. Thery C, Zitvogel L, Amigorena S. Exosomes: composition, biogenesis and function. *Nat Rev Immunol.* (2002) 2:569–79. doi: 10.1038/nri855

221. de Carvalho JV, de Castro RO, da Silva EZ, Silveira PP, da Silva-Januario ME, Arruda E, et al. Nef neutralizes the ability of exosomes from CD4$^+$ T cells to act as decoys during HIV-1 infection. *PLoS ONE.* (2014) 9:e113691. doi: 10.1371/journal.pone.0113691

222. Naslund TI, Paquin-Proulx D, Paredes PT, Vallhov H, Sandberg JK, Gabrielsson S. Exosomes from breast milk inhibit HIV-1 infection of dendritic cells and subsequent viral transfer to CD4$^+$ T cells. *AIDS.* (2014) 28:171–80. doi: 10.1097/QAD.0000000000000159

223. Dias MVS, Costa CS, daSilva LLP. The ambiguous roles of extracellular vesicles in HIV replication and pathogenesis. *Front Microbiol.* (2018) 9:2411. doi: 10.3389/fmicb.2018.02411

224. Mfunyi CM, Vaillancourt M, Vitry J, Nsimba Batomene TR, Posvandzic A, Lambert AA, et al. Exosome release following activation of the dendritic cell

225. Popov S, Chenine AL, Gruber A, Li PL, Ruprecht RM. Long-term productive human immunodeficiency virus infection of CD1a-sorted myeloid dendritic cells. *J Virol.* (2005) 79:602–8. doi: 10.1128/JVI.79.1.602-608.2005

226. Sol-Foulon N, Moris A, Nobile C, Boccaccio C, Engering A, Abastado JP, et al. HIV-1 Nef-induced upregulation of DC-SIGN in dendritic cells promotes lymphocyte clustering and viral spread. *Immunity.* (2002) 16:145–55. doi: 10.1016/S1074-7613(02)00260-1

227. St. Gelais C, Coleman CM, Wang JH, Wu L. HIV-1 Nef enhances dendritic cell-mediated viral transmission to CD4$^+$ T cells and promotes T-cell activation. *PLoS ONE.* (2012) 7:e34521. doi: 10.1371/journal.pone.0034521

228. Coleman CM, Spearman P, Wu L. Tetherin does not significantly restrict dendritic cell-mediated HIV-1 transmission and its expression is upregulated by newly synthesized HIV-1 Nef. *Retrovirology.* (2011) 8:26. doi: 10.1186/1742-4690-8-26

229. Neil SJ, Zang T, Bieniasz PD. Tetherin inhibits retrovirus release and is antagonized by HIV-1 Vpu. *Nature.* (2008) 451:425–30. doi: 10.1038/nature06553

230. Perez-Caballero D, Zang T, Ebrahimi A, McNatt MW, Gregory DA, Johnson MC, et al. Tetherin inhibits HIV-1 release by directly tethering virions to cells. *Cell.* (2009) 139:499–511. doi: 10.1016/j.cell.2009.08.039

231. Venkatesh S, Bieniasz PD. Mechanism of HIV-1 virion entrapment by tetherin. *PLoS Pathog.* (2013) 9:e1003483. doi: 10.1371/journal.ppat.1003483

232. Kuhl BD, Sloan RD, Donahue DA, Bar-Magen T, Liang C, Wainberg MA. Tetherin restricts direct cell-to-cell infection of HIV-1. *Retrovirology.* (2010) 7:115. doi: 10.1186/1742-4690-7-115

233. Turville SG, Aravantinou M, Stossel H, Romani N, Robbiani M. Resolution of de novo HIV production and trafficking in immature dendritic cells. *Nat Methods.* (2008) 5:75–85. doi: 10.1038/nmeth1137

234. Cote SC, Plante A, Tardif MR, Tremblay MJ. Dectin-1/TLR2 and NOD2 agonists render dendritic cells susceptible to infection by X4-using HIV-1 and promote cis-infection of CD4$^+$ T cells. *PLoS ONE.* (2013) 8:e67735. doi: 10.1371/journal.pone.0067735

235. Huber A, Dammeijer F, Aerts J, Vroman H. Current state of dendritic cell-based immunotherapy: opportunities for *in vitro* antigen loading of different DC subsets? *Front Immunol.* (2018) 9:2804. doi: 10.3389/fimmu.2018.02804

236. Tel J, Aarntzen EH, Baba T, Schreibelt G, Schulte BM, Benitez-Ribas D, et al. Natural human plasmacytoid dendritic cells induce antigen-specific T-cell responses in melanoma patients. *Cancer Res.* (2013) 73:1063–75. doi: 10.1158/0008-5472.CAN-12-2583

237. Prue RL, Vari F, Radford KJ, Tong H, Hardy MY, D'Rozario R, et al. A phase I clinical trial of CD1c (BDCA-1)$^+$ dendritic cells pulsed with HLA-A*0201 peptides for immunotherapy of metastatic hormone refractory prostate cancer. *J Immunother.* (2015) 38:71–6. doi: 10.1097/CJI.0000000000000063

238. Schreibelt G, Bol KF, Westdorp H, Wimmers F, Aarntzen EH, Duivemande Boer T, et al. Effective clinical responses in metastatic melanoma patients after vaccination with primary myeloid dendritic cells. *Clin Cancer Res.* (2016) 22:2155–66. doi: 10.1158/1078-0432.CCR-15-2205

239. Porkolab V, Chabrol E, Varga N, Ordanini S, Sutkeviciu Te I, Thepaut M, et al. Rational-differential design of highly specific glycomimetic ligands: targeting DC-SIGN and excluding langerin recognition. *ACS Chem Biol.* (2018) 13:600–8. doi: 10.1021/acschembio.7b00958

240. Ochyl LJ, Moon JJ. Dendritic cell membrane vesicles for activation and maintenance of antigen-specific T cells. *Adv Healthc Mater.* (2019) 8:e1801091. doi: 10.1002/adhm.201801091

immunoreceptor: a potential role in HIV-1 pathogenesis. *Virology.* (2015) 484:103–12. doi: 10.1016/j.virol.2015.05.013

2

What's in a name? Some early and current issues in dendritic cell nomenclature

David Vremec[1] and Ken Shortman[1,2,3]*

[1] *The Walter and Eliza Hall Institute, Melbourne, VIC, Australia,* [2] *Department of Medical Biology, The University of Melbourne, Melbourne, VIC, Australia,* [3] *Burnet Institute, Melbourne, VIC, Australia*

Keywords: dendritic cells, DC subsets, monocytes, macrophages, nomenclature

Edited by:
Martin Guilliams,
Ghent University – VIB, Belgium

Reviewed by:
Elodie Segura,
Institut Curie, France
Steffen Jung,
Weizmann Institute of Science, Israel

*Correspondence:
Ken Shortman
shortman@wehi.edu.au

The name dendritic cell (DC) was given by Steinman to describe the unusual cell type he saw in spleen cell suspensions. This morphological description is not sufficient to specify the cell of so much interest to immunologists; many cells can adopt a similar form. A useful functional definition evolved as Steinman and colleagues explored the immunological properties of this novel cell type (1). DCs were considered as antigen collecting and processing cells able to present antigen on MHC molecules and efficiently activate even primary T-cells. Nowadays, immunologists would likely add to this definition, a capacity to sense the context in which the antigen was collected, via receptors for pathogen or damaged cell-derived material. Why might we need to go beyond the name "dendritic cell" for cells with these well-understood functions? Some limitations of this single name arose early in DC research. This article surveys some problems of definition encountered in past work from our own laboratory. The problems we encountered arose from two sources, the first the discovery of different DC subsets and the need to determine whether these represented different maturation states or separate sub-lineages. The second was the difficulty in distinguishing these DC subsets from macrophages.

Our first hint that there could be distinct types of DCs came from our studies with Wu and Ardavin on thymic T and DC development (2). We were surprised to find that a high proportion of mouse thymic DCs stained with antibodies against characteristic T-cell markers, such as CD8α; it was a relief to find they did not stain with antibodies against CD3 or the T-cell receptor! Pickup of material from thymocytes was eliminated as an explanation. We then found a similar but less frequent DC subset staining for surface CD8α among the DCs in mouse spleen and these DCs were shown to express mRNA for CD8α (3). Others had already reported some staining of DCs with anti-CD8; our work emphasized that these CD8$^+$ DCs were a distinct population, CD8α expression being positively correlated with expression of DEC205 but inversely correlated with expression of other markers such as CD4, CD11b, and, as illustrated in **Figure 1**, SIRPα (4, 5).

Immunological interest in the CD8$^+$ and CD8$^-$ DC subsets increased when it became apparent from the work of many laboratories that these DCs differed in immunological functions. Differences were apparent in the expression of toll-like and other microbial pattern recognition receptors, in the cytokines produced on activation, in the fate of the T-cells they stimulated, in their capacity to phagocytize dead cells, and in the processing of antigens for MHC class I versus MHC class II presentation [reviewed in Ref. (7)]. The key findings from our laboratory were that the CD8$^+$ DCs, when appropriately stimulated, were the most potent producers of IL12p70 (8), and that the CD8$^+$ DCs have a strong bias to cross-presenting exogenous antigens, both soluble and particulate, for MHC class I presentation (9, 10).

An important issue became whether these functionally distinct DC types represented different lineages, or were simply different maturation states within one very plastic lineage. There was direct evidence, confirmed by us, that some CD8$^-$ DCs could on adoptive transfer, produce CD8$^+$ DCs. However, these CD8$^+$ DCs proved to be generated from a small number of early members of the

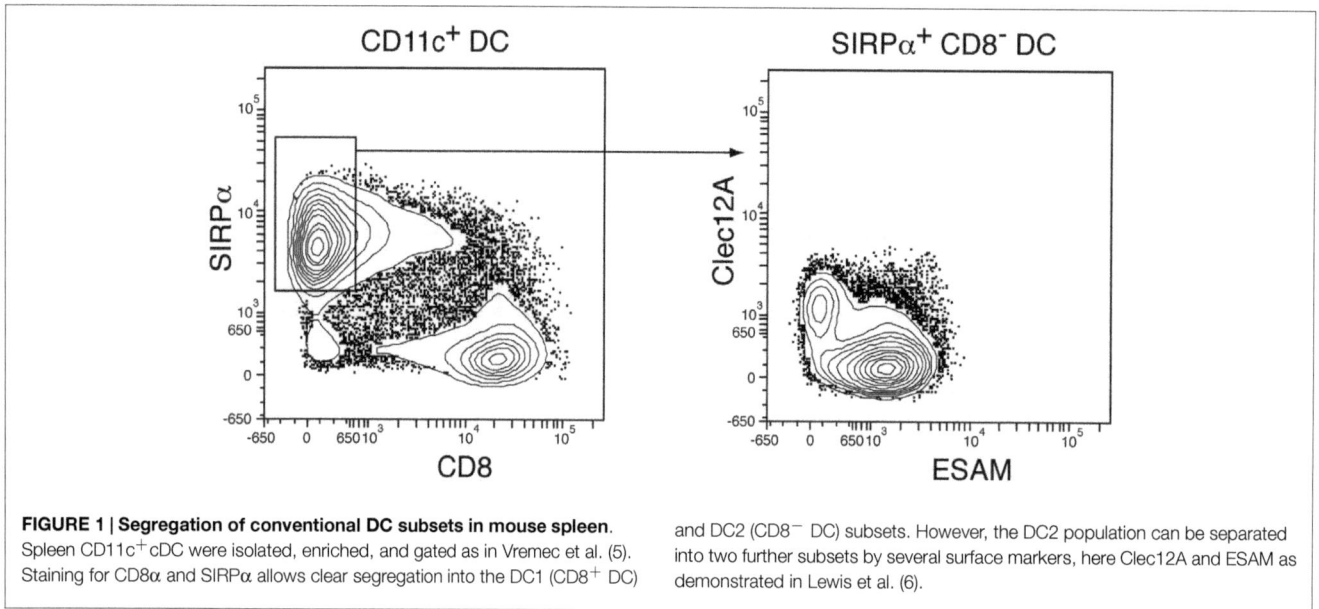

FIGURE 1 | Segregation of conventional DC subsets in mouse spleen. Spleen CD11c$^+$cDC were isolated, enriched, and gated as in Vremec et al. (5). Staining for CD8α and SIRPα allows clear segregation into the DC1 (CD8$^+$ DC) and DC2 (CD8$^-$ DC) subsets. However, the DC2 population can be separated into two further subsets by several surface markers, here Clec12A and ESAM as demonstrated in Lewis et al. (6).

CD8$^+$ DC lineage that had not yet acquired CD8α expression; the bulk of CD8$^-$ DCs did not give rise to CD8$^+$ DCs (11). Although sequential maturation states were found within the CD8$^+$ DC lineage, with early forms lacking CD8α expression (12) and the earliest stages lacking capacity for antigen cross-presentation (13), there was a clear developmental separation from the bulk of CD8$^-$ DCs. A further distinct DC type in mouse spleen became evident when the mouse equivalent of the human type 1 interferon-producing plasmacytoid dendritic cell (pDC) was identified (14). Although some pDCs expressed CD8α (15), they were clearly a separate lineage from the CD8$^+$ conventional DCs (cDCs). Our subsequent work with Naik showed that spleen CD8$^+$ DCs and CD8$^-$ DCs represented separate cDC sub-lineages derived via pre-DC populations from a bone marrow pro-DC or common dendritic cell precursor (CDP) (16–18). Thus, because of differences in surface phenotype, immunological functions, and developmental pathways, these two spleen cDC populations had to be distinguished, and the terms CD8$^+$ cDC and CD8$^-$ cDC became current.

In line with the pioneering work of Salomon et al. (19) and Anjuere et al. (20), we extended our analysis of DC subtypes from mouse spleen to mouse lymph nodes (LNs) (4, 21). Spleen should contain only what we termed the "lymphoid tissue resident" cDCs whereas LNs should contain both these and the "migratory" cDC type arriving via the lymph from other tissues. It was immediately apparent that the level of DC heterogeneity was greater than in spleen. One source of heterogeneity was the existence of different activation states within even one DC lineage. In particular, the DC that had migrated even in steady state from peripheral tissues such as skin into LNs were more activated than those remaining in skin, and more activated than their lymphoid tissue resident counterparts; the DCs that had migrated expressed higher surface levels of MHC class II and of co-stimulator molecules such as CD86. Although they were first called "mature" DCs they proved to be not fully activated but "semi-mature"; they were not producing

cytokines and were likely tolerogenic (22). A similar transformation termed "spontaneous activation" occurred when spleen cDCs were isolated and placed in culture (23). In both cases, further signals, such as given by microbial products interacting with TLR ligands, were required to produce a fully active, cytokine secreting immunogenic DC. However, even when these different activation states were considered, further cDC subsets not found in spleen were apparent, such as the epidermal Langerhans cell-derived LN DCs. The full lineage complexity of LN has now been well delineated by other laboratories, a notable finding being the existence of a migratory form of the CD8$^+$ DC lineage but lacking CD8α expression, commonly termed as the CD103$^+$ cDCs (24–26).

Our second problem with DC nomenclature arose as we attempted to distinguish DCs from macrophages, a particularly difficult exercise in inflamed tissues. It was also difficult to relate the DC populations we isolated from steady state mouse spleen with the DCs produced by culture of monocytes with granulocyte-macrophage colony stimulating factor (GM-CSF), a well-established model of DC generation (27). At that time, it was generally assumed that all DCs and macrophages would be bone marrow derived cells, with monocytes as the common late precursor. Some questioned whether it was valid even to consider DCs as a separate entity rather than as a macrophage variant (28). We had some sympathy with this view, since in experiments with Metcalf we had difficulty in segregating DCs from macrophages in the peritoneal fluid from mice expressing high levels of GM-CSF (29). Although cells with DC function and surface phenotype could be segregated from macrophages at the extremes of the distribution, there appeared to be a continuum of phenotypes rather than two discrete populations. For us the clarification came when, with Naik, the immediate precursor of the spleen cDC was isolated and shown to be distinct from monocytes and unable to produce macrophages (16). We termed these pre-DCs. This led to the view that there were two different routes to cells with DC antigen presenting function, one via monocytes and more often found under conditions of inflammation, the other derived from CDP/pro-DC precursors in bone marrow then via pre-DC to the types of DC found in steady state lymphoid tissue (17). The culture model finally developed for generation of the type of DCs found in steady state became bone marrow stimulated with Flt3 ligand, rather that with GM-CSF (30, 31). Thus, the developmental pathway leading to DC functions became a major criterion for segregating and naming DC subtypes.

It was then possible to segregate DCs derived from monocytes from the cDCs found in steady state spleen. However, it is evident from the account above that our previous nomenclature of the subsets of spleen cDCs based on CD8α expression was inadequate. Certain pDC subsets also expressed CD8α. Early DCs of "CD8$^+$ cDC" lineage in spleen did not express CD8α. The migratory version of the same lineage, the CD103$^+$ DCs, did not express it. And finally, CD8 was not expressed by human DCs. A major advance was the demonstration in several laboratories of an equivalent of the mouse "CD8$^+$ cDC" lineage within human DCs, and the finding that the chemokine receptor XCR1 and the C-type lectin-like molecule Clec9A, rather than CD8, served as common DC surface markers crossing this species barrier [commentary in Ref. (32)]. The proposed designation of this DC subtype as DC1 overcomes the previous nomenclature problems (33).

In contrast to these advances in understanding the DC1 subset, the CD8$^-$ CD11b^{high} SIRPα^{high} cDCs (designated as DC2) have been less studied and still present nomenclature issues. We had already separated spleen CD8$^-$ DCs into two subsets based on CD4 expression (5), but the significance of this remains obscure. A more meaningful separation can now be made based on surface expression of Clec12A (DCAL2, MICL) versus DCIR2 or ESAM (6, 34, 35). An example of such segregation is shown in **Figure 1**. Importantly, these DC subsets differ in both developmental requirements and immunological characteristics; formation of DCIR2$^+$ ESAM^{high} Clec12A$^-$ DCs requires Notch2 signaling and this subset selectively responds to flagellin and induces Th2 responses. Will these differences demand a further division into DC2 and DC3 subtypes? Or will one of these, particularly the

Clec12A$^+$ subset, prove to be part of the monocyte-derived group? These questions require further work.

It is notable that ontogeny has led to a better understanding and provided one logical basis for DC classification (33). Will ontogeny be the best guide for DC nomenclature in future? We can foresee one area where it may cause confusion. A proportion of mouse pDCs and the CD8α-expressing subset of cDCs in the mouse thymus have a potential route of development from lymphoid rather than myeloid precursors (36, 37). These DC types have D–J rearrangements in their Ig heavy chain genes, a characteristic of lymphoid-origin cells (38). The extent to which a lymphoid route contributes to their development in steady state is still unclear, but the potential is there. Yet, the thymic CD8$^+$ DCs are similar to the splenic CD8$^+$ DCs of myeloid origin, and pDCs developing from myeloid or lymphoid precursors have similar surface phenotype and immunological functions. Should they have separate names according to their developmental origin, or should this "convergent" development lead to cells with the same name? There may yet be fine differences in function that eventually will be important to specify, but at present they are called by the same name. One resolution of this paradox comes from the likelihood that, despite the differences in bone marrow precursor surface markers, a common molecular program for pDC or for CD8$^+$ cDC formation has been initiated, with transcription factors that override any previous precursor orientation. Considering ontological origin in terms of these final molecular programs, rather than by the surface markers on the precursor cells, should overcome the paradox resulting from apparent convergent differentiation.

Acknowledgments

Our research was supported by the National Health and Medical Research Council, Australia and was made possible through the Victorian State Government Operational Infrastructure Support and Australian Government NHMRC IRIISS.

References

1. Steinman RM. Decisions about dendritic cells: past, present and future. *Annu Rev Immunol* (2011) **30**:1–22. doi:10.1146/annurev-immunol-100311-102839

2. Shortman K, Wu L, Ardavin C, Vremec D, Stozik F, Winkel K, et al. Thymic dendritic cells: surface phenotype, developmental origin and function. In: Banchereau J, Schmitt D, editors. *Dendritic Cells in Fundamental and Clinical Immunology*. New York, NY: Plenum Publishing Corporation (1995). p. 21–9.

3. Vremec D, Zorbas M, Scollay R, Saunders DJ, Ardavin CF, Wu L, et al. The surface phenotype of dendritic cells purified from mouse thymus and spleen: investigation of the CD8 expression by a subpopulation of dendritic cells. *J Exp Med* (1992) **176**:47–58. doi:10.1084/jem.176.1.47

4. Vremec D, Shortman K. Dendritic cell subtypes in mouse lymphoid organs: cross-correlation of surface markers, changes with incubation, and differences among thymus, spleen, and lymph nodes. *J Immunol* (1997) **159**:565–73.

5. Vremec D, Pooley J, Hochrein H, Wu L, Shortman K. CD4 and CD8 expression by dendritic cell subtypes in mouse thymus and spleen. *J Immunol* (2000) **164**:2978–86. doi:10.4049/jimmunol.164.6.2978

6. Lewis KL, Caton ML, Bogunovic M, Greter M, Grajkowska LT, Ng D, et al. Notch2 receptor signalling controls functional differentiation of dendritic cells in the spleen and intestine. *Immunity* (2011) **35**:780–91. doi:10.1016/j.immuni.2011.08.013

7. Shortman K, Heath WR. The CD8$^+$ dendritic cell subset. *Immunol Rev* (2010) **234**:18–31. doi:10.1111/j.0105-2896.2009.00870.x

8. Hochrein H, Shortman K, Vremec D, Scott B, Hertzog P, O'Keeffe M. Differential production of IL-12, IFN-alpha, and IFN-gamma by mouse dendritic cell subsets. *J Immunol* (2001) **166**:5448–55. doi:10.4049/jimmunol.166.9.5448

9. Pooley JL, Heath WR, Shortman K. Cutting edge: intravenous soluble antigen is presented to CD4 T cells by CD8$^-$ dendritic cells, but cross-presented to CD8 T cells by CD8$^+$ dendritic cells. *J Immunol* (2001) **166**:5327–30. doi:10.4049/jimmunol.166.9.5327

10. Schnorrer P, Behrens GM, Wilson NS, Pooley JL, Smith CM, El-Sukkari D, et al. The dominant role of CD8$^+$ dendritic cells in cross-presentation is not dictated by antigen capture. *Proc Natl Acad Sci U S A* (2006) **103**:10729–34. doi:10.1073/pnas.0601956103

11. Naik S, Vremec D, Wu L, O'Keeffe M, Shortman K. CD8α$^+$ mouse spleen dendritic cells do not originate from the CD8α$^-$ dendritic cell subset. *Blood* (2003) **102**:601–4. doi:10.1182/blood-2002-10-3186

12. Bedoui S, Prato S, Mintern J, Gebhardt T, Zhan Y, Lew A, et al. Characterisation of an intermediate splenic precursor of CD8$^+$ dendritic cells capable of inducing antiviral T cell responses. *J Immunol* (2009) **182**:4200–7. doi:10.4049/jimmunol.0802286

13. Sathe P, Pooley J, Vremec D, Mintern J, Jin JO, Wu L, et al. The acquisition of antigen cross-presentation function by newly formed dendritic cells. *J Immunol* (2011) **186**:5184–92. doi:10.4049/jimmunol.1002683

14. Asselin-Paturel C, Boonstra A, Dalod M, Durand I, Yessaad N, Dezutter-Dambuyant C, et al. Mouse type I IFN-producing cells are immature APCs with plasmacytoid morphology. *Nat Immunol* (2001) **2**:1144–50. doi:10.1038/ni736

15. O'Keeffe M, Hochrein H, Vremec D, Caminschi I, Miller JL, Anders EM, et al. Mouse plasmacytoid cells: long-lived cells, heterogeneous in surface phenotype and function, that differentiate into CD8α⁺ dendritic cells only after microbial stimulus. *J Exp Med* (2002) **196**:1307–19. doi:10.1084/jem.20021031

16. Naik SH, Metcalf D, van Nieuwenhuijze A, Wicks I, Wu L, O'Keeffe M, et al. Intrasplenic steady-state dendritic cell precursors that are distinct from monocytes. *Nat Immunol* (2006) **7**:663–71. doi:10.1038/ni1340

17. Shortman K, Naik SH. Steady-state and inflammatory dendritic-cell development. *Nat Rev Immunol* (2007) **7**:19–30. doi:10.1038/nri1996

18. Naik SH, Sathe P, Park H-Y, Metcalf D, Proietto AI, Dakic A, et al. Development of plasmacytoid and conventional dendritic cell subtypes from single in vitro and in vivo-derived precursors. *Nat Immunol* (2007) **8**:1217–26. doi:10.1038/ni1522

19. Salomon B, Cohen JL, Masurier C, Klatzmann D. Three populations of mouse lymph node dendritic cells with different origins and dynamics. *J Immunol* (1998) **160**:708–17.

20. Anjuere F, Martin P, Ferraro I, Frage GM, del Hoyo GM, Wright N, et al. Definition of dendritic cell populations present in the spleen, Peyer's patches, lymph nodes, and skin of the mouse. *Blood* (1999) **93**:590–8.

21. Henri S, Vremec D, Kamath A, Waithman J, Williams S, Benoist C, et al. The dendritic cell populations of mouse lymph nodes. *J Immunol* (2001) **167**:741–8. doi:10.4049/jimmunol.167.2.741

22. Wilson NS, El-Sukkari D, Belz GT, Smith CM, Steptoe RJ, Heath WR, et al. Most lymphoid organ dendritic cell types are phenotypically and functionally immature. *Blood* (2003) **102**:2187–94. doi:10.1182/blood-2003-02-0513

23. Vremec D, O'Keeffe M, Wilson A, Ferrero I, Koch U, Radtke F, et al. Factors determining the spontaneous activation of splenic dendritic cells in culture. *Innate Immun* (2011) **17**:338–52. doi:10.1177/1753425910371396

24. Poulin LF, Henri S, de Bovis B, Devilard E, Kissenpfennig A, Malissen B. The dermis contains langerin+ dendritic cells that develop and function independently of epidermal Langerhans cells. *J Exp Med* (2007) **24**:3119–31. doi:10.1084/jem.20071724

25. Ginhoux F, Collin MP, Bogunovic M, Abel M, Leboeuf M, Helft J, et al. Blood-derived dermal langerin+ dendritic cells survey the skin in the steady state. *J Exp Med* (2007) **204**:3133–46. doi:10.1084/jem.20071733

26. Bursch LS, Wang L, Igyarto B, Kissenpfennig A, Malissen B, Kaplan DH, et al. Identification of a novel population of langerin⁺ dendritic cells. *J Exp Med* (2007) **204**:3147–56. doi:10.1084/jem.20071966

27. Sallusto F, Lanzavecchia A. Efficient presentation of soluble antigen by cultured human dendritic cells is maintained by granulocyte/macrophage colony-stimulating factor plus interleukin 4 and downregulated by tumor necrosis factor alpha. *J Exp Med* (1994) **179**:1109–18. doi:10.1084/jem.179.4.1109

28. Hume DA. Macrophages as APC and the dendritic cell myth. *J Immunol* (2008) **181**:5829–35. doi:10.4049/jimmunol.181.9.5829

29. Metcalf D, Shortman K, Vremec D, Mifsud S, Di Rago L. Effects of excess GM-CSF levels on hematopoiesis and leukemia development in GM-CSF/max 41 double transgenic mice. *Leukemia* (1996) **10**:713–9.

30. Brasel K, De Smedt T, Smith JL, Maliszewski CR. Generation of murine dendritic cells from flt-3-ligand-supplemented bone marrow cultures. *Blood* (2000) **96**:3029–39.

31. Naik SH, Proietto AI, Wilson NS, Dakic A, Schnorrer P, Fuchsberger M, et al. Cutting edge: generation of splenic CD8⁺ and CD8⁻ dendritic cell equivalents in Fms-like tyrosine kinase 3 ligand bone marrow cultures. *J Immunol* (2005) **174**:6592–7. doi:10.4049/jimmunol.174.11.6592

32. Villadangos JA, Shortman K. Found in translation: the human equivalent of mouse CD8⁺ dendritic cells. *J Exp Med* (2010) **207**:1131–4. doi:10.1084/jem.20100985

33. Guilliams M, Ginhoux F, Jakubzinck C, Naik S, Onai N, Schraml BU, et al. Dendritic cells, monocytes and macrophages – a proposal for a unifying nomenclature based on ontogeny. *Nat Rev Immunol* (2014) **14**:571–8. doi:10.1038/nri3712

34. Lahoud MH, Proietto A, Ahmet F, Kitsoulis S, Eidsmo L, Wu L, et al. The C-type lectin Clec12A present on mouse and human dendritic cells can serve as a target for antigen delivery and enhancement of antibody responses. *J Immunol* (2009) **182**:7587–94. doi:10.4049/jimmunol.0900464

35. Kasahara S, Clark EA. Dendritic cell associated lectin 2 (DCAL2) defines a distinct CD8α dendritic cell subset. *J Leukoc Biol* (2012) **91**:418–37. doi:10.1189/jlb.0711384

36. Shortman K, Sathe P, Vremec D, Naik S, O'Keeffe M. Plasmacytoid dendritic cell development. *Adv Immunol* (2013) **120**:105–26. doi:10.1016/B978-0-12-417028-5.00004-1

37. Ardavin C, Wu L, Li CL, Shortman K. Thymic dendritic cells and T cells develop simultaneously within the thymus from a common precursor population. *Nature* (1993) **362**:761–3. doi:10.1038/362761a0

38. Corcoran L, Ferrero I, Vremec D, Lucas K, Waithman J, O'Keeffe M, et al. The lymphoid past of mouse plasmacytoid cells and thymic dendritic cells. *J Immunol* (2003) **170**:4926–32. doi:10.4049/jimmunol.170.10.4926

3

Human and mouse mononuclear phagocyte networks: a tale of two species?

Gary Reynolds[1,2] and Muzlifah Haniffa[1]*

[1] Human Dendritic Cell Laboratory, Institute of Cellular Medicine, Newcastle University, Newcastle upon Tyne, UK,
[2] Musculoskeletal Research Group, Institute of Cellular Medicine, Newcastle University, Newcastle upon Tyne, UK

Edited by:
Shalin Naik,
Walter and Eliza Hall Institute,
Australia

Reviewed by:
Kristen J. Radford,
University of Queensland, Australia
Simon Yona,
University College London, UK

*Correspondence:
Muzlifah Haniffa,
Human Dendritic Cell Laboratory,
Institute of Cellular Medicine, The
Medical School, Newcastle
University, Framlington Place,
Newcastle upon Tyne NE2 4HH, UK
m.a.haniffa@ncl.ac.uk

Dendritic cells (DCs), monocytes, and macrophages are a heterogeneous population of mononuclear phagocytes that are involved in antigen processing and presentation to initiate and regulate immune responses to pathogens, vaccines, tumor, and tolerance to self. In addition to their afferent sentinel function, DCs and macrophages are also critical as effectors and coordinators of inflammation and homeostasis in peripheral tissues. Harnessing DCs and macrophages for therapeutic purposes has major implications for infectious disease, vaccination, transplantation, tolerance induction, inflammation, and cancer immunotherapy. There has been a paradigm shift in our understanding of the developmental origin and function of the cellular constituents of the mononuclear phagocyte system. Significant progress has been made in tandem in both human and mouse mononuclear phagocyte biology. This progress has been accelerated by comparative biology analysis between mouse and human, which has proved to be an exceptionally fruitful strategy to harmonize findings across species. Such analyses have provided unexpected insights and facilitated productive reciprocal and iterative processes to inform our understanding of human and mouse mononuclear phagocytes. In this review, we discuss the strategies, power, and utility of comparative biology approaches to integrate recent advances in human and mouse mononuclear phagocyte biology and its potential to drive forward clinical translation of this knowledge. We also present a functional framework on the parallel organization of human and mouse mononuclear phagocyte networks.

Keywords: mononuclear phagocyte system, dendritic cells, macrophages, monocytes, comparative genomics

Introduction

The mononuclear phagocyte system (MPS) is a branch of the immune system comprising dendritic cells (DCs), macrophages, and monocytes (1–3). The many functions of the MPS include tissue maintenance and healing, innate immunity and pathogen clearance, and the induction of adaptive immune responses (1–3). Manipulating these functions could lead to clinical benefit, such as modulating DCs to develop antigen-specific anti-tumor immunity or suppressing peripheral autoreactive T cell responses in autoimmunity (4, 5). Several factors need to be considered in designing immunotherapy targeting the MPS, including cellular or pathway target choice and the relevant disease and tissue context. Diversity and plasticity of the MPS, two core features that are paramount for directing the quantity and quality of specific immune responses, have frustrated attempts to develop successful focused therapies. The additional variable of local tissue environment, which also heavily influences the composition and function of resident and infiltrating mononuclear phagocytes (MPs), also requires careful consideration (1–3).

The MPS was conceived in the 1960s by van Furth to encompass a family of phagocytic mononuclear leukocytes regarded as functional variations of monocytes (6). DCs were embraced as members of the MPS several years later (7). The revolutionary discovery that human monocytes and CD34+ hematopoietic stem cells (HSCs) could be differentiated into DC (mo-DC) and macrophage-like (mo-Mac) cells provided a convenient *in vitro* model to study human MP biology (8–10). However, murine studies have demonstrated the independence of many DCs, macrophages, and Langerhans cells (LCs) from blood monocytes questioning the accuracy of human *in vitro* monocyte-derived cells in recapitulating *in vivo* populations (11–16). Conventional DCs arise from HSCs along a lineage that does not go through a monocyte stage and are dependent on the growth factor receptor FLT3 (11). In contrast, the majority of tissue macrophages arise from prenatally seeded precursors that can survive into adulthood and are dependent on CSF1-R (12–16).

The constituents of MPS share overlapping surface markers, which poses a challenge in parsing functionally distinct populations. A rewarding approach to unravel this complexity has been comparative biology analysis (17–28). In essence, comparative biology relies on the concept that core developmental programs and functions such as differential CD4 and CD8 T cell priming, cross-presentation, migration, and cytokine production are likely to be non-redundant and conserved between species. In support of this, around 99% of murine genes have human analogs and around 96% are syntenic, despite the two species having 80 million years of divergent evolution (29). Comparative transcriptomic mapping has revealed conserved gene expression profiles in the two species allowing parallels to be drawn between DC and macrophage subsets (17–28). This approach places comparative analysis as the central fulcrum facilitating the integration of fundamental immunology to fertilize clinical translational strands (**Figure 1**). Integrating this workflow with

cutting-edge technologies including single-cell genomics and proteomics approaches has the potential to accelerate discovery in basic MP biology and its clinical applicability (**Figure 1**). Comparative biology has revealed further insights into the origin and function of human and mouse mononuclear phagocyte populations (17–28) and generated new hypotheses to be tested in both species.

The concept of functional specialization as an inherent property imprinted by MP ontogeny and tissue anatomy has been well demonstrated in many murine studies [reviewed in Ref. (1, 3, 30)]. However, the MPS possesses an additional layer of complexity in the form of dynamic mobility, plasticity, and adaptability to tissue/local microenvironment both in steady state and in inflammation (1, 3, 31). These issues have been particularly difficult to dissect in human, where the temporal resolution to observe these kinetics is constrained by snapshot analysis during inflammation and disease without adequate recourse to their onset and evolution (**Figure 2**). Snapshot observations during inflammation may be confounded by temporal variations in MPS composition and function resulting in highly variable biological data. This variability may account for the biological noise inherently observed with outbred humans in contrast to inbred mice in specific pathogen free (SPF) facilities.

Mononuclear phagocytes and their progenitors are in dynamic equilibrium between peripheral tissue, blood, and bone marrow (1, 3, 31, 32). The distinction between MPs within peripheral interstitial tissue and blood can be difficult to establish in highly vascularized organs such as liver and spleen, where large sinusoids are present adjacent to discontinuous endothelial lining that enables greater mobility of leukocytes within these organs. In addition, inflammatory perturbations affect the dynamic equilibrium between tissue, blood, and bone marrow compartments favoring the relative expansion and egress of specific lineages in response to distinct stimuli (33–35). Expansion of monocyte-derived cells dominates the response to inflammatory stimuli in tissue but little is known regarding their fate upon resolution of inflammation (35). Peripheral tissue DCs migrate to the lymph node where they mediate their potent functions upon inflammatory stimuli. Whether they play a prominent role in local tissue immune regulation and how migratory DCs are repopulated during inflammation and its resolution has been poorly characterized.

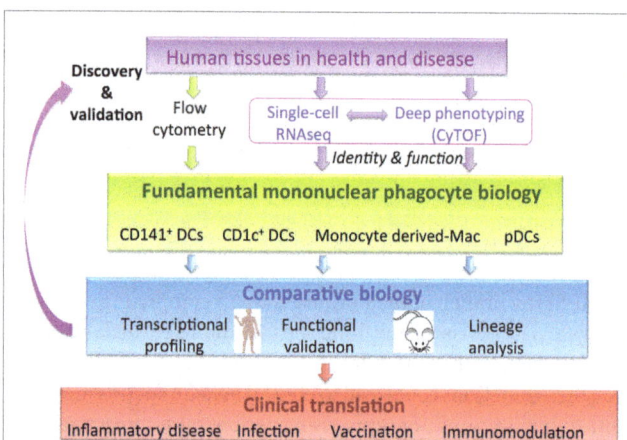

FIGURE 1 | Comparative biology is a validation and discovery tool to pull-through fundamental knowledge in MPS biology to clinical translation. Incorporation of new genomics and proteomics methodologies will accelerate discovery.

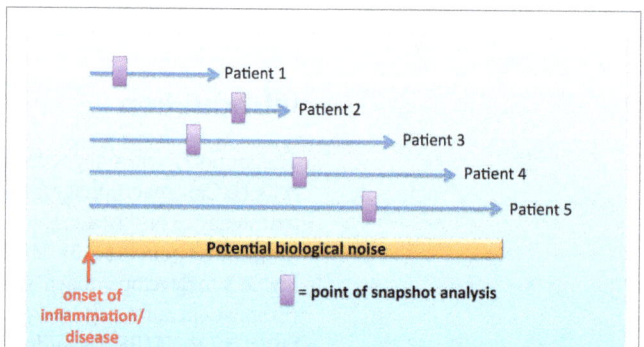

FIGURE 2 | Biological noise with snapshot analysis during temporal course of inflammation and disease.

Comparative Biology to Interrogate Human and Mouse MP Networks

Identifying homology between mice and humans in other hematopoietic cells such as T and B cells has been relatively simple at phenotype and practical levels because of shared lymphocyte surface markers (CD3/CD4/CD8 and CD19, respectively) as well as the relative ease of isolating lymphocytes, which form 90% of human peripheral blood mononuclear cells (c.f. <1% being DCs). Nevertheless, there are functional differences in lymphocytes between the two species, such as differentiation requirements for IL-17 (36) and GM-CSF (37, 38) secreting CD4[+] T cells, the specificity of granzyme and FOXP3 expression to define natural Tregs (39), the distinct classes of immunoglobulin (40) and human CD1a, 1b, and 1c-restricted responses to lipid molecules (41). Unfortunately, components of the human and mouse MPS lack overlapping phenotypic markers, hampering initial progress in identifying homologous populations between species.

A range of –omics technologies such as transcriptomics, metabolomics, proteomics, and epigenomics could potentially be employed to assess proximity between species. Of these approaches, transcriptomics is technically most tractable and generates enough complexity to achieve good definition between populations (*n*-dimensions where *n* is the number of genes analyzed) (42, 43). Transcriptome-based comparison of various hematopoietic lineages between human and mouse shows broad conservation but also highlighted specific differences and transcriptional divergence due to gene duplication (43).

Transcriptomics

The hypothesis underlying comparative transcriptomics is that the identified MP populations were present in a shared ancestor and that these same subsets are present in modern animals. Furthermore, despite divergent evolution over time, cells from each subset will have a conserved transcriptomic signature similar to that of its equivalent in the other species. Two approaches are generally used to measuring this similarity: (1) unsupervised hierarchical clustering and principal component analysis (PCA), which assigns samples a point in *n*-dimensional space (*n* corresponding to the number of genes analyzed) and applying a distance metric with greater proximity suggesting a developmental relationship, or (2) supervised assessment of defined transcriptome signature enrichment between populations of interest exemplified by gene set enrichment analysis (GSEA) (44) and its later variations (45).

In hierarchical clustering, the Euclidean distance is calculated between samples. In PCA, the same Euclidean metric is used after the *n*-dimensional data are projected on to the two or three dimensions over which the most variation occurs. This approach has the disadvantages inherent in using large sets of gene data, large number of variables/genes, and high inter-sample variability when testing a limited number of samples. The consistent finding that tissue-specific genes predominate in DC microarray transcriptomes highlights the first point. As a result, microarray data of DC subsets from the same tissue tend to cluster together rather than with their equivalent in blood or another tissue (46). This can be corrected for by techniques such as excluding genes that are differentially expressed between pooled cells from each

tissue (and classifying these "tissue-specific") (23, 26) or through using an abbreviated gene panel that is enriched for genes that are known to give good definition between DC subsets (17). An important corollary of this finding is that, while the relative contribution of ontogeny and environment to DC function remains to be determined, the list of genes that define ontogeny is a small fraction of the genes that are modulated by the environment and highlights a potential drawback of using blood DCs as a proxy for tissue DCs.

The use of GSEA derives from large-scale microarray data in which it was recognized that groups of co-regulated functionally linked genes may be more relevant than the few genes that are most significantly differentially regulated but functionally unrelated. This approach is dependent upon an *a priori* understanding of gene function and this can introduce bias. When GSEA has been used in aligning DC subsets between species, a "query signature" is produced that defines the subset of interest. Samples in the test population can then be interrogated for whether they are enriched for this query signature. The underlying analysis is based on the non-parametric goodness-of-fit Kolmogorov–Smirnov test statistic with the reference probability distribution that of the query signature. GSEA and its later variant connectivity map analysis (CMAP) have been successfully used to identify homologous MP populations between species and the developmental origin of human inflammatory DCs (17, 23, 25, 26, 47). Steady state homologous MP populations in human and mouse blood, lymph node, and peripheral tissues are illustrated in **Figure 3**.

Most transcriptomics studies thus far on MPs have involved ensemble or bulk-population analysis. This introduces an inherent bias, as cell populations have to be defined *a priori* based on expression of specific markers. More recently, the application of single-cell RNA-sequencing (sc-RNA-Seq) with unbiased analysis potential has been successfully used to interrogate cellular heterogeneity to

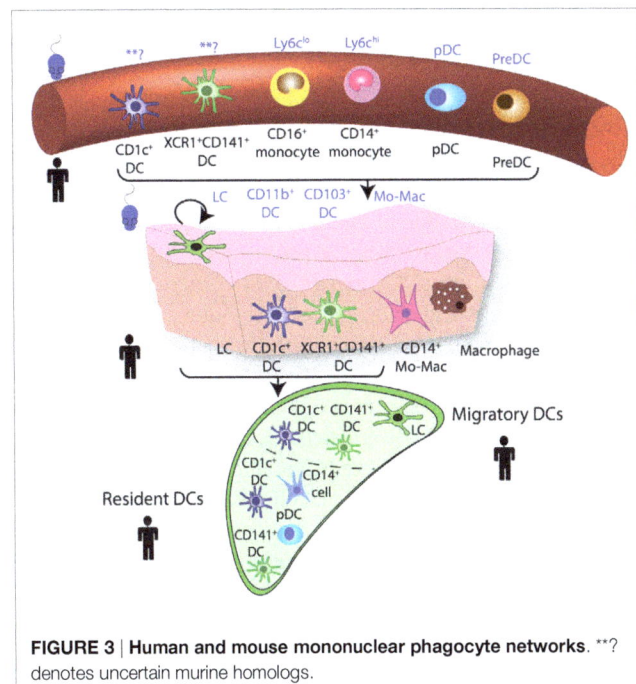

FIGURE 3 | Human and mouse mononuclear phagocyte networks. **? denotes uncertain murine homologs.

uncover new cell populations, functional immune states, and to establish cellular lineage hierarchies and lymphocyte differentiation programs (48–53). These technical advances combined with novel computational approaches have the potential to revolutionize our understanding of MPS biology by unraveling predicted and unexpected functional heterogeneity, which underpins the dynamic repertoire of our immune system in health and disease.

Proteomics

Proteomics analysis has revealed differences in viral sensing pathways between murine splenic DC subsets (54) and identified the murine common monocyte progenitor (cMOP), an intermediate cell-type between the monocyte/macrophage and DC precursor (MDP) and monocyte (55). However, current large-scale proteomics approaches require high cell numbers for robust analysis and are impractical for rare populations, especially from limited human tissue material. Protein expression on a more limited scale has been the mainstay of conventional flow cytometry to define populations and assess MP functions at single-cell resolution. Although the number of parameters that can be analyzed simultaneously is limited (17–18 parameters using commercial instruments), the application of new unbiased probabilistic analysis to define populations could reveal new insights to MP heterogeneity (56). Mass cytometry (CyTOF) provides additional parameters (up to 100) and combined with unbiased population assignment has enormous discovery potential. This combined analysis on mouse myeloid cell populations has revealed far greater population heterogeneity than previously appreciated (57).

Functional Validation

Comparative functional analysis between mouse and human MPs has resulted in variable findings [reviewed in Ref. (30, 58)]. It is unknown if this is due to true biological differences or experimental factors which are not comparable within and between species, including the common use of murine *in vivo* models in contrast to human *in vitro* assays to assess MP functions. Conserved functions are detailed in **Figure 4**.

Lineage Analysis

The power and utility of comparative biology to identify homologous MP populations is beginning to be applied to MP lineage analysis. The recent identification of the successive downstream progenies of human MDP; the Common DC precursor (CDP) and precursor of myeloid DCs (pre-cDCs) exploited the conserved dependency on growth factors and cytokines between human and mouse DC precursors (27, 28). Similarly, comparative analysis suggested the monocyte-origin of human dermal CD14+ cells (25) and inflammatory DCs (47). The preservation of LCs and dermal macrophages in GATA2 and biallelic IRF8 deficiencies show that they are bone marrow independent in the steady state and similar to their murine counterparts, also arise from prenatally seeded precursors (59, 60).

DC, Monocyte, and Macrophage Subsets in Mice and Humans

This approach of using ontogeny and by extension transcription factor dependence to define MPS populations was formalized

Cell	Human	Common specialisations	Mouse
cDC1	CD141 XCR1 CLEC9A CADM1	• Cross-presentation • IL-12 and IFNλ production • Expression of TLR3 • Generate Th1/Th2 responses	CD103/CD8 XCR1 Clec9A CADM1
cDC2	CD1c CD11b SIRPα	• CD4+ T cell responses • IL-1β, IL-6 and IL-23 production • Expression of all TLRs apart from TLR3(mouse) and TLR9 (human) • Generate Th2/Th17 responses	CD24 CD11b SIRPα
pDC	CD123 CD303 CD304	• Anti-viral responses • IFNα production • Expression of TLR7 and TLR9	SiglecH Bst2 Ly6c
CD14+ monocyte / Ly6chi monocyte	CD14 CCR2hi CD62L	• Inflammatory responses • DC- and macrophage-like differentiation in vitro • High CCR2 expression	Ly6chi CCR2hi
CD16+ monocyte / Ly6clo monocyte	CD16 CCR2lo CX3CR1hi	• Patrol endothelium • DC- and macrophage-like differentiation in vitro • Low CCR2 expression	Ly6clo CCR2lo CX3CR1hi
LC	Langerin CD1a+++ CD11clo	• Maintain epidermal integrity • Maintain Tregs • Generate Th17 responses	Langerin CD24 CD11b F4/80
Macrophage	DC-SIGN FXIIIa LYVE-1 CD206	• Tissue-resident specialisation conserved between species (e.g. Kupffer cells, microglia, osteoclasts)	F4/80 CD64 MerTK

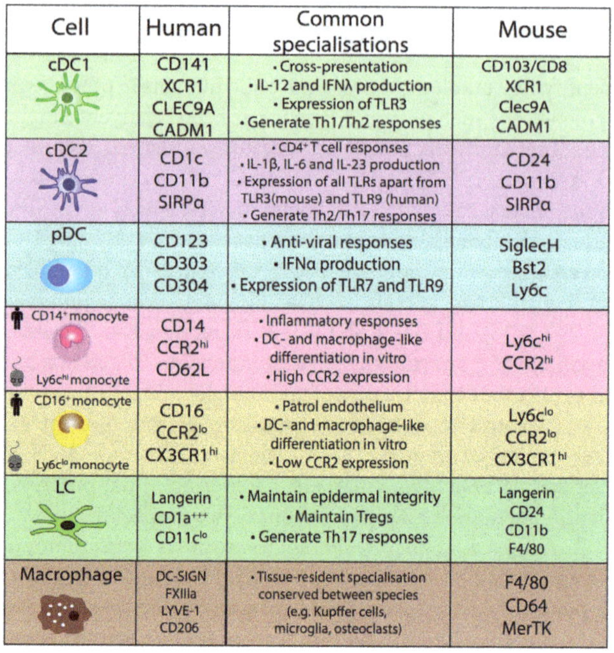

FIGURE 4 | Conserved specializations between human and mouse mononuclear phagocytes.

recently in a proposed nomenclature (61). In this scheme, four adult HSC-derived MP populations are described in mice: two conventional/classical DC subsets (cDC1 and cDC2), plasmacytoid DCs (pDCs), and monocyte-derived cells (61). Both cDCs and pDCs are derived from murine CDP (62, 63). The CDP-derived cells are defined by their dependence on specific transcription factors (TFs): cDC1 are Batf3-dependent, cDC2 are Irf4-dependent, and pDC are E2-2-dependent (61). This definition is unambiguous and avoids using surface markers that can vary between tissues and in inflammation. While the ontogeny approach aids definition of murine populations, it cannot be easily transferred to human DC nomenclature, due to inherent logistical difficulties of human ontogeny studies. However, with the aid of comparative biology approaches, homologous populations between human and mouse MP subsets can be identified and inferences between species on ontogeny and function can be made (**Figures 3 and 4**).

cDC1
Phenotype

This subset is identified in mouse by the expression of CD8α in the spleen and CD103 in non-lymphoid tissues (NLT). Its human equivalent in blood and NLT were initially defined by their high expression of CD141 (thrombomodulin, BDCA-3) (19–23). However, this antigen can be upregulated on blood monocytes and expressed promiscuously by other DC subsets in human tissue (23). The cell adhesion molecule CADM1 (NECL2), C-type lectin, CLEC9A (which recognizes damaged cells), and the chemokine receptor XCR1 are expressed on human and mouse cDC1 (19, 22, 64–66). However, CADM1 expression is not restricted to leukocytes and CLEC9A is also expressed on murine DC precursors

(67, 68). Although cDC1 is the only leukocyte expressing XCR1, a commercial antibody against it is currently unavailable. Notably, langerin is expressed on murine but not human cDC1 (23, 69, 70).

Homology

Homology between human (XCR1+CD141+DCs) and mouse (CD8+/CD103+) cDC1 was demonstrated by comparative transcriptomics, phenotype, and functional analyses (17, 19–21, 23, 71). Furthermore, blood and skin CD141+ DCs cluster together separately from CD1c+ DCs, CD14+ and CD16+ monocytes, and pDCs, suggesting that skin XCR1+CD141+ DCs are the tissue equivalents of blood XCR1+CD141+ DCs (23).

Transcription Factors

In addition to Batf3 (72), murine cDC1 differentiation requires Irf8 (73), Id2 (74, 75), and NFIL3 (76). In human, shRNA knockdown of *BATF3* in cord blood HSCs inhibits their differentiation into cDC1 *in vitro* (22). However, cDC1 were detectable in humanized mice reconstituted with *BATF3* knockdown CD34+ HSCs (22). A possible explanation for this seeming contradiction was shown in mice, where in inflammatory conditions (specifically in the presence of IL-12), other members of the Batf family of TFs appear to be able to compensate for loss of Batf3 (77). *ID2* mRNA is expressed at low amounts in human CD34+ HSCs but upregulated during DC differentiation in the presence of GM-CSF and IL-4 (74). Its role is potentially in suppressing B cell differentiation from a common precursor. Definitive evidence for the requirement of ID2 and NFIL3 in cDC1 development in humans is lacking and highlights the potential difficulties of translating TF-based definitions of DC subsets from mice to humans.

Function

The cDC1 subset is thought to be able to efficiently prime CD8+ T cells through functional specializations such as cross-presentation of antigens and the production of IL-12p70 (78–80). This process is important in the induction of tumor immunity and the control of viral and bacterial infections when DCs are not the malignant cells or directly infected. The expression of Clec9A and XCR1 by both murine and human cDC1s supports this notion. cDC1s express a more limited TLR profile than cDC2s with high expression of TLR3 and TLR10 but without TLR4, −5, −7 and −9 (54, 81). TLR3 senses viral dsRNA but the role of TLR10 is currently unknown. Human cDC1s do not produce large amounts of IL-12p70 in response to TLR ligands alone but do following the combination of TLR ligands and CD40-CD40L signaling through activated T cells (71), in common with the finding in mice (82). IFN-λ is produced by murine and human cDC1 upon stimulation with the TLR3 agonist, poly I:C (83).

Murine cDC1s have an advantage over other subsets at cross-presentation of antigens by being able to (1) maintain optimal phagosomal pH for antigen processing (84) and (2) enhance the transfer of proteins from the endosome in to the cytosol so they can be loaded on to MHC Class I (85). This advantage is apparent when assessing cross-presentation of dead cell-derived antigens and upon stimulation with TLR3. However, recent data showed that murine cDC2 are also able to cross-present and cross-prime antigen upon stimulation with R848, a TLR7/8 agonist (86). In human, cDC1

appears to be superior at cross-presenting cell-derived antigen, particularly upon polyI:C stimulation (19–21, 71) and when antigens are delivered to late endosomes and lysosomes (87). However, in common with mice, cDC2 are also able to cross-present soluble antigen and long-peptide particularly upon R848 stimulation (88, 89). The variable findings reported may also be due to type of antigens used in the cross-presentation assays and the validity of comparing murine *in vivo* models with human *in vitro* assays.

In mouse, cDC1 preferentially induce Th1 immune response through IL-12p70 production (90, 91), although Th2 induction has also been reported (92). In human, both cDC1 and cDC2 have been shown to induce Th1 and Th2 responses (93). cDC1s were also shown to promote enhanced Th2 differentiation in response to TSLP in an influenza infection humanized mouse model (94). As most human experiments are performed using blood DCs and *in vitro*, it has been logistically difficult to establish pathogen and tissue-specific effects relevant for driving Th priming *in vivo*.

cDC2
Phenotype

cDC2s in mice are lin−MHCIIhiCD11c+CD11b+. However, this fraction also includes monocyte-derived cells and macrophages (95). This is demonstrated by the variable depletion of cells from this fraction in Flt3 or Csf1r KO mice suggesting contamination by Flt3-independent cells (75). This is in contrast to the near complete absence of cDC1 in Flt3 KO mice (74).

Genetic tracing using Clec9A-reporter mouse to identify all CDP-derived cells demonstrated near-complete labeling of cDC1s but variable labeling of CD11b+ DCs in NLT (68). Although this is in keeping with the presence of monocyte-derived cells and macrophages within CD11b+ cells, it does not exclude the possibility of an alternative DC differentiation program that does not undergo a monocyte or CDP intermediate stage. Splenic CD11b+ DCs are divided into an ESAMhi population that requires Notch2-, Flt3, and LTβ-signaling for its development and a monocyte-like ESAMloClec12A+CX3CR1+ population that is Flt3-independent and expresses high levels of CD14, TNFα, CCR2, and Lyz2 (96). In murine lung, it has been possible to divide the MHCII+CD11c+CD11b+ fraction into CD11b+CD64+ monocyte-macrophage cells and CD11b+CD24+ cDC2s (24). In murine skin, the MHCII+CD11b+Langerin− fraction comprises cDC2, monocyte-derived cells, and macrophages (97).

There is evidence that similar heterogeneity may be present within human cDC2. Only 170 genes characterized human cDC2, in comparison to 1020 for cDC1 and 1065 genes for pDCs (23). This limited list of differentially expressed genes predicts heterogeneity within the boundaries of the phenotype parameters used to define human cDC2, specifically a subpopulation derived from or closely related to another mononuclear phagocyte such as CD14+ monocytes.

Human cDC2 (CD1c+ DCs) are defined as lin−MHCII+CD14−CD16−CD11c+CD1c+ cells, a definition they share with *in vitro* monocyte-derived DCs. Although human peripheral blood and murine cDC2 additionally express CD11b, CX3CR1, and SIRPα, these antigens do not distinguish them from monocyte-derived cells (24, 98). Uniquely in the small intestine, cDC2s co-express CD103 and SIRPα (24, 26). *In vitro* human mo-DCs express

CD206/MMR and CD1a but peripheral blood cDC2 do not (47, 99). However, tissue CD1c⁺ DCs express CD206 and CD1a (100, 101). In addition, some tissue CD1c⁺ DCs co-express CD14 particularly during inflammation (47).

Homology

The transcriptional signatures of human blood CD1c⁺ DCs are enriched with that of mouse spleen CD4⁺/CD11b⁺ DCs (17, 23). In NLT, the transcriptional signatures of human small intestine CD103⁺SIRPα⁺ DCs and dermal CD1c⁺ DCs are enriched with that of murine spleen and mesenteric lymph node CD11b⁺ DCs and dermal CD11b⁺ DCs, respectively (25, 26). A similar relationship was also observed between murine lung CD11b⁺ DCs with human blood CD1c⁺ DCs (24).

Transcription Factors

cDC2 development has been shown to be dependent on the TFs Irf4, PU.1, RelB, and RBPJ (24, 96, 102–108). Irf4 directly supports MHC class II antigen presentation to promote CD4⁺ T cell responses (109). In humans, CD1c⁺ DCs express high amounts of IRF4 (24). Interestingly, IRF4 is also required for mo-DC differentiation, suggesting a shared differentiation program between cDC2 and mo-DC. PU.1 interacts with Irf4 but also upregulates Flt3 expression critical for early DC differentiation in mice (110, 111). The PU.1 binding site in the Flt3 promoter is conserved in mice and humans, and so it is thought to be similarly required for DC differentiation in humans (111). Administration of Flt3 results in expansion of DC subsets in lymphoid and non-lymphoid tissue (112). PU.1 mutations in humans and mice are associated with myeloid leukemias (113). Biallelic human IRF8 K108E mutation resulted in complete loss of monocytes, pDCs, cDC1, and cDC2 in the peripheral blood (60). Surprisingly, human autosomal dominant IRF8 T108A mutation results in selective loss of the cDC2 subset and IL-12 production (60). It is now apparent from studies on Irf8ᴿ²⁹⁴ᶜ(BXH2) and Irf8⁻/⁻ mice that in addition to cDC1, pDCs and monocytes are also dependent on IRF8 (73, 75, 114–116). However, cDC2 frequency in mice with Irf8⁻/⁻ and the hypomorphic mutation Irf8ᴿ²⁹⁴ᶜ are unaffected, in contrast to the findings in humans (73, 75, 114).

Function

The transcriptome of cDC2s is enriched for genes related to antigen processing such as LAMP1, LAMP2, and cathepsins (117). Murine cDC2s have been shown to be able to promote Th17, Th2, and regulatory T cell responses depending upon the pathogen and antigen stimulus (24, 108, 118–120). This may be a consequence of their innate plasticity but could also relate to unresolved heterogeneity within murine cDC2. In human, cDC2 have been shown to induce Th17 differentiation (24).

Both human and mouse cDC2 share many transcriptional and functional similarities with monocyte-derived cells (24, 25, 47, 97). Both cDC2 and mo-DC are capable of promoting naïve CD4⁺ and CD8⁺ T cell proliferation and in mice cDC2 appear to be superior at trafficking to lymph nodes (97, 98), leading to the hypothesis that mo-DCs specialize in activating tissue-tropic T cells. Mo-DCs also produce higher levels of monocyte-attracting

chemokines (CCL2, CCL7, CCL12) than cDC2s (98). Human blood cDC2 have a TLR expression profile that is close to murine lymphoid cDC2 with significantly higher levels of TLRs 2, 4, and 5 than other DC subsets (81), a profile it also shares with in vitro mo-DCs (121). The pathogenic role of cDC2 in human disease is not clear but they have been shown to accumulate in conditions such as RA (122), chronic kidney disease (123), and atopic airway inflammation (124), although their distinction from inflammatory mo-DCs is unclear. Human cDC2 are also implicated in the accumulation of CD103⁺CD8⁺ mucosal T cells in the lung and promote fibrosis in the kidney through production of TGFβ (123, 125). Finally, human and mouse cDC2 share a similar cytokine production profile which includes IL-6, IL-23, and IL-1β (24, 81, 126, 127). In addition, unlike murine cDC2, human blood cDC2 can secrete high amounts of IL-12p70 upon in vitro stimulation with R848 and LPS, which was augmented in the presence of IFNγ and CD40L (89).

Plasmacytoid DCs (pDCs)
Phenotype

Plasmacytoid DCs are specialized IFNα producing cells that were first described in human peripheral blood and tonsil (128–131). In blood, their morphology resembles that of lymphocytes but upon in vitro culture with IL-3 and CD40L, they acquire dendrites resembling myeloid DCs (129). pDCs are identified in mice by expression of CD11cⁱⁿᵗCD11b⁻B220⁺ in combination with markers such as SiglecH and CD317 (BST2) to exclude a subset of NK cells and precursors of cDCs (132). In humans, they are identified by expression of CD123, CD303, and CD304. CD123 is the IL-3 receptor alpha chain and is also expressed on precursor cells, basophils, and eosinophils (133, 134). CD303 (BDCA-2) is a C-type lectin that is specifically expressed by human pDCs (135). Functionally, it has a role in antigen capture and when ligated it inhibits IFNα production (136). CD304 (BDCA-4) is uniquely expressed by pDC in peripheral blood but is also expressed by other cells such as endothelial cells (137).

Homology

The relative distance of the pDC transcriptome from other leukocyte subsets and its conservation directly aligns murine and human pDCs (17). However, a subset of murine pDCs also appears to have cDC differentiation potential (138, 139), which has not been observed in human.

Transcription Factors

Plasmacytoid DC development in humans and mice is dependent on the transcription factor E2-2 (140). E2-2 opposes default differentiation of precursors into cDCs and controls expression of a range of pDC-associated TFs, including SpiB, Irf7, and Irf8 (140, 141). In humans, haploinsufficiency of E2-2 results in Pitt-Hopkins syndrome, a condition with a range of features including developmental delay and characteristic facial features but without known clinical immunodeficiency (142). A population of CD45RA⁺CD123⁺ cells is present in the blood of patients with Pitt-Hopkins syndrome but these cells fail to express CD303 and have severely reduced expression of IFNα, indicating that loss of E2-2 blocks full pDC differentiation (140). The transcription factor

SpiB is required for IFNα production by pDCs in mice (143). SpiB-knockdown in human CD34+ HSCs inhibits pDC differentiation *in vitro* (144).

Function

Plasmacytoid DCs have a functional program that is well-conserved between mice and humans (145). In contrast to cDCs, pDCs express a narrow range of pattern recognition (146). Both mouse and human pDCs express TLR7 and TLR9 (146). TLR8 is expressed at very low amounts if any by human pDCs (81, 147, 148) and appears to have a different function in mice (146, 149, 150). pDCs in both mice and humans are specialized in the production of IFNα and thought to be important in viral immunity but also human autoimmunity such as SLE (151, 152).

Monocytes and Monocyte-Derived Cells
Phenotype

Two subsets of monocytes exist in mice and can be distinguished by the differential expression of Ly6C, CCR2, and CX3CR1. Similarly in humans, there are two monocyte subsets in peripheral blood identified by expression of CD14 and CD16 (CD14++CD16- and CD14+CD16+) as well as an intermediate phenotype (CD14++CD16+). In addition to these antigens, human monocytes are also heterogeneous for the expression of the angiopoietin receptor, Tie2, and 6-sulfoLanNAc(Slan), a carbohydrate modification of the P-selectin glycoprotein ligand-1 (PSGL-1) (153, 154).

Homology

Homology between peripheral blood monocyte subsets has been demonstrated by the extensive transcript enrichment between Ly6ChiCX3CR1lo and CD14++CD16- monocytes and between Ly6CloCX3CR1hi and CD16+ monocytes (18, 23, 155).

Transcription Factors

The TFs that regulate the sequential differentiation of HSCs into MDP in mice include PU.1, Irf8, and Klf4 [reviewed in Ref. (156)]. PU.1 is required at each developmental bifurcation including HSC maintenance (157) and the generation of early myeloid progenitors (16, 158–160). Similarly in humans, PU.1 is required for monocyte differentiation from CD34+ cord blood precursors (161). In murine monopoiesis, Irf8 and Klf4 act together to skew differentiation toward monocytes by antagonizing the granulocyte-supporting TF C/EBPα (115, 162). Consistent with this, human autosomal recessive Irf8 deficiency results in complete loss of circulating monocytes and DCs in the presence of neutrophilia (60).

The TFs that control cell-fate decisions downstream of MDP are less well defined. In mice, Irf5 and TCFEB are implicated during MDP to CMoP differentiation (55). The TF Nur77 has been implicated in Ly6CloCX3CR1hi monocyte generation (163).

PU.1 and MafB act antagonistically to support human monocyte differentiation into mo-DC and mo-Mac, respectively *in vitro* (164). Irf4 was also implicated in human *in vitro* mo-DC differentiation (165). Irf5 promotes the differentiation of classical/M1 macrophages from human monocytes *in vitro* (166). In contrast, Irf4 activates transcription of the alternative/M2 macrophage markers in mice (167) and humans (168, 169).

Function

CD14+ human and Ly6ChiCX3CR1lo murine monocytes can exhibit considerable functional plasticity as demonstrated by their acquisition of DC-like and macrophage-like characteristics *in vitro* and *in vivo*. Recent fate mapping studies have demonstrated that monocytes do not contribute to tissue-resident macrophages in the steady state (12, 14, 15), with the notable exception of gut and dermal macrophages (14, 97, 170). However, monocytes can give rise to tissue macrophage-like cells in inflammation (35, 171). Monocytes can also differentiate into DC-like cells in the steady state in mucosal tissues and skin (97, 172). This process is enhanced during inflammation (97, 98, 173), including infections with Leishmania (34), Influenza (174), Trypanosoma (175), Listeria (33), and pulmonary *Aspergillus* (176). Alternatively, rather than DC-like or macrophage-like differentiation, monocytes may remain as tissue monocytes upon extravasation (177).

CD14+CD16+ intermediate and CD16+ non-classical monocytes are expanded in multiple disease, infection, and inflammatory states (178). CD16+ monocytes "patrol" the endothelium *in vivo*, are weak phagocytes, and sense nucleic acids and viruses via TLR7 and 8 receptors (155). Additional heterogeneity has been reported within human monocytes. Tie2+ monocytes are associated with angiogenesis and Slan+CD16+ cells, which are also present in inflamed skin, are potent producers of TNF α, IL-1β, and IL-12 (179, 180). Monocyte-derived dermal CD14+ cells express IL-1α (25) have been shown to induce differentiation of follicular helper T cells (126) and provide direct B cell help (181).

Langerhans Cells

Langerhans cells are located in epidermal surfaces such as skin and are characterized by the presence of cytoplasmic organelles containing Langerin called Birbeck granules (182). The function of these organelles is unclear but their absence does not affect their capacity to process and present antigen (183). LCs form a dynamic network with adjacent keratinocytes and protrude dendrites through tight junctions to pick up antigens that have passed the stratum corneum barrier (184). The easy accessibility of LCs and their functional plasticity has generated significant interest in targeting them for vaccination strategies (185).

In the steady state, LCs are maintained independently of the bone marrow through local self-renewal (186–188). Human LCs can proliferate *in situ* and have been shown to remain donor in origin up to 10 years after limb transplant (189–191). During inflammation, LCs can be replaced by circulating precursors. The identity of the circulating LC precursor remains unclear. In mice, there appears to be two waves of replenishment with monocytes in the first wave giving rise to short-term LCs that retain some monocyte features and an as yet unknown CD34+ HSC-derived precursor that gives rise to long-term LCs (186, 187). In humans, CD1c+ DCs are able to upregulate langerin and CD1a, a phenotype resembling LCs, upon *in vitro* culture with TSLP and TGFβ or GM-CSF and BMP7, but the relevance of this to *in vivo* LC differentiation is uncertain (192, 193). Although human LCs can self-renew locally after BMT, they are replaced by donor-derived cells, even after non-myeloablative transplant conditioning (194–196).

Langerhans cells are developmentally independent of Flt3 but dependent on Csf1r. However, it is IL-34 signaling through Csf1r, rather than Csf1, that is critical for LC development and maintenance (197). IL-34 is also expressed in human skin but the dependence of human LCs on this cytokine remains untested.

Phenotype
Human and murine LCs are CD11c^{lo}, langerin^{hi}, EPCAM$^+$, and also characterized by the presence of cytoplasmic Birbeck granules (198). In human, LCs are additionally CD1a^{hi} and CD1c$^+$ (23, 199).

Homology
The homology between LCs in humans and mouse is obvious given their exclusive anatomical occupancy and shared expression of langerin, EPCAM, and presence of Birbeck granules. Comparative transcriptomic analysis of human and mouse LCs has never been performed.

Transcription Factors
Langerhans cell development is dependent on PU.1, Runx3, and Id2, although the latter may be dispensable for bone marrow-derived LCs (74, 188, 200, 201).

Function
Langerhans cells are able to induce different immune responses depending on the context. Depletion of murine LCs can either exacerbate or suppress contact hypersensitivity immune response [reviewed in Ref. (202)]. In a mouse model of graft versus host disease (GVHD), LCs neither primed CD8$^+$ T cells nor programed their homing to the epidermis but were required for their effector function *in situ* (203). This is consistent with their inability to cross-present antigen *in vivo* (80, 204), although cross-presentation has been reported using *in vitro* assays (205). In mice, LCs appear to be critical for Th17 response against the yeast form of *Candida albicans* in the epidermis through engagement of Dectin-1 and their subsequent production of IL-6 (206). In humans, failure to generate effective Th17 responses (as a result of a range of mutations in, for example, IL-17RA, IL-17F, STAT1 genes) can result in chronic mucocutaneous candidiasis (CMC) (207). However, it is unclear if immunity against *Candida* infections in the skin in healthy individuals is dependent upon LCs. Notably, human LCs do not appear to express Dectin-1, which is important for *Candida* recognition (208). *In vitro* human LCs appear versatile and are capable of generating Th1, Th2 (209), Th17 (210), Th22 (211), and Treg (212) responses depending on the experimental conditions used.

Macrophages
Macrophages are a diverse population of tissue-resident cells with roles in inflammation, tissue homeostasis, and repair. Macrophage identity and function can be influenced by three variables: (1) resident tissue environment; (2) exposure to activation signals; and (3) ontogeny (monocyte- vs. prenatal precursor-derived) [reviewed in Ref. (3)].

The nomenclature of macrophages is based upon their tissue of origin [for example, Kupffer cells (liver), osteoclasts (bone), and microglia (CNS)]. This is in recognition of the central influence of environment on their phenotype and function. Examples of these functional specializations include breakdown of RBCs (Kupffer cells and splenic macrophages), bone resorption (osteoclasts), gut peristalsis (muscularis macrophages), and neural network development and maintenance (microglia) (213–215). Although macrophages in the vast majority of tissues, except dermis and the lamina propria, are prenatally derived, their preservation into adulthood by self-renewal is variable by site and in the presence of inflammation [(15, 216, 217) and reviewed in Ref. (218)]. The relative preservation of dermal macrophages and LCs in patients lacking circulating blood monocytes and DCs due to heterozygote GATA2 and biallelic IRF8 deficiencies supports a prenatal origin of some human macrophages (59, 60).

Microarray transcriptome analysis has identified several thousand transcripts with greater than twofold difference in expression between macrophages from different sites in mice (219), supporting unique local microenvironment-related characteristics. These tissue specific transcripts are more prominent within macrophages than DCs (219) and may reflect the tissue-resident nature of macrophages. The impact and underlying mechanisms of environmental regulation on macrophages was elegantly demonstrated by the unique epigenetic modulation of macrophage in distinct tissues and the ability of macrophages from one environment to develop the characteristics of their counterparts in another tissue (220, 221).

Phenotype
Murine macrophages express the antigens CD11b, CD68, CSF1R, and F4/80 (215). With the exception of F4/80 which is predominantly expressed on eosinophils (222), these antigens are also expressed on human macrophages (223). Furthermore, human alveolar macrophages were shown to express many antigens, which are conserved at transcript level with murine bone marrow-derived macrophages (163).

Homology
Comparative analysis between human and mouse macrophage populations has been poorly studied. In skin, homologous monocyte-derived dermal macrophage populations have been identified (25) but the murine counterparts of human dermal macrophages containing melanin-granules (melanophages) remain uncertain. While a range of transcriptional analyses of human macrophage populations in health and disease have been performed, comparisons between human tissues and across species have not been rigorously undertaken (224).

Transcription Factors
The transcriptional requirements of murine YS-derived macrophages differ to those of HSC-derived macrophages. YS-derived microglia require PU.1 and Irf8 but are independent of Myb, Id2, Batf3, and Klf4 (2, 12, 225). Consistent with macrophage tissue specializations, additional TFs such NFATc1 and Spi-C have been shown to be required for osteoclasts and splenic and bone marrow macrophage differentiation, respectively (226–228).

Function
The M1/M2 paradigm has been described to model the diverse programs of macrophage activation but has largely relied on

in vitro generated macrophages. This has provided a useful tool to examine macrophage activation in the absence of tissue-specific effects. More recently, a spectrum of responses, with M1 and M2 being two poles of a continuum that is transcriptionally apparent, were identified (229). It is unclear how closely human and murine macrophages are aligned in response to a similar range of stimuli. There are inter-species differences in the response to a single stimulus (LPS) between human and mouse *in vitro* derived macrophages; INOS transcript is preferentially induced in mouse but human macrophages characteristically upregulate CCL20, CXCL13, IL-7R, P2RX7, and STAT4 (230).

Mononuclear Phagocytes in Inflammation

Classical Ly6C^{hi}CX3CR1^{lo} monocytes infiltrate inflamed tissues where they can acquire either DC or macrophage properties (33, 231). This *in vivo* process (thought to be analogous to *in vitro* mo-Mac and mo-DC differentiation) can be influenced by local microbiota (97, 98, 170, 171). In infection and disease, monocyte-derived cells accumulate in greater numbers in a broad range of tissues [reviewed in Ref. (232)]. In many such models of infection, they are non-redundant and required for clearance of pathogens by promoting protective Th1 and Th17 responses (34, 233, 234). This suggests that despite shared functions with resident conventional DCs, there are important differences that require the presence of monocyte-derived cells to overcome infection. In murine experimental autoimmune encephalomyelitis, monocytes infiltrate the CNS but are not long-lived and following resolution do not contribute to the microglial pool (231). Analysis of murine Kupffer cells suggests functional heterogeneity between resident and recruited populations (235).

Snapshot analysis of inflamed human tissue similarly reveals additional subsets that are not present in health [(47, 99, 179, 236, 237) and reviewed in Ref. (31)]. These include inflammatory dendritic epidermal cells (IDECs) found in atopic dermatitis, TNF, and iNOS producing DCs (Tip DCs) and slan DCs, found in psoriasis (99, 179, 236, 237). In rheumatoid arthritis synovial fluid and malignant ascites, there is an accumulation of cells that express overlapping markers with blood CD1c$^+$ DCs but additionally express CD1a, CD206, SIRPα, and CD14 (47). Monocytes can acquire DC characteristics when cultured with *ex vivo* GM-CSF-primed synovial T cells, which potentially suggests a mechanism for their generation (238). Histiocytes are pathological MPs expressing CD68 and CD163. It is unknown if these cells, often found in granulomas, arise from resident macrophages or are monocyte-derived. Further studies are required to establish the *in vivo* differentiation requirements of inflammatory MP populations and how they contribute to disease.

Conclusion

In this review, we have discussed the parallel organization of the MPS between humans and mice. We demonstrate the use of comparative biology approaches as both a validation and discovery tool to dissect the development and functional heterogeneity of mononuclear phagocytes in a reciprocal manner across the two species. The incorporation of high-dimensional unbiased single-cell genomics and proteomics technologies will facilitate the interrogation of functionally relevant populations with indiscrete phenotypes and validate current definitions of cell-types based on limited antigen expression profile particularly during inflammation. This combined strategy will accelerate the translation of fundamental MPS biology to clinical benefit through enhanced understanding of the pathomechanisms of disease and facilitate the development of novel approaches in vaccination and cancer immunotherapy.

Acknowledgments

MH and GR are funded by The Wellcome Trust, UK [Intermediate Clinical Fellowship; WT088555A (MH) and Clinical Research Training Fellowship; WT098914MA (GR)]. The authors wish to acknowledge Matthew Collin for assistance with illustration.

References

1. Merad M, Sathe P, Helft J, Miller J, Mortha A. The dendritic cell lineage: ontogeny and function of dendritic cells and their subsets in the steady state and the inflamed setting. *Annu Rev Immunol* (2013) 31:563–604. doi:10.1146/annurev-immunol-020711-074950

2. Ginhoux F, Jung S. Monocytes and macrophages: developmental pathways and tissue homeostasis. *Nat Rev Immunol* (2014) 14:392–404. doi:10.1038/nri3671

3. Varol C, Mildner A, Jung S. Macrophages: development and tissue specialization. *Annu Rev Immunol* (2015) 33:643–75. doi:10.1146/annurev-immunol-032414-112220

4. Bancherau J, Steinman RM. Dendritic cells and the control of immunity. *Nature* (1998) 392:245–52. doi:10.1038/32588

5. Steinman RM, Bancherau J. Taking dendritic cells into medicine. *Nature* (2007) 449:419–26. doi:10.1038/nature06175

6. van Furth R, Cohn ZA, Hirsch JG, Humphrey JH, Spector WG, Langevoort HL. The mononuclear phagocyte system: a new classification of macrophages, monocytes, and their precursor cells. *Bull World Health Organ* (1972) 46:845–52.

7. van Furth R. Current view on the mononuclear phagocyte system. *Immunobiology* (1982) 161:178–85. doi:10.1016/S0171-2985(82)80072-7

8. Cline MJ. Bactericidal activity of human macrophages: analysis of factors influencing the killing of *Listeria* monocytogenes. *Infect Immun* (1970) 2:156–61.

9. Sallusto F, Lanzavecchia A. Efficient presentation of soluble antigen by cultured human dendritic cells is maintained by granulocyte/macrophage colony-stimulating factor plus interleukin 4 and downregulated by tumor necrosis factor alpha. *J Exp Med* (1994) 179:1109–18. doi:10.1084/jem.179.4.1109

10. Caux C, Vanbervliet B, Massacrier C, Dezutter-Dambuyant C, de Saint-Vis B, Jacquet C, et al. CD34+ hematopoietic progenitors from human cord blood differentiate along two independent dendritic cell pathways in response to GM-CSF+TNF alpha. *J Exp Med* (1996) 184:695–706. doi:10.1084/jem.184.2.695

11. Liu K, Victora GD, Schwickert TA, Guermonprez P, Meredith MM, Yao K, et al. In vivo analysis of dendritic cell development and homeostasis. *Science* (2009) 324:392–7. doi:10.1126/science.1170540

12. Schulz C, Gomez Perdiguero E, Chorro L, Szabo-Rogers H, Cagnard N, Kierdorf K, et al. A lineage of myeloid cells independent of Myb and hematopoietic stem cells. *Science* (2012) 336:86–90. doi:10.1126/science.1219179

13. Hoeffel G, Wang Y, Greter M, See P, Teo P, Malleret B, et al. Adult Langerhans cells derive predominantly from embryonic fetal liver monocytes with a minor contribution of yolk sac-derived macrophages. *J Exp Med* (2012) 209:1167–81. doi:10.1084/jem.20120340

14. Yona S, Kim KW, Wolf Y, Mildner A, Varol D, Breker M, et al. Fate mapping reveals origins and dynamics of monocytes and tissue macrophages under homeostasis. *Immunity* (2013) 38:79–91. doi:10.1016/j.immuni.2012.12.001

15. Perdiguero EG, Klapproth K, Schulz C, Busch K, Azzoni E, Crozet L, et al. Tissue-resident macrophages originate from yolk-sac-derived erythro-myeloid progenitors. *Nature* (2014) 518(7540):547–51. doi:10.1038/nature13989

16. Hoeffel G, Chen J, Lavin Y, Low D, Almeida FF, See P, et al. C-myb(+) erythro-myeloid progenitor-derived fetal monocytes give rise to adult tissue-resident macrophages. *Immunity* (2015) 42:665–78. doi:10.1016/j.immuni.2015.03.011

17. Robbins SH, Walzer T, Dembele D, Thibault C, Defays A, Bessou G, et al. Novel insights into the relationships between dendritic cell subsets in human and mouse revealed by genome-wide expression profiling. *Genome Biol* (2008) 9:R17. doi:10.1186/gb-2008-9-1-r17

18. Ingersoll MA, Spanbroek R, Lottaz C, Gautier EL, Frankenberger M, Hoffmann R, et al. Comparison of gene expression profiles between human and mouse monocyte subsets. *Blood* (2010) 115:e10–9. doi:10.1182/blood-2009-07-235028

19. Bachem A, Guttler S, Hartung E, Ebstein F, Schaefer M, Tannert A, et al. Superior antigen cross-presentation and XCR1 expression define human CD11c+CD141+ cells as homologues of mouse CD8+ dendritic cells. *J Exp Med* (2010) 207:1273–81. doi:10.1084/jem.20100348

20. Crozat K, Guiton R, Contreras V, Feuillet V, Dutertre CA, Ventre E, et al. The XC chemokine receptor 1 is a conserved selective marker of mammalian cells homologous to mouse CD8alpha+ dendritic cells. *J Exp Med* (2010) 207:1283–92. doi:10.1084/jem.20100223

21. Jongbloed SL, Kassianos AJ, McDonald KJ, Clark GJ, Ju X, Angel CE, et al. Human CD141+ (BDCA-3)+ dendritic cells (DCs) represent a unique myeloid DC subset that cross-presents necrotic cell antigens. *J Exp Med* (2010) 207:1247–60. doi:10.1084/jem.20092140

22. Poulin LF, Reyal Y, Uronen-Hansson H, Schraml BU, Sancho D, Murphy KM, et al. DNGR-1 is a specific and universal marker of mouse and human Batf3-dependent dendritic cells in lymphoid and nonlymphoid tissues. *Blood* (2012) 119:6052–62. doi:10.1182/blood-2012-01-406967

23. Haniffa M, Shin A, Bigley V, McGovern N, Teo P, See P, et al. Human tissues contain CD141(hi) cross-presenting dendritic cells with functional homology to mouse CD103(+) nonlymphoid dendritic cells. *Immunity* (2012) 37:60–73. doi:10.1016/j.immuni.2012.04.012

24. Schlitzer A, McGovern N, Teo P, Zelante T, Atarashi K, Low D, et al. IRF4 transcription factor-dependent CD11b+ dendritic cells in human and mouse control mucosal IL-17 cytokine responses. *Immunity* (2013) 38:970–83. doi:10.1016/j.immuni.2013.04.011

25. McGovern N, Schlitzer A, Gunawan M, Jardine L, Shin A, Poyner E, et al. Human dermal CD14(+) cells are a transient population of monocyte-derived macrophages. *Immunity* (2014) 41:465–77. doi:10.1016/j.immuni.2014.08.006

26. Watchmaker PB, Lahl K, Lee M, Baumjohann D, Morton J, Kim SJ, et al. Comparative transcriptional and functional profiling defines conserved programs of intestinal DC differentiation in humans and mice. *Nat Immunol* (2014) 15:98–108. doi:10.1038/ni.2768

27. Lee J, Breton G, Oliveira TY, Zhou YJ, Aljoufi A, Puhr S, et al. Restricted dendritic cell and monocyte progenitors in human cord blood and bone marrow. *J Exp Med* (2015) 212:385–99. doi:10.1084/jem.20141442

28. Breton G, Lee J, Zhou YJ, Schreiber JJ, Keler T, Puhr S, et al. Circulating precursors of human CD1c+ and CD141+ dendritic cells. *J Exp Med* (2015) 212:401–13. doi:10.1084/jem.20141441

29. Waterston RH, Lindblad-Toh K, Birney E, Rogers J, Abril JF, Agarwal P, et al. Initial sequencing and comparative analysis of the mouse genome. *Nature* (2002) 420:520–62. doi:10.1038/nature01262

30. Haniffa M, Collin M, Ginhoux F. Ontogeny and functional specialization of dendritic cells in human and mouse. *Adv Immunol* (2013) 120:1–49. doi:10.1016/B978-0-12-417028-5.00001-6

31. Haniffa M, Gunawan M, Jardine L. Human skin dendritic cells in health and disease. *J Dermatol Sci* (2015) 77(2):85–92. doi:10.1016/j.jdermsci.2014.08.012

32. Collin M, Bigley V, Haniffa M, Hambleton S. Human dendritic cell deficiency: the missing ID? *Nat Rev Immunol* (2011) 11:575–83. doi:10.1038/nri3046

33. Serbina NV, Salazar-Mather TP, Biron CA, Kuziel WA, Pamer EG. TNF/iNOS-producing dendritic cells mediate innate immune defense against bacterial infection. *Immunity* (2003) 19:59–70. doi:10.1016/S1074-7613(03)00171-7

34. Leon B, Lopez-Bravo M, Ardavin C. Monocyte-derived dendritic cells formed at the infection site control the induction of protective T helper 1 responses against Leishmania. *Immunity* (2007) 26:519–31. doi:10.1016/j.immuni.2007.01.017

35. Serbina NV, Hohl TM, Cherny M, Pamer EG. Selective expansion of the monocytic lineage directed by bacterial infection. *J Immunol* (2009) 183:1900–10. doi:10.4049/jimmunol.0900612

36. de Jong E, Suddason T, Lord GM. Translational mini-review series on Th17 cells: development of mouse and human T helper 17 cells. *Clin Exp Immunol* (2010) 159:148–58. doi:10.1111/j.1365-2249.2009.04041.x

37. Noster R, Riedel R, Mashreghi MF, Radbruch H, Harms L, Haftmann C, et al. IL-17 and GM-CSF expression are antagonistically regulated by human T helper cells. *Sci Transl Med* (2014) 6:241ra80. doi:10.1126/scitranslmed.3008706

38. Codarri L, Gyulveszi G, Tosevski V, Hesske L, Fontana A, Magnenat L, et al. RORgammat drives production of the cytokine GM-CSF in helper T cells, which is essential for the effector phase of autoimmune neuroinflammation. *Nat Immunol* (2011) 12:560–7. doi:10.1038/ni.2027

39. Roncarolo MG, Battaglia M. Regulatory T-cell immunotherapy for tolerance to self antigens and alloantigens in humans. *Nat Rev Immunol* (2007) 7:585–98. doi:10.1038/nri2138

40. Mestas J, Hughes CC. Of mice and not men: differences between mouse and human immunology. *J Immunol* (2004) 172:2731–8. doi:10.4049/jimmunol.172.5.2731

41. Van Rhijn I, Ly D, Moody DB. CD1a, CD1b, and CD1c in immunity against mycobacteria. *Adv Exp Med Biol* (2013) 783:181–97. doi:10.1007/978-1-4614-6111-1_10

42. van de Mortel JE, Aarts MG. Comparative transcriptomics – model species lead the way. *New Phytol* (2006) 170:199–201. doi:10.1111/j.1469-8137.2006.01708.x

43. Shay T, Jojic V, Zuk O, Rothamel K, Puyraimond-Zemmour D, Feng T, et al. Conservation and divergence in the transcriptional programs of the human and mouse immune systems. *Proc Natl Acad Sci U S A* (2013) 110:2946–51. doi:10.1073/pnas.1222738110

44. Subramanian A, Tamayo P, Mootha VK, Mukherjee S, Ebert BL, Gillette MA, et al. Gene set enrichment analysis: a knowledge-based approach for interpreting genome-wide expression profiles. *Proc Natl Acad Sci U S A* (2005) 102:15545–50. doi:10.1073/pnas.0506580102

45. Lamb J, Crawford ED, Peck D, Modell JW I, Blat C, Wrobel MJ, et al. The connectivity map: using gene-expression signatures to connect small molecules, genes, and disease. *Science* (2006) 313:1929–35. doi:10.1126/science.1132939

46. Lindstedt M, Lundberg K, Borrebaeck CA. Gene family clustering identifies functionally associated subsets of human in vivo blood and tonsillar dendritic cells. *J Immunol* (2005) 175:4839–46. doi:10.4049/jimmunol.175.8.4839

47. Segura E, Touzot M, Bohineust A, Cappuccio A, Chiocchia G, Hosmalin A, et al. Human inflammatory dendritic cells induce th17 cell differentiation. *Immunity* (2013) 38:336–48. doi:10.1016/j.immuni.2012.10.018

48. Jaitin DA, Kenigsberg E, Keren-Shaul H, Elefant N, Paul F, Zaretsky I, et al. Massively parallel single-cell RNA-seq for marker-free decomposition of tissues into cell types. *Science* (2014) 343:776–9. doi:10.1126/science.1247651

49. Mahata B, Zhang X, Kolodziejczyk AA, Proserpio V, Haim-Vilmovsky L, Taylor AE, et al. Single-cell RNA sequencing reveals T helper cells synthesizing steroids de novo to contribute to immune homeostasis. *Cell Rep* (2014) 7:1130–42. doi:10.1016/j.celrep.2014.04.011

50. Trapnell C, Cacchiarelli D, Grimsby J, Pokharel P, Li S, Morse M, et al. The dynamics and regulators of cell fate decisions are revealed by pseudotemporal ordering of single cells. *Nat Biotechnol* (2014) 32:381–6. doi:10.1038/nbt.2859

51. Treutlein B, Brownfield DG, Wu AR, Neff NF, Mantalas GL, Espinoza FH, et al. Reconstructing lineage hierarchies of the distal lung epithelium using single-cell RNA-seq. *Nature* (2014) 509:371–5. doi:10.1038/nature13173

52. Durruthy-Durruthy R, Gottlieb A, Hartman BH, Waldhaus J, Laske RD, Altman R, et al. Reconstruction of the mouse otocyst and early neuroblast lineage at single-cell resolution. *Cell* (2014) 157:964–78. doi:10.1016/j.cell.2014.03.036

53. Arsenio J, Kakaradov B, Metz PJ, Kim SH, Yeo GW, Chang JT. Early specification of CD8+ T lymphocyte fates during adaptive immunity revealed by single-cell gene-expression analyses. *Nat Immunol* (2014) 15:365–72. doi:10.1038/ni.2842

54. Luber CA, Cox J, Lauterbach H, Fancke B, Selbach M, Tschopp J, et al. Quantitative proteomics reveals subset-specific viral recognition in dendritic cells. *Immunity* (2010) 32:279–89. doi:10.1016/j.immuni.2010.01.013

55. Hettinger J, Richards DM, Hansson J, Barra MM, Joschko AC, Krijgsveld J, et al. Origin of monocytes and macrophages in a committed progenitor. *Nat Immunol* (2013) 14:821–30. doi:10.1038/ni.2638

56. Amir, el- AD, Davis KL, Tadmor MD, Simonds EF, Levine JH, Bendall SC, et al. viSNE enables visualization of high dimensional single-cell data and reveals phenotypic heterogeneity of leukemia. *Nat Biotechnol* (2013) 31:545–52. doi:10.1038/nbt.2594

57. Becher B, Schlitzer A, Chen J, Mair F, Sumatoh HR, Teng KW, et al. High-dimensional analysis of the murine myeloid cell system. *Nat Immunol* (2014) 15(12):1181–9. doi:10.1038/ni.3006

Human and mouse mononuclear phagocyte networks: a tale of two...

35

58. Segura E, Amigorena S. Cross-presentation by human dendritic cell subsets. *Immunol Lett* (2014) **158**:73–8. doi:10.1016/j.imlet.2013.12.001

59. Bigley V, Haniffa M, Doulatov S, Wang XN, Dickinson R, McGovern N, et al. The human syndrome of dendritic cell, monocyte, B and NK lymphoid deficiency. *J Exp Med* (2011) **208**:227–34. doi:10.1084/jem.20101459

60. Hambleton S, Salem S, Bustamante J, Bigley V, Boisson-Dupuis S, Azevedo J, et al. IRF8 mutations and human dendritic-cell immunodeficiency. *N Engl J Med* (2011) **365**(2):127–38. doi:10.1056/NEJMoa1100066

61. Guilliams M, Ginhoux F, Jakubzick C, Naik SH, Onai N, Schraml BU, et al. Dendritic cells, monocytes and macrophages: a unified nomenclature based on ontogeny. *Nat Rev Immunol* (2014) **14**(8):571–8. doi:10.1038/nri3712

62. Naik SH, Sathe P, Park HY, Metcalf D, Proietto AI, Dakic A, et al. Development of plasmacytoid and conventional dendritic cell subtypes from single precursor cells derived in vitro and in vivo. *Nat Immunol* (2007) **8**:1217–26. doi:10.1038/ni1522

63. Onai N, Kurabayashi K, Hosoi-Amaike M, Toyama-Sorimachi N, Matsushima K, Inaba K, et al. A clonogenic progenitor with prominent plasmacytoid dendritic cell developmental potential. *Immunity* (2013) **38**:943–57. doi:10.1016/j.immuni.2013.04.006

64. Galibert L, Diemer GS, Liu Z, Johnson RS, Smith JL, Walzer T, et al. Nectin-like protein 2 defines a subset of T-cell zone dendritic cells and is a ligand for class-I-restricted T-cell-associated molecule. *J Biol Chem* (2005) **280**:21955–64. doi:10.1074/jbc.M502095200

65. Bachem A, Hartung E, Guttler S, Mora A, Zhou X, Hegemann A, et al. Expression of XCR1 characterizes the Batf3-dependent lineage of dendritic cells capable of antigen cross-presentation. *Front Immunol* (2012) **3**:214. doi:10.3389/fimmu.2012.00214

66. Contreras V, Urien C, Guiton R, Alexandre Y, Vu Manh TP, Andrieu T, et al. Existence of CD8alpha-like dendritic cells with a conserved functional specialization and a common molecular signature in distant mammalian species. *J Immunol* (2010) **185**:3313–25. doi:10.4049/jimmunol.1000824

67. Tatsumi K, Taatjes DJ, Wadsworth MP, Bouchard BA, Bovill EG. Cell adhesion molecule 1 (CADM1) is ubiquitously present in the endothelium and smooth muscle cells of the human macro- and micro-vasculature. *Histochem Cell Biol* (2012) **138**:815–20. doi:10.1007/s00418-012-1024-2

68. Schraml BU, van Blijswijk J, Zelenay S, Whitney PG, Filby A, Acton SE, et al. Genetic tracing via DNGR-1 expression history defines dendritic cells as a hematopoietic lineage. *Cell* (2013) **154**:843–58. doi:10.1016/j.cell.2013.07.014

69. Ginhoux F, Collin MP, Bogunovic M, Abel M, Leboeuf M, Helft J, et al. Blood-derived dermal langerin+ dendritic cells survey the skin in the steady state. *J Exp Med* (2007) **204**:3133–46. doi:10.1084/jem.20071733

70. Bigley V, McGovern N, Milne P, Dickinson R, Pagan S, Cookson S, et al. Langerin-expressing dendritic cells in human tissues are related to CD1c+ dendritic cells and distinct from Langerhans cells and CD141high XCR1+ dendritic cells. *J Leukoc Biol* (2014) **97**(4):627–34. doi:10.1189/jlb.1HI0714-351R

71. Poulin LF, Salio M, Griessinger E, Anjos-Afonso F, Craciun L, Chen JL, et al. Characterization of human DNGR-1+ BDCA3+ leukocytes as putative equivalents of mouse CD8alpha+ dendritic cells. *J Exp Med* (2010) **207**:1261–71. doi:10.1084/jem.20092618

72. Hildner K, Edelson BT, Purtha WE, Diamond M, Matsushita H, Kohyama M, et al. Batf3 deficiency reveals a critical role for CD8alpha+ dendritic cells in cytotoxic T cell immunity. *Science* (2008) **322**:1097–100. doi:10.1126/science.1164206

73. Schiavoni G, Mattei F, Sestili P, Borghi P, Venditti M, Morse HC, et al. ICSBP is essential for the development of mouse type I interferon-producing cells and for the generation and activation of CD8alpha(+) dendritic cells. *J Exp Med* (2002) **196**:1415–25. doi:10.1084/jem.20021263

74. Hacker C, Kirsch RD, Ju XS, Hieronymus T, Gust TC, Kuhl C, et al. Transcriptional profiling identifies Id2 function in dendritic cell development. *Nat Immunol* (2003) **4**:380–6. doi:10.1038/ni903

75. Ginhoux F, Liu K, Helft J, Bogunovic M, Greter M, Hashimoto D, et al. The origin and development of nonlymphoid tissue CD103+ DCs. *J Exp Med* (2009) **206**:3115–30. doi:10.1084/jem.20091756

76. Kashiwada M, Pham NL, Pewe LL, Harty JT, Rothman PB. NFIL3/E4BP4 is a key transcription factor for CD8alpha(+) dendritic cell development. *Blood* (2011) **117**:6193–7. doi:10.1182/blood-2010-07-295873

77. Tussiwand R, Lee WL, Murphy TL, Mashayekhi M, KC W, Albring JC, et al. Compensatory dendritic cell development mediated by BATF-IRF interactions. *Nature* (2012) **490**:502–7. doi:10.1038/nature11531

78. den Haan JM, Lehar SM, Bevan MJ. CD8(+) but not CD8(−) dendritic cells cross-prime cytotoxic T cells in vivo. *J Exp Med* (2000) **192**:1685–96. doi:10.1084/jem.192.12.1685

79. Iyoda T, Shimoyama S, Liu K, Omatsu Y, Akiyama Y, Maeda Y, et al. The CD8+ dendritic cell subset selectively endocytoses dying cells in culture and in vivo. *J Exp Med* (2002) **195**:1289–302. doi:10.1084/jem.20020161

80. Bedoui S, Whitney PG, Waithman J, Eidsmo L, Wakim L, Caminschi I, et al. Cross-presentation of viral and self antigens by skin-derived CD103+ dendritic cells. *Nat Immunol* (2009) **10**:488–95. doi:10.1038/ni.1724

81. Hemont C, Neel A, Heslan M, Braudeau C, Josien R. Human blood mDC subsets exhibit distinct TLR repertoire and responsiveness. *J Leukoc Biol* (2013) **93**:599–609. doi:10.1189/jlb.0912452

82. Schulz O, Edwards AD, Schito M, Aliberti J, Manickasingham S, Sher A, et al. CD40 triggering of heterodimeric IL-12 p70 production by dendritic cells in vivo requires a microbial priming signal. *Immunity* (2000) **13**:453–62. doi:10.1016/S1074-7613(00)00045-5

83. Lauterbach H, Bathke B, Gilles S, Traidl-Hoffmann C, Luber CA, Fejer G, et al. Mouse CD8alpha+ DCs and human BDCA3+ DCs are major producers of IFN-lambda in response to poly IC. *J Exp Med* (2010) **207**:2703–17. doi:10.1084/jem.20092720

84. Savina A, Peres A, Cebrian I, Carmo N, Moita C, Hacohen N, et al. The small GTPase Rac2 controls phagosomal alkalinization and antigen crosspresentation selectively in CD8(+) dendritic cells. *Immunity* (2009) **30**:544–55. doi:10.1016/j.immuni.2009.01.013

85. Lin ML, Zhan Y, Proietto AI, Prato S, Wu L, Heath WR, et al. Selective suicide of cross-presenting CD8+ dendritic cells by cytochrome c injection shows functional heterogeneity within this subset. *Proc Natl Acad Sci U S A* (2008) **105**:3029–34. doi:10.1073/pnas.0712394105

86. Desch AN, Gibbings SL, Clambey ET, Janssen WJ, Slansky JE, Kedl RM, et al. Dendritic cell subsets require cis-activation for cytotoxic CD8 T-cell induction. *Nat Commun* (2014) **5**:4674. doi:10.1038/ncomms5674

87. Cohn L, Chatterjee B, Esselborn F, Smed-Sorensen A, Nakamura N, Chalouni C, et al. Antigen delivery to early endosomes eliminates the superiority of human blood BDCA3+ dendritic cells at cross presentation. *J Exp Med* (2013) **210**:1049–63. doi:10.1084/jem.20121251

88. Segura E, Durand M, Amigorena S. Similar antigen cross-presentation capacity and phagocytic functions in all freshly isolated human lymphoid organ-resident dendritic cells. *J Exp Med* (2013) **210**:1035–47. doi:10.1084/jem.20121103

89. Nizzoli G, Krietsch J, Weick A, Steinfelder S, Facciotti F, Gruarin P, et al. Human CD1c+ dendritic cells secrete high levels of IL-12 and potently prime cytotoxic T-cell responses. *Blood* (2013) **122**:932–42. doi:10.1182/blood-2013-04-495424

90. Mashayekhi M, Sandau MM I, Dunay R, Frickel EM, Khan A, Goldszmid RS, et al. CD8alpha(+) dendritic cells are the critical source of interleukin-12 that controls acute infection by *Toxoplasma gondii* tachyzoites. *Immunity* (2011) **35**:249–59. doi:10.1016/j.immuni.2011.08.008

91. Martinez-Lopez M, Iborra S, Conde-Garrosa R, Sancho D. Batf3-dependent CD103+ dendritic cells are major producers of IL-12 that drive local Th1 immunity against *Leishmania* major infection in mice. *Eur J Immunol* (2015) **45**:119–29. doi:10.1002/eji.201444651

92. Nakano H, Free ME, Whitehead GS, Maruoka S, Wilson RH, Nakano K, et al. Pulmonary CD103(+) dendritic cells prime Th2 responses to inhaled allergens. *Mucosal Immunol* (2012) **5**:53–65. doi:10.1038/mi.2011.47

93. Segura E, Valladeau-Guilemond J, Donnadieu M-H, Sastre-Garau X, Soumelis V, Amigorena S. Characterization of resident and migratory dendritic cells in human lymph nodes. *J Exp Med* (2012) **209**:653–60. doi:10.1084/jem.20111457

94. Yu CI, Becker C, Metang P, Marches F, Wang Y, Toshiyuki H, et al. Human CD141+ dendritic cells induce CD4+ T cells to produce type 2 cytokines. *J Immunol* (2014) **193**(9):4335–43. doi:10.4049/jimmunol.1401159

95. Ginhoux F, Schlitzer A. CD11b+ DCs rediscovered: implications for vaccination. *Expert Rev Vaccines* (2014) **13**:445–7. doi:10.1586/14760584.2014.893196

96. Lewis KL, Caton ML, Bogunovic M, Greter M, Grajkowska LT, Ng D, et al. Notch2 receptor signaling controls functional differentiation of dendritic cells in the spleen and intestine. *Immunity* (2011) **35**:780–91. doi:10.1016/j.immuni.2011.08.013

97. Tamoutounour S, Guilliams M, Montanana Sanchis F, Liu H, Terhorst D, Malosse C, et al. Origins and functional specialization of macrophages and of conventional and monocyte-derived dendritic cells in mouse skin. *Immunity* (2013) **39**:925–38. doi:10.1016/j.immuni.2013.10.004

98. Plantinga M, Guilliams M, Vanheerswynghels M, Deswarte K, Branco-Madeira F, Toussaint W, et al. Conventional and monocyte-derived CD11b(+) dendritic cells initiate and maintain T helper 2 cell-mediated immunity to house dust mite allergen. *Immunity* (2013) 38:322–35. doi:10.1016/j.immuni.2012.10.016

99. Wollenberg A, Mommaas M, Oppel T, Schottdorf EM, Gunther S, Moderer M. Expression and function of the mannose receptor CD206 on epidermal dendritic cells in inflammatory skin diseases. *J Invest Dermatol* (2002) 118:327–34. doi:10.1046/j.0022-202x.2001.01665.x

100. Zaba LC, Fuentes-Duculan J, Steinman RM, Krueger JG, Lowes MA. Normal human dermis contains distinct populations of CD11c+BDCA-1+ dendritic cells and CD163+FXIIIA+ macrophages. *J Clin Invest* (2007) 117:2517–25. doi:10.1172/JCI32282

101. Ochoa MT, Loncaric A, Krutzik SR, Becker TC, Modlin RL. Dermal dendritic cells" comprise two distinct populations: CD1(+) dendritic cells and CD209(+) macrophages. *J Invest Dermatol* (2008) 128(9):2225–31. doi:10.1038/jid.2008.56

102. Wu L, D'Amico A, Winkel KD, Suter M, Lo D, Shortman K. RelB is essential for the development of myeloid-related CD8alpha- dendritic cells but not of lymphoid-related CD8alpha+ dendritic cells. *Immunity* (1998) 9:839–47. doi:10.1016/S1074-7613(00)80649-4

103. Guerriero A, Langmuir PB, Spain LM, Scott EW. PU.1 is required for myeloid-derived but not lymphoid-derived dendritic cells. *Blood* (2000) 95:879–85.

104. Suzuki S, Honma K, Matsuyama T, Suzuki K, Toriyama K, Akitoyo I, et al. Critical roles of interferon regulatory factor 4 in CD11bhighCD8alpha- dendritic cell development. *Proc Natl Acad Sci U S A* (2004) 101:8981–6. doi:10.1073/pnas.0402139101

105. Tamura T, Tailor P, Yamaoka K, Kong HJ, Tsujimura H, O'Shea JJ, et al. IFN regulatory factor-4 and -8 govern dendritic cell subset development and their functional diversity. *J Immunol* (2005) 174:2573–81. doi:10.4049/jimmunol.174.5.2573

106. Caton ML, Smith-Raska MR, Reizis B. Notch-RBP-J signaling controls the homeostasis of CD8- dendritic cells in the spleen. *J Exp Med* (2007) 204:1653–64.

107. Satpathy AT, Briseno CG, Lee JS, Ng D, Manieri NA, Kc W, et al. Notch2-dependent classical dendritic cells orchestrate intestinal immunity to attaching-and-effacing bacterial pathogens. *Nat Immunol* (2013) 14:937–48. doi:10.1038/ni.2679

108. Williams JW, Tjota MY, Clay BS, Vander Lugt B, Bandukwala HS, Hrusch CL, et al. Transcription factor IRF4 drives dendritic cells to promote Th2 differentiation. *Nat Commun* (2013) 4:2990. doi:10.1038/ncomms3990

109. Vander Lugt B, Khan AA, Hackney JA, Agrawal S, Lesch J, Zhou M, et al. Transcriptional programming of dendritic cells for enhanced MHC class II antigen presentation. *Nat Immunol* (2014) 15:161–7. doi:10.1038/ni.2795

110. Brass AL, Kehrli E, Eisenbeis CF, Storb U, Singh H. Pip, a lymphoid-restricted IRF, contains a regulatory domain that is important for autoinhibition and ternary complex formation with the Ets factor PU.1. *Genes Dev* (1996) 10:2335–47. doi:10.1101/gad.10.18.2335

111. Carotta S, Dakic A, D'Amico A, Pang SH, Greig KT, Nutt SL, et al. The transcription factor PU.1 controls dendritic cell development and Flt3 cytokine receptor expression in a dose-dependent manner. *Immunity* (2010) 32:628–41. doi:10.1016/j.immuni.2010.05.005

112. Maraskovsky E, Daro E, Roux E, Teepe M, Maliszewski CR, Hoek J, et al. In vivo generation of human dendritic cell subsets by Flt3 ligand. *Blood* (2000) 96:878–84.

113. Dakic A, Wu L, Nutt SL. Is PU.1 a dosage-sensitive regulator of haemopoietic lineage commitment and leukaemogenesis? *Trends Immunol* (2007) 28:108–14. doi:10.1016/j.it.2007.01.006

114. Tailor P, Tamura T, Morse HCR, Ozato K. The BXH2 mutation in IRF8 differentially impairs dendritic cell subset development in the mouse. *Blood* (2008) 111:1942–5. doi:10.1182/blood-2007-07-100750

115. Kurotaki D, Osato N, Nishiyama A, Yamamoto M, Ban T, Sato H, et al. Essential role of the IRF8-KLF4 transcription factor cascade in murine monocyte differentiation. *Blood* (2013) 121:1839–49. doi:10.1182/blood-2012-06-437863

116. Yanez A, Ng MY, Hassanzadeh-Kiabi N, Goodridge HS. IRF8 acts in lineage-committed rather than oligopotent progenitors to control neutrophil vs monocyte production. *Blood* (2015) 125:1452–9. doi:10.1182/blood-2014-09-600833

117. Dudziak D, Kamphorst AO, Heidkamp GF, Buchholz VR, Trumpfheller C, Yamazaki S, et al. Differential antigen processing by dendritic cell subsets in vivo. *Science* (2007) 315:107–11. doi:10.1126/science.1136080

118. Persson EK, Uronen-Hansson H, Semmrich M, Rivollier A, Hagerbrand K, Marsal J, et al. IRF4 transcription-factor-dependent CD103(+)CD11b(+) dendritic cells drive mucosal T helper 17 cell differentiation. *Immunity* (2013) 38:958–69. doi:10.1016/j.immuni.2013.03.009

119. Gao Y, Nish SA, Jiang R, Hou L, Licona-Limon P, Weinstein JS, et al. Control of T helper 2 responses by transcription factor IRF4-dependent dendritic cells. *Immunity* (2013) 39:722–32. doi:10.1016/j.immuni.2013.08.028

120. Zhou Q, Ho AW, Schlitzer A, Tang Y, Wong KH, Wong FH, et al. GM-CSF-licensed CD11b+ lung dendritic cells orchestrate Th2 immunity to blomia tropicalis. *J Immunol* (2014) 193:496–509. doi:10.4049/jimmunol.1303138

121. Iwasaki A, Medzhitov R. Toll-like receptor control of the adaptive immune responses. *Nat Immunol* (2004) 5:987–95. doi:10.1038/ni1112

122. Moret FM, Hack CE, van der Wurff-Jacobs KM, de Jager W, Radstake TR, Lafeber FP, et al. Intra-articular CD1c-expressing myeloid dendritic cells from rheumatoid arthritis patients express a unique set of T cell-attracting chemokines and spontaneously induce Th1, Th17 and Th2 cell activity. *Arthritis Res Ther* (2013) 15:R155. doi:10.1186/ar4338

123. Kassianos AJ, Wang X, Sampangi S, Muczynski K, Healy H, Wilkinson R. Increased tubulointerstitial recruitment of human CD141(hi) CLEC9A(+) and CD1c(+) myeloid dendritic cell subsets in renal fibrosis and chronic kidney disease. *Am J Physiol Renal Physiol* (2013) 305:F1391–401. doi:10.1152/ajprenal.00318.2013

124. Jahnsen FL, Moloney ED, Hogan T, Upham JW, Burke CM, Holt PG. Rapid dendritic cell recruitment to the bronchial mucosa of patients with atopic asthma in response to local allergen challenge. *Thorax* (2001) 56:823–6. doi:10.1136/thorax.56.11.823

125. Yu CI, Becker C, Wang Y, Marches F, Helft J, Leboeuf M, et al. Human CD1c dendritic cells drive the differentiation of CD103 CD8 mucosal effector T cells via the cytokine TGF-beta. *Immunity* (2013) 38(4):818–30. doi:10.1016/j.immuni.2013.03.004

126. Klechevsky E, Morita R, Liu M, Cao Y, Coquery S, Thompson-Snipes L, et al. Functional specializations of human epidermal Langerhans cells and CD14+ dermal dendritic cells. *Immunity* (2008) 29:497–510. doi:10.1016/j.immuni.2008.07.013

127. Haniffa M, Ginhoux F, Wang XN, Bigley V, Abel M, Dimmick I, et al. Differential rates of replacement of human dermal dendritic cells and macrophages during hematopoietic stem cell transplantation. *J Exp Med* (2009) 206:371–85. doi:10.1084/jem.20081633

128. O'Doherty U, Peng M, Gezelter S, Swiggard WJ, Betjes M, Bhardwaj N, et al. Human blood contains two subsets of dendritic cells, one immunologically mature and the other immature. *Immunology* (1994) 82:487–93.

129. Grouard G, Rissoan MC, Filgueira L, Durand I, Banchereau J, Liu YJ. The enigmatic plasmacytoid T cells develop into dendritic cells with interleukin (IL)-3 and CD40-ligand. *J Exp Med* (1997) 185:1101–11. doi:10.1084/jem.185.6.1101

130. Cella M, Jarrossay D, Facchetti F, Alebardi O, Nakajima H, Lanzavecchia A, et al. Plasmacytoid monocytes migrate to inflamed lymph nodes and produce large amounts of type I interferon. *Nat Med* (1999) 5:919–23. doi:10.1038/11360

131. Siegal FP, Kadowaki N, Shodell M, Fitzgerald-Bocarsly PA, Shah K, Ho S, et al. The nature of the principal type 1 interferon-producing cells in human blood. *Science* (1999) 284:1835–7. doi:10.1126/science.284.5421.1835

132. Satpathy AT, Wu X, Albring JC, Murphy KM. Re(de)fining the dendritic cell lineage. *Nat Immunol* (2012) 13:1145–54. doi:10.1038/ni.2467

133. Taussig DC, Pearce DJ, Simpson C, Rohatiner AZ, Lister TA, Kelly G, et al. Hematopoietic stem cells express multiple myeloid markers: implications for the origin and targeted therapy of acute myeloid leukemia. *Blood* (2005) 106:4086–92. doi:10.1182/blood-2005-03-1072

134. Toba K, Koike T, Shibata A, Hashimoto S, Takahashi M, Masuko M, et al. Novel technique for the direct flow cytofluorometric analysis of human basophils in unseparated blood and bone marrow, and the characterization of phenotype and peroxidase of human basophils. *Cytometry* (1999) 35:249–59. doi:10.1002/(SICI)1097-0320(19990301)35:3<249::AID-CYTO8>3.3.CO;2-F

135. Dzionek A, Fuchs A, Schmidt P, Cremer S, Zysk M, Miltenyi S, et al. BDCA-2, BDCA-3, and BDCA-4: three markers for distinct subsets of dendritic cells in human peripheral blood. *J Immunol* (2000) 165:6037–46. doi:10.4049/jimmunol.165.11.6037

136. Dzionek A, Sohma Y, Nagafune J, Cella M, Colonna M, Facchetti F, et al. BDCA-2, a novel plasmacytoid dendritic cell-specific type II C-type lectin, mediates antigen capture and is a potent inhibitor of interferon alpha/beta induction. *J Exp Med* (2001) 194:1823–34. doi:10.1084/jem.194.12.1823

137. Lebre MC, Jongbloed SL, Tas SW, Smeets TJ I, McInnes B, Tak PP. Rheumatoid arthritis synovium contains two subsets of CD83-DC-LAMP- dendritic cells

Human and mouse mononuclear phagocyte networks: a tale of two...

37

with distinct cytokine profiles. *Am J Pathol* (2008) **172**:940–50. doi:10.2353/ajpath.2008.070703

138. Schlitzer A, Loschko J, Mair K, Vogelmann R, Henkel L, Einwachter H, et al. Identification of CCR9- murine plasmacytoid DC precursors with plasticity to differentiate into conventional DCs. *Blood* (2011) **117**:6562–70. doi:10.1182/blood-2010-12-326678

139. Schlitzer A, Heiseke AF, Einwachter H, Reindl W, Schiemann M, Manta CP, et al. Tissue-specific differentiation of a circulating CCR9- pDC-like common dendritic cell precursor. *Blood* (2012) **119**:6063–71. doi:10.1182/blood-2012-03-418400

140. Cisse B, Caton ML, Lehner M, Maeda T, Scheu S, Locksley R, et al. Transcription factor E2-2 is an essential and specific regulator of plasmacytoid dendritic cell development. *Cell* (2008) **135**:37–48. doi:10.1016/j.cell.2008.09.016

141. Ghosh HS, Cisse B, Bunin A, Lewis KL, Reizis B. Continuous expression of the transcription factor e2-2 maintains the cell fate of mature plasmacytoid dendritic cells. *Immunity* (2010) **33**:905–16. doi:10.1016/j.immuni.2010.11.023

142. Peippo M, Ignatius J. Pitt-Hopkins syndrome. *Mol Syndromol* (2012) **2**:171–80.

143. Sasaki I, Hoshino K, Sugiyama T, Yamazaki C, Yano T, Iizuka A, et al. Spi-B is critical for plasmacytoid dendritic cell function and development. *Blood* (2012) **120**:4733–43. doi:10.1182/blood-2012-06-436527

144. Schotte R, Nagasawa M, Weijer K, Spits H, Blom B. The ETS transcription factor Spi-B is required for human plasmacytoid dendritic cell development. *J Exp Med* (2004) **200**:1503–9. doi:10.1084/jem.20041231

145. Hochrein H, O'Keeffe M, Wagner H. Human and mouse plasmacytoid dendritic cells. *Hum Immunol* (2002) **63**:1103–10. doi:10.1016/S0198-8859(02)00748-6

146. Fuchsberger M, Hochrein H, O'Keeffe M. Activation of plasmacytoid dendritic cells. *Immunol Cell Biol* (2005) **83**:571–7. doi:10.1111/j.1440-1711.2005.01392.x

147. Jarrossay D, Napolitani G, Colonna M, Sallusto F, Lanzavecchia A. Specialization and complementarity in microbial molecule recognition by human myeloid and plasmacytoid dendritic cells. *Eur J Immunol* (2001) **31**:3388–93. doi:10.1002/1521-4141(200111)31:11<3388::AID-IMMU3388>3.0.CO;2-Q

148. Kadowaki N, Ho S, Antonenko S, Malefyt RW, Kastelein RA, Bazan F, et al. Subsets of human dendritic cell precursors express different toll-like receptors and respond to different microbial antigens. *J Exp Med* (2001) **194**:863–9. doi:10.1084/jem.194.6.863

149. Demaria O, Pagni PP, Traub S, de Gassart A, Branzk N, Murphy AJ, et al. TLR8 deficiency leads to autoimmunity in mice. *J Clin Invest* (2010) **120**:3651–62. doi:10.1172/JCI42081

150. Desnues B, Macedo AB, Roussel-Queval A, Bonnardel J, Henri S, Demaria O, et al. TLR8 on dendritic cells and TLR9 on B cells restrain TLR7-mediated spontaneous autoimmunity in C57BL/6 mice. *Proc Natl Acad Sci U S A* (2014) **111**:1497–502. doi:10.1073/pnas.1314121111

151. Gilliet M, Cao W, Liu YJ. Plasmacytoid dendritic cells: sensing nucleic acids in viral infection and autoimmune diseases. *Nat Rev Immunol* (2008) **8**:594–606. doi:10.1038/nri2358

152. Garcia-Romo GS, Caielli S, Vega B, Connolly J, Allantaz F, Xu Z, et al. Netting neutrophils are major inducers of type I IFN production in pediatric systemic lupus erythematosus. *Sci Transl Med* (2011) **3**:73ra20. doi:10.1126/scitranslmed.3001201

153. Murdoch C, Tazzyman S, Webster S, Lewis CE. Expression of tie-2 by human monocytes and their responses to angiopoietin-2. *J Immunol* (2007) **178**:7405–11. doi:10.4049/jimmunol.178.11.7405

154. Schakel K, Kannagi R, Kniep B, Goto Y, Mitsuoka C, Zwirner J, et al. 6-Sulfo LacNAc, a novel carbohydrate modification of PSGL-1, defines an inflammatory type of human dendritic cells. *Immunity* (2002) **17**:289–301. doi:10.1016/S1074-7613(02)00393-X

155. Cros J, Cagnard N, Woollard K, Patey N, Zhang SY, Senechal B, et al. Human CD14dim monocytes patrol and sense nucleic acids and viruses via TLR7 and TLR8 receptors. *Immunity* (2010) **33**:375–86. doi:10.1016/j.immuni.2010.08.012

156. Rosenbauer F, Tenen DG. Transcription factors in myeloid development: balancing differentiation with transformation. *Nat Rev Immunol* (2007) **7**:105–17. doi:10.1038/nri2024

157. Staber PB, Zhang P, Ye M, Welner RS, Nombela-Arrieta C, Bach C, et al. Sustained PU.1 levels balance cell-cycle regulators to prevent exhaustion of adult hematopoietic stem cells. *Mol Cell* (2013) **49**:934–46. doi:10.1016/j.molcel.2013.01.007

158. Scott EW, Simon MC, Anastasi J, Singh H. Requirement of transcription factor PU.1 in the development of multiple hematopoietic lineages. *Science* (1994) **265**:1573–7. doi:10.1126/science.8079170

159. Kueh HY, Champhekar A, Nutt SL, Elowitz MB, Rothenberg EV. Positive feedback between PU.1 and the cell cycle controls myeloid differentiation. *Science* (2013) **341**:670–3. doi:10.1126/science.1240831

160. Laslo P, Spooner CJ, Warmflash A, Lancki DW, Lee HJ, Sciammas R, et al. Multilineage transcriptional priming and determination of alternate hematopoietic cell fates. *Cell* (2006) **126**:755–66. doi:10.1016/j.cell.2006.06.052

161. Rosa A, Ballarino M, Sorrentino A, Sthandier O, De Angelis FG, Marchioni M, et al. The interplay between the master transcription factor PU.1 and miR-424 regulates human monocyte/macrophage differentiation. *Proc Natl Acad Sci U S A* (2007) **104**:19849–54. doi:10.1073/pnas.0706963104

162. Kurotaki D, Yamamoto M, Nishiyama A, Uno K, Ban T, Ichino M, et al. IRF8 inhibits C/EBPalpha activity to restrain mononuclear phagocyte progenitors from differentiating into neutrophils. *Nat Commun* (2014) **5**:4978. doi:10.1038/ncomms5978

163. Hanna RN, Carlin LM, Hubbeling HG, Nackiewicz D, Green AM, Punt JA, et al. The transcription factor NR4A1 (Nur77) controls bone marrow differentiation and the survival of Ly6C- monocytes. *Nat Immunol* (2011) **12**:778–85. doi:10.1038/ni.2063

164. Bakri Y, Sarrazin S, Mayer UP, Tillmanns S, Nerlov C, Boned A, et al. Balance of MafB and PU.1 specifies alternative macrophage or dendritic cell fate. *Blood* (2005) **105**:2707–16. doi:10.1182/blood-2004-04-1448

165. Lehtonen A, Veckman V, Nikula T, Lahesmaa R, Kinnunen L, Matikainen S, et al. Differential expression of IFN regulatory factor 4 gene in human monocyte-derived dendritic cells and macrophages. *J Immunol* (2005) **175**:6570–9. doi:10.4049/jimmunol.175.10.6570

166. Krausgruber T, Blazek K, Smallie T, Alzabin S, Lockstone H, Sahgal N, et al. IRF5 promotes inflammatory macrophage polarization and TH1-TH17 responses. *Nat Immunol* (2011) **12**:231–8. doi:10.1038/ni.1990

167. Satoh T, Takeuchi O, Vandenbon A, Yasuda K, Tanaka Y, Kumagai Y, et al. The Jmjd3-Irf4 axis regulates M2 macrophage polarization and host responses against helminth infection. *Nat Immunol* (2010) **11**:936–44. doi:10.1038/ni.1920

168. Derlindati E, Dei Cas A, Montanini B, Spigoni V, Curella V, Aldigeri R, et al. Transcriptomic analysis of human polarized macrophages: more than one role of alternative activation? *PLoS One* (2015) **10**:e0119751. doi:10.1371/journal.pone.0119751

169. Martinez FO, Helming L, Milde R, Varin A, Melgert BN, Draijer C, et al. Genetic programs expressed in resting and IL-4 alternatively activated mouse and human macrophages: similarities and differences. *Blood* (2013) **121**:e57–69. doi:10.1182/blood-2012-06-436212

170. Bain CC, Bravo-Blas A, Scott CL, Gomez Perdiguero E, Geissmann F, Henri S, et al. Constant replenishment from circulating monocytes maintains the macrophage pool in the intestine of adult mice. *Nat Immunol* (2014) **15**:929–37. doi:10.1038/ni.2967

171. Bain CC, Scott CL, Uronen-Hansson H, Gudjonsson S, Jansson O, Grip O, et al. Resident and pro-inflammatory macrophages in the colon represent alternative context-dependent fates of the same Ly6Chi monocyte precursors. *Mucosal Immunol* (2013) **6**:498–510. doi:10.1038/mi.2012.89

172. Varol C, Landsman L, Fogg DK, Greenshtein L, Gildor B, Margalit R, et al. Monocytes give rise to mucosal, but not splenic, conventional dendritic cells. *J Exp Med* (2007) **204**:171–80. doi:10.1084/jem.20061011

173. Cheong C, Matos I, Choi JH, Dandamudi DB, Shrestha E, Longhi MP, et al. Microbial stimulation fully differentiates monocytes to DC-SIGN/CD209(+) dendritic cells for immune T cell areas. *Cell* (2010) **143**:416–29. doi:10.1016/j.cell.2010.09.039

174. Nakano H, Lin KL, Yanagita M, Charbonneau C, Cook DN, Kakiuchi T, et al. Blood-derived inflammatory dendritic cells in lymph nodes stimulate acute T helper type 1 immune responses. *Nat Immunol* (2009) **10**:394–402. doi:10.1038/ni.1707

175. Bosschaerts T, Guilliams M, Stijlemans B, Morias Y, Engel D, Tacke F, et al. Tip-DC development during parasitic infection is regulated by IL-10 and requires CCL2/CCR2, IFN-gamma and MyD88 signaling. *PLoS Pathog* (2010) **6**:e1001045. doi:10.1371/journal.ppat.1001045

176. Hohl TM, Rivera A, Lipuma L, Gallegos A, Shi C, Mack M, et al. Inflammatory monocytes facilitate adaptive CD4 T cell responses during respiratory fungal infection. *Cell Host Microbe* (2009) **6**:470–81. doi:10.1016/j.chom.2009.10.007

177. Jakubzick C, Gautier EL, Gibbings SL, Sojka DK, Schlitzer A, Johnson TE, et al. Minimal differentiation of classical monocytes as they survey steady-state

tissues and transport antigen to lymph nodes. *Immunity* (2013) **39**:599–610. doi:10.1016/j.immuni.2013.08.007

178. Wong KL, Yeap WH, Tai JJ, Ong SM, Dang TM, Wong SC. The three human monocyte subsets: implications for health and disease. *Immunol Res* (2012) **53**:41–57. doi:10.1007/s12026-012-8297-3

179. Hansel A, Gunther C, Ingwersen J, Starke J, Schmitz M, Bachmann M, et al. Human slan (6-sulfo LacNAc) dendritic cells are inflammatory dermal dendritic cells in psoriasis and drive strong TH17/TH1 T-cell responses. *J Allergy Clin Immunol* (2011) **127**:.e1–9. doi:10.1016/j.jaci.2010.12.009

180. Brumeanu TD, Swiggard WJ, Steinman RM, Bona CA, Zaghouani H. Efficient loading of identical viral peptide onto class II molecules by antigenized immunoglobulin and influenza virus. *J Exp Med* (1993) **178**:1795–9. doi:10.1084/jem.178.5.1795

181. Matthews K, Chung NP, Klasse PJ, Moore JP, Sanders RW. Potent induction of antibody-secreting B cells by human dermal-derived CD14+ dendritic cells triggered by dual TLR ligation. *J Immunol* (2012) **189**:5729–44. doi:10.4049/jimmunol.1200601

182. Birbeck MS, Breathnach AS, Everall JD. An electron microscope study of basal melanocytes and high-level clear cells (Langerhans cells) in vitiligo. *J Invest Dermatol* (1961) **37**:51–64. doi:10.1038/jid.1961.7

183. Kissenpfennig A, Ait-Yahia S, Clair-Moninot V, Stossel H, Badell E, Bordat Y, et al. Disruption of the langerin/CD207 gene abolishes birbeck granules without a marked loss of Langerhans cell function. *Mol Cell Biol* (2005) **25**:88–99. doi:10.1128/MCB.25.1.88-99.2005

184. Kubo A, Nagao K, Yokouchi M, Sasaki H, Amagai M. External antigen uptake by Langerhans cells with reorganization of epidermal tight junction barriers. *J Exp Med* (2009) **206**:2937–46. doi:10.1084/jem.20091527

185. Teunissen MB, Haniffa M, Collin MP. Insight into the immunobiology of human skin and functional specialization of skin dendritic cell subsets to innovate intradermal vaccination design. *Curr Top Microbiol Immunol* (2012) **351**:25–76. doi:10.1007/82_2011_169

186. Sere K, Baek JH, Ober-Blobaum J, Muller-Newen G, Tacke F, Yokota Y, et al. Two distinct types of Langerhans cells populate the skin during steady state and inflammation. *Immunity* (2012) **37**:905–16. doi:10.1016/j.immuni.2012.07.019

187. Nagao K, Kobayashi T, Moro K, Ohyama M, Adachi T, Kitashima DY, et al. Stress-induced production of chemokines by hair follicles regulates the trafficking of dendritic cells in skin. *Nat Immunol* (2012) **13**:744–52. doi:10.1038/ni.2353

188. Chopin M, Seillet C, Chevrier S, Wu L, Wang H, Morse HCR, et al. Langerhans cells are generated by two distinct PU.1-dependent transcriptional networks. *J Exp Med* (2013) **210**:2967–80. doi:10.1084/jem.20130930

189. Chorro L, Sarde A, Li M, Woollard KJ, Chambon P, Malissen B, et al. Langerhans cell (LC) proliferation mediates neonatal development, homeostasis, and inflammation-associated expansion of the epidermal LC network. *J Exp Med* (2009) **206**:3089–100. doi:10.1084/jem.20091586

190. Kanitakis J, Petruzzo P, Dubernard JM. Turnover of epidermal Langerhans' cells. *N Engl J Med* (2004) **351**:2661–2. doi:10.1056/NEJM200412163512523

191. Kanitakis J, Morelon E, Petruzzo P, Badet L, Dubernard JM. Self-renewal capacity of human epidermal Langerhans cells: observations made on a composite tissue allograft. *Exp Dermatol* (2011) **20**:145–6. doi:10.1111/j.1600-0625.2010.01146.x

192. Martinez-Cingolani C, Grandclaudon M, Jeanmougin M, Jouve M, Zollinger R, Soumelis V. Human blood BDCA-1 dendritic cells differentiate into Langerhans-like cells with thymic stromal lymphopoietin and TGF-beta. *Blood* (2014) **124**:2411–20. doi:10.1182/blood-2014-04-568311

193. Milne P, Bigley V, Gunawan M, Haniffa M, Collin M. CD1c+ blood dendritic cells have Langerhans cell potential. *Blood* (2014) **125**(3):470–3. doi:10.1182/blood-2014-08-593582

194. Merad M, Manz MG, Karsunky H, Wagers A, Peters W, Charo I, et al. Langerhans cells renew in the skin throughout life under steady-state conditions. *Nat Immunol* (2002) **3**:1135–41. doi:10.1038/ni852

195. Collin MP, Hart DN, Jackson GH, Cook G, Cavet J, Mackinnon S, et al. The fate of human Langerhans cells in hematopoietic stem cell transplantation. *J Exp Med* (2006) **203**:27–33. doi:10.1084/jem.20051787

196. Mielcarek M, Kirkorian AY, Hackman RC, Price J, Storer BE, Wood BL, et al. Langerhans cell homeostasis and turnover after nonmyeloablative and myeloablative allogeneic hematopoietic cell transplantation. *Transplantation* (2014) **98**(5):563–8. doi:10.1097/TP.0000000000000097

197. Greter M, Lelios I, Pelczar P, Hoeffel G, Price J, Leboeuf M, et al. Stroma-derived interleukin-34 controls the development and maintenance of Langerhans cells and the maintenance of microglia. *Immunity* (2012) **37**:1050–60. doi:10.1016/j.immuni.2012.11.001

198. Kashihara M, Ueda M, Horiguchi Y, Furukawa F, Hanaoka M, Imamura S. A monoclonal antibody specifically reactive to human Langerhans cells. *J Invest Dermatol* (1986) **87**:602–7. doi:10.1111/1523-1747.ep12455849

199. Stoitzner P, Romani N, McLellan AD, Tripp CH, Ebner S. Isolation of skin dendritic cells from mouse and man. *Methods Mol Biol* (2010) **595**:235–48. doi:10.1007/978-1-60761-421-0_16

200. Fainaru O, Woolf E, Lotem J, Yarmus M, Brenner O, Goldenberg D, et al. Runx3 regulates mouse TGF-beta-mediated dendritic cell function and its absence results in airway inflammation. *EMBO J* (2004) **23**:969–79. doi:10.1038/sj.emboj.7600085

201. Chopin M, Nutt SL. Establishing and maintaining the Langerhans cell network. *Semin Cell Dev Biol* (2014). doi:10.1016/j.semcdb.2014.02.001

202. Kaplan DH, Kissenpfennig A, Clausen BE. Insights into Langerhans cell function from Langerhans cell ablation models. *Eur J Immunol* (2008) **38**:2369–76. doi:10.1002/eji.200838397

203. Bennett CL, Fallah-Arani F, Conlan T, Trouillet C, Goold H, Chorro L, et al. Langerhans cells regulate cutaneous injury by licensing CD8 effector cells recruited to the skin. *Blood* (2011) **117**(26):7063–9. doi:10.1182/blood-2011-01-329185

204. Henri S, Poulin LF, Tamoutounour S, Ardouin L, Guilliams M, de Bovis B, et al. CD207+ CD103+ dermal dendritic cells cross-present keratinocyte-derived antigens irrespective of the presence of Langerhans cells. *J Exp Med* (2010) **207**:189–206. doi:10.1084/jem.20091964

205. Stoitzner P, Tripp CH, Eberhart A, Price KM, Jung JY, Bursch L, et al. Langerhans cells cross-present antigen derived from skin. *Proc Natl Acad Sci U S A* (2006) **103**:7783–8. doi:10.1073/pnas.0509307103

206. Kashem SW, Igyarto BZ, Gerami-Nejad M, Kumamoto Y, Mohammed J, Jarrett E, et al. Candida albicans morphology and dendritic cell subsets determine T helper cell differentiation. *Immunity* (2015) **42**:356–66. doi:10.1016/j.immuni.2015.01.008

207. Puel A, Cypowyj S, Marodi L, Abel L, Picard C, Casanova JL. Inborn errors of human IL-17 immunity underlie chronic mucocutaneous candidiasis. *Curr Opin Allergy Clin Immunol* (2012) **12**:616–22. doi:10.1097/ACI.0b013e328358cc0b

208. Ni L, Gayet I, Zurawski S, Duluc D, Flamar AL, Li XH, et al. Concomitant activation and antigen uptake via human dectin-1 results in potent antigen-specific CD8+ T cell responses. *J Immunol* (2010) **185**:3504–13. doi:10.4049/jimmunol.1000999

209. Furio L, Briotet I, Journeaux A, Billard H, Peguet-Navarro J. Human Langerhans cells are more efficient than CD14(-)CD1c(+) dermal dendritic cells at priming naive CD4(+) T cells. *J Invest Dermatol* (2010) **130**:1345–54. doi:10.1038/jid.2009.424

210. Mathers AR, Janelsins BM, Rubin JP, Tkacheva OA, Shufesky WJ, Watkins SC, et al. Differential capability of human cutaneous dendritic cell subsets to initiate Th17 responses. *J Immunol* (2009) **182**:921–33. doi:10.4049/jimmunol.182.2.921

211. Fujita H, Nograles KE, Kikuchi T, Gonzalez J, Carucci JA, Krueger JG. Human Langerhans cells induce distinct IL-22-producing CD4+ T cells lacking IL-17 production. *Proc Natl Acad Sci U S A* (2009) **106**:21795–800. doi:10.1073/pnas.0911472106

212. Seneschal J, Clark RA, Gehad A, Baecher-Allan CM, Kupper TS. Human epidermal Langerhans cells maintain immune homeostasis in skin by activating skin resident regulatory t cells. *Immunity* (2012) **36**(5):873–84. doi:10.1016/j.immuni.2012.03.018

213. Muller PA, Koscso B, Rajani GM, Stevanovic K, Berres ML, Hashimoto D, et al. Crosstalk between muscularis macrophages and enteric neurons regulates gastrointestinal motility. *Cell* (2014) **158**:300–13. doi:10.1016/j.cell.2014.04.050

214. Paolicelli RC, Bolasco G, Pagani F, Maggi L, Scianni M, Panzanelli P, et al. Synaptic pruning by microglia is necessary for normal brain development. *Science* (2011) **333**:1456–8. doi:10.1126/science.1202529

215. Wynn TA, Chawla A, Pollard JW. Macrophage biology in development, homeostasis and disease. *Nature* (2013) **496**:445–55. doi:10.1038/nature12034

216. Epelman S, Lavine KJ, Beaudin AE, Sojka DK, Carrero JA, Calderon B, et al. Embryonic and adult-derived resident cardiac macrophages are maintained through distinct mechanisms at steady state and during inflammation. *Immunity* (2014) **40**:91–104. doi:10.1016/j.immuni.2013.11.019

Human and mouse mononuclear phagocyte networks: a tale of two...

39

217. Guilliams M, De Kleer I, Henri S, Post S, Vanhoutte L, De Prijck S, et al. Alveolar macrophages develop from fetal monocytes that differentiate into long-lived cells in the first week of life via GM-CSF. *J Exp Med* (2013) **210**:1977–92. doi:10.1084/jem.20131199

218. Sieweke MH, Allen JE. Beyond stem cells: self-renewal of differentiated macrophages. *Science* (2013) **342**:1242974. doi:10.1126/science.1242974

219. Gautier EL, Shay T, Miller J, Greter M, Jakubzick C, Ivanov S, et al. Gene-expression profiles and transcriptional regulatory pathways that underlie the identity and diversity of mouse tissue macrophages. *Nat Immunol* (2012) **13**:1118–28. doi:10.1038/ni.2419

220. Lavin Y, Winter D, Blecher-Gonen R, David E, Keren-Shaul H, Merad M, et al. Tissue-resident macrophage enhancer landscapes are shaped by the local microenvironment. *Cell* (2014) **159**:1312–26. doi:10.1016/j.cell.2014.11.018

221. Gosselin D, Link VM, Romanoski CE, Fonseca GJ, Eichenfield DZ, Spann NJ, et al. Environment drives selection and function of enhancers controlling tissue-specific macrophage identities. *Cell* (2014) **159**:1327–40. doi:10.1016/j.cell.2014.11.023

222. Hamann J, Koning N, Pouwels W, Ulfman LH, van Eijk M, Stacey M, et al. EMR1, the human homolog of F4/80, is an eosinophil-specific receptor. *Eur J Immunol* (2007) **37**:2797–802. doi:10.1002/eji.200737553

223. Murray PJ, Wynn TA. Protective and pathogenic functions of macrophage subsets. *Nat Rev Immunol* (2011) **11**:723–37. doi:10.1038/nri3073

224. Gordon S, Pluddemann A, Martinez Estrada F. Macrophage heterogeneity in tissues: phenotypic diversity and functions. *Immunol Rev* (2014) **262**:36–55. doi:10.1111/imr.12223

225. Kierdorf K, Erny D, Goldmann T, Sander V, Schulz C, Perdiguero EG, et al. Microglia emerge from erythromyeloid precursors via Pu.1- and Irf8-dependent pathways. *Nat Neurosci* (2013) **16**:273–80. doi:10.1038/nn.3318

226. Takayanagi H, Kim S, Koga T, Nishina H, Isshiki M, Yoshida H, et al. Induction and activation of the transcription factor NFATc1 (NFAT2) integrate RANKL signaling in terminal differentiation of osteoclasts. *Dev Cell* (2002) **3**:889–901. doi:10.1016/S1534-5807(02)00369-6

227. Kohyama M, Ise W, Edelson BT, Wilker PR, Hildner K, Mejia C, et al. Role for Spi-C in the development of red pulp macrophages and splenic iron homeostasis. *Nature* (2009) **457**:318–21. doi:10.1038/nature07472

228. Haldar M, Kohyama M, So AY, Kc W, Wu X, Briseno CG, et al. Heme-mediated SPI-C induction promotes monocyte differentiation into iron-recycling macrophages. *Cell* (2014) **156**:1223–34. doi:10.1016/j.cell.2014.01.069

229. Xue J, Schmidt SV, Sander J, Draffehn A, Krebs W, Quester I, et al. Transcriptome-based network analysis reveals a spectrum model of human macrophage activation. *Immunity* (2014) **40**:274–88. doi:10.1016/j.immuni.2014.01.006

230. Schroder K, Irvine KM, Taylor MS, Bokil NJ, Le Cao KA, Masterman KA, et al. Conservation and divergence in toll-like receptor 4-regulated gene expression in primary human versus mouse macrophages. *Proc Natl Acad Sci U S A* (2012) **109**:E944–53. doi:10.1073/pnas.1110156109

231. Ajami B, Bennett JL, Krieger C, McNagny KM, Rossi FM. Infiltrating monocytes trigger EAE progression, but do not contribute to the resident microglia pool. *Nat Neurosci* (2011) **14**:1142–9. doi:10.1038/nn.2887

232. Segura E, Amigorena S. Inflammatory dendritic cells in mice and humans. *Trends Immunol* (2013) **34**:440–5. doi:10.1016/j.it.2013.06.001

233. Zhan Y, Xu Y, Seah S, Brady JL, Carrington EM, Cheers C, et al. Resident and monocyte-derived dendritic cells become dominant IL-12 producers under different conditions and signaling pathways. *J Immunol* (2010) **185**:2125–33. doi:10.4049/jimmunol.0903793

234. Fei M, Bhatia S, Oriss TB, Yarlagadda M, Khare A, Akira S, et al. TNF-alpha from inflammatory dendritic cells (DCs) regulates lung IL-17A/IL-5 levels and neutrophilia versus eosinophilia during persistent fungal infection. *Proc Natl Acad Sci U S A* (2011) **108**:5360–5. doi:10.1073/pnas.1015476108

235. Ikarashi M, Nakashima H, Kinoshita M, Sato A, Nakashima M, Miyazaki H, et al. Distinct development and functions of resident and recruited liver Kupffer cells/macrophages. *J Leukoc Biol* (2013) **94**:1325–36. doi:10.1189/jlb.0313144

236. Lowes MA, Chamian F, Abello MV, Fuentes-Duculan J, Lin SL, Nussbaum R, et al. Increase in TNF-alpha and inducible nitric oxide synthase-expressing dendritic cells in psoriasis and reduction with efalizumab (anti-CD11a). *Proc Natl Acad Sci U S A* (2005) **102**:19057–62. doi:10.1073/pnas.0509736102

237. Fujita H, Shemer A, Suarez-Farinas M, Johnson-Huang LM, Tintle S, Cardinale I, et al. Lesional dendritic cells in patients with chronic atopic dermatitis and psoriasis exhibit parallel ability to activate T-cell subsets. *J Allergy Clin Immunol* (2011) **128**:e1–12. doi:10.1016/j.jaci.2011.05.016

238. Reynolds G, Gibbon JR, Pratt AG, Wood MJ, Coady D, Raftery G, et al. Synovial CD4+ T-cell-derived GM-CSF supports the differentiation of an inflammatory dendritic cell population in rheumatoid arthritis. *Ann Rheum Dis* (2015). doi:10.1136/annrheumdis-2014-206578

4

Guardians of the gut – murine intestinal macrophages and dendritic cells

Mor Gross[1,2], Tomer-Meir Salame[1,2] and Steffen Jung[1,2]*

[1] *Department of Immunology, Weizmann Institute of Science, Rehovot, Israel,* [2] *Biological Services, Weizmann Institute of Science, Rehovot, Israel*

Edited by:
Martin Guilliams,
Ghent University – VIB, Belgium

Reviewed by:
Masaaki Murakami,
Hokkaido University, Japan
Amir Ghaemmaghami,
The University of Nottingham, UK

*Correspondence:
Steffen Jung,
Weizmann Institute of Science,
Department of Immunology, Rehovot
76100, Israel
s.jung@weizmann.ac.il

Intestinal mononuclear phagocytes find themselves in a unique environment, most prominently characterized by its constant exposure to commensal microbiota and food antigens. This anatomic setting has resulted in a number of specializations of the intestinal mononuclear phagocyte compartment that collectively contribute the unique steady state immune landscape of the healthy gut, including homeostatic innate lymphoid cells, B, and T cell compartments. As in other organs, macrophages and dendritic cells (DCs) orchestrate in addition the immune defense against pathogens, both in lymph nodes and mucosa-associated lymphoid tissue. Here, we will discuss origins and functions of intestinal DCs and macrophages and their respective subsets, focusing largely on the mouse and cells residing in the lamina propria.

Keywords: gut, dendritic cells, macrophages, homeostasis, inflammation, IBD

The Unique Characteristics of the Gut Landscape

Intestinal mononuclear phagocytes are located in a unique anatomic environment that necessitated the evolution of special functional adaptations of these cells. Exposure to commensal bacteria and harmful pathogens, as well as nutrients and food antigens, in the intestinal lumen force the immune system to continuously weigh tolerogenic and protective immune response. Disruption of this critical and delicate balance can result in devastating inflammatory reactions, e.g., hyper-reactivity to food components (1) or inflammatory bowel diseases (IBD), such as Crohn's disease or ulcerative colitis (2).

Both dendritic cells (DC) and macrophages are found spread throughout the connective tissue that underlies the epithelial layer of the gut, the lamina propria. Moreover, representatives of the two main mononuclear phagocyte families are also located in mucosa-associated lymphoid tissue (MALT), including Peyers' Patches and isolated lymphoid follicles (ILFs) (3). DC and macrophages have distinct, yet complementary roles in maintaining gut homeostasis and immune defense. In keeping with their migratory capacity, DC translocate from the lamina propria via the lymphatics to the gut-draining mesenteric lymph nodes (MsnLNs), where they present antigens to naïve T cells, polarize them toward effector fates, and thus establish the adaptive branch of the immune system (4).

Macrophages, on the other hand, are believed to contribute to the local clearance of bacteria from the tissue, translate alert signals to other immune cells, secrete cytokines to establish the local homeostatic immune cell network, and participate in T cell re-stimulation and maintenance within the lamina propria (5).

DC and macrophages can, as discussed in detail below, be divided into several subpopulations with defined origins, overlapping and distinct surface marker profiles, functions and

TABLE 1 | Mononuclear phagocytes and their respective subsets in the lamina propria of the mouse intestine.

Intestinal mononuclear phagocyte	Main markers (additional markers)	Location	Precursor	Growth/transcription/ environmental factor dependence	Functional specialization	Additional comments	Selected references SI, LI indicate organ of study: small or large intestine
DC	CD103+ CD11b– (CD24+, XCR1+)	Lamina propria, MALT	preDC	Flt-3L Irf8, Id2, Batf-3	Cross-presentation	Equivalent of splenic XCR1+ CD8a+ DC	Edelson et al. (6) SI Ginhoux et al. (7) SI Becker et al. (8) SI Crozat et al. (9) SI Schlitzer et al. (10) SI
	CD103+ CD11b+ (CD24+, Sirpα+)	Lamina propria, MALT	preDC	Flt-3L (partially) Csf-2 (GM-CSF), Irf-4, Notch2, Retinoic acid (ileum)	Required for generation and priming of TH17 cells	More prevalent in ileum	Bogunovic et al. (11) SI, LI Lewis et al. (12) SI, LI Welty et al. (13) SI, LI Schlitzer et al. (10) SI Persson et al. (14) SI, LI Klebanoff et al. (15) SI
	CD103– CD11b+		preDC	Flt-3L, Csf-1 (M-CSF)	Priming of IL-17 and INFγ-producing T cells		Bogunovic et al. (11) SI, LI Cerovic et al. (16) SI Scott et al. (17) SI, LI
	CD103– CD11b–		preDC	Ftl3L	Priming of TH17 (in vitro)		Cerovic et al. (16)
Macrophages	CD64+ CX3CR1+ CD11c+ (F4/80+ CD11b+)	Lamina propria	Ly6C+ monocytes	Csf-1 (M-CSF) Csf-2 (GM-CSF) (in colon)			Niess et al. (18) SI Varol et al. (19) SI Bogunovic et al. (11) SI Mortha et al. (20) Cecchini et al. (38)
	CD64+ CX3CR1+ CD11c– (F4/80+ CD11b+)	Lamina propria	Ly6C+ monocytes	Csf-1 (M-CSF) Notch 1/2			Ishifune et al. (21) SI Cecchini et al. (38), SI LI
	CD64+ CX3CR1+ CD169+ (F4/80+ CD11b+)	Crypt proximity	Ly6C+ monocytes	Csf-1 (M-CSF)			Hiemstra et al. (22) LI Cecchini et al. (38), SI LI
	CD64+ CX3CR1+ (F4/80+ CD11b+)	Muscularis layer	Ly6C+ monocytes	Csf-1 (M-CSF)	Communication with neurons		Muller et al. (23) SI, LI Cecchini et al. (38), SI LI

locations. The best characterized DC and macrophage subsets and their key features are summarized in **Table 1**.

With this review, we provide an overview on the characteristics and function of intestinal macrophages and DC in the mouse, including specific roles of their subpopulations. We will discuss distinct origins, roles in maintaining gut homeostasis, and the interactions between these cells and other immune cells. Finally, we will review their communication with their non-immune microenvironment and elaborate on emerging roles of macrophages and DC in inflammation.

Intestinal Macrophages

Macrophages are the most abundant mononuclear phagocytes in the steady-state gut lamina propria (3, 24). Intestinal macrophages are currently best characterized by their expression of CD64, the Fcγ receptor 1 (FcγRI) (25), and the chemokine receptor CX$_3$CR1 (18), as well as the F4/80 antigen (EGF-like module containing mucin-like hormone receptor-like 1-EMR1) and the integrins CD11b and CD11c (26). Due to the high surface expression levels of the chemokine receptor CX$_3$CR1 by gut macrophages, these cells can also be readily detected, isolated, and studied *in situ* using intra-vital microscopy on mice harboring a GFP reporter gene inserted into the CX$_3$CR1 locus (27).

Ontogeny

Like other tissue macrophages (28), also intestinal macrophages are first established before birth from precursors originating in the yolk sac or fetal liver (29). However, unlike macrophages in most other tissues, these embryo-derived cells are replaced in the gut shortly after birth by cells that derive from Ly6C$^+$ blood monocytes (29). The adult monocyte-derived cells display a uniquely short half-life for macrophages (30) indicating their continuous renewal. The monocytic origin of intestinal macrophages was first established in adoptive transfer experiments, involving the transfer of CX3CR1^{gfp} monocyte-precursors and monocytes into CD11c-DTR transgenic mice, whose CD11c-expressing cells, including intestinal macrophages, were depleted by a diphtheria toxin challenge (11, 19, 31). During their differentiation into gut macrophages, monocytes lose Ly6C expression, while other surface markers, such as MHCII, F4/80, CD64, CD11c, and CX3CR1 are up-regulated (25, 32, 33). Moreover, the cells acquire a characteristic anti-inflammatory gene expression profile (32, 34), whose timely establishment and maintenance are critical for gut homeostasis (35). This includes the expression of IL-10, TREM-2, IRAK-M, and tumor necrosis factor (TNF)AIP3 genes, but also of TNFα, which has both pro- and -anti-inflammatory activity (32). Of note, this expression profile is robust, as it seems to withstand acute challenges, such as the ones associated with oral dextran sulfate sodium (DSS) exposure (32). The molecular cues that drive the "education" of the macrophages in various regions of the gut remain to be defined, but the epithelium is likely to play a role in this process. Epithelial cells could control macrophage differentiation by secretion of immune-regulatory factors, such as thymic stromal lymphopoietin (TSLP), transforming growth factor-β (TGF-β), and prostaglandin E-2 (PGE-2) (36). In addition, recent findings suggested that semaphorin 7A, which is secreted by epithelial cells, contributes to the induction of IL-10 expression by CX_3CR1^+ intestinal macrophages (37). Also, colony-stimulating factor 2 (Csf-1; previously named macrophage colony-stimulating factor, M-CSF) and colony-stimulating factor 2 (Csf-2; previously named granulocyte-macrophage colony-stimulating factor, GM-CSF) play a role in the development of macrophages. Csf-1 is a crucial factor for monocyte development, as Csf-1-deficient osteopetrotic (op/op) mice display reduced levels of $F4/80^+$ cells in the small and large intestine after the first few days of life (28, 38, 39). Csf-2-depleted mice were shown have reduced numbers of $CD11c^+$ colonic macrophages (20).

Of note, $Ly6C^+$ monocytes fail to acquire the characteristic macrophage quiescence during intestinal inflammation, but under this condition respond to local factors that trigger pattern recognition receptors, such as TLRs and NLRs, giving rise to pro-inflammatory macrophages (32). These pro-inflammatory cells, which in acute inflammation outnumber the resident macrophage population, secrete IL-12, IL-23, TNF-α, and inducible nitric oxide synthase (iNOS) (32).

A key suppressor of macrophage-associated inflammation is the IL-10/IL-10 receptor (IL-10R) axis, as mice bearing mutations in IL10-Ra in intestinal $CX3CR1^+$ macrophages developed severe colitis (35) comparable to the pathology reported for IL-10-deficient animals (40). This central critical role of IL-10 in maintaining the non-inflammatory state of macrophages, and thereby, gut homeostasis is also supported by research conducted on samples from humans with loss of function mutations in IL-10R (41). The latter provides an explanation for the severe early onset of colitis observed in pediatric patients harboring nonsense and missense mutations in IL-10R, which reduce IL-10R expression and hamper its signaling cascades (42). Interestingly though, IL-10 production by intestinal macrophages, although also prominent, seems to be redundant for the maintenance of gut homeostasis (35); rather the system seems to rely on alternative IL-10 sources, such as Treg cells (43).

Homeostatic monocyte recruitment to the gut is thought to depend on the chemokine receptor CCR2, as CCR2-deficient mice display less intestinal macrophages and CCR2-deficient intestinal macrophages are underrepresented in mixed bone marrow chimeras (24, 25). The exact factors and mechanisms that ensure homeostatic $Ly6C^+$ monocyte recruitment to the steady state gut are, however, still unknown. While they are likely related to the microbiota exposure of the tissue, analysis of germ-free animals has yielded conflicting results (29, 34, 44, 45). The latter could be due to intestinal embryo-derived macrophages that might persist in the absence of arising competition by an adult monocyte influx.

Macrophage Heterogeneity

Interestingly, emerging evidence suggests that intestinal macrophages are more heterogeneous than previously thought. Monocyte-derived $CD11b^+$ CX_3CR1^+ cells in the gut comprise both $CD11c^+$ and $CD11c^-$ cells. While differential functions of these cells remain to be established, studies into this matter might profit from the recent finding that generation of $CD11c^+$, but not $CD11c^-$ CX_3CR1^+ intestinal macrophages requires Notch signaling (21). A subpopulation of CD169-expressing CX_3CR1^+ macrophages has been reported to be associated with the intestinal crypts (22), although these cells will require further functional characterization. Bogunovic and colleagues recently reported an intriguing $CX3CR1^+$ macrophage subpopulation that resides in the muscularis layer and communicates with enteric neurons to regulate gastrointestinal motility (23). Importantly, we and others have recently shown that macrophages isolated from distinct tissues, such as the liver, lung, brain, and peritoneum, differ considerably with respect to their gene expression profile (46, 47). As expected, this diversity is also prominently reflected in the differential enhancer usage of these cells, as inferred from highly divergent histone modifications (47). Moreover, given that the number of regulatory elements by far exceeds the number of genes (48, 49), this heterogeneity is even more pronounced, including both active and poised enhancer states (47). This applies, albeit to a lesser extent, also to macrophages located in proximal and distal segments of the gut (47). Epigenetic heterogeneity of intestinal macrophages likely reflects monocyte exposure to distinct environmental cues in ileum and colon during their local differentiation (32, 47). In-depth understanding of how these macrophage identities are established, including the hierarchy of induced transcription factors, could yield valuable insights into monocyte differentiation that might be applicable to other tissues and inflammatory settings. PU.1 is a pioneering factor, which induces c-fms transcription and is hence required for macrophage differentiation (50). Intestinal macrophages are furthermore characterized by prominent expression of the Runt-related transcription factor 3 (Runx-3) (47). Interestingly, mice that harbor Runx3 deficiency develop spontaneous colitis (51). Other candidates that might be involved in the establishment of the intestinal macrophage signature are the interferon regulatory factors 4 and 5 (Irf-4, Irf-5), shown to be associated with classical and alternative macrophage activation, respectively (52–54).

Macrophage Interactions with Their Environment

Macrophage Communication with the Epithelial Cell Layer

Pioneering studies by Rescignio and colleagues revealed that certain intestinal mononuclear phagocytes can penetrate the intestinal epithelium by virtue of expression of tight junction proteins and formations of dendritic projections (55). These structures, later termed trans-epithelial dendrites (TEDs) (56), were subsequently ascribed to macrophages expressing CX_3CR1 (18)

and allegedly allow these non-migratory cells to sense, and potentially sample, the luminal content (18, 56). TED formation by macrophages in the terminal region of the ileum was found to be dependent on expression of both CX$_3$CR1 macrophages and its membrane-tethered ligand CX$_3$CL1/Fractalkine by selected epithelial cells (57). CX$_3$CR1-deficient and CX$_3$CL1-deficient mice were reported to be relatively protected from acute, DSS-induced colitis (58) – a phenotype that might be related to TED formation (57). Likewise, CX$_3$CR1-deficient mice were shown to display impaired oral tolerance, which was related to impaired IL-10 production by intestinal macrophages, though not their TED formation (59). Finally, there is evidence for a potential role of CX$_3$CR1$^+$ macrophages in the capture of luminal bacteria (60) and even the transport of the latter to lymph nodes, at least under conditions of dysbiosis (61). However, the exact definition of macrophage contributions in their native tissue context remains challenging, because it requires their accurate discrimination from closely related and phenotypically similar monocyte-derived DC.

Apart from their role in maintaining intestinal immune homeostasis, gut macrophages also contribute critically to epithelial wound healing. Macrophages associated with the crypts of Lieberkuehn in the colon were reported to assist, following tissue damage, the proliferation and survival of epithelial progenitor cells in a Myd88-dependent manner (62–64). Moreover, in a murine model of acute epithelial regeneration in the colon, activated macrophages supported tissue repair by up-regulating expression of IL-3 and IL-4, while inhibiting secretion of TNF and interferon-γ (IFN-γ) in the lamina propria (3, 65). Macrophages also appear to be able to influence the permeability of the epithelium barrier via the secretion of IL-6 and NO, thereby potentially increasing the invasion of pathogens (66).

Communication with Immune Cells

Macrophages are inferior to DC in their ability to prime naïve T cells (67). This might be due to their rapid degradation of ingested proteins, which impairs their ability to retain antigens for presentation (68). Moreover, at least in steady state, intestinal CX$_3$CR1$^+$ macrophages lack expression of CCR7, i.e., the chemokine receptor required for migration to the MsnLNs (25, 69). Rather, the cells that reside in the lamina propria have been proposed to maintain the functionality of FoxP-3$^+$ T regulatory cells that migrated back from the MsnLNs into the tissue (59). Thus, while Treg cell generation of CX$_3$CR1-deficient mice is unimpaired, these animals harbor reduced Treg cell numbers in the lamina propria, a phenotype that is associated with impaired oral tolerance (59). In light of other data (70), the authors of this study linked the reduced FoxP-3$^+$ Treg cell numbers to impaired production of IL-10 by CX$_3$CR1$^+$ macrophages (59). However, the latter might have to be revised, since CX$_3$CR1^{Cre}:IL10$^{fl/fl}$ mice were shown to harbor unimpaired FoxP-3$^+$ Treg cell numbers (35). Also, interactions between CX$_3$CR1$^+$ macrophages and Th17 cells, which are rarely found in intestinal lymphoid tissues and, though primed in the MsnLN, might terminally differentiate in the lamina propria, remain incompletely defined. On one hand, it was shown that intestinal CD70^{hi} CX$_3$CR1$^+$ macrophages are activated by commensal-derived ATP and drive the *in vitro*

differentiation of Th17 cells (71, 72). On the other hand, intestinal macrophages were reported to counteract Th17 generation that is promoted by CD103$^+$CD11b$^+$ DC (73, 74). Of note, CD103$^+$CD11b$^+$ DCs and Th17 cells co-localize in the intestinal tract, as the number of both cells drop from the duodenum to the ileum, and they are scarce in the colon. By contrast, CX$_3$CR1$^+$ macrophages and FoxP3$^+$ Treg cells are most abundant in the colon (74).

Recent findings revealed an intriguing cross-talk between intestinal macrophages and innate lymphoid cells (ILC). Thus, in response to luminal stimuli and using a signaling pathway involving the TLR adaptor Myd88, macrophages were shown to secrete IL-1β and in turn induce production of csf-2 by RORγt$^+$ type 3 ILC (20). Mice lacking Csf-2 display reduced numbers of colonic macrophages and DC, associated with a hampered Treg cell compartment (20). Moreover, in a *Citrobacter* infection model CX$_3$CR1$^+$ macrophages were shown to promote ILC production of IL-22 via secretion of IL-23 (75), in line with another report (76). Interestingly, CX$_3$CR1$^+$ macrophage-derived IL-23 not only induces IL-22 but also seems to concomitantly suppress IL-12 production by CD103$^+$ CD11b$^-$ DC and thereby prevents otherwise detrimental immunopathology (77). Notably, the latter finding provides first evidence for the existence of a direct cross-talk among intestinal mononuclear phagocytes in tissue context, a topic that clearly deserves further study.

Intestinal Dendritic Cells

Dendritic cells are specialized in communicating with T cells, curbing autoreactivity and activating T cell immunity in response to threats. Specifically, DC provide T cells with antigenic peptides that are presented in MHC context, co-stimulation and instructing cytokines that govern T cell polarization into effector cells (67). In order to maintain homeostasis and avoid inflammatory responses toward innocuous antigens, gut DC employ tolerogenic mechanisms that allow them to dampen adaptive immunity. MsnLN- and lamina propria-resident CD103$^+$ DC secrete, for example, retinoic acid (RA) and transforming growth factor-β (TGF-β), which promote the generation of Foxp3$^+$ Treg cells and contribute to the differentiation of plasma cells, which secrete IgA (78, 79).

Classification and Ontogeny

Intestinal DC in mice are characterized by the surface expression of the integrins CD11c (α_X) and CD103 ($\alpha_E\beta7$) (11, 19, 69). More recently, CD24 and Sirpα have been introduced for the better discrimination of DC from macrophages (8, 10). CD103$^+$ DC in the gut arise from dedicated DC precursors, or preDC, and accordingly, mice deficient for fms-related tyrosine kinase-3 receptor (Flt-3) or its ligand Flt-3L have significantly decreased levels of intestinal DC (7, 19). Other, currently though less well-characterized DC progenitors are α4β7$^+$ so-called "pre-μDC," which are generated in the bone marrow and were shown to give rise to classical CD103$^+$ DC and CCR9$^+$ plasmacytoid DC (80).

Classical CD103$^+$ DC are divided into two major subpopulations according to their expression of CD11b (α_M) (81). CD103$^+$ CD11b$^+$ DC and CD103$^+$ CD11b$^-$ DC display distinct abundance in small and large intestine, present different additional surface markers, and require different growth factors for their development (82, 83).

CD103$^+$ CD11b$^+$ DC are developmentally related to CD11b$^+$ CD8α$^-$ splenic DCs (15) and found in the lamina propria of the small and large intestine. They can migrate in CCR7-dependent manner (84) to the MsnLNs, where they present luminal antigens to T cells. CD103$^+$ CD11b$^+$ DCs likely represent a heterogeneous population, as a fraction of them is Csf-2-dependent (3). Development of CD103$^+$ CD11b$^+$ DC, but not of CD103$^+$ CD11b$^-$ DC, is hampered in Csf-2R-deficient mice (85) and when expression of Notch-2 (12, 76) or IRF-4 (14) is impaired. Moreover, CD103$^+$ CD11b$^+$ DC numbers are also reduced in absence of RA and under conditions of vitamin A deprivation (15).

CD103$^+$ CD11b$^-$ DC are more prevalent in lymphoid organs – the Peyer's Patches, MsnLNs, and ILFs (7, 69). However, they can be found also in animals lacking these structures, and are hence not limited to lymphoid tissues (3). Similar to classical CD8α$^+$ DC in the spleen, CD103$^+$ CD11b$^-$ DC depend on the expression of the transcription factors BatF-3 and Irf-8 (6, 15). Like the former, they also express the chemokine receptor XCR1 that has emerged as a universal marker for this DC subset in mouse and human (8, 9). The connection between CD103$^+$ CD11b$^-$ DC and CD8α$^+$ DC is also supported by the fact that the number of CD103$^+$ CD11b$^-$ DC was shown to increase, alongside with splenic CD8α$^+$ DC, in mice that display constitutive β-catenin activation (86). Moreover, like splenic CD8α$^+$ DC (87), also CD103$^+$ CD11b$^-$ DC are specialized in cross-presentation (88).

The exact definition of intestinal DC is complicated, since monocyte-derived cells can acquire phenotypic and functional DC hallmarks. Studies have described a population of CD103$^-$CX3CR1$^+$CD11b$^+$ DC, which resides in the lamina propria (11, 16). These cells are CSFR-1 dependent and appear to be derived from Ly6C^{high} monocytes (11). Recent studies also reported that under inflammatory conditions, these CD103$^-$CX3CR1$^+$CD11b$^+$ DC expressed CCR7 and migrated in the intestinal lymph, similar to classical intestinal DC, and induced the differentiation of IL-17 and IFN-γ producing T cells (16, 17).

Antigen Sensing and Uptake

CD103$^+$ DC, present in the lamina propria and associated with the intestinal epithelium lining the villi, provide surveillance of the luminal environment (30). They detect foreign and inflammatory signals, acquire and present antigens and interact with T cells by migrating to secondary lymphoid organs (3). Located deep in the core of the villous lamina propria, CD103$^+$CD11b$^+$ DC would seemingly have limited access to luminal signals, unless antigens or bacteria cross the epithelium or are imported into the lamina propria by other cells, e.g., macrophages, epithelial M cells, or small intestine goblet cells (36, 89, 90). However, lamina propria-resident CD103$^+$ DC were shown to migrate into the epithelial cell layer and capture bacterial antigens (90).

DC Migration

Mucosal T cell priming, arguably one of the primary roles of gut DC, is believed to be restricted to lymphoid tissues (3).

Intestinal DC are hence bound to migrate from the lamina propria to the MsnLNs, or within Peyer's Patches into T cell zones. Indeed, CD103$^+$ DC were detected in the intestinal lymph under homeostatic conditions (69, 84). In addition, after systemic BrdU administration, labeled CD103$^+$ DC were found in the lamina propria before they could be discerned in the MsnLNs (30). LN-resident CD103$^+$ DC are thus derived from the tissue and constantly immigrate (30, 91). Interestingly, steady state migration of intestinal CD103$^+$ DC does not appear to be induced by the microbiota or by TLR signaling (92), but may rather depend on a low, tonic release of inflammatory cytokines, or result from spontaneous DC maturation. Nevertheless, entry of CD103$^+$ DC into the MsnLNs is of course considerably enhanced by pro-inflammatory cytokines or TLR ligands (93, 94). Migration of intestinal DC depends on CCR7, both in steady state and under inflammatory conditions. Accordingly, CCR7 expression is up-regulated in DC before their migration from the tissue into the MsnLN (84) and CCR7 deficient DC fail to migrate (69, 84, 95). Moreover, it was recently shown that DC can also migrate from the lamina propria into the epithelial layer (90) and can thus gain direct access to antigen and luminal bacteria. Hence, following challenge with *Salmonella*, accumulation of the bacteria was first observed in DC of the epithelial fraction and only subsequently in DC in the lamina propria (90).

DC and the Epithelium

DC intimately interact with the epithelial layer of the intestine by a variety of mechanisms. Small intestinal goblet cells were shown to transfer small soluble antigens from the intestinal lumen to CD103$^+$ DC (89). Chemokines secreted by enterocytes in response to TLR ligand exposure can induce the above-mentioned relocation of lamina propria DC to the epithelium (90). In addition, it is becoming more and more evident that epithelial cells play a critical role in maintaining DC in a tolerogenic state, compatible with gut homeostasis. Epithelial and stromal cells secrete factors, which are thought to induce DC tolerance, such as RA, TGF-β, PGE-2, and TSLP (3, 82, 96–99). In parallel to ILC (20), intestinal epithelial cells regulate retinal dehydrogenase (RALDH) expression by CD103$^+$ DC that the cells need to metabolize retinoids. Specifically, epithelial cells express a critical cytosolic retinoid chaperone, the cellular retinol binding protein II, which is required for *in vivo* imprinting of gut DC by lumenal retinoids (99, 100). Supporting this notion, the *in vitro* co-culture of bone marrow- or spleen-derived DC with epithelial cells results in the up-regulation of CD103 and RALDH, together with TGF-β imprinted homing potential on T cells (101–103). These data establish the potential of intestinal epithelial cells to educate intestinal DC, although further *in vivo* studies and higher resolution, with respect to cell subsets, are required to better elucidate the underlying mechanisms.

DC Communication with Intestinal T Cells

Intestinal CD103$^+$ DC, found in lamina propria, Peyer's Patches, and the MsnLNs program T cells to express the gut-homing factors CCR9 and α4β7 integrin (101, 104, 105). Concomitantly, DC can also induce the development of FoxP-3$^+$ and IL-10 producing

Treg cells (106) and prime Th17 cells (17, 107, 108). The majority of these DC-governed priming events require TGF-β signaling and RA, which are generated in the DC by enzymatic conversion of all-trans-retinal, a derivative of vitamin A, using RALDH2 (101, 109, 110). Indeed, RA has emerged as the critical conditioning factor for intestinal DC, as vitamin A is crucial for the activity of the enzyme RALDH in DC. Without RALDH, the ability of DC to imprint T cells is hampered, and restored only after vitamin A administration (111). The balance between RA and TGF-β levels seems to determine the fate of Treg cells primed by DC, as presence of both RA and TGF-β favor the development of FoxP-3$^+$ cells, while RA induces the generation of IL-10 producing T cells (106).

Other enzymes that influence the outcome of T cell priming are indoleamine 2,3 dioxygenase (IDO) and TSLP. IDO is expressed also by DC in other tissues and was shown to inhibit the development of effector T cells and promote Treg cell generation (112, 113). TSLP is, as mentioned above, secreted by epithelial cells, but also by the intestinal DC, themselves. In the presence

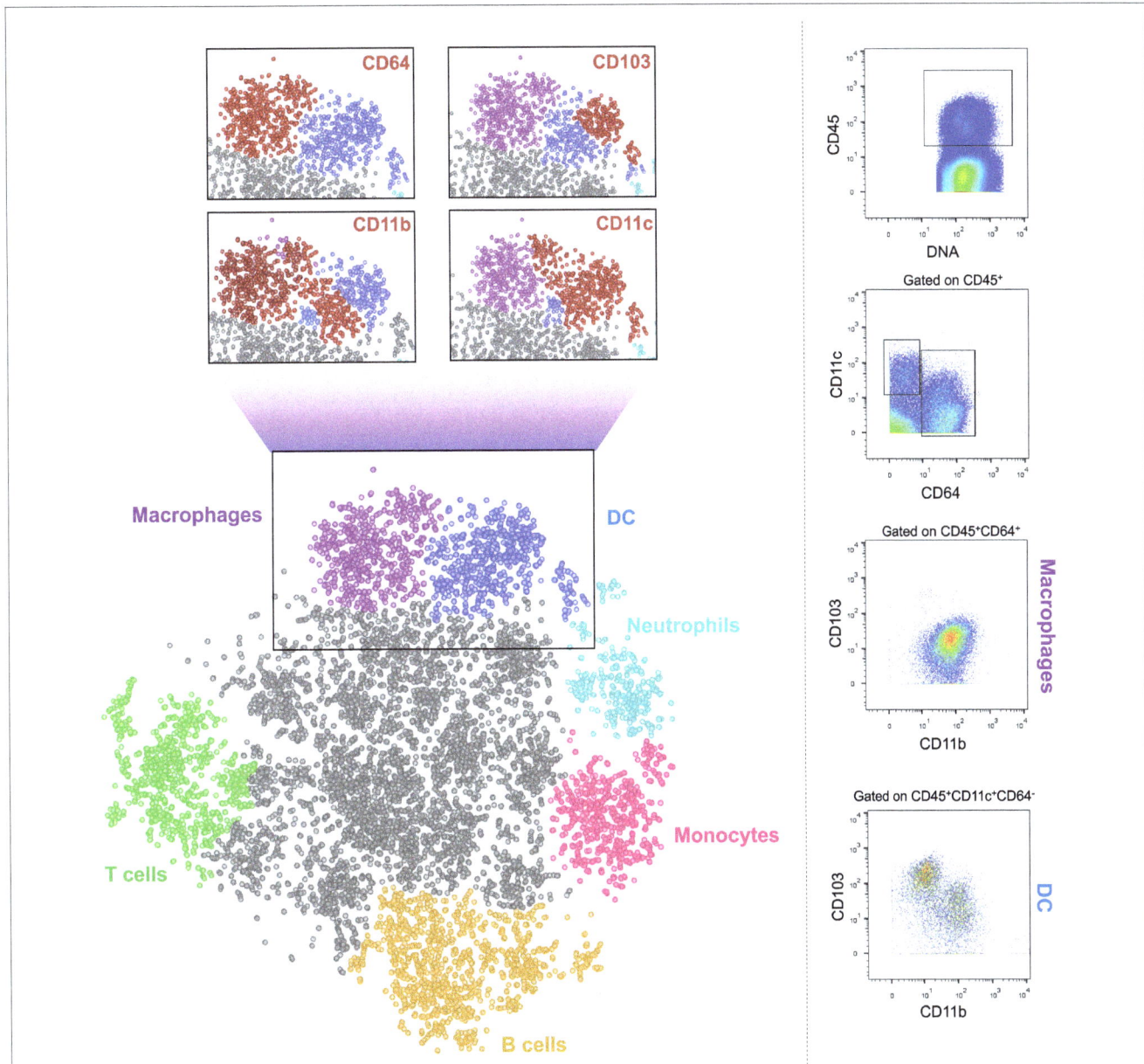

FIGURE 1 | CyTOF analysis of CD45$^+$ cells from murine colon. Cells were isolated from colon of 6–9 weeks old WT female C57Bl/6 mice and stained with a panel of 26 cell surface markers. The results were gated for live, single, CD45$^+$ cells. Bh-SNE analysis and clustering were performed by Accense (http://www.cellaccense.com/) and the results were processed by GIMP. Colors indicate high levels of the following markers: green – TCRβ, CD3e (T cells), Orange – B220 (B cells), light blue – Ly6G (granulocytes), pink –Ly6C (monocytes), purple – CD64, F4/80 (macrophages), blue – clustered by Accence, different DC populations, gray – non-identified or non-specific cells. Red populations in zoom-in black squares indicate high levels of the marker written. Representative of at least four separate, independent experiments.

of TSLP, Th17 responses are restricted due to a reduced ability to produce IL-17, and Treg cell differentiation is up-regulated (107). The ability of the intestinal DC compartment to generate Th17 cells seems to be associated with CD103$^+$ CD11b$^+$ DC, as the frequency of Th17 cells is reduced in mice lacking these DC due to either IRF-4 or Notch-deficiency (10, 12, 14), or as a result of conditional ablation of this DC subset (13). Interestingly though, a recent study showed that also another subpopulation of DC, i.e. CCR2$^+$ CD103$^-$ CD11b$^+$ DC can induce IL-17a production in CD4$^+$ T cells and effectively prime Th17 cells, probably via IL-12/IL-23p40 secretion (17).

Intestinal DC, Inflammation, and Immune Response

In steady state, intestinal DC are probably mainly tolerogenic. Under inflammatory conditions, however, they can become highly effective T cell activators (114). Induction of experimental colitis results in the accumulation of CD103$^+$ DC with an inflammatory profile in the MsnLNs (114). These DC express less RALDH and TGF-β and instead of promoting Treg cell formation, now induce Th1 inflammatory responses (114). While Th17 polarization might be carried out by CD103$^+$ CD11b$^+$ DC (12), differentiation of CD8$^+$ effector T cells under inflammatory conditions seems to be dependent on CD103$^+$ CD11b$^-$ CD8α$^+$ DC that migrated into the lymph (88).

Flagellin stimulation causes TLR-5$^+$ CD103$^+$ DC in the small intestine to promote differentiation of Th17 cells and secrete IL-23, which in turn induces IL-22 production by ILC3 and subsequent epithelial up-regulation of antibacterial peptides (115).

In summary, DC are major players in maintaining homeostasis in the intestine. While tolerogenic at steady state, under inflammatory conditions they tip the scales and activate the immune system. They can migrate between different compartments of the intestine – from the lamina propria to the epithelium and into the MsnLNs – and execute different immune responses in each tissue. Further research regarding the location of DC, their functions and characteristics should shed new light on the role of these cells in the intestine.

Concluding Remarks and a Glimpse to the Future

In summary, macrophages and DC critically contribute to intestinal homeostasis and immune defense. Both cellular

compartments have been subdivided into discrete subpopulations, which though currently mainly phenotypically defined, in some cases have been assigned distinct activities. The challenge ahead is to better define precise roles of these subsets both in health and under inflammatory conditions, first in the mouse but then also in the human. This task is complicated by the fact that many of the used markers used to distinguish between subpopulations of DC and macrophages are shared by the two types of mononuclear phagocytes. Moreover, under inflammatory conditions monocyte-derived cells further blur the picture. Collectively, this highlights the need to define cells by multiple parameters, including both surface and intracellular markers. Single cell transcriptome analysis is likely to help with this task (116, 117). However, classic flow cytometry analysis using fluorescent dye-coupled antibodies allows only a very limited simultaneous panel of markers due to the few dyes available and the spectral overlap of their emission. This problem might, in the near future, be solved by spectral cytometry systems that use ultrafast optical spectroscopy combined with flow cytometry to differentiate between the emission curves of different fluorophores, thus enabling the use of dozens of antibodies in one sample (118). Moreover, a new cell analyzer has been introduced, which uses mass cytometry instead of flow cytometry and is termed cytometry by Time-Of-Flight, or CyTOF (119). Instead of conjugations to fluorophores, this machine uses conjugations to heavy metal isotopes. Such metals do not exist naturally in the cells, so background is insignificant. The stained cells are injected into the CyTOF and are evaporated in a plasma chamber. The metals are ionized, hit the TOF detector, and their mass is measured, allowing the machine to determine the expression levels of the markers on each cell. This multiple-parameter approach enables to explore entire immune cell populations and subpopulations from the same tissue. As exemplified in **Figure 1**, such global analysis methods might well hold the key for the better definition and understanding of the cellular make-up of the intestine. No doubt, that with the recent development in the fields of cell cytometry and RNA sequencing, more pieces of this complex puzzle of the characteristics and roles of mononuclear phagocytes in the gut will be detected and put in place.

Acknowledgments

This work was supported by the European Research Council (340345).

References

1. Meresse B, Malamut G, Cerf-Bensussan N. Celiac disease: an immunological jigsaw. *Immunity* (2012) **36**:907–19. doi:10.1016/j.immuni.2012.06.006

2. Xavier RJ, Podolsky DK. Unravelling the pathogenesis of inflammatory bowel disease. *Nature* (2007) **448**:427–34. doi:10.1038/nature06005

3. Farache J, Zigmond E, Shakhar G, Jung S. Contributions of dendritic cells and macrophages to intestinal homeostasis and immune defense. *Immunol Cell Biol* (2013) **91**:232–9. doi:10.1038/icb.2012.79

4. Bekiaris V, Persson EK, Agace WW. Intestinal dendritic cells in the regulation of mucosal immunity. *Immunol Rev* (2014) **260**:86–101. doi:10.1111/imr.12194

5. Zigmond E, Jung S. Intestinal macrophages: well educated exceptions from the rule. *Trends Immunol* (2013) **34**:162–8. doi:10.1016/j.it.2013.02.001

6. Edelson BT, Kc W, Juang R, Kohyama M, Benoit LA, Klekotka PA, et al. Peripheral CD103+ dendritic cells form a unified subset developmentally related to CD8alpha+ conventional dendritic cells. *J Exp Med* (2010) **207**:823–36. doi:10.1084/jem.20091627

7. Ginhoux F, Liu K, Helft J, Bogunovic M, Greter M, Hashimoto D, et al. The origin and development of nonlymphoid tissue CD103+ DCs. *J Exp Med* (2009) **206**:3115–30. doi:10.1084/jem.20091756

8. Becker M, Guttler S, Bachem A, Hartung E, Mora A, Jakel A, et al. Ontogenic, phenotypic, and functional characterization of XCR1(+) dendritic cells leads to a consistent classification of intestinal dendritic cells based on the expression of

XCR1 and SIRPalpha. *Front Immunol* (2014) **5**:326. doi:10.3389/fimmu.2014.00326

9. Crozat K, Guiton R, Contreras V, Feuillet V, Dutertre CA, Ventre E, et al. The XC chemokine receptor 1 is a conserved selective marker of mammalian cells homologous to mouse CD8alpha+ dendritic cells. *J Exp Med* (2010) **207**:1283–92. doi:10.1084/jem.20100223

10. Schlitzer A, Mcgovern N, Teo P, Zelante T, Atarashi K, Low D, et al. IRF4 transcription factor-dependent CD11b+ dendritic cells in human and mouse control mucosal IL-17 cytokine responses. *Immunity* (2013) **38**:970–83. doi:10.1016/j.immuni.2013.04.011

11. Bogunovic M, Ginhoux F, Helft J, Shang L, Hashimoto D, Greter M, et al. Origin of the lamina propria dendritic cell network. *Immunity* (2009) **31**:513–25. doi:10.1016/j.immuni.2009.08.010

12. Lewis KL, Caton ML, Bogunovic M, Greter M, Grajkowska LT, Ng D, et al. Notch2 receptor signaling controls functional differentiation of dendritic cells in the spleen and intestine. *Immunity* (2011) **35**:780–91. doi:10.1016/j.immuni.2011.08.013

13. Welty NE, Staley C, Ghilardi N, Sadowsky MJ, Igyarto BZ, Kaplan DH. Intestinal lamina propria dendritic cells maintain T cell homeostasis but do not affect commensalism. *J Exp Med* (2013) **210**:2011–24. doi:10.1084/jem.20130728

14. Persson EK, Uronen-Hansson H, Semmrich M, Rivollier A, Hagerbrand K, Marsal J, et al. IRF4 transcription-factor-dependent CD103(+)CD11b(+) dendritic cells drive mucosal T helper 17 cell differentiation. *Immunity* (2013) **38**:958–69. doi:10.1016/j.immuni.2013.03.009

15. Klebanoff CA, Spencer SP, Torabi-Parizi P, Grainger JR, Roychoudhuri R, Ji Y, et al. Retinoic acid controls the homeostasis of pre-cDC-derived splenic and intestinal dendritic cells. *J Exp Med* (2013) **210**:1961–76. doi:10.1084/jem.20122508

16. Cerovic V, Houston SA, Scott CL, Aumeunier A, Yrlid U, Mowat AM, et al. Intestinal CD103(-) dendritic cells migrate in lymph and prime effector T cells. *Mucosal Immunol* (2013) **6**:104–13. doi:10.1038/mi.2012.53

17. Scott CL, Bain CC, Wright PB, Sichien D, Kotarsky K, Persson EK, et al. CCR2CD103 intestinal dendritic cells develop from DC-committed precursors and induce interleukin-17 production by T cells. *Mucosal Immunol* (2014) **8**(2):327–39. doi:10.1038/mi.2014.70

18. Niess JH, Brand S, Gu X, Landsman L, Jung S, Mccormick BA, et al. CX3CR1-mediated dendritic cell access to the intestinal lumen and bacterial clearance. *Science* (2005) **307**:254–8. doi:10.1126/science.1102901

19. Varol C, Vallon-Eberhard A, Elinav E, Aychek T, Shapira Y, Luche H, et al. Intestinal lamina propria dendritic cell subsets have different origin and functions. *Immunity* (2009) **31**:502–12. doi:10.1016/j.immuni.2009.06.025

20. Mortha A, Chudnovskiy A, Hashimoto D, Bogunovic M, Spencer SP, Belkaid Y, et al. Microbiota-dependent crosstalk between macrophages and ILC3 promotes intestinal homeostasis. *Science* (2014) **343**:1249288. doi:10.1126/science.1249288

21. Ishifune C, Maruyama S, Sasaki Y, Yagita H, Hozumi K, Tomita T, et al. Differentiation of CD11c+ CX3CR1+ cells in the small intestine requires notch signaling. *Proc Natl Acad Sci U S A* (2014) **111**:5986–91. doi:10.1073/pnas.1401671111

22. Hiemstra IH, Beijer MR, Veninga H, Vrijland K, Borg EG, Olivier BJ, et al. The identification and developmental requirements of colonic CD169(+) macrophages. *Immunology* (2014) **142**:269–78. doi:10.1111/imm.12251

23. Muller PA, Koscso B, Rajani GM, Stevanovic K, Berres ML, Hashimoto D, et al. Crosstalk between muscularis macrophages and enteric neurons regulates gastrointestinal motility. *Cell* (2014) **158**:300–13. doi:10.1016/j.cell.2014.04.050

24. Bain CC, Mowat AM. Intestinal macrophages – specialised adaptation to a unique environment. *Eur J Immunol* (2011) **41**:2494–8. doi:10.1002/eji.201141714

25. Tamoutounour S, Henri S, Lelouard H, De Bovis B, De Haar C, Van Der Woude CJ, et al. CD64 distinguishes macrophages from dendritic cells in the gut and reveals the Th1-inducing role of mesenteric lymph node macrophages during colitis. *Eur J Immunol* (2012) **42**:3150–66. doi:10.1002/eji.201242847

26. Pabst O, Bernhardt G. The puzzle of intestinal lamina propria dendritic cells and macrophages. *Eur J Immunol* (2010) **40**:2107–11. doi:10.1002/eji.201040557

27. Jung S, Aliberti J, Graemmel P, Sunshine MJ, Kreutzberg GW, Sher A, et al. Analysis of fractalkine receptor CX(3)CR1 function by targeted deletion and green fluorescent protein reporter gene insertion. *Mol Cell Biol* (2000) **20**:4106–14. doi:10.1128/MCB.20.11.4106-4114.2000

28. Ginhoux F, Jung S. Monocytes and macrophages: developmental pathways and tissue homeostasis. *Nat Rev Immunol* (2014) **14**:392–404. doi:10.1038/nri3671

29. Bain CC, Bravo-Blas A, Scott CL, Gomez Perdiguero E, Geissmann F, Henri S, et al. Constant replenishment from circulating monocytes maintains the macrophage pool in the intestine of adult mice. *Nat Immunol* (2014) **15**:929–37. doi:10.1038/ni.2967

30. Jaensson E, Uronen-Hansson H, Pabst O, Eksteen B, Tian J, Coombes JL, et al. Small intestinal CD103+ dendritic cells display unique functional properties that are conserved between mice and humans. *J Exp Med* (2008) **205**:2139–49. doi:10.1084/jem.20080414

31. Varol C, Landsman L, Fogg DK, Greenshtein L, Gildor B, Margalit R, et al. Monocytes give rise to mucosal, but not splenic, conventional dendritic cells. *J Exp Med* (2007) **204**:171–80. doi:10.1084/jem.20061011

32. Zigmond E, Varol C, Farache J, Elmaliah E, Satpathy AT, Friedlander G, et al. Ly6C hi monocytes in the inflamed colon give rise to proinflammatory effector cells and migratory antigen-presenting cells. *Immunity* (2012) **37**:1076–90. doi:10.1016/j.immuni.2012.08.026

33. Bain CC, Scott CL, Uronen-Hansson H, Gudjonsson S, Jansson O, Grip O, et al. Resident and pro-inflammatory macrophages in the colon represent alternative context-dependent fates of the same Ly6Chi monocyte precursors. *Mucosal Immunol* (2013) **6**:498–510. doi:10.1038/mi.2012.89

34. Rivollier A, He J, Kole A, Valatas V, Kelsall BL. Inflammation switches the differentiation program of Ly6Chi monocytes from antiinflammatory macrophages to inflammatory dendritic cells in the colon. *J Exp Med* (2012) **209**:139–55. doi:10.1084/jem.20101387

35. Zigmond E, Bernshtein B, Friedlander G, Walker CR, Yona S, Kim KW, et al. Macrophage-restricted interleukin-10 receptor deficiency, but not IL-10 deficiency, causes severe spontaneous colitis. *Immunity* (2014) **40**:720–33. doi:10.1016/j.immuni.2014.03.012

36. Artis D. Epithelial-cell recognition of commensal bacteria and maintenance of immune homeostasis in the gut. *Nat Rev Immunol* (2008) **8**:411–20. doi:10.1038/nri2316

37. Kang S, Okuno T, Takegahara N, Takamatsu H, Nojima S, Kimura T, et al. Intestinal epithelial cell-derived semaphorin 7A negatively regulates development of colitis via alphavbeta1 integrin. *J Immunol* (2012) **188**:1108–16. doi:10.4049/jimmunol.1102084

38. Cecchini MG, Dominguez MG, Mocci S, Wetterwald A, Felix R, Fleisch H, et al. Role of colony stimulating factor-1 in the establishment and regulation of tissue macrophages during postnatal development of the mouse. *Development* (1994) **120**:1357–72.

39. Dai XM, Ryan GR, Hapel AJ, Dominguez MG, Russell RG, Kapp S, et al. Targeted disruption of the mouse colony-stimulating factor 1 receptor gene results in osteopetrosis, mononuclear phagocyte deficiency, increased primitive progenitor cell frequencies, and reproductive defects. *Blood* (2002) **99**:111–20. doi:10.1182/blood.V99.1.111

40. Kuhn R, Lohler J, Rennick D, Rajewsky K, Muller W. Interleukin-10-deficient mice develop chronic enterocolitis. *Cell* (1993) **75**:263–74. doi:10.1016/0092-8674(93)80068-P

41. Shouval DS, Biswas A, Goettel JA, Mccann K, Conaway E, Redhu NS, et al. Interleukin-10 receptor signaling in innate immune cells regulates mucosal immune tolerance and anti-inflammatory macrophage function. *Immunity* (2014) **40**:706–19. doi:10.1016/j.immuni.2014.03.011

42. Glocker EO, Kotlarz D, Boztug K, Gertz EM, Schaffer AA, Noyan F, et al. Inflammatory bowel disease and mutations affecting the interleukin-10 receptor. *N Engl J Med* (2009) **361**:2033–45. doi:10.1056/NEJMoa0907206

43. Rubtsov YP, Rasmussen JP, Chi EY, Fontenot J, Castelli L, Ye X, et al. Regulatory T cell-derived interleukin-10 limits inflammation at environmental interfaces. *Immunity* (2008) **28**:546–58. doi:10.1016/j.immuni.2008.02.017

44. Niess JH, Adler G. Enteric flora expands gut lamina propria CX3CR1+ dendritic cells supporting inflammatory immune responses under normal and inflammatory conditions. *J Immunol* (2010) **184**:2026–37. doi:10.4049/jimmunol.0901936

45. Ueda Y, Kayama H, Jeon SG, Kusu T, Isaka Y, Rakugi H, et al. Commensal microbiota induce LPS hyporesponsiveness in colonic macrophages via the production of IL-10. *Int Immunol* (2010) **22**:953–62. doi:10.1093/intimm/dxq449

46. Gautier EL, Shay T, Miller J, Greter M, Jakubzick C, Ivanov S, et al. Gene-expression profiles and transcriptional regulatory pathways that underlie the

identity and diversity of mouse tissue macrophages. *Nat Immunol* (2012) **13**:1118–28. doi:10.1038/ni.2419

47. Lavin Y, Winter D, Blecher-Gonen R, David E, Keren-Shaul H, Merad M, et al. Tissue-resident macrophage enhancer landscapes are shaped by the local microenvironment. *Cell* (2014) **159**:1312–26. doi:10.1016/j.cell.2014.11.018

48. Ghisletti S, Barozzi I, Mietton F, Polletti S, De Santa F, Venturini E, et al. Identification and characterization of enhancers controlling the inflammatory gene expression program in macrophages. *Immunity* (2010) **32**:317–28. doi:10.1016/j.immuni.2010.02.008

49. Heinz S, Benner C, Spann N, Bertolino E, Lin YC, Laslo P, et al. Simple combinations of lineage-determining transcription factors prime cis-regulatory elements required for macrophage and B cell identities. *Mol Cell* (2010) **38**:576–89. doi:10.1016/j.molcel.2010.05.004

50. Sasmono RT, Oceandy D, Pollard JW, Tong W, Pavli P, Wainwright BJ, et al. A macrophage colony-stimulating factor receptor-green fluorescent protein transgene is expressed throughout the mononuclear phagocyte system of the mouse. *Blood* (2003) **101**:1155–63. doi:10.1182/blood-2002-02-0569

51. Brenner O, Levanon D, Negreanu V, Golubkov O, Fainaru O, Woolf E, et al. Loss of Runx3 function in leukocytes is associated with spontaneously developed colitis and gastric mucosal hyperplasia. *Proc Natl Acad Sci U S A* (2004) **101**:16016–21. doi:10.1073/pnas.0407180101

52. Satoh T, Takeuchi O, Vandenbon A, Yasuda K, Tanaka Y, Kumagai Y, et al. The Jmjd3-Irf4 axis regulates M2 macrophage polarization and host responses against helminth infection. *Nat Immunol* (2010) **11**:936–44. doi:10.1038/ni.1920

53. Krausgruber T, Blazek K, Smallie T, Alzabin S, Lockstone H, Sahgal N, et al. IRF5 promotes inflammatory macrophage polarization and TH1-TH17 responses. *Nat Immunol* (2011) **12**:231–8. doi:10.1038/ni.1990

54. Okabe Y, Medzhitov R. Tissue-specific signals control reversible program of localization and functional polarization of macrophages. *Cell* (2014) **157**:832–44. doi:10.1016/j.cell.2014.04.016

55. Rescigno M, Urbano M, Valzasina B, Francolini M, Rotta G, Bonasio R, et al. Dendritic cells express tight junction proteins and penetrate gut epithelial monolayers to sample bacteria. *Nat Immunol* (2001) **2**:361–7. doi:10.1038/86373

56. Vallon-Eberhard A, Landsman L, Yogev N, Verrier B, Jung S. Transepithelial pathogen uptake into the small intestinal lamina propria. *J Immunol* (2006) **176**:2465–9. doi:10.4049/jimmunol.176.4.2465

57. Kim KW, Vallon-Eberhard A, Zigmond E, Farache J, Shezen E, Shakhar G, et al. In vivo structure/function and expression analysis of the CX3C chemokine fractalkine. *Blood* (2011) **118**:e156–67. doi:10.1182/blood-2011-04-348946

58. Kostadinova FI, Baba T, Ishida Y, Kondo T, Popivanova BK, Mukaida N. Crucial involvement of the CX3CR1-CX3CL1 axis in dextran sulfate sodium-mediated acute colitis in mice. *J Leukoc Biol* (2010) **88**:133–43. doi:10.1189/jlb.1109768

59. Hadis U, Wahl B, Schulz O, Hardtke-Wolenski M, Schippers A, Wagner N, et al. Intestinal tolerance requires gut homing and expansion of FoxP3+ regulatory T cells in the lamina propria. *Immunity* (2011) **34**:237–46. doi:10.1016/j.immuni.2011.01.016

60. Hapfelmeier S, Muller AJ, Stecher B, Kaiser P, Barthel M, Endt K, et al. Microbe sampling by mucosal dendritic cells is a discrete, MyD88-independent step in DeltainvG S. Typhimurium colitis. *J Exp Med* (2008) **205**:437–50. doi:10.1084/jem.20070633

61. Diehl GE, Longman RS, Zhang JX, Breart B, Galan C, Cuesta A, et al. Microbiota restricts trafficking of bacteria to mesenteric lymph nodes by CX3CR1(hi) cells. *Nature* (2013) **494**:116–20. doi:10.1038/nature11809

62. Slavin J, Nash JR, Kingsnorth AN. Effect of transforming growth factor beta and basic fibroblast growth factor on steroid-impaired healing intestinal wounds. *Br J Surg* (1992) **79**:69–72. doi:10.1002/bjs.1800790124

63. Fukata M, Michelsen KS, Eri R, Thomas LS, Hu B, Lukasek K, et al. Toll-like receptor-4 is required for intestinal response to epithelial injury and limiting bacterial translocation in a murine model of acute colitis. *Am J Physiol Gastrointest Liver Physiol* (2005) **288**:G1055–65. doi:10.1152/ajpgi.00328.2004

64. Pull SL, Doherty JM, Mills JC, Gordon JI, Stappenbeck TS. Activated macrophages are an adaptive element of the colonic epithelial progenitor niche necessary for regenerative responses to injury. *Proc Natl Acad Sci U S A* (2005) **102**:99–104. doi:10.1073/pnas.0405979102

65. Seno H, Miyoshi H, Brown SL, Geske MJ, Colonna M, Stappenbeck TS. Efficient colonic mucosal wound repair requires Trem2 signaling. *Proc Natl Acad Sci U S A* (2009) **106**:256–61. doi:10.1073/pnas.0803343106

66. Du Plessis J, Vanheel H, Janssen CE, Roos L, Slavik T, Stivaktas PI, et al. Activated intestinal macrophages in patients with cirrhosis release NO and IL-6 that may disrupt intestinal barrier function. *J Hepatol* (2013) **58**:1125–32. doi:10.1016/j.jhep.2013.01.038

67. Mildner A, Jung S. Development and function of dendritic cell subsets. *Immunity* (2014) **40**:642–56. doi:10.1016/j.immuni.2014.04.016

68. Delamarre L, Pack M, Chang H, Mellman I, Trombetta ES. Differential lysosomal proteolysis in antigen-presenting cells determines antigen fate. *Science* (2005) **307**:1630–4. doi:10.1126/science.1108003

69. Schulz O, Jaensson E, Persson EK, Liu X, Worbs T, Agace WW, et al. Intestinal CD103+, but not CX3CR1+, antigen sampling cells migrate in lymph and serve classical dendritic cell functions. *J Exp Med* (2009) **206**:3101–14. doi:10.1084/jem.20091925

70. Murai M, Turovskaya O, Kim G, Madan R, Karp CL, Cheroutre H, et al. Interleukin 10 acts on regulatory T cells to maintain expression of the transcription factor Foxp3 and suppressive function in mice with colitis. *Nat Immunol* (2009) **10**:1178–84. doi:10.1038/ni.1791

71. Atarashi K, Nishimura J, Shima T, Umesaki Y, Yamamoto M, Onoue M, et al. ATP drives lamina propria T(H)17 cell differentiation. *Nature* (2008) **455**:808–12. doi:10.1038/nature07240

72. Medina-Contreras O, Geem D, Laur O, Williams IR, Lira SA, Nusrat A, et al. CX3CR1 regulates intestinal macrophage homeostasis, bacterial translocation, and colitogenic Th17 responses in mice. *J Clin Invest* (2011) **121**:4787–95. doi:10.1172/JCI59150

73. Denning TL, Wang YC, Patel SR, Williams IR, Pulendran B. Lamina propria macrophages and dendritic cells differentially induce regulatory and interleukin 17-producing T cell responses. *Nat Immunol* (2007) **8**:1086–94. doi:10.1038/ni1511

74. Denning TL, Norris BA, Medina-Contreras O, Manicassamy S, Geem D, Madan R, et al. Functional specializations of intestinal dendritic cell and macrophage subsets that control Th17 and regulatory T cell responses are dependent on the T cell/APC ratio, source of mouse strain, and regional localization. *J Immunol* (2011) **187**:733–47. doi:10.4049/jimmunol.1002701

75. Longman RS, Diehl GE, Victorio DA, Huh JR, Galan C, Miraldi ER, et al. CX(3)CR1(+) mononuclear phagocytes support colitis-associated innate lymphoid cell production of IL-22. *J Exp Med* (2014) **211**:1571–83. doi:10.1084/jem.20140678

76. Satpathy AT, Briseno CG, Lee JS, Ng D, Manieri NA, Kc W, et al. Notch2-dependent classical dendritic cells orchestrate intestinal immunity to attaching-and-effacing bacterial pathogens. *Nat Immunol* (2013) **14**:937–48. doi:10.1038/ni.2679

77. Aychek T, Mildner A, Yona S, Kim KW, Lampl N, Reich-Zeliger S, et al. IL-23-mediated mononuclear phagocyte crosstalk protects mice from *Citrobacter rodentium*-induced colon immunopathology. *Nat Commun* (2015) **6**:6525. doi:10.1038/ncomms7525

78. Sun CM, Hall JA, Blank RB, Bouladoux N, Oukka M, Mora JR, et al. Small intestine lamina propria dendritic cells promote de novo generation of Foxp3 T reg cells via retinoic acid. *J Exp Med* (2007) **204**:1775–85. doi:10.1084/jem.20070602

79. Coombes JL, Powrie F. Dendritic cells in intestinal immune regulation. *Nat Rev Immunol* (2008) **8**:435–46. doi:10.1038/nri2335

80. Zeng R, Oderup C, Yuan R, Lee M, Habtezion A, Hadeiba H, et al. Retinoic acid regulates the development of a gut-homing precursor for intestinal dendritic cells. *Mucosal Immunol* (2012) **6**:847–56. doi:10.1038/mi.2012.123

81. Milling S, Yrlid U, Cerovic V, Macpherson G. Subsets of migrating intestinal dendritic cells. *Immunol Rev* (2010) **234**:259–67. doi:10.1111/j.0105-2896.2009.00866.x

82. Pulendran B, Tang H, Denning TL. Division of labor, plasticity, and crosstalk between dendritic cell subsets. *Curr Opin Immunol* (2008) **20**:61–7. doi:10.1016/j.coi.2007.10.009

83. Mowat AM, Agace WW. Regional specialization within the intestinal immune system. *Nat Rev Immunol* (2014) **14**:667–85. doi:10.1038/nri3738

84. Jang MH, Sougawa N, Tanaka T, Hirata T, Hiroi T, Tohya K, et al. CCR7 is critically important for migration of dendritic cells in intestinal lamina propria to mesenteric lymph nodes. *J Immunol* (2006) **176**:803–10. doi:10.4049/jimmunol.176.2.803

85. Cerovic V, Bain CC, Mowat AM, Milling SW. Intestinal macrophages and dendritic cells: what's the difference? *Trends Immunol* (2014) **35**:270–7. doi:10.1016/j.it.2014.04.003

86. Cohen SB, Smith NL, Mcdougal C, Pepper M, Shah S, Yap GS, et al. Beta-catenin signaling drives differentiation and proinflammatory function of IRF8-dependent dendritic cells. *J Immunol* (2014) **194**(1):210–22. doi:10.4049/jimmunol.1402453

87. den Haan JM, Lehar SM, Bevan MJ. CD8(+) but not CD8(-) dendritic cells cross-prime cytotoxic T cells in vivo. *J Exp Med* (2000) **192**:1685–96. doi:10.1084/jem.192.12.1685

88. Cerovic V, Houston SA, Westlund J, Utriainen L, Davison ES, Scott CL, et al. Lymph-borne CD8alpha dendritic cells are uniquely able to cross-prime CD8 T cells with antigen acquired from intestinal epithelial cells. *Mucosal Immunol* (2014) **8**(1):38–48. doi:10.1038/mi.2014.40

89. McDole JR, Wheeler LW, Mcdonald KG, Wang B, Konjufca V, Knoop KA, et al. Goblet cells deliver luminal antigen to CD103+ dendritic cells in the small intestine. *Nature* (2012) **483**:345–9. doi:10.1038/nature10863

90. Farache J, Koren I, Milo I, Gurevich I, Kim KW, Zigmond E, et al. Luminal bacteria recruit CD103+ dendritic cells into the intestinal epithelium to sample bacterial antigens for presentation. *Immunity* (2013) **38**:581–95. doi:10.1016/j.immuni.2013.01.009

91. Bimczok D, Sowa EN, Faber-Zuschratter H, Pabst R, Rothkotter HJ. Site-specific expression of CD11b and SIRPalpha (CD172a) on dendritic cells: implications for their migration patterns in the gut immune system. *Eur J Immunol* (2005) **35**:1418–27. doi:10.1002/eji.200425726

92. Wilson NS, Young LJ, Kupresanin F, Naik SH, Vremec D, Heath WR, et al. Normal proportion and expression of maturation markers in migratory dendritic cells in the absence of germs or toll-like receptor signaling. *Immunol Cell Biol* (2008) **86**:200–5. doi:10.1038/sj.icb.7100125

93. Sierro F, Dubois B, Coste A, Kaiserlian D, Kraehenbuhl JP, Sirard JC. Flagellin stimulation of intestinal epithelial cells triggers CCL20-mediated migration of dendritic cells. *Proc Natl Acad Sci U S A* (2001) **98**:13722–7. doi:10.1073/pnas.241308598

94. Yrlid U, Milling SW, Miller JL, Cartland S, Jenkins CD, Macpherson GG. Regulation of intestinal dendritic cell migration and activation by plasmacytoid dendritic cells, TNF-alpha and type 1 IFNs after feeding a TLR7/8 ligand. *J Immunol* (2006) **176**:5205–12. doi:10.4049/jimmunol.176.9.5205

95. Forster R, Davalos-Misslitz AC, Rot A. CCR7 and its ligands: balancing immunity and tolerance. *Nat Rev Immunol* (2008) **8**:362–71. doi:10.1038/nri2297

96. Rimoldi M, Chieppa M, Salucci V, Avogadri F, Sonzogni A, Sampietro GM, et al. Intestinal immune homeostasis is regulated by the crosstalk between epithelial cells and dendritic cells. *Nat Immunol* (2005) **6**:507–14. doi:10.1038/ni1192

97. Zeuthen LH, Fink LN, Frokiaer H. Epithelial cells prime the immune response to an array of gut-derived commensals towards a tolerogenic phenotype through distinct actions of thymic stromal lymphopoietin and transforming growth factor-beta. *Immunology* (2008) **123**:197–208. doi:10.1111/j.1365-2567.2007.02687.x

98. Iliev ID, Spadoni I, Mileti E, Matteoli G, Sonzogni A, Sampietro GM, et al. Human intestinal epithelial cells promote the differentiation of tolerogenic dendritic cells. *Gut* (2009) **58**:1481–9. doi:10.1136/gut.2008.175166

99. Stock A, Booth S, Cerundolo V. Prostaglandin E2 suppresses the differentiation of retinoic acid-producing dendritic cells in mice and humans. *J Exp Med* (2011) **208**:761–73. doi:10.1084/jem.20101967

100. McDonald KG, Leach MR, Brooke KW, Wang C, Wheeler LW, Hanly EK, et al. Epithelial expression of the cytosolic retinoid chaperone cellular retinol binding protein II is essential for in vivo imprinting of local gut dendritic cells by lumenal retinoids. *Am J Pathol* (2012) **180**:984–97. doi:10.1016/j.ajpath.2011.11.009

101. Johansson-Lindbom B, Svensson M, Pabst O, Palmqvist C, Marquez G, Forster R, et al. Functional specialization of gut CD103+ dendritic cells in the regulation of tissue-selective T cell homing. *J Exp Med* (2005) **202**:1063–73. doi:10.1084/jem.20051100

102. Edele F, Molenaar R, Gutle D, Dudda JC, Jakob T, Homey B, et al. Cutting edge: instructive role of peripheral tissue cells in the imprinting of T cell homing receptor patterns. *J Immunol* (2008) **181**:3745–9. doi:10.4049/jimmunol.181.6.3745

103. Iliev ID, Mileti E, Matteoli G, Chieppa M, Rescigno M. Intestinal epithelial cells promote colitis-protective regulatory T-cell differentiation through dendritic cell conditioning. *Mucosal Immunol* (2009) **2**:340–50. doi:10.1038/mi.2009.13

104. Mora JR, Bono MR, Manjunath N, Weninger W, Cavanagh LL, Rosemblatt M, et al. Selective imprinting of gut-homing T cells by Peyer's patch dendritic cells. *Nature* (2003) **424**:88–93. doi:10.1038/nature01726

105. Annacker O, Coombes JL, Malmstrom V, Uhlig HH, Bourne T, Johansson-Lindbom B, et al. Essential role for CD103 in the T cell-mediated regulation of experimental colitis. *J Exp Med* (2005) **202**:1051–61. doi:10.1084/jem.20040662

106. Bakdash G, Vogelpoel LT, Van Capel TM, Kapsenberg ML, De Jong EC. Retinoic acid primes human dendritic cells to induce gut-homing, IL-10-producing regulatory T cells. *Mucosal Immunol* (2014) **8**(2):265–78. doi:10.1038/mi.2014.64

107. Spadoni I, Iliev ID, Rossi G, Rescigno M. Dendritic cells produce TSLP that limits the differentiation of Th17 cells, fosters Treg development, and protects against colitis. *Mucosal Immunol* (2012) **5**:184–93. doi:10.1038/mi.2011.64

108. Goto Y, Panea C, Nakato G, Cebula A, Lee C, Diez MG, et al. Segmented filamentous bacteria antigens presented by intestinal dendritic cells drive mucosal Th17 cell differentiation. *Immunity* (2014) **40**:594–607. doi:10.1016/j.immuni.2014.03.005

109. Coombes JL, Siddiqui KR, Arancibia-Carcamo CV, Hall J, Sun CM, Belkaid Y, et al. A functionally specialized population of mucosal CD103+ DCs induces Foxp3+ regulatory T cells via a TGF-beta and retinoic acid-dependent mechanism. *J Exp Med* (2007) **204**:1757–64. doi:10.1084/jem.20070590

110. Manicassamy S, Ravindran R, Deng J, Oluoch H, Denning TL, Kasturi SP, et al. Toll-like receptor 2-dependent induction of vitamin A-metabolizing enzymes in dendritic cells promotes T regulatory responses and inhibits autoimmunity. *Nat Med* (2009) **15**:401–9. doi:10.1038/nm.1925

111. Molenaar R, Knippenberg M, Goverse G, Olivier BJ, De Vos AF, O'Toole T, et al. Expression of retinaldehyde dehydrogenase enzymes in mucosal dendritic cells and gut-draining lymph node stromal cells is controlled by dietary vitamin A. *J Immunol* (2011) **186**:1934–42. doi:10.4049/jimmunol.1001672

112. Cherayil BJ. Indoleamine 2,3-dioxygenase in intestinal immunity and inflammation. *Inflamm Bowel Dis* (2009) **15**:1391–6. doi:10.1002/ibd.20910

113. Matteoli G, Mazzini E, Iliev ID, Mileti E, Fallarino F, Puccetti P, et al. Gut CD103+ dendritic cells express indoleamine 2,3-dioxygenase which influences T regulatory/T effector cell balance and oral tolerance induction. *Gut* (2010) **59**:595–604. doi:10.1136/gut.2009.185108

114. Laffont S, Siddiqui KR, and Powrie F. (2010). Intestinal inflammation abrogates the tolerogenic properties of MLN CD103+ dendritic cells. *Eur J Immunol* **40**:1877–1883. doi:10.1002/eji.200939957

115. Kinnebrew MA, Buffie CG, Diehl GE, Zenewicz LA, Leiner I, Hohl TM, et al. Interleukin 23 production by intestinal CD103(+)CD11b(+) dendritic cells in response to bacterial flagellin enhances mucosal innate immune defense. *Immunity* (2012) **36**:276–87. doi:10.1016/j.immuni.2011.12.011

116. Jaitin DA, Kenigsberg E, Keren-Shaul H, Elefant N, Paul F, Zaretsky I, et al. Massively parallel single-cell RNA-seq for marker-free decomposition of tissues into cell types. *Science* (2014) **343**:776–9. doi:10.1126/science.1247651

117. Zeisel A, Munoz-Manchado AB, Codeluppi S, Lonnerberg P, La Manno G, Jureus A, et al. Cell types in the mouse cortex and hippocampus revealed by single-cell RNA-seq. *Science* (2015) **347**:1138–42. doi:10.1126/science.aaa1934

118. Gregori G, Rajwa B, Patsekin V, Jones J, Furuki M, Yamamoto M, et al. Hyperspectral cytometry. *Curr Top Microbiol Immunol* (2014) **377**:191–210. doi:10.1007/82_2013_359

119. Bendall SC, Nolan GP, Roederer M, Chattopadhyay PK. A deep profiler's guide to cytometry. *Trends Immunol* (2012) **33**:323–32. doi:10.1016/j.it.2012.02.010

5

A systematic approach to identify markers of distinctly activated human macrophages

Bayan Sudan[1], Mark A. Wacker[1], Mary E. Wilson[1,2] and Joel W. Graff[1,2]*

[1] Infectious Diseases Division, Department of Internal Medicine, University of Iowa, Iowa City, IA, USA, [2] Iowa City VA Medical Center, Iowa City, IA, USA

Edited by:
Martin Guilliams,
VIB Inflammation Research Center,
Belgium

Reviewed by:
Elodie Segura,
Institut Curie, France
Sam Basta,
Queen's University, Canada

*Correspondence:
Joel W. Graff,
Infectious Diseases Division,
Department of Internal Medicine,
University of Iowa, 400 EMRB, Iowa
City, IA 52242, USA
joel-graff@uiowa.edu

Polarization has been a useful concept for describing activated macrophage pheno-types and gene expression profiles. However, macrophage activation status within tumors and other settings are often inferred based on only a few markers. Complicating matters for relevance to human biology, many macrophage activation markers have been best characterized in mice and sometimes are not similarly regulated in human macrophages. To identify novel markers of activated human macrophages, gene expression profiles for human macrophages of a single donor subjected to 33 distinct activating conditions were obtained and a set of putative activation markers were subsequently evaluated in macrophages from multiple donors using integrated fluidic circuit (IFC)-based RT-PCR. Using unsupervised hierarchical clustering of the microarray screen, highly altered transcripts (>4-fold change in expression) sorted the macrophage transcription profiles into two major and 13 minor clusters. Among the 1874 highly altered transcripts, over 100 were uniquely altered in one major or two related minor clusters. IFC PCR-derived data confirmed the microarray results and determined the kinetics of expression of potential macrophage activation markers. Transcripts encoding chemokines, cytokines, and cell surface were prominent in our analyses. The activation markers identified by this study could be used to better characterize tumor-associated macrophages from biopsies as well as other macrophage populations collected from human clinical samples.

Keywords: human macrophages, activation markers, microarray, integrated fluidic circuit RT-PCR, macrophage polarization

Introduction

Macrophages assume critical roles in almost every tissue and disease state through their ability to assume distinct functional capacities in different microenvironments. Macrophages respond to a variety of external stimuli to assume different polarized activation states. Distinctly polarized macrophages, modeled *in vitro* using specific activating conditions, can be defined by functional attributes such as microbicidal activity, and by unique gene expression profiles. An early study contrasting functional and gene expression differences between IFNγ- and IL-4-treated macrophages proposed that the latter phenotype be described as alternative activation (1), a very different macrophage phenotype from IFNγ- or classically activated macrophages. Since that time, many additional polarized macrophage types, induced by different stimuli, have been proposed.

Several competing systems have been proposed in an attempt to provide a framework that describes the complexity of macrophage polarization. The first system describes macrophage phenotypes as a linear continuum with M1 (classically activated) and M2 (alternatively activated) macrophages at opposite ends (2, 3). The second system describes macrophage phenotypes as a spectrum akin to a color wheel, with classically activated, wound healing, and regulatory macrophages used as examples of unique polarized phenotypes that do not fit well within a linear continuum (4). A modified version of the M1–M2 system acknowledged the diversity of macrophage phenotypes with descriptions such as M1a, M1b, M2a, M2b, and M2c (5, 6). Additions to the M1–M2 nomenclature system have proposed naming macrophages differentiated in the presence of CXCL4 as "M4" (7) and IL-17-treated macrophages "M17" (8). To standardize the burgeoning descriptions of polarized macrophage types, it has been suggested that the activation condition be defined in the name of the polarized macrophage [M(IL-4), M(IL-10), M(LPS), M(IFNγ), and so forth (9)]. To preserve clarity, we have employed this descriptive nomenclature system to describe the activated macrophages in the current report (**Table 1**).

Macrophages are often very abundant within tumors (12, 13). There is evidence that macrophages can promote tumorigenesis, tumor growth, and metastasis (14). Despite macrophage pro-tumor activities, tumor-associated macrophages (TAMs) display a wide range of phenotypic diversity within a tumor due to ontogeny, activation signals, and localization (15). The plasticity of macrophage phenotypes is well known (16, 17) and this characteristic has provided a therapeutic target whereby macrophages are encouraged to switch functionally from pro-tumor to anti-tumor. Clinical approaches that modify macrophage activation in this way include blockade of M-CSF, low-dose irradiation, and combinational therapies (18–21). What is lacking is a thoroughly characterized and reliable set of macrophage activation markers that would allow for improved characterization of activation patterns, and monitoring of the therapeutic efficacy of macrophage-targeted treatments.

Gene expression profiles using microarrays have been used to analyze activation of primary human monocytes and monocyte-derived macrophages (MDMs) (7, 22–32). Until very recently (33), most transcriptome-based approaches to characterize polarized macrophages contrasted two macrophage-activating conditions in each study. Using a blood sample from a single human donor, we surveyed gene expression profiles in primary macrophages activated with 33 different activating conditions. This data set served as a rich resource for identifying putative human macrophage activation markers. As a follow-up approach, integrated fluidic circuit (IFC)-based RT-PCR was used to examine a panel of transcripts to verify the reproducibility of the gene expression changes from multiple donors. This latter assay was also used to determine the expression kinetics of previously described markers of human macrophages as well as novel markers identified by the microarray-based screen.

TABLE 1 | Macrophage-activating conditions and nomenclature used in this study.

Single stimulus treatments	Previous nomenclature	Current nomenclature
1. GM-CSF (100 ng/ml)	M1	M(GM-CSF)
2. IFNβ (20 ng/ml)		M(IFNβ)
3. IFNγ (20 ng/ml)	M1, classical	M(IFNγ)
4. IL-1β (100 ng/ml)		M(IL-1β)
5. IL-4 (20 ng/ml)	M2, M2a, alternative, wound healing	M(IL-4)
6. IL-10 (50 ng/ml)	M2c	M(IL-10)
7. TGFβ (5 ng/ml)	M2c	M(TGFβ)
8. TNFα (100 ng/ml)		M(TNFα)
9. Curdlan (20 μg/ml)[a]		M(Curdlan)
10. TDB (20 μg/ml)		M(TDB)
11. PolyI:C (2 μg/ml)		M(PolyI:C)
12. LPS (10 ng/ml)	M1, classical	M(LPS10)
13. LPS (100 ng/ml)	M1, classical	M(LPS100)
14. Adenosine (100 μM)		M(Ado)
15. IgG-OVA immune complexes (IC)[a]		M(IC)
16. Dexamethasone (100 nM)	M2c	M(Dex)

Combinational treatments	Previous nomenclature	Current nomenclature
17. TDB + IFNγ		M(TDB + IFNγ)
18. TDB + IC[a]		M(TDB + IC)
19. TDB + IL-4		M(TDB + IL-4)
20. TDB + IL-10		M(TDB + IL-10)
21. LPS (10 ng/ml) + IFNγ	M1	M(LPS + IFNγ)
22. LPS (100 ng/ml) + IC[a]	M2b, regulatory	M(LPS + IC)
23. LPS (10 ng/ml) + IL-4		M(LPS + IL-4)
24. LPS (10 ng/ml) + IL-10		M(LPS + IL-10)
25. Adenosine + IFNγ		M(Ado + IFNγ)
26. Adenosine + IC[a]		M(Ado + IC)
27. Adenosine + IL-10		M(Ado + IL-10)
28. TGFβ + GM-CSF		M(TGFβ + GM-CSF)
29. TGFβ + IL-1β		M(TGFβ + IL-1β)
30. TGFβ + LPS (100 ng/ml)		M(LPS + TGFβ)
31. Dexamethasone + GM-CSF		M(Dex + GM-CSF)
32. Dexamethasone + IL-1β		M(Dex + IL-1β)
33. Dexamethasone + LPS (100 ng/ml)		M(LPS + Dex)

[a] Treatments with chicken ovalbumin and with Curdlan likely had endotoxin contamination due to the extraction processes used to obtain these reagents (10, 11).

Materials and Methods

Human Subjects

Human subject protocols were approved by Institutional Review Boards of the University of Iowa and the Iowa City Veterans Affairs Medical Center. Peripheral blood samples from anonymous, healthy donors were acquired through the DeGowin Blood Bank at the University of Iowa.

Integrated Fluidic Circuit-Based RT-PCR

RNA purified from MDMs using TRIzol was reverse transcribed in random hexamer-primed reactions with SuperScript III RT (Invitrogen). The cDNA was pre-amplified for 14 PCR cycles in reactions primed by a master mix of 48 TaqMan Gene Expression Assays (Applied Biosystems) using PreAmp Master Mix (Applied

Biosystems) with a modified protocol according to recommendations by Fluidigm. Following a 1:5 dilution of pre-amplified product in water, 48 samples and 48 TaqMan Gene Expression Assays were loaded onto 48.48 Dynamic Array IFC plates (Fluidigm) using the 48.48 MX IFC Controller (Fluidigm). Real-time PCR was performed using the BioMark System for Genetic Analysis (Fluidigm). Cycle threshold (Ct) values were determined using real-time PCR analysis v3.1.3 software (Fluidigm). Ct values corresponding to transcripts encoding ACTB, B2M, and TBP were used as endogenous controls. Changes in transcript expression were calculated using the $\Delta\Delta Ct$ method and converted to $\log_2$ scale using Excel 2010. Line graphs of time course experiments were generated using Prism 6 (GraphPad). Heat maps were generated using Partek Genomic Suite software.

Cell Purification and Culture

Peripheral blood mononuclear cells (PBMCs) were isolated from the blood by density sedimentation using Ficoll-Paque PLUS (GE Healthcare) and maintained in Petri dishes at a density of 5e7 cells/dish in 10 ml RP-10 medium [RPMI 1640 medium (Gibco) supplemented with 10% fetal bovine serum (Gibco), 100 U/ml penicillin, 50 µg/ml gentamicin, and 5 ng/ml M-CSF (eBioscience)]. After 10 days, non-adherent cells were removed by rinsing and the adherent MDMs were dislodged with cell scraping following incubation at 37°C for 10 min in 0.25% Trypsin/1 mM EDTA solution (Gibco). MDMs were seeded in 12-well tissue culture-treated plates (Corning) at 5e5 cells/well in 2 ml RP-10 and allowed to rest for 2 days at 37°C. Before treatment with macrophage-activating conditions, the culture medium was replaced with 1 ml fresh RP-10 per well. At 24 h post-treatment, RNA was purified from MDMs using TRIzol Reagent (Invitrogen).

Macrophage-Activating Stimuli

All stock solutions were stored at −80°C unless otherwise noted. The sources of human recombinant cytokines were as follows: IL-1β (eBioscience), IL-4 (PeproTech), IL-10 (R&D Systems), IFNβ (PeproTech), IFNγ (PeproTech), GM-CSF (eBioscience), TNFα (PeproTech), and TGFβ (R&D Systems). These cytokines were stored at concentrations recommended by the manufacturers and were subjected to no more than two freeze–thaw cycles. Dexamethasone powder (Sigma-Aldrich) was suspended in 1 part ethanol and subsequently diluted in 49 parts medium to a stock concentration of 50 µM. Phenol-extracted *Escherichia coli* 055:B5 LPS (Sigma-Aldrich) and polyinosinic:polycytidylic acid sodium salt (PolyI:C) (Sigma-Aldrich) were stored at a stock concentration of 1 mg/ml in RP-10. Adenosine (Sigma-Aldrich) was suspended in RP-10 at a stock concentration of 10 mM.

Chicken ovalbumin (MP Biomedicals) was suspended at 2 mg/ml in PBS lacking Ca^{++} or Mg^{++} (Gibco) and goat anti-chicken ovalbumin (MP Biomedicals) was suspended in water at 16 mg/ml. Immune complexes (IC) were prepared fresh for each experiment by combining ~10:1M excess of antibody to antigen and incubating with end-over-end rotation at room temperature for 30 min. Curdlan (InvivoGen) was also freshly prepared for each experiment by suspension in RP-10 at a concentration of 1 mg/ml.

Trehalose-6,6-dibehenate (TDB) (InvivoGen) was suspended at a concentration of 10 mg/ml in DMSO and heated to 60°C for 30 s. After vortexing, the TDB/DMSO solution was diluted to 1 mg/ml by the addition of PBS. This stock solution was heated to 60°C for 15 min and stored at 4°C.

Microarrays

RNA sample preparation for microarrays and the subsequent hybridization to the Illumina beadchips were performed at the University of Iowa DNA Facility. Three Human HT-12 v4 BeadChips (Illumina) were processed individually in this experiment with 1 sample from an untreated control and 11 samples from polarized macrophages loaded onto each array. Briefly, 100 ng total RNA from each of the 36 samples was amplified and converted to biotin-cRNA using the Epicenter TargetAmp-Nano Labeling Kit for Illumina Expression Bead-Chip (Illumina). The biotin-aRNA product was purified using the RNeasy MinElute Cleanup Kit (Qiagen) according to modifications from Epicenter. Seven hundred fifty nanograms of this product were mixed with Illumina hybridization buffer, placed onto each beadchip array, and incubated with rocking at 58°C for 17 h in an Illumina Hybridization Oven. Following hybridization, the arrays were washed, blocked, and stained with streptavidin-Cy3 using the Whole-Genome Gene Expression Direct Hybridization Assay (Illumina). Beadchip arrays were scanned with the iScan System (Illumina) and data were collected using the GenomeStudio software v2011.1 (Illumina). The expression data has been deposited in NCBI Geo repository (GSE68854).

Transcript Expression Analysis

Partek Genomic Suite v6.5 (Partek) was used to perform robust multi-array averaging and to calculate gene expression changes. A data set comprising of 1874 transcripts with changes in expression of more than fourfold relative to untreated controls was submitted to unsupervised hierarchical clustering and principal components analysis using default settings in Partek Genomic Suite software. Briefly, for unsupervised hierarchical clustering, agglomerative clustering was used to determine Euclidean dissimilarity distances using an average linkage method. For principal components analysis, a dispersion matrix based on correlations was normalized using Eigenvector scaling. Contribution of individual transcripts to each of the principal components was determined using the FactoMineR package in R. After principal component analysis was completed the contribution of each transcript to each of the components was extracted and ranked using Excel 2010.

Correlation coefficients were calculated for each pairwise combination of the 33 activated macrophage expression profiles for the 1874 regulated transcripts using the corandPvalue function of the WGCNA package in R. The data were then converted to heat maps using Excel (Microsoft).

For gene ontology (GO) analysis, the STRING database (version 9.05; string-db.org) was used to identify the 1615 protein coding RNAs in our set of 1874 regulated transcripts. Also, within the STRING database website, the GO categories enriched in the set of 1615 regulated transcripts identified as protein coding RNAs were determined.

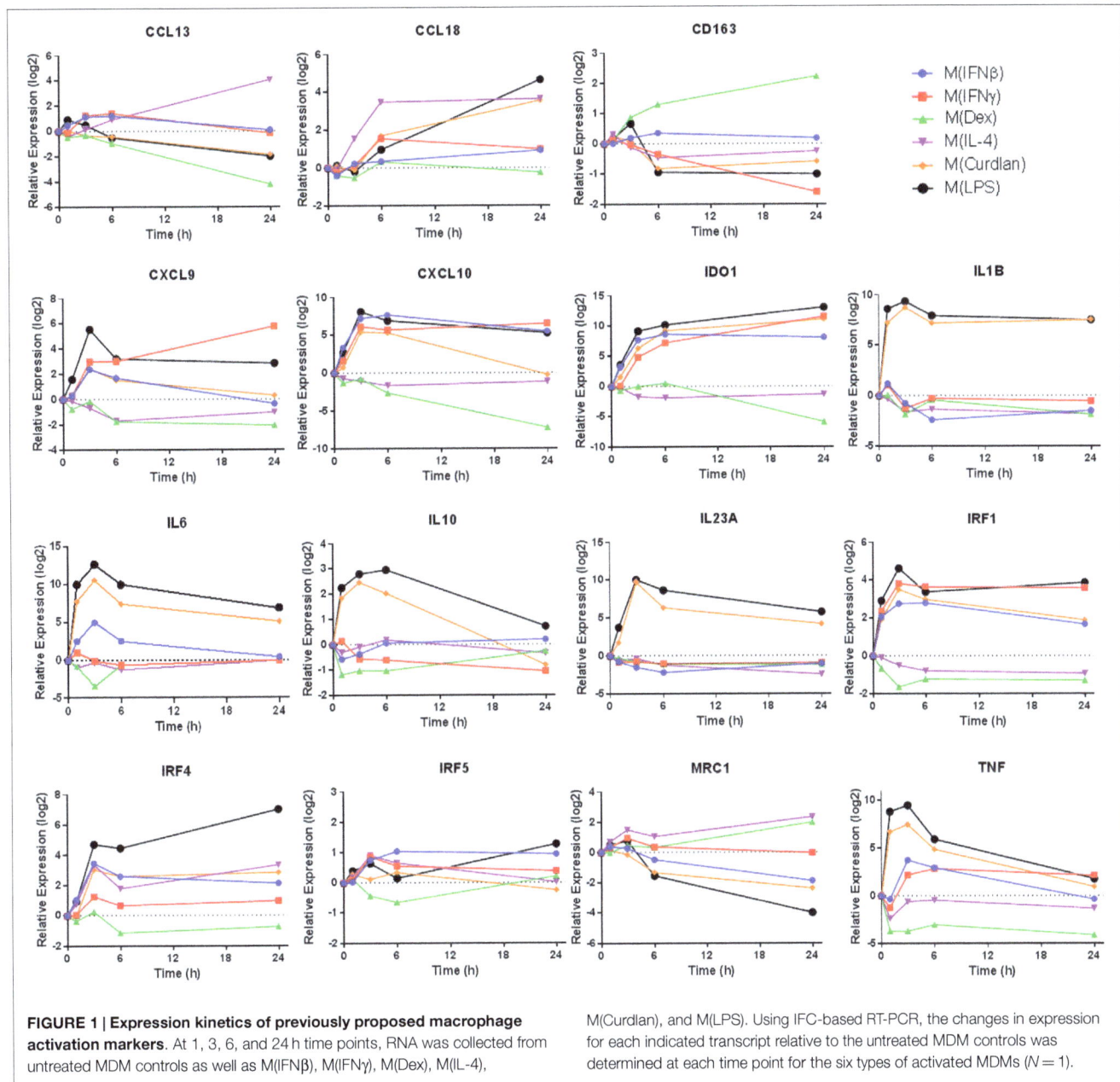

FIGURE 1 | Expression kinetics of previously proposed macrophage activation markers. At 1, 3, 6, and 24 h time points, RNA was collected from untreated MDM controls as well as M(IFNβ), M(IFNγ), M(Dex), M(IL-4), M(Curdlan), and M(LPS). Using IFC-based RT-PCR, the changes in expression for each indicated transcript relative to the untreated MDM controls was determined at each time point for the six types of activated MDMs (N = 1).

Results

Survey of Proposed Human Macrophage Activation Marker Expression in MDMs Responding to Six Distinct Activation Conditions

Transcripts used as markers of polarized human macrophages should change expression in response to one stimulus or a limited number of related activation stimuli. Additionally, macrophage activation markers should have sustained, rather than transient, changes in expression. In primary human macrophages responding to a variety of activation conditions, we evaluated the expression kinetics of transcripts that encode 11 proposed activation markers (4, 6, 9) over a 24-h period (**Figure 1**). Several observations from this survey were notable. First, some commonly assessed transcripts, TNF and IL-10, were rapidly induced in M(LPS) and M(Curdlan) but returned to near basal expression by the 24-h time point. Second, although many genes were similarly regulated in M(IFNβ) and M(IFNγ), the expression patterns of CD163 and CXCL9 were distinct in response to these two interferon types. Third, most markers have been noted because of their increased expression in response to macrophage activation conditions but many transcripts in this panel showed a remarkable reduction in expression. Finally, the expression level of many activation marker transcripts was either continuing to change or was sustained at high levels at the 24-h time point. Together, these observations revealed there is a need for a systematic attempt to identify reliable activation markers whose expression was either up- or down-regulated in human macrophages, and which

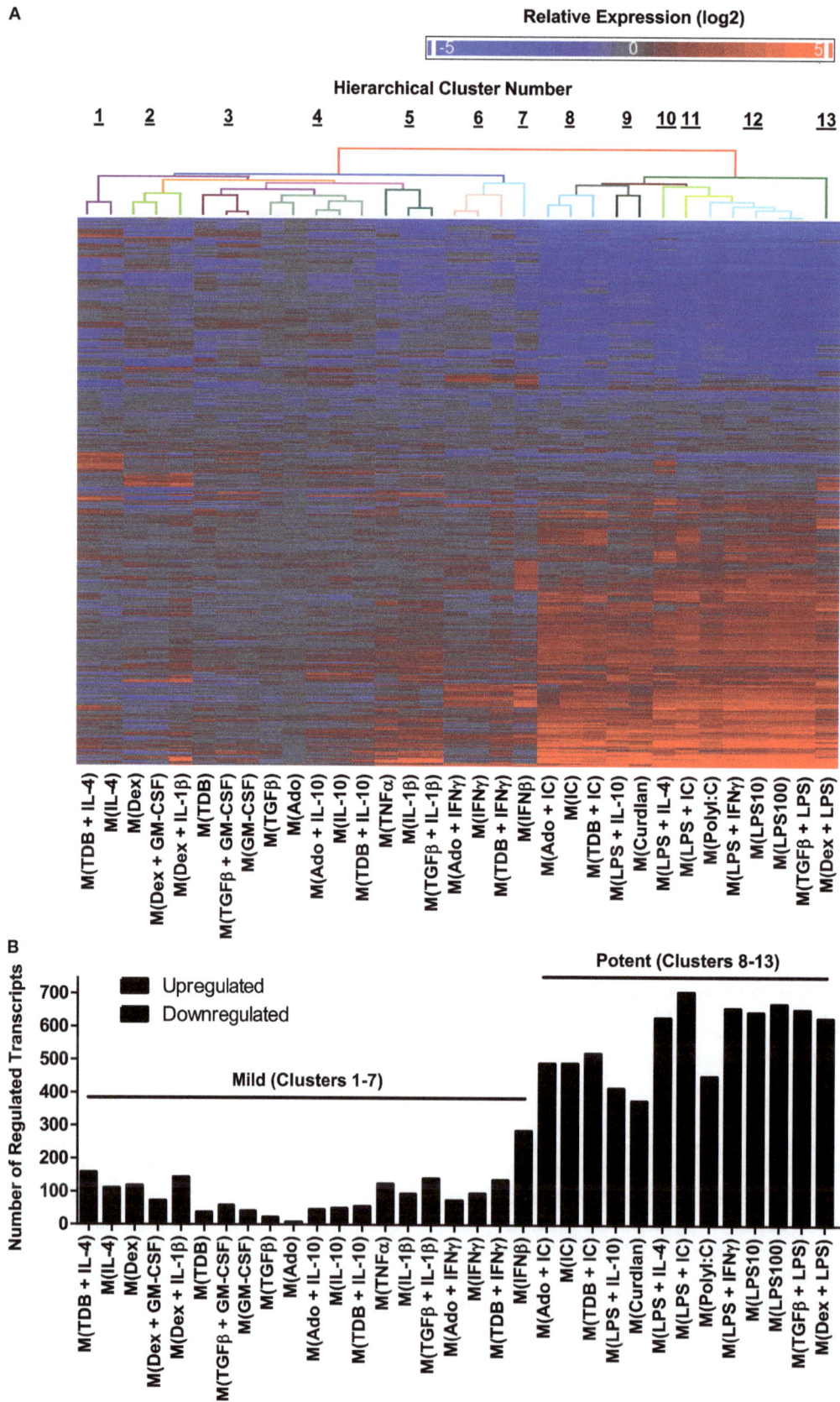

FIGURE 2 | Continued

FIGURE 2 | Continued
Hierarchical clustering of gene expression profiles from activated human MDMs separated into 2 major clusters and 13 minor clusters. Microarrays were performed using RNA collected from MDMs at 24 h post-treatment with 33 distinct activation conditions ($N = 1$). **(A)** A set of 1874 regulated transcripts defined as having >4-fold change in expression levels relative to untreated controls was compiled and displayed as a heat map ($\log_2$ scale). Gene expression profiles were sorted according to unsupervised hierarchical clustering of genes and treatments. Dissimilarity distances between gene expression profiles are displayed using a color-coded dendrogram to indicate 13 hierarchical clusters. See Section "Results" for dissimilarity distance cut-off rationale. Arranged in the same order as shown here, transcript names and quantitation of expression level changes are available in Table S1 in Supplementary Material. **(B)** Number of upregulated and down-regulated transcripts within each gene expression profile. Potent and mild macrophage activation conditions are indicated.

exhibited sustained expression level changes in response to an array of activation conditions.

Expression Profiling of a Diverse Array of Activated Human Macrophages

To screen for transcripts representing putative human macrophage activation markers, microarrays were performed using samples collected from human MDMs derived from a single donor, subjected to 33 unique activating conditions (**Table 1**) for 24 h. Sixteen of the conditions were composed of a single activating stimulus. Eight cytokines comprised the largest category of macrophage-activating stimuli used in this study and represent a spectrum of pro- and anti-inflammatory molecules that are abundantly expressed in sites where MDMs would be recruited such as infections or wounds. Pathogen-associated molecular patterns (PAMPs) recognized by C-type lectin receptors (CLRs) or toll-like receptors (TLRs) were the second largest set of macrophage-activating stimuli in this study and consisted of Curdlan (dectin-1 agonist), TDB (trehalose-6,6-dibehenate; mincle agonist), polyI:C (TLR3 agonist), or one of two concentrations of LPS (TLR4 agonist). Another set of stimuli, IgG–OVA IC and adenosine, were selected for their ability to reprogram inflammatory macrophages to become non-inflammatory (34, 35). Finally, we selected the glucocorticoid, dexamethasone, as an immunosuppressive stimulus. The remaining 17 conditions consisted of pairs of the above macrophage-activating stimuli (**Table 1**). The macrophage-activating conditions were selected with the expectation that they would lead to diverse gene expression profiles providing insights into the potential diversity of macrophage gene expression programs.

We first focused our attention on regulated transcripts that had changes in abundance of over fourfold relative to untreated controls changes. A data set of 1874 regulated transcripts that were differentially expressed in MDMs responding to one or more of the macrophage-activating conditions was compiled. Unsupervised hierarchical clustering was performed to evaluate the expression profiles of the regulated transcripts; this is summarized in a heat map that includes a dendrogram indicating relative dissimilarity distances between gene expression profiles of each polarized macrophage type (**Figure 2A**). Official gene names of the regulated transcripts and calculated expression changes are provided as supplemental material (Table S1 in Supplementary Material).

We considered whether the clustering analysis results separated the gene expression profiles corresponding to previously studied macrophage activation states as denoted in **Table 1**. Consistent with the previous reports (29, 33), gene expression profiles of M(LPS + IFNγ) (previously named "M1") macrophages were quite different from that of M(IL-4) (previously named "M2a") macrophages. By contrast, the profile of M(LPS + IC) macrophages (previously named "M2b") was very similar to the profiles of M(LPS + IFNγ), separated only by the profile of M(PolyI:C). Since M(LPS + IFNγ) and M(LPS + IC) are known to have different biological activities (6), we divided the 33 macrophage expression profiles into 13 clusters, the lowest dissimilarity distance cut-off that successfully separated these profiles (**Figure 2A**).

Microarray Results were Confirmed Using IFC Arrays

The IFC array-based real-time RT-PCR platform provided a high-throughput mechanism to accurately verify the expression of a large set of transcripts in samples from multiple human donors. We used several strategies to select a panel of transcripts with diverse expression patterns out of the 1874 regulated transcripts, which were re-assessed on multiple samples using IFC arrays. First, we included the 11 transcripts analyzed in **Figure 1**. Next, we used the STRING database (version 9.05) to identify enriched GOs for the 1615 protein coding RNAs in our set of 1874 regulated transcripts. Among the GOs categories that were enriched in our data set, we chose to focus on chemokine activity, cell surface, and cytokine activity because these GO categories were highly enriched (Table S2 in Supplementary Material). Finally, we selected transcripts that were uniquely regulated in one or two minor clusters. The final panel included a combination of transcripts that represented changes occurring in each of the 33 macrophage-activating conditions. We also mined the data set for reliable endogenous controls to include in the panel. Among the potential endogenous controls we considered, the expression levels of TBP (define) and B2M (define) transcripts appeared to be the least affected by the 33 macrophage-activating conditions (Table S1 in Supplementary Material).

The samples obtained from activated MDMs of a single donor that were analyzed by microarray were re-assessed using IFC arrays. Approximately 10 transcripts were not detected when using a pre-established Ct cut-off. Strong linear correlation for 15 representative transcripts was observed when comparing expression levels determined by microarray and by IFC arrays (Figure S1 in Supplementary Material). The remaining detectable transcripts in our panel had expression levels that also showed strong linear correlation when comparing microarray and RT-PCR results (data not shown). Overall, these results confirmed the microarray measurements using an independent approach and provide convincing evidence that IFC arrays was a dependable method for measuring transcript expression.

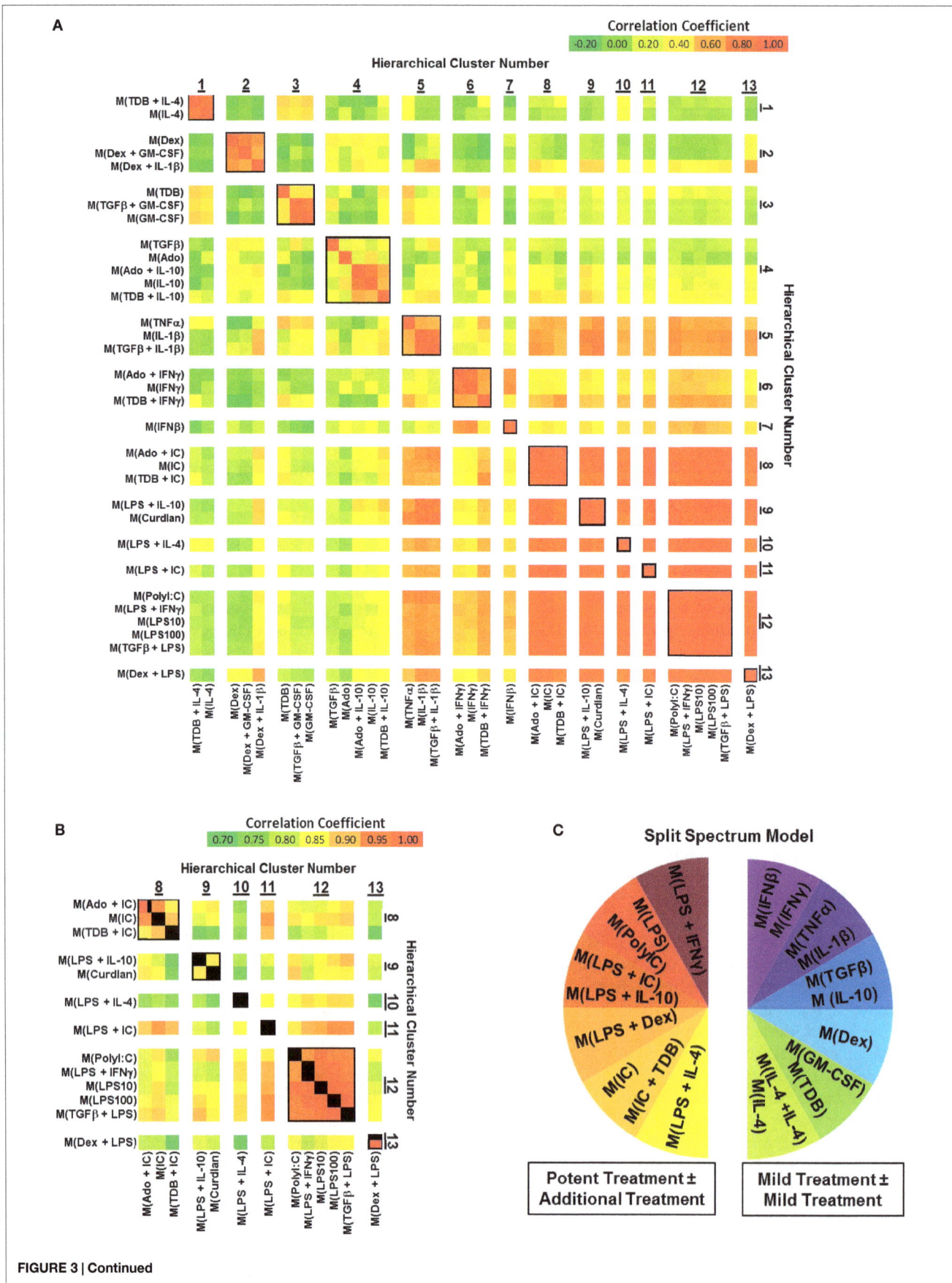

FIGURE 3 | Continued

FIGURE 3 | Continued
Comparing correlation coefficients supported the separation of gene expression profiles into two clusters, which can be modeled as a "split spectrum." Correlation coefficients were calculated for using the 1874 regulated transcript data set (N = 1). **(A)** Each pairwise combination of the 33 gene expression profiles are displayed as a heat map with a range of

coefficients of −0.2 to 1.0. **(B)** Each pairwise combination of the 13 "potent" gene expression profiles in clusters 8–13 are displayed as a heat map with a restricted range of coefficients from 0.7 to 1.0. **(C)** A "Split Spectrum" model of macrophage activation can be used to emphasize the high degree of correlation between treatments with at least one potent macrophage-activating stimulus.

Macrophage-Activating Conditions can be Categorized as "Mild" or "Potent" Based on the Number of Transcripts regulated in Response to the Stimuli

There was a large range in the number of highly regulated transcripts in each activated macrophage expression profile (**Figure 2B**). Specifically, the 20 activated macrophage types within clusters 1–7 had relatively few regulated transcripts (93 ± 14), whereas the 13 activated macrophage types within clusters 8–13 had large numbers of regulated transcripts (564 ± 31). We propose that macrophage-activating conditions can be categorized as "mild" or "potent" based on the number of transcripts the treatment alters.

When considering the mild and potent clusters of the unsupervised hierarchical clustering, we noted that the gene expression profiles did not segregate along previously described M1–M2 divisions (**Table 1**). Polarized macrophage types, M(IFNγ) and M(LPS), which have each been considered "M1" macrophage types sorted into the mild and potent clusters, respectively. Similarly, macrophage types formerly named "M2," M(IL-4), and M(LPS + IC) were categorized as mild and potent, respectively.

We considered the possibility that the wide discrepancy in the number of transcripts regulated in MDMs responding to mild and potent activating conditions was due to suboptimal concentrations of the "mild" stimulus. To address this, MDMs were treated with each of the 11 single treatment macrophage-activating conditions that were categorized as mild at concentrations ranging from 4-fold higher to 16-fold lower those used in the microarray-based experiments. In general, modest dose responses were observed. In response to the majority of the mild stimuli tested (IFNβ, IFNγ, IL-1β, IL-4, IL-10, and TNFα), the amplitude of change in expression for any given transcript was routinely <4-fold between the lowest and highest concentrations for the activating stimulus tested (Figure S2 in Supplementary Material). This suggests that the window of activity is wide for these stimuli and further suggests that use of higher concentrations of these stimuli would be unlikely to revise their macrophage-activating categorization from "mild" to "potent."

Evaluation of Correlation Coefficients Between Activated MDM Gene Expression Profiles Supports Conclusions Drawn from Hierarchical Clustering Analyses

Correlation coefficients were determined for each pairwise combination of activated MDM gene expression profiles in the set of 1874 regulated transcripts (**Figure 3**). This analysis further substantiated the categorization of gene expression profiles into mild and potent categories as shown by unsupervised hierarchical clustering (**Figure 2**). As an example, there was a consistently higher

gene expression profile correlation when the profiles of potently activated macrophages (clusters 8–13) were paired with profiles from potently activated (clusters 1–7), macrophages (**Figure 3A**). Also, we note that, when using a different color scale (**Figure 3B**), correlations between profiles in clusters 8–13 were noticeable and supported the division of the gene expression profiles of the potently activated macrophage gene expression profiles into minor clusters.

In a recent microarray-based study (33), at least 9 clusters of activated macrophages in a data set derived from human MDMs activated with 28 distinct stimuli. In agreement with the level of clustering as the previous study, we now show using unsupervised hierarchical clustering and correlation coefficient analyses that human MDMs activated with the 33 macrophage activation conditions used in this study form at least 13 clusters. Both studies support a spectrum model of macrophage activation. Because of the strong "mild" and "potent" macrophage-activating condition categories described here, we propose that macrophage activation may best be described using a "split spectrum" model (**Figure 3C**).

Verifying Transcripts that Serve as Markers for the "Potent" Macrophage Activation Conditions

The first principal component (PC1) explains (42.9%) of the variance in the data set of 1874 regulated genes while PC2 and PC3 each contributed to ~10% of the variance and the remaining principal components each accounted for <5% of the variance (**Figure 4A** and data not shown). A scatterplot of regulated gene expression profiles based on the first two principal components segregated profiles in clusters 1–7 from those in clusters 8–13 along the PC1 axis (**Figure 4B**). The expression profiles of the 50 transcripts that contributed the most to PC1 were subjected to unsupervised hierarchical clustering and displayed as a heat map (**Figure 4C**). There was an obvious distinction between gene expression responses between the profiles within the mild and potent major clusters; the transcripts robustly regulated by potent macrophage-activating conditions and relatively unaltered by mild macrophage-activating conditions underlie PC1 and account for the major source of variance for the diverse spectrum of polarized macrophage gene expression profiles in this study.

There were many transcripts in addition to the 50 noted in **Figure 4C** that contributed to PC1. Using samples collected for analysis in **Figure 1**, we monitored the change in expression of four transcripts (CCL5, IRG1, MT1G, and S100A8) that contributed to PC1 over 24 h (**Figure 4D**). CCL5 and IRG1 transcripts showed immediate increased expression levels that were sustained through the 24-h time point. By contrast, delayed increases were seen for the expression levels of MT1G and S100A8 transcripts. These four transcripts, in addition to IL1B, IL6, and IL23A that were previously seen to have sustained high expression levels in

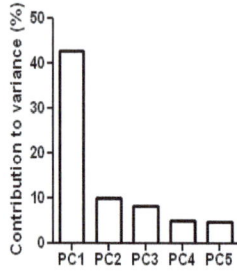

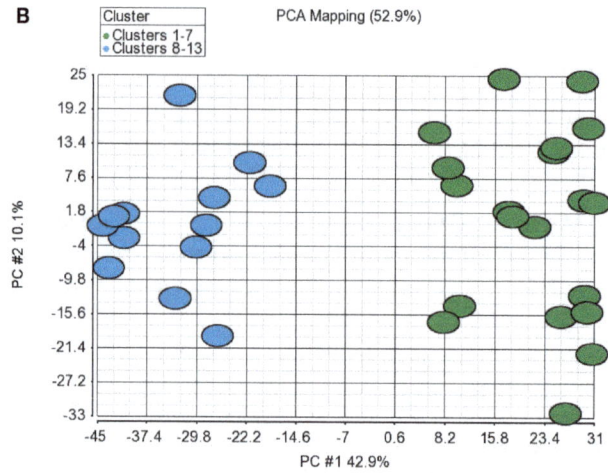

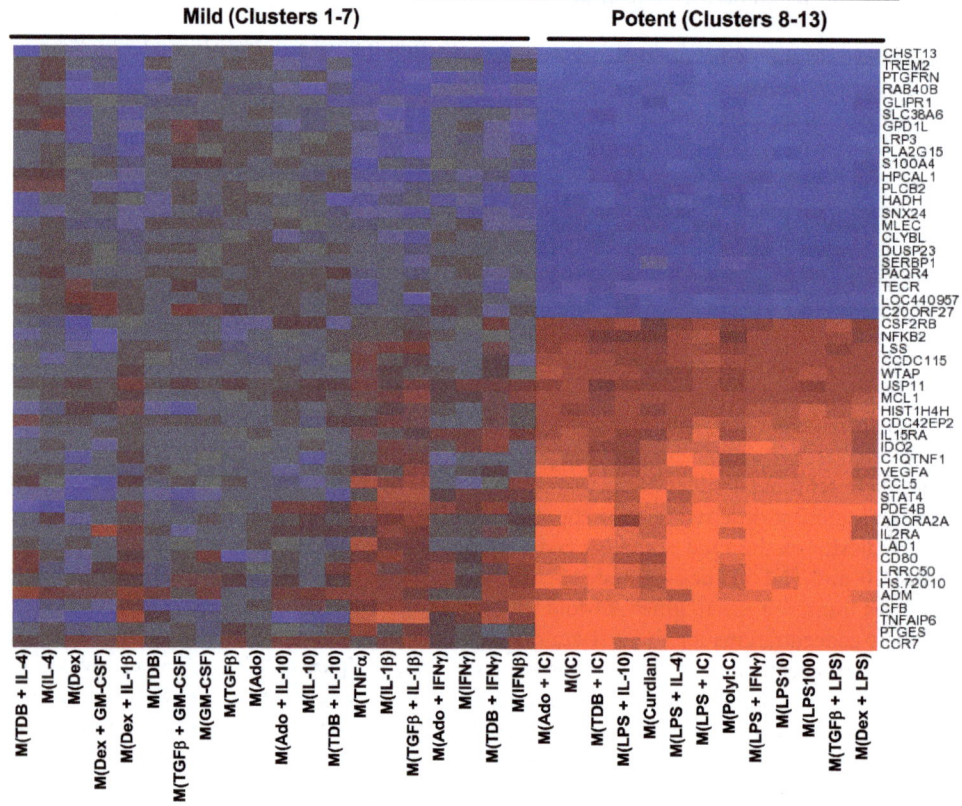

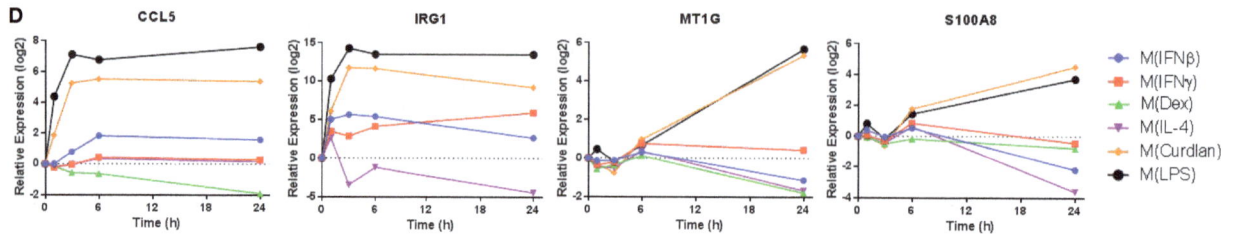

FIGURE 4 | Continued

FIGURE 4 | Continued
Transcripts universally regulated by "potent" macrophage-activating conditions in clusters 8–13 were the largest source of variance in the polarized MDM gene expression profiles. Principal components analysis was performed using the data set of 1874 regulated transcripts from the 33 gene expression profiles. **(A)** The contribution of PC1–PC5 to the variance is shown. **(B)** Scatterplot displays gene expression profiles according to PC1 and PC2 scores with color coding based on "mild" (clusters 1–7) and

"potent" (clusters 8–13) categorization. **(C)** The 50 transcripts that contributed the most to PC1 were sorted according to unsupervised hierarchical clustering results. Changes in transcript expression levels relative to untreated MDMs are depicted as a heat map (log$_2$ scale). **(D)** Using IFC-based RT-PCR and samples from **Figure 1**, the changes in expression for each indicated transcript relative to the untreated MDM controls was determined at the 1, 3, 6, and 24 h time points for the six types of activated MDMs (N = 1).

M(LPS) and M(Curdlan) macrophages, suggesting that numerous transcripts that can be reliably used as markers of "potent" activation conditions.

Evaluating the Use of Chemokine Transcripts as Macrophage Activation Markers

Chemokines not only play an important functional role in macrophage activity but also include some of the earliest proposed markers of macrophage polarization (2, 5). We generated a heat map of transcript expression changes for chemokines from the C–C and C–X–C subfamilies from the 1874 transcripts (**Figure 5A**). Since IL-4 treated macrophages have been well characterized, the chemokines were sorted according to their average expression in the two activated macrophage types that form cluster 1, M(IL-4) and M(TDB + IL-4). Among the remaining 12 clusters, the chemokine expression profiles from macrophages in cluster 3, comprised of M(TDB), M(TGFβ + GM-CSF), and M(GM-CSF) macrophages, appeared to have the most similar trend in chemokine expression. The overall chemokine expression patterns from all other profiles shared little resemblance to those in cluster 1.

Transcripts for two chemokines, CCL13 and CCL22, accumulated in macrophages treated with IL-4 for 24 h (**Figure 5A**). Interestingly, the upregulation of these chemokines in response to IL-4 was delayed relative to other treatments that induced transient upregulation: interferons for CCL13 (**Figure 1**) and PAMPs for CCL22 (**Figure 5B**). These observations suggests that CCL13 and CCL22 can be used as specific markers for M(IL-4) as long as enough time has elapsed since the activation occurred.

We noted that nearly all chemokines had reduced expression in M(Dex) macrophages according to the microarray results (**Figure 5A**). This observation was confirmed when monitoring the kinetics of expression for five chemokine transcripts described above (CCL5, CCL13, CCL18, CXCL9, and CXCL10) (**Figures 1** and **4D**) and in four additional chemokines (CCL2, CCL3, CCL22, and CXCL5) (**Figure 5B**). The general trend of repressing chemokine production in M(Dex) macrophages may hint at a mechanism by which dexamethasone acts as an immunosuppressive molecule.

Donor-to-Donor Variability in Gene Expression Regulation was Minimal in Most Circumstances but was Occasionally seen in some Minor Clusters

A caveat to the results described until this point is that they were based on MDMs derived from two human donors: one donor for monitoring transcript expression kinetics and one donor for transcriptional profiling. Since donor-to-donor variability among human MDM responses was a concern, the expression profiles for many transcripts was determined in samples derived from the microarray experiment and from two additional donors whose MDMs were treated with all 33 macrophage activation conditions (**Figure 6**).

The strong correlation between the microarray results and the IFC PCR results for the first donor was discussed above (Figure S1 in Supplementary Material). Importantly, the last two rows for each transcript, which show the results for the two additional donors, indicated that the MDM responses were, in general, similar to those of the first donor (**Figure 6**). There were a few transcripts (CCL22, CXCL10, IL10, ITGB7, and TGM2) that had strong opposing changes in expression from one donor to the next (**Figure 6**). It is noteworthy that in these instances, the difference in expression was restricted to a limited number of clusters. For example, CCL22 expression regulation tended to be similar in response to all 33 macrophages activation conditions for all 3 donors; the notable exception was seen in the 5 macrophages activation conditions within cluster 12 for the second donor (**Figure 6**). This result is unlikely due to the polyIC and LPS treatments being suboptimal in the experiment involving MDMs from donor 2 since other transcripts, such as CCL5, were regulated similarly in all three donors for the macrophage-activating conditions that make up cluster 12.

A recent mass cytometry-based study produced a high-dimension data set from a panel of 38 antibodies to effectively identify signature expression patterns of myeloid cell populations in mice from a number of tissues (36). Since the dimensionality of data sets produced by mass cytometry and IFC PCR are similar, we tested whether the 13 clusters originally defined by unsupervised hierarchical clustering of the 1874 regulated transcripts (**Figure 2**) could be effectively identified using IFC PCR results (**Figure 7A**). The majority of the 13 clusters remained clusters for each of the three donors (**Figure 7B**). Even the "clusters" composed of a single type of activated macrophage type [i.e., M(IFNβ)] maintained their distinctness relative to the other activated macrophage types. We conclude that gene expression platforms such as IFC PCR monitor a large enough set of macrophage activation marker transcripts to identify an overall macrophage population's type/cluster while still allowing for detection of subtle donor-to-donor differences.

Putative Activation Markers were Identified for Specific Clusters of Polarized Human Macrophages

Macrophage activation markers would ideally have large expression changes in a single cluster or polarized macrophage type. We therefore queried the gene expression profiles in the current

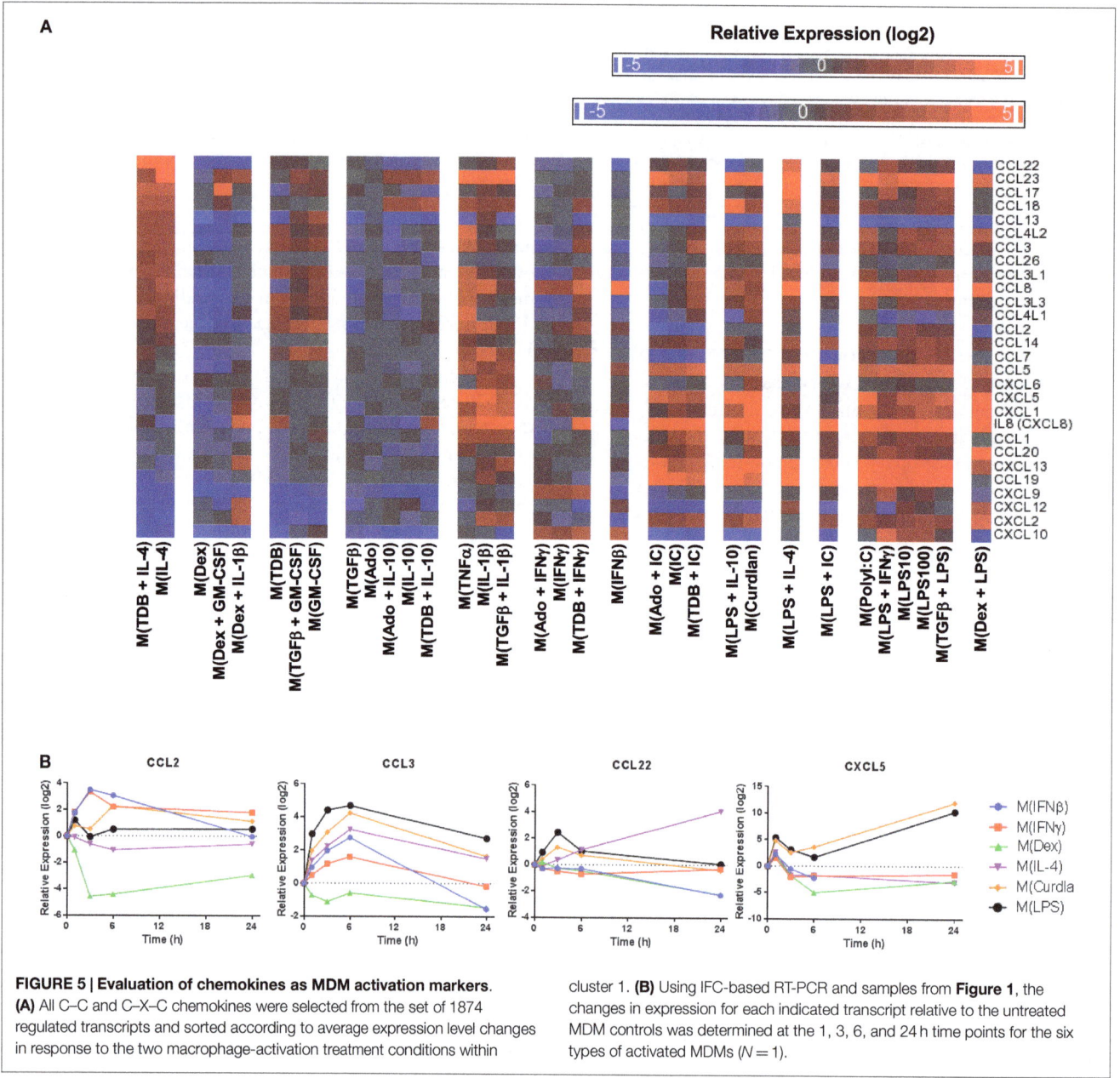

FIGURE 5 | Evaluation of chemokines as MDM activation markers.
(A) All C–C and C–X–C chemokines were selected from the set of 1874 regulated transcripts and sorted according to average expression level changes in response to the two macrophage-activation treatment conditions within cluster 1. **(B)** Using IFC-based RT-PCR and samples from **Figure 1**, the changes in expression for each indicated transcript relative to the untreated MDM controls was determined at the 1, 3, 6, and 24 h time points for the six types of activated MDMs ($N = 1$).

study to identify activation markers specific to each of the 13 clusters formed by the unsupervised hierarchical clustering analysis. Many putative activation markers were identified in macrophages activated with IL-4, dexamethasone, or IFNβ (**Figures 8–10**).

IL-4 was used as an activation condition for gene expression profiles in clusters 1 and 10. We identified transcripts that were strongly upregulated only within cluster 1, within both cluster 1 and cluster 10, or only within cluster 10. Examples of transcripts that fit these gene expression profiles were readily detected within our data set (**Figure 8A**). Analysis of the kinetics of expression for three of the transcripts identified by this screening approach showed that while ALOX15 and CD1B each appear to be good markers for M(IL-4), although the increase in CD1B was delayed until the 24 h time point, FABP4 was not robustly induce in

M(IL-4) but could still be a valuable marker as this transcript was potently down-regulated in response to several macrophage-activating conditions (**Figure 8B**). This latter observation was consistent with the microarray data (**Figure 8A**).

Given the relative ease of finding IL-4-associated activation markers in our data set, we switched our attention to identifying additional activation markers. Dexamethasone-associated activation markers were identified that were specifically upregulated in macrophage-activating conditions from only within cluster 2, within both clusters 2 and 13, and only within cluster 13 (**Figure 9A**). The expression kinetics was determined for three of the transcripts identified by the microarray screen as dexamethasone responsive (**Figure 9B**). Of these, ALOX15B and MFGE8 appear to be a markers for M(Dex) at early and late time points, respectively.

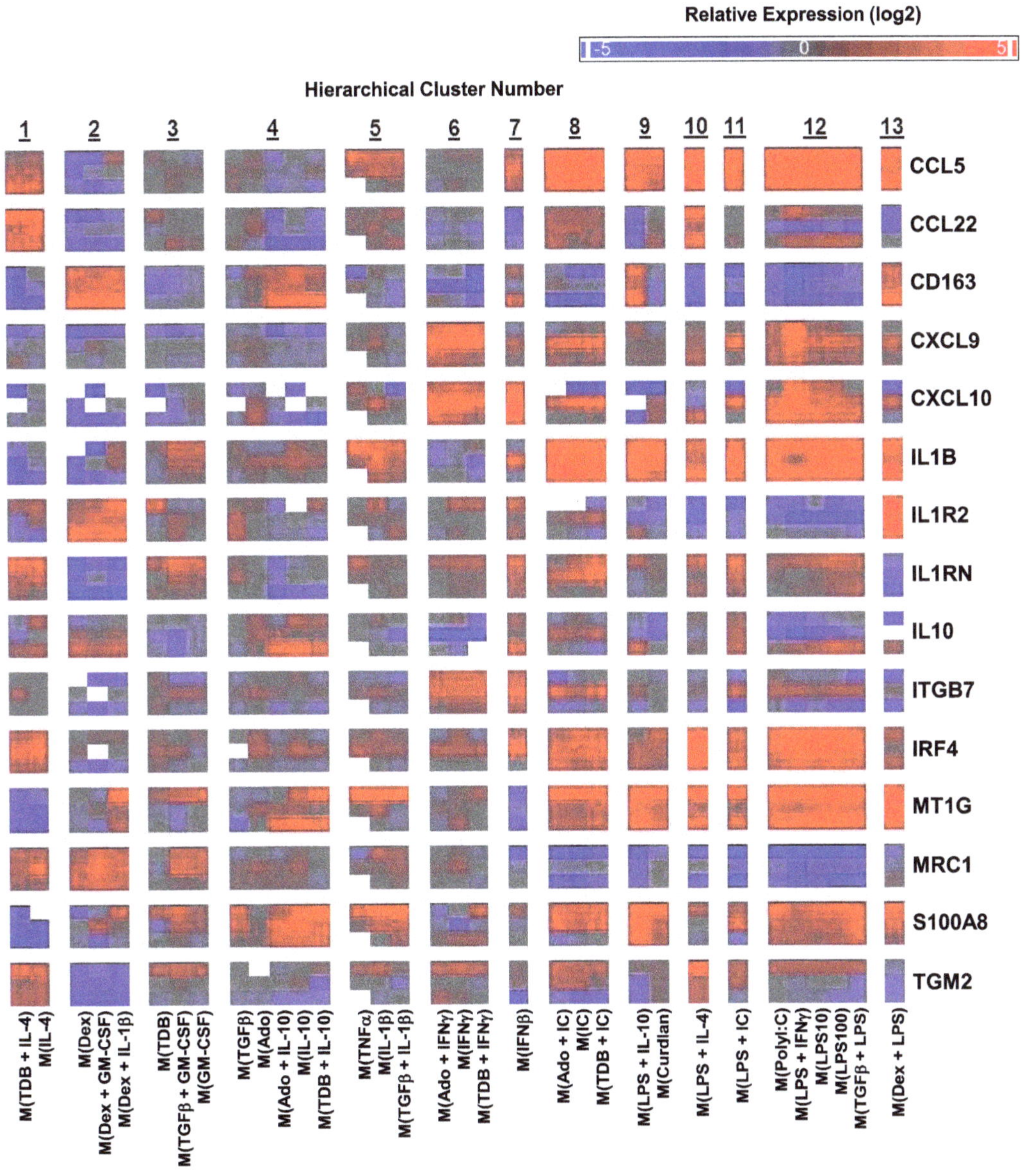

FIGURE 6 | Variability in donor-to-donor MDM gene expression responses was often limited to specific clusters. IFC-based RT-PCR was used to determine the expression of 48 transcripts (45 putative macrophage activation markers and 3 endogenous controls) in MDMs at 24 h post-treatment with 33 distinct activation conditions (columns) (N = 3). Shown here are the results for 15 of the activation marker transcripts. The RNA collected from the first donor (first row for each indicated transcript) had been used in the microarray studies and the RNA from two additional donors (second and third row for each indicated transcript) was collected in independent experiments. Blank areas within clusters represent samples did not meet the Ct cut-off of 25 or, in the case of the third M(TNFα) sample, did not load properly into the IFC device.

Next, potential activation markers or IFNβ-treated macrophages were identified within cluster 7 (**Figure 10A**). Further analysis showed that AXL, IFIT, and ZBP1 were all induced rapidly in M(IFNβ) and with delayed kinetics in M(LPS) (**Figure 10B**). This observation may be explained by indirect induction of these genes by LPS-induced IFNβ production.

Discussion

Characterization of TAMs has shifted from quantifying macrophage density in and around tumors to evaluating markers of activation (15, 37). It is important to note that macrophage activation markers have been used to categorize macrophage activation, typically using the M1–M2 nomenclature, yet the regulation patterns of these markers in macrophages responding to a wide variety of activation conditions are not well understood. Using a combined microarray- and IFC array-based approach in this study, previously proposed markers of macrophage activation were better characterized and novel markers of macrophage activation were identified.

In the earliest report using M1–M2 nomenclature, the authors stated that "M-1 and M-2, while useful for conceptualizing

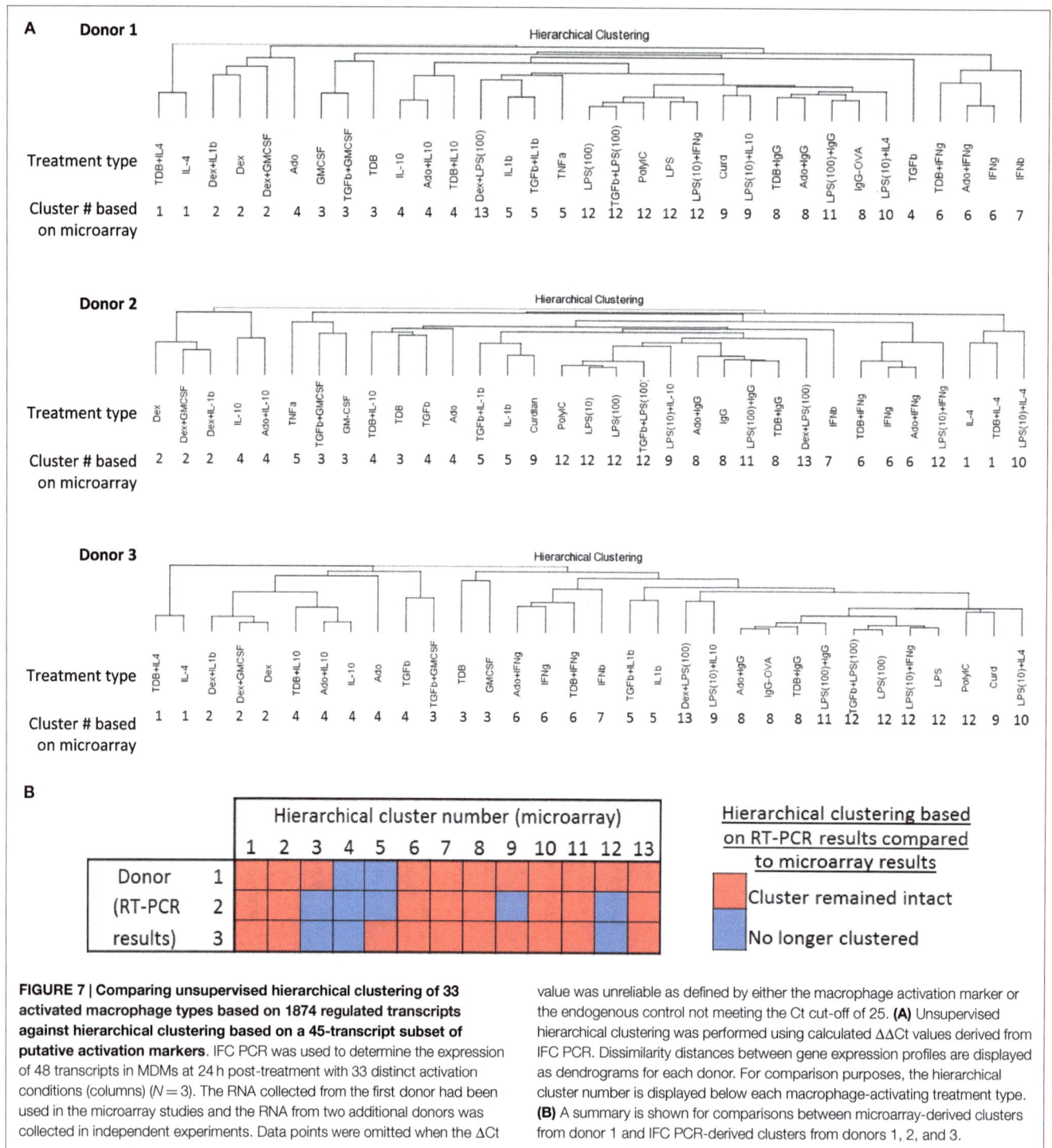

FIGURE 7 | Comparing unsupervised hierarchical clustering of 33 activated macrophage types based on 1874 regulated transcripts against hierarchical clustering based on a 45-transcript subset of putative activation markers. IFC PCR was used to determine the expression of 48 transcripts in MDMs at 24 h post-treatment with 33 distinct activation conditions (columns) (N = 3). The RNA collected from the first donor had been used in the microarray studies and the RNA from two additional donors was collected in independent experiments. Data points were omitted when the ΔCt value was unreliable as defined by either the macrophage activation marker or the endogenous control not meeting the Ct cut-off of 25. **(A)** Unsupervised hierarchical clustering was performed using calculated ΔΔCt values derived from IFC PCR. Dissimilarity distances between gene expression profiles are displayed as dendrograms for each donor. For comparison purposes, the hierarchical cluster number is displayed below each macrophage-activating treatment type. **(B)** A summary is shown for comparisons between microarray-derived clusters from donor 1 and IFC PCR-derived clusters from donors 1, 2, and 3.

immune responses, certainly could be an oversimplification" and that "there may be a continuum of phenotypes between M-1 and M-2 macrophages" (3). A recently proposed framework argued against using the M1–M2 nomenclature yet upheld the linear model concept that suggested M(IFNγ) and M(IL-4) to represent the polar extremes (9). However, both the results of the current study and those reported by Xue et al. (33) support a spectrum model of macrophage activation rather than a linear model.

Unsupervised hierarchical clustering, correlation coefficient analysis, and principal components analysis of the regulated transcripts each support the concept that macrophage polarized states in this study can be sorted into two major clusters. We designated these clusters "mild" and "potent" to convey the number transcripts altered in response to each specific macrophage-activating condition. It is important to note that, although we have evaluated more macrophage activation conditions in a macrophage

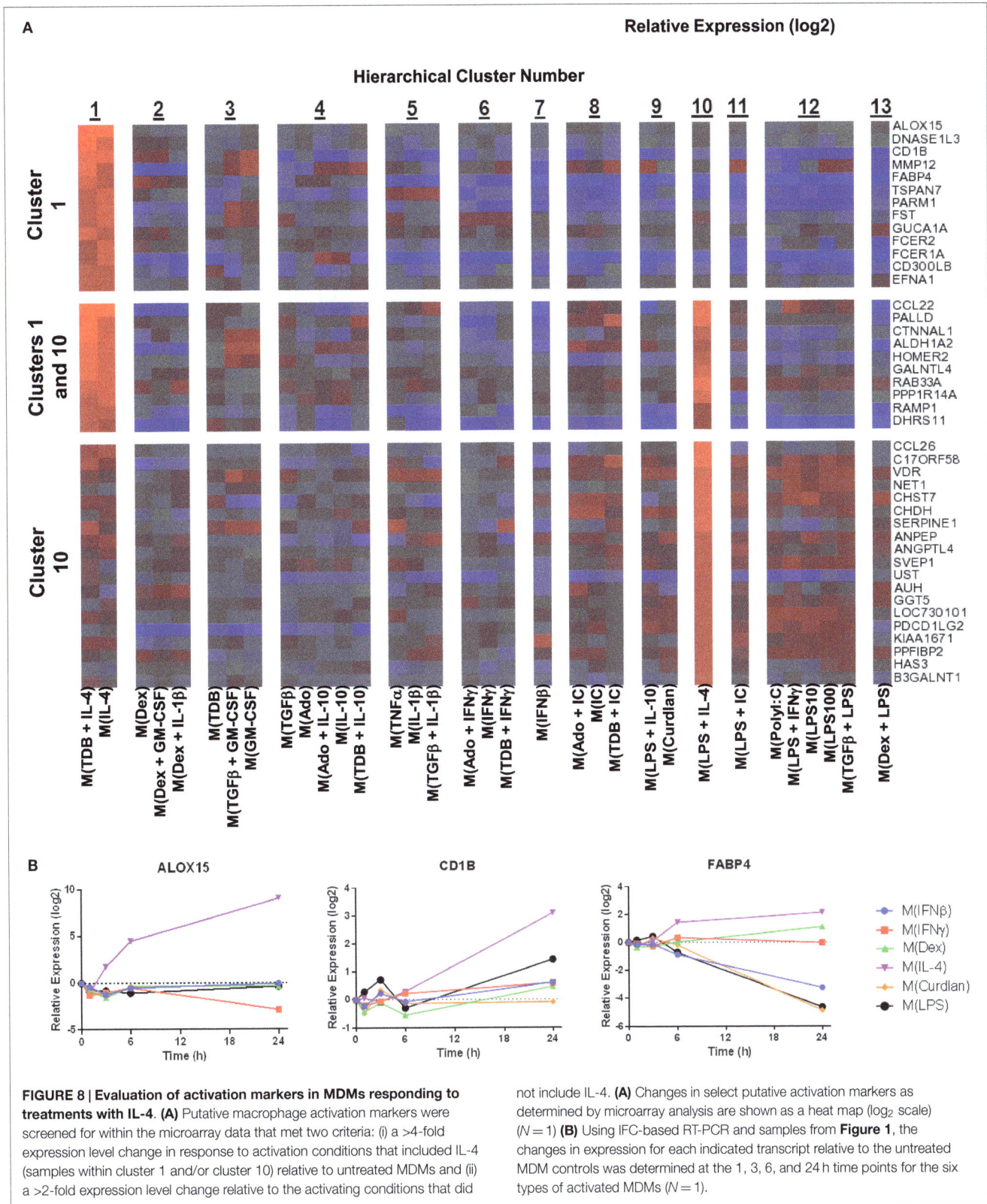

FIGURE 8 | Evaluation of activation markers in MDMs responding to treatments with IL-4. (A) Putative macrophage activation markers were screened for within the microarray data that met two criteria: (i) a >4-fold expression level change in response to activation conditions that included IL-4 (samples within cluster 1 and/or cluster 10) relative to untreated MDMs and (ii) a >2-fold expression level change relative to the activating conditions that did not include IL-4. **(A)** Changes in select putative activation markers as determined by microarray analysis are shown as a heat map (log$_2$ scale) ($N = 1$) **(B)** Using IFC-based RT-PCR and samples from **Figure 1**, the changes in expression for each indicated transcript relative to the untreated MDM controls was determined at the 1, 3, 6, and 24 h time points for the six types of activated MDMs ($N = 1$).

activation study that has previously been published, there could be activation conditions that will have an intermediate number of regulated transcripts making our split spectrum model potentially incorrect. Indeed, Xue et al. (33) studied macrophage responses to 28 activation conditions and we found that the free fatty acid conditions from their study may represent an "intermediate"

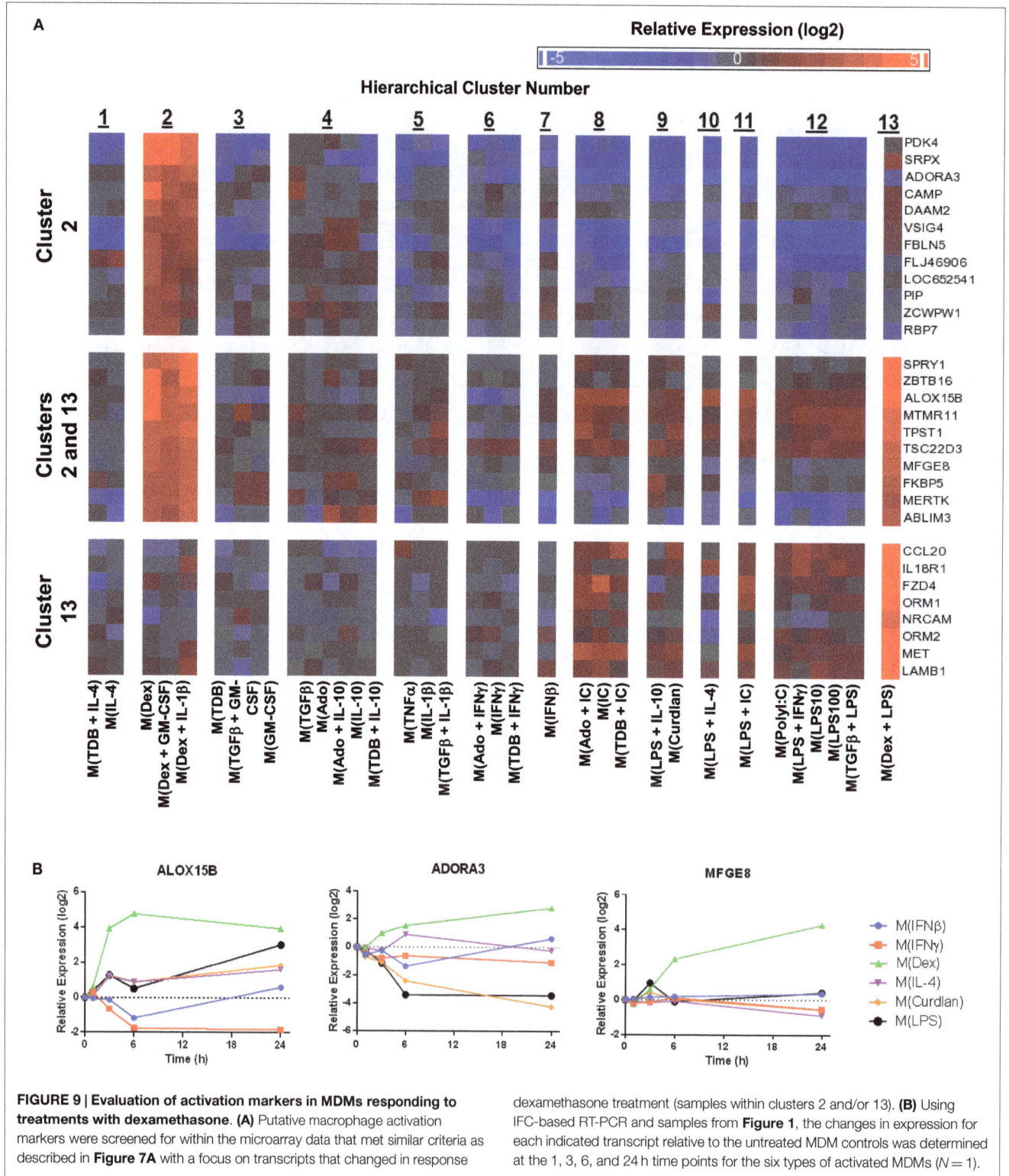

FIGURE 9 | Evaluation of activation markers in MDMs responding to treatments with dexamethasone. (A) Putative macrophage activation markers were screened for within the microarray data that met similar criteria as described in **Figure 7A** with a focus on transcripts that changed in response dexamethasone treatment (samples within clusters 2 and/or 13). **(B)** Using IFC-based RT-PCR and samples from **Figure 1**, the changes in expression for each indicated transcript relative to the untreated MDM controls was determined at the 1, 3, 6, and 24 h time points for the six types of activated MDMs ($N = 1$).

cluster (analysis not shown). While our "split spectrum" model may not represent the entirety of the spectrum, it raises the idea that strength of macrophage activation may be worth considering in future attempts to accurately describe macrophage activation/polarization.

In the analysis of the principal components, special attention was warranted for PC1 because it accounted for four times more of the variance than any other principal component. The single treatment macrophage-activating conditions that contributed to PC1 were immune complexes, Curdlan, polyIC, and LPS.

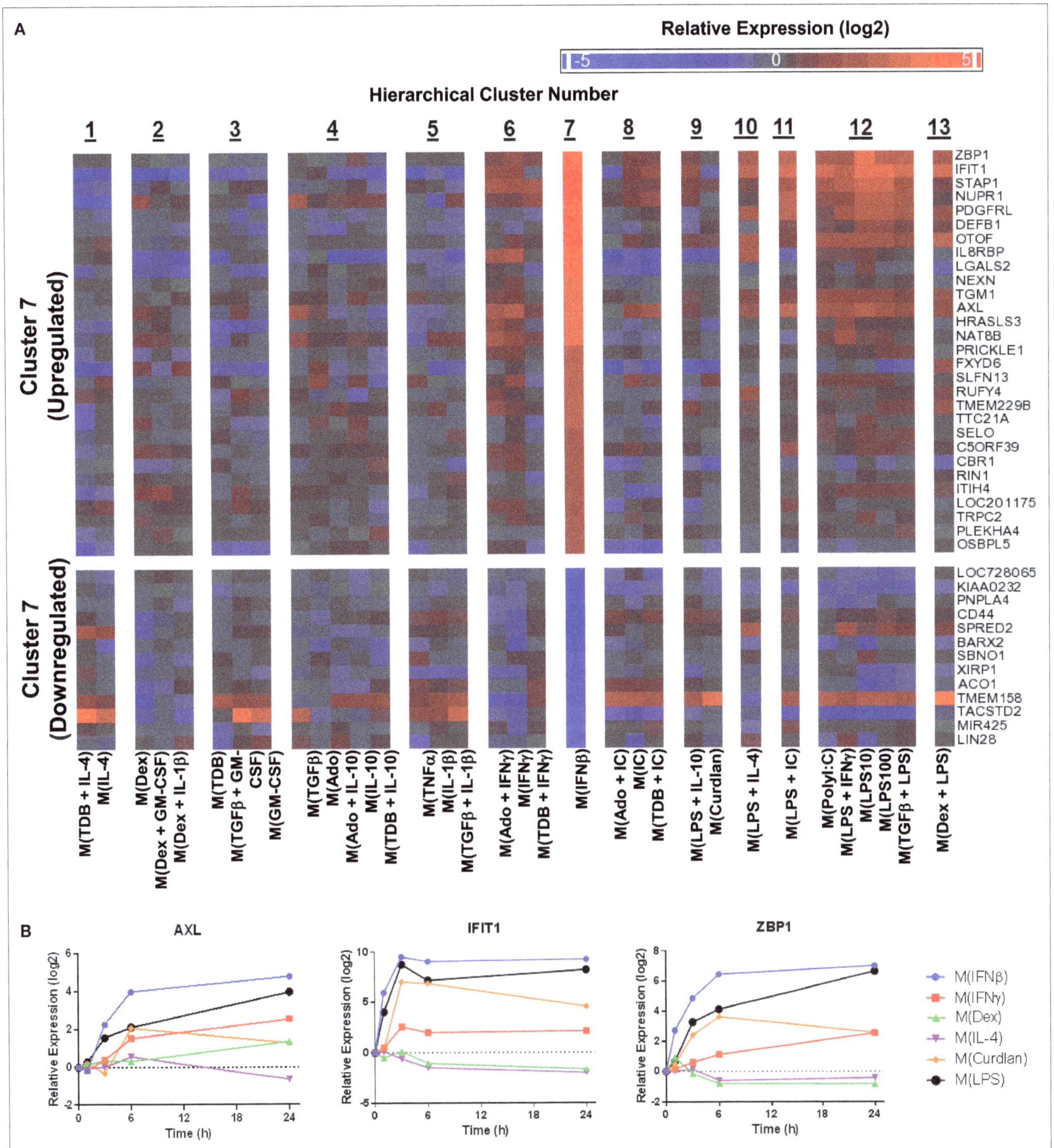

FIGURE 10 | Evaluation of activation markers in MDMs responding to treatment with IFNβ. (A) Putative macrophage activation markers were screened for within the microarray data that met similar criteria as described in **Figure 7A** with a focus on transcripts that changed in response IFNβ treatment (cluster 7). **(B)** Using IFC-based RT-PCR and samples from **Figure 1**, the changes in expression for each indicated transcript relative to the untreated MDM controls was determined at the 1, 3, 6, and 24 h time points for the six types of activated MDMs ($N = 1$).

All combinational treatments that contributed to PC1 contained one or two of these potent stimuli. Treatment of macrophages with immune complexes and Curdlan initially signal through Fcγ receptor/Syk/Card9 pathways while treatment with polyIC and LPS signal through TRIF and/or MyD88 pathways. Despite these initial differences, there is substantial overlap triggered by the potent stimuli further downstream pathway signaling. For example, activation of pathways such as NF-κB and MAPK may directly and indirectly account for the regulated expression of many transcripts that contributed to PC1. Importantly, as noted in **Table 1**, chicken ovalbumin and Curdlan are often contaminated with substantial levels of endotoxin, so the "potent" activation conditions may be mostly or in part a consequence of TLR-initiated signaling (10, 11). Future studies will assess the extent that TLR signaling may have contributed to the alterations in the M(Curdlan) and M(IC) macrophage gene expression profiles.

There was substantial evidence, both gene expression and functional, that the mild and potent polarized macrophage types of our data set should be divided into smaller clusters. To define these clusters, we chose to separate our gene expression profiles based on known differences that occur in response macrophage-activating conditions rather than using a statistically based dissimilarity cut-off in the unsupervised hierarchical clustering. Specifically, we noted that M(LPS + IFNγ) and M(LPS + IC) were situated close to each other according to unsupervised hierarchical clustering analysis (**Figure 2**). Important functional differences in macrophages treated with these two distinct activating conditions such as cytokine production (IL-12 vs. IL-10) and ability to skew CD4$^+$ T cell responses (Th1 vs. Th2) (35, 38–40) supported the segregation of these gene expression profiles into separate clusters. Therefore, the dissimilarity distance between these two gene expression profiles served as our cut-off to rationally sort the 33 gene expression profiles into 13 clusters.

It is notable that if the gene expression profiles had been segregated based on dissimilarity distances into 14 clusters instead of 13, the 5 gene expression profiles currently grouped within "cluster 4" would have been split into 2 clusters. Furthermore, correlation coefficients within cluster 4 were markedly higher when comparing gene expression profiles from MDMs activated with conditions that included IL-10 (**Figure 3**). Finally, hierarchical clustering based on IFC PCR results (**Figure 7**) failed to retain the integrity of cluster 4 in any of the three donors. These observations suggest that subdividing the 33 gene expression profiles into more than 13 clusters may have been warranted starting with subdividing cluster 4. Future functional studies will be useful for supporting or modifying our current classification of 13 clusters for these 33 macrophage-activating conditions.

In order for macrophage activation markers to be useful, it is critical to know whether each marker is regulated by a wide variety or a limited number of stimuli. In our initial time course analysis survey of previously proposed macrophage activation markers, few of the 11 transcripts were found to be highly specific for a specific type of activated macrophage. Therefore, microarrays were performed and then surveyed to identify novel macrophage activation markers. This approach proved to be useful for identifying markers differentially expressed by activated macrophages in all the potent conditions used in this study (**Figure 4**) and in many of the minor clusters (**Figures 7–9**).

Our approach of screening for macrophage activation markers by surveying microarray results of a single donor's macrophage responses to 33 different activation conditions and following up with IFC arrays proved effective. Also, use of unbiased, bottom-up analyses of the microarray results argue against previously proposed top-down linear frameworks describing macrophage activation states, such as the M1–M2 system (3, 9). We note that our results are in line with the spectrum model proposed by Xue et al. from their microarray data set (33). There are likely to be more clusters of activated macrophages than the 13 described here and the 9 described by Xue et al. (33). Taken together, we conclude that measuring the expression changes in a panel of well-characterized markers would provide a useful tool to accurately differentiate various activation states associated with functional activity of TAMs or other macrophage populations.

Acknowledgments

We would like to thank Dr. Fayyaz Sutterwala and Dr. John Janczy for advice on preparing immune complexes and Dr. Robert Philibert for use of the Fluidigm BioMark IFC real-time PCR machine. These studies were supported in part by grants 1IK2BX001627 (JG), 5I01BX001983 (MEW), and 5I01BX000536 (MEW) from the Department of Veterans Affairs, Veterans Health Administration, Office of Research and Development, Biomedical Laboratory Research and Development and grants AI076233 and AI045540 from the National Institutes of Health (MEW). Studies were conducted during a period of support for MAW on NIH training grant AI007511 (MW). The content of this manuscript are solely the responsibility of the authors and do not necessarily represent the official views of the granting agencies.

Supplementary Material

Table S1 | Gene expression profiles in the 33 activated macrophage types for the 1874 regulated transcripts.

Table S2 | GO enrichment for 1615 gene names recognized by STRING v. 9.05.

Figure S1 | IFC PCR-calculated transcript expression changes correlated well with results from the microarrays. Scatterplots of gene expression level changes of the indicated transcripts as determined by microarray and by Fluidigm IFC-based RT-PCR. RNA samples collected at 24 h post-treatment from activated MDMs of a single donor were used as template for both assays.

Figure S2 | Dose-dependent changes in MDM transcript expression levels were minimal across a broad range of concentrations for most of the mild, single stimulus treatments. IFC-based RT-PCR was used to monitor the expression of 48 transcripts in MDMs from a single donor treated with four different concentrations of 11 indicated mild treatment stimuli. The concentrations tested were 4×, 1×, 1/4×, and 1/16× relative to the concentration described for each stimulus in **Table 1**. For each treatment, transcripts that had at least a fourfold change in expression (>2 or <−2 on log$_2$ scale) in any of the four tested concentrations were selected for display.

References

1. Stein M, Keshav S, Harris N, Gordon S. Interleukin 4 potently enhances murine macrophage mannose receptor activity: a marker of alternative immunologic macrophage activation. *J Exp Med* (1992) **176**:287–92. doi:10.1084/jem.176.1.287

2. Mantovani A, Sozzani S, Locati M, Allavena P, Sica A. Macrophage polarization: tumor-associated macrophages as a paradigm for polarized M2 mononuclear

phagocytes. *Trends Immunol* (2002) **23**:549–55. doi:10.1016/S1471-4906(02)02302-5

3. Mills CD, Kincaid K, Alt JM, Heilman MJ, Hill AM. M-1/M-2 macrophages and the Th1/Th2 paradigm. *J Immunol* (2000) **164**:6166–73. doi:10.4049/jimmunol.164.12.6166

4. Mosser DM, Edwards JP. Exploring the full spectrum of macrophage activation. *Nat Rev Immunol* (2008) **8**:958–69. doi:10.1038/nri2448

5. Mantovani A, Sica A, Sozzani S, Allavena P, Vecchi A, Locati M. The chemokine system in diverse forms of macrophage activation and polarization. *Trends Immunol* (2004) **25**:677–86. doi:10.1016/j.it.2004.09.015

6. Martinez FO, Sica A, Mantovani A, Locati M. Macrophage activation and polarization. *Front Biosci* (2008) **13**:453–61. doi:10.2741/2692

7. Gleissner CA, Shaked I, Little KM, Ley K. CXC chemokine ligand 4 induces a unique transcriptome in monocyte-derived macrophages. *J Immunol* (2010) **184**:4810–8. doi:10.4049/jimmunol.0901368

8. Zizzo G, Cohen PL. IL-17 stimulates differentiation of human anti-inflammatory macrophages and phagocytosis of apoptotic neutrophils in response to IL-10 and glucocorticoids. *J Immunol* (2013) **190**:5237–46. doi:10.4049/jimmunol.1203017

9. Murray PJ, Allen JE, Biswas SK, Fisher EA, Gilroy DW, Goerdt S, et al. Macrophage activation and polarization: nomenclature and experimental guidelines. *Immunity* (2014) **41**:14–20. doi:10.1016/j.immuni.2014.06.008

10. Sonck E, Devriendt B, Goddeeris B, Cox E. Varying effects of different beta-glucans on the maturation of porcine monocyte-derived dendritic cells. *Clin Vaccine Immunol* (2011) **18**:1441–6. doi:10.1128/CVI.00080-11

11. Watanabe J, Miyazaki Y, Zimmerman GA, Albertine KH, McIntyre TM. Endotoxin contamination of ovalbumin suppresses murine immunologic responses and development of airway hyper-reactivity. *J Biol Chem* (2003) **278**:42361–8. doi:10.1074/jbc.M307752200

12. Allavena P, Sica A, Solinas G, Porta C, Mantovani A. The inflammatory micro-environment in tumor progression: the role of tumor-associated macrophages. *Crit Rev Oncol Hematol* (2008) **66**:1–9. doi:10.1016/j.critrevonc.2007.07.004

13. Coussens LM, Werb Z. Inflammation and cancer. *Nature* (2002) **420**:860–7. doi:10.1038/nature01322

14. Pollard JW. Tumour-educated macrophages promote tumour progression and metastasis. *Nat Rev Cancer* (2004) **4**:71–8. doi:10.1038/nrc1256

15. Van Overmeire E, Laoui D, Keirsse J, Van Ginderachter JA, Sarukhan A. Mechanisms driving macrophage diversity and specialization in distinct tumor microenvironments and parallelisms with other tissues. *Front Immunol* (2014) **5**:127. doi:10.3389/fimmu.2014.00127

16. Stout RD, Jiang C, Matta B, Tietzel I, Watkins SK, Suttles J. Macrophages sequentially change their functional phenotype in response to changes in microenvironmental influences. *J Immunol* (2005) **175**:342–9. doi:10.4049/jimmunol.175.1.342

17. Stout RD, Suttles J. Functional plasticity of macrophages: reversible adaptation to changing microenvironments. *J Leukoc Biol* (2004) **76**:509–13. doi:10.1189/jlb.0504272

18. Klug F, Prakash H, Huber PE, Seibel T, Bender N, Halama N, et al. Low-dose irradiation programs macrophage differentiation to an iNOS(+)/M1 phenotype that orchestrates effective T cell immunotherapy. *Cancer Cell* (2013) **24**:589–602. doi:10.1016/j.ccr.2013.09.014

19. Pallasch CP, Leskov I, Braun CJ, Vorholt D, Drake A, Soto-Feliciano YM, et al. Sensitizing protective tumor microenvironments to antibody-mediated therapy. *Cell* (2014) **156**:590–602. doi:10.1016/j.cell.2013.12.041

20. Pyonteck SM, Akkari L, Schuhmacher AJ, Bowman RL, Sevenich L, Quail DF, et al. CSF-1R inhibition alters macrophage polarization and blocks glioma progression. *Nat Med* (2013) **19**:1264–72. doi:10.1038/nm.3337

21. Ries CH, Cannarile MA, Hoves S, Benz J, Wartha K, Runza V, et al. Targeting tumor-associated macrophages with anti-CSF-1R antibody reveals a strategy for cancer therapy. *Cancer Cell* (2014) **25**:846–59. doi:10.1016/j.ccr.2014.05.016

22. Bosco MC, Puppo M, Santangelo C, Anfosso L, Pfeffer U, Fardin P, et al. Hypoxia modifies the transcriptome of primary human monocytes: modulation of novel immune-related genes and identification of CC-chemokine ligand 20 as a new hypoxia-inducible gene. *J Immunol* (2006) **177**:1941–55. doi:10.4049/jimmunol.177.3.1941

23. Bostrom P, Magnusson B, Svensson PA, Wiklund O, Boren J, Carlsson LM, et al. Hypoxia converts human macrophages into triglyceride-loaded foam cells. *Arterioscler Thromb Vasc Biol* (2006) **26**:1871–6. doi:10.1161/01.ATV.0000229665.78997.0b

24. Chang YC, Chen TC, Lee CT, Yang CY, Wang HW, Wang CC, et al. Epigenetic control of MHC class II expression in tumor-associated macrophages by decoy receptor 3. *Blood* (2008) **111**:5054–63. doi:10.1182/blood-2007-12-130609

25. Hu X, Park-Min KH, Ho HH, Ivashkiv LB. IFN-gamma-primed macrophages exhibit increased CCR2-dependent migration and altered IFN-gamma responses mediated by Stat1. *J Immunol* (2005) **175**:3637–47. doi:10.4049/jimmunol.175.6.3637

26. Irvine KM, Andrews MR, Fernandez-Rojo MA, Schroder K, Burns CJ, Su S, et al. Colony-stimulating factor-1 (CSF-1) delivers a proatherogenic signal to human macrophages. *J Leukoc Biol* (2009) **85**:278–88. doi:10.1189/jlb.0808497

27. Jura J, Wegrzyn P, Korostynski M, Guzik K, Oczko-Wojciechowska M, Jarzab M, et al. Identification of interleukin-1 and interleukin-6-responsive genes in human monocyte-derived macrophages using microarrays. *Biochim Biophys Acta* (2008) **1779**:383–9. doi:10.1016/j.bbagrm.2008.04.006

28. Locati M, Deuschle U, Massardi ML, Martinez FO, Sironi M, Sozzani S, et al. Analysis of the gene expression profile activated by the CC chemokine ligand 5/RANTES and by lipopolysaccharide in human monocytes. *J Immunol* (2002) **168**:3557–62. doi:10.4049/jimmunol.168.7.3557

29. Martinez FO, Gordon S, Locati M, Mantovani A. Transcriptional profiling of the human monocyte-to-macrophage differentiation and polarization: new molecules and patterns of gene expression. *J Immunol* (2006) **177**:7303–11. doi:10.4049/jimmunol.177.10.7303

30. Scotton CJ, Martinez FO, Smelt MJ, Sironi M, Locati M, Mantovani A, et al. Transcriptional profiling reveals complex regulation of the monocyte IL-1 beta system by IL-13. *J Immunol* (2005) **174**:834–45. doi:10.4049/jimmunol.174.2.834

31. Sironi M, Martinez FO, D'Ambrosio D, Gattorno M, Polentarutti N, Locati M, et al. Differential regulation of chemokine production by Fcgamma receptor engagement in human monocytes: association of CCL1 with a distinct form of M2 monocyte activation (M2b, Type 2). *J Leukoc Biol* (2006) **80**:342–9. doi:10.1189/jlb.1005586

32. Tassiulas I, Hu X, Ho H, Kashyap Y, Paik P, Hu Y, et al. Amplification of IFN-alpha-induced STAT1 activation and inflammatory function by Syk and ITAM-containing adaptors. *Nat Immunol* (2004) **5**:1181–9. doi:10.1038/ni1126

33. Xue J, Schmidt SV, Sander J, Draffehn A, Krebs W, Quester I, et al. Transcriptome-based network analysis reveals a spectrum model of human macrophage activation. *Immunity* (2014) **40**:274–88. doi:10.1016/j.immuni.2014.01.006

34. Cohen HB, Briggs KT, Marino JP, Ravid K, Robson SC, Mosser DM. TLR stimulation initiates a CD39-based autoregulatory mechanism that limits macrophage inflammatory responses. *Blood* (2013) **122**:1935–45. doi:10.1182/blood-2013-04-496216

35. Sutterwala FS, Noel GJ, Salgame P, Mosser DM. Reversal of proinflammatory responses by ligating the macrophage Fcgamma receptor type I. *J Exp Med* (1998) **188**:217–22. doi:10.1084/jem.188.1.217

36. Becher B, Schlitzer A, Chen J, Mair F, Sumatoh HR, Teng KW, et al. High-dimensional analysis of the murine myeloid cell system. *Nat Immunol* (2014) **15**:1181–9. doi:10.1038/ni.3006

37. Quatromoni JG, Eruslanov E. Tumor-associated macrophages: function, phenotype, and link to prognosis in human lung cancer. *Am J Transl Res* (2012) **4**:376–89.

38. Anderson CF, Mosser DM. A novel phenotype for an activated macrophage: the type 2 activated macrophage. *J Leukoc Biol* (2002) **72**:101–6.

39. Edwards JP, Zhang X, Frauwirth KA, Mosser DM. Biochemical and functional characterization of three activated macrophage populations. *J Leukoc Biol* (2006) **80**:1298–307. doi:10.1189/jlb.0406249

40. Gerber JS, Mosser DM. Reversing lipopolysaccharide toxicity by ligating the macrophage Fc gamma receptors. *J Immunol* (2001) **166**:6861–8. doi:10.4049/jimmunol.166.11.6861

6

Ontogeny of tissue-resident macrophages

Guillaume Hoeffel* and Florent Ginhoux*

Singapore Immunology Network (SIgN), Agency for Science, Technology and Research (A*STAR), Singapore, Singapore

Edited by:
Peter M. Van Endert,
Université Paris Descartes, France

Reviewed by:
Meredith O'Keeffe,
Burnet Institute for Medical Research,
Australia
Jean M. Davoust,
Institut National de la Santé et la
Recherche Médicale, France

*Correspondence:
Guillaume Hoeffel and
Florent Ginhoux,
Singapore Immunology Network
(SIgN), Agency for Science,
Technology and Research (A*STAR),
8A Biomedical Grove, IMMUNOS
Building #3-4, Biopolis,
138648 Singapore
guillaumehoeffel1@gmail.com;
florent_ginhoux@immunol.
a-star.edu.sg

The origin of tissue-resident macrophages, crucial for homeostasis and immunity, has remained controversial until recently. Originally described as part of the mononuclear phagocyte system, macrophages were long thought to derive solely from adult blood circulating monocytes. However, accumulating evidence now shows that certain macrophage populations are in fact independent from monocyte and even from adult bone marrow hematopoiesis. These tissue-resident macrophages derive from sequential seeding of tissues by two precursors during embryonic development. Primitive macrophages generated in the yolk sac (YS) from early erythro-myeloid progenitors (EMPs), independently of the transcription factor c-Myb and bypassing monocytic intermediates, first give rise to microglia. Later, fetal monocytes, generated from c-Myb$^+$ EMPs that initially seed the fetal liver (FL), then give rise to the majority of other adult macrophages. Thus, hematopoietic stem cell-independent embryonic precursors transiently present in the YS and the FL give rise to long-lasting self-renewing macrophage populations.

Keywords: macrophages, monocytes, fetal liver, yolk sac, C-Myb, erythro-myeloid progenitors, hematopoiesis, hematopoietic stem cells

Introduction

Ilya (Elie) Metchnikoff first described the mechanism of phagocytosis and the cells responsible for this process over a century ago. These professional phagocytic cells were named "macrophages" (from the Greek derivation macro = large and phage = devouring, "large devouring cells"). These were separate from "microphages," which included polymorphonuclear phagocytes (1). Determining the role of macrophages in pathogenic infections was one of the fundamental observations leading to the concept of cellular immunity (2). Through this seminal work, Metchnikoff anticipated the central role of macrophages in tissue inflammation and homeostasis. We recommend an elegant historical review for more details about Metchnikoff's work by Yona and Gordon in this issue (3).

Since then, the definition of the phagocyte system has been continuously refined, and our understanding of the wide-ranging functions of macrophages has been substantially expanded. It is now clear that, in addition to their classical function in the activation and resolution of tissue inflammation, macrophages also play roles in tissue-specific functions, tissue remodeling during angiogenesis and organogenesis, and wound healing, to name a few (4). Macrophages are exquisitely adapted to their local environment, acquiring organ-specific functionalities during developmental stages and the steady state (4). Macrophages are able to support multiple tissue functions, integrating cues from both the outside environment and their microenvironment to act as rheostatic cells of tissue function. Thus, tissue-resident macrophages represent an attractive target for modern medicine to treat a wide spectrum of diseases in which they have been implicated, including atherosclerosis, autoimmune diseases, neurodegenerative and metabolic disorders, and tumor growth (5–8). Understanding the origin and developmental pathways of macrophages will help to design novel intervention strategies targeting these cells in tissue-specific sites.

A number of observations now indicate that certain macrophage populations derive from embryonic precursors sequentially seeding tissues during development (9–13). Two macrophage progenitors, yolk sac (YS) macrophages and fetal monocytes, have been described in the embryo, but their exact nature and origin were not fully understood until recently (14, 15). Here, we discuss recent developments in our understanding of the origin of adult tissue-resident macrophages, exploring the sequence of progenitors generated during embryonic and adult hematopoiesis. We focus on the relative contributions of YS macrophages and fetal or adult monocytes, including a discussion of our own recent data exploring the heterogeneity of fetal monocyte developmental pathways.

Early Concepts

Macrophages form part of the mononuclear phagocyte system (MPS), which also includes circulating monocytes and dendritic cells (16). Until recently, our vision of macrophage origin and homeostasis was largely based on seminal studies that used *in vivo* radioisotope labeling and radiation chimera experiments. These studies led to the early dogma that resident macrophages were constantly replenished from circulating bone marrow (BM)-derived monocytes as a continuum of differentiation (17–19).

In agreement with that concept, studying the ontogeny of the MPS revealed that monocytes and macrophages derived from macrophage and dendritic cell progenitors (MDPs) present in the BM, which are phenotypically defined as lineage-c-kit$^+$CX3CR1$^+$Flt3$^+$CD115$^+$ (20). MDPs further differentiate through a newly described common monocyte precursor (cMoP), phenotypically defined as lineage$^-$c-kit$^+$CX3CR1$^+$Flt3$^-$CD115$^+$ (21), that gives rise to the two main subsets of circulating monocytes distinguished by the expression of Ly6C (22).

Specific tissue macrophages, such as dermal, gut, and heart macrophages, seemed to follow the model of Van Furth, that macrophages are derived from monocytes (23–25). However, this model did not fit in all cases and evidence also emerged indicating that macrophages were long-lived cells, able to self-renew locally. Hashimoto was the first to speculate that Langerhans cells (LCs) represented a self-perpetuating "intraepithelial phagocytic system" (26). Performing a human skin transplantation assay onto nude mice, Krueger et al. described the remarkable longevity of LCs, which were able to persist in the grafts for more than 2 months (27). Their ability to self-renew through proliferation was later described using DNA densitometry (28). Similar conclusions were drawn soon after regarding alveolar macrophages (29). The dominant concept of "the monocytic origin" of tissue macrophages was also challenged through experiments in animals with prolonged monocytopenia following strontium-89 monocyte depletion, in which liver Kupffer cells were shown to maintain cell numbers by increasing local proliferation (30, 31).

More recently, the use of long-term parabiotic mice and subsequent fate-mapping models have challenged the MPS paradigm and revealed that, unlike all other hematopoietic cells, which rely on hematopoietic stem cell (HSC)-derived BM progenitors, certain macrophage populations possess the unique ability to self-renew locally independently of circulating precursors (32–36). Initial studies describing the presence of macrophages in embryonic tissues suggested that tissue macrophages derived from embryonic progenitors. In rodents, macrophage-like cells first described in the brain rudiment and in the developing skin (37, 38) were named "fetal macrophages" and found to exhibit a high capacity for proliferation (39). These observations suggested that adult macrophages derive from fetal macrophages established during early development. However, whether these fetal macrophages were maintained until adulthood or were replaced postnatally was not addressed until recently. In addition, the exact nature and the origin of fetal macrophage progenitors remained unclear.

Embryonic Hematopoiesis

Mammalian embryos produce several transient waves of hematopoietic cells before the establishment of HSCs in the BM during late gestation (40, 41). The multiple embryonic waves are differentially regulated in time and space and exhibit distinct lineage potentials. Importantly, they contribute to hematopoietic populations that persist until adulthood. These waves include primitive hematopoiesis in the YS, and definitive hematopoiesis, which comprises a transient definitive stage, generating multi-lineage erythro-myeloid progenitors (EMPs) and lympho-myeloid progenitors (LMPs), and a definitive stage characterized by the production of HSCs in the aorta-gonad-mesonephros (AGM). These transient progenitors establish themselves transiently in the fetal liver (FL) during the mid to late stages of hematopoiesis. The sequential waves of hematopoiesis can overlap in time and space (**Figure 1**) and remain difficult to separate clearly, even with the most recent fate-mapping tools available.

Primitive Hematopoiesis

In mice, the first hematopoietic progenitors appear in the extra-embryonic YS blood islands at around embryonic age 7.25 (E7.25), where primitive hematopoiesis is initiated, producing mainly nucleated erythrocytes. This observation linked the myeloid progenitors observed in the YS at E7 with the emergence of YS macrophages after E9.0 [**Figure 2**; Ref. (42–45)]. Primitive hematopoiesis was also shown to produce megakaryocyte progenitors (46). The denomination "primitive" was given to reflect the production of embryonic erythroblasts, like those observed in lower species such as fish, amphibians, and birds, and remaining nucleated throughout their life span (47–49). This denomination was extended to macrophages in the YS due to their concomitant development prior to FL hematopoiesis. Interestingly, no clear evidence of monocytic intermediates was reported at this stage, although the seminal study of Cline and Moore did mention the existence of local intermediate cells between progenitors and functional macrophages (43). Studies by Naito and Takahashi et al. clarified the emergence of primitive macrophages in the YS blood islands in the mouse and rat, observing an absence of endogenous peroxidase activity as a surrogate marker for an absence of monocytic intermediates, such as those found in the

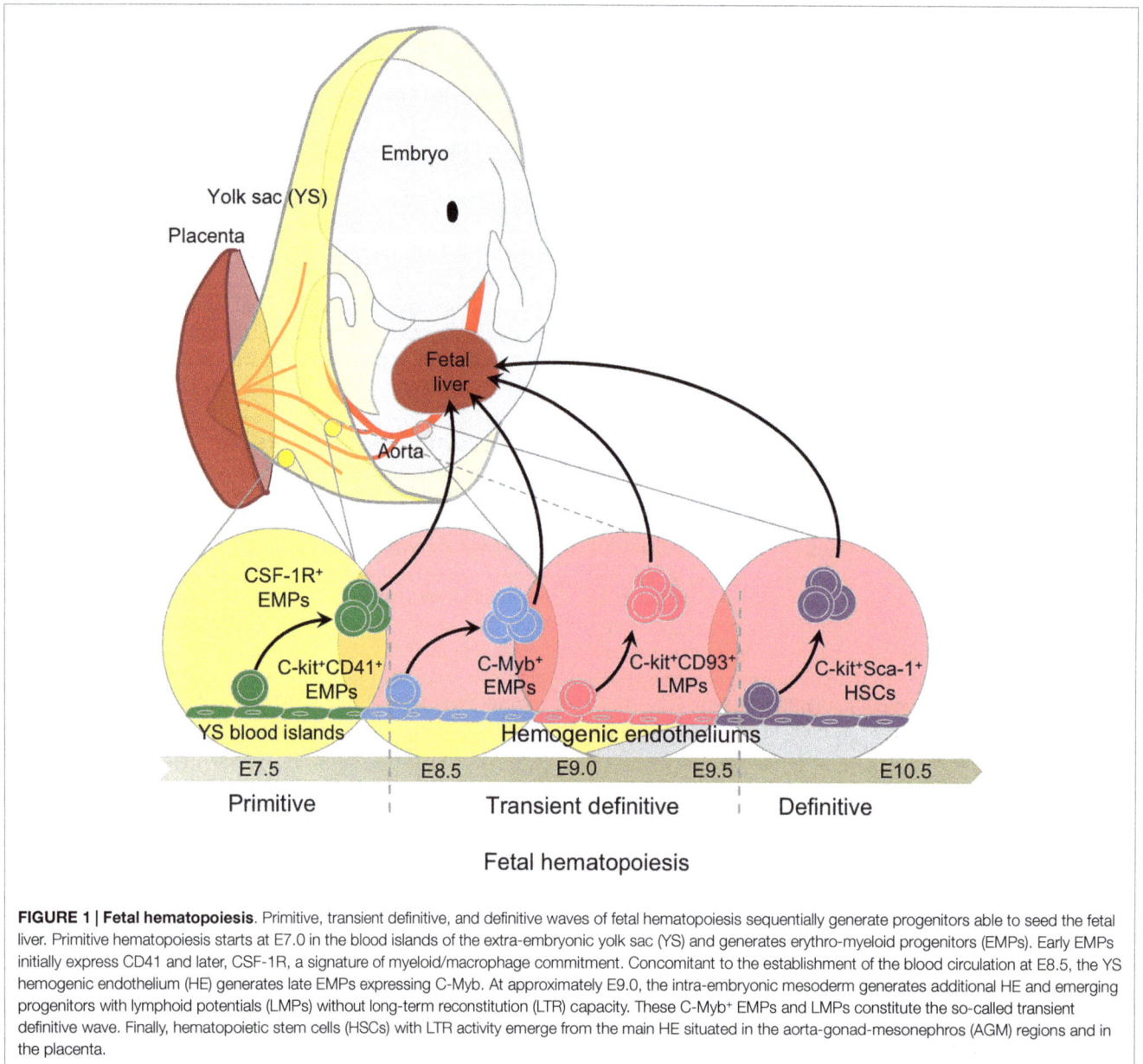

FIGURE 1 | Fetal hematopoiesis. Primitive, transient definitive, and definitive waves of fetal hematopoiesis sequentially generate progenitors able to seed the fetal liver. Primitive hematopoiesis starts at E7.0 in the blood islands of the extra-embryonic yolk sac (YS) and generates erythro-myeloid progenitors (EMPs). Early EMPs initially express CD41 and later, CSF-1R, a signature of myeloid/macrophage commitment. Concomitant to the establishment of the blood circulation at E8.5, the YS hemogenic endothelium (HE) generates late EMPs expressing C-Myb. At approximately E9.0, the intra-embryonic mesoderm generates additional HE and emerging progenitors with lymphoid potentials (LMPs) without long-term reconstitution (LTR) capacity. These C-Myb+ EMPs and LMPs constitute the so-called transient definitive wave. Finally, hematopoietic stem cells (HSCs) with LTR activity emerge from the main HE situated in the aorta-gonad-mesonephros (AGM) regions and in the placenta.

BM (50–52), suggesting a unique developmental pathway for YS macrophages (53, 54).

Transient Definitive Hematopoiesis

The quest to elucidate the origins of embryonic HSCs led to the discovery of earlier lineage-restricted HSC-independent progenitors seeding the FL at E10.5. These progenitors arise concurrently with the transition of primitive to definitive erythropoiesis and were thus considered to form a transient stage of definitive hematopoiesis (45, 47, 55). Transient definitive hematopoiesis consists of progenitors sequentially acquiring myeloid, then lymphoid potential, without exhibiting the long-term reconstitution potential of HSCs. Seminal work from Palis and colleagues on embryonic erythropoiesis in the YS described the parallel emergence of multiple myeloid lineage potential progenitors from E8.25 in the YS (45). Palis et al. first observed the emergence of definitive progenitors for mast cells and a bipotential granulocyte/macrophage progenitor. These progenitors then migrated to the FL through the bloodstream after E8.5, once circulation was established (44, 56). From this pattern of development, the authors concluded that definitive hematopoietic progenitors arise in the YS, migrate through the bloodstream, and seed the FL to rapidly initiate the first phase of intra-embryonic hematopoiesis. Similarly, primitive and definitive erythropoiesis, associated with myelopoiesis, was also shown to emerge prior to HSC in the zebrafish embryo (57). Bertrand et al. showed that definitive hematopoiesis initiates in the posterior blood island with only transient proliferative potential. Because these HSC-independent definitive progenitors were observed to produce definitive erythroid and myeloid

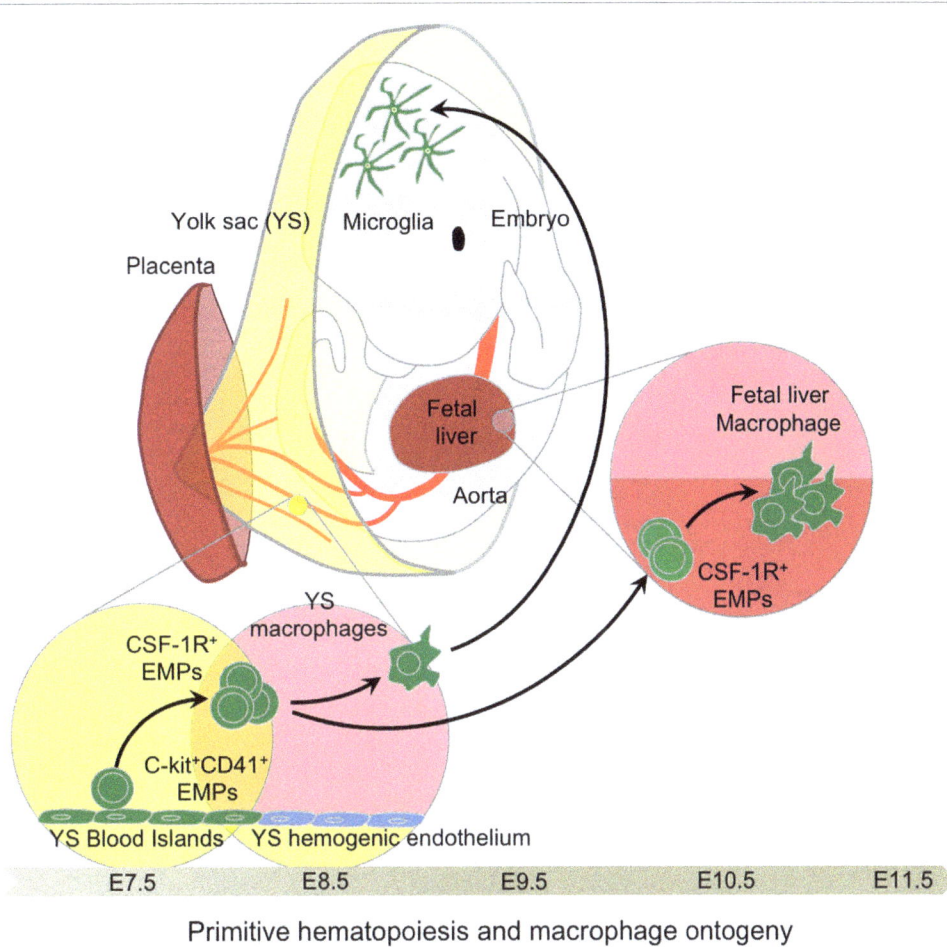

FIGURE 2 | Primitive hematopoiesis and yolk sac macrophage ontogeny. Early EMPs emerge in the YS around E7.5 before establishment of the blood circulation. They express CD41 and CSF-1R and are independent of the transcription factor C-Myb. Upon establishment of the blood circulation around E8.5, EMPs differentiate into primitive macrophages as well as primitive erythrocytes and granulocytes. Primitive macrophages seed all fetal tissues, in particular the head where they will give rise to future brain microglia that are able to continuously self-renew throughout adulthood. EMPs seeding the fetal liver briefly expand to generate a local macrophage population, likely important for sustaining enucleation of primitive erythrocytes passing through the sinusoid prior to the establishment of definitive hematopoiesis and the generation of fetal monocyte-derived macrophages in the fetal liver.

cell types, but not to colonize the zebrafish thymus (implying that they are devoid of lymphoid potential), this population was termed EMPs (57). Interestingly, EMPs can also emerge from the hemogenic endothelium (HE) located in the placenta and umbilical cord (58) and colonize the FL from E9.5 (55) to participate in definitive hematopoiesis. Further studies advanced the field significantly by identifying CD41 as an early marker (pre-CD45) for defining hematopoietic progenitors, including EMPs, emerging from the YS (59, 60). Altogether, these important studies provided phenotypic and functional analyses of the first hematopoietic progenitors and demonstrated that definitive hematopoiesis proceeds through two distinct waves during embryonic development (**Figure 3**).

In parallel, several groups have also identified other multipotential progenitors with lymphoid- or myelo-lymphoid-restricted potential in the YS and the developing para-aortic splanchnopleura (P-Sp) prior to HSCs (61, 62). Lacaud et al. also described AA4.1 (CD93)$^+$ multipotential progenitors present in the E14.5

FL with T cell, B cell, and macrophage potential (63), although the precise origin of these progenitors was not addressed. At the same time, the team of Jacobssen identified Flt3$^+$ lympho-myeloid progenitors (LMPs) devoid of erythrocyte and megakaryocyte capacity (64). Later, cells with myelo-erythroid and lymphoid lineage potential, such as B-1 cells present in the adult spleen, were associated with E9.5 YS progenitors expressing AA4.1 and CD19 (65). A year later, the same team also identified T cell potential within the E9.5 YS progenitors (66). Using a Rag-1-Cre fate-mapping model, Boiers et al. confirmed that these LMPs emerged at approximately E9.5 in the YS, seeding the FL by E11.5 to give rise to T and B cells, as well as granulocytes and monocytes in the E14.5 FL, prior to HSCs (67). Finally, lympho-myeloid progenitors isolated from the dorsal aorta at E9.0 were shown to acquire long-term reconstitution capacity after a few days of *in vitro* culture with stromal cells, and were called immature HSCs (68). Without preculture, these multipotential progenitors can only engraft natural killer (NK)-deficient Rag2γc$^{-/-}$ mice.

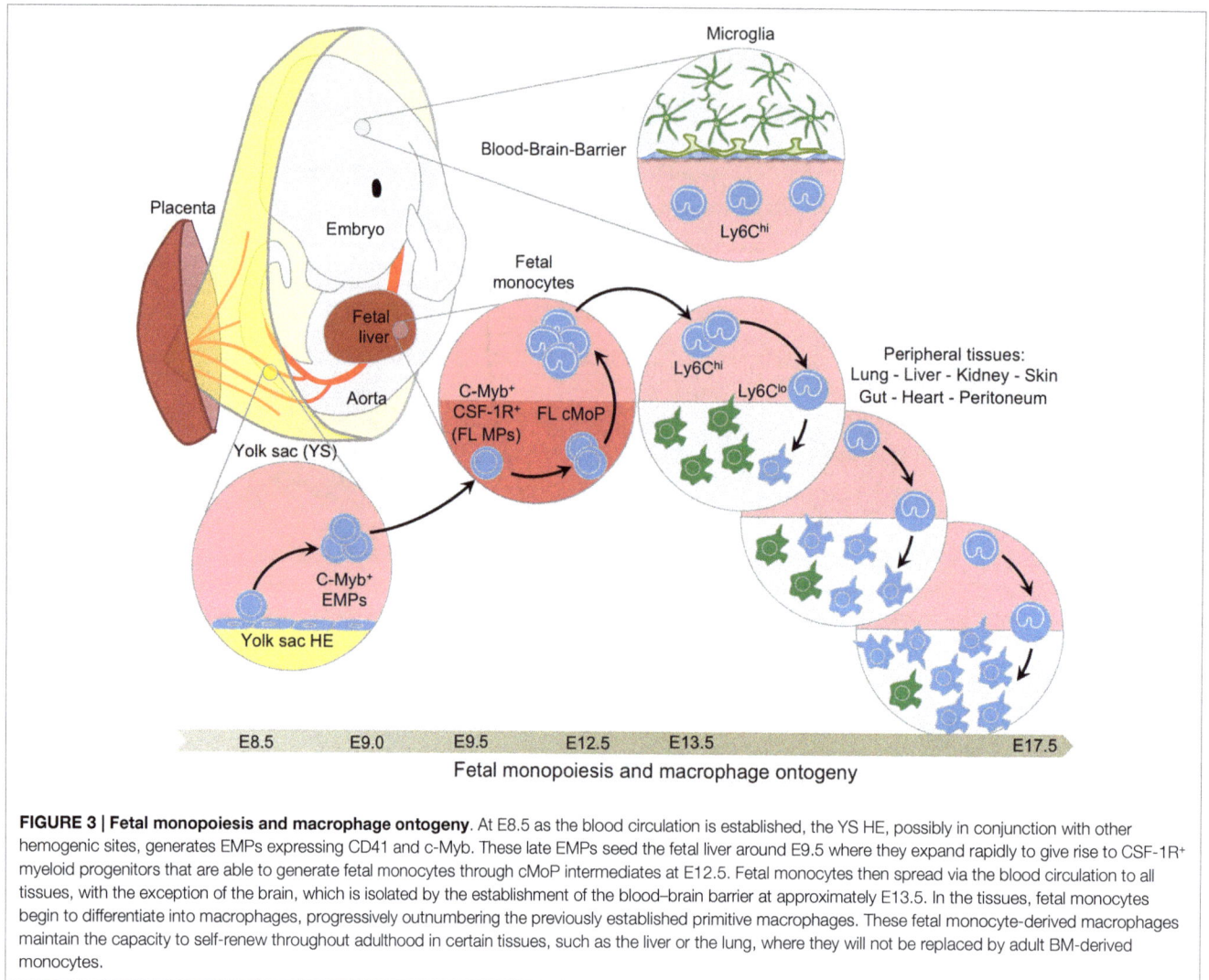

FIGURE 3 | Fetal monopoiesis and macrophage ontogeny. At E8.5 as the blood circulation is established, the YS HE, possibly in conjunction with other hemogenic sites, generates EMPs expressing CD41 and c-Myb. These late EMPs seed the fetal liver around E9.5 where they expand rapidly to give rise to CSF-1R+ myeloid progenitors that are able to generate fetal monocytes through cMoP intermediates at E12.5. Fetal monocytes then spread via the blood circulation to all tissues, with the exception of the brain, which is isolated by the establishment of the blood–brain barrier at approximately E13.5. In the tissues, fetal monocytes begin to differentiate into macrophages, progressively outnumbering the previously established primitive macrophages. These fetal monocyte-derived macrophages maintain the capacity to self-renew throughout adulthood in certain tissues, such as the liver or the lung, where they will not be replaced by adult BM-derived monocytes.

Whether these immature HSCs arise from LMPs or represent a distinct wave of progenitors remains to be clarified. However, these seminal studies provided strong evidence that lymphoid potential can emerge from the YS, prior to HSC-budding from the AGM (69).

Because the emergence of EMPs and LMPs overlaps in time and space, they could not be distinguished clearly until recently. Previous reports had suggested that lymphoid potential was restricted to the CD41-negative cell fraction (59, 65). However, CD41 is also expressed in a sub-fraction of FL HSCs, and so this phenotypical distinction spread some confusion (47). Finally, a recent report from the group of Palis clarified this point by showing that co-expression of c-kit, CD41, and CD16/32 defines EMPs and allows their separation from other progenitors with lymphoid potential, such as those giving rise to the B-1 cell (70). McGrath et al. extended the notion of EMPs by showing their potential to generate neutrophils, megakaryocytes, macrophages, and erythrocytes. Finally, transplantation of EMPs in immune-compromised adult mice can also provide transient adult red blood cell reconstitution (70).

To conclude, commitment to hematopoietic fates begins during gastrulation in the YS, which represents the only site of primitive erythropoiesis and also serves as the first source of transient definitive hematopoietic progenitors. HE develops from the YS to various intra-embryonic sites, and acquires myeloid and then lymphoid lineage potentials in overlapping waves, highlighting the complexity of the hematopoietic output. Whether some of these progenitors arise from independent sources or represent different maturation stages of a shared hematopoietic wave, culminating with the generation of HSCs, needs to be further clarified. However, it is tempting to speculate that the clear contrasts in differentiation/lineage potential do not reside in their intrinsic potential, but rather in the extrinsic signals provided by the local environment.

Definitive Hematopoiesis

The complex hierarchy of stem and progenitor cells in the BM is first established during embryonic development starting with the emergence of small numbers of HSCs from the AGM at E10.5 in murine embryos or at 5 weeks in human embryos

(71, 72). After E9.5 in the mouse, with the determination of the intra-embryonic mesoderm toward a hematopoietic lineage, new waves of hematopoietic progenitors emerge within the HE of the embryo proper (**Figure 4**), first in the P-Sp region and the umbilical and vitelline arterial regions of the embryo, then in the AGM region and the placenta (55, 73, 74). The hematopoietic activities of the P-Sp and AGM first generate immature HSCs and then mature HSCs, which are defined by their capacity to reconstitute adult conventional mice (long-term reconstitution; LTR). Both immature and mature HSCs seed the FL at approximately E10.5 (68, 71, 75, 76) to establish definitive hematopoiesis (40, 77, 78). A maturation step seems necessary for immature HSCs to express their LTR activity in full, which is then maintained until adulthood (68). However, further investigations using a fate-mapping system would be necessary to confirm this model.

The FL becomes the major hematopoietic organ after E11.5, generating all hematopoietic lineages. Importantly, the FL itself does not produce progenitors *de novo*, but rather recruits progenitors derived from the YS and other hemogenic sites, to initiate definitive hematopoiesis (79) in parallel with the expansion of the definitive HSC population before their migration to the spleen and BM (80).

The contribution of HSCs to FL hematopoiesis is complex to evaluate, partly because of the lack of specific fate-mapping models, and also the relatively limited knowledge regarding embryonic HSC maintenance and homeostasis in this environment. The capacity for long-term reconstitution, which defines functional HSCs, is present in the AGM by E10.5 (76). However, lineage-specific commitment may not occur *in vivo* immediately after reaching the FL environment. A number of other progenitors generated during transient definitive hematopoiesis, as discussed above, are already present and able to give rise to almost all cell lineages, which could prevent HSC consumption and differentiation (**Figure 5**). Evaluation of HSC contribution has long been based on the assumption that all hematopoietic cells in the FL were derived from HSCs as is the case in the BM (81). Many multipotential progenitors share the same phenotype with pre-HSC and HSCs, such as the expression of CD41 and AA4.1 (60), adding to this confusion. The combination of the marker Sca-1 and new markers such as those from the SLAM family (82) have greatly helped to clarify the characterization of HSCs, defined now as Lin⁻ckit⁺Sca-1⁺CD150⁺CD48⁻CD244⁻. However, no specific fate-mapping model exists to characterize embryonic HSC progeny with the exception of the Flt3-Cre

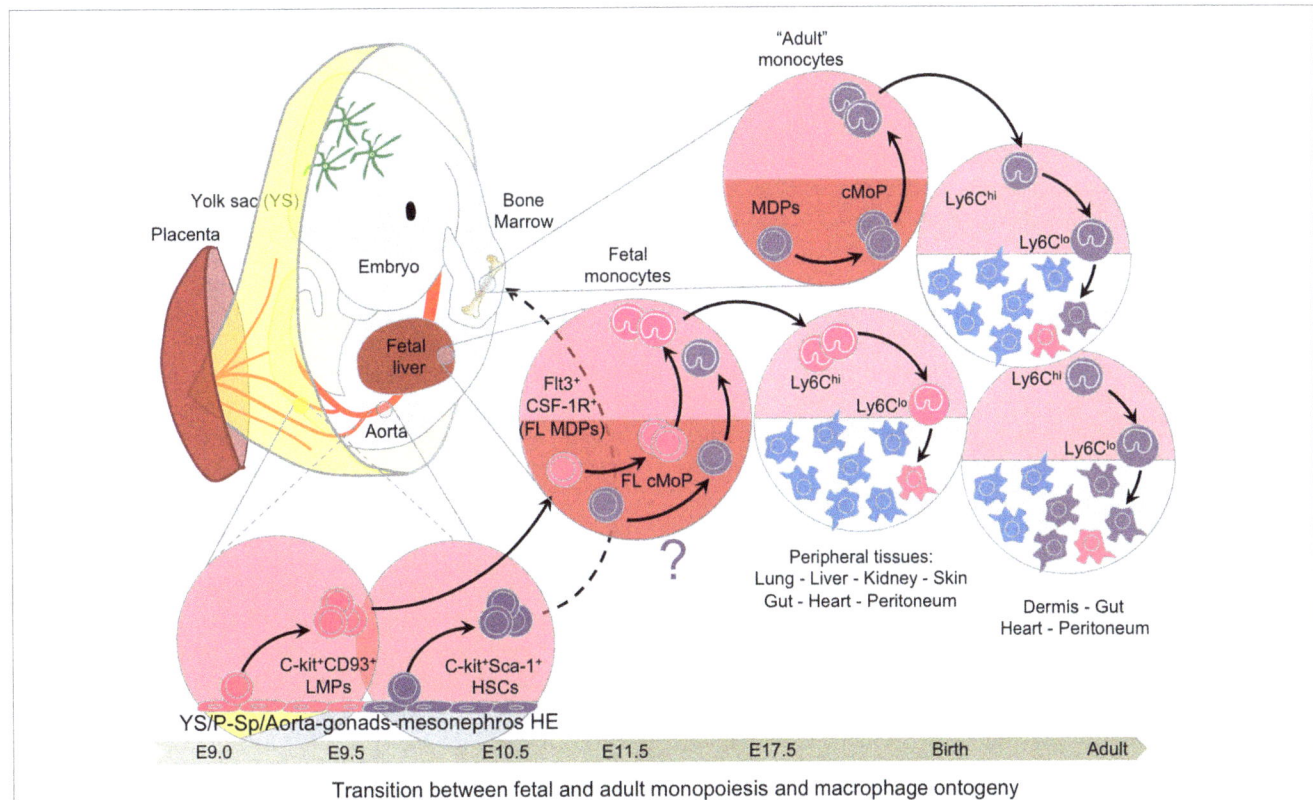

Transition between fetal and adult monopoiesis and macrophage ontogeny

FIGURE 4 | Transition between fetal and adult hematopoiesis. Hemogenic endothelial cells from extra and intra-embryonic hematopoietic tissues generate C-Myb-dependent multipotential progenitors, such as LMPs and pre-HSCs, between E9.0 and E10.5, culminating with the emergence of mature HSCs with long-term reconstitution-bearing potential. CD93 (AA4.1) expression is associated with the emergence of lymphoid potential, whereas Sca-1 is the hallmark of HSCs. These progenitors seed the fetal liver around E10/E11, expanding and giving rise to the various lineages of the hematopoietic system, including fetal monocytes. These late fetal monocytes continue to participate in the tissue-resident macrophage network until hematopoiesis switches completely from the fetal liver to the bone marrow. Once adult hematopoiesis begins to take place in the bone marrow generating monocytes, certain tissues, such as the dermis, heart peritoneum, and the gut, continue to recruit adult monocytes to generate resident macrophages and replace with time the embryonic-derived macrophages.

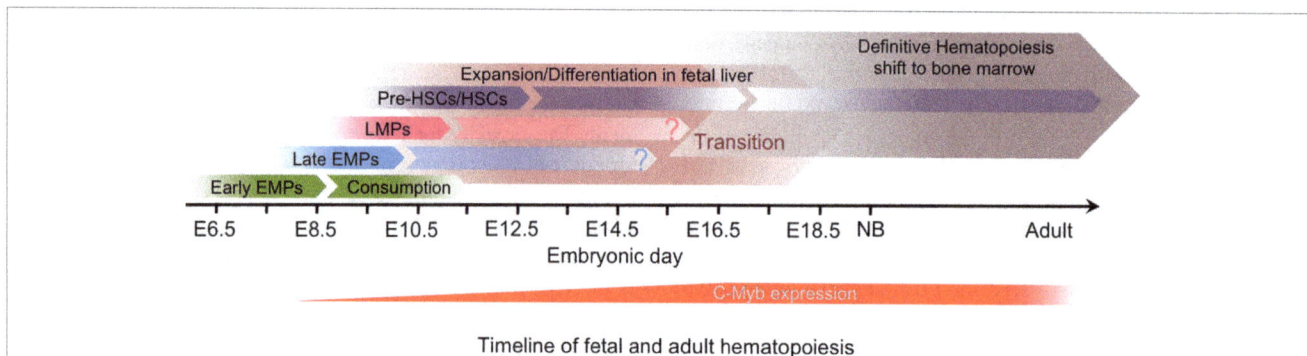

FIGURE 5 | Timeline of fetal and adult hematopoiesis. The primitive hematopoiesis is initiated in the yolk sac independently of C-Myb activity, and generates early CSF-1R⁺ EMPs that give rise to YS macrophages without monocytic intermediates during a short time window and will establish the brain microglia. The transient definitive hematopoiesis and then the definitive hematopoiesis are both dependent on C-Myb activity and generate progenitors that differentiate in the fetal liver. The transient definitive wave, which include EMPs and then LMPs, give rise in particular to fetal monocytes that seed the tissues prior to birth to establish the self-renewing tissue-resident macrophage network. Although only HSCs, which result from the definitive hematopoiesis, seem to be maintained in the bone marrow in adults, the relative contribution of the transient definitive wave to the adult immune system remains unclear.

model (83), which was used until now with the assumption that embryonic and adult HSCs follow similar differentiation pathways. Our recent report suggests that the Flt3-Cre model can also be used to follow the progeny of LMPs (15). Furthermore, in the nascent BM, the long-term repopulation (LTR) capacity that characterizes functional HSCs is only observed at around E17.5 (84). Considering the time required to initiate full HSC differentiation, these data suggest that proper adult HSC-derived hematopoiesis does not take place in the BM until a few days after birth. Characterization of the functional specificities and regulatory pathways of HE that give rise to HSCs versus those that generate EMPs and other multipotential progenitors could aid the development of new fate-mapping models and improve our understanding of this process (85). Use of other fate-mapping models such as the Runx1-Mer-Cre-Mer (Runx1-iCre) (86), Tie2-Mer-Cre-Mer mice (14), and the c-kit-Mer-Cre-Mer mice (87) provided complementary results, although a careful analysis of the targeted cells in time and space is not yet fully available for the last two models. We present here our best interpretation of the data provided in these two recent studies that have used these models in light of the literature and our own results and experience using the Runx1-iCre model (**Figure 6**).

Embryonic and Adult Precursors of Adult Tissue-Resident Macrophages

Yolk Sac Macrophages

Yolk sac macrophages first appear in the YS blood islands at E9 (albeit in small numbers) with a unique pattern of differentiation that bypasses the monocytic intermediate stage seen in adult macrophages (50, 52). YS-derived primitive macrophages spread into the embryo proper through the blood as soon as the circulatory system is fully established (from E8.5 to E10) (56), and migrate to various tissues, including the brain. Importantly, this occurs before the onset of fetal monocyte production by the FL, which starts around E11.5/E12.5 (92). These primitive macrophages retain the high proliferative potential observed in the YS as they

colonize various tissues (52, 93–95). Primitive macrophages may contribute to many fundamental processes during mid and late embryogenesis, such as clearance of dead cells or tissue maturation. In this regard, the developmental process of interdigital cell death removal during the mouse footplate remodeling that occurs between E12.5 and E14.5 is of interest as the interdigit regions become heavily populated by macrophages and most of the dead cells were shown to be rapidly engulfed by macrophages (96). However, mouse models devoid of primitive macrophages such as the colony-stimulating factor 1 receptor (CSF-1R) KO (Florent Ginhoux, unpublished data,) and PU.1 KO (97) appears to exhibit a normal interdigit web tissue. Wood et al. observed that interdigit web tissue in PU.1 KO was only slightly retarded, suggesting that other cell type such as neighboring mesenchymal cells were compensating (97). In addition, we recently showed that depletion of primitive macrophages and hence of embryonic microglia, affected the progression of dopaminergic axons in the forebrain and the laminar positioning of subsets of neocortical interneurons, likely through phagocytic mechanisms (98).

Schulz et al. highlighted further differences between primitive and definitive hematopoiesis, showing that the latter relies on the transcription factor Myb, while YS-derived macrophages are Myb-independent, and are instead dependent on PU.1 (12). This again reinforces the view that YS-derived macrophages constitute an independent lineage, distinct from the progeny of definitive HSCs. Schulz et al. exploited the differential dependence of primitive versus definitive hematopoiesis on the transcription factor c-Myb and reported that E16.5 tissue macrophage populations were not affected by the loss of c-Myb. Using a CSF-1R-iCre fate-mapping model of YS macrophages, they also reported the persistence of YS macrophages progeny in adult tissue-resident macrophage populations (lung, liver, and pancreas, as well as in the brain and skin), although the level of labeling was minimal (below 3–5%) and decreased with time. The authors concluded that tissue-resident macrophages were therefore derived from a c-Myb-independent lineage via YS macrophages (12), data supporting the initial report showing that microglia arise from

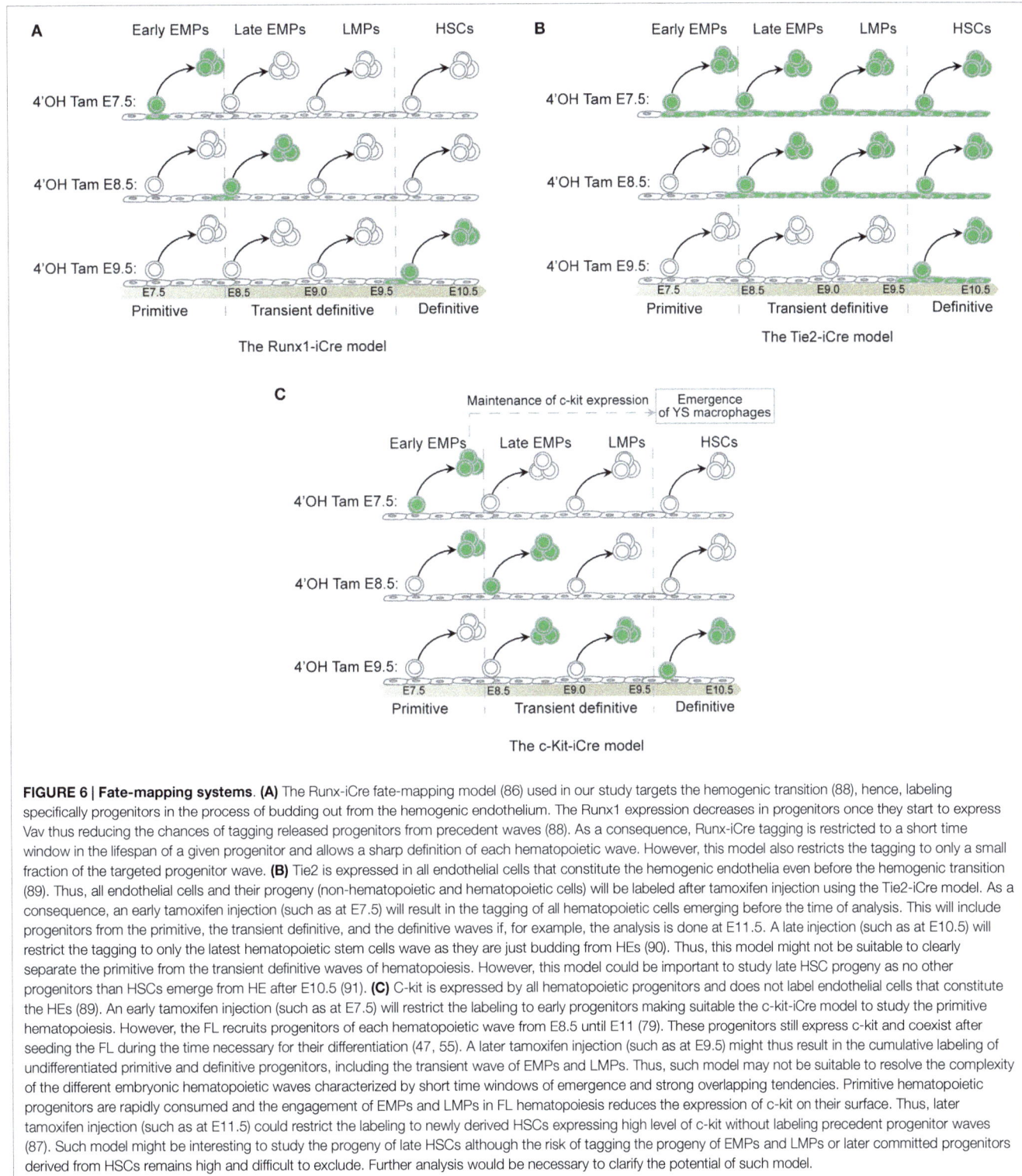

FIGURE 6 | Fate-mapping systems. **(A)** The Runx-iCre fate-mapping model (86) used in our study targets the hemogenic transition (88), hence, labeling specifically progenitors in the process of budding out from the hemogenic endothelium. The Runx1 expression decreases in progenitors once they start to express Vav thus reducing the chances of tagging released progenitors from precedent waves (88). As a consequence, Runx-iCre tagging is restricted to a short time window in the lifespan of a given progenitor and allows a sharp definition of each hematopoietic wave. However, this model also restricts the tagging to only a small fraction of the targeted progenitor wave. **(B)** Tie2 is expressed in all endothelial cells that constitute the hemogenic endothelia even before the hemogenic transition (89). Thus, all endothelial cells and their progeny (non-hematopoietic and hematopoietic cells) will be labeled after tamoxifen injection using the Tie2-iCre model. As a consequence, an early tamoxifen injection (such as at E7.5) will result in the tagging of all hematopoietic cells emerging before the time of analysis. This will include progenitors from the primitive, the transient definitive, and the definitive waves if, for example, the analysis is done at E11.5. A late injection (such as at E10.5) will restrict the tagging to only the latest hematopoietic stem cells wave as they are just budding from HEs (90). Thus, this model might not be suitable to clearly separate the primitive from the transient definitive waves of hematopoiesis. However, this model could be important to study late HSC progeny as no other progenitors than HSCs emerge from HE after E10.5 (91). **(C)** C-kit is expressed by all hematopoietic progenitors and does not label endothelial cells that constitute the HEs (89). An early tamoxifen injection (such as at E7.5) will restrict the labeling to early progenitors making suitable the c-kit-iCre model to study the primitive hematopoiesis. However, the FL recruits progenitors of each hematopoietic wave from E8.5 until E11 (79). These progenitors still express c-kit and coexist after seeding the FL during the time necessary for their differentiation (47, 55). A later tamoxifen injection (such as at E9.5) might thus result in the cumulative labeling of undifferentiated primitive and definitive progenitors, including the transient wave of EMPs and LMPs. Thus, such model may not be suitable to resolve the complexity of the different embryonic hematopoietic waves characterized by short time windows of emergence and strong overlapping tendencies. Primitive hematopoietic progenitors are rapidly consumed and the engagement of EMPs and LMPs in FL hematopoiesis reduces the expression of c-kit on their surface. Thus, later tamoxifen injection (such as at E11.5) could restrict the labeling to newly derived HSCs expressing high level of c-kit without labeling precedent progenitor waves (87). Such model might be interesting to study the progeny of late HSCs although the risk of tagging the progeny of EMPs and LMPs or later committed progenitors derived from HSCs remains high and difficult to exclude. Further analysis would be necessary to clarify the potential of such model.

YS macrophages (9). Embryonic origin of macrophages was further supported by the work of Yona et al. and Hashimoto et al. showing that adult monocytes do not substantially contribute to tissue macrophages under steady-state conditions (13, 33). Furthermore, Yona et al. suggested the existence of a CX3CR1+ precursor for some of the monocyte-independent macrophages,

although the exact nature of this precursor was not elucidated. In fact, both YS macrophages and fetal monocytes express CX3CR1 (9, 11, 15) and could therefore correspond to the unidentified precursors suggested by Yona et al. However, using a CSF-1R-iCre fate-mapping model, also used by Schulz et al. (12), another study noted that the YS macrophage contribution in the brain, the adult liver, and the heart was maintained although at a minimal level that decreased with time (99). Interestingly, the level of labeling was always higher in microglia than that in the liver or the heart, suggesting that the level of YS macrophage contribution may differ between tissues and that YS macrophages may be differently replaced over time by later waves of progenitors, which follow tissue-distinct kinetics (discussed below). Our own report using the Runx1-iCre fate-mapping model (86) indicated that only microglia, specialized macrophages of the central nervous system, were derived solely from primitive macrophages while all other tissue macrophages derived from definitive hematopoiesis (9).

To understand whether YS macrophages might be the sole progenitors of every other adult macrophages, we asked what impact their *in utero* depletion would have on the subsequent generation of fetal tissue macrophages. CSF-1R is expressed on YS macrophages and fetal monocytes, but only the development of the former is actively dependent on CSF-1R (9, 11). Thus, we attempted to deplete YS macrophages by transiently inhibiting the CSF-1R signaling pathway using a blocking anti-CSF-1R antibody, as recently described (98). Importantly, after complete depletion of primitive YS macrophages in E10.5 embryos and thus of most macrophages in treated embryos at E14.5, tissue macrophages (including microglia) were able to repopulate to normal levels before birth. These data suggest that YS macrophages are dispensable for the generation of tissue-resident macrophages in the embryo, and that another CSF-1R-independent embryonic precursor can functionally replace YS macrophages during development (15, 98). Using a combination of both the CSF-1R-iCre and the Runx1-iCre fate-mapping models, we noted that although YS macrophages infiltrate all tissues (including lung, liver, kidney, skin, gut, heart, pancreas, and stomach) until E13.5, a second wave of precursors, with a monocytic morphology and phenotype, supersedes them after E14.5 with the exception of the brain where YS macrophages are maintained until adulthood (15). A fuller understanding of this process may help to resolve some of the earlier discrepancies regarding the contribution of YS macrophages.

Fetal Monocytes

Fetal monocytes were described by Naito et al. (92). Focusing their study on liver Kupffer cells (the resident macrophages of the liver) during embryonic development, they exploited the endogenous peroxidase activity of monocytes and pro-monocytes granules described earlier by van Furth et al. (18, 92). Naito et al. observed the transient appearance of peroxidase activity, a signature for monocyte and pro-monocyte granule activity, during the *in vitro* generation of macrophages from a preparation of FL-dissociated cells (92). In the YS and at early stages of FL development, no peroxidase activity was observed, suggesting that primitive macrophages first seed the FL. At a later stage, the peroxidase activity increased, suggesting the presence of monocytic intermediates. *In vitro* clonal expansion assays confirmed the existence of two types of colonies, those containing fetal monocytes and those devoid of them. This provided early evidence for the existence of two distinct developmental pathways leading to the generation of Kupffer cells, although at this

stage, direct differentiation of fetal monocytes into macrophages *in vivo* had not been demonstrated (100).

To investigate the developmental event leading to the emergence of tissue-specific macrophages, we initially focused on the LC, the specialized myeloid population of the epidermis. While YS macrophages seed the embryonic skin before E13.5, we discovered that the major fraction of adult LCs is in fact derived from fetal monocytes that are generated in the FL from E12.5 and are then recruited into fetal skin at E14.5 (11). These cells share a similar phenotype to their adult counterparts; however, they are generated independently of CSF-1R expression (9, 11). They possess high proliferative potential, and, in contrast to their adult counterparts, express few genes related to pathogen recognition and immune activation (15). Further studies should clarify whether such differences reflect monocyte immaturity imposed by a sterile fetal environment, or rather dedicated functional specializations that have yet to be unraveled. *In utero* adoptive transfers combined with fate-mapping studies unequivocally confirmed *in situ* differentiation of fetal monocytes into adult LCs (11). Fetal monocytes were then demonstrated to be the precursor of adult macrophages in lung alveoli by intranasal injection (10, 101). Fetal monocytes were also shown to be involved in the generation of adult macrophages of the heart (99). In fact, fetal monocytes become the major leukocyte within the blood circulation after E13.5, spreading to all tissues. This occurred independently of the CCL2/CCR2 axis (15), suggesting an alternative mechanism of exit from the FL and/or recruitment by fetal tissues. Moreover, we were able to fate-map, from before birth to adulthood, the local differentiation of fetal monocytes into resident macrophages, by taking advantage of the specific expression of S100a4 in fetal monocytes compared to YS macrophages (15). Only the brain remained free from fetal monocyte infiltration, possibly resulting from the isolation of the brain by the nascent blood–brain barrier as early as E13.5 (15, 102). Thus, these data now reveal that fetal monocytes are the major circulating embryonic precursor for all macrophages, with the exception of the brain. The absence of monocyte precursor contribution to the microglial pool could result from a lack of intrinsic potential or a lack of access to the developing brain due to the nascent blood–brain barrier. Interestingly, we observed a major influx of monocytes in the brain at E14.5 in our YS macrophage depletion model, and preliminary data using our fetal monocyte S100a4-Cre/WT fate-mapping model combined with *in utero* depletion of YS macrophages suggest that fetal monocytes are capable of giving rise to microglia under certain conditions (Hoeffel & Ginhoux, personal communication). Whether this atypical fetal monocyte infiltration reflects a compensatory mechanism to fulfill an empty niche in the brain or results from a disruption of the blood–brain barrier remains to be investigated.

Adult Monocytes

BM-derived circulating monocytes were considered the only precursors for all tissue-resident macrophages since the seminal work of van Furth et al. (17–18). Although this dogma was entirely revisited recently with the emergence of sophisticated fate-mapping tools as well as parabiotic models, the physiological contribution of circulating adult monocytes to the adult macrophage network remains valid at least in certain tissues. The continuous recruitment of circulating monocytes to the dermis has been shown to shape the adult dermal macrophage network (25). Although this study did not employ fate-mapping techniques, Tamoutounour's data suggest the existence in the dermis of both a prenatal pool of macrophages and a second pool derived from adult blood monocytes. The authors argue that the dermis, in contrast to the epidermis, continues to recruit circulating monocytes in adulthood, most likely facilitated by its high level of vascularization. The macrophage network in the intestine follows a similar model. Data from Bain et al. suggest that embryonic macrophages do not persist in adulthood in the gut, and are replaced constantly by circulating adult monocytes (23), convincingly showing that adult monocytes are the source of intestine-resident macrophages. The role of commensal microbiota in this process is supported by the observation that the use of germ-free animals or treatment with broad-spectrum antibiotics results in a significant reduction in the recruitment of Ly6C$^+$ monocytes to the colon (23). The macrophage network of the heart has also been shown to contain a component of YS macrophages and fetal monocyte-derived macrophages, both of which are maintained in adulthood (99). However, similar to the dermis and the gut, adult monocytes seem to replace embryonic macrophages progressively over time (24). The decreasing capacity for self-renewal of embryonic macrophages with age observed by Molawi et al. may explain the requirement for continuous recruitment of monocyte-derived macrophages to the heart in the absence of inflammation. It remains to be clarified whether this phenomenon occurs in other tissues as a result of aging. In agreement, proliferation of YS macrophages and fetal monocytes is very high during development (20–40% before E14.5) but decreases progressively to 10% few days after birth in most tissues and decreases to almost undetectable levels in adults (15). Interestingly, macrophage turnover seems different from one tissue to another. Following BrdU incorporation at steady state, almost no proliferation was observed in adult gut macrophages (23), while 2–5% was measured in adult heart macrophages (24). Macrophage proliferation activity can also be mobilized upon inflammation. For example, peritoneal macrophages can increase their proliferation rate from 1 to 9% in response to parasite infection or in response to IL-4 stimulation (35), while enhanced local proliferation of macrophages in atherosclerotic lesions sustain disease progression (103). The characterization of local signals regulating macrophage proliferation as well as the presence of specialized tissue niches that sustain macrophage survival, proliferation, or even "stemness" will be fundamental to better understand their tissue homeostasis.

The macrophage network of the lymphoid system seems to follow a similar pattern than in the gut and dermis. Although the lymph nodes (LN) start to develop very early in the embryo (104), they become functionally active only within the first week after birth recruiting and organizing B and T cell areas when follicles start to shape with connections to afferent lymphatics via the subcapsular sinus (105). Although macrophages are known to participate in lymphangiogenesis during development, notably by the production of VEGF (106, 107), the precise origin of the different LN macrophage populations remain poorly understood

(108). The high level of foreign antigens passing through the LN during the lifespan, support the model of a constant replenishment of the local macrophage pool by circulating adult monocytes. However, the work of Jakubzick et al. suggests otherwise as tissue-patrolling monocytes at steady state seem to enter the LN without any sign of local differentiation to macrophages or dendritic cells (34). Further studies using fate-mapping systems should be addressed to clarify this point. Spleen macrophages are generated prenatally (13, 33). However, red pulp macrophages and marginal zone macrophages seem highly dependent, respectively, on the transcription factor SPI-C (109) and on the nuclear receptor LXR (110), also expressed by circulating monocytes and suggest again that embryonic-derived macrophages are replaced over time by adult monocytes-derived macrophages. The use of the S100a4-Cre fate-mapping model in our hands supports these observations and similar conclusions were obtained for BM and peritoneal macrophages (15). Although tissue microenvironment shapes certain macrophage functional specificities (111), through an ontogenic point of view, the composition of each tissue-resident macrophage pool evolves throughout life and the respective origins of each macrophage population may account for some of their key functions and cellular behaviors in a given tissue. Hence, a new challenge is to understand if an embryonic or adult origin matters for the function and the activation states of tissue-resident macrophages.

Origin and Development of YS Macrophages and Fetal Monocytes

Origin of YS Macrophages

Bertrand et al., in line with the seminal work of Palis (45), described two sequential myeloid waves within the early YS (42). Using an *in vitro* culture reporter system, Bertrand et al. observed a first wave of monopotent progenitors that gave rise only to macrophages, followed by a second wave that gave rise to a mix of granulocytes, monocytes, and macrophages. More recently, Kierdorf et al. revisited the work of Bertrand et al. exploiting organotypic embryonic brain slices to demonstrate that microglial cells derived from YS EMPs (112). Kierdorf et al. also showed that these EMPs did not express the transcription factor c-Myb, associating them with the progenitors reported by Schulz et al. (12), although a direct link with the generation of microglia *in vivo* in adulthood was not conclusively demonstrated. More recently, Perdiguero et al. used the CSF-1R-iCre fate-mapping model to show that YS macrophages are derived from CSF-1R$^+$ EMPs (14). Hence, these two studies suggest that YS macrophages, and thus microglia, would originate from c-Myb-independent CSF-1R$^+$ EMPs. Furthermore, Perdiguero et al. demonstrated that CSF-1R$^+$ EMPs were able to seed the FL by E10.5, suggesting that these progenitors could later populate other tissue niches and produce YS-like macrophage later during development in others tissues. Nevertheless, these data do not explain the low percentage of labeled adult macrophages observed by Schulz et al. using the CSF-1R-iCre fate-mapping model (12). Later observations by Epelman et al. (99), and more recently by our group using the same fate-mapping model (15), indicated that the ability of CSF-1R$^+$ EMP to reach the FL could explain the surprising maintenance of primitive macrophages until E16.5 in c-Myb null embryos, where primitive macrophages generated in the YS as well as in the FL would be able to fulfill the empty niche left by the absence of c-Myb-dependent myeloid cells, that include fetal monocytes. However, this may not reflect the physiological situation and may instead result from a compensatory mechanism to ensure the presence of macrophages in all tissues in the absence of c-Myb activity and fetal monocytes. Using the same CSF-1R-iCre fate-mapping model (15), we were able to follow the maintenance of microglia in the brain by self-renewal from E10.5 until adulthood, linking them with CSF-1R$^+$ EMPs and confirming the previous observations of Perdiguero et al. (14). However, for all other macrophage populations, the reduction of fate-mapping reporter labeling after E13.5 confirmed the progressive replacement of YS macrophages by another unlabeled precursor arising from a different hematopoietic wave.

We previously showed that Runx1$^+$ YS progenitors that emerged at E7.5 give rise to YS macrophages and microglia (9, 11). Using both the Runx1-iCre and the CSF-1R-iCre fate-mapping models, we showed that these E7.5 Runx1$^+$ YS progenitors were in fact the same CSF-1R$^+$ EMPs described by Perdiguero et al. and Kierdorf et al., which contributed to the generation of YS macrophages and, to a lesser extent, those seeding the FL (14, 15, 112). However, we also observed their disappearance from the FL after E11.5 indicative of a rapid local consumption/differentiation rather than long-term maintenance. Our results also suggest that these early CSF-1R$^+$ EMPs are able to contribute to a short-term maintenance of macrophages in the FL (**Figure 2**), but do not contribute to other tissue macrophages as evidenced by their rapid disappearance from the blood circulation after E14.5 (15). This transient population in the FL may be due to a local immediate requirement for macrophages, at least during the onset of FL hematopoiesis, to perform efficient enucleation of primitive erythrocytes passing through the FL sinusoids (100, 113). Combining historical evidences showing their direct lineage connection with the emergence of YS macrophages and recent findings showing their independence with c-Myb activity, we propose that CSF-1R$^+$ EMPs should be designated as primitive EMPs.

Origin of Fetal Monocytes

Because adult monocytes are derived from HSCs in the BM, it would be reasonable to assume that embryonic HSCs might also give rise to fetal monocytes in the developing liver. In agreement with this hypothesis, we have identified a population in the FL similar to adult MDPs that have the potential to generate fetal cMoPs and monocytes following *in vitro* culture (15). Exploiting the Flt3-Cre tomato fate-mapping model (83), we then followed the progeny of embryonic HSCs. However, the poor labeling observed between E14.5 and E17.5 in FL monocytes and macrophages contrasted with the strong labeling of FL MDPs, suggesting that HSCs had limited involvement in the generation of fetal monocytes (15). Nonetheless, the limited but significant labeling in fetal monocytes and macrophages at birth suggested an increasing derivation from fetal HSCs, assuming that fetal HSCs follow a similar Flt3-dependent differentiation pathway as adult HSCs. In parallel, gene array analysis highlighted a strong lymphoid signature within fetal MDPs (15), indicative of their

derivation from the recently described YS-derived LMPs (67). Thus, LMPs may be important for the generation of a small but significant proportion of fetal monocytes prior to the expansion of mature HSCs (**Figure 4**). Further investigations using more specific fate-mapping models will be necessary to elucidate the exact contribution of LMPs as well as the hematopoietic transition between the FL and the BM.

Importantly, we observed that fetal monocytes were not tagged with the CSF-1R-Cre model that label early CSF-1R$^+$ EMPs, suggesting that fetal monopoiesis is not dependent on CSF-1R$^+$ EMPs, consistent with our previous data (9, 11) and with our YS macrophage depletion results (15, 98). Furthermore, the Runx1-iCre fate-mapping model allowed us to identify two waves of EMPs that arise sequentially before LMPs in the YS. These included an early wave, arising at E7.5 that differentiates locally into YS macrophages; and a later wave tagged at E8.5, that migrates and seeds the FL following the establishment of the blood circulation before E9.0. Early EMPs tagged at E7.5 were therefore related to those described previously by Kierdorf and Perdiguero (14, 112). The late EMPs tagged at E8.5, however, expressed c-Myb, expanded more efficiently in the FL, and differentiated *in vivo* into fetal cMoPs, constituting the major component of the fetal monocyte population as well as the fetal monocyte-derived macrophage population (**Figure 3**), which was able to maintain itself in all tissues tested (15).

The existence of two distinct EMP waves is in agreement with Bertrand et al. who reported an early wave of macrophage progenitors restricted to the YS, and a second wave that was able to reach the FL to participate in definitive hematopoiesis (42). The differential expression of c-Myb between early and late EMPs is in agreement with previous reports indicating that primitive hematopoiesis can occur in the absence of c-Myb, especially for the generation of monopotent macrophage progenitors (114), whereas EMPs from definitive hematopoiesis express and are dependent on c-Myb activity (45, 62, 115).

Notably, a previous study showed that c-Myb ablation strongly compromises definitive hematopoiesis (116). Palis et al. observed that c-Myb is expressed prior to and during the early development of definitive erythrocyte progenitors (45). Thus, late EMPs and LMPs, as well as HSCs, express c-Myb (15, 45, 61, 62), suggesting that the entire fetal monopoiesis machinery is reliant on this transcription factor. In agreement, the CD11b^{hi}F480^{lo} population, which in our hands contains fetal monocytes, was completely absent in the c-Myb-deficient embryo (12, 116). As a consequence, the contribution of c-Myb-dependent progenitors to tissue-resident macrophage populations could not be evaluated in c-Myb-deficient embryos, where c-Myb-independent YS macrophages maintain themselves as a compensatory mechanism due to the absence of c-Myb-dependent fetal monocytes that normally outcompete them. Because c-Myb expression is upregulated during the successive steps of fetal monopoiesis (15), the switch in EMP localization between the YS and the FL may indeed be orchestrated by c-Myb. As a consequence, most tissue-resident macrophages derived from fetal monocytes would therefore rely on c-Myb activity. Altogether we propose that c-Myb$^+$ EMPs giving rise to the first circulating monocytes should be designated as definitive EMPs.

Conclusion

Recent reports have drastically changed the view of the development of the MPS and shed light on the multiple layers that define fetal hematopoiesis. It is now evident that fetal monocytes form the major precursors of most adult tissue-resident macrophages, and further investigations are now necessary to clarify how they shape macrophage heterogeneity. Examining how tissues imprint specific fates in these circulating precursors will aid our understanding of the mechanisms that control the tissue-specific functions of macrophages in the steady state, and thus may uncover new therapeutic opportunities in diverse pathological settings such as metabolic diseases, fibrosis, and carcinogenesis.

References

1. Cavaillon JM. The historical milestones in the understanding of leukocyte biology initiated by Elie Metchnikoff. *J Leukoc Biol* (2011) **90**:413–24. doi:10.1189/jlb.0211094

2. Tauber AI. Metchnikoff and the phagocytosis theory. *Nat Rev Mol Cell Biol* (2003) **4**:897–901. doi:10.1038/nrm1244

3. Yona S, Gordon S. From the reticuloendothelial to mononuclear phagocyte system – the unaccounted years. *Front Immunol* (2015) **6**:328. doi:10.3389/fimmu.2015.00328

4. Wynn TA, Chawla A, Pollard JW. Macrophage biology in development, homeostasis and disease. *Nature* (2013) **496**:445–55. doi:10.1038/nature12034

5. McNelis JC, Olefsky JM. Macrophages, immunity, and metabolic disease. *Immunity* (2014) **41**:36–48. doi:10.1016/j.immuni.2014.05.010

6. Moore KJ, Sheedy FJ, Fisher EA. Macrophages in atherosclerosis: a dynamic balance. *Nat Rev Immunol* (2013) **13**:709–21. doi:10.1038/nri3520

7. Noy R, Pollard JW. Tumor-associated macrophages: from mechanisms to therapy. *Immunity* (2014) **41**:49–61. doi:10.1016/j.immuni.2014.06.010

8. Prinz M, Priller J. Microglia and brain macrophages in the molecular age: from origin to neuropsychiatric disease. *Nat Rev Neurosci* (2014) **15**:300–12. doi:10.1038/nrn3722

9. Ginhoux F, Greter M, Leboeuf M, Nandi S, See P, Gokhan S, et al. Fate mapping analysis reveals that adult microglia derive from primitive macrophages. *Science* (2010) **330**:841–5. doi:10.1126/science.1194637

10. Guilliams M, De Kleer I, Henri S, Post S, Vanhoutte L, De Prijck S, et al. Alveolar macrophages develop from fetal monocytes that differentiate into long-lived cells in the first week of life via GM-CSF. *J Exp Med* (2013) **210**:1977–92. doi:10.1084/jem.20131199

11. Hoeffel G, Wang Y, Greter M, See P, Teo P, Malleret B, et al. Adult Langerhans cells derive predominantly from embryonic fetal liver monocytes with a minor contribution of yolk sac-derived macrophages. *J Exp Med* (2012) **209**:1167–81. doi:10.1084/jem.20120340

12. Schulz C, Gomez Perdiguero E, Chorro L, Szabo-Rogers H, Cagnard N, Kierdorf K, et al. A lineage of myeloid cells independent of Myb and hematopoietic stem cells. *Science* (2012) **336**:86–90. doi:10.1126/science.1219179

13. Yona S, Kim KW, Wolf Y, Mildner A, Varol D, Breker M, et al. Fate mapping reveals origins and dynamics of monocytes and tissue macrophages under homeostasis. *Immunity* (2013) **38**:79–91. doi:10.1016/j.immuni.2012.12.001

14. Gomez Perdiguero E, Klapproth K, Schulz C, Busch K, Azzoni E, Crozet L, et al. Tissue-resident macrophages originate from yolk-sac-derived erythro-myeloid progenitors. *Nature* (2015) **518**:547–51. doi:10.1038/nature13989

15. Hoeffel G, Chen J, Lavin Y, Low D, Almeida FF, See P, et al. C-myb(+) erythro-myeloid progenitor-derived fetal monocytes give rise to adult tissue-resident macrophages. *Immunity* (2015) **42**:665–78. doi:10.1016/j.immuni.2015.03.011

16. Chow A, Brown BD, Merad M. Studying the mononuclear phagocyte system in the molecular age. *Nat Rev Immunol* (2011) **11**:788–98. doi:10.1038/nri3087

17. Van Furth R, Cohn ZA. The origin and kinetics of mononuclear phagocytes. *J Exp Med* (1968) **128**:415–35. doi:10.1084/jem.128.3.415

18. Van Furth R, Cohn ZA, Hirsch JG, Humphrey JH, Spector WG, Langevoort HL. The mononuclear phagocyte system: a new classification of macrophages, monocytes, and their precursor cells. *Bull World Health Organ* (1972) **46**:845–52.

19. Virolainen M. Hematopoietic origin of macrophages as studied by chromosome markers in mice. *J Exp Med* (1968) **127**:943–52. doi:10.1084/jem.127.5.943

20. Auffray C, Fogg DK, Narni-Mancinelli E, Senechal B, Trouillet C, Saederup N, et al. CX3CR1+ CD115+ CD135+ common macrophage/DC precursors and the role of CX3CR1 in their response to inflammation. *J Exp Med* (2009) **206**:595–606. doi:10.1084/jem.20081385

21. Hettinger J, Richards DM, Hansson J, Barra MM, Joschko AC, Krijgsveld J, et al. Origin of monocytes and macrophages in a committed progenitor. *Nat Immunol* (2013) **14**:821–30. doi:10.1038/ni.2638

22. Geissmann F, Jung S, Littman DR. Blood monocytes consist of two principal subsets with distinct migratory properties. *Immunity* (2003) **19**:71–82. doi:10.1016/S1074-7613(03)00174-2

23. Bain CC, Bravo-Blas A, Scott CL, Gomez Perdiguero E, Geissmann F, Henri S, et al. Constant replenishment from circulating monocytes maintains the macrophage pool in the intestine of adult mice. *Nat Immunol* (2014) **15**:929–37. doi:10.1038/ni.2967

24. Molawi K, Wolf Y, Kandalla PK, Favret J, Hagemeyer N, Frenzel K, et al. Progressive replacement of embryo-derived cardiac macrophages with age. *J Exp Med* (2014) **211**:2151–8. doi:10.1084/jem.20140639

25. Tamoutounour S, Guilliams M, Montanana Sanchis F, Liu H, Terhorst D, Malosse C, et al. Origins and functional specialization of macrophages and of conventional and monocyte-derived dendritic cells in mouse skin. *Immunity* (2013) **39**:925–38. doi:10.1016/j.immuni.2013.10.004

26. Hashimoto K, Tarnowski WM. Some new aspects of the Langerhans cell. *Arch Dermatol* (1968) **97**:450–64. doi:10.1001/archderm.1968.01610100090015

27. Krueger GG, Daynes RA, Emam M. Biology of Langerhans cells: selective migration of Langerhans cells into allogeneic and xenogeneic grafts on nude mice. *Proc Natl Acad Sci U S A* (1983) **80**:1650–4. doi:10.1073/pnas.80.6.1650

28. Czernielewski J, Vaigot P, Prunieras M. Epidermal Langerhans cells – a cycling cell population. *J Invest Dermatol* (1985) **84**:424–6. doi:10.1111/1523-1747.ep12265523

29. Tarling JD, Lin HS, Hsu S. Self-renewal of pulmonary alveolar macrophages: evidence from radiation chimera studies. *J Leukoc Biol* (1987) **42**:443–6.

30. Naito M, Takahashi K. The role of Kupffer cells in glucan-induced granuloma formation in the liver of mice depleted of blood monocytes by administration of strontium-89. *Lab Invest* (1991) **64**:664–74.

31. Yamada M, Naito M, Takahashi K. Kupffer cell proliferation and glucan-induced granuloma formation in mice depleted of blood monocytes by strontium-89. *J Leukoc Biol* (1990) **47**:195–205.

32. Ajami B, Bennett JL, Krieger C, Tetzlaff W, Rossi FM. Local self-renewal can sustain CNS microglia maintenance and function throughout adult life. *Nat Neurosci* (2007) **10**:1538–43. doi:10.1038/nn2014

33. Hashimoto D, Chow A, Noizat C, Teo P, Beasley MB, Leboeuf M, et al. Tissue-resident macrophages self-maintain locally throughout adult life with minimal contribution from circulating monocytes. *Immunity* (2013) **38**:792–804. doi:10.1016/j.immuni.2013.04.004

34. Jakubzick C, Gautier EL, Gibbings SL, Sojka DK, Schlitzer A, Johnson TE, et al. Minimal differentiation of classical monocytes as they survey steady-state tissues and transport antigen to lymph nodes. *Immunity* (2013) **39**:599–610. doi:10.1016/j.immuni.2013.08.007

35. Jenkins SJ, Ruckerl D, Cook PC, Jones LH, Finkelman FD, van Rooijen N, et al. Local macrophage proliferation, rather than recruitment from the

blood, is a signature of TH2 inflammation. *Science* (2011) **332**:1284–8. doi:10.1126/science.1204351

36. Merad M, Manz MG, Karsunky H, Wagers A, Peters W, Charo I, et al. Langerhans cells renew in the skin throughout life under steady-state conditions. *Nat Immunol* (2002) **3**:1135–41. doi:10.1038/ni852

37. Romani N, Schuler G, Fritsch P. Ontogeny of Ia-positive and Thy-1-positive leukocytes of murine epidermis. *J Invest Dermatol* (1986) **86**:129–33. doi:10.1111/1523-1747.ep12284135

38. Takahashi K, Takahashi H, Naito M, Sato T, Kojima M. Ultrastructural and functional development of macrophages in the dermal tissue of rat fetuses. *Cell Tissue Res* (1983) **232**:539–52. doi:10.1007/BF00216427

39. Mizoguchi S, Takahashi K, Takeya M, Naito M, Morioka T. Development, differentiation, and proliferation of epidermal Langerhans cells in rat ontogeny studied by a novel monoclonal antibody against epidermal Langerhans cells, RED-1. *J Leukoc Biol* (1992) **52**:52–61.

40. Orkin SH, Zon LI. Hematopoiesis: an evolving paradigm for stem cell biology. *Cell* (2008) **132**:631–44. doi:10.1016/j.cell.2008.01.025

41. Tavian M, Peault B. Embryonic development of the human hematopoietic system. *Int J Dev Biol* (2005) **49**:243–50. doi:10.1387/ijdb.041957mt

42. Bertrand JY, Jalil A, Klaine M, Jung S, Cumano A, Godin I. Three pathways to mature macrophages in the early mouse yolk sac. *Blood* (2005) **106**:3004–11. doi:10.1182/blood-2005-02-0461

43. Cline MJ, Moore MA. Embryonic origin of the mouse macrophage. *Blood* (1972) **39**:842–9.

44. Palis J, Chan RJ, Koniski A, Patel R, Starr M, Yoder MC. Spatial and temporal emergence of high proliferative potential hematopoietic precursors during murine embryogenesis. *Proc Natl Acad Sci U S A* (2001) **98**:4528–33. doi:10.1073/pnas.071002398

45. Palis J, Robertson S, Kennedy M, Wall C, Keller G. Development of erythroid and myeloid progenitors in the yolk sac and embryo proper of the mouse. *Development* (1999) **126**:5073–84.

46. Tober J, Koniski A, McGrath KE, Vemishetti R, Emerson R, de Mesy-Bentley KK, et al. The megakaryocyte lineage originates from hemangioblast precursors and is an integral component both of primitive and of definitive hematopoiesis. *Blood* (2007) **109**:1433–41. doi:10.1182/blood-2006-06-031898

47. Frame JM, McGrath KE, Palis J. Erythro-myeloid progenitors: "definitive" hematopoiesis in the conceptus prior to the emergence of hematopoietic stem cells. *Blood Cells Mol Dis* (2013) **51**:220–5. doi:10.1016/j.bcmd.2013.09.006

48. Gulliver G. Observations on the sizes and shapes of red corpuscles of the blood of vertebrates, with drawings of them to a uniform scale, and extended and revised tables of measurements. *Proc Zool Soc Lond* (1875):474–95.

49. Palis J. Primitive and definitive erythropoiesis in mammals. *Front Physiol* (2014) **5**:3. doi:10.3389/fphys.2014.00003

50. Naito M, Yamamura F, Nishikawa S, Takahashi K. Development, differentiation, and maturation of fetal mouse yolk sac macrophages in cultures. *J Leukoc Biol* (1989) **46**:1–10.

51. Takahashi K, Naito M. Development, differentiation, and proliferation of macrophages in the rat yolk sac. *Tissue Cell* (1993) **25**:351–62. doi:10.1016/0040-8166(93)90077-X

52. Takahashi K, Yamamura F, Naito M. Differentiation, maturation, and proliferation of macrophages in the mouse yolk sac: a light-microscopic, enzyme-cytochemical, immunohistochemical, and ultrastructural study. *J Leukoc Biol* (1989) **45**:87–96.

53. Faust N, Huber MC, Sippel AE, Bonifer C. Different macrophage populations develop from embryonic/fetal and adult hematopoietic tissues. *Exp Hematol* (1997) **25**:432–44.

54. Morioka Y, Naito M, Sato T, Takahashi K. Immunophenotypic and ultrastructural heterogeneity of macrophage differentiation in bone marrow and fetal hematopoiesis of mouse in vitro and in vivo. *J Leukoc Biol* (1994) **55**:642–51.

55. Lin Y, Yoder MC, Yoshimoto M. Lymphoid progenitor emergence in the murine embryo and yolk sac precedes stem cell detection. *Stem Cells Dev* (2014) **23**:1168–77. doi:10.1089/scd.2013.0536

56. McGrath KE, Koniski AD, Malik J, Palis J. Circulation is established in a stepwise pattern in the mammalian embryo. *Blood* (2003) **101**:1669–76. doi:10.1182/blood-2002-08-2531

57. Bertrand JY, Kim AD, Violette EP, Stachura DL, Cisson JL, Traver D. Definitive hematopoiesis initiates through a committed erythromyeloid progenitor in the zebrafish embryo. *Development* (2007) **134**:4147–56. doi:10.1242/dev.012385

58. Dzierzak E, Speck NA. Of lineage and legacy: the development of mammalian hematopoietic stem cells. *Nat Immunol* (2008) **9**:129–36. doi:10.1038/ni1560

59. Ferkowicz MJ, Starr M, Xie X, Li W, Johnson SA, Shelley WC, et al. CD41 expression defines the onset of primitive and definitive hematopoiesis in the murine embryo. *Development* (2003) **130**:4393–403. doi:10.1242/dev.00632

60. Mikkola HK, Orkin SH. The journey of developing hematopoietic stem cells. *Development* (2006) **133**:3733–44. doi:10.1242/dev.02568

61. Godin I, Dieterlen-Lievre F, Cumano A. Emergence of multipotent hemopoietic cells in the yolk sac and paraaortic splanchnopleura in mouse embryos, beginning at 8.5 days postcoitus. *Proc Natl Acad Sci U S A* (1995) **92**:773–7. doi:10.1073/pnas.92.23.10815b

62. Yoder MC, Hiatt K, Dutt P, Mukherjee P, Bodine DM, Orlic D. Characterization of definitive lymphohematopoietic stem cells in the day 9 murine yolk sac. *Immunity* (1997) **7**:335–44. doi:10.1016/S1074-7613(00)80355-6

63. Lacaud G, Carlsson L, Keller G. Identification of a fetal hematopoietic precursor with B cell, T cell, and macrophage potential. *Immunity* (1998) **9**:827–38. doi:10.1016/S1074-7613(00)80648-2

64. Adolfsson J, Mansson R, Buza-Vidas N, Hultquist A, Liuba K, Jensen CT, et al. Identification of Flt3+ lympho-myeloid stem cells lacking erythro-megakaryocytic potential a revised road map for adult blood lineage commitment. *Cell* (2005) **121**:295–306. doi:10.1016/j.cell.2005.02.013

65. Yoshimoto M, Montecino-Rodriguez E, Ferkowicz MJ, Porayette P, Shelley WC, Conway SJ, et al. Embryonic day 9 yolk sac and intra-embryonic hemogenic endothelium independently generate a B-1 and marginal zone progenitor lacking B-2 potential. *Proc Natl Acad Sci U S A* (2011) **108**:1468–73. doi:10.1073/pnas.1015841108

66. Yoshimoto M, Porayette P, Glosson NL, Conway SJ, Carlesso N, Cardoso AA, et al. Autonomous murine T-cell progenitor production in the extra-embryonic yolk sac before HSC emergence. *Blood* (2012) **119**:5706–14. doi:10.1182/blood-2011-12-397489

67. Boiers C, Carrelha J, Lutteropp M, Luc S, Green JC, Azzoni E, et al. Lymphomyeloid contribution of an immune-restricted progenitor emerging prior to definitive hematopoietic stem cells. *Cell Stem Cell* (2013) **13**:535–48. doi:10.1016/j.stem.2013.08.012

68. Kieusseian A, Brunet de la Grange P, Burlen-Defranoux O, Godin I, Cumano A. Immature hematopoietic stem cells undergo maturation in the fetal liver. *Development* (2012) **139**:3521–30. doi:10.1242/dev.079210

69. Cumano A, Dieterlen-Lievre F, Godin I. Lymphoid potential, probed before circulation in mouse, is restricted to caudal intraembryonic splanchnopleura. *Cell* (1996) **86**:907–16. doi:10.1016/S0092-8674(00)80166-X

70. McGrath KE, Frame JM, Fegan KH, Bowen JR, Conway SJ, Catherman SC, et al. Distinct sources of hematopoietic progenitors emerge before HSCs and provide functional blood cells in the mammalian embryo. *Cell Rep* (2015) **11**:1892–904. doi:10.1016/j.celrep.2015.05.036

71. Muller AM, Medvinsky A, Strouboulis J, Grosveld F, Dzierzak E. Development of hematopoietic stem cell activity in the mouse embryo. *Immunity* (1994) **1**:291–301. doi:10.1016/1074-7613(94)90081-7

72. Tavian M, Robin C, Coulombel L, Peault B. The human embryo, but not its yolk sac, generates lympho-myeloid stem cells: mapping multipotent hematopoietic cell fate in intraembryonic mesoderm. *Immunity* (2001) **15**:487–95. doi:10.1016/S1074-7613(01)00193-5

73. Gekas C, Dieterlen-Lievre F, Orkin SH, Mikkola HK. The placenta is a niche for hematopoietic stem cells. *Dev Cell* (2005) **8**:365–75. doi:10.1016/j.devcel.2004.12.016

74. Mikkola HK, Gekas C, Orkin SH, Dieterlen-Lievre F. Placenta as a site for hematopoietic stem cell development. *Exp Hematol* (2005) **33**:1048–54. doi:10.1016/j.exphem.2005.06.011

75. de Bruijn MF, Speck NA, Peeters MC, Dzierzak E. Definitive hematopoietic stem cells first develop within the major arterial regions of the mouse embryo. *EMBO J* (2000) **19**:2465–74. doi:10.1093/emboj/19.11.2465

76. Kumaravelu P, Hook L, Morrison AM, Ure J, Zhao S, Zuyev S, et al. Quantitative developmental anatomy of definitive haematopoietic stem cells/long-term repopulating units (HSC/RUs): role of the aorta-gonad-mesonephros (AGM) region and the yolk sac in colonisation of the mouse embryonic liver. *Development* (2002) **129**:4891–9.

77. Golub R, Cumano A. Embryonic hematopoiesis. *Blood Cells Mol Dis* (2013) **51**:226–31. doi:10.1016/j.bcmd.2013.08.004

78. Medvinsky A, Rybtsov S, Taoudi S. Embryonic origin of the adult hematopoietic system: advances and questions. *Development* (2011) **138**:1017–31. doi:10.1242/dev.040998

79. Swiers G, Rode C, Azzoni E, de Bruijn MF. A short history of hemogenic endothelium. *Blood Cells Mol Dis* (2013) **51**:206–12. doi:10.1016/j.bcmd.2013.09.005

80. Christensen JL, Wright DE, Wagers AJ, Weissman IL. Circulation and chemotaxis of fetal hematopoietic stem cells. *PLoS Biol* (2004) **2**:E75. doi:10.1371/journal.pbio.0020075

81. Kawamoto H, Katsura Y. A new paradigm for hematopoietic cell lineages: revision of the classical concept of the myeloid-lymphoid dichotomy. *Trends Immunol* (2009) **30**:193–200. doi:10.1016/j.it.2009.03.001

82. Oguro H, Ding L, Morrison SJ. SLAM family markers resolve functionally distinct subpopulations of hematopoietic stem cells and multipotent progenitors. *Cell Stem Cell* (2013) **13**:102–16. doi:10.1016/j.stem.2013.05.014

83. Boyer SW, Schroeder AV, Smith-Berdan S, Forsberg EC. All hematopoietic cells develop from hematopoietic stem cells through Flk2/Flt3-positive progenitor cells. *Cell Stem Cell* (2011) **9**:64–73. doi:10.1016/j.stem.2011.04.021

84. Coskun S, Chao H, Vasavada H, Heydari K, Gonzales N, Zhou X, et al. Development of the fetal bone marrow niche and regulation of HSC quiescence and homing ability by emerging osteolineage cells. *Cell Rep* (2014) **9**:581–90. doi:10.1016/j.celrep.2014.09.013

85. Chen MJ, Li Y, De Obaldia ME, Yang Q, Yzaguirre AD, Yamada-Inagawa T, et al. Erythroid/myeloid progenitors and hematopoietic stem cells originate from distinct populations of endothelial cells. *Cell Stem Cell* (2011) **9**:541–52. doi:10.1016/j.stem.2011.10.003

86. Samokhvalov IM, Samokhvalova NI, Nishikawa S. Cell tracing shows the contribution of the yolk sac to adult haematopoiesis. *Nature* (2007) **446**:1056–61. doi:10.1038/nature05725

87. Sheng J, Ruedl C, Karjalainen K. Most tissue-resident macrophages except microglia are derived from fetal hematopoietic stem cells. *Immunity* (2015) **43**:382–93. doi:10.1016/j.immuni.2015.07.016

88. Chen MJ, Yokomizo T, Zeigler BM, Dzierzak E, Speck NA. Runx1 is required for the endothelial to haematopoietic cell transition but not thereafter. *Nature* (2009) **457**:887–91. doi:10.1038/nature07619

89. Cumano A, Godin I. Ontogeny of the hematopoietic system. *Annu Rev Immunol* (2007) **25**:745–85. doi:10.1146/annurev.immunol.25.022106.141538

90. Boisset JC, van Cappellen W, Andrieu-Soler C, Galjart N, Dzierzak E, Robin C. In vivo imaging of haematopoietic cells emerging from the mouse aortic endothelium. *Nature* (2010) **464**:116–20. doi:10.1038/nature08764

91. Busch K, Klapproth K, Barile M, Flossdorf M, Holland-Letz T, Schlenner SM, et al. Fundamental properties of unperturbed haematopoiesis from stem cells in vivo. *Nature* (2015) **518**:542–6. doi:10.1038/nature14242

92. Naito M, Takahashi K, Nishikawa S. Development, differentiation, and maturation of macrophages in the fetal mouse liver. *J Leukoc Biol* (1990) **48**:27–37.

93. Lichanska AM, Hume DA. Origins and functions of phagocytes in the embryo. *Exp Hematol* (2000) **28**:601–11. doi:10.1016/S0301-472X(00)00157-0

94. Naito M, Umeda S, Yamamoto T, Moriyama H, Umezu H, Hasegawa G, et al. Development, differentiation, and phenotypic heterogeneity of murine tissue macrophages. *J Leukoc Biol* (1996) **59**:133–8.

95. Sorokin SP, McNelly NA, Blunt DG, Hoyt RF Jr. Macrophage development: III. Transformation of pulmonary macrophages from precursors in fetal lungs and their later maturation in organ culture. *Anat Rec* (1992) **232**:551–71. doi:10.1002/ar.1092320411

96. Hopkinson-Woolley J, Hughes D, Gordon S, Martin P. Macrophage recruitment during limb development and wound healing in the embryonic and foetal mouse. *J Cell Sci* (1994) **107**(Pt 5):1159–67.

97. Wood W, Turmaine M, Weber R, Camp V, Maki RA, McKercher SR, et al. Mesenchymal cells engulf and clear apoptotic footplate cells in macrophage-less PU.1 null mouse embryos. *Development* (2000) **127**:5245–52.

98. Squarzoni P, Oller G, Hoeffel G, Pont-Lezica L, Rostaing P, Low D, et al. Microglia modulate wiring of the embryonic forebrain. *Cell Rep* (2014) **8**:1271–9. doi:10.1016/j.celrep.2014.07.042

99. Epelman S, Lavine KJ, Beaudin AE, Sojka DK, Carrero JA, Calderon B, et al. Embryonic and adult-derived resident cardiac macrophages are maintained through distinct mechanisms at steady state and during inflammation. *Immunity* (2014) **40**:91–104. doi:10.1016/j.immuni.2013.11.019

100. Naito M, Hasegawa G, Takahashi K. Development, differentiation, and maturation of Kupffer cells. *Microsc Res Tech* (1997) **39**:350–64. doi:10.1002/(SICI)1097-0029(19971115)39:4<350::AID-JEMT5>3.3.CO;2-V

101. Schneider C, Nobs SP, Kurrer M, Rehrauer H, Thiele C, Kopf M. Induction of the nuclear receptor PPAR-gamma by the cytokine GM-CSF is critical for the differentiation of fetal monocytes into alveolar macrophages. *Nat Immunol* (2014) **15**:1026–37. doi:10.1038/ni.3005

102. Daneman R, Zhou L, Kebede AA, Barres BA. Pericytes are required for blood-brain barrier integrity during embryogenesis. *Nature* (2010) **468**:562–6. doi:10.1038/nature09513

103. Robbins CS, Hilgendorf I, Weber GF, Theurl I, Iwamoto Y, Figueiredo JL, et al. Local proliferation dominates lesional macrophage accumulation in atherosclerosis. *Nat Med* (2013) **19**:1166–72. doi:10.1038/nm.3258

104. van de Pavert SA, Mebius RE. New insights into the development of lymphoid tissues. *Nat Rev Immunol* (2010) **10**:664–74. doi:10.1038/nri2832

105. Randall TD, Carragher DM, Rangel-Moreno J. Development of secondary lymphoid organs. *Annu Rev Immunol* (2008) **26**:627–50. doi:10.1146/annurev.immunol.26.021607.090257

106. Kubota Y, Takubo K, Shimizu T, Ohno H, Kishi K, Shibuya M, et al. M-CSF inhibition selectively targets pathological angiogenesis and lymphangiogenesis. *J Exp Med* (2009) **206**:1089–102. doi:10.1084/jem.20081605

107. Gordon EJ, Rao S, Pollard JW, Nutt SL, Lang RA, Harvey NL. Macrophages define dermal lymphatic vessel calibre during development by regulating lymphatic endothelial cell proliferation. *Development* (2010) **137**:3899–910. doi:10.1242/dev.050021

108. Gray EE, Cyster JG. Lymph node macrophages. *J Innate Immun* (2012) **4**:424–36. doi:10.1159/000337007

109. Haldar M, Kohyama M, So AY, Kc W, Wu X, Briseno CG, et al. Heme-mediated SPI-C induction promotes monocyte differentiation into iron-recycling macrophages. *Cell* (2014) **156**:1223–34. doi:10.1016/j.cell.2014.01.069

110. A-Gonzalez N, Guillen JA, Gallardo G, Diaz M, de la Rosa JV, Hernandez IH, et al. The nuclear receptor LXRalpha controls the functional specialization of splenic macrophages. *Nat Immunol* (2013) **14**:831–9. doi:10.1038/ni.2622

111. Lavin Y, Winter D, Blecher-Gonen R, David E, Keren-Shaul H, Merad M, et al. Tissue-resident macrophage enhancer landscapes are shaped by the local microenvironment. *Cell* (2014) **159**:1312–26. doi:10.1016/j.cell.2014.11.018

112. Kierdorf K, Erny D, Goldmann T, Sander V, Schulz C, Perdiguero EG, et al. Microglia emerge from erythromyeloid precursors via Pu.1- and Irf8-dependent pathways. *Nat Neurosci* (2013) **16**:273–80. doi:10.1038/nn.3318

113. McGrath KE, Kingsley PD, Koniski AD, Porter RL, Bushnell TP, Palis J. Enucleation of primitive erythroid cells generates a transient population of "pyrenocytes" in the mammalian fetus. *Blood* (2008) **111**:2409–17. doi:10.1182/blood-2007-08-107581

114. Clarke D, Vegiopoulos A, Crawford A, Mucenski M, Bonifer C, Frampton J. In vitro differentiation of c-myb(-/-) ES cells reveals that the colony forming capacity of unilineage macrophage precursors and myeloid progenitor commitment are c-Myb independent. *Oncogene* (2000) **19**:3343–51. doi:10.1038/sj.onc.1203661

115. Sumner R, Crawford A, Mucenski M, Frampton J. Initiation of adult myelopoiesis can occur in the absence of c-Myb whereas subsequent development is strictly dependent on the transcription factor. *Oncogene* (2000) **19**:3335–42. doi:10.1038/sj.onc.1203660

116. Mucenski ML, McLain K, Kier AB, Swerdlow SH, Schreiner CM, Miller TA, et al. A functional c-myb gene is required for normal murine fetal hepatic hematopoiesis. *Cell* (1991) **65**:677–89. doi:10.1016/0092-8674(91)90099-K

7

From the reticuloendothelial to mononuclear phagocyte system – the unaccounted years

Simon Yona[1]* and Siamon Gordon[2]

[1] University College London, London, UK, [2] Sir William Dunn School of Pathology, The University of Oxford, Oxford, UK

Edited by:
Peter M. Van Endert,
Université Paris Descartes, France

Reviewed by:
Edda Fiebiger,
Harvard Medical School, USA
Richard A. Kroczek,
Robert Koch-Institute, Germany
Jean M. Davoust,
Institut National de la Santé et la
Recherche Médicale, France

*Correspondence:
Simon Yona,
University College London, 5
University Street,
London WC1E 6JF, UK
s.yona@ucl.ac.uk

It is over 125 years since Ilya Metchnikoff described the significance of phagocytosis. In this review, we examine the early origins and development of macrophage research continuing after his death in 1916, through the period of the reticuloendothelial system. Studies on these cells resulted in a substantial literature spanning immunology, hematology, biochemistry, and pathology. Early histological studies on morphology and in situ labeling laid the foundations to appreciate the diversity and functional capacity of these cells in the steady state and during pathology. We complete this phagocyte retrospective with the establishment of the mononuclear phagocyte system nomenclature half a century ago.

Keywords: macrophage, monocyte, Metchnikoff, phagocytosis, dendritic cells, inflammation

Introduction

The earliest accounts of macrophage research are closely linked with the widespread introduction of the microscope in the mid-nineteenth century, 300 years following the seminal microscopic observations of Antony van Leeuwenhoek (1700) (1). In the histological accounts, von Kölliker (1847) detected cells in the spleen containing particles; later Preyer (1867) observed the internalization of erythrocytes by splenic cells and proposed that this occurred by an active process (2, 3). However, investigators at the time did not associate such observations with a defense mechanism. In fact, Klebs (1872) believed just the opposite, proposing that these cells assist the transport of bacteria to lymphatic tissue (4). Koch (1878) also concluded that these cells provide a suitable microenvironment for bacilli to multiply and disseminate to other tissues, – the so-called Trojan horse theory – after observing numerous bacilli within leukocytes, while studying frogs treated with anthrax (5). Therefore, although cytological observations of the mid-nineteenth century recognized the ability of leukocytes to devour (*fressen*) erythrocytes and microorganisms, opinion at the time did not associate this event with host defense, nor was there a consensus that the process was active or merely the penetration of foreign material into cells to aid infection.

By the late nineteenth century, Metchnikoff (1892), the Russian zoologist and forefather of cellular immunity, established the idea of the phagocyte (6–8). Metchnikoff was the first to fully appreciate the capabilities and purpose of phagocytosis, by performing a series of classical studies spanning from simple unicellular organisms to complex vertebrates. The description of Metchnikoff's discovery of phagocytosis documented by his wife Olga, now rests in the pantheon of immunology legend.

> ...One day when the family had gone to see some performing apes at the circus, Metchnikoff with his microscope introduced a rose thorn into the transparent body of a starfish larva, Metchnikoff observed the accumulation of phagocytes surrounding the foreign material and attempting to devour the splinter... (9).

It is important to remember that Metchnikoff started his career as an evolutionary developmental embryologist, influenced by Darwin's recent publication On the Origin of Species in 1859. As an embryologist, Metchnikoff modeled the early formation of the embryo in primitive organisms, such as sponges, and proposed that an inner *"parenchymella"* contained wandering cells of mesodermal origin capable of taking up particulate matter during embryogenesis. These studies may have been the foundation for Metchnikoff's phagocytosis theory. Later, Metchnikoff recognized the multiple tasks performed by phagocytosis; as an embryologist, the reabsorption of tissue during embryogenesis, as a zoologist, a common feeding mechanism of unicellular organisms and as a pathologist, its role in host defense. Therefore, when Metchnikoff performed his most notable study, the rose thorn experiment at Messina culminated in the phagocytic process we know today. Metchnikoff was one of, if not the, earliest to demonstrate the evolutionary functional adaptation of a particular biological process, in this case phagocytosis, from a simple feeding mechanism for unicellular organisms, to a developmental requirement during embryogenesis and finally as a necessity for host defense (3, 10).

Metchnikoff's phagocytes comprised two populations he termed macrophages (large eaters) and microphages (small eaters, later known as polymorphonuclear leukocytes). Contrary to Rudolf Virchow's impression that inflammation is a continuous life threatening menace, Metchnikoff regarded it as a healing or salutary reaction as postulated 100 years earlier by the Scottish surgeon and collector John Hunter (1794) (11). Therefore, Metchnikoff concluded that the ability of cells to engulf foreign microorganisms acts as an active defense mechanism, giving rise to the concept of cellular innate immunity. At the time, this triggered extensive debate between humoral and cellular schools of thought. Two major events at the turn of the twentieth century helped to reconcile this dispute. First, in 1908, the Nobel Prize in Physiology or Medicine was awarded jointly to Metchnikoff, advocate of the cellularists and to Ehrlich, the champion of humoralist dogmas *"in recognition of their work in immunity"*. Second, in 1903, Wright and Douglas proposed the concept of "opsonization" as a humoral mechanism to increase the susceptibility of bacteria to phagocytosis. These investigators claimed that humoral and cellular functions were not mutually exclusive, rather interdependent (12). This theory was spoofed by George Bernard Shaw, in the introduction to his play "The Doctor's Dilemma" in 1906.

> . . . Sir Almroth Wright, following up one of Metchnikoff's most suggestive biological romances, discovered that the white corpuscles or phagocytes, which attack and devour disease germs for us, do their work only when we butter the disease germs appetizingly for them with natural sauce which Sir Almroth Wright named opsonin. . . (13).

The Reticuloendothelial System

By the early decades of the twentieth century descriptions of the phagocyte system had become chaotic, not least since the term macrophage had become synonymous with adventitia cell, anode cell, clasmatocyte, dictocyte, erythrophagocyte, histiocyte, polyblast, pyrrhol cell, and rhagiocrine cell; the many terms bestowed on these cells (>30 different names) (14) revealed that the divergence of opinion at the time as to the relationship of these cells to one another and from tissue-to-tissue. Not only were tissue phagocytes given a variety of bewildering names but also their origin remained unknown. From time-to-time, historic discoveries are lost in the ether of *a priori* thought; this is certainly true for histological techniques that assisted in the classification of blood cytology. Until Ehrlich's early effort to develop leukocyte cytological staining, scholars of blood operated solely on fresh samples. Ehrlich's aim was to take advantage of the known chemical structures of dyes and their interaction with cellular bodies to map and characterize the anatomy of blood cells. By using aniline dyes in combination with neutral dyes and the morphology of the nucleus, he was able to divide cells of the blood into mononucleated lymphocytes, some of which were large, large mononuclears with indented nuclei (now known as monocytes) and polymorphonuclear cells that were neutrophilic (neutrophils), acidophilic (eosinophils), or basophilic granules (basophils/mast cells) (15). By the early twentieth century, Ribbert (1904) had performed studies with lithium carmine solution injected into the peripheral circulation and observed the specific uptake and storage by a group of cells, which became *vitally stained* (16). These were subsequently demonstrated to be mononuclear cells phagocytosing particulate matter. Clark and Clark (17) described these large mononuclear cells in tissues to be the same as "clasmatocytes," described by Louis-Antoine Ranvier, the "Polyblasts" of Alexander Maximow and the "Histiocytes" defined by Kenji Kiyono. Following these early observations, it became apparent that a large number of histological dyes including trypan blue, neutral red, isamine blue, and other colloids discriminated phagocytes from fibroblasts. The systematic analysis of tissues and dyes led Karl Albert Ludwig Aschoff (1924) to coin the term "reticuloendothelial system" (RES) to describe this group of cells, with their ability to incorporate vital dyes from the circulation (18). Reticulo refers to the propensity of these large phagocytic cells in various organs to form a network or a *reticulum* by cytoplasmic extensions; *endothelial* refers to their proximity to the vascular endothelium (19), from which they were sometimes believed to arise, these cells formed Aschoff's unified system throughout the organism. The capture and clearance of unwanted particulate material from blood and lymph were considered to be the major function of the RES. Although opinions about the origin of cells of the RES will be discussed later in this series; at the time, Aschoff considered that the cells of the RES were derived locally and that both histiocytes and reticulum cells shared a common origin.

Cells of the Reticuloendothelial System

Metchnikoff had previously described the dissemination of macrophages throughout the organism and Aschoff's system implied a common function of the cells of the RES even in the absence of inflammation. In the next section, we highlight some of the tissue locations and possible functions assigned at the time.

Kupffer Cells

The macrophages of the liver are located within the sinusoids, which is composed of four cell types, each with its own morphology and function. Karl Wilhelm von Kupffer (1876) observed "*Sternzellen*" (star cells) in the liver and believed them to be an integral part of the hepatic endothelium. Later, Tadeusz Browicz (1899) identified Kupffer's cells as the phagocytes of the liver (20) (sometimes known as Browicz–Kupffer cells) and observed that they could take-up a large percentage of vital stain. In the early 1930s, Peyton Rous developed an ingenious method to isolate Kupffer cells of the liver. Rous and Beard injected a suspension of gamma ferrous oxide i.v., light in weight but highly magnetic particles, Kupffer cells efficiently phagocytosed these particles. They then perfused and processed the liver and the phagocytes were then selected by magnetic force, to the best of our knowledge the first description of magnetic cell sorting (21, 22), enabling the extraction of macrophages from a solid tissue for examination *in vitro*. The origin of Kupffer cells like all cells of the RES at the time remained a source of continued confusion and debate. At the American Association of Anatomists conference in 1925, M. R. Lewis presented a paper comparing Kupffer cells isolated from frogs, thought to be derived from endothelial cells, side-by-side with an examination of clasmatocytes and concluded these cells were identical in morphology and function (23).

In 1950s, Baruch Benacerraf, Nobel laureate in 1980 for his work on MHC with George Snell and Jean Dausset (24), teamed up with Guido Biozzi, a young Italian in the Halpern laboratory in Paris in a productive collaboration. They developed techniques to study clearance of particulates from blood and formulated equations that govern this in mammals. In subsequent work, Biozzi bred strains of mice differing in the quantitative antibody response to various antigens. Biozzi mice are still in use to study autoimmune inflammatory neurological disease (25). These studies in mice and guinea pigs helped to introduce genetic approaches to the role of macrophages in innate and adaptive immunity.

Microglia and the Origin of the RES

Virchow (1858) acknowledged that the central nervous system (CNS) was composed of both neurones and interstitial cells, which he termed neuroglia (26, 27). By the end of the nineteenth century, the Scottish pathologist William Ford Robertson confirmed that neuroglia were indeed composed of multiple cell types (28). Robertson continued to investigate this heterogeneous population of cells; with the aid of platinum staining techniques he was able to distinguish a novel cell type in the brain he termed mesoglia (as he believed that they were mesodermal in origin). Finally, Robertson deduced that mesoglia possessed phagocytic properties (29). In fact, Robertson had identified oligodendroglia and mesodermal derived cells, under the term mesoglia. In 1913, Santiago Ramon y Cajal described a group of cells derived from the mesoderm as the "*third element*" of the CNS, the first element being neurones and the second element the astrocytes, derived from ectoderm. It was the Spanish pathologist Pio Del Rio-Hortega who revolutionized our understanding of neuroglia from a series of detailed studies using silver carbonate impregnation staining. He uncovered a homogeneous group of cells within the CNS with tree like projections and predicted that they possess phagocytic functions within the CNS; he termed these cells as microglia (26, 27). Hortega laid the groundwork for microglia research in a lecture given at University of Oxford, microglia enter the CNS during development from mesodermal origin where they disseminate throughout the CNS and take-up a branched ramified cytological appearance. He went on to explain that they remain evenly spaced in the steady state, while pathological insults cause microglia to take on an amoeboid morphology, express the ability to phagocytose and to migrate (30). These studies confirmed that the microglia of the CNS belonged to the RES. The account he gave in Oxford remains accurate to this day. Interestingly, although microglia were unable to take-up vital dyes because of the blood brain barrier, they were known to be highly phagocytic, reticuloendothelial cells readily stained by silver carbonate. Although Hortega described microglia to be derived from cells of the mesoderm during embryogenesis, this still remained a matter of great debate, until recently. Early observations in the late nineteenth and several studies in the early twentieth described microglia during neurodegenerative diseases without a clear understanding of their origin.

Osteoclasts

The histological identification of a cell that resorbs bone can be traced back to the early 1850s. Tomes and De Morgan (31) described within a section of diseased femur, cavities that were associated with an increase in nucleated cells (31). By 1873, Kölliker described multinucleated giant cells involved in bone absorption that he termed *Ostoklast* and anticipated that these cells are involved in homeostatic and pathological bone degradation (32, 33). The notion of a bone-resorbing cell became widely accepted (34). Over the next 50 years, the morphology of the osteoclast was refined and interestingly these large multinucleated cells showed variation in size and nuclear content; in pigs, they contained as many as 100 nuclei (35) while human osteoclasts could contain up to 50 nuclei (32). John Loutit an Australian born pioneer in radiation biology from the late 1940s studied not only osteoclast origin from blood precursors but also the biology of bone marrow transplantation after irradiation (36) in a long and productive career at Harwell MRC laboratories.

Alveolar Macrophages and Phagocytosis

The lung also contains many mononuclear phagocytes, which are associated with the alveoli and the alveolar space (37). The macrophages within the alveolar space were initially known as "dust cells" because of their content of intracellular carbon particles. There is a constant requirement to keep the alveolar space free of particles and pathogens allowing for optimal oxygen transfer, the major role of these cells. As the lung occupies a unique accessible position among internal organs, it is constantly in contact with the external world. In the late 1950s, Karrer observed the efficient phagocytosis of India ink exclusively by free alveolar macrophages, similar to the previous observations of increased carbon particles in these cells of city dwellers (38, 39). The question of the origin of the macrophage troubled cytologists and immunologists for most of the twentieth century; this was no different in the lung.

One of the best-studied pathologies in the first half of the twentieth century in relation to macrophages was pulmonary tuberculosis. The initial stage of tuberculosis displays the transient influx of neutrophils described by Maximow in 1925 (40). However, these cells are unable to destroy the bacilli and monocyte/macrophages remain the most prominent infected host cells. From 1920s, Sabin, the first full female Member of the Rockefeller Institute and first female elected to the National Academy of Sciences, considered the monocyte response to tuberculosis the most significant process, "*cellular and immunological reactions in tuberculosis center around the monocyte,*" when she first proposed this she was mocked by her peers. Sabin observed monocytes to become epithelioid cells that develop into macrophage giant cells (41), previously described by the German pathologist Theodor Langhans as a hallmark of tuberculous granulomata already in the nineteenth century. In 1930s, Max Lurie, an advocate of the monocyte theory, used inbred rabbits to study their susceptibility to bovine tuberculosis. Lurie observed resistant inbred rabbits went on to develop cavitary tuberculosis while susceptible families went on to develop disseminated tuberculosis (42). The Australian immunologist, George Mackaness studied the role of anti-TB drugs on infected macrophages when a student at the Sir William Dunn School of Pathology, University of Oxford with Howard Florey in the early 1950s. His subsequent studies in the sixties at the Trudeau Institute in Saranac Lake defined T lymphocyte-dependent activation of macrophages by BCG and *Listeria* infection (43, 44), Mackaness coined the term macrophage activation, the so-called "angry" macrophages (45). Dannenberg has been another pioneer of macrophage research in experimental and clinical tuberculosis, especially in the characterization of the granuloma (46).

Other resident phagocytic populations were described in many tissues during this period, for example, in the skin (Langerhans cells), gut, lympho-hemopoietic tissues, reproductive and endocrine organs, and placenta (Hoffbauer cells). We draw attention to specialized macrophages in bone marrow stroma, where they appear at the center of hematopoietic islands, first described in detail by Marcel Bessis and his collaborators (47). John Humphrey drew attention to the marginal metallophilic macrophages located in a zone between the red and white pulp of spleen, especially in rodents (48, 49). They line a sinusoidal space where they sample circulating blood for viruses, for example, and play an important role in clearance of T cell-independent immunogenic polysaccharides. Tingible body macrophages (TBM) were identified by Walther Flemming in 1885; located in germinal centers. TBM contain phagocytosed apoptotic cell debris (tingible bodies) and are involved in the clearance of apoptotic lymphocytes (50), these observations were confirmed by electron microscopy in the early 1960s (51). Finally, the peritoneal macrophages of the mouse, responsible for much of our knowledge of macrophage immunobiology, were first described as a tractable cell population by Cohn only in 1962 (52).

The Origin of Macrophages

As Aschoff was formulating the requirements of the RES, a number of research groups were searching for the origin of macrophages. In 1914, Awrorow and Timofejewskij concluded from the outgrowth of leukocytes from leukemic blood that the lymphocyte is the progenitor from which macrophages arise (53). A few years later, several *in vitro* studies described the differentiation of blood monocytes into macrophages (53–55); Carrel and Ebeling (55) and Lewis and Lewis (23) observed that blood cultures over time developed into macrophages that had phagocytosed the debris of other blood cells, concluding that these monocyte-derived cells became actively phagocytic and were indistinguishable from macrophages in staining with neutral red (54, 55). At the same time, in 1925, Sabin took a cytological approach using neutral red staining to examine resident macrophage populations in connective tissue (clasmatocytes), concluding that a proportion of macrophages were derived from bone marrow-derived monocytes (56). However, the first *in vivo* study to examine how mammalian blood monocytes behave during an acute pathological insult was performed by Ebert and Florey (57) at the University of Oxford, using the rabbit ear chamber, observing diapedesis of blood monocytes toward the site of tissue injury. These monocytes transformed into macrophages during the inflammatory process; they concluded "*The cells originating from monocytes eventually became cells which we are calling histiocytes, which are indistinguishable from the so-called resting wandering tissue-cell of Maximow*" (57). Twenty-five years later, Volkman and Gowans (58) confirmed these findings using thymidine autoradiography and parabiosis inferring that macrophage precursors are rapidly dividing cells derived from a remote site during inflammation (58). Takahashi mentions in his comprehensive review on macrophage ontology how the Japanese pathologist Amano with the aid of supravital staining at the inflammatory foci observed blood monocytes to be precursors of the macrophage (59). Finally, Marchesi and Florey employing electron microscopy were able to distinguish the earliest phase of monocyte migration during mild inflammation, which occurred during the maximal efflux of neutrophils (14, 60). The conclusion from these studies suggests that macrophages derived from circulating monocytes were based on inflammatory models. Therefore, a more accurate conclusion would be that during inflammation monocytes become effector cells by concentrating at the site of injury with the ability to produce large quantities of inflammatory mediators (61).

Another important and well-studied population of recruited monocyte-derived cells are the foam cells, a hallmark of atheromatous pathology. A pupil of Maximow in St. Petersburg and later the student of Aschoff in Freiburg (62), Anitschkow (63, 64) showed that simply by feeding rabbits purified cholesterol caused vascular changes leading to the formation of lesions similar to those seen during atherosclerosis in humans (63, 64). Anitschkow decided that these cholesterol-laden cells were of leukocyte origin (65). Anitschkow's work on lipid storage was compared to the work by Robert Koch on the tubercle bacillus (66). It was mainly as a result of the work by Russell Ross in 1970s that these monocyte-derived cells have been categorized as part of a fat modified inflammatory process (67). Recruited monocytes can also give rise to multinucleated giant cells, not only a feature of tuberculosis. They are found in many granulomatous inflammatory diseases, including viral and parasitic infections, and in responses to foreign bodies and fat necrosis (Touton cells), named after the German dermatologist Karl Touton (1885) (68).

The accumulation of data and the introduction of new techniques highlighted that the cells of the RES differ in morphology, function, and origin (14). In addition, the underlying processes involved in these functional and morphologic alterations remained unknown. Is there a proliferating mononuclear phagocyte population within the RES, constantly differentiating in the steady state? These questions continued to puzzle scientists throughout the twentieth century.

The Mononuclear Phagocyte System

> ... The most immature cell of the mononuclear phagocyte system ... is the promonocyte ... that by dividing gives rise to monocytes ... Monocytes in the circulation constitute a mobile pool of relatively immature cells on their way from the place of origin to the tissues. At sites where conditions are favourable for phagocytosis, these cells become macrophages... (69).

As knowledge accumulated, the term RES was regarded as insufficient to describe resident phagocytes and their antecedents. At a scientific meeting in Leiden in 1969, a group of prominent pathologists/immunologists proposed the term "mononuclear phagocyte system" (MPS) as a more accurate term (**Figure 1**) (69). The MPS at the time comprised monocytes and macrophages derived from the bone marrow derived monocytes. Nevertheless, little evidence existed to suggest that monocytes differentiate into resident macrophages under steady state conditions. Interestingly, Maximow proposed on the basis of embryonic studies on amphibians and mammals that macrophages and leukocytes may arise from distinct lineages (70).

While the MPS was being formulated in 1960s, immunologists were in pursuit of the "*third cell*" (71) a requirement for adaptive-immune responses. Steinman and Cohn in their landmark study (1973) identified and characterized the dendritic cell (DC) as distinct from macrophages (72, 73); over time, the DC became accepted as the third arm in the trinity we know today as the

MPS (74). Monocytes, macrophages, and DCs are distinguished on the basis of morphology, function, and origin, yet collectively constitute the MPS.

As more data accumulated in the early decades of the twenty-first century, it emerged that most tissue macrophage populations in adults in the steady state are maintained independent of the bone marrow and rely almost exclusively on self-renewal (75, 59, 76–86). These data facilitated the reexamination of the concept of the MPS (87).

Conclusion

We have highlighted only a few of the many milestones of macrophage biology from its early origins to the establishment of the MPS nomenclature in 1968. Studies during this period resulted in a substantial literature spanning immunology, hematology, and pathology. A number of important issues emerge from a retrospective analysis of the literature. First, we learn that rarely in science do revolutions occur from a single Eureka moment rather years of observation culminate in new findings. While Metchnikoff's phagocytosis theory seems to have emerged from his experiments on starfish larvae, he had previously observed cells capable of taking up particulate matter during embryogenesis (10). Second, the first half of the twentieth century, blighted by two World Wars, had profound impacts on science, resulting in a geographical and common language shift of scientific research. Third, the techniques used routinely in the laboratory shifted from the pathologists' tool box of the microscope and microbiology to the immunologists' introduction of cell transfer, thymidine autoradiography, immunohistochemistry, parabiosis, electron microscopy, later flow cytometry, cell, and molecular biology. However, if one was able to transport Metchnikoff, Aschoff, or Cohn to a conference in 2015 on mononuclear phagocytes they would perhaps not appreciate the cytokines, chemokines, blots, and plots; however, the fundamental questions and discussions remain familiar; what is the origin of these cells? How do they phagocytose? Do macrophages proliferate *in situ*? How is particulate material recognized and cleared? This is why it is important to examine the history of our field since our research questions today are more closely linked to the past than we may appreciate.

The macrophage story is not over. In recent months, further strides have been made in examining the molecular signatures, characterizing the MPS in the steady state and upon enhanced recruitment of monocytes during inflammation (88–94). These studies highlight collective attributes of the macrophages; however, they also show significant local adaptations associated with particular functions within a specific organ. The next stage on this journey will include recreating *in vitro* the phenotypes of these specific populations using induced pluripotent stem cells, and gaining a greater insight into how these cells behave under steady state conditions *in situ*, as well as during and after the inflammatory response. Finally, the role of the circulating monocyte is also undergoing a re-evaluation; previously, monocytes were viewed as the bridge from bone marrow progenitors to fully differentiated tissue macrophages not only after inflammation, injury, and infection but also for resident populations in the

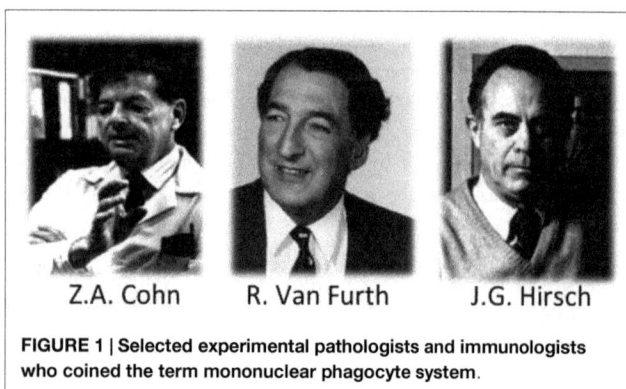

Z.A. Cohn   R. Van Furth   J.G. Hirsch

FIGURE 1 | Selected experimental pathologists and immunologists who coined the term mononuclear phagocyte system.

absence of inflammation, as stated in Van Furth's description of the MPS "*Monocytes in the circulation constitute a mobile pool of relatively immature cells on their way from the place of origin to the tissues*" (69). Moreover, monocytes should now be further investigated as distinct precursors of only newly recruited monocyte-derived cells and as effector cells in their own right.

References

A number of the citations in this review are historical and may be unavailable to readers. If you are interested in obtaining a copy of a particular publication please contact the corresponding author. In addition, please be aware the authors have cited each paper as it appeared at the time of publication, many journal names have changed overtime papers are cited as they appeared at the time of print.

1. Wootton D. *Bad Medicine: Doctors Doing Harm Since Hippocrates*. Oxford: Oxford University Press (2006).
2. Preyer W. Ueber amöboide blutkörperchen. *Arch Pathol Anat Physiol Klin Med* (1867) **30**:417–41.
3. Tauber AI, Chernyak L. *Metchnikoff and the Origins of Immunology: From Metaphor to Theory*. Oxford: Oxford University Press (1991).
4. Klebs E. *Beiträge zur Pathologischen Anatomie der Schusswunden*. Leipzig: von F.C.W. Vogel (1872).
5. Koch R. *Untersuchungen Über Die Aetiologie der Wundinfektionskrankheiten*. Leipzig: von F.C.W. Vogel (1878).
6. Metchnikoff E. *Leçons sur la Pathologie Comparée de L'inflammation Faites à l'Institut Pasteur en Avril et mai 1891*. Paris: G. Masson Libraire de l'academie de Medecine (1892).
7. Metchnikoff E. Ueber den kampf der zellen gegen erysipel-kokken, ein beitrag zur phagocytenlehre. *Arch Pathol Anat Physiol Klin Med* (1887) **107**: 209–49.
8. Gordon S. Elie Metchnikoff: father of natural immunity. *Eur J Immunol* (2008) **38**:3257–64. doi:10.1002/eji.200838855
9. Metchnikoff O. *Life of Elie Metchnikoff 1845-1916*. Boston, NY: Houghton Mifflin (1921).
10. Tauber AI. Ilya Metchnikoff: from evolutionist to immunologist, and back again. In: Harman O, Dietrich MR, editors. *Outsider Scientists Routes to Innovation in Biology*. Chicago, IL: The University of Chicago Press (2013). p. 259–74.
11. Hunter J. *A Treatise on the Blood, Inflammation and Gunshot Wounds*. London: George Nicol (1794).
12. Wright AE, Douglas SR. An experimental investigation of the role of the blood fluids in connection with phagocytosis. *Proc R Soc Lond* (1903) **72**:357–70. doi:10.1098/rspl.1903.0062
13. Shaw GB. *The Doctor's Dilemma (play)* (1926). London: Constable.
14. Gall E. The cytological identity and interrelation of mesenchymal cells of lymphoid tissue. *Ann N Y Acad Sci* (1958) **73**:120–30. doi:10.1111/j.1749-6632.1959.tb40796.x
15. Ehrlich P. Methodologische beiträge zur physiologie und pathologie der verschiedenen formen der leukocyten. *Z Klin Med* (1880) **1**:553–60.
16. Florey HW. *General Pathology*. London: Lloyd-Luke (Medical Books) (1970).
17. Clark ER, Clark EL. Relation of monocytes of the blood to the tissue macrophages. *Am J Anat* (1930) **46**:149–85. doi:10.1002/aja.1000460105
18. Aschoff L. Das reticulo-endotheliale system. *Ergeb Inn Med Kinderheilkd* (1924) **26**:1–118.
19. Wintrobe MW. *Blood, Pure and Eloquent*. New York: McGraw-Hill Book Company (1980).
20. Sródka A, Gryglewski RW, Szczepariski W. Browicz or Kupffer cells? *Pol J Pathol* (2006) **57**:183–5.
21. Rous P, Beard JW. Selection with the magnet and cultivation of "reticuloendothelial" cells. *Science* (1933) **77**:92. doi:10.1126/science.77.1986.92
22. Rous P, Beard JW. Selection with the magnet and cultivation of reticuloendothelial cells (Kupffer cells). *J Exp Med* (1934) **59**:577–91. doi:10.1084/jem.59.5.577
23. Lewis MR, Lewis WH. Monocytes, macrophages, epithelioid cells, and giant cells. In: Sabin FR, Mayer AW, editors. *American Association of Anatomists*. Cleveland, OH: Western Reserve University (1925). p. 391.
24. Odelberg W. Baruch benacerraf les prix nobel. *Les Prix Nobel. The Nobel Prizes 1980*. Stockholm: Almqvist & Wiksell (1981).
25. Amor S, Smith PA, Hart B, Baker D. Biozzi mice: of mice and human neurological diseases. *J Neuroimmunol* (2005) **165**:1–10. doi:10.1016/j.jneuroim.2005.04.010
26. Rezaie P, Male D. Mesoglia & microglia – a historical review of the concept of mononuclear phagocytes within the central nervous system. *J Hist Neurosci* (2002) **11**:325–74. doi:10.1076/jhin.11.4.325.8531
27. Kettenmann H, Hanisch UK, Noda M, Verkhratsky A. Physiology of microglia. *Physiol Rev* (2011) **91**:461–553. doi:10.1152/physrev.00011.2010
28. Robertson WF. The normal histology and pathology of neuroglia. *J Ment Sci* (1897) **43**:733–52.
29. Robertson WF. *A Text-Book of Pathology in Relation to Mental Diseases*. Edinburgh: William F Clay (1900).
30. del Rio-Hortega P. The microglia. *Lancet* (1939) **233**:1023–6. doi:10.1016/S0140-6736(00)60571-8
31. Tomes TW, De Morgan C. Observations on the structure and development of bone. *Philos Trans* (1853) **143**:109–39. doi:10.1098/rstl.1853.0004
32. Kölliker A. *Die Normale Resorption des Knochengewebes und Ihre Bedeutung für Die Entstehung der Typischen Knochenformen*. Leipzig: von F.C.W. Vogel (1873).
33. Hancox NM. The osteoclast. *Biol Rev Camb Philos Soc* (1949) **24**:448–71. doi:10.1111/j.1469-185X.1949.tb00583.x
34. Wegner G. Myeloplaxen und knochenresorption. *Arch Pathol Anat Physiol Klin Med* (1872) **56**:523–33.
35. Arey LB. The origin, growth and fate of osteoclasts and their relation to bone resorption. *Am J Anat* (1920) **26**:314–45. doi:10.1002/aja.1000260302
36. Lyon M, Mollison PL. John Freeman Loutit 19 February 1910-11 June 1992. *Biogr Mems Fell R Soc* (1994) **40**:237–52. doi:10.1098/rsbm.1994.0037
37. Lang FJ. The reaction of lung tissue to tuberculous infection in vitro – two plates. *J Infect Dis* (1925) **37**:430–42. doi:10.1093/infdis/37.5.431
38. Robertson OH. Phagocytosis of foreign material in the lung. *Physiol Rev* (1941) **21**:112–39.
39. Karrer HE. The ultrastructure of mouse lung: the alveolar macrophage. *J Biophys Biochem Cytol* (1958) **4**:693–700. doi:10.1083/jcb.4.6.693
40. Maximow A. Role of the nongranular blood leukocytes in the formation of the tubercle. *J Infect Dis* (1925) **37**:418–29. doi:10.1093/infdis/37.5.418
41. McMaster PD, Heidelberger M. Florence Rena Sabin. *Natl Acad Sci Biograph Memoirs* (1960) **34**:270–319.
42. Lurie MB. Heredity, constitution and tuberculosis, an experimental study. *Am Rev Tuberc Suppl* (1941) **44**:1–125.
43. Mackaness GB. The growth of tubercle bacilli in monocytes from normal and vaccinated rabbits. *Am Rev Tuberc* (1954) **69**:495–504.
44. Miki K, Mackaness GB. The passive transfer of acquired resistance to *Listeria monocytogenes*. *J Exp Med* (1964) **120**:93–103. doi:10.1084/jem.120.1.93
45. North RJ. George B. Mackaness, M.D., D.Phil., F.R.S. *Tuberculosis* (2007) **87**:391. doi:10.1016/j.tube.2007.04.001
46. Dannenberg AM. *Pathogenesis of Human Pulmonary Tuberculosis: Insights from the Rabbit Model*. Washington, DC: ASM Press (2006).
47. Manwani D, Bieker JJ. The erythroblastic island. *Curr Top Dev Biol* (2008) **82**:23–53. doi:10.1016/S0070-2153(07)00002-6
48. Humphrey JL. Marginal zone and marginal sinus macrophages in the mouse are distinct populations. *Adv Exp Med Biol* (1979) **114**:381–8.
49. Humphrey JH. Splenic macrophages: antigen presenting cells for T1-2 antigens. *Immunol Lett* (1985) **11**:149–52. doi:10.1016/0165-2478(85)90161-0
50. Flemming W. Studien über regeneration der gewebe. *Arch mikroskopische Anat* (1885) **24**:50–91. doi:10.1007/BF02960374

Acknowledgments

We have cited only a fraction of investigations during this period. We dedicate this review to all the scientists who have contributed to the field especially E. Metchnikoff and Z. A. Cohn, whose contributions continue to impact research today.

51. Swartzendruber DC, Congdon CC. Electron microscope observations on tingible body macrophages in mouse spleen. *J Cell Biol* (1963) **19**:641–6. doi:10.1083/jcb.19.3.641

52. Cohn ZA. Determinants of infection in the peritoneal cavity. II. Factors influencing the fate of *Staphylococcus aureus* in the mouse. *Yale J Biol Med* (1962) **35**:29–47.

53. Awrorow PP, Timofejewskij AD. Kultivierungsversuche von leukämischem blute. *Virchows Arch A Pathol Pathol Anat* (1914) **216**:184–214.

54. Lewis MR, Lewis WH. The transformation of white blood cells – into clasmatocytes (macrophages), epithelioid cells, and giant cells. *J Am Med Assoc* (1925) **84**:798–9. doi:10.1001/jama.1925.02660370008002

55. Carrel A, Ebeling AH. The fundamental properties of the fibroblast and the macrophage: II. The macrophage. *J Exp Med* (1926) **44**:285–305. doi:10.1084/jem.44.2.261

56. Sabin FR, Doan CA, Cunningham RS. Discrimination of two types of phagocytic cells in the connective tissues by the supravital technique. *Contrib Embryol* (1925) **16**:125–62.

57. Ebert RH, Florey HW. The extravascular development of the monocyte observed in vivo. *Br J Exp Pathol* (1939) **20**:342–56.

58. Volkman A, Gowans JL. The origin of macrophages from the bone marrow in the rat. *Br J Exp Pathol* (1965) **46**:62–70.

59. Takahashi K. Development and differentiation of macrophages and related cell: historical review and current concepts. *J Clin Exp Hematop* (2001) **41**:1–33. doi:10.3960/jslrt.41.1

60. Marchesi VT, Florey HW. Electron micrographic observations on the emigration of leucocytes. *Q J Exp Physiol Cogn Med Sci* (1960) **45**:343–8.

61. Mildner A, Yona S, Jung S. A close encounter of the third kind: monocyte-derived cells. *Adv Immunol* (2013) **120**:69–103. doi:10.1016/B978-0-12-417028-5.00003-X

62. Konstantinov IE, Mejevoi N, Anichkov NM. Nikolai N. Anichkov and his theory of atherosclerosis. *Tex Heart Inst J* (2006) **33**:417–23.

63. Anitschkow N. Über die histogenese der myokardveränderungen bei einigen intoxikationen. *Virchows Arch A Pathol Pathol Anat* (1913) **211**:193–237.

64. Anitschkow N. Über die veränderung der kaninchenaorta bei experimenteller cholesterinsteatose. *Beitr Pathol Anat* (1913) **56**:379–404.

65. Steinberg D. *The Cholesterol Wars: The Skeptics vs the Preponderance of Evidence.* London: Academic Press (2007).

66. Dock W. Research in arteriosclerosis; the first fifty years. *Ann Intern Med* (1958) **49**:699–705. doi:10.7326/0003-4819-49-3-699

67. Willerson JT, Majesky MW, Fuster V. Russell Ross, PhD – visionary basic scientist in cardiovascular medicine – in memoriam. *Circulation* (2001) **103**:478–9. doi:10.1161/01.CIR.103.4.478

68. Aterman K, Remmele W, Smith M. Karl Touton and his "xanthelasmatic giant cell." A selective review of multinucleated giant cells. *Am J Dermatopathol* (1988) **10**:257–69.

69. van Furth R, Cohn ZA, Hirsch JG, Humphrey JH, Spector WG, Langevoort HL. [Mononuclear phagocytic system: new classification of macrophages, monocytes and of their cell line]. *Bull World Health Organ* (1972) **47**:651–8.

70. Maximow A. Über die entwicklung der blut-und bindegewebszellen beim säugetier-embryo. *Folia Haematol* (1907) **4**:611–26.

71. Mosier DE, Coppleson LW. A three-cell interaction required for the induction of the primary immune response in vitro. *Proc Natl Acad Sci U S A* (1968) **61**:542–7. doi:10.1073/pnas.61.2.542

72. Steinman RM, Cohn ZA. Identification of a novel cell type in peripheral lymphoid organs of mice. I. Morphology, quantitation, tissue distribution. *J Exp Med* (1973) **137**:1142–62. doi:10.1084/jem.137.5.1142

73. Nussenzweig MC, Steinman RM. Contribution of dendritic cells to stimulation of the murine syngeneic mixed leukocyte reaction. *J Exp Med* (1980) **151**:1196–212. doi:10.1084/jem.151.5.1196

74. van Furth R. Identification of mononuclear phagocytes: overview and definitions. In: Adams DO, Edelson PJ, Koren H, editors. *Methods for Studying Mononuclear Phagocytes.* London: Academic Press (1980). p. 243–52.

75. Ginhoux F, Greter M, Leboeuf M, Nandi S, See P, Gokhan S, et al. Fate mapping analysis reveals that adult microglia derive from primitive macrophages. *Science* (2010) **330**:841–5. doi:10.1126/science.1194637

76. Yamada M, Naito M, Takahashi K. Kupffer cell proliferation and glucan-induced granuloma formation in mice depleted of blood monocytes by strontium-89. *J Leukoc Biol* (1990) **47**:195–205.

77. Naito M, Hasegawa G, Takahashi K. Development, differentiation, and maturation of Kupffer cells. *Microsc Res Tech* (1997) **39**:350–64. doi:10.1002/(SICI)1097-0029(19971115)39:4<350::AID-JEMT5>3.3.CO;2-V

78. Alliot F, Godin I, Pessac B. Microglia derive from progenitors, originating from the yolk sac, and which proliferate in the brain. *Brain Res Dev Brain Res* (1999) **117**:145–52. doi:10.1016/S0165-3806(99)00113-3

79. Lichanska AM, Hume DA. Origins and functions of phagocytes in the embryo. *Exp Hematol* (2000) **28**:601–11. doi:10.1016/S0301-472X(00)00157-0

80. Bigley V, Haniffa M, Doulatov S, Wang XN, Dickinson R, Mcgovern N, et al. The human syndrome of dendritic cell, monocyte, B and NK lymphoid deficiency. *J Exp Med* (2011) **208**:227–34. doi:10.1084/jem.20101459

81. Schulz C, Gomez Perdiguero E, Chorro L, Szabo-Rogers H, Cagnard N, Kierdorf K, et al. A lineage of myeloid cells independent of Myb and hematopoietic stem cells. *Science* (2012) **336**:86–90. doi:10.1126/science.1219179

82. Hashimoto D, Chow A, Noizat C, Teo P, Beasley MB, Leboeuf M, et al. Tissue-resident macrophages self-maintain locally throughout adult life with minimal contribution from circulating monocytes. *Immunity* (2013) **38**:792–804. doi:10.1016/j.immuni.2013.04.004

83. Yona S, Kim KW, Wolf Y, Mildner A, Varol D, Breker M, et al. Fate mapping reveals origins and dynamics of monocytes and tissue macrophages under homeostasis. *Immunity* (2013) **38**:79–91. doi:10.1016/j.immuni.2012.12.001

84. Epelman S, Lavine KJ, Randolph GJ. Origin and functions of tissue macrophages. *Immunity* (2014) **41**:21–35. doi:10.1016/j.immuni.2014.06.013

85. Gomez Perdiguero E, Klapproth K, Schulz C, Busch K, Azzoni E, Crozet L, et al. Tissue-resident macrophages originate from yolk-sac-derived erythro-myeloid progenitors. *Nature* (2015) **518**:547–51. doi:10.1038/nature13989

86. Hoeffel G, Chen J, Lavin Y, Low D, Almeida FF, See P, et al. C-myb(+) erythro-myeloid progenitor-derived fetal monocytes give rise to adult tissue-resident macrophages. *Immunity* (2015) **42**:665–78. doi:10.1016/j.immuni.2015.03.011

87. Guilliams M, Ginhoux F, Jakubzick C, Naik SH, Onai N, Schraml BU, et al. Dendritic cells, monocytes and macrophages: a unified nomenclature based on ontogeny. *Nat Rev Immunol* (2014) **14**:571–8. doi:10.1038/nri3712

88. Mildner A, Chapnik E, Manor O, Yona S, Kim KW, Aychek T, et al. Mononuclear phagocyte miRNome analysis identifies miR-142 as critical regulator of murine dendritic cell homeostasis. *Blood* (2013) **121**:1016–27. doi:10.1182/blood-2012-07-445999

89. Becher B, Schlitzer A, Chen JM, Mair F, Sumatoh HR, Teng KWW, et al. High-dimensional analysis of the murine myeloid cell system. *Nat Immunol* (2014) **15**:1181–9. doi:10.1038/ni.3006

90. Gautier EL, Ivanov S, Williams JW, Huang SCC, Marcelin G, Fairfax K, et al. Gata6 regulates aspartoacylase expression in resident peritoneal macrophages and controls their survival. *J Exp Med* (2014) **211**:1525–31. doi:10.1084/jem.20140570

91. Gosselin D, Link VM, Romanoski CE, Fonseca GJ, Eichenfield DZ, Spann NJ, et al. Environment drives selection and function of enhancers controlling tissue-specific macrophage identities. *Cell* (2014) **159**:1327–40. doi:10.1016/j.cell.2014.11.023

92. Lavin Y, Winter D, Blecher-Gonen R, David E, Keren-Shaul H, Merad M, et al. Tissue-resident macrophage enhancer landscapes are shaped by the local microenvironment. *Cell* (2014) **159**:1312–26. doi:10.1016/j.cell.2014.11.018

93. Okabe Y, Medzhitov R. Tissue-specific signals control reversible program of localization and functional polarization of macrophages. *Cell* (2014) **157**:832–44. doi:10.1016/j.cell.2014.04.016

94. Rosas M, Davies LC, Giles PJ, Liao CT, Kharfan B, Stone TC, et al. The transcription factor Gata6 links tissue macrophage phenotype and proliferative renewal. *Science* (2014) **344**:645–8. doi:10.1126/science.1251414

8

A hitchhiker's guide to myeloid cell subsets: practical implementation of a novel mononuclear phagocyte classification system

Martin Guilliams [1,2]* and Lianne van de Laar [1,2]*

[1] Laboratory of Immunoregulation, VIB Inflammation Research Center, Ghent University, Ghent, Belgium, [2] Department of Respiratory Medicine, University Hospital Ghent, Ghent, Belgium

Edited by:
Ken J. Ishii,
National Institute of Biomedical
Innovation, Japan

Reviewed by:
Richard A. Kroczek,
Robert Koch-Institute, Germany
Kazuhiro Suzuki,
Osaka University, Japan

*Correspondence:
Martin Guilliams and
Lianne van de Laar,
Laboratory of Immunoregulation, VIB
Inflammation Research Center, Ghent
University, Technologiepark 927,
Zwijnaarde, 9052 Ghent, Belgium
martin.guilliams@irc.vib-ugent.be;
lianne.vandelaar@irc.vib-ugent.be

The classification of mononuclear phagocytes as either dendritic cells or macrophages has been mainly based on morphology, the expression of surface markers, and assumed functional specialization. We have recently proposed a novel classification system of mononuclear phagocytes based on their ontogeny. Here, we discuss the practical application of such a classification system through a number of prototypical examples we have encountered while hitchhiking from one subset to another, across species and between steady-state and inflammatory settings. Finally, we discuss the advantages and drawbacks of such a classification system and propose a number of improvements to move from theoretical concepts to concrete guidelines.

Keywords: nomenclature, dendritic cells, macrophages, monocytes, classification

Introduction

In the science fiction series created by Douglas Adams (1), the Hitchhiker's Guide to the Galaxy starts as follows: "*Space is big. Really big. You just won't believe how vastly hugely mind-boggling big it is. You may think it's a long way down the road to the chemist's, but that's just peanuts to space.*" Given the complexity of the mononuclear phagocyte (Star)system (MPS), one could easily give a similar warning to readers who are trying to make some sense of the huge number of hypothetically distinct dendritic cell (DC) and macrophage (MΦ) subsets. At the last International DC Symposium (DC2014, Tours – France), we counted at least 28 different DC subsets that were described using various surface markers and nomenclature systems in distinct species. If one would add the different MΦ subsets and the Cytof technology allowing to measure the expression of more than 30 different surface markers per cell, one could with a bit of luck end up with "42" as answer to the ultimate myeloid question of how many mononuclear phagocyte subsets exist in life, the universe, and everything. Although this would be great for fans of Douglas Adams, without Babel Fish to help us make some sense of so many different subsets, this evolution will not be beneficial for communication among myeloid cell experts, let alone for the communication toward pharmaceutical companies, scientific editors, medical doctors, or graduate students. We will here try to simplify this apparent complexity through a number of practical examples and theoretical concepts. Having hitchhiked from MΦ to DC labs studying myeloid cells in various tissues and in distinct inflammatory conditions, we would, in accordance with the Hitchhiker's Guide to the Galaxy, advise the following: do not panic and bring your towel along.

Members of the Mononuclear Phagocyte System

In the original MPS model proposed by Ralph van Furth, James Hirsch, and Zanvil Cohn, MΦs were proposed to derive from circulating monocytes (2). A couple of years later, Ralph Steinman and Zanvil Cohn identified DCs, which were also included in the MPS (3). The fact that DCs could be derived from human and mouse monocytes in GM-CSF-driven *in vitro* cultures (4–8) and *in vivo* upon inflammation or in barrier tissues (9–15) supported this concept. For a historical overview of the MPS field, we redirect the readers to the review of Simon Yona and Siamon Gordon in this issue (16). The identification of mouse hematopoietic precursors committed to the DC lineage called the common DC progenitors (CDPs – giving rise to pDCs and cDCs) and pre-cDCs (giving rise to cDCs) that are distinct from monocytes and can give rise to the so-called conventional DCs (cDCs) induced a first conceptual revolution in the field (12, 17–20). Moreover, Flt3-L, and not GM-CSF, was shown to be critically involved in the development of cDCs *in vitro* (8, 21–23) and *in vivo* (24–28). Recently, two additional committed precursors were identified in mice: the pre-pDC precursor that preferentially differentiates into pDCs (29), and the monocyte-committed common monocyte progenitor (cMop) (30). Importantly, the human equivalent of the pre-cDC, CDP, and cMop was recently identified (31, 32). A second conceptual revolution in the field was driven by the finding that most tissue-resident MΦs do not derive from circulating HSC-derived monocytes but develop from embryonic precursors, i.e., the yolk-sac MΦs (YS MΦs) or fetal liver (FL) monocytes (33–39). The relative contribution of YS MΦ-derived

and FL monocyte-derived MΦs seems to vary from one organ to another (40–42). It was recently demonstrated that almost all MΦs have a YS origin [either directly from YS MΦs or through YS-derived EMPs (39)]. This may seem in contradiction with the proposed partial origin from FL monocytes (35, 43). However, it is now clear that YS-derived EMPs seed the FL and go through a FL monocyte intermediate before differentiating into most tissue-resident MΦs (44), reconciling most of the apparent discrepancies in the field. Together, these findings have challenged the MPS dogma and revealed that most DCs and MΦs derive from distinct committed precursors rather than from circulating HSC-derived monocytes (**Figure 1**).

Revisiting the Classification of Mononuclear Phagocytes

Historically, mononuclear phagocytes were classified as DCs or MΦs based on a restricted set of surface markers (CD11c and MHCII for DCs versus F4/80 for MΦs), proposed functional specialization (antigen-presentation and migration to lymph nodes for DCs versus phagocytosis for MΦs) and/or morphological features (dendritic-shaped cells for DCs versus large vacuolar cells for MΦs). However, these features are often not mutually exclusive. For example, although CD11c and MHCII are typically associated with DCs, alveolar MΦ are CD11c^{hi} and MHCII is expressed by intestinal MΦs (35, 45). Ideal surface markers allowing identification of the distinct myeloid cell subsets across tissues and species are still incomplete. Markers typically associated with some myeloid cell subsets can be lost or acquired by other subsets. The monocyte-associated

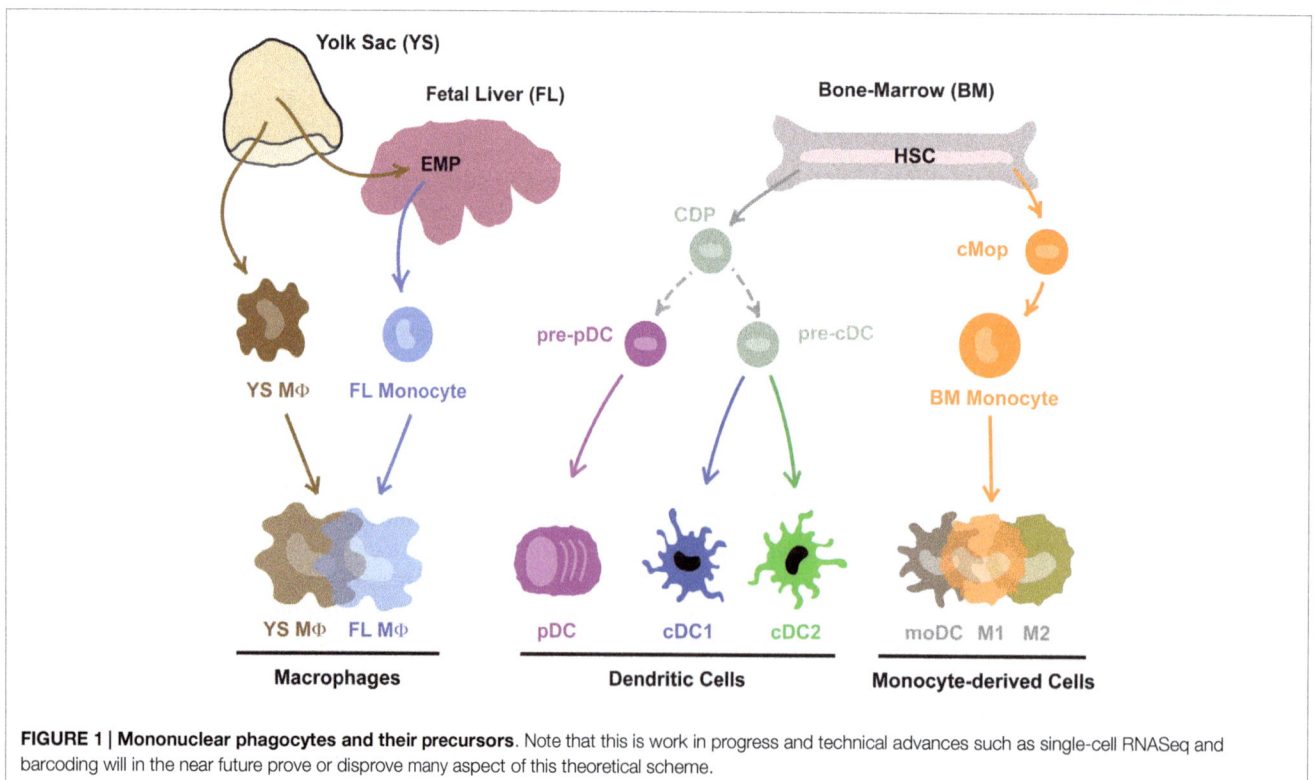

FIGURE 1 | Mononuclear phagocytes and their precursors. Note that this is work in progress and technical advances such as single-cell RNASeq and barcoding will in the near future prove or disprove many aspect of this theoretical scheme.

marker Ly-6C is rapidly down-regulated on many monocyte-derived cells (MCs) upon entrance in the tissues (45–48) and is expressed on pDCs (and lowly expressed on some cDCs). The pDC-associated marker mPDCA1 (stained with 120G8) can be acquired by MCs during inflammation (49). Alveolar MΦs (50) and Kupffer Cells (unpublished data) can upregulate CD11b during inflammation. Finally, BDCA3 is expressed on both human cDC1s and MCs (51). Thus, the inability to consistently identify myeloid cell subsets irrespective of tissue, species, or inflammatory state makes surface markers unattractive as basis for classification.

We would also propose to avoid a classification based primarily on functional specialization. First, each myeloid cell subset can perform more than one prototypical function. MΦs are often linked to phagocytosis of dead cells and pathogens but also have important immunomodulatory and metabolic functions. Second, subsets can acquire or lose functional capacities during inflammation as recently demonstrated for cDC2s that acquire cross-presentation capacities upon TLR stimulation (52). Therefore, we propose to disregard function as a basis for classifying cells.

Instead of surface markers, functional specialization, or morphology, we have recently suggested to classify cells based on their cellular origin, which could allow a more robust classification system (53). This would yield three big groups of cells (**Figure 1**): (i) embryonic progenitor-derived MΦs, (ii) CDP-derived DCs (that would be subdivided into cDC1s, cDC2s, and pDCs), and (iii) MCs. As these precursors have now been identified in both the mouse and the human, this allows one classification system across tissues and species.

Although precursor-based classification would provide a robust and species-conserved system, at the end of the day the function of the cells is what really matters for converting our knowledge into therapeutic advances for patients. Regrouping all the DCs into three big subsets of cDC1s, cDC2s, and pDCs will thus have the disadvantage of lumping together cells that may be in very different functional activation states. Similarly, MCs have been shown to be particularly plastic cells (54). Therefore, we propose to add a second classification level to the fixed ontogeny-based Level1 (**Figure 2**). Addition of a Level2 allows specification of the cellular activation state, the micro-anatomical localization or simply the surface markers utilized to identify the cells in a particular study. Of note, when defining the Level2 it will be important to avoid generalizations as a given function is often performed by only a fraction of the cells studied. We would thus propose to restrict the Level2 to objective criteria that can be measured at the single-cell level.

Practical Implementation for DCs

Historically, DCs were divided into subsets based on surface markers that differed between tissues and species, such as CD207 (Langerin) in the skin, CD103 (IntegrinαE) in the intestine, CD11b (IntegrinαM) in the lungs, CD4/CD8α in the spleen, and CD24/CD172α for *in vitro* differentiated DCs (**Figure 3**). Human DCs, on the other hand, have been divided into CD141$^+$ (BDCA3) and CD1c$^+$ (BDCA1) DCs. pDCs are identified by the expression of BDCA4 and BDCA2 in human, but by B220, mPDCA1 (BST2, recognized by 120G8), or Siglec-H in mice (53). Technical advances in multi-color flow cytometry have made matters worse with evermore "novel DC subsets" based on the expression of additional surface markers. By comparing the gene-expression profile of DCs isolated from various tissues and species, one can appreciate three big clusters of DCs (55–61). The pDC cluster includes mouse PDCA1$^+$ pDCs and human BDCA2$^+$BDCA4$^+$ pDCs. The cDC1 cluster comprises dermal CD207$^+$CD103$^+$ cDC1s, lung CD103$^+$CD11b$^-$cDC1s, splenic CD8a$^+$CD4$^-$ cDC1s, intestinal CD103$^+$CD11b$^-$ cDC1s and human blood BDCA3$^+$ cDC1s. Dermal CD207$^-$CD11b$^+$ cDC2s, lung CD103$^-$CD11b$^+$ cDC2s, splenic CD8a$^-$CD4$^+$ cDC2s, intestinal CD103$^+$CD11b$^+$ cDC2s and human blood BDCA1$^+$ cDC2s form the cDC2 cluster (62, 63). This bio-IT-driven analysis also revealed that within the cDC population, XCR1 and Sirpα are, respectively, expressed by all cDC1s and cDC2s across tissues, allowing an improved identification of these cells (51, 64–70). Note, however, that Sirpα is also expressed by other myeloid cells than cDC2s, showing the need for correct cDC identification prior to using this marker to distinguish cDC2s from cDC1s. Strikingly, this gene-expression-based division is supported by the existence of distinct pre-committed precursors (29, 71, 72) and by differential developmental transcription factor requirement of cDC1s,

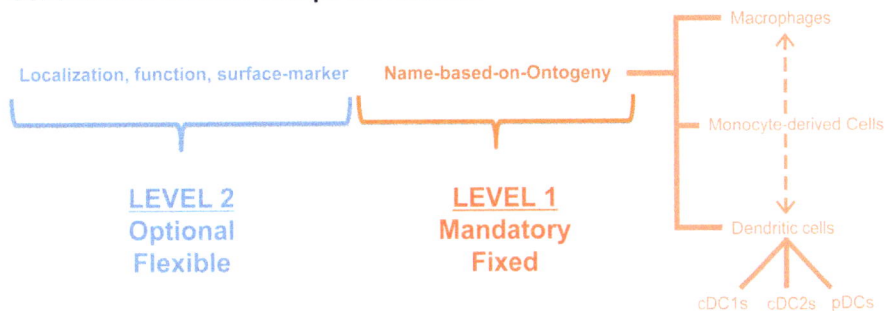

FIGURE 2 | A nomenclature system in two levels would have the advantage that cells can be first classified based on a restricted set of names (in this proposition according to their cellular origin: MΦ, MC, cDC1, cDC2, pDC) that would be applicable across species and across tissues, but the second level would still allow some flexibility to denote a distinct activation state or localization.

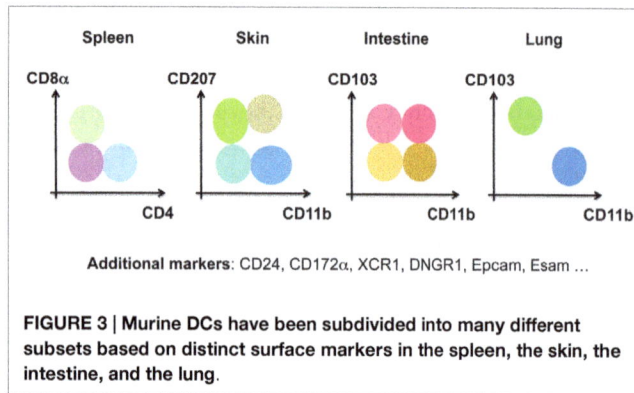

FIGURE 3 | Murine DCs have been subdivided into many different subsets based on distinct surface markers in the spleen, the skin, the intestine, and the lung.

cDC2s, and pDCs in the mouse. cDC1s, but not cDC2s, require BATF3 (71, 73, 74), ID2 (28, 75, 76), NFIL3 (77), and IRF8 (28, 71, 78–80) for their development, while cDC2s, but not cDC1s, are dependent on RELB (81), RBPJ (82), and IRF4 (79, 83–85). pDC development has been shown to be driven by E2-2 (86, 87).

The subdivision of DCs in three distinct Level1 groups is thus supported by their gene-expression profiles, cellular origin, and transcription factor requirement. However, these cells can acquire a distinct functional activation state from one tissue to another and in distinct inflammatory settings, underlining the need for a Level2 system. This can be illustrated by the capacity of intestinal cDCs to produce retinoic acid and promote the generation of induced regulatory T cells (iT$_{REG}$s) (88–91). Identification of DCs with superior iT$_{REG}$ inducing ability is clinically relevant as the prevalence of food allergies, celiac disease and inflammatory bowel diseases is currently rising throughout the western world. Originally, it was described that CD103$^+$ but not CD103$^-$ intestinal DCs excel in iT$_{REG}$ generation in a retinoic acid-dependent manner (89, 90). It is now clear that CD103$^+$ intestinal DCs comprise two ontogenically distinct subsets, CD103$^+$CD11b$^-$ cDC1s and CD103$^+$CD11b$^+$ cDC2s (74). Interestingly, rather than being associated with either of the two subsets, about half of the intestinal CD103$^+$CD11b$^-$ cDC1s were shown to possess the capacity to produce retinoic acid, while only one-third of the cDC2s do (91). Moreover, on a per cell basis retinoic acid producing CD103$^+$ cDC1s were the best at inducing iT$_{REG}$s (**Figure 4**). These data reveal that CD103$^+$ cDC1s, although broadly considered as a homogeneous subset, consist of 50% cells that are very efficient at inducing iT$_{REG}$s and 50% cells that are not. Interestingly, dermal cDC2s have higher retinoic acid-dependent iT$_{REG}$ induction activity than dermal cDC1s (91). We hypothesize that this functional heterogeneity may be explained by the existence of distinct micro-environments within organs, inducing diverse functional modules on DCs. The finding that important functional modules can be acquired by only a fraction of cDC1s and/or cDC2s, which can moreover differ from one organ to another, illustrates the need for a Level2 nomenclature for DCs.

Another example of functional heterogeneity within DCs concerns the cDC2s. Splenic cDC2s contain a subpopulation that expresses CD4 and is specifically localized in the bridging channels (92). This localization has been shown to be EBI2-driven and essential to drive antibody production by B cells. The development of these CD4$^+$ cDC2s is Notch2 dependent. Note also that

Notch2 deficiency is associated with defects in T$_H$17 induction (93, 94). In addition, it was found that KLF4 controls the development of a subpopulation of CD24^{lo}CD11b^{lo}Sirpα^{hi} cDC2s in the dermis (95). Importantly, loss of KLF4 was associated with loss of T$_H$2 induction. Thus, although cDC2s have been proposed to excel at both the induction of T$_H$2 (47, 96) and T$_H$17 responses (84, 85, 93), it may well be that these functional modules are in fact expressed by subpopulations of cDC2s (controlled by KLF4 and Notch2, respectively). In conclusion, although the current knowledge of early DC development in the bone-marrow seems to support only three big groups of DCs (cDC1s, cDC2s, and pDCs), it appears that a second layer of tissue-specific signals imprint operative gene modules on a fraction of DCs. Depending on their micro-localization, subpopulations of cDC1s, cDC2s, or pDCs will acquire distinct functional properties, requiring a flexible Level2 to classify and describe functionally distinct subpopulations.

A final example of a need for a Level2 classification involves inflammation-induced changes of surface marker expression. When mice are infected with the influenza virus, there is a transient change in the CD103 versus CD11b expression profile of lung cDCs, yielding four instead of two lung DC subsets (**Figure 5**). If one considers these as four distinct DC subsets, one could conclude that influenza infection disrupts hematopoiesis in the bone-marrow, as has been shown for Toxoplasma infection (97). Alternatively, these novel CD103/CD11b expression patterns may represent distinct local activation states of cDC1s or cDC2s. We have studied the cellular origin of the "novel" DC subsets arising during influenza infection (Neyt et al., manuscript in preparation). Our preliminary data suggest that CD103$^+$CD11b$^+$ cells are cDC2s that acquire CD103 expression during inflammation rather than a completely new subset. In conclusion, although we cannot rule out the existence of additional DC subsets that specifically develop during inflammation, when in doubt we propose to first evaluate whether cells with a novel surface receptor expression profile represent a different activation state of cDC1s or cDC2s before assuming the existence of a novel cDC3.

Practical Implementation for Embryonic Macrophages

In our classification system based on ontogeny all mononuclear phagocytes of embryonic origin are grouped together under a single Level1 as "macrophages" (**Figure 1**). This includes liver-resident Kupffer cells, brain-resident microglial cells, lung-resident alveolar MΦs but also epidermis-resident Langerhans cells. In effect, this would thus classify Langerhans cells as MΦs and not as DCs, based on the fact that these cells derive from embryonic precursors that seed the epidermis around birth and then self-maintain throughout life (43, 98). We propose to keep the historical names for MΦs with undisputed identities. Mouse Kupffer cells, for example, do not require a different nomenclature since these cells have a well-defined cellular origin [embryonic (34, 38, 39)], localization (i.e., the liver sinusoids), and gene-expression profile (99, 100). However, we would like to emphasize that not any F4/80$^+$ cell in the liver should be categorized as Kupffer cell. MCs infiltrating

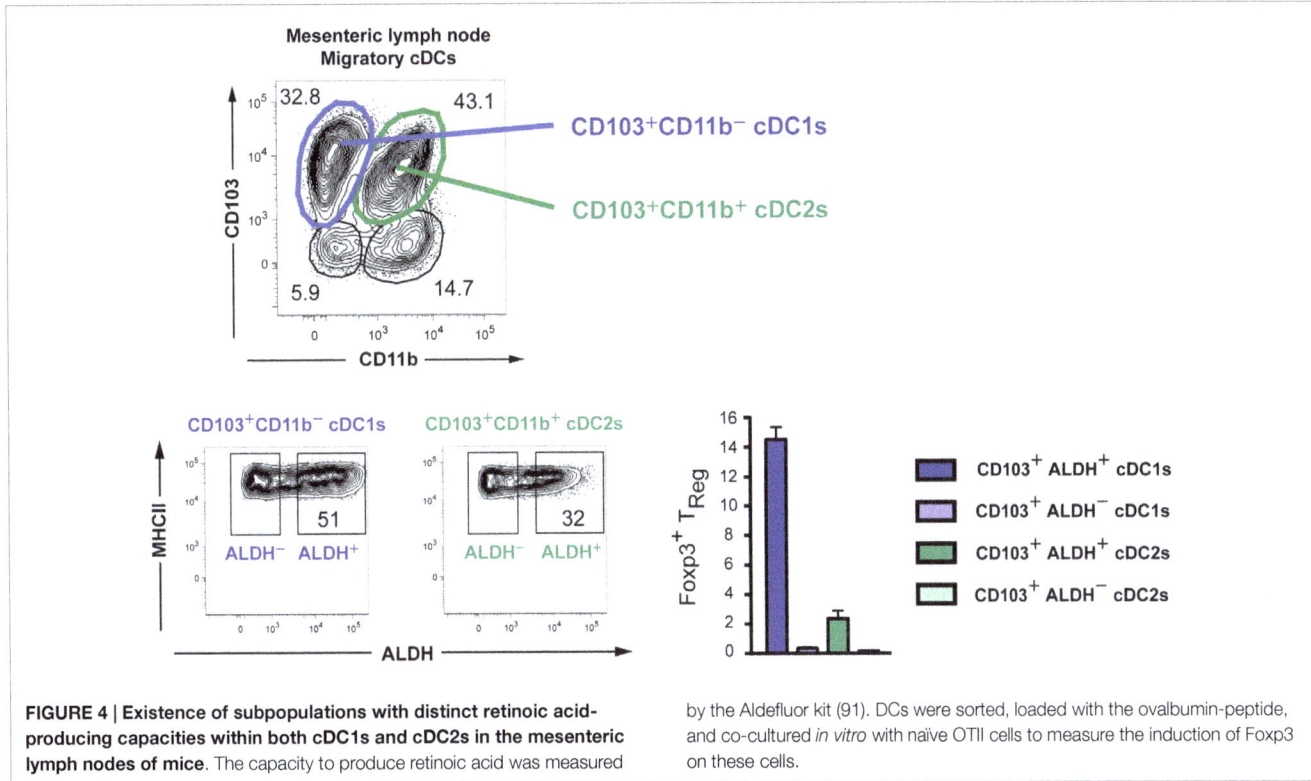

FIGURE 4 | Existence of subpopulations with distinct retinoic acid-producing capacities within both cDC1s and cDC2s in the mesenteric lymph nodes of mice. The capacity to produce retinoic acid was measured by the Aldefluor kit (91). DCs were sorted, loaded with the ovalbumin-peptide, and co-cultured *in vitro* with naïve OTII cells to measure the induction of Foxp3 on these cells.

the liver during acetaminophen-induced injury also express F4/80 but are short-lived and acquire a gene-expression profile that is strikingly different from resident Kupffer cells (100). Similarly, MCs infiltrating the central nervous system during inflammation are short-lived and do not acquire the specific gene-expression profile of embryonic microglia (101–103). As such, any MΦ-like cell in the liver or the brain should not be classified as Kupffer cell or microglia, respectively, as is often the case. Unfortunately, tools to correctly distinguish MCs from resident MΦs have long been lacking. In a way, this is surprising given the huge difference in gene-expression profile between resident embryonic MΦs and recruited MCs in these disease models. We have now identified several surface markers that are expressed by Kupffer cells but not MCs recruited during liver injury (Scott et al. manuscript in preparation) and we expect that given the striking heterogeneity of tissue-resident MΦs (104, 105) many of these MΦ-specific markers will be found. This will facilitate the correct classification of these cells and pave the way toward unraveling the functional differences between recruited MCs and tissue-resident embryonic MΦs during inflammation.

Practical Implementation for Monocyte-Derived Cells

Monocytes are particularly plastic cells. This can be appreciated using *in vitro* culture systems. Monocytes cultured with GM-CSF express some DC-like characteristics and have therefore long been referred to as moDCs. By contrast,

culturing monocytes with M-CSF induces their differentiation into MΦ-like cells (moMΦs). Adding IL-4 or IFN-γ to M-CSF cultures further polarizes MCs into the so-called "classically activated MΦs/M1s" or "alternatively activated MΦs/M2s" (106), with strikingly different gene-expression profiles and metabolic modules (107). In a nomenclature system based on ontogeny, moDCs, M1s, or M2s are however first classified as MCs (Level1). In theory, this does not prevent further Level2 classification as "dendritic MCs," "classically activated MCs," or "alternatively activated MCs." However, we feel this polarized classification implies functional characteristics that are often not assessed experimentally. For example, MCs classified as M1s are typically associated with pathogen killing, M2s with wound healing, and moDCs with antigen-presentation (**Figure 6**). However, the identification of "dendritic MCs/moDCs," "classically activated MCs/M1s," or "alternatively activated MCs/M2s" *in vivo* turned out to be very challenging. In fact, profiling of MCs isolated from various inflamed tissues or *in vitro* culture systems reveals that monocytes can acquire a much broader transcriptional repertoire than suggested by the three-way M1/M2/moDC model. In recent efforts to further characterize the heterogeneity of MC activation states, Schultze and colleagues compared the gene-expression profile of MCs stimulated with a vast array of cytokines and TLR ligands. Instead of yielding a polarized model, the unbiased bio-informatics-driven clustering approach revealed a spectrum model (54). In our view, this spectrum model can be taken one step further and include the "dendritic MCs/moDCs" derived from GM-CSF-induced bone-marrow cultures as yet another extreme of the continuum of cellular faiths that can be acquired by monocytes. Rather

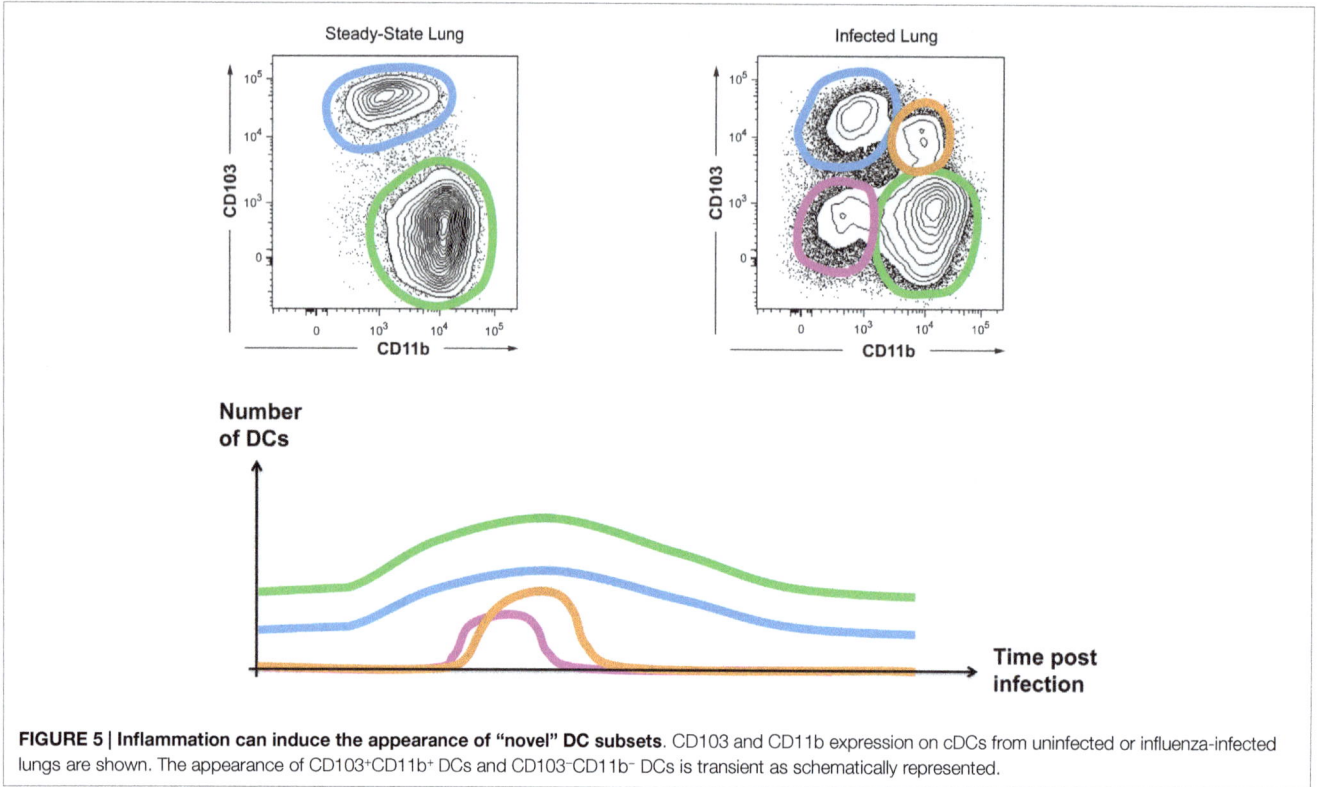

FIGURE 5 | Inflammation can induce the appearance of "novel" DC subsets. CD103 and CD11b expression on cDCs from uninfected or influenza-infected lungs are shown. The appearance of CD103+CD11b+ DCs and CD103−CD11b− DCs is transient as schematically represented.

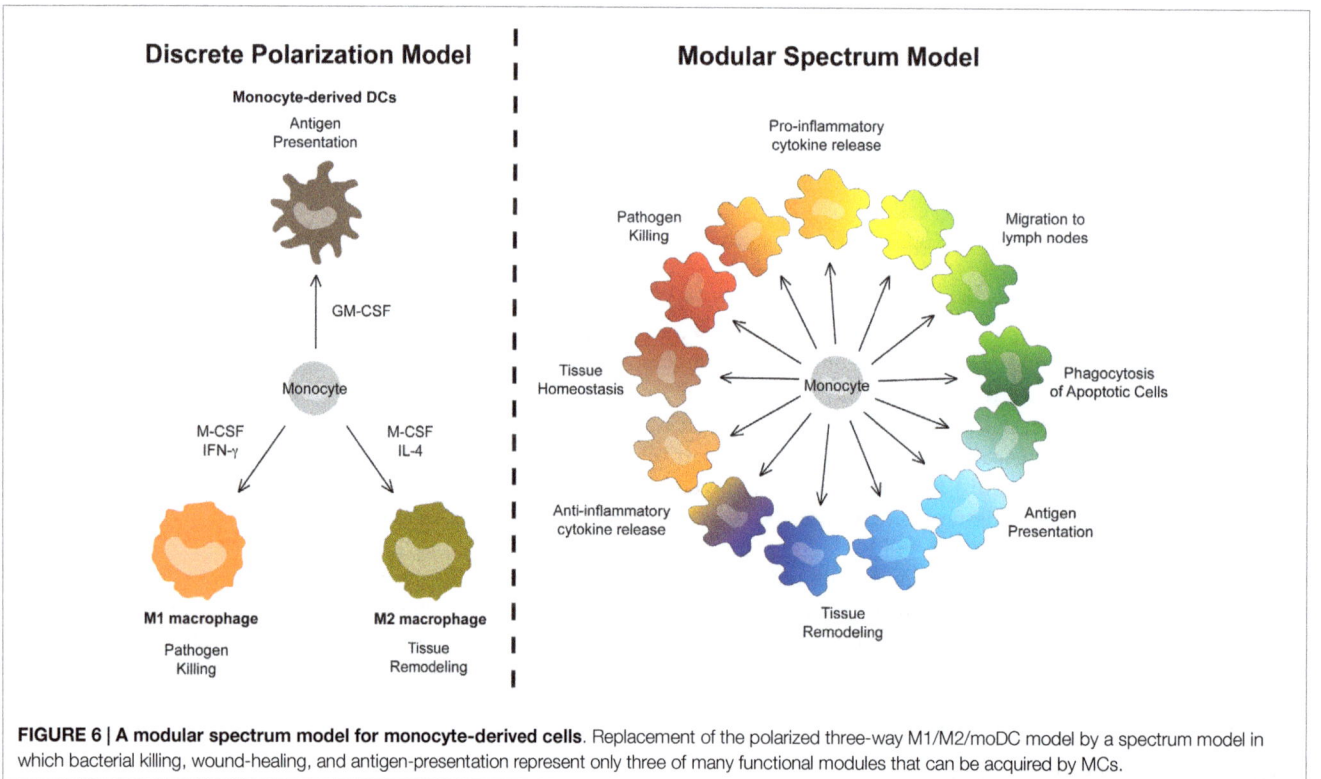

FIGURE 6 | A modular spectrum model for monocyte-derived cells. Replacement of the polarized three-way M1/M2/moDC model by a spectrum model in which bacterial killing, wound-healing, and antigen-presentation represent only three of many functional modules that can be acquired by MCs.

than unique end points, bacterial killing, wound-healing, and antigen-presentation represent three of many functional modules that can be acquired by MCs in a spectrum model that can be graphically represented by a continuous circle (**Figure 6**).

One important consequence of the herein-described classification would be the regrouping of moDCs and moMΦs under a single MC Level1. We feel this will represent an improvement for the field due to the lack of clear, mutually exclusive features

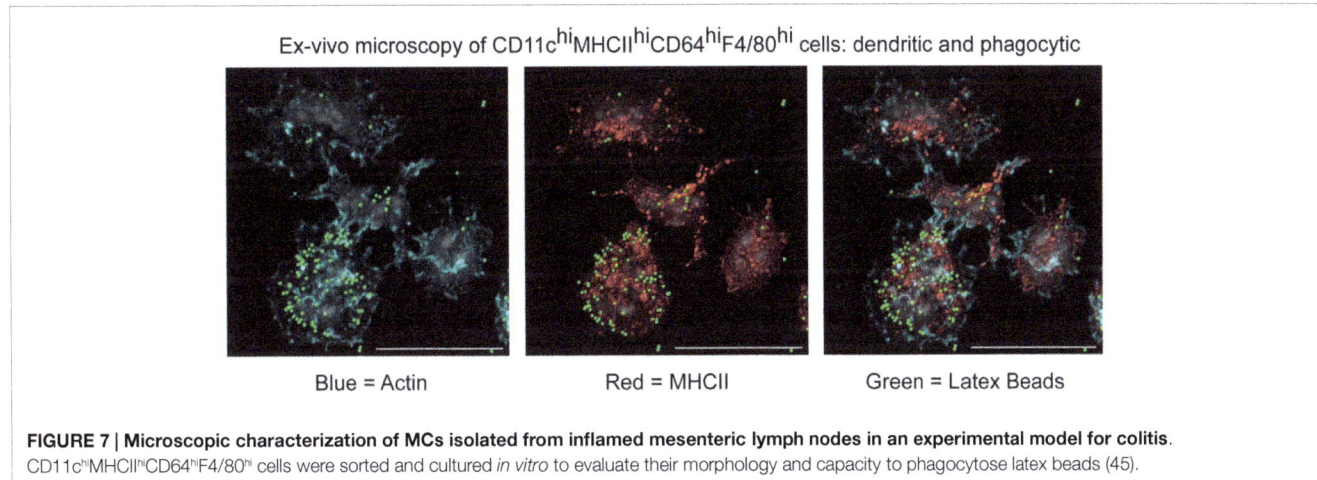

Ex-vivo microscopy of CD11c^{hi}MHCII^{hi}CD64^{hi}F4/80^{hi} cells: dendritic and phagocytic

Blue = Actin   Red = MHCII   Green = Latex Beads

FIGURE 7 | Microscopic characterization of MCs isolated from inflamed mesenteric lymph nodes in an experimental model for colitis. CD11c^{hi}MHCII^{hi}CD64^{hi}F4/80^{hi} cells were sorted and cultured *in vitro* to evaluate their morphology and capacity to phagocytose latex beads (45).

that can be used to objectively separate moDCs from moMΦs (53). This problem can be illustrated by the MCs present within inflamed mesenteric lymph nodes during an experimental model for colitis (45). In this study, we called these cells moMΦs because they were CD64^{hi}F4/80^{hi} and excelled at phagocytosis (**Figure 7**). However, we could perfectly, like the Powrie group (108), have classified these cells as moDCs based on their CD11c^{hi}MHCII^{hi} profile, their localization within the T cell zone, their antigen-presentation capacity, or their dendritic morphology (**Figure 7**). By classifying these cells as CD64^{hi}F4/80^{hi} and/or CD11c^{hi}MHCII^{hi} MCs, they are recognized as one lineage, which will promote understanding and simplify communication between different research groups without preventing the study of DC-like or MΦ-like properties of specific MCs.

Finally, we do not expect MCs to be homogeneous in inflamed tissues. We and others found iNOS, the enzyme that is used by MΦs to produce NO and that is classically associated with an M1 activation state, to be typically expressed by only 10% of MCs (11, 45, 109). Since NO is bactericidal, suppressive for T cells, and can induce serious tissue damage, it makes perfect sense to study the factors that induce the expression of iNOS on a fraction of MCs. But by classifying these cells as "iNOS$^+$ MCs" instead of "M1 macrophages" or "TIP-DCs" one avoids to associate functions that have not been proven for these cells such as the antigen-presentation activity typically expected from DCs [in fact mice lacking monocytes showed identical T cell priming suggesting that TIP-DCs are not essential for this function (11)].

The Tough Cases Part I: moDCs as Fourth DC Subset?

MCs fitting the complete list of characteristics attributed to moDC, including migratory and antigen-presentation capacities comparable to cDCs, are not easily identified *in vivo*. We have described migratory MCs upon house-dust mite (HDM) exposure in the lungs (47), but their migration is much less efficient as compared to cDCs and required very high (and non-physiological) doses of HDM. In fact, we found that the majority of HDM-induced

pulmonary MCs are not migratory cells but instead play an important role in the secretion of inflammatory chemokines that orchestrate the local immune responses. Similarly, a low-grade migration of CCR2$^+$CD64^{int} MCs was described upon DSS inflammation in the skin but this was minor as compared to cDC migration (48). Moreover, compared to cDCs these dermal CCR2$^+$CD64^{int} MCs displayed a rather modest antigen–antigen presentation capacity.

The most convincing pieces of evidence for cDC-like features of MCs come from *in vitro* culture systems. Bone-marrow cells cultured with GM-CSF yield cells with excellent antigen-presentation capacity that acquire CCR7, the chemokine receptor controlling migration of cDCs from tissues to lymph nodes, upon TLR stimulation and can migrate to the lymph nodes upon *in vivo* transfer (110, 111). This culture system has been used in many labs and is globally accepted to yield a homogeneous population of moDCs. This concept was first challenged by a study using single-cell transcriptomics (112). Among LPS-stimulated GM-CSF-induced bone-marrow-derived moDCs, the majority of cells were found to show high expression of inflammatory genes such as TNF, IL1, and CXCL10, while a smaller subset had much lower expression of these genes but displayed a signature reminiscent of "mature DCs," including high expression of CCR7 (113, 114). This was originally interpreted as functional heterogeneity among moDCs. However, in what we consider a landmark paper, Reis e Sousa and colleagues now demonstrate that this minor "mature" population in fact represents cDC2s that contaminate these cultures. These cDC2s displayed lower production of inflammatory cytokines but much better CCR7-ligand-induced migration and antigen-presentation as compared to GM-CSF-induced MCs (115, 116). This implies that many of the DC-like features of GM-CSF-induced moDCs should in fact not be attributed to MCs, but to a minor contaminating cDC2 population. All in all, both *in vitro* and *in vivo* data thus point toward a lower migration and antigen-presentation capacity of MCs as compared to cDCs, but conversely a higher production of inflammatory cytokines and chemokines. We therefore propose that in an inflamed organ the core business of cDCs will be migration to the draining lymph nodes and activation of naïve T cells, whereas MCs will primarily orchestrate local inflammatory responses. Note that this has important consequences for DC-based vaccination strategies as this may explain why MC-based vaccines have only yielded modest clinical responses (117). The recent advances in cDC culture systems and the proper identification of committed circulating DC-precursors (31, 32, 118) may therefore pave the way toward more efficient cDC-based vaccination strategies.

The Tough Cases Part II: Steady-State MCs Versus Embryonic Macrophages

Most MΦ-like cells present in steady-state tissues are of embryonic origin (33–36, 38, 39, 44). However, puzzling exceptions have been reported. Although embryonic MΦs colonize the intestine and the heart before birth, these cells are thereafter progressively replaced by MCs. Importantly, these cells are relatively short-lived and continuous monocyte-recruitment is required to maintain the MC pool in these tissues (45, 46, 119–121). Similarly, monocytes are continuously recruited to the steady-state dermis (48). Therefore, while in some steady-state tissues, including the lung and the spleen, monocytes have been proposed to remain in an undifferentiated state (37, 122); in others, including the intestine, the skin, and the heart, they acquire a MΦ-like phenotype. The classification of MCs that differentiate in these steady-state organs and that replace the embryonic MΦs is difficult. They do not fit the profile of the MCs that are recruited to inflamed tissues, including pulmonary infection (36), auto-immune brain inflammation (101, 102), and acute liver injury (100), since in these inflammatory settings MCs do not replace the embryonic MΦs and display a very different gene-expression profile. Future research will be required to compare the functional properties and gene-expression profile of the embryonic MΦs present in the intestine, the skin, and the heart to the ones from the MCs that replace them with time. It will be interesting to compare the influence of tissue-imprinting to the intrinsic differences associated with their distinct cellular origin. Embryonic MΦs were recently compared to their bone-marrow-derived counterparts that replace them after irradiation-induced depletion. It was found that both cells share between 50 and 90% of the tissue-specific epigenetic landscape (105). This emphasizes the importance of tissue-imprinting, but at the same time implies that between 10 and 50% of the epigenetic landscape could be governed by the cellular origin of the cells. Future research will be required to assess the functional relevance of these findings. In the meantime, the classification of MΦ-like MCs in steady-state tissues remains difficult.

The Way Forward

The Level1 that forms the scaffold of the herein proposed classification system is in part based on elegant murine fate-mapping systems developed to study the cellular origin of MΦs (33, 34, 38, 39, 43, 44) and DCs [(123) and (124) in this issue]. Although the recent identification of committed DC-precursors distinct from monocytes in humans suggests that many of the principles identified in mice apply to the human immune system, this remains to be formally proven. Moreover, many of these murine fate-mapping systems label only a small fraction of the cells per population, rendering functional studies difficult. In cases where classification as cDC1, cDC2, pDC, MC, or MΦ is not obvious, a core set of signature genes that are specific for each cell type could facilitate correct Level1 identification. However, such signature genes are not easily identified. In addition, identification based on surface receptors would be most practical since it would allow the sorting of living cells through flow cytometry for functional assays. Ideally, such markers would be conserved across species. We are currently data-mining the gene-expression profiles of cells from various tissues and species to try to identify such markers. This can however represent a catch22. To find markers specifically expressed by the different populations, one requires pure gene-expression profiles, but correct sorting of the cells without contamination by other populations for RNA profiling requires the very markers we are looking for. Recent technological advances in single-cell RNA sequencing will allow to profile the gene expression of mixed populations. This may at last disentangle mixed myeloid populations and will hopefully provide the field with new markers that can then be validated with the current fate-mapping systems.

Although the current classification systems should thus be seen as work in progress, we are confident that in the near future better markers will be found which faithfully translate the cellular origin of cells and will form a practical base for the Level1 classification of myeloid cells. The Level2 classification should in our view be kept as flexible as possible to allow researchers to focus on one particular functional attribute of their cells of interest without implying too many additional features that have not been studied. Finally, it is noteworthy that in parallel to our proposition a nomenclature system for MΦs was proposed (106). In this proposition, terms implying functional specialization such as "classically activated macrophages" (pro-inflammatory) or "alternatively activated macrophages" (anti-inflammatory) were replaced by an objective description of how a MΦ is cultured *in vitro* [e.g., MΦ (IL-10)] or identified *in vivo* (e.g., "Relma^hi MΦ"). Thus, the common core Level1 would be MΦ and the added description provides the Level2. This and our classification system thus share three important principles: (i) elimination of terms that imply functional specialization as much as possible, (ii) introduction of a fixed Level1 system across species and tissues, and (iii) permitting flexibility through a Level2 system. Irrespective of which system is used to define the Level1 [ontogeny as we propose, or gene-expression profile as proposed by the Dalod group in this issue (60, 61)], we feel these three principles should be maintained for a future and hopefully definitive classification system.

Acknowledgments

We thank Bernard Malissen, Hugues Lelouard, Bart Lambrecht, and Katrijn Neyt for the use of unpublished data. MG is supported by a Marie Curie Reintegration grant, an Odysseus grant, and several FWO grants of the Flemish Government. LL is supported by an EMBO long-term fellowship and a Marie Curie intra-European fellowship.

References

1. Adams D. *The Hitch Hiker's Guide to the Galaxy*. London: Pan Books (1979).

2. van Furth R, Cohn ZA, Hirsch JG, Humphrey JH, Spector WG, Langevoort HL. [Mononuclear phagocytic system: new classification of macrophages, monocytes and of their cell line]. *Bull World Health Organ* (1972) 47:651–8.

3. van Furth R. "Identification of Mononuclear Phagocytes: Overview and Definitions," in *Methods for Studying Mononuclear Phagocytes*. London: Academic Press (1980). p. 243–52.

4. Inaba K, Inaba M, Romani N, Aya H, Deguchi M, Ikehara S, et al. Generation of large numbers of dendritic cells from mouse bone marrow cultures supplemented with granulocyte/macrophage colony-stimulating factor. *J Exp Med* (1992) 176:1693–702. doi:10.1084/jem.176.6.1693

5. Sallusto F, Lanzavecchia A. Efficient presentation of soluble antigen by cultured human dendritic cells is maintained by granulocyte/macrophage colony-stimulating factor plus interleukin 4 and downregulated by tumor necrosis factor alpha. *J Exp Med* (1994) 179:1109–18. doi:10.1084/jem.179.4.1109

6. Caux C, Vanbervliet B, Massacrier C, Dezutter-Dambuyant C, De Saint-Vis B, Jacquet C, et al. CD34+ hematopoietic progenitors from human cord blood differentiate along two independent dendritic cell pathways in response to GM-CSF+TNF alpha. *J Exp Med* (1996) 184:695–706. doi:10.1084/jem.184.2.695

7. Palucka KA, Taquet N, Sanchez-Chapuis F, Gluckman JC. Dendritic cells as the terminal stage of monocyte differentiation. *J Immunol* (1998) 160:4587–95.

8. Xu Y, Zhan Y, Lew AM, Naik SH, Kershaw MH. Differential development of murine dendritic cells by GM-CSF versus Flt3 ligand has implications for inflammation and trafficking. *J Immunol* (2007) 179:7577–84. doi:10.4049/jimmunol.179.11.7577

9. Randolph GJ, Inaba K, Robbiani DF, Steinman RM, Muller WA. Differentiation of phagocytic monocytes into lymph node dendritic cells in vivo. *Immunity* (1999) 11:753–61. doi:10.1016/S1074-7613(00)80149-1

10. Geissmann F, Jung S, Littman DR. Blood monocytes consist of two principal subsets with distinct migratory properties. *Immunity* (2003) 19:71–82. doi:10.1016/S1074-7613(03)00174-2

11. Serbina NV, Salazar-Mather TP, Biron CA, Kuziel WA, Pamer EG. TNF/iNOS-producing dendritic cells mediate innate immune defense against bacterial infection. *Immunity* (2003) 19:59–70. doi:10.1016/S1074-7613(03)00171-7

12. Naik SH, Metcalf D, Van Nieuwenhuijze A, Wicks I, Wu L, O'Keeffe M, et al. Intrasplenic steady-state dendritic cell precursors that are distinct from monocytes. *Nat Immunol* (2006) 7:663–71. doi:10.1038/ni1340

13. Varol C, Landsman L, Fogg DK, Greenshtein L, Gildor B, Margalit R, et al. Monocytes give rise to mucosal, but not splenic, conventional dendritic cells. *J Exp Med* (2007) 204:171–80. doi:10.1084/jem.20061011

14. Bogunovic M, Ginhoux F, Helft J, Shang L, Hashimoto D, Greter M, et al. Origin of the lamina propria dendritic cell network. *Immunity* (2009) 31(3):513–25. doi:10.1016/j.immuni.2009.08.010

15. Varol C, Vallon-Eberhard A, Elinav E, Aychek T, Shapira Y, Luche H, et al. Intestinal lamina propria dendritic cell subsets have different origin and functions. *Immunity* (2009) 31:502–12. doi:10.1016/j.immuni.2009.06.025

16. Yona S, Gordon S. From the reticulo-endothelial to mononuclear phagocyte system – the unaccounted years. *Front Immunol* (2015) 6:328. doi:10.3389/fimmu.2015.00328

17. Diao J, Winter E, Cantin C, Chen W, Xu L, Kelvin D, et al. In situ replication of immediate dendritic cell (DC) precursors contributes to conventional DC homeostasis in lymphoid tissue. *J Immunol* (2006) 176:7196–206. doi:10.4049/jimmunol.176.12.7196

18. Naik SH, Sathe P, Park HY, Metcalf D, Proietto AI, Dakic A, et al. Development of plasmacytoid and conventional dendritic cell subtypes from single precursor cells derived in vitro and in vivo. *Nat Immunol* (2007) 8:1217–26. doi:10.1038/ni1522

19. Onai N, Obata-Onai A, Schmid MA, Ohteki T, Jarrossay D, Manz MG. Identification of clonogenic common Flt3+M-CSFR+ plasmacytoid and conventional dendritic cell progenitors in mouse bone marrow. *Nat Immunol* (2007) 8:1207–16. doi:10.1038/ni1518

20. Merad M, Sathe P, Helft J, Miller J, Mortha A. The dendritic cell lineage: ontogeny and function of dendritic cells and their subsets in the steady state

and the inflamed setting. *Annu Rev Immunol* (2013) 31:563–604. doi:10.1146/annurev-immunol-020711-074950

21. Saunders D, Lucas K, Ismaili J, Wu L, Maraskovsky E, Dunn A, et al. Dendritic cell development in culture from thymic precursor cells in the absence of granulocyte/macrophage colony-stimulating factor. *J Exp Med* (1996) 184:2185–96. doi:10.1084/jem.184.6.2185

22. Pulendran B, Banchereau J, Burkeholder S, Kraus E, Guinet E, Chalouni C, et al. Flt3-ligand and granulocyte colony-stimulating factor mobilize distinct human dendritic cell subsets in vivo. *J Immunol* (2000) 165:566–72. doi:10.4049/jimmunol.165.1.566

23. Naik SH, Proietto AI, Wilson NS, Dakic A, Schnorrer P, Fuchsberger M, et al. Cutting edge: generation of splenic CD8+ and CD8- dendritic cell equivalents in Fms-like tyrosine kinase 3 ligand bone marrow cultures. *J Immunol* (2005) 174:6592–7. doi:10.4049/jimmunol.174.11.6592

24. Salomon B, Cohen JL, Masurier C, Klatzmann D. Three populations of mouse lymph node dendritic cells with different origins and dynamics. *J Immunol* (1998) 160:708–17.

25. McKenna HJ, Stocking KL, Miller RE, Brasel K, De Smedt T, Maraskovsky E, et al. Mice lacking flt3 ligand have deficient hematopoiesis affecting hematopoietic progenitor cells, dendritic cells, and natural killer cells. *Blood* (2000) 95:3489–97.

26. Karsunky H, Merad M, Cozzio A, Weissman IL, Manz MG. Flt3 ligand regulates dendritic cell development from Flt3+ lymphoid and myeloid-committed progenitors to Flt3+ dendritic cells in vivo. *J Exp Med* (2003) 198:305–13. doi:10.1084/jem.20030323

27. Waskow C, Liu K, Darrasse-Jeze G, Guermonprez P, Ginhoux F, Merad M, et al. The receptor tyrosine kinase Flt3 is required for dendritic cell development in peripheral lymphoid tissues. *Nat Immunol* (2008) 9:676–83. doi:10.1038/ni.1615

28. Ginhoux F, Liu K, Helft J, Bogunovic M, Greter M, Hashimoto D, et al. The origin and development of nonlymphoid tissue CD103+ DCs. *J Exp Med* (2009) 206:3115–30. doi:10.1084/jem.20091756

29. Onai N, Kurabayashi K, Hosoi-Amaike M, Toyama-Sorimachi N, Matsushima K, Inaba K, et al. A clonogenic progenitor with prominent plasmacytoid dendritic cell developmental potential. *Immunity* (2013) 38:943–57. doi:10.1016/j.immuni.2013.04.006

30. Hettinger J, Richards DM, Hansson J, Barra MM, Joschko AC, Krijgsveld J, et al. Origin of monocytes and macrophages in a committed progenitor. *Nat Immunol* (2013) 14:821–30. doi:10.1038/ni.2638

31. Breton G, Lee J, Zhou YJ, Schreiber JJ, Keler T, Puhr S, et al. Circulating precursors of human CD1c+ and CD141+ dendritic cells. *J Exp Med* (2015) 212:401–13. doi:10.1084/jem.20141441

32. Lee J, Breton G, Oliveira TY, Zhou YJ, Aljoufi A, Puhr S, et al. Restricted dendritic cell and monocyte progenitors in human cord blood and bone marrow. *J Exp Med* (2015) 212:385–99. doi:10.1084/jem.20141442

33. Ginhoux F, Greter M, Leboeuf M, Nandi S, See P, Gokhan S, et al. Fate mapping analysis reveals that adult microglia derive from primitive macrophages. *Science* (2010) 330:841–5. doi:10.1126/science.1194637

34. Schulz C, Gomez Perdiguero E, Chorro L, Szabo-Rogers H, Cagnard N, Kierdorf K, et al. A lineage of myeloid cells independent of Myb and hematopoietic stem cells. *Science* (2012) 336:86–90. doi:10.1126/science.1219179

35. Guilliams M, De Kleer I, Henri S, Post S, Vanhoutte L, De Prijck S, et al. Alveolar macrophages develop from fetal monocytes that differentiate into long-lived cells in the first week of life via GM-CSF. *J Exp Med* (2013) 210:1977–92. doi:10.1084/jem.20131199

36. Hashimoto D, Chow A, Noizat C, Teo P, Beasley MB, Leboeuf M, et al. Tissue-resident macrophages self-maintain locally throughout adult life with minimal contribution from circulating monocytes. *Immunity* (2013) 38:792–804. doi:10.1016/j.immuni.2013.04.004

37. Jakubzick C, Gautier EL, Gibbings SL, Sojka DK, Schlitzer A, Johnson TE, et al. Minimal differentiation of classical monocytes as they survey steady-state tissues and transport antigen to lymph nodes. *Immunity* (2013) 39:599–610. doi:10.1016/j.immuni.2013.08.007

38. Yona S, Kim KW, Wolf Y, Mildner A, Varol D, Breker M, et al. Fate mapping reveals origins and dynamics of monocytes and tissue macrophages under homeostasis. *Immunity* (2013) 38:79–91. doi:10.1016/j.immuni.2012.12.001

39. Gomez Perdiguero E, Klapproth K, Schulz C, Busch K, Azzoni E, Crozet L, et al. Tissue-resident macrophages originate from yolk-sac-derived erythro-myeloid progenitors. *Nature* (2015) 518:547–51. doi:10.1038/nature13989

40. Ginhoux F, Jung S. Monocytes and macrophages: developmental pathways and tissue homeostasis. *Nat Rev Immunol* (2014) **14**:392–404. doi:10.1038/nri3671

41. Scott CL, Henri S, Guilliams M. Mononuclear phagocytes of the intestine, the skin, and the lung. *Immunol Rev* (2014) **262**:9–24. doi:10.1111/imr.12220

42. Varol C, Mildner A, Jung S. Macrophages: development and tissue specialization. *Annu Rev Immunol* (2015) **33**:643–75. doi:10.1146/annurev-immunol-032414-112220

43. Hoeffel G, Wang Y, Greter M, See P, Teo P, Malleret B, et al. Adult Langerhans cells derive predominantly from embryonic fetal liver monocytes with a minor contribution of yolk sac-derived macrophages. *J Exp Med* (2012) **209**(6):1167–81. doi:10.1084/jem.20120340

44. Hoeffel G, Chen J, Lavin Y, Low D, Almeida FF, See P, et al. C-myb(+) erythro-myeloid progenitor-derived fetal monocytes give rise to adult tissue-resident macrophages. *Immunity* (2015) **42**:665–78. doi:10.1016/j.immuni.2015.03.011

45. Tamoutounour S, Henri S, Lelouard H, De Bovis B, De Haar C, Van Der Woude CJ, et al. CD64 distinguishes macrophages from dendritic cells in the gut and reveals the Th1-inducing role of mesenteric lymph node macrophages during colitis. *Eur J Immunol* (2012) **42**(12):3150–66. doi:10.1002/eji.201242847

46. Bain CC, Scott CL, Uronen-Hansson H, Gudjonsson S, Jansson O, Grip O, et al. Resident and pro-inflammatory macrophages in the colon represent alternative context-dependent fates of the same Ly6C(hi) monocyte precursors. *Mucosal Immunol* (2012) **6**(3):498–510. doi:10.1038/mi.2012.89

47. Plantinga M, Guilliams M, Vanheerswynghels M, Deswarte K, Branco-Madeira F, Toussaint W, et al. Conventional and monocyte-derived CD11b(+) dendritic cells initiate and maintain T helper 2 cell-mediated immunity to house dust mite allergen. *Immunity* (2013) **38**:322–35. doi:10.1016/j.immuni.2012.10.016

48. Tamoutounour S, Guilliams M, Montanana Sanchis F, Liu H, Terhorst D, Malosse C, et al. Origins and functional specialization of macrophages and of conventional and monocyte-derived dendritic cells in mouse skin. *Immunity* (2013) **39**:925–38. doi:10.1016/j.immuni.2013.10.004

49. Geurtsvankessel CH, Bergen IM, Muskens F, Boon L, Hoogsteden HC, Osterhaus AD, et al. Both conventional and interferon killer dendritic cells have antigen-presenting capacity during influenza virus infection. *PLoS One* (2009) **4**:e7187. doi:10.1371/journal.pone.0007187

50. Janssen WJ, Barthel L, Muldrow A, Oberley-Deegan RE, Kearns MT, Jakubzick C, et al. Fas determines differential fates of resident and recruited macrophages during resolution of acute lung injury. *Am J Respir Crit Care Med* (2011) **184**:547–60. doi:10.1164/rccm.201011-1891OC

51. Balan S, Ollion V, Colletti N, Chelbi R, Montanana-Sanchis F, Liu H, et al. Human XCR1+ dendritic cells derived in vitro from CD34+ progenitors closely resemble blood dendritic cells, including their adjuvant responsiveness, contrary to monocyte-derived dendritic cells. *J Immunol* (2014) **193**:1622–35. doi:10.4049/jimmunol.1401243

52. Desch AN, Gibbings SL, Clambey ET, Janssen WJ, Slansky JE, Kedl RM, et al. Dendritic cell subsets require cis-activation for cytotoxic CD8 T-cell induction. *Nat Commun* (2014) **5**:4674. doi:10.1038/ncomms5674

53. Guilliams M, Ginhoux F, Jakubzick C, Naik SH, Onai N, Schraml BU, et al. Dendritic cells, monocytes and macrophages: a unified nomenclature based on ontogeny. *Nat Rev Immunol* (2014) **14**:571–8. doi:10.1038/nri3712

54. Xue J, Schmidt SV, Sander J, Draffehn A, Krebs W, Quester I, et al. Transcriptome-based network analysis reveals a spectrum model of human macrophage activation. *Immunity* (2014) **40**:274–88. doi:10.1016/j.immuni.2014.01.006

55. Robbins SH, Walzer T, Dembele D, Thibault C, Defays A, Bessou G, et al. Novel insights into the relationships between dendritic cell subsets in human and mouse revealed by genome-wide expression profiling. *Genome Biol* (2008) **9**:R17. doi:10.1186/gb-2008-9-1-r17

56. Crozat K, Guiton R, Guilliams M, Henri S, Baranek T, Schwartz-Cornil I, et al. Comparative genomics as a tool to reveal functional equivalences between human and mouse dendritic cell subsets. *Immunol Rev* (2010) **234**:177–98. doi:10.1111/j.0105-2896.2009.00868.x

57. Guilliams M, Henri S, Tamoutounour S, Ardouin L, Schwartz-Cornil I, Dalod M, et al. From skin dendritic cells to a simplified classification of human and mouse dendritic cell subsets. *Eur J Immunol* (2010) **40**:2089–94. doi:10.1002/eji.201040498

58. Miller MA, Feng XJ, Li G, Rabitz HA. Identifying biological network structure, predicting network behavior, and classifying network state with high dimensional model representation (HDMR). *PLoS One* (2012) **7**:e37664. doi:10.1371/journal.pone.0037664

59. Vu Manh TP, Marty H, Sibille P, Le Vern Y, Kaspers B, Dalod M, et al. Existence of conventional dendritic cells in gallus gallus revealed by comparative gene expression profiling. *J Immunol* (2014) **192**(10):4510–7. doi:10.4049/jimmunol.1303405

60. Vu Manh TP, Bertho N, Hosmalin A, Schwartz-Cornil I, Dalod M. Investigating evolutionary conservation of dendritic cell subset identity and functions. *Front Immunol* (2015) **6**:260. doi:10.3389/fimmu.2015.00260

61. Vu Manh TP, Elhmouzi-Younes J, Urien C, Ruscanu S, Jouneau L, Bourge M, et al. Defining mononuclear phagocyte subset homology across several distant warm-blooded vertebrates through comparative transcriptomics. *Front Immunol* (2015) **6**:299. doi:10.3389/fimmu.2015.00299

62. Durand M, Segura E. The known unknowns of the human dendritic cell network. *Front Immunol* (2015) **6**:129. doi:10.3389/fimmu.2015.00129

63. Reynolds G, Haniffa M. Human and mouse mononuclear phagocyte networks: a tale of two species? *Front Immunol* (2015) **6**:330. doi:10.3389/fimmu.2015.00330

64. Bachem A, Guttler S, Hartung E, Ebstein F, Schaefer M, Tannert A, et al. Superior antigen cross-presentation and XCR1 expression define human CD11c+CD141+ cells as homologues of mouse CD8+ dendritic cells. *J Exp Med* (2010) **207**:1273–81. doi:10.1084/jem.20100348

65. Contreras V, Urien C, Guiton R, Alexandre Y, Vu Manh TP, Andrieu T, et al. Existence of CD8alpha-like dendritic cells with a conserved functional specialization and a common molecular signature in distant mammalian species. *J Immunol* (2010) **185**:3313–25. doi:10.4049/jimmunol.1000824

66. Crozat K, Guiton R, Contreras V, Feuillet V, Dutertre CA, Ventre E, et al. The XC chemokine receptor 1 is a conserved selective marker of mammalian cells homologous to mouse CD8alpha+ dendritic cells. *J Exp Med* (2010) **207**:1283–92. doi:10.1084/jem.20100223

67. Crozat K, Tamoutounour S, Vu Manh TP, Fossum E, Luche H, Ardouin L, et al. Cutting edge: expression of XCR1 defines mouse lymphoid-tissue resident and migratory dendritic cells of the CD8alpha+ type. *J Immunol* (2011) **187**:4411–5. doi:10.4049/jimmunol.1101717

68. Bachem A, Hartung E, Guttler S, Mora A, Zhou X, Hegemann A, et al. Expression of XCR1 characterizes the Batf3-dependent lineage of dendritic cells capable of antigen cross-presentation. *Front Immunol* (2012) **3**:214. doi:10.3389/fimmu.2012.00214

69. Becker M, Guttler S, Bachem A, Hartung E, Mora A, Jakel A, et al. Ontogenic, phenotypic, and functional characterization of XCR1(+) dendritic cells leads to a consistent classification of intestinal dendritic cells based on the expression of XCR1 and SIRPalpha. *Front Immunol* (2014) **5**:326. doi:10.3389/fimmu.2014.00326

70. Gurka S, Hartung E, Becker M, Kroczek RA. Mouse conventional dendritic cells can be universally classified based on the mutually exclusive expression of XCR1 and SIRPalpha. *Front Immunol* (2015) **6**:35. doi:10.3389/fimmu.2015.00035

71. Grajales-Reyes GE, Iwata A, Albring J, Wu X, Tussiwand R, Kc W, et al. Batf3 maintains autoactivation of Irf8 for commitment of a CD8alpha conventional DC clonogenic progenitor. *Nat Immunol* (2015) **16**(7):708–17. doi:10.1038/ni.3197

72. Schlitzer A, Sivakamasundari V, Chen J, Sumatoh HR, Schreuder J, Lum J, et al. Identification of cDC1- and cDC2-committed DC progenitors reveals early lineage priming at the common DC progenitor stage in the bone marrow. *Nat Immunol* (2015) **16**(7):718–28. doi:10.1038/ni.3200

73. Hildner K, Edelson BT, Purtha WE, Diamond M, Matsushita H, Kohyama M, et al. Batf3 deficiency reveals a critical role for CD8alpha+ dendritic cells in cytotoxic T cell immunity. *Science* (2008) **322**:1097–100. doi:10.1126/science.1164206

74. Edelson BT, Kc W, Juang R, Kohyama M, Benoit LA, Klekotka PA, et al. Peripheral CD103+ dendritic cells form a unified subset developmentally related to CD8alpha+ conventional dendritic cells. *J Exp Med* (2010) **207**:823–36. doi:10.1084/jem.20091627

75. Hacker C, Kirsch RD, Ju XS, Hieronymus T, Gust TC, Kuhl C, et al. Transcriptional profiling identifies Id2 function in dendritic cell development. *Nat Immunol* (2003) **4**:380–6. doi:10.1038/ni903

76. Jackson JT, Hu Y, Liu R, Masson F, D'Amico A, Carotta S, et al. Id2 expression delineates differential checkpoints in the genetic program of CD8alpha+ and CD103+ dendritic cell lineages. *EMBO J* (2011) **30**:2690–704. doi:10.1038/emboj.2011.163

77. Kashiwada M, Pham NL, Pewe LL, Harty JT, Rothman PB. NFIL3/E4BP4 is a key transcription factor for CD8alpha(+) dendritic cell development. *Blood* (2011) **117**:6193–7. doi:10.1182/blood-2010-07-295873

78. Schiavoni G, Mattei F, Sestili P, Borghi P, Venditti M, Morse HC III, et al. ICSBP is essential for the development of mouse type I interferon-producing cells and for the generation and activation of CD8alpha(+) dendritic cells. *J Exp Med* (2002) **196**:1415–25. doi:10.1084/jem.20021263

79. Tamura T, Tailor P, Yamaoka K, Kong HJ, Tsujimura H, O'Shea JJ, et al. IFN regulatory factor-4 and -8 govern dendritic cell subset development and their functional diversity. *J Immunol* (2005) **174**:2573–81. doi:10.4049/jimmunol.174.5.2573

80. Tailor P, Tamura T, Morse HC III, Ozato K. The BXH2 mutation in IRF8 differentially impairs dendritic cell subset development in the mouse. *Blood* (2008) **111**:1942–5. doi:10.1182/blood-2007-07-100750

81. Wu L, D'Amico A, Winkel KD, Suter M, Lo D, Shortman K. RelB is essential for the development of myeloid-related CD8alpha- dendritic cells but not of lymphoid-related CD8alpha+ dendritic cells. *Immunity* (1998) **9**:839–47. doi:10.1016/S1074-7613(00)80649-4

82. Caton ML, Smith-Raska MR, Reizis B. Notch-RBP-J signaling controls the homeostasis of CD8- dendritic cells in the spleen. *J Exp Med* (2007) **204**:1653–64. doi:10.1084/jem.20062648

83. Suzuki S, Honma K, Matsuyama T, Suzuki K, Toriyama K, Akitoyo I, et al. Critical roles of interferon regulatory factor 4 in CD11bhighCD8alpha- dendritic cell development. *Proc Natl Acad Sci U S A* (2004) **101**:8981–6. doi:10.1073/pnas.0402139101

84. Persson EK, Uronen-Hansson H, Semmrich M, Rivollier A, Hagerbrand K, Marsal J, et al. IRF4 transcription-factor-dependent CD103CD11b dendritic cells drive mucosal T helper 17 cell differentiation. *Immunity* (2013) **38**(5):958–69. doi:10.1016/j.immuni.2013.03.009

85. Schlitzer A, Mcgovern N, Teo P, Zelante T, Atarashi K, Low D, et al. IRF4 transcription factor-dependent CD11b+ dendritic cells in human and mouse control mucosal IL-17 cytokine responses. *Immunity* (2013) **38**:970–83. doi:10.1016/j.immuni.2013.04.011

86. Cisse B, Caton ML, Lehner M, Maeda T, Scheu S, Locksley R, et al. Transcription factor E2-2 is an essential and specific regulator of plasmacytoid dendritic cell development. *Cell* (2008) **135**:37–48. doi:10.1016/j.cell.2008.09.016

87. Ghosh HS, Cisse B, Bunin A, Lewis KL, Reizis B. Continuous expression of the transcription factor e2-2 maintains the cell fate of mature plasmacytoid dendritic cells. *Immunity* (2010) **33**:905–16. doi:10.1016/j.immuni.2010.11.023

88. Benson MJ, Pino-Lagos K, Rosemblatt M, Noelle RJ. All-trans retinoic acid mediates enhanced T reg cell growth, differentiation, and gut homing in the face of high levels of co-stimulation. *J Exp Med* (2007) **204**:1765–74. doi:10.1084/jem.20070719

89. Coombes JL, Siddiqui KR, Arancibia-Carcamo CV, Hall J, Sun CM, Belkaid Y, et al. A functionally specialized population of mucosal CD103+ DCs induces Foxp3+ regulatory T cells via a TGF-beta and retinoic acid-dependent mechanism. *J Exp Med* (2007) **204**:1757–64. doi:10.1084/jem.20070590

90. Sun CM, Hall JA, Blank RB, Bouladoux N, Oukka M, Mora JR, et al. Small intestine lamina propria dendritic cells promote de novo generation of Foxp3 T reg cells via retinoic acid. *J Exp Med* (2007) **204**:1775–85. doi:10.1084/jem.20070602

91. Guilliams M, Crozat K, Henri S, Tamoutounour S, Grenot P, Devilard E, et al. Skin-draining lymph nodes contain dermis-derived CD103(-) dendritic cells that constitutively produce retinoic acid and induce Foxp3(+) regulatory T cells. *Blood* (2010) **115**:1958–68. doi:10.1182/blood-2009-09-245274

92. Gatto D, Wood K, Caminschi I, Murphy-Durland D, Schofield P, Christ D, et al. The chemotactic receptor EBI2 regulates the homeostasis, localization and immunological function of splenic dendritic cells. *Nat Immunol* (2013) **14**:446–53. doi:10.1038/ni.2555

93. Lewis KL, Caton ML, Bogunovic M, Greter M, Grajkowska LT, Ng D, et al. Notch2 receptor signaling controls functional differentiation of dendritic cells in the spleen and intestine. *Immunity* (2011) **35**:780–91. doi:10.1016/j.immuni.2011.08.013

94. Satpathy AT, Briseno CG, Lee JS, Ng D, Manieri NA, Kc W, et al. Notch2-dependent classical dendritic cells orchestrate intestinal immunity to attaching-and-effacing bacterial pathogens. *Nat Immunol* (2013) **14**:937–48. doi:10.1038/ni.2679

95. Tussiwand R, Everts B, Grajales-Reyes GE, Kretzer NM, Iwata A, Bagaitkar J, et al. Klf4 expression in conventional dendritic cells is required for T helper 2 cell responses. *Immunity* (2015) **42**:916–28. doi:10.1016/j.immuni.2015.04.017

96. Gao Y, Nish SA, Jiang R, Hou L, Licona-Limon P, Weinstein JS, et al. Control of T helper 2 responses by transcription factor IRF4-dependent dendritic cells. *Immunity* (2013) **39**:722–32. doi:10.1016/j.immuni.2013.08.028

97. Chou DB, Sworder B, Bouladoux N, Roy CN, Uchida AM, Grigg M, et al. Stromal-derived IL-6 alters the balance of myeloerythroid progenitors during *Toxoplasma gondii* infection. *J Leukoc Biol* (2012) **92**:123–31. doi:10.1189/jlb.1011527

98. Chorro L, Sarde A, Li M, Woollard KJ, Chambon P, Malissen B, et al. Langerhans cell (LC) proliferation mediates neonatal development, homeostasis, and inflammation-associated expansion of the epidermal LC network. *J Exp Med* (2009) **206**:3089–100. doi:10.1084/jem.20091586

99. Jaitin DA, Kenigsberg E, Keren-Shaul H, Elefant N, Paul F, Zaretsky I, et al. Massively parallel single-cell RNA-seq for marker-free decomposition of tissues into cell types. *Science* (2014) **343**:776–9. doi:10.1126/science.1247651

100. Zigmond E, Samia-Grinberg S, Pasmanik-Chor M, Brazowski E, Shibolet O, Halpern Z, et al. Infiltrating monocyte-derived macrophages and resident Kupffer cells display different ontogeny and functions in acute liver injury. *J Immunol* (2014) **193**(1):344–53. doi:10.4049/jimmunol.1400574

101. Ajami B, Bennett JL, Krieger C, Tetzlaff W, Rossi FM. Local self-renewal can sustain CNS microglia maintenance and function throughout adult life. *Nat Neurosci* (2007) **10**:1538–43. doi:10.1038/nn2014

102. Ajami B, Bennett JL, Krieger C, Mcnagny KM, Rossi FM. Infiltrating monocytes trigger EAE progression, but do not contribute to the resident microglia pool. *Nat Neurosci* (2011) **14**:1142–9. doi:10.1038/nn.2887

103. Yamasaki R, Lu H, Butovsky O, Ohno N, Rietsch AM, Cialic R, et al. Differential roles of microglia and monocytes in the inflamed central nervous system. *J Exp Med* (2014) **211**:1533–49. doi:10.1084/jem.20132477

104. Gautier EL, Shay T, Miller J, Greter M, Jakubzick C, Ivanov S, et al. Gene-expression profiles and transcriptional regulatory pathways that underlie the identity and diversity of mouse tissue macrophages. *Nat Immunol* (2012) **13**:1118–28. doi:10.1038/ni.2419

105. Lavin Y, Winter D, Blecher-Gonen R, David E, Keren-Shaul H, Merad M, et al. Tissue-resident macrophage enhancer landscapes are shaped by the local microenvironment. *Cell* (2014) **159**:1312–26. doi:10.1016/j.cell.2014.11.018

106. Murray PJ, Allen JE, Biswas SK, Fisher EA, Gilroy DW, Goerdt S, et al. Macrophage activation and polarization: nomenclature and experimental guidelines. *Immunity* (2014) **41**:14–20. doi:10.1016/j.immuni.2014.06.008

107. Jha AK, Huang SC, Sergushichev A, Lampropoulou V, Ivanova Y, Loginicheva E, et al. Network integration of parallel metabolic and transcriptional data reveals metabolic modules that regulate macrophage polarization. *Immunity* (2015) **42**:419–30. doi:10.1016/j.immuni.2015.02.005

108. Siddiqui KR, Laffont S, Powrie F. E-cadherin marks a subset of inflammatory dendritic cells that promote T cell-mediated colitis. *Immunity* (2010) **32**:557–67. doi:10.1016/j.immuni.2010.03.017

109. Guilliams M, Movahedi K, Bosschaerts T, Vandendriessche T, Chuah MK, Herin M, et al. IL-10 dampens TNF/inducible nitric oxide synthase-producing dendritic cell-mediated pathogenicity during parasitic infection. *J Immunol* (2009) **182**:1107–18. doi:10.4049/jimmunol.182.2.1107

110. Kuipers H, Soullie T, Hammad H, Willart M, Kool M, Hijdra D, et al. Sensitization by intratracheally injected dendritic cells is independent of antigen presentation by host antigen-presenting cells. *J Leukoc Biol* (2009) **85**:64–70. doi:10.1189/jlb.0807519

111. Cheong C, Matos I, Choi JH, Dandamudi DB, Shrestha E, Longhi MP, et al. Microbial stimulation fully differentiates monocytes to DC-SIGN/CD209(+) dendritic cells for immune T cell areas. *Cell* (2010) **143**:416–29. doi:10.1016/j.cell.2010.09.039

112. Shalek AK, Satija R, Adiconis X, Gertner RS, Gaublomme JT, Raychowdhury R, et al. Single-cell transcriptomics reveals bimodality in expression and splicing in immune cells. *Nature* (2013) **498**:236–40. doi:10.1038/nature12172

113. Ohl L, Mohaupt M, Czeloth N, Hintzen G, Kiafard Z, Zwirner J, et al. CCR7 governs skin dendritic cell migration under inflammatory and steady-state conditions. *Immunity* (2004) **21**:279–88. doi:10.1016/j.immuni.2004.06.014

114. Jang MH, Sougawa N, Tanaka T, Hirata T, Hiroi T, Tohya K, et al. CCR7 is critically important for migration of dendritic cells in intestinal lamina propria to mesenteric lymph nodes. *J Immunol* (2006) **176**:803–10. doi:10.4049/jimmunol.176.2.803

115. Guilliams M, Malissen B. A death notice for in-vitro-generated GM-CSF dendritic cells? *Immunity* (2015) **42**:988–90. doi:10.1016/j.immuni.2015.05.020

116. Helft J, Bottcher J, Chakravarty P, Zelenay S, Huotari J, Schraml BU, et al. GM-CSF mouse bone marrow cultures comprise a heterogeneous population of CD11c(+)MHCII(+) macrophages and dendritic cells. *Immunity* (2015) **42**:1197–211. doi:10.1016/j.immuni.2015.05.018

117. Wimmers F, Schreibelt G, Skold AE, Figdor CG, De Vries IJ. Paradigm shift in dendritic cell-based immunotherapy: from in vitro generated monocyte-derived DCs to naturally circulating DC subsets. *Front Immunol* (2014) **5**:165. doi:10.3389/fimmu.2014.00165

118. Poulin LF, Salio M, Griessinger E, Anjos-Afonso F, Craciun L, Chen JL, et al. Characterization of human DNGR-1+ BDCA3+ leukocytes as putative equivalents of mouse CD8alpha+ dendritic cells. *J Exp Med* (2010) **207**:1261–71. doi:10.1084/jem.20092618

119. Bain CC, Bravo-Blas A, Scott CL, Gomez Perdiguero E, Geissmann F, Henri S, et al. Constant replenishment from circulating monocytes maintains the macrophage pool in the intestine of adult mice. *Nat Immunol* (2014) **15**:929–37. doi:10.1038/ni.2967

120. Epelman S, Lavine KJ, Beaudin AE, Sojka DK, Carrero JA, Calderon B, et al. Embryonic and adult-derived resident cardiac macrophages are maintained through distinct mechanisms at steady state and during inflammation. *Immunity* (2014) **40**:91–104. doi:10.1016/j.immuni.2013.11.019

121. Molawi K, Wolf Y, Kandalla PK, Favret J, Hagemeyer N, Frenzel K, et al. Progressive replacement of embryo-derived cardiac macrophages with age. *J Exp Med* (2014) **211**:2151–8. doi:10.1084/jem.20140639

122. Swirski FK, Nahrendorf M, Etzrodt M, Wildgruber M, Cortez-Retamozo V, Panizzi P, et al. Identification of splenic reservoir monocytes and their deployment to inflammatory sites. *Science* (2009) **325**:612–6. doi:10.1126/science.1175202

123. Schraml BU, Van Blijswijk J, Zelenay S, Whitney PG, Filby A, Acton SE, et al. Genetic tracing via DNGR-1 expression history defines dendritic cells as a hematopoietic lineage. *Cell* (2013) **154**:843–58. doi:10.1016/j.cell.2013.07.014

124. Poltorak MP, Schraml BU. Fate mapping of dendritic cells. *Front Immunol* (2015) **6**:199. doi:10.3389/fimmu.2015.00199

9

Transcriptional regulation of mononuclear phagocyte development

Roxane Tussiwand[1]* and Emmanuel L. Gautier[2]

[1] *Department of Biomedicine, University of Basel, Basel, Switzerland,* [2] *INSERM UMR_S 1166, Sorbonne Universités, UPMC Univ Paris 06, Pitié-Salpêtrière Hospital, Paris, France*

Edited by:
Martin Guilliams,
Ghent University, Belgium

Reviewed by:
Elodie Segura,
Institut Curie, France
Vuk Cerovic,
RWTH Aachen, Germany

*Correspondence:
Roxane Tussiwand
r.tussiwand@unibas.ch

Mononuclear phagocytes (MP) are a quite unique subset of hematopoietic cells, which comprise dendritic cells (DC), monocytes as well as monocyte-derived and tissue-resident macrophages. These cells are extremely diverse with regard to their origin, their phenotype as well as their function. Developmentally, DC and monocytes are constantly replenished from a bone marrow hematopoietic progenitor. The ontogeny of macrophages is more complex and is temporally linked and specified by the organ where they reside, occurring early during embryonic or perinatal life. The functional heterogeneity of MPs is certainly a consequence of the tissue of residence and also reflects the diverse ontogeny of the subsets. In this review, we will highlight the developmental pathways of murine MP, with a particular emphasis on the transcriptional factors that regulate their development and function. Finally, we will discuss and point out open questions in the field.

Keywords: transcription factors, development, dendritic cells, macrophages, immunity

INTRODUCTION

The mononuclear-phagocyte system (MPS), which comprises dendritic cells (DCs), macrophages, and monocytes, is a heterogeneous group of myeloid cells. The complexity of the MPS is equally reflected by the plasticity in function and phenotype that characterizes each subset depending on their location and activation state. Specialized subsets of mononuclear phagocytes (MP) reside in defined anatomical locations, are critical for the homeostatic maintenance of tissues, and provide the link between innate and adaptive immune responses during infections. The ability of MP to maintain or to induce the correct tolerogenic or inflammatory milieu also resides in their complex subset specialization. Such subset heterogeneity is obtained through lineage diversification and specification, which is controlled by defined transcriptional networks and programs. Understanding the MP biology means to define their transcriptional signature, which is required during lineage commitment, and which characterizes each subset's features. This review will focus on the transcriptional regulation of the MPS; in particular, what determines lineage commitment and functional identity; we will emphasize recent advances in the field of single-cell analysis and highlight unresolved questions in the field.

THE MONONUCLEAR-PHAGOCYTE SYSTEM NETWORK

As summarized in **Table 1**, the MPS is a rather heterogeneous group of myeloid cells, which includes DC, monocytes, and macrophages (1). DCs are mostly short lived and characterized by a half-life that varies between few days up to few weeks (2). This subset of MPs is equipped with pattern recognition receptors (PRR) and is specialized in antigen capture and presentation to T cells (3). At least three different DC subsets have been identified: plasmacytoid DCs (pDCs), and two common or conventional DC (cDC) subsets; cDC1, which express CD24, and CD8α in lymphoid tissues, or CD103 in peripheral organs; and cDC2, which express CD4, CD11b, and CD172 (1). This latter subset of cDCs is heterogeneous and seems to comprise also monocyte-derived DCs and activated macrophages, which have acquired a DC phenotype and most likely function (4).

Steady-state monocytes are short-lived MPs. They are sub-divided into two major subsets: patrolling and inflammatory monocytes, which are characterized by low and high expression of Ly-6C, respectively (5). Inflammatory monocytes are recruited and extravasate into infected tissues. They play a role in main-taining the correct inflammatory milieu, are important in the resolution of inflammation and in certain tissues monocytes will replenish the pool of resident macrophages (5–7). The role of patrolling monocytes is less clear but they are certainly involved in the homeostasis of the endothelium (8, 9).

The last subset of MP comprises the mostly long-lived tissue-resident macrophages (10). This subset is present in every devel-oping as well as mature tissue, which is highly heterogeneous in terms of phenotype and function, reflecting the physiological needs of the organ of origin (11). Macrophages are thought to be required for the correct development and maintenance of tissues. This topological-related feature is possibly the reason for their extreme heterogeneity and their tissue specialization (12).

Collectively, MPs are highly plastic myeloid cells, which can perform very diverse functions. **Table 1** summarizes the mostly used surface markers in mice and the function attributed to the different MP subsets.

TRANSCRIPTIONAL REGULATION OF DENDRITIC CELLS DEVELOPMENT

As shown in **Figure 1**, lineage development of hematopoietic progenitor cells along DC lineage occurs through an orchestrated expression pattern of transcription factors (TF), yet the precise molecular mechanisms of lineage restriction and determination remains largely unexplained (2, 13–17). The analysis of gene-targeted mice has revealed the functional importance of a few

TABLE 1 | Summarized are the three major murine MPs: dendritic cells, monocytes, and macrophages.

MPs	Subset	Surface MK	Functions
Dendritic cells	pDCs	SiglecH, Bst2	Production of type 1 IFN (antiviral response)
	cDC1	XCR1, CD103/CD8, Clec9a	Th1 and CTL immunity, cross-presentation, IL-12 production
	cDC2	CD11b, Sirp-α	Th2 and Th17 immunity, production of IL-23 and IL-6
Monocytes	Ly6C high inflammatory	Ly6C hi CCR2 hi	Differentiate into DCs and tissue macrophages during inflammation
	Ly6C low patrolling	Ly6C low CCR2 low Cx3Cr1	Endothelial integrity
Macrophages	Tissue specific	F4/80, MerTK, CD64 CD11b	Tissue specific

Each MP subset can be further subdivided into different subsets based on surface marker expression and function and as indicated.

critical TFs in DC development, with some of them affecting all DCs and some affecting specific subsets (18). DC progenitors are present within the fms-related tyrosine kinase 3 (Flt3)-expressing bone marrow fraction and sustained Flt3 signaling can be con-sidered as instructive for DC development (19–22). Consistently, Flt3-ligand (Flt3L) supports the *in vitro* differentiation of progeni-tor cells into both pDCs and cDCs (23, 24). Genetic deletion of Flt3L, its receptor, or treatment of mice with Flt3 inhibitors leads to a 10-fold reduction of lymphoid-organ pDCs and cDCs (25, 26). Moreover, Flt3L injection or overexpression of Flt3L results in the expansion of both pDCs and cDCs in all lymphoid and non-lymphoid organs (27, 28). Engagement of Flt3 by Flt3L induces Stat3 phosphorylation and activation, identifying Stat3 as the critical checkpoint of Flt3-induced DC development and proliferation (29, 30). Mirroring Flt3 deficiency, Stat3-deficient mice have severely reduced DC progenitors and mature cells (29). Similarly, deletion of the transcriptional repressor growth factor independent 1 (Gfi1) results in impaired DC development (31). Gfi1-deficient mice show reduced Stat3 phosphorylation and nuclear translocation, with increased expression levels of the Stat3 negative regulators SOCS3 and PIAS3 suggesting that Gfi1 is downstream of Stat3 signaling in the Flt3-Flt3L-induced DC developmental pathway (31). However, the role of Gfi1 is more complex since mice deficient for this repressor show multiple hematopoietic impairments (32, 33). The defects related to Gfi1 deficiency can partially be related to dysregulation of Id2 expression (34–36). However, further studies using subset-specific deletion models will be instrumental to precisely dissect specific transcrip-tional requirements within the MP lineage. Similarly, despite the experimental evidence of DC expansion following sustained Flt3 signaling, the instructive mechanism promoting DC development is still unclear, given the broad expression of Flt3 on all short-term uncommitted hematopoietic progenitors (ST-HSC) (37, 38). A long non-coding RNA (lncRNA), named lnc-DC, was recently suggested to be the missing key element regulating Stat3 activity exclusively in DCs (39). lnc-DC RNA is expressed by mature DCs and by monocyte-derived DCs and seems to directly interact with Stat3 preventing its de-phosphorylation by SHP1. Furthermore, knockdown experiments of lnc-DC *in vitro* showed impaired DC development from mouse BM progenitors. The conservation of this lnc-DC in terms of function and of its consensus elements at the promoter region across species supports the hypothesis of a new level of regulation present in DC development. However, in mice the transcript seems translated into a highly expressed protein in adipose tissue (40). Further studies are therefore needed to understand potential species-specificities as well as its require-ment *in vivo* under steady-state conditions.

Proceeding along the DC developmental pathway, three major branches of mature DCs are identified: pDCs, CD24+ cDC1, and CD11b+ cDC2 (3, 16). pDCs and cDC1 both express and depend on the transcription factor interferon regulatory factor 8 (Irf8), while cDC2 express and are partially dependent on Irf4 (1, 18, 41–44). Despite major advances in our understanding of the transcriptional requirement during DC development, we are still unable to draw a clear developmental map (**Figure 2**) (13, 18). This may reflect subset heterogeneity as well as the plasticity, which characterizes DCs. Also, the expression of the different TFs

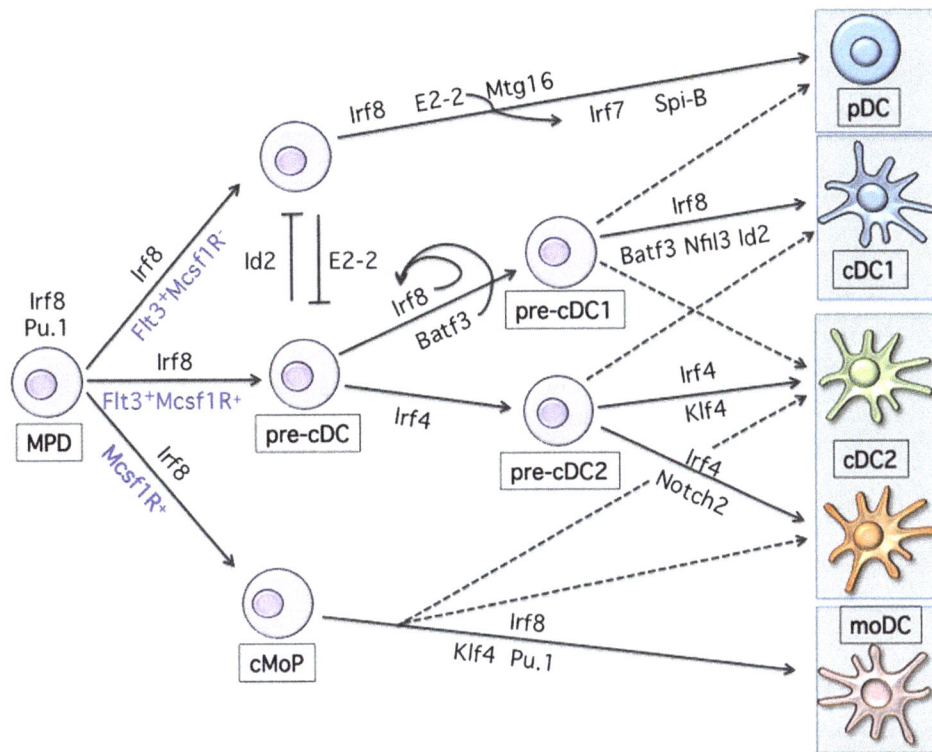

FIGURE 1 | Transcriptional development of dendritic cells. Shown are the major transcription factors known to be involved in DC lineage commitment. Development occurs from a Flt3-, Irf8-expressing hematopoietic progenitor. Progressive acquisition of one or more TFs will result in differentiation toward a specific MP subset. Loss or reduction of one or more TFs can, to some extent, redirect commitment to another lineage.

is not unique and can change during differentiation and activation further complicating the picture.

During early stages of DC development, a progenitor that expresses high levels of Irf8 and shows developmental potential toward all DCs can be identified (42). It is likely that the first branching choice will determine whether pDCs or cDCs commitment occurs. The balance of the E-protein transcription factor 4 (Tcf4), also known as E2-2, and the E-protein inhibitor of DNA binding 2 (Id2) seems to determine lineage development toward pDCs or cDCs, respectively (45–50). Constitutive or inducible deletion of E2-2 in CD11c-expressing cells blocks the development of pDCs but not cDCs, while overexpression of Id2 inhibits pDC development (47, 48). E2-2 is required not only during development but also for lineage maintenance of pDCs (47, 51). Several targets of E2-2 have been identified such as SpiB, Irf8, and Irf7, and all contribute to pDCs lineage specification (47, 51). Despite the requirement for E2-2 during pDCs commitment, how Id2 and E2-2 are conversely induced and regulated is still an open question. Recently, the eight-twenty-one (ETO) protein 2 or Mtg16 (also referred as core-binding factor, runt domain, alpha subunit 2, translocated to 3 Cbfa2t3) was suggested to target and repress Id2 together with E2-2 and inhibit Irf8-expressing cDC1 development, while favoring pDC commitment (52). Consistently, Id2 and Mtg16 double-deficient mice show restored pDC potential (52). However, Mtg16 seems to act together with E2-2 leaving the question on how lineage determination toward E2-2- or Id2-expressing progenitors

occurs, still open. On the other side, one other candidate, which could be involved in reinforcing lineage fate toward Irf8-expressing cDC1 at the expenses of pDCs could be the leucine zipper transcription factor E4BP4, also referred as Nfil3 (53). Mice deficient for this TF show increased pDC and reduced cDC1 development (53). The mechanism of action remains to be elucidated since Id2 expression does not appear to be perturbed and only the basic leucine zipper transcription factor ATF-like 3 (Batf3) expression was shown to be reduced (53). Phenotypically, a bias toward pDC development has been observed within the macrophage-colony-stimulating factor receptor (M-CSFR) negative progenitors, whereas cDCs precursors are enriched within the M-CSFR expressing BM fraction (54, 55). These results may suggest that under sustained M-CSF stimulation uncommitted progenitors may lose the potential toward pDCs. Alternatively, as recently suggested, the absence of GM-CSF signaling, which induces STAT5 phosphorylation, could be the permissive condition to promote pDC development (56). Accumulation and/or withdrawal of specific cytokines during proliferation and differentiation as well as regulation of TF levels through division of progenitor cells could partially explain how BM niches influence development and lineage commitment (57, 58).

Proceeding along DC development, a common cDC progenitor able to differentiate *in vivo* into both CD24[+] cDC1 and CD11b[+] cDC2 was identified (55, 59–61). And recently, lineage-tracing studies allowed further dissection of cDC commitment and resulted fundamental to establish the transcriptional

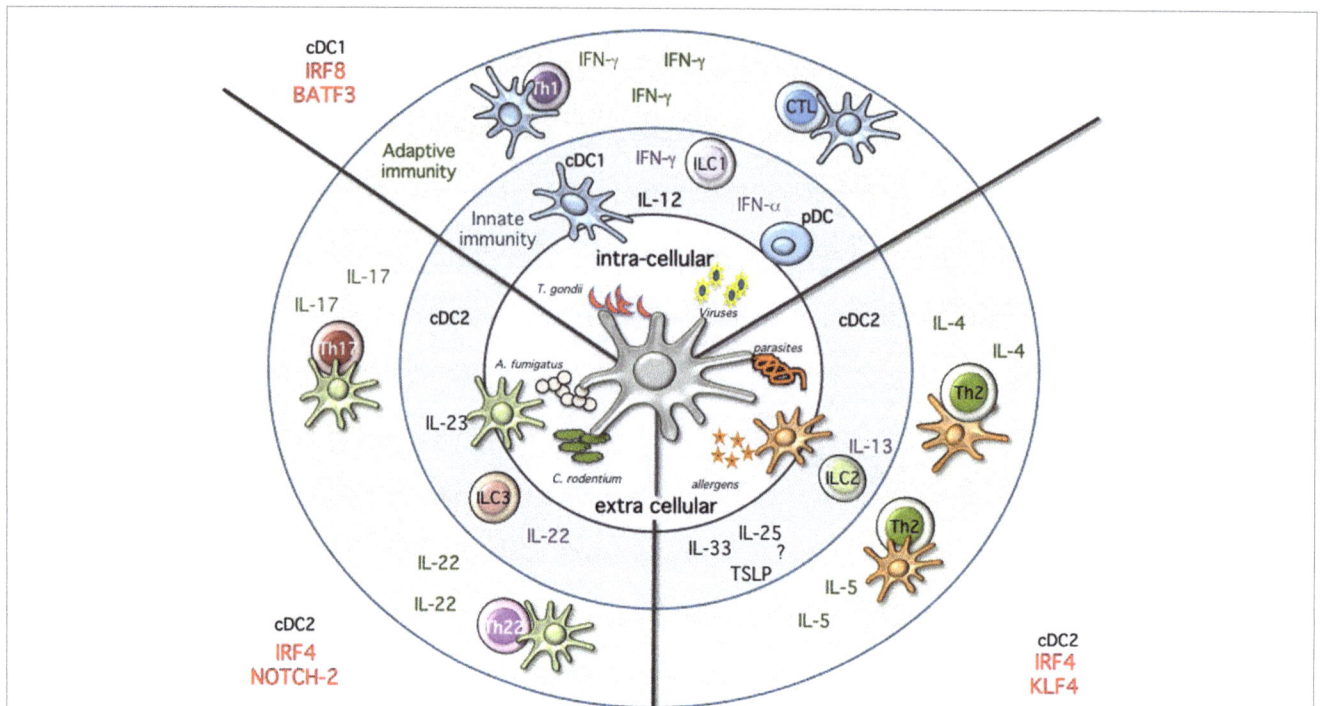

FIGURE 2 | Immune modules dendritic cells will sense the environment and start the immune response by producing cytokines, activating innate immune cells, and priming T cells. Intracellular pathogens, such as *Toxoplasma gondii*, or viral infections will activate cDC1 and pDCs. These subsets are specialized in the production of IL-12 and type 1 IFNs, respectively. High amounts of IL-12 will activate innate lymphoid cells 1 (ILC1) to produce IFN-g and ultimately leading to the priming of Th1 immunity and sustain IFN-g secretion. Following viral infections, pDCs will produce high amounts of type 1 IFNs, while cDC1 will prime CTL response through cross-presentation of infected cells. Immunity against fungi and extracellular pathogens is mostly mediated by Irf4/Notch2 cDC2, which produce high amounts of IL-23, leading to the activation of ILC3 and IL-22 production as well as priming of Th17 and Th22 T cells. Th2 immunity during allergic reactions and following parasitic infections requires KLf4-dependent cDC2. In this case, the mechanism seems to be more complex and may require ILC2, but ultimately results in the activation of Th2 cells and the production of IL-4 and IL-5.

requirements during development of clonogenic cDC progenitors (62, 63). The expression pattern of the zinc finger and BTB domain containing 46 transcription factor Zbtb46 (also called Btbd4) can be considered as cDC-lineage specific within hematopoietic cells (64–66). This TF is not present on pDCs and is induced on monocyte-derived DCs, supporting on the one hand early divergence of pDCs during DC commitment, and on the other hand suggesting a developmental convergence between cDCs and monocyte-derived DCs (4, 66). Similarly, lineage-tracing experiments were performed using mice expressing Cre recombinase under the control of Clec9a also referred as Dngr-1 (67). Although labeling is not absolute on all cDCs subsets, it seems to be restricted to pre-cDC progeny, without marking inflammatory-derived DCs (67). The use of these reporter mouse models will help us better characterize the ontogeny of specific cDCs subsets also depending on the tissue of origin and whether under steady-state or inflammatory conditions.

The CD24+ cDC1 branch of cDCs depends on the transcription factors Irf8, Id2, Nfil3, and Batf3 (68). The generation of mice deficient for Batf3 has revealed the common origin and the lineage identity of Irf8-expressing cDC1 cells, also referred as CD8a+ or CD103+ across all lymphoid and peripheral organs (69, 70). However, while only Irf8 was shown to be necessary for commitment, Id2, Nfil3, and Batf3 are dispensable under certain conditions (71, 72). Lineage choice seems influenced by high and

sustained levels of Irf8 during cDC1 commitment. Binding of Batf3 and Irf8 to an AP1-IRF composite element (AICE) within the Irf8 super-enhancer in CD24- or Zbtb46-gfp-expressing immediate progenitors leads to sustained Irf8 expression and cDC1 development (62). In the absence of Batf3, reduced Irf8 levels, redirect commitment of a CD24-expressing cDC1 progenitor toward the Irf4-expressing cDC2 lineage (62).

Despite the recent advances, how the branching of cDC1 and cDC2 occurs is still an open question. The recent identification of a committed cDC2 progenitor might help to identify the key factors involved in this process: we still need to understand how expression of Irf4 progressively replaces Irf8, and how those two TF determine the identity of these subsets. Furthermore, the cDC2 lineage, as already mentioned, is highly heterogeneous and possibly contains multiple subsets (1, 2, 11, 16, 18). Mature CD11b-expressing cDC2 express high levels of Irf4, suggesting an important role for this TF within this lineage. And indeed, absence of Irf4 impairs the development as well as the function of cDC2 (42, 44, 73–77). In mice lacking IRF4 in CD11c-expressing cells, cDC2 numbers are reduced in lung and small intestinal DCs, while no difference is reported for skin (44, 74). However, reduction in lung and lamina propria cDCs is only observed upon deletion of Irf4 in early progenitors (44; 74). Despite, normal numbers of skin DCs in Irf4-deficient animals, migration to draining lymph nodes is impaired as a consequence of defective induction of CCR7 (78). Furthermore, reduced up-regulation of MHC-II and co-stimulatory molecules is also associated with Irf4 deficiency (75, 77, 78). Collectively, Irf4 shows a broad action across different tissues and potentially subsets, and further studies are required to be able to understand the specific requirement of this TF during development.

Other TFs reported to display a reduction of cDC2 are RelB, Notch2, RbpJ, and the Kruppel-like Factor 4 (Klf4) (79–85). Notch2 is required for terminal differentiation of endothelial cell-selective adhesion molecule (ESAM)-expressing splenic cDC (81, 83). Similar to Notch2 deficiency, mice compromised in Runx3 (86, 87) and in the alternative NF-kB pathway show a reduction in the development of ESAM+ cDCs (80, 88). However, a survival disadvantage in competitive settings appears to be present in mice with compromised NF-kB signaling, suggesting caution in proposing the requirement for NF-kB during DCs development (80, 88). Klf4 deficiency results in impaired development of the so-called "double negative" DCs in skin draining LN and a partial reduction of Sirp-α but not splenic ESAM-expressing cDC2 across all the organs (84). In these mice, cDC progenitors are impaired in their ability to down-regulate Irf8 and up-regulate Irf4. However, the *in vitro* differentiation potential of Irf4-expressing cDCs as well as expression of Irf4 on peripheral cDCs is not compromised. This can be explained by the existence of at least two cDC2 subsets, where only the Klf4/Irf4-dependent one is developmentally impaired. Alternatively, a different maturation/activation state, which requires Klf4, may exist within the Irf4-expressing cDC2 subset.

Collectively, a partial reduction associated with the lack of one or the other TF confirms the developmental, and supports the subset-specific heterogeneity observed in single-cell sequencing experiments for the Irf4 and CD11b-expressing cDC2 cells (89–92). The transcriptional diversity, which characterizes these cDC2 cells, results and reflects a functional heterogeneity (**Figure 2**). Notch2 cDC2 are required for anti bacterial Th17/IL-22 immunity, while Klf4 deficiency results in impaired Th2 immunity (83, 84, 93). Expression of Irf4 in cDCs is necessary for both Th17 and Th2 responses further highlighting the complexity of this TF in DC biology (44, 73, 74, 77). Understanding whether the absence of a subset or a functional defect caused by a transcriptional deficiency on the remaining subset could account for the observed phenotypes will require subset-specific deletion. Furthermore, we also need to explore more in detail the influence of tissues on the different subsets. Are tissue-specific cues driving the expression of a transcriptional signature in a similar way as recently revealed for macrophages? (12) Are the differences reflecting a developmental or a functional heterogeneity? Is a developmental convergence between cDCs and monocyte-derived DCs creating the confusion within this branch of cDCs. We need a better characterization of the different subsets, which fall under the broad umbrella of CD11b or Irf4-expressing cDC2 and some progress has certainly been made with the introduction of new reporter mice as previously discussed as well as the recently identified committed progenitor. Teasing this heterogeneous pool of Irf4-expressing cDC2 apart is currently an active field of investigation (90, 91, 94). And new technologies will be instrumental to improve our comprehension of the molecular clues, which regulate lineage commitment. A recent report analyzed stage and subset-specific expression of mi-RNAs during DC development and miR-142 was identified as

a key regulator of cDC2 differentiation, further adding additional complexity to our current understanding of DC development (91).

Better genetic models are needed and will possibly be soon developed as a result of the recently published single-cell analysis (16, 89). Identifying TFs or surface markers, which would compromise or trace the development of one lineage independently of the anatomical localization, as previously done in Batf3$^{-/-}$ mice for Irf8-dependent cDC1 would be of great advantage (69, 70).

TRANSCRIPTIONAL REGULATION OF MONOCYTE DEVELOPMENT

The molecular regulation, which defines monocyte differentiation and lineage commitment, is poorly understood (95). Most of the identified TFs, that result in impaired monocyte development, also show an effect on other hematopoietic lineages. The transcription factors Irf8, Sfpi1 (PU.1), Egr-1, Stat3, Gfi1, Gata2, Gbx2, Nur77, retinoic acid receptors, C/EBPα and C/EBPβ, Klf4, and c-Maf as well as members of the NF-κB family members are all involved in monocyte differentiation, however their function is often redundant, certainly not limited to monocytes and in some cases mediating proliferative and/or survival rather than instructive cues (96). Most of the TFs involved in monocyte differentiation are shared within the myelo-monocytic branch. Some of them were already mentioned as important during DC development; others are involved in macrophage and or granulocyte commitment; we are therefore aware that we can only provide here a simplified transcriptional path, which leads to monocyte development and that more efforts are required to better understand.

Expression of the ETS family transcription factor Sfpi1 or PU.1 at early stages is suggested to antagonize on the one hand key regulators of other developmental pathways, such as GATA-1 for erythroid lineage, and on the other hand activate myeloid-specific factors such as Irf8, Klf4, and Erg1 (95). A critical step in monocyte differentiation is the induction of Csf1R expression at the cell surface. This seems to be regulated by Klf4 and Irf8, however both factors are also involved in cDC development, as previously discussed, (85). Furthermore, Csf1R is also needed for macrophage development.

The identification of a committed progenitor with monocyte-restricted potential called cMoP confirmed high expression levels of the above-mentioned TF (97). However, none of those is unique to monocyte differentiation and potentially complex genetic models will be required to unravel the transcriptional map required for monocyte lineage specification.

ORIGIN OF TISSUE-RESIDENT MACROPHAGES

As discussed above for DCs, similar questions arise considering tissue-resident macrophage origin and development. Lineage-tracing studies recently revisited their origin and revealed how their maintenance in adult tissues is mostly independent from monocytes and adult definitive hematopoiesis (10). Indeed, tissue-resident macrophages were proposed to develop from a Myb-independent but Sfpi1 (PU.1)-dependent fetal progenitor present in the yolk sac (YS) (5, 6, 15, 98–100) and capable of seeding the developing embryo and self-renewing during adulthood. This developmental path was first described for microglia, the brain-resident macrophages (98, 99), but still remained elusive for a number of other macrophage populations. Using similar tools, the contribution of YS progenitors to a number of adult tissue-resident macrophage populations was next assessed and only very limited input was found in most tissues tested (101). In parallel, other studies conducted in the lung and skin found that resident alveolar macrophages and Langerhans cells originated from fetal monocytes (99, 102). A recently described hypothesis is now trying to bridge these findings by proposing the existence of erythro-myeloid progenitors (EMP) distinct from hematopoietic stem cells (HSCs), which develop in YS (E8.5) and colonize the fetal liver at E16.5 giving rise to fetal erythrocytes, macrophages, granulocytes, and monocyte (103, 104). Such progenitors would generate microglia early during fetal development and participate to Kupffer cells and Langerhans cells development, but its definitive participation to the generation of other tissue-resident populations, as well as its long-term persistence, still remains to be firmly established. Indeed, a very recent study is now arguing that fetal HSCs, and not YS progenitors or EMPs, give rise to most tissue macrophage populations, except microglia known to originate from YS progenitors other than HSCs (105). This study also highlighted that while most tissue macrophages subsets maintain by self-renewal in the adult, peritoneal, dermal, and colonic residents macrophages needed continuous HSCs input to be maintained during lifetime. Accordingly, gut macrophages, most likely a specific population of macrophage residing in the serosa (106), were shown to derive from HSC-derived circulating monocytes (107). Moreover, blood monocytes can participate to the maintenance of heart macrophages in the adult (108). During inflammation, in addition to tissue-resident macrophages, some macrophages found in tissue differentiate from locally recruited Ly-6C^{hi} monocyte. Such monocyte-derived macrophages reside only for a short period of time in the tissue until inflammation resolves, and are cleared through local cell death (109). Overall, these studies suggest that there is probably more than a single developmental pathway to generate tissue macrophages and to support their self-renewal potential and unique long-term maintenance ability (**Figure 3**).

PATHWAYS ALLOWING FOR TISSUE-RESIDENT MACROPHAGE DEVELOPMENT AND MAINTENANCE

Proceeding along development, Runt-related transcription factor 1 (RUNX1) is required at early stages of myeloid lineage specification and regulates the expression of Sfpi1 (PU.1) which has to be expressed at high levels to allow for development and maintenance of macrophage differentiation (57). One of the most crucial target genes of PU.1 during macrophage development is *Csf1r*, which encodes the receptor for M-CSF and IL-34 (110). Signaling of M-CSFR through either M-CSF or IL-34 allows for the maintenance of tissue-resident macrophages (111, 112). Other TFs required for macrophage development, which co-operate with PU.1 in lineage determination, are AML1 and CCAAT enhancer-binding proteins (C/EBP) (113, 114). Overall, our understanding of the molecular pathways controlling tissue

macrophage development in general, as well as their maintenance, remains poorly defined and further studies are need to better characterize how their development is regulated. While for DCs, deletion of a subset might result in minor consequences, macrophages are thought to be critical for the organogenesis and organ homeostasis, therefore deletion of a subset could be deleterious for the life of the individual or only compatible with compensation through alternative subsets or pathways.

TISSUE-RESIDENT MACROPHAGES DIVERSITY AND TISSUE-SPECIFIC TRANSCRIPTION FACTORS CONTROLLING RESIDENT MACROPHAGE DEVELOPMENT AND MAINTENANCE

Our analysis of the transcriptional landscape of tissue-resident macrophages revealed wide heterogeneity across tissues, leading to the definition of population-specific signatures. These specific signatures were recently shown to rely on distinct enhancer landscapes shaped by the tissue microenvironment (115, 116). Using the Immunological Genome database, the reconstruction of lineage-specific regulation from gene-expression profiles across lineages (117) revealed gene modules selectively associated with a single tissue macrophage population (12). Additionally, TFs were predicted to regulate these modules, and thus could potentially influence the development of resident macrophages in a tissue-specific manner (12). Among others, predicted regulators included Spi-C for red pulp macrophages, which confirmed precedent findings (118) and thus validated the predictive power of the algorithm. Indeed, Spi-C is a TF closely related to Sfpi1 and highly expressed in spleen red pulp macrophages compared to other phagocytes (12, 118). Mice deficient for Spi-C lack splenic red pulp macrophages (118), leading to defective red blood cells recycling and iron accumulation in the spleen. At which levels Spi-C acts to control the differentiation and/or survival of red pulp macrophage remains to be determined. LXRα is another TF needed for splenic red pulp macrophage development (119), and whether Spi-C and LXRα interact together in this process is not known. Interestingly, intracellular heme accumulation following erythrocytes uptake induced Spi-C expression by stimulating the degradation of its transcriptional inhibitor Bach1 (118). Thus, heme-induced Spi-C controls the functionality of splenic red pulp macrophages, but also their maintenance albeit by an undetermined mechanism.

Similarly, PPARγ was identified as a regulator for lung macrophages (120). It is a ligand-controlled TF of the nuclear receptor family known for its role in lipid metabolism (121). Previous work has shown that PPARγ expression is important to maintain lung macrophages functionality and surfactant catabolism (122). We reported that conditional deletion of PPARγ in lung macrophages strikingly altered their transcriptome (120). Dysregulated expression of a number of genes involved in lipid metabolism was observed (120), and many of these genes were known targets of the sterol-responsive transcription factor LXR. Accordingly, increased sterol accumulation was observed in lung macrophages lacking PPARγ, as well as decreased expression of genes involved in inflammation and immunity (120). Using a different gene deletion

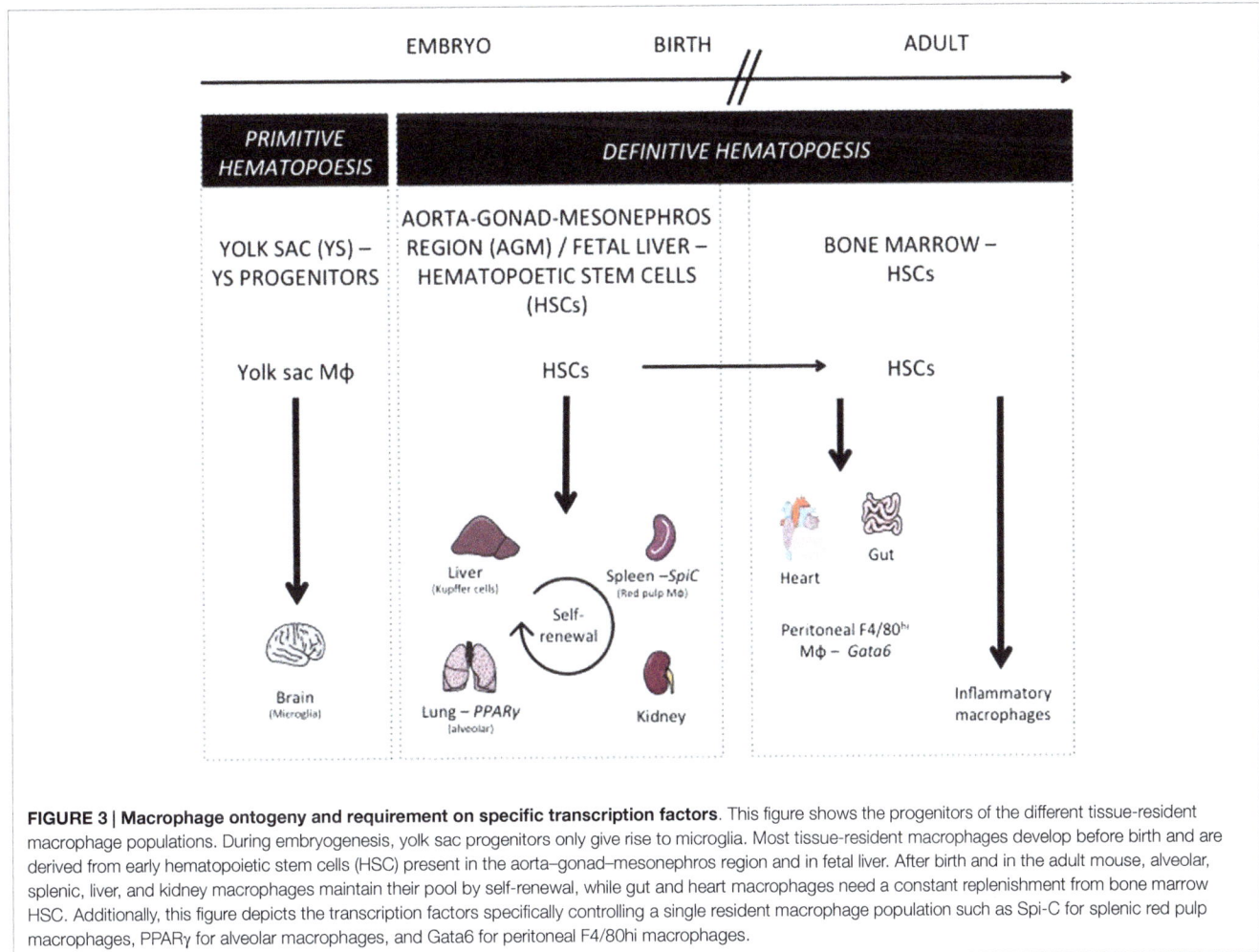

FIGURE 3 | Macrophage ontogeny and requirement on specific transcription factors. This figure shows the progenitors of the different tissue-resident macrophage populations. During embryogenesis, yolk sac progenitors only give rise to microglia. Most tissue-resident macrophages develop before birth and are derived from early hematopoietic stem cells (HSC) present in the aorta–gonad–mesonephros region and in fetal liver. After birth and in the adult mouse, alveolar, splenic, liver, and kidney macrophages maintain their pool by self-renewal, while gut and heart macrophages need a constant replenishment from bone marrow HSC. Additionally, this figure depicts the transcription factors specifically controlling a single resident macrophage population such as Spi-C for splenic red pulp macrophages, PPARγ for alveolar macrophages, and Gata6 for peritoneal F4/80hi macrophages.

approach, it was recently shown that PPARγ could also be key in controlling the development of this subset. Such discrepancy between models might relate to the different temporal induction of the cre expressing strains used in these two studies (123).

Finally, GATA6 was identified as a specific peritoneal macrophage regulator and we observed that its expression was selectively found in F4/80+ peritoneal macrophages across many lineages tested (12), suggesting that it may represent the master regulator of tissue-resident peritoneal macrophages. Interestingly, GATA6 expression by resident peritoneal macrophages was dependent on retinoic acid signaling *in vivo* (12) and mice lacking GATA6 in macrophages, generated by crossing Gata6$^{fl/fl}$ mice with Lyz2-cre, showed a strong reduction in F4/80+ peritoneal macrophages (115–117). Additionally, Th2 inflammation following parasitic infection failed to increase peritoneal macrophage numbers in Lyz2-cre × Gata6$^{fl/fl}$ mice (124), as described for wild-type mice (125). Impaired steady-state numbers of peritoneal macrophages in the absence of GATA6 was accompanied by impaired self-renewal, marked increased in S/G2-M cell cycle phases and accumulation of multinucleated macrophages due to impaired cytokinesis (126). While reduced survival of peritoneal macrophage already explains the strong contraction in their number, impaired cytokinesis will likely further exacerbate the phenotype. GATA6 deficiency

in F4/80+ peritoneal macrophages led to the down-regulation of Aspa mRNA, which encodes an aspartoacylase generating acetyl-CoA, a central cellular metabolite, from *N*-acetylaspartate (124). Interestingly, mice lacking Aspa showed reduced F4/80+ peritoneal macrophages. Overall, a tissue-specific transcriptional network driven by GATA6 controls multiple pathways all required for the maintenance of F4/80+ peritoneal macrophages.

CONCLUDING REMARKS

In the past few years, major advances have been made in our understanding how the development of myeloid cells occurs. DNA, RNA protein sequencing and characterization on entire tissues and populations is now a more accessible technology. This combined with improved multicolor flow cytometry and CyTOF technology has allowed us to better understand which TFs identify specific subsets and developmental stages during hematopoietic development (89, 90, 124, 127). However, as it is often the case, the better our analysis tools become, the more complex the picture appears. And despite these advances, we are now starting to perceive how many more gaps need to be filled in order to be able to draw a definitive road map for every MP subset. Significant progress has been made in defining, which TF are needed during DC and macrophage development in specific tissues. Monocyte development, however, is still elusive and most of the factors identified rather compromise their survival, making hard to discriminate between developmental and survival defects. Moreover, in the past few years, it has become obvious how tissues are able to influence not only the phenotype but also the function of the different subsets. This observation translates in changes in the transcriptional signature, which identifies each subset in a given tissue. Tissue-associated hallmarks have been mostly studied in macrophages, however profound consequences appear to matter also within DC subsets. It will therefore be important to discriminate between tissue- versus subset-specific transcriptional identity, to define intrinsic properties, and functional potential for every subset across and within the different tissues. On the one hand, it is attractive to think that subset specialization similar as for T and innate lymphoid cells is also present within the myeloid compartment. On the other hand, we are aware that myeloid cells are characterized by an elevated intrinsic functional plasticity. Anatomical compartments, pathogen and antigen dose as well as small micro-environmental cues might drastically influence the phenotype, the transcriptional landscape as well as the function of the different subsets during immune responses and we are just starting to explore in depth the complexity of the different subset in response to an immunological insult (90, 127). For DCs, a model has been recently suggested which takes into account the development of a subset with its immunological function. As shown in **Figure 2**, expression of Irf8 and Batf3 is needed in response to pathogens or immunological conditions where IFN-γ is required. On the other side, Irf4 is essential to stimulate Th17, Th2, and IL-22 responses. Within the Irf4 response, Notch2 and Klf4 are specifically required for Th17/IL-22 or Th2 immunity, respectively. The scenario, which appears, is consistent with functional modules of transcription across different cell types, i.e., Klf4 is also necessary for goblet cell development and polarization of M2 macrophages, whereas Notch is required for ILC3 development. Similarly, Nfil3 is important not only for cDC1 but also for NK and ILC1 cells development.

Several TFs that are required for MP development have been characterized; we can draw a map for their temporal requirement along development but for most of them the precise mechanism of action and their targets still need to be identified. Furthermore, since the developmental as well as the functional requirements for a transcriptional pathway are most often shared, caution is necessary to ascribe a specific role to a TF. Recently, a Waddington landscape was suggested to explain the plasticity in DC development (92). A similar concept may reflect and be applied to the entire MP system, where lineage commitment, specific functions, as well as subset identity could depend on the achievement of a threshold of a pool of TFs, rather than a unique master regulator. This concept would explain the so-called "graded-commitment" obtained from barcoding individual progenitors and performing lineage-tracing experiments (128). A second level of complexity is characterized by the fact that multiple subsets share the same TF, though the functional requirements are different. For example, Irf4 seems to regulate migration in skin DCs but not in other peripheral tissues, such as lungs. The functional outcome might be shared; such as in both cases antigen presentation is impaired, however it is important to understand the different requirements depending on the tissue of origin.

The study of MP is characterized by blurry phenotypic boundaries, which do not allow for unequivocal identification of the different subsets. The absence of specific markers leads to the absence of specific genetic tools and sometimes conflicting or unclear results are present in the literature. For lineage-specific deletion within cDC, we still relay on CD11c–cre mice, despite the evident limitations of this model. For monocytes as well as macrophages we lack genetic models, which would allow for selective and specific depletion of inflammatory or patrolling monocytes as well as tissue macrophages. Efforts to generate better lineage-deleter mouse models are therefore required and should be a priority in the next future to better understand the development as well as the contribution of MPs during an immune response.

ACKNOWLEDGMENTS

RT is supported by the Swiss National Fund with an SNF Professorship (NBM 1570). EG is funded by grants (LIPOCAMD and MACLEAR) from the National Agency of Research (ANR) and by a grant from the Fondation de France (00056835).

REFERENCES

1. Guilliams M, Ginhoux F, Jakubzick C, Naik SH, Onai N, Schraml BU, et al. Dendritic cells, monocytes and macrophages: a unified nomenclature based on ontogeny. *Nat Rev Immunol* (2014) **14**:571–8. doi:10.1038/nri3712
2. Merad M, Sathe P, Helft J, Miller J, Mortha A. The dendritic cell lineage: ontogeny and function of dendritic cells and their subsets in the steady state and the inflamed setting. *Annu Rev Immunol* (2013) **31**:563–604. doi:10.1146/annurev-immunol-020711-074950
3. Satpathy AT, Wu X, Albring JC, Murphy KM. Re(de)fining the dendritic cell lineage. *Nat Immunol* (2012) **13**:1145–54. doi:10.1038/ni.2467
4. Schlitzer A, McGovern N, Ginhoux F. Dendritic cells and monocyte-derived cells: two complementary and integrated functional systems. *Semin Cell Dev Biol* (2015) **41**:9–22. doi:10.1016/j.semcdb.2015.03.011
5. Yona S, Kim KW, Wolf Y, Mildner A, Varol D, Breker M, et al. Fate mapping reveals origins and dynamics of monocytes and tissue macrophages under homeostasis. *Immunity* (2013) **38**:79–91. doi:10.1016/j.immuni.2012.12.001
6. Varol C, Yona S, Jung S. Origins and tissue-context-dependent fates of blood monocytes. *Immunol Cell Biol* (2009) **87**:30–8. doi:10.1038/icb.2008.90
7. Zigmond E, Varol C, Farache J, Elmaliah E, Satpathy AT, Friedlander G, et al. Ly6C hi monocytes in the inflamed colon give rise to proinflammatory effector cells and migratory antigen-presenting cells. *Immunity* (2012) **37**:1076–90. doi:10.1016/j.immuni.2012.08.026
8. Geissmann F, Auffray C, Palframan R, Wirrig C, Ciocca A, Campisi L, et al. Blood monocytes: distinct subsets, how they relate to dendritic cells, and their possible roles in the regulation of T-cell responses. *Immunol Cell Biol* (2008) **86**:398–408. doi:10.1038/icb.2008.19
9. Auffray C, Fogg D, Garfa M, Elain G, Join-Lambert O, Kayal S, et al. Monitoring of blood vessels and tissues by a population of monocytes with patrolling behavior. *Science* (2007) **317**:666–70. doi:10.1126/science.1142883
10. Hashimoto D, Chow A, Noizat C, Teo P, Beasley MB, Leboeuf M, et al. Tissue-resident macrophages self-maintain locally throughout adult life with minimal contribution from circulating monocytes. *Immunity* (2013) **38**:792–804. doi:10.1016/j.immuni.2013.04.004
11. Malissen B, Tamoutounour S, Henri S. The origins and functions of dendritic cells and macrophages in the skin. *Nat Rev Immunol* (2014) **14**:417–28. doi:10.1038/nri3683
12. Gautier EL, Shay T, Miller J, Greter M, Jakubzick C, Ivanov S, et al. Gene-expression profiles and transcriptional regulatory pathways that underlie the identity and diversity of mouse tissue macrophages. *Nat Immunol* (2012) **13**:1118–28. doi:10.1038/ni.2419

13. Belz GT, Nutt SL. Transcriptional programming of the dendritic cell network. *Nat Rev Immunol* (2012) **12**:101–13. doi:10.1038/nri3149

14. Miller JC, Brown BD, Shay T, Gautier EL, Jojic V, Cohain A, et al. Deciphering the transcriptional network of the dendritic cell lineage. *Nat Immunol* (2012) **13**:888–99. doi:10.1038/ni.2370

15. Hashimoto D, Miller J, Merad M. Dendritic cell and macrophage heterogeneity in vivo. *Immunity* (2011) **35**:323–35. doi:10.1016/j.immuni.2011.09.007

16. Mildner A, Jung S. Development and function of dendritic cell subsets. *Immunity* (2014) **40**:642–56. doi:10.1016/j.immuni.2014.04.016

17. Shortman K, Naik SH. Steady-state and inflammatory dendritic-cell development. *Nat Rev Immunol* (2007) **7**:19–30. doi:10.1038/nri1996

18. Murphy KM. Transcriptional control of dendritic cell development. *Adv Immunol* (2013) **120**:239–67. doi:10.1016/B978-0-12-417028-5.00009-0

19. Maraskovsky E, Brasel K, Teepe M, Roux ER, Lyman SD, Shortman K, et al. Dramatic increase in the numbers of functionally mature dendritic cells in Flt3 ligand-treated mice: multiple dendritic cell subpopulations identified. *J Exp Med* (1996) **184**:1953–62. doi:10.1084/jem.184.5.1953

20. Maraskovsky E, Pulendran B, Brasel K, Teepe M, Roux ER, Shortman K, et al. Dramatic numerical increase of functionally mature dendritic cells in FLT3 ligand-treated mice. *Adv Exp Med Biol* (1997) **417**:33–40. doi:10.1007/978-1-4757-9966-8_6

21. Karsunky H, Merad M, Cozzio A, Weissman IL, Manz MG. Flt3 ligand regulates dendritic cell development from Flt3+ lymphoid and myeloid-committed progenitors to Flt3+ dendritic cells in vivo. *J Exp Med* (2003) **198**:305–13. doi:10.1084/jem.20030323

22. Naik SH, O'Keeffe M, Proietto A, Hochrein H, Shortman K, Wu L. CD8+, CD8-, and plasmacytoid dendritic cell generation in vitro using flt3 ligand. *Methods Mol Biol* (2010) **595**:167–76. doi:10.1007/978-1-60761-421-0_10

23. Brasel K, De Smedt T, Smith JL, Maliszewski CR. Generation of murine dendritic cells from flt3-ligand-supplemented bone marrow cultures. *Blood* (2000) **96**:3029–39.

24. Gilliet M, Boonstra A, Paturel C, Antonenko S, Xu XL, Trinchieri G, et al. The development of murine plasmacytoid dendritic cell precursors is differentially regulated by Flt3-ligand and granulocyte/macrophage colony-stimulating factor. *J Exp Med* (2002) **195**:953–8. doi:10.1084/jem.20020045

25. McKenna HJ, Stocking KL, Miller RE, Brasel K, De Smedt T, Maraskovsky E, et al. Mice lacking Flt3 ligand have deficient hematopoiesis affecting hematopoietic progenitor cells, dendritic cells, and natural killer cells. *Blood* (2000) **95**:3489–97.

26. Tussiwand R, Onai N, Mazzucchelli L, Manz MG. Inhibition of natural type I IFN-producing and dendritic cell development by a small molecule receptor tyrosine kinase inhibitor with Flt3 affinity. *J Immunol* (2005) **175**:3674–80. doi:10.4049/jimmunol.175.6.3674

27. Manfra DJ, Chen SC, Jensen KK, Fine JS, Wiekowski MT, Lira SA. Conditional expression of murine Flt3 ligand leads to expansion of multiple dendritic cell subsets in peripheral blood and tissues of transgenic mice. *J Immunol* (2003) **170**:2843–52. doi:10.4049/jimmunol.170.6.2843

28. Waskow C, Liu K, Darrasse-Jeze G, Guermonprez P, Ginhoux F, Merad M, et al. The receptor tyrosine kinase Flt3 is required for dendritic cell development in peripheral lymphoid tissues. *Nat Immunol* (2008) **9**:676–83. doi:10.1038/ni.1615

29. Laouar Y, Welte T, Fu XY, Flavell RA. STAT3 is required for Flt3L-dependent dendritic cell differentiation. *Immunity* (2003) **19**:903–12. doi:10.1016/S1074-7613(03)00332-7

30. Onai N, Obata-Onai A, Tussiwand R, Lanzavecchia A, Manz MG. Activation of the Flt3 signal transduction cascade rescues and enhances type I interferon-producing and dendritic cell development 1. *J Exp Med* (2006) **203**:227–38. doi:10.1084/jem.20051645

31. Rathinam C, Geffers R, Yucel R, Buer J, Welte K, Moroy T, et al. The transcriptional repressor Gfi1 controls STAT3-dependent dendritic cell development and function. *Immunity* (2005) **22**:717–28. doi:10.1016/j.immuni.2005.04.007

32. Hock H, Hamblen MJ, Rooke HM, Schindler JW, Saleque S, Fujiwara Y, et al. Gfi-1 restricts proliferation and preserves functional integrity of haematopoietic stem cells. *Nature* (2004) **431**:1002–7. doi:10.1038/nature02994

33. Karsunky H, Mende I, Schmidt T, Moroy T. High levels of the onco-protein Gfi-1 accelerate T-cell proliferation and inhibit activation induced T-cell death in Jurkat T-cells. *Oncogene* (2002) **21**:1571–9. doi:10.1038/sj.onc.1205216

34. Yucel R, Karsunky H, Klein-Hitpass L, Moroy T. The transcriptional repressor Gfi1 affects development of early, uncommitted c-Kit+ T cell progenitors and CD4/CD8 lineage decision in the thymus. *J Exp Med* (2003) **197**:831–44. doi:10.1084/jem.20021417

35. Kim W, Klarmann KD, Keller JR. Gfi-1 regulates the erythroid transcription factor network through Id2 repression in murine hematopoietic progenitor cells. *Blood* (2014) **124**:1586–96. doi:10.1182/blood-2014-02-556522

36. Li H, Ji M, Klarmann KD, Keller JR. Repression of Id2 expression by Gfi-1 is required for B-cell and myeloid development. *Blood* (2010) **116**:1060–9. doi:10.1182/blood-2009-11-255075

37. Mackarehtschian K, Hardin JD, Moore KA, Boast S, Goff SP, Lemischka IR. Targeted disruption of the flk2/flt3 gene leads to deficiencies in primitive hematopoietic progenitors. *Immunity* (1995) **3**:147–61. doi:10.1016/1074-7613(95)90167-1

38. Adolfsson J, Mansson R, Buza-Vidas N, Hultquist A, Liuba K, Jensen CT, et al. Identification of Flt3+ lympho-myeloid stem cells lacking erythro-megakaryocytic potential a revised road map for adult blood lineage commitment. *Cell* (2005) **121**:295–306. doi:10.1016/j.cell.2005.02.013

39. Wang P, Xue Y, Han Y, Lin L, Wu C, Xu S, et al. The STAT3-binding long noncoding RNA lnc-DC controls human dendritic cell differentiation. *Science* (2014) **344**:310–3. doi:10.1126/science.1251456

40. Wu Y, Smas CM. Wdnm1-like, a new adipokine with a role in MMP-2 activation. *Am J Physiol Endocrinol Metab* (2008) **295**:E205–15. doi:10.1152/ajpendo.90316.2008

41. Tamura T, Tailor P, Yamaoka K, Kong HJ, Tsujimura H, O'Shea JJ, et al. IFN regulatory factor-4 and -8 govern dendritic cell subset development and their functional diversity. *J Immunol* (2005) **174**:2573–81. doi:10.4049/jimmunol.174.5.2573

42. Suzuki S, Honma K, Matsuyama T, Suzuki K, Toriyama K, Akitoyo I, et al. Critical roles of interferon regulatory factor 4 in CD11bhighCD8alpha-dendritic cell development. *Proc Natl Acad Sci U S A* (2004) **101**:8981–6. doi:10.1073/pnas.0402139101

43. Yamamoto M, Kato T, Hotta C, Nishiyama A, Kurotaki D, Yoshinari M, et al. Shared and distinct functions of the transcription factors IRF4 and IRF8 in myeloid cell development. *PLoS One* (2011) **6**:e25812. doi:10.1371/journal.pone.0025812

44. Williams JW, Tjota MY, Clay BS, Vander LB, Bandukwala HS, Hrusch CL, et al. Transcription factor IRF4 drives dendritic cells to promote Th2 differentiation. *Nat Commun* (2013) **4**:2990. doi:10.1038/ncomms3990

45. Li HS, Yang CY, Nallaparaju KC, Zhang H, Liu YJ, Goldrath AW, et al. The signal transducers STAT5 and STAT3 control expression of Id2 and E2-2 during dendritic cell development. *Blood* (2012) **120**:4363–73. doi:10.1182/blood-2012-07-441311

46. Jackson JT, Hu Y, Liu R, Masson F, D'Amico A, Carotta S, et al. Id2 expression delineates differential checkpoints in the genetic program of CD8alpha+ and CD103+ dendritic cell lineages. *EMBO J* (2011) **30**:2690–704. doi:10.1038/emboj.2011.163

47. Nagasawa M, Schmidlin H, Hazekamp MG, Schotte R, Blom B. Development of human plasmacytoid dendritic cells depends on the combined action of the basic helix-loop-helix factor E2-2 and the Ets factor Spi-B. *Eur J Immunol* (2008) **38**:2389–400. doi:10.1002/eji.200838470

48. Cisse B, Caton ML, Lehner M, Maeda T, Scheu S, Locksley R, et al. Transcription factor E2-2 is an essential and specific regulator of plasmacytoid dendritic cell development 1. *Cell* (2008) **135**:37–48. doi:10.1016/j.cell.2008.09.016

49. Spits H, Couwenberg F, Bakker AQ, Weijer K, Uittenbogaart CH. Id2 and Id3 inhibit development of CD34(+) stem cells into predendritic cell (pre-DC)2 but not into pre-DC1. Evidence for a lymphoid origin of pre-DC2. *J Exp Med* (2000) **192**:1775–84. doi:10.1084/jem.192.12.1775

50. Hacker C, Kirsch RD, Ju XS, Hieronymus T, Gust TC, Kuhl C, et al. Transcriptional profiling identifies Id2 function in dendritic cell development. *Nat Immunol* (2003) **4**:380–6. doi:10.1038/ni903

51. Ghosh HS, Cisse B, Bunin A, Lewis KL, Reizis B. Continuous expression of the transcription factor e2-2 maintains the cell fate of mature plasmacytoid dendritic cells. *Immunity* (2010) **33**:905–16. doi:10.1016/j.immuni.2010.11.023

52. Ghosh HS, Ceribelli M, Matos I, Lazarovici A, Bussemaker HJ, Lasorella A, et al. ETO family protein Mtg16 regulates the balance of dendritic cell subsets by repressing Id2. *J Exp Med* (2014) **211**:1623–35. doi:10.1084/jem.20132121

53. Kashiwada M, Pham NL, Pewe LL, Harty JT, Rothman PB. NFIL3/E4BP4 is a key transcription factor for CD8{alpha}+ dendritic cell development. *Blood* (2011) **117**:6193–7. doi:10.1182/blood-2010-07-295873

54. Onai N, Kurabayashi K, Hosoi-Amaike M, Toyama-Sorimachi N, Matsushima K, Inaba K, et al. A clonogenic progenitor with prominent plasmacytoid dendritic cell developmental potential. *Immunity* (2013) 38:943–57. doi:10.1016/j.immuni.2013.04.006

55. Liu K, Victora GD, Schwickert TA, Guermonprez P, Meredith MM, Yao K, et al. In vivo analysis of dendritic cell development and homeostasis. *Science* (2009) 324:392–7. doi:10.1126/science.1170540

56. Kurotaki D, Yamamoto M, Nishiyama A, Uno K, Ban T, Ichino M, et al. IRF8 inhibits C/EBPalpha activity to restrain mononuclear phagocyte progenitors from differentiating into neutrophils. *Nat Commun* (2014) 5:4978. doi:10.1038/ncomms5978

57. Kueh HY, Champhekar A, Nutt SL, Elowitz MB, Rothenberg EV. Positive feedback between PU.1 and the cell cycle controls myeloid differentiation. *Science* (2013) 341:670–3. doi:10.1126/science.1240831

58. Heissig B, Hattori K, Dias S, Friedrich M, Ferris B, Hackett NR, et al. Recruitment of stem and progenitor cells from the bone marrow niche requires MMP-9 mediated release of kit-ligand. *Cell* (2002) 109:625–37. doi:10.1016/S0092-8674(02)00754-7

59. Naik SH, Proietto AI, Wilson NS, Dakic A, Schnorrer P, Fuchsberger M, et al. Cutting edge: generation of splenic CD8+ and CD8- dendritic cell equivalents in Fms-like tyrosine kinase 3 ligand bone marrow cultures. *J Immunol* (2005) 174:6592–7. doi:10.4049/jimmunol.174.11.6592

60. Naik SH, Metcalf D, van Nieuwenhuijze A, Wicks I, Wu L, O'Keeffe M, et al. Intrasplenic steady-state dendritic cell precursors that are distinct from monocytes. *Nat Immunol* (2006) 7:663–71. doi:10.1038/ni1340

61. Liu K, Waskow C, Liu X, Yao K, Hoh J, Nussenzweig M. Origin of dendritic cells in peripheral lymphoid organs of mice. *Nat Immunol* (2007) 8:578–83. doi:10.1038/ni1462

62. Grajales-Reyes GE, Iwata A, Albring J, Wu X, Tussiwand R, Kc W, et al. Batf3 maintains autoactivation of Irf8 for commitment of a CD8alpha(+) conventional DC clonogenic progenitor. *Nat Immunol* (2015) 16:708–17. doi:10.1038/ni.3197

63. Schlitzer A, Sivakamasundari V, Chen J, Sumatoh HR, Schreuder J, Lum J, et al. Identification of cDC1- and cDC2-committed DC progenitors reveals early lineage priming at the common DC progenitor stage in the bone marrow. *Nat Immunol* (2015) 16:718–28. doi:10.1038/ni.3200

64. Meredith MM, Liu K, Kamphorst AO, Idoyaga J, Yamane A, Guermonprez P, et al. Zinc finger transcription factor zDC is a negative regulator required to prevent activation of classical dendritic cells in the steady state. *J Exp Med* (2012) 209:1583–93. doi:10.1084/jem.20121003

65. Reizis B. Classical dendritic cells as a unique immune cell lineage. *J Exp Med* (2012) 209:1053–6. doi:10.1084/jem.20121038

66. Satpathy AT, Kc W, Albring JC, Edelson BT, Kretzer NM, Bhattacharya D, et al. Zbtb46 expression distinguishes classical dendritic cells and their committed progenitors from other immune lineages. *J Exp Med* (2012) 209:1135–52. doi:10.1084/jem.20120030

67. Schraml BU, van Blijswijk J, Zelenay S, Whitney PG, Filby A, Acton SE, et al. Genetic tracing via DNGR-1 expression history defines dendritic cells as a hematopoietic lineage. *Cell* (2013) 154:843–58. doi:10.1016/j.cell.2013.07.014

68. Hildner K, Edelson BT, Purtha WE, Diamond M, Matsushita H, Kohyama M, et al. Batf3 deficiency reveals a critical role for CD8alpha+ dendritic cells in cytotoxic T cell immunity. *Science* (2008) 322:1097–100. doi:10.1126/science.1164206

69. Edelson BT, Kc W, Juang R, Kohyama M, Benoit LA, Klekotka PA, et al. Peripheral CD103+ dendritic cells form a unified subset developmentally related to CD8alpha+ conventional dendritic cells. *J Exp Med* (2010) 207:823–36. doi:10.1084/jem.20091627

70. Ginhoux F, Liu K, Helft J, Bogunovic M, Greter M, Hashimoto D, et al. The origin and development of nonlymphoid tissue CD103+ DCs. *J Exp Med* (2009) 206:3115–30. doi:10.1084/jem.20091756

71. Seillet C, Jackson JT, Markey KA, Hill GR, Macdonald KP, Nutt SL, et al. CD8alpha+ DCs can be induced in the absence of transcription factors Id2, Nfil3 and Batf3. *Blood* (2013) 121:1574–83. doi:10.1182/blood-2012-07-445650

72. Tussiwand R, Lee WL, Murphy TL, Mashayekhi M, Wumesh KC, Albring JC, et al. Compensatory dendritic cell development mediated by BATF-IRF interactions. *Nature* (2012) 490:502–7. doi:10.1038/nature11531

73. Persson EK, Uronen-Hansson H, Semmrich M, Rivollier A, Hagerbrand K, Marsal J, et al. IRF4 transcription factor-dependent CD103(+)CD11b(+)

74. dendritic cells drive mucosal T helper 17 cell differentiation. *Immunity* (2013) 38:958–69. doi:10.1016/j.immuni.2013.03.009

74. Schlitzer A, McGovern N, Teo P, Zelante T, Atarashi K, Low D, et al. IRF4 transcription factor-dependent CD11b(+) dendritic cells in human and mouse control mucosal IL-17 cytokine responses. *Immunity* (2013) 38:970–83. doi:10.1016/j.immuni.2013.04.011

75. Vander LB, Khan AA, Hackney JA, Agrawal S, Lesch J, Zhou M, et al. Transcriptional programming of dendritic cells for enhanced MHC class II antigen presentation. *Nat Immunol* (2014) 15:161–7. doi:10.1038/ni.2795

76. Zhou Q, Ho AW, Schlitzer A, Tang Y, Wong KH, Wong FH, et al. CD11b+ lung dendritic cells orchestrate Th2 immunity to *Blomia tropicalis*. *J Immunol* (2014) 193:496–509. doi:10.4049/jimmunol.1303138

77. Gao Y, Nish SA, Jiang R, Hou L, Licona-Limon P, Weinstein JS, et al. Control of T helper 2 responses by transcription factor IRF4-dependent dendritic cells. *Immunity* (2013) 39:722–32. doi:10.1016/j.immuni.2013.08.028

78. Bajana S, Roach K, Turner S, Paul J, Kovats S. IRF4 promotes cutaneous dendritic cell migration to lymph nodes during homeostasis and inflammation. *J Immunol* (2012) 189:3368–77. doi:10.4049/jimmunol.1102613

79. Burkly L, Hession C, Ogata L, Reilly C, Marconi LA, Olson D, et al. Expression of relB is required for the development of thymic medulla and dendritic cells. *Nature* (1995) 373:531–6. doi:10.1038/373531a0

80. Wu L, D'Amico A, Winkel KD, Suter M, Lo D, Shortman K. RelB is essential for the development of myeloid-related CD8alpha– dendritic cells but not of lymphoid-related CD8alpha+ dendritic cells. *Immunity* (1998) 9:839–47. doi:10.1016/S1074-7613(00)80649-4

81. Caton ML, Smith-Raska MR, Reizis B. Notch-RBP-J signaling controls the homeostasis of CD8- dendritic cells in the spleen. *J Exp Med* (2007) 204:1653–64. doi:10.1084/jem.20062648

82. Cheng P, Nefedova Y, Miele L, Osborne BA, Gabrilovich D. Notch signaling is necessary but not sufficient for differentiation of dendritic cells. *Blood* (2003) 102:3980–8. doi:10.1182/blood-2003-04-1034

83. Satpathy AT, Briseno CG, Lee JS, Ng D, Manieri NA, Kc W, et al. Notch2-dependent classical dendritic cells orchestrate intestinal immunity to attaching-and-effacing bacterial pathogens. *Nat Immunol* (2013) 14:937–48. doi:10.1038/ni.2679

84. Tussiwand R, Everts B, Grajales-Reyes GE, Kretzer NM, Iwata A, Bagaitkar J, et al. Klf4 expression in conventional dendritic cells is required for T helper 2 cell responses. *Immunity* (2015) 42:916–28. doi:10.1016/j.immuni.2015.04.017

85. Kurotaki D, Osato N, Nishiyama A, Yamamoto M, Ban T, Sato H, et al. Essential role of the IRF8-KLF4 transcription factor cascade in murine monocyte differentiation. *Blood* (2013) 121:1839–49. doi:10.1182/blood-2012-06-437863

86. Dicken J, Mildner A, Leshkowitz D, Touw IP, Hantisteanu S, Jung S, et al. Transcriptional reprogramming of CD11b+Esam(hi) dendritic cell identity and function by loss of Runx3. *PLoS One* (2013) 8:e77490. doi:10.1371/journal.pone.0077490

87. Satpathy AT, Briseño CG, Cai X, Michael DG, Chou C. Hsiung S, et al. Runx1 and Cbfβ regulate the development of dendritic cell precursors by restricting granulocyte/macrophage lineages. *Blood* (2014) 123:2968–77. doi:10.1182/blood-2013-11-539643

88. Kobayashi T, Walsh PT, Walsh MC, Speirs KM, Chiffoleau E, King CG, et al. TRAF6 is a critical factor for dendritic cell maturation and development. *Immunity* (2003) 19:353–63. doi:10.1016/S1074-7613(03)00230-9

89. Amit I, Garber M, Chevrier N, Leite AP, Donner Y, Eisenhaure T, et al. Unbiased reconstruction of a mammalian transcriptional network mediating pathogen responses. *Science* (2009) 326:257–63. doi:10.1126/science.1179050

90. Jaitin DA, Kenigsberg E, Keren-Shaul H, Elefant N, Paul F, Zaretsky I, et al. Massively parallel single-cell RNA-seq for marker-free decomposition of tissues into cell types. *Science* (2014) 343:776–9. doi:10.1126/science.1247651

91. Mildner A, Chapnik E, Manor O, Yona S, Kim KW, Aychek T, et al. Mononuclear phagocyte miRNome analysis identifies miR-142 as critical regulator of murine dendritic cell homeostasis. *Blood* (2013) 121:1016–27. doi:10.1182/blood-2012-07-445999

92. Paul F, Amit I. Plasticity in the transcriptional and epigenetic circuits regulating dendritic cell lineage specification and function. *Curr Opin Immunol* (2014) 30:1–8. doi:10.1016/j.coi.2014.04.004

93. Lewis KL, Caton ML, Bogunovic M, Greter M, Grajkowska LT, Ng D, et al. Notch2 receptor signaling controls functional differentiation of dendritic

cells in the spleen and intestine. *Immunity* (2011) **35**:780–91. doi:10.1016/j.immuni.2011.08.013

94. Watchmaker PB, Lahl K, Lee M, Baumjohann D, Morton J, Kim SJ, et al. Comparative transcriptional and functional profiling defines conserved programs of intestinal DC differentiation in humans and mice. *Nat Immunol* (2014) **15**:98–108. doi:10.1038/ni.2768

95. Terry RL, Miller SD. Molecular control of monocyte development. *Cell Immunol* (2014) **291**:16–21. doi:10.1016/j.cellimm.2014.02.008

96. Friedman AD. Transcriptional control of granulocyte and monocyte development. *Oncogene* (2007) **26**:6816–28. doi:10.1038/sj.onc.1210764

97. Hettinger J, Richards DM, Hansson J, Barra MM, Joschko AC, Krijgsveld J, et al. Origin of monocytes and macrophages in a committed progenitor. *Nat Immunol* (2013) **14**:821–30. doi:10.1038/ni.2638

98. Schulz C, Gomez PE, Chorro L, Szabo-Rogers H, Cagnard N, Kierdorf K, et al. A lineage of myeloid cells independent of Myb and hematopoietic stem cells. *Science* (2012) **336**:86–90. doi:10.1126/science.1219179

99. Ginhoux F, Greter M, Leboeuf M, Nandi S, See P, Gokhan S, et al. Fate mapping analysis reveals that adult microglia derive from primitive macrophages. *Science* (2010) **330**:841–5. doi:10.1126/science.1194637

100. Geissmann F, Manz MG, Jung S, Sieweke MH, Merad M, Ley K. Development of monocytes, macrophages, and dendritic cells. *Science* (2010) **327**:656–61. doi:10.1126/science.1178331

101. Epelman S, Lavine KJ, Beaudin AE, Sojka DK, Carrero JA, Calderon B, et al. Embryonic and adult-derived resident cardiac macrophages are maintained through distinct mechanisms at steady state and during inflammation. *Immunity* (2014) **40**:91–104. doi:10.1016/j.immuni.2013.11.019

102. Guilliams M, De Kleer I, Henri S, Post S, Vanhoutte L, De Prijck S, et al. Alveolar macrophages develop from fetal monocytes that differentiate into long-lived cells in the first week of life via GM-CSF. *J Exp Med* (2013) **210**:1977–92. doi:10.1084/jem.20131199

103. Gomez Perdiguero E, Klapproth K, Schulz C, Busch K, Azzoni E, Crozet L, et al. Tissue-resident macrophages originate from yolk-sac-derived erythro-myeloid progenitors. *Nature* (2015) **518**:547–51. doi:10.1038/nature13989

104. Hoeffel G, Wang Y, Greter M, See P, Teo P, Malleret B, et al. Adult Langerhans cells derive predominantly from embryonic fetal liver monocytes with a minor contribution of yolk sac-derived macrophages. *J Exp Med* (2012) **209**:1167–81. doi:10.1084/jem.20120340

105. Sheng J, Ruedl C, Karjalainen K. Most tissue-resident macrophages except microglia are derived from fetal hematopoietic stem cells. *Immunity* (2015) **43**:382–93. doi:10.1016/j.immuni.2015.07.016

106. Muller PA, Koscso B, Rajani GM, Stevanovic K, Berres ML, Hashimoto D, et al. Crosstalk between muscularis macrophages and enteric neurons regulates gastrointestinal motility. *Cell* (2014) **158**:300–13. doi:10.1016/j.cell.2014.04.050

107. Bain CC, Bravo-Blas A, Scott CL, Gomez Perdiguero E, Geissmann F, Henri S, et al. Constant replenishment from circulating monocytes maintains the macrophage pool in the intestine of adult mice. *Nat Immunol* (2014) **15**:929–37. doi:10.1038/ni.2967

108. Molawi K, Wolf Y, Kandalla PK, Favret J, Hagemeyer N, Frenzel K, et al. Progressive replacement of embryo-derived cardiac macrophages with age. *J Exp Med* (2014) **211**:2151–8. doi:10.1084/jem.20140639

109. Gautier EL, Ivanov S, Lesnik P, Randolph GJ. Local apoptosis mediates clearance of macrophages from resolving inflammation in mice. *Blood* (2013) **122**:2714–22. doi:10.1182/blood-2013-01-478206

110. Chihara T, Suzu S, Hassan R, Chutiwitoonchai N, Hiyoshi M, Motoyoshi K, et al. IL-34 and M-CSF share the receptor Fms but are not identical in biological activity and signal activation. *Cell Death Differ* (2010) **17**:1917–27. doi:10.1038/cdd.2010.60

111. Greter M, Lelios I, Pelczar P, Hoeffel G, Price J, Leboeuf M, et al. Stroma-derived interleukin-34 controls the development and maintenance of Langerhans cells and the maintenance of microglia. *Immunity* (2012) **37**:1050–60. doi:10.1016/j.immuni.2012.11.001

112. Wang Y, Szretter KJ, Vermi W, Gilfillan S, Rossini C, Cella M, et al. IL-34 is a tissue-restricted ligand of CSF1R required for the development of Langerhans cells and microglia. *Nat Immunol* (2012) **13**:753–60. doi:10.1038/ni.2360

113. Zhang P, Iwasaki-Arai J, Iwasaki H, Fenyus ML, Dayaram T, Owens BM, et al. Enhancement of hematopoietic stem cell repopulating capacity and self-renewal in the absence of the transcription factor C/EBP alpha. *Immunity* (2004) **21**:853–63. doi:10.1016/j.immuni.2004.11.006

114. Zhang DE, Hetherington CJ, Meyers S, Rhoades KL, Larson CJ, Chen HM, et al. CCAAT enhancer-binding protein (C/EBP) and AML1 (CBF alpha2) synergistically activate the macrophage colony-stimulating factor receptor promoter. *Mol Cell Biol* (1996) **16**:1231–40.

115. Gosselin D, Link VM, Romanoski CE, Fonseca GJ, Eichenfield DZ, Spann NJ, et al. Environment drives selection and function of enhancers controlling tissue-specific macrophage identities. *Cell* (2014) **159**:1327–40. doi:10.1016/j.cell.2014.11.023

116. Lavin Y, Winter D, Blecher-Gonen R, David E, Keren-Shaul H, Merad M, et al. Tissue-resident macrophage enhancer landscapes are shaped by the local microenvironment. *Cell* (2014) **159**:1312–26. doi:10.1016/j.cell.2014.11.018

117. Jojic V, Shay T, Sylvia K, Zuk O, Sun X, Kang J, et al. Identification of transcriptional regulators in the mouse immune system. *Nat Immunol* (2013) **14**:633–43. doi:10.1038/ni.2587

118. Haldar M, Kohyama M, So AY, Kc W, Wu X, Briseno CG, et al. Heme-mediated SPI-C induction promotes monocyte differentiation into iron-recycling macrophages. *Cell* (2014) **156**:1223–34. doi:10.1016/j.cell.2014.01.069

119. A-Gonzalez N, Guillen JA, Gallardo G, Diaz M, de la Rosa JV, Hernandez IH, et al. The nuclear receptor LXRalpha controls the functional specialization of splenic macrophages. *Nat Immunol* (2013) **14**:831–9. doi:10.1038/ni.2622

120. Gautier EL, Chow A, Spanbroek R, Marcelin G, Greter M, Jakubzick C, et al. Systemic analysis of PPARgamma in mouse macrophage populations reveals marked diversity in expression with critical roles in resolution of inflammation and airway immunity. *J Immunol* (2012) **189**:2614–24. doi:10.4049/jimmunol.1200495

121. Tontonoz P, Spiegelman BM. Fat and beyond: the diverse biology of PPARgamma. *Annu Rev Biochem* (2008) **77**:289–312. doi:10.1146/annurev.biochem.77.061307.091829

122. Baker AD, Malur A, Barna BP, Ghosh S, Kavuru MS, Malur AG, et al. Targeted PPAR{gamma} deficiency in alveolar macrophages disrupts surfactant catabolism. *J Lipid Res* (2010) **51**:1325–31. doi:10.1194/jlr.M001651

123. Schneider C, Nobs SP, Kurrer M, Rehrauer H, Thiele C, Kopf M. Induction of the nuclear receptor PPAR-gamma by the cytokine GM-CSF is critical for the differentiation of fetal monocytes into alveolar macrophages. *Nat Immunol* (2014) **15**:1026–37. doi:10.1038/ni.3005

124. Gautier EL, Ivanov S, Williams JW, Huang SC, Marcelin G, Fairfax K, et al. Gata6 regulates aspartoacylase expression in resident peritoneal macrophages and controls their survival. *J Exp Med* (2014) **211**:1525–31. doi:10.1084/jem.20140570

125. Jenkins SJ, Ruckerl D, Cook PC, Jones LH, Finkelman FD, Van Rooijen N, et al. Local macrophage proliferation, rather than recruitment from the blood, is a signature of TH2 inflammation. *Science* (2011) **332**:1284–8. doi:10.1126/science.1204351

126. Rosas M, Davies LC, Giles PJ, Liao CT, Kharfan B, Stone TC, et al. The transcription factor Gata6 links tissue macrophage phenotype and proliferative renewal. *Science* (2014) **344**:645–8. doi:10.1126/science.1251414

127. Becher B, Schlitzer A, Chen J, Mair F, Sumatoh HR, Teng KW, et al. High-dimensional analysis of the murine myeloid cell system. *Nat Immunol* (2014) **15**:1181–9. doi:10.1038/ni.3006

128. Naik SH, Perie L, Swart E, Gerlach C, van Rooij N, de Boer RJ, et al. Diverse and heritable lineage imprinting of early haematopoietic progenitors. *Nature* (2013) **496**:229–32. doi:10.1038/nature12013

10

Defining mononuclear phagocyte subset homology across several distant warm-blooded vertebrates through comparative transcriptomics

Thien-Phong Vu Manh[1,2,3]*‡, Jamila Elhmouzi-Younes[4]†‡, Céline Urien[4], Suzana Ruscanu[4], Luc Jouneau[4], Mickaël Bourge[5], Marco Moroldo[6], Gilles Foucras[7,8], Henri Salmon[9,10], Hélène Marty[9,10], Pascale Quéré[9,10], Nicolas Bertho[4], Pierre Boudinot[4], Marc Dalod[1,2,3]*§ and Isabelle Schwartz-Cornil[4]*§

[1] UM2, Centre d'Immunologie de Marseille-Luminy, Aix Marseille Université, Marseille, France, [2] U1104, INSERM, Marseille, France, [3] UMR7280, CNRS, Marseille, France, [4] UR892, Virologie et Immunologie Moléculaires, INRA, Domaine de Vilvert, Jouy-en-Josas, France, [5] IFR87 La Plante et son Environnement, IMAGIF CNRS, Gif-sur-Yvette, France, [6] CRB GADIE, Génétique Animale et Biologie Intégrative, INRA, Domaine de Vilvert, Jouy-en-Josas, France, [7] UMR1225, Université de Toulouse, INPT, ENVT, Toulouse, France, [8] UMR1225, Interactions Hôtes-Agents Pathogènes, INRA, Toulouse, France, [9] UMR1282, Infectiologie et Santé Publique, INRA, Nouzilly, France, [10] UMR1282, Université François Rabelais de Tours, Tours, France

Edited by:
Shalin Naik,
Walter and Eliza Hall Institute, Australia

Reviewed by:
Christophe Jean Desmet,
University of Liege, Belgium
Richard A. Kroczek,
Robert Koch-Institute, Germany

*Correspondence:
Thien-Phong Vu Manh and Marc Dalod,
Centre d'Immunologie de Marseille-Luminy, Parc Scientifique et Technologique de Luminy, Case 906, Marseille Cedex 9 F-13288, France
vumanh@ciml.univ-mrs.fr;
dalod@ciml.univ-mr.fr
Isabelle Schwartz-Cornil,
UR892, Virologie et Immunologie Moléculaires, INRA, Domaine de Vilvert, Jouy-en-Josas Cedex 78352, France
isabelle.schwartz@jouy.inra.fr

†**Present address:**
Jamila Elhmouzi-Younes, CEA, Division of Immuno-Virology, IDMIT Center, Institute for Emerging Diseases and Innovative Therapies (iMETI), DSV, Fontenay-aux-Roses, France

‡Thien-Phong Vu Manh and Jamila Elhmouzi-Younes have contributed equally to this work.

§Senior co-authorship

Mononuclear phagocytes are organized in a complex system of ontogenetically and functionally distinct subsets, that has been best described in mouse and to some extent in human. Identification of homologous mononuclear phagocyte subsets in other vertebrate species of biomedical, economic, and environmental interest is needed to improve our knowledge in physiologic and physio-pathologic processes, and to design intervention strategies against a variety of diseases, including zoonotic infections. We developed a streamlined approach combining refined cell sorting and integrated comparative transcriptomics analyses which revealed conservation of the mononuclear phagocyte organization across human, mouse, sheep, pigs and, in some respect, chicken. This strategy should help democratizing the use of omics analyses for the identification and study of cell types across tissues and species. Moreover, we identified conserved gene signatures that enable robust identification and universal definition of these cell types. We identified new evolutionarily conserved gene candidates and gene interaction networks for the molecular regulation of the development or functions of these cell types, as well as conserved surface candidates for refined subset phenotyping throughout species. A phylogenetic analysis revealed that orthologous genes of the conserved signatures exist in teleost fishes and apparently not in Lamprey.

Keywords: comparative biology, immunology, dendritic cells, monocytes, macrophages, genomic and bio-informatic methods

Introduction

Reaching the global health objective requires to improve disease prevention and treatments in humans and in a wide variety of animal species. To achieve that goal, knowledge of the immune system, and particularly of the mononuclear phagocyte system that orchestrates the immune response, needs to be translated across species in order to develop better vaccines and immune response-targeting therapies in relevant species.

The mononuclear phagocytes encompass three main functional cell types: monocytes (Mo), macrophages (MP), and DC. The main functions of Mo are to patrol the body to detect infections and to produce microbicidal compounds including TNF, superoxide, or nitric oxide intermediates, or to differentiate into MP. The main function of MP is to preserve tissue homeostasis through trophic and scavenger functions. DCs are professional antigen-presenting cells that are key instructors of immunity, controlling tolerance to self and immune defense against pathogens. However, beyond these generic definitions, each of these mononuclear phagocyte category encompasses a complex array of different subtypes with distinct ontogeny and functions, as described extensively in mice and to some extent in humans. Mo include at least two main subsets, classical Mo (cMo) and non-classical Mo (ncMo) (1), that express different innate immune recognition receptors and mediate distinct functions, with ncMo showing the original property of patrolling blood vessels (2). Adult MP are derived either from embryonic precursors and self-renew in tissues, or in some cases are replenished from circulating Mo (2–6). The MP subtypes populating different tissues show distinct molecular and functional characteristics which are in a large part determined by their anatomical microenvironment (7, 8). Two cell types with morphologic and functional features of DC derive from the Mo/MP lineage, namely monocyte-derived DC (MoDC) and Langerhans cells (9). MoDC are generated (i) upon inflammatory stimuli *in vivo* (10), (ii) at steady-state in the skin (3), and (iii) upon culture of purified Mo or of total bone marrow cells with GM-CSF ± IL-4 *in vitro* (11, 12). Langerhans cells derive from embryonic monocytic precursors upon IL-34 signaling and populate the outer layer of epithelia (13). Finally, three types of *bona fide* DC exist, the plasmacytoid DC (pDC) and the conventional DC (cDC) cDC1 and cDC2 types which derive from a bone marrow common DC precursor and are present both in lymphoid organs and as interstitial DC in the parenchyma of non-lymphoid tissues such as skin, lung, gut, and liver (14). Comparative transcriptomic analyses pioneered by us and used by other groups, as well as functional studies, have demonstrated the existence of similar mononuclear phagocytes and DC subsets between human and mice (15–20). DC subset candidates have also been described in other mammals such as in ruminants and pigs. However, no systematic study has demonstrated the existence of a framework of homologous DC subsets throughout distant species [for review see Ref. (21)]. Overall, it remains unknown whether a similar diversity in mononuclear phagocyte subsets exists across distant mammals and vertebrates, and when during evolution this complex organization of the mononuclear phagocyte system arose.

The combination of phenotypic, functional, and ontogenic studies used in the mouse model cannot be used to define cell subsets in most other species of interest due to technical, financial, or ethical limitations. As the ontogeny and functions of cell types are instructed by specific gene-expression modules, cell type identity can be defined by its molecular fingerprinting (22). We thus reasoned that mononuclear phagocyte subset identity could be defined by gene-expression profiling, whatever the species. In addition, cell types that are homologous between species must exhibit closer molecular fingerprints and gene-expression programs than non-homologous cell types, based on the definition of homologous cell types as "those cells that evolved from the same precursor cell type in the last common ancestor" (23).

In this paper, we developed a streamlined approach (see Figure S1 in Supplementary Material) to identify homologous mononuclear phagocyte subsets in distant species with reference to the mouse, consisting in (i) designing antibody panels for sorting candidate cell subsets to high marker-based purity, (ii) generating genome expression profiling of the sorted cell subsets, and (iii) performing computational transcriptomic analyses to establish gene signatures and compare them to the transcriptomic fingerprints of the well-characterized immune cell types of the mouse referent species. Our analysis was extended to chicken cell subsets, showing that it is amenable to establish mononuclear phagocyte subset homology throughout vertebrates. We also derived gene-expression signatures and gene interaction networks that are selectively expressed in mononuclear phagocyte subsets in a conserved manner throughout distant mammals and that can be used to identify homologous subsets throughout species. The conserved gene-expression signatures and networks not only encompassed genes with known functions in mononuclear phagocyte subsets but also pointed out novel candidate genes likely involved in the ontogeny or functional specialization of these cell types. Finally, we conducted a phylogenetic analysis to examine the presence in bony fishes and in Lamprey of orthologs of genes from the transcriptomic signatures identified in mammals.

Materials and Methods

Pigs and Sheep for Blood Collection

All animal experiments were carried out under licenses issued by the Direction of the Veterinary Services of Versailles (accreditation numbers B78-93) and under approval of the Committee on the Ethics of Animal Experiments of AgroParisTech and INRA-Jouy-en-Josas (COMETHEA, authorization number 00604.01). The eight pigs (blood) used in this study (four males, four females) were around 2 years old and weighted between 60 and 85 kg. Down-sized pigs were kept at the Centre d'Imagerie Interventionnelle (Jouy-en-Josas). «Prealpe» female sheep (total 37, 50–80 kg), originate from and were raised in the «Unité Commune d'Expérimentation Animale» in Jouy-en-Josas, France. Blood (<400 ml/animal) was collected by venous puncture on sodium citrate.

Isolation of DC Subset Candidates, B Lymphocytes, and Mo from Pig Blood

PBMC were obtained from pig peripheral blood buffy coat samples by 1.076 g/ml density Percoll (GE Healthcare) gradient centrifugation (24). For B cell sorting, PBMC were surface-labeled with 2 µg/ml primary monoclonal antibody (mAb) against IgL (K139 3E1, IgG2a) followed by Alexa647-conjugated goat anti-mouse

IgG2a antibodies (Invitrogen). For pDC sorting, PBMC were surface-labeled with 2 µg/ml primary mAb anti-pig CD4 (PT90A, IgG2a), CD3 (8E6, IgG1), CD14 (CAM36, IgG1), and CD172A (74-22-15, IgG2b) followed by Alexa488, phycoerythrin (PE), or Alexa647-conjugated goat anti-mouse isotype-specific antibodies (Invitrogen). Blood pDC candidates were sorted as CD3⁻ CD14⁻ CD4⁺ CD172^{int} cells, based on previously published indicative data (25). For cDC candidates and Mo sorting, PBMC were surface-labeled with 2 µg/ml mAb anti-pig IgL (K139 3E1, IgG2a), anti-pig IgG (K138 4C2, IgM), anti-pig IgM (PG145A, IgM), anti-pig CD4 (PT90A, IgG2a), anti-human and pig cross-reacting CD14 (TUK4, IgG2a), anti-pig CD172A (74-22-15, IgG1), anti-artiodactyl MHC class II (Th21A, IgG2b), and chicken anti-human and artiodactyl cross-reacting CADM1 (3E1, IgY). The primary antibodies were revealed with Alexa488, PE, or Alexa647-conjugated goat anti-mouse isotype-specific antibodies and with donkey anti-chicken IgY Peridinin Chlorophyll Protein Complex (PerCP)-conjugated IgG. The cDC2 candidates were isolated as FSC^{hi} IgL⁻ IgG⁻ IgM⁻ CD4⁻ CD14⁻ MHC class II⁺ CADM1⁻ CD172^{hi} or CD172^{int} cells. The cDC1 candidates were isolated as FSC^{hi} IgL⁻ IgG⁻ IgM⁻ CD4⁻ CD14⁻ MHC class II⁺ CADM1⁺ CD172^{lo} cells. Mo candidates were sorted as MHC class II⁻ CD172^{hi} cells. Non-relevant antibodies (IgG1, IgG2a, IgG2b, and IgM) were systematically used as controls to measure the level of non-specific background signal caused by primary antibodies. The cell subsets were sorted by flow cytometry on the ImaGif Cytometry platform using the analyzer-sorter MoFlo XDP cytometer and the Summit 5.2 software from Beckman Coulter (cytometric assessment of post-sort purity >98%). The numbers of DCs that were collected per pig lay between 2 and 3×10^5 for pDC, 25 and 47×10^3 for cDC1, 20 and 40×10^5 for cDC2 candidates.

Isolation of DC Subset Candidates from Sheep Blood and B Lymphocytes and Macrophages from Sheep Spleen

Sheep PBMC were loaded on 1.065 density iodixanol gradient (Optiprep, Nycomed Pharma) to isolate low density cells from blood. Sheep pDC candidates were isolated by flow cytometry as previously described (26). For isolating sheep cDC candidates, the low density PBMC from several sheep were reacted with anti-CD11c mAb (2 µg/ml, OM1 clone, IgG1) followed by a saturating concentration of pacific blue-labeled anti-mouse IgG donkey Fab (50 µg/ml). After extensive wash, cells were further incubated anti-CD172A mAb (2 µg/ml, ILA24, IgG1) followed by a saturating concentration of Alexa488-labeled anti-mouse IgG donkey Fab (50 µg/ml). After extensive wash, cells were incubated with 2 µg/ml primary mAbs anti-ruminant B cells (DU-204, IgM), CD11b (ILA130, IgG2a), TCR1γ/δ receptor (CC15, IgG2a), CD45RB (CC76, IgG1), and chicken anti-human and artiodactyl cross-reacting CADM1 (3E1, IgY). The IgM and IgG2a primary antibodies were revealed with PE-conjugated goat anti-mouse isotype-specific antibodies, the IgG1 anti-CD45RB primary antibody was revealed with Alexa647-conjugated goat anti-mouse IgG1 antibody, and the anti-CADM1 with anti-IgY PerCP-conjugated IgG. The cDC2 candidates were isolated by flow cytometry as B⁻ CD11b⁻ TCR1⁻ CD45RB⁻ CD11c⁺ CADM1^{lo} CD172^{hi} FSC^{hi} cells. The cDC1 candidates were isolated by flow cytometry as B⁻ CD11b⁻ TCR1⁻ CD45RB⁻ CD11c⁺ CADM1^{hi} CD172^{lo} FSC^{hi} cells. The numbers of DCs that were collected per sheep lay between 1 and 2×10^5 for pDC, around 600 for cDC1, and around 4000 for cDC2. The far lower amounts of collected blood cDCs from sheep as compared to pig may probably originate from the multiple staining steps due to the necessity to separately identify several IgG1 as primary antibodies. B cells and MP were sorted by flow cytometry from isolated sheep spleen cells using the anti-ruminant B cell (DU-204, IgM) and anti-CD14 (CAM36, IgG1), respectively.

Production of Sheep MoDC

Three independent cultures of sheep MoDC were produced with GM-CSF as previously described (27).

RNA Extraction and Hybridization on Microarrays

Total RNA from subsets was extracted using the Arcturus PicoPure RNA Isolation Kit (Arcturus Life Technologies) and checked for quality with an Agilent 2100 Bioanalyzer using RNA 6000 Nano or Pico Kits (Agilent Technologies). All RNA samples had an RNA integrity number (RIN) above 8.5. When insufficient total RNA amounts for hybridization were obtained (<25 ng for sheep DNA chips, <50 ng for pig DNA chips), the RNAs from the sorted subsets of distinct animal were mixed. RNA amplification and labeling was performed using the one-color Low Input Quick Amp Labeling kit (Agilent Technologies) following the manufacturer's recommendations. Each RNA sample (25 ng for sheep and 50 ng for pig) was amplified and cyanin 3 (Cy3) labeled, and subsequently the complementary RNA (cRNA) was checked for quality on a Nanodrop and on an Agilent 2100 Bioanalyzer. The cRNAs (600 ng) were fragmented and used for hybridization on custom-designed Agilent ovine and porcine arrays. Our arrays for sheep and pig were custom-designed based on the commercial ovine Agilent arrays for these two species, as previously described (28, 29). In brief, the commercial probes with poor Sigreannot scores (30) were replaced with new probes designed using the e-array software from Agilent Technologies and including ovine or porcine orthologs of genes known to be selectively expressed in human and mouse DC subsets (15). After hybridization of the cRNAs on the custom-designed ovine array, the chips were washed according to the manufacturer's protocol and scanned using a G2565CA scanner (Agilent Technologies) at the resolution of 3 µm. The resulting .tiff images were extracted using the Feature Extraction software v10.7.3.1 (Agilent Technologies), using the GE1_107_Sep09 protocol. All the protocols used can be obtained by contacting the CRB GADIE facility[1]. The transcriptomic data from the chicken immune subsets were obtained from a previous study (31). All microarray data have been deposited in the Gene Expression Omnibus (GEO) database under reference numbers GSE9810, GSE53500, GSE55642 which have already been released and GSE66311 which is under embargo until publication of the present study.

---

[1]http://crb-gadie.inra.fr/

Computational Pipeline to Assess Cell Subset Homology Across Species

We have designed a computational pipeline in order to define cell subset homology across species, based on the analysis of gene-expression microarray data. In the current study, it has been applied to identify homology relationships between mononuclear phagocyte cells in mammalian species and then extended to the comparison with a more distant species (chicken). However, it can be applied to any cell type and to any species, provided that the annotations of the genes for each species are sufficiently well documented to allow the retrieval of the orthologous genes. In order to perform the comparison of expression profiles of cells coming from different species, thus from different platforms, we have designed two independent procedures. The first procedure (Figure S2A in Supplementary Material) is based on the assessment of the conservation of cell-specific fingerprints/signatures, as assessed by performing gene set enrichment analyses (GSEA, see below) between pairs of cell types. The second procedure (Figure S2B in Supplementary Material, see below) consists in cross-normalizing the expression datasets coming from the different species, in order to simultaneously examine the relationships between all cell types together.

Cross-Species Transcriptome Comparison by Pairwise Gene Set Enrichment Analyses

The methodological pipeline is depicted in Figure S2A in Supplementary Material, based on an example with comparison of three different species (A, B, and C). Species A is the reference species, i.e., the species for which the cell types are the most accurately described and generally also for which gene orthologous relationships can be retrieved from (mouse or human here). Species B and C are the test species, i.e., the species for which the identity of the cell types has to be established. Coming from three different platforms, the expression datasets have different numbers of probes, illustrated by boxes of different size. In brief, the strategy is to examine by GSEA whether the transcriptomic fingerprint of a given cell type (X) from the referent species A is enriched in one cell type (Y) of a test species (B for example) as compared to all other cell types of the same species. If this is the case, this would support the hypothesis that the cell type Y from the test species B is homologous to the cell type X of the referent species A. To perform these high-throughput GSEA in a processive way that could be easily reproduced and interpreted by other researchers devoid of bio-informatics expertise, we designed and implemented a dedicated software, called Bubble GUM (manuscript in preparation in which an extensive description of the software will be provided)[2]. Bubble GUM encompasses two main modules, GeneSign and BubbleMap, respectively, dedicated to the generation of gene sets and to their use for GSEA applied to multiple pairwise comparisons of samples integrated together into a simple graphical output that helps in the interpretation of the results. The first step consists in extracting from the reference species the transcriptomic fingerprints of each cell type. A cell-specific transcriptomic fingerprint can be defined as the list of genes that

are more highly expressed in the cell type of interest than in all other cell types. These fingerprints were extracted using the "Min (test) vs. Max (ref)" method [(minimum expression among all replicates for all samples for which the transcriptomic fingerprint is defined/maximum expression among all replicates used as reference) ≥1.5-fold] (15, 32), using the GeneSign module of Bubble GUM. These transcriptomic fingerprints, in gene symbol format, will be assessed for enrichment on the expression datasets of species B and C. Thus, it is necessary to convert the probe annotations from the arrays of species B and C into the gene symbol of their orthologous counterparts in species A. For this purpose, we used the orthology relationships defined by the Sigenae pipeline which annotated the pig and sheep genes with their human and mouse orthologous gene symbols (30). The genes present on the gene chips of species B and C that were not associated to an orthologous counterpart in species A remained annotated with the gene symbol corresponding to their species of origin. The statistical enrichment of the cell-specific transcriptomic fingerprints extracted from the reference species A were then calculated between pairwise comparisons of cell types from species B or C with the GSEA methodology, using gene set permutations for computing the p-values and false-discovery rates (FDR) (33). This was achieved, and the results graphically represented, by using the BubbleMap module of Bubble GUM.

Cross-Normalization of the Species-Specific Expression Datasets

Using the same starting expression datasets, this is an alternative strategy which is complementary to the pairwise GSEA of the species-specific expression datasets, since it allows clustering all cell types together based on the overall evaluation of the proximity of their expression patterns of hundreds to thousands of orthologous genes. The first step (Orthology Filter) consists in aligning the genes across the species (A, B, and C). It requires retaining only one representative probe per gene for each species/platform. This is needed since, in microarray designs, many genes are often each represented several times by a number of individual probes having each a different signal-to-noise ratio. However, probes have no equivalence across species, whereas genes do. In our experience, the signal-to-noise ratio is generally better for probes that have the strongest signal in positive control samples, while certain probes that have a low signal-to-noise ratio can give misleading high fold changes across conditions when using a limited number of replicates. Hence, we computed for each probe in each platform the sum of normalized expression values across all samples and kept for each gene the probe that had the highest computed value. Then, for the genes of species B and C, we retrieved the gene symbol of their orthologous counterparts in species A (reference species). In the example illustrated in Figure S2B in Supplementary Material, species A is the reference: the genes of species B and C are thus annotated using the gene symbol of the orthologous genes in the species A. The genes not represented in each of the gene chip platform used were removed from the analysis. This Orthology Filter yielded a filtered expression dataset for each species, where the number of genes and their associated symbols were similar between all species, as illustrated by boxes of the same size (Figure S2B in Supplementary Material). In order to be able

---

[2] http://www.ciml.univ-mrs.fr/applications/BubbleGUM/index.html

to rigorously merge the different datasets together, the dynamic ranges of expression values for each gene across all species must be homogenized by setting for each dataset and for each gene across all samples the mean expression to 0 and the variance to 1, a process called data centering and reduction. To prevent this mathematical transformation of the expression data to introduce noise by forcing artifactual expression changes for genes that were unregulated in the initial datasets, it is mandatory to remove all the genes that are not regulated in at least one of the datasets. This thus requires keeping only the genes that are differentially expressed between at least two cell types for each of the species studied. This was achieved in the second and third steps of the data processing. The second step (Differentially Expressed Gene, DEG, Filter) consisted in identifying independently in each dataset the genes that are differentially expressed between at least two cell types. The identification of DEG was performed by calculating the minimal ratio between each pairwise comparison of cell types and by selecting only the genes for which this minimal ratio was higher than twofold. The third step (DEG intersection) consisted in keeping only the genes that were common to all filtered DEG lists, i.e., the orthologous genes which expression was modulated across samples in each of the species studied. The fourth step consisted in data centering and reduction for each dataset, which was performed using the R statistical environment. This step consists in setting, for each dataset, the mean to 0 and the variance to 1, so that all datasets are comparable. In the fifth and final step, the different datasets were merged together simply by aligning their rows based on the common gene symbol extracted from species A. The final cross-normalized expression dataset including the data for all species was then used to perform canonical analyses for classification of samples, namely here hierarchical clustering.

Generation of Conserved Cross-Species Cell Type-Specific Signatures

For each species (human, mouse, sheep, and pig), the transcriptomic fingerprint of each cell type was generated by selecting the genes more highly expressed in the cell type of interest, as compared to all other studied cell types of the same species in the case of "absolute" transcriptomic signatures, or as compared to selected cell populations of the same species in the case of "relative" transcriptomic signatures, using the "Min (test) vs. Max (ref) $\geq$1-fold" method. Once the fingerprints had been obtained for each species for a given cell type, the gene identifiers were all converted into their corresponding official human gene symbol using BioMart and we selected the intersection of these four lists as the final conserved cross-species transcriptomic signature specific of that cell type, with the following exceptions. First, for certain cell types such as MoDC, data were available from only three, and not four, species. Second, in order to avoid removing putatively relevant signature genes, we kept in these signatures the genes found in all species but one, when their absence in the signature of that given species was due to absence or non-functionality of corresponding ProbeSets on the array of that species.

Real-Time PCR

For relative quantitation of gene expression in subsets, RNA was reverse transcribed using random primers and the Multiscribe reverse transcriptase (Applied Biosystems). Real-time PCR (qPCR) was carried out with 300 nM primers in a final reaction volume of 25 µl of 1 X SYBR Green PCR Master Mix (Applied Biosystems). The primers used to amplify ovine and porcine cDNA were designed with the Primer Express software (v2.0) using publically available GenBank sequences (Table S1 in Supplementary Material). PCR cycling conditions were 95°C for 10 min, linked to 40 cycles of 95°C for 15 s and 60°C for 1 min. Real-time qPCR data were collected by the Mastercycler® e0p realplex-Eppendorf system and $2^{-\Delta Ct}$ calculations for the relative expression of the different genes (arbitrary units) were performed with the Realplex software using GAPDH for normalization. All qPCR reactions showed >95% efficacy.

Results

Isolation of Mononuclear Phagocyte Subset Candidates from Artiodactyl Blood or Spleen Using a Set of Surface Markers

In order to establish a framework of homologous mononuclear phagocyte subsets across different species, we selected two mammalian artiodactyl species, sheep and pig, belonging to the Laurasatherians, a phylogenetically distant order from the Euarchontoglires that include the human and mouse species (**Figure 1**). A set of available antibodies exist to isolate cell subsets in these species of interest as food animals, hosts of zoonotic diseases, and biomedical models. We focused on blood or spleen immune subsets, because (i) a large source of transcriptomic data is available from this compartment in the human and mouse reference species, (ii) they are readily accessible with a minimum of technical biases in all species, and (iii) their gene-expression profiles are not expected to be influenced by peripheral tissue imprinting. We designed antibody panels to sort the subsets. In human and mice, cDC lack expression of T and B lymphocyte and Mo/MP markers and they abundantly express CD11c and MHC class II. Independent groups identified SIRPα as a conserved marker suitable to distinguish cDC2 from cDC1 across species (17, 34). Whereas XCR1 stands as the best marker for identifying cDC1 (34–41), appropriate reagents are not yet available in species outside human and mouse, and CADM1, whose sequence is highly conserved in evolution (42), can be used as an alternative (43, 44). cDC1 and cDC2 candidates were isolated from sheep and pig low density blood cells after exclusion of irrelevant cells (**Figure 2A** for sheep and **Figure 2B** for pigs, see Material and Methods section). The «candidate» nature of a sorted cell subset is marked by a star before the considered subset name in this paper. Due to restricted reagent availability, CD11c and MHC2 class II markers were used to isolate sheep and pig cDC, respectively.

In the case of pig, two populations being CADM1$^-$ CD172$^+$ or CD172^{int} were identified and selected as potential candidates and designated as *cDC2 and **cDC2, respectively (**Figure 2B**). We previously published the marker phenotype, morphology, and type I IFN production properties of sheep lymph and blood *pDC as CD45RB$^+$ FSC^{high} TCRγ/δ$^-$ B$^-$ CD11b$^-$ cells (26, 48). The sorted cells were very potent at type I IFN production upon viral-type stimulation, demonstrating at the functional level that they were highly enriched in pDC. Moreover, the sorted cells

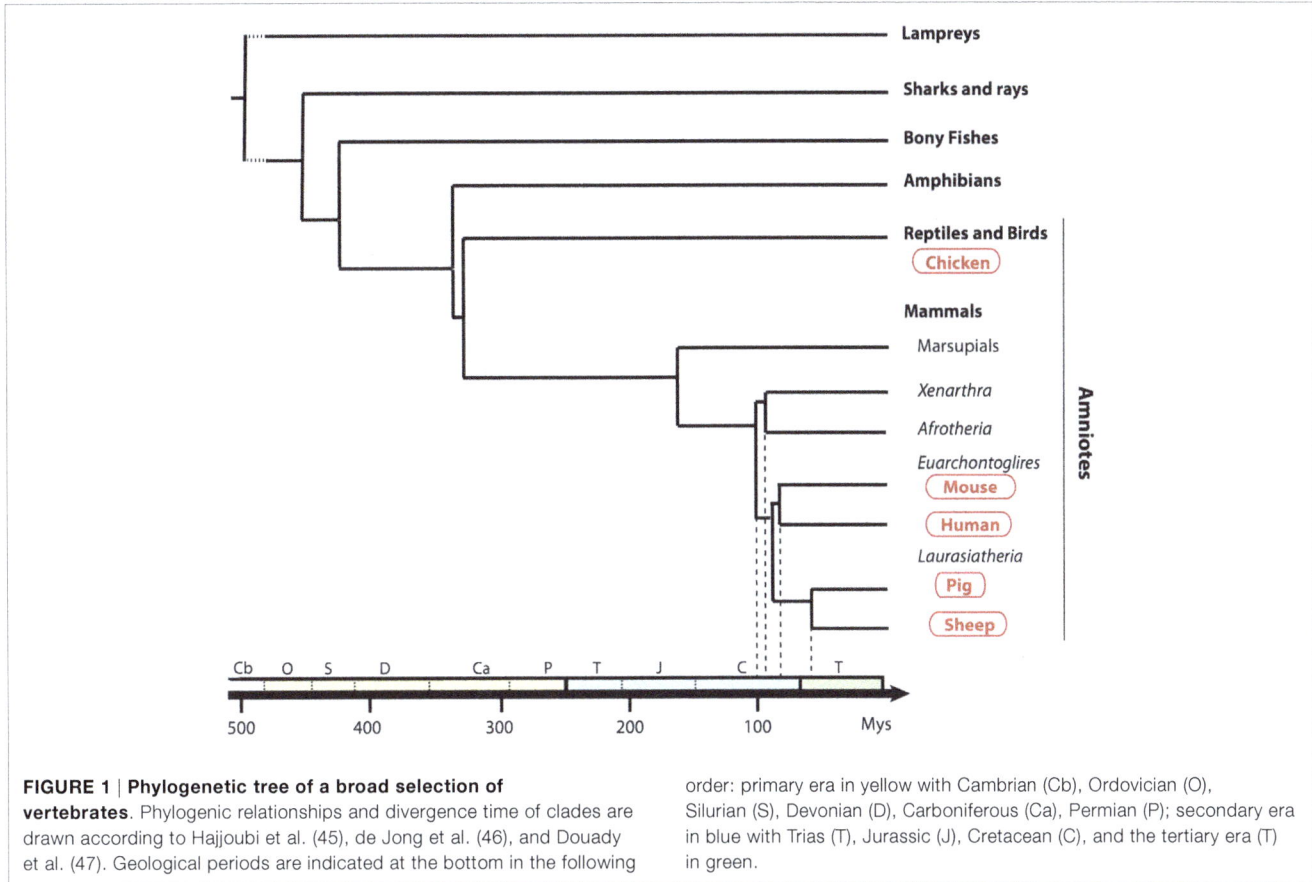

FIGURE 1 | Phylogenetic tree of a broad selection of vertebrates. Phylogenic relationships and divergence time of clades are drawn according to Hajjoubi et al. (45), de Jong et al. (46), and Douady et al. (47). Geological periods are indicated at the bottom in the following order: primary era in yellow with Cambrian (Cb), Ordovician (O), Silurian (S), Devonian (D), Carboniferous (Ca), Permian (P); secondary era in blue with Trias (T), Jurassic (J), Cretacean (C), and the tertiary era (T) in green.

had the expected size and plasmacytoid morphology, indicating that they were not contaminated by other types of myeloid cells (48). Blood *pDC from pigs were sorted as CD3− CD4+ CD172^{dim} cells based on marker phenotype, morphology, and type I IFN production properties established by others (25, 49). Pig *Mo were sorted as CD172^{high} MHC class II− cells and sheep splenic *MP as CD14+ cells.

To decrease the risk of improper identification of sorted cell subsets, we performed a quality control consisting in examining the expression of a few control genes by qRT-PCR (**Figure 2C**) prior to performing genome-wide transcriptomic analyses. Control genes were chosen based on their high selective expression in a given subset of mononuclear phagocytes in a conserved manner between mouse and human (15, 36) and encompassed *TCF4* for pDC, *CD14* for Mo/MP, *FLT3* for cDC and pDC, *ZBTB46* for cDC, *BATF3* and *XCR1* for cDC1. As expected, *TCF4* was expressed to much higher levels in sheep (26) and pig *pDC as compared to all other cell types examined except for pig **cDC2. *CD14* was expressed at much higher levels in sheep and pig *mono/MP as compared to all other cell types examined except one of the two replicates of pig *cDC2. *FLT3* was expressed at much higher levels in sheep and pig *cDC1 and in sheep *cDC2 as compared to all other cell types examined. *BATF3* and *XCR1* were expressed at higher levels in sheep and pig *cDC1 as compared to all other cell types examined. Importantly, these control analyses have allowed us to improve

our initial strategy for sheep *cDC1 and *cDC2 sorting. In fact, in our initial sorting (Figure S3 in Supplementary Material), the CD45RB+ cells were not excluded to sort cDC candidates, and the *cDC1 were found to express high levels of *TCF4* mRNA, leading us to refine the sheep cDC sorting as presented in **Figure 2**. Thus, overall these control analyses validated our strategy for phenotypic identification and flow cytometry purification of sheep and pig *pDC, *Mo/MP and *cDC1 DC, and of sheep *cDC2. In the case of pig cell subsets, the nature of **cDC2 and *cDC2 was not clear since the former expressed high levels of *TCF4* and *XCR1*, and the latter expressed relatively high levels of *CD14* in one out of two replicates. Because pig **cDC2 presented a relatively high expression level of both *TCF4* and *XCR1*, we concluded that they were significantly contaminated by pDC and cDC1. Therefore, we excluded these cells from further analyses and assumed that pig *cDC2 cells were the proper candidate.

Use of Pairwise Gene Set Enrichment Analyses for Assessment of the Similarity Between Mononuclear Phagocyte Subsets Across Distant Mammal Species

As a first approach to establish mononuclear phagocyte subset homology across species, we determined the level of similarity between artiodactyl, mouse, and human mononuclear phagocytes using pairwise GSEA, as previously performed to characterize

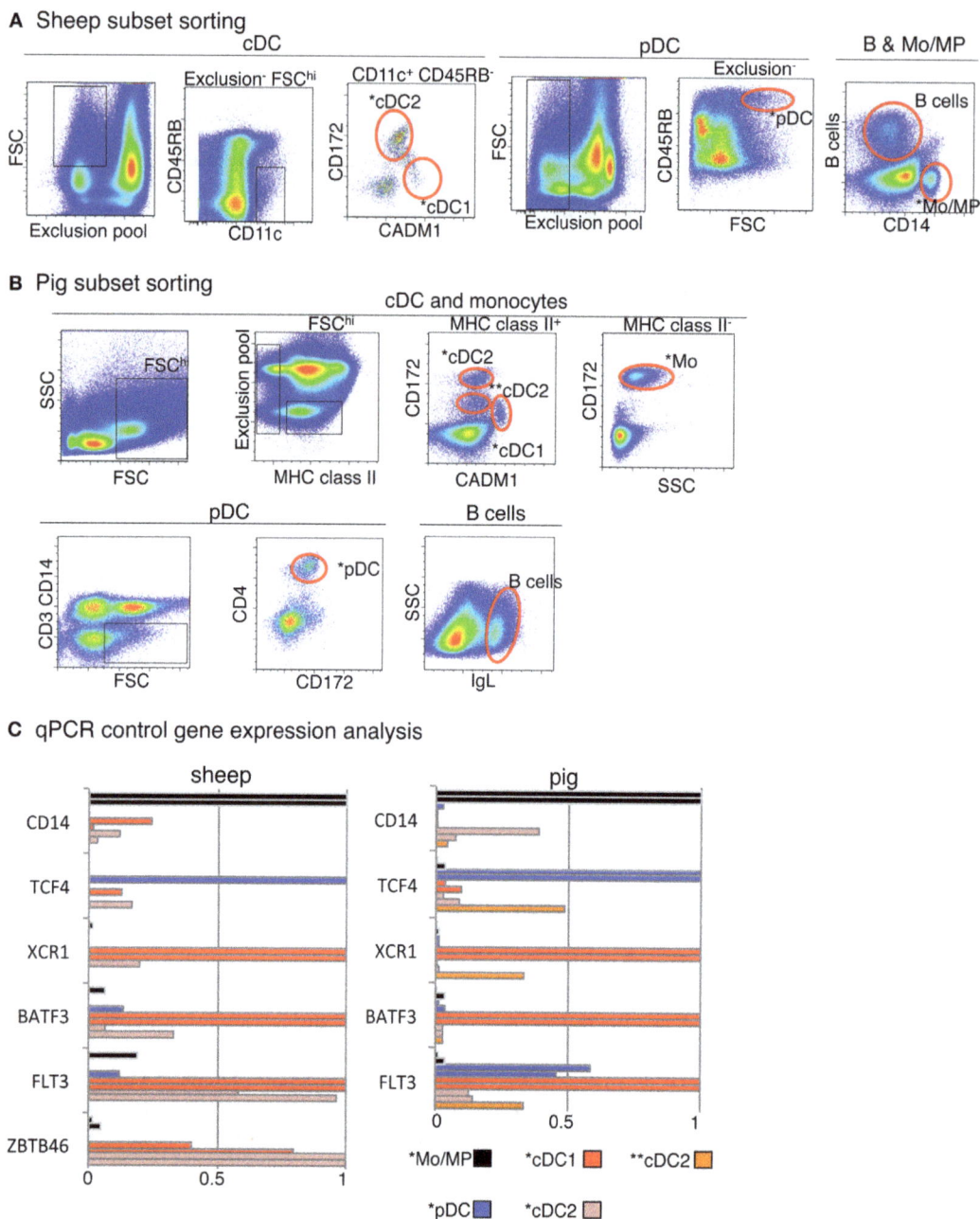

FIGURE 2 | Sorting of B cells, DC subset candidates, and Mo/MP candidates from pig and sheep blood or spleen and analysis of their expression of control genes. (A) Sheep cell subset sorting from blood and spleen. For sorting of blood cDC subset candidates, low density blood cells were gated on FSChi CD11c+ B- CD11b- TCR1- CD45RB- cells and analyzed for CADM1 and CD172 expression, based on isotype control references for each staining. The CADM1hi CD172lo (*cDC1) and CADM1lo CD172+ (*cDC2) cells were sorted. Blood pDC candidates (*pDC) were sorted as low density FSChi B- CD11b- TCR1- CD8- CD11c- CD45RB+ cells. Splenic candidate *Mo/MP were sorted as CD14+ cells. Splenic B cells were identified as DU-2-104+ cells. **(B)** Pig cell subset sorting from blood. For cDC candidate sorting, low density PBMC were gated on FSChi MHC class II+ B- CD14- CD4- cells and analyzed for CADM1 and CD172 expression. One cDC1 candidate population was identified and sorted, as

CADM1+ CD172lo (*cDC1). Two cDC2 candidate populations were identified and sorted, as CADM1- CD172hi (*cDC2) and CADM1- CD172int (**cDC2). Candidate Mo were sorted as CD172+ MHC2- cells (*Mo). Candidate pDC were sorted as CD3- CD14- CD4+ CD172int cells (*pDC). B cells were identified and sorted as IgL+ cells. **(C)** qPCR analysis of the expression of control genes in sorted candidates from one or two animals. RNA from candidate cell subsets (left, sheep; right, pig) were subjected to detection of control transcripts by qPCR. Control transcripts were chosen based on their high selective expression in specific subsets of mononuclear phagocytes in a conserved manner between mouse and man, i.e., *TCF4* for pDC, *FLT3*, *BATF3* and *ZBTB46* for cDC, *XCR1* for cDC1, and CD14 for Mo/MP. Data are represented as relative expression levels normalized to maximal expression across cell types, each bar corresponding to a distinct animal.

human immune cell subsets (19) and chicken cDC (31). To that aim, we used publicly available transcriptomic data from a selection of human and mouse immune cell types (Data Sheet S1 in Supplementary Material). We established human and mouse transcriptomic fingerprints for B cells, pDC, cDC1, Mo/MP, MoDC, cMo, and ncMo as the list of genes that are expressed at least 1.5-fold higher in the index cell population than in a large number of other immune cell types (Data Sheet S2 in Supplementary Material). B lymphocytes were chosen in all species as a reference cell subset, because their phenotypic identification in each species and their homology across species are already well established, and because they are expected to share with mononuclear phagocytes a genetic program underlying their common function of antigen-presenting cells. We generated a common fingerprint for Mo and tissue MP because their gene program is very close in the mouse (9), even though tissue MP generally derive from embryonic precursors rather than from circulating blood Mo. We could not establish a human or mouse cDC2 transcriptomic fingerprint with a sufficiently large number of genes for subsequent reliable statistical analysis. We also defined relative transcriptomic signatures for cDC vs. Mo/MP as the list of genes that are 1.5-fold higher in all cDC relatively to Mo and MP from different tissues, and reciprocally (Data Sheet S2 in Supplementary Material). Finally, we identified transcriptomic fingerprints from human and mouse MoDC (12). We then tested whether the transcriptomic signatures of mouse and human immune cell types were enriched between sheep or pig candidate cell subsets using GSEA (33) (**Figures 3** and **4**). As control, since homologies between mouse and human cell subsets have been previously demonstrated by other methods of transcriptional analyses (15, 16, 18), mouse fingerprints were also used for GSEA analysis on human cells (Figure S4 in Supplementary Material) and reciprocally (Figure S5 in Supplementary Material).

As expected, sheep B cells were significantly enriched for the expression of both human and mouse B cell transcriptomic fingerprints as compared to all other sheep cell subsets examined (**Figure 3, ❶**). The sheep *pDC were enriched for the human and mouse pDC fingerprints in most comparisons (**Figure 3, ❷**), suggesting that sheep *pDC correspond to homologs of human and mouse pDC. However, both mouse and human pDC fingerprints were not significantly enriched in the comparison of sheep *pDC with *cDC2 (NES = 1.29 and 1.24, and FDR = 1.0 and 1.0, respectively), indicating that sheep pDC probably contaminate sheep *cDC2 despite exclusion of CD45RB+ cells for their purification. The sheep *pDC were also enriched for the human B cell fingerprint in most comparisons (except with sheep B cells), what can be partly explained by the known overlap between the gene-expression program of pDC and B (15, 50–52); however, the human pDC fingerprint is not enriched in the sheep *pDC comparison with B cells and the extent of the human B cell fingerprint enrichment in sheep *pDC is above the expectations provided by similar analyses in the human and mouse reference species (Figures S4 and S5 in Supplementary Material), all of this indicating that B cells are likely to contaminate sheep *pDC despite exclusion with a pan-B cell marker for *pDC selection. Finally, *cDC2 did not show a clear enrichment for any human

and mouse signatures (**Figure 3, ❸**). However, it is also the case when examining enrichment of mouse cell subset fingerprints in human cDC2 (Figure S4 in Supplementary Material, **❸**) and reciprocally (Figure S5 in Supplementary Material, **❸**). Hence, this GSEA approach is not very informative for identification of cDC2, due to the lack of robust human or mouse fingerprints that are specific of this cell type as mentioned earlier. Sheep *cDC1 were significantly enriched in the human and mouse cDC1 fingerprints in all comparisons (**Figure 3, ❹**). They were also enriched systematically in the mouse cDC vs. Mo/MP fingerprints. This suggested that sheep *cDC1 correspond to true homologs to human and mouse cDC1. Sheep splenic *Mo/MP were strongly enriched for the human and mouse Mo/MP vs. cDC fingerprints except when compared to MoDC, and not for the human and mouse fingerprints of B lymphocytes, pDC, or cDC (**Figure 3, ❺**). This confirmed that sheep splenic *Mo/MP belong to the monocytic lineage and not to the B nor DC lineages. However, their precise identity remained unclear as they were enriched for the mouse cMo fingerprint but not for the human cMo or ncMo fingerprints. When mouse fingerprints were applied on human immune cell subsets comparisons and vice versa, there was also no consistent alignment of ncMo between the two species (Figures S4 and S5 in Supplementary Material, highlights **❺** and **❻**). Finally, sheep *MoDC that were derived from bone marrow cells in GM-CSF (27), were systematically and strongly enriched in the human and mouse MoDC signatures (**Figure 3, ❻**), confirming the homology between these three populations.

A similar analysis for pig candidate cell subsets also clearly established similarities with their putative human and mouse equivalents for B cells (**Figure 4, ❶**), pDC (**Figure 4, ❷**), and cDC1 (**Figure 4, ❹**) but not for cDC2 (**Figure 4, ❸**). Pig *Mo were clearly enriched for human and mouse fingerprints of cells of the monocytic lineage, and not for human and mouse signatures of B lymphocytes, pDC, or cDC (**Figure 4, ❺**).

Thus, altogether, GSEA analysis of the sheep and pig data for the fingerprints of human and mouse immune cell subsets gave results as informative as those obtained when comparing together human and mouse cell types, and clearly established similarities between sheep and pig cell subset candidates and their putative human and mouse equivalents for B cells, pDC, cDC1, and MoDC. Further analyses are necessary to precisely identify the nature of sheep and pig *cDC2 and *Mo/MP subsets.

Confirmation and Extension of the Conclusions on the Similarity Between Mononuclear Phagocyte Subsets Through Global and Simultaneous Analysis of the Gene-Expression Profiling of All Cell Types from Mammalian Species Using Hierarchical Clustering

In order to confirm the identification of homologous mononuclear phagocytes across species as deduced from GSEA analyses, and to potentially gain more insights into the exact nature of pig and sheep *cDC2 and *Mo/MP, we next processed all the data together for global analysis by hierarchical clustering (**Figure 5**). Only the genes that showed significant variation in their expression across subsets in each species were selected

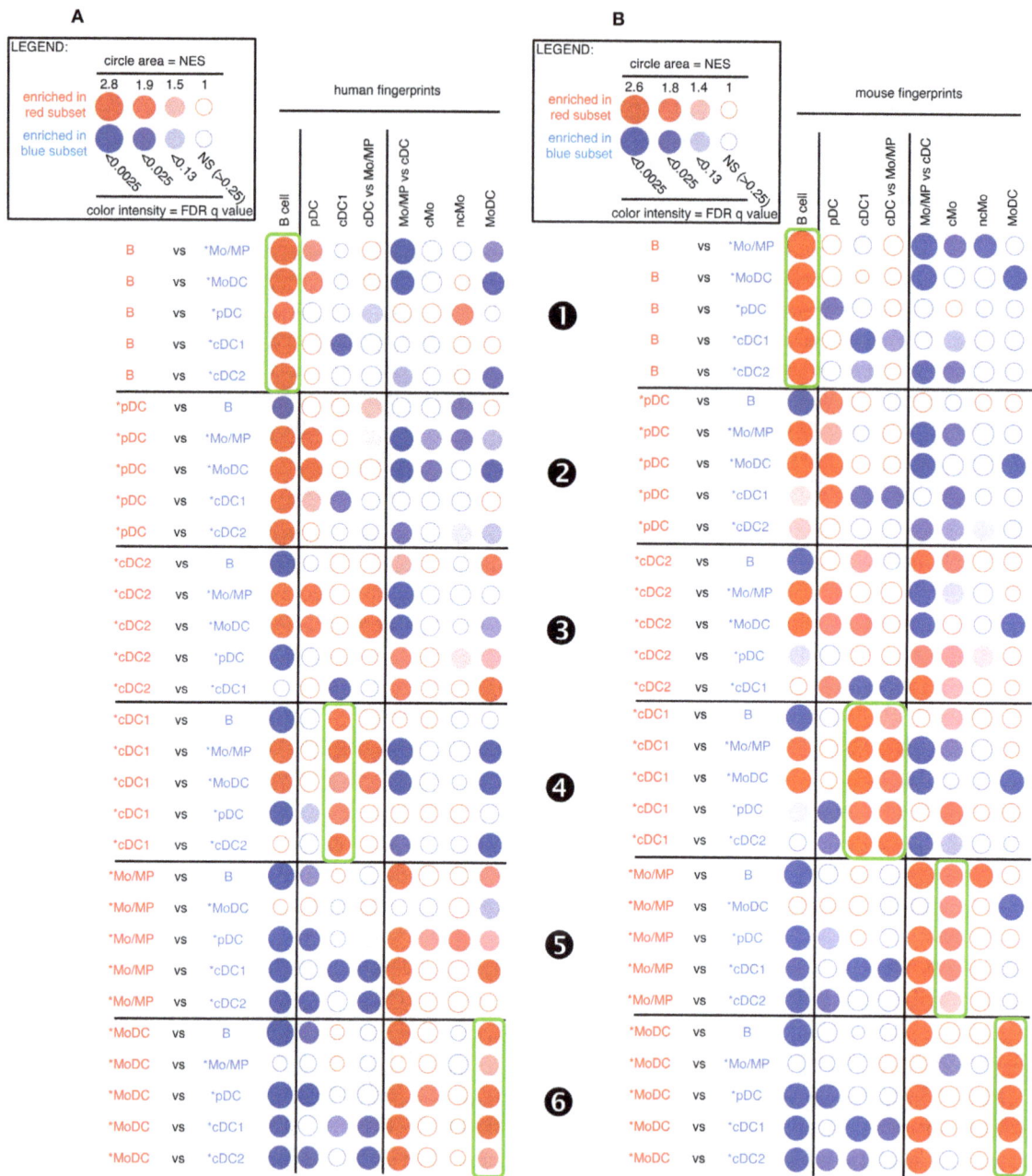

FIGURE 3 | GSEA-based assessment of the identity of sheep cell subset candidates by comparison with well-defined human and mouse mononuclear phagocytes. Candidate sheep cell subsets were compared to one another for their relative enrichment in transcriptomic fingerprints (GeneSets) specific of human **(A)** or mouse **(B)** mononuclear phagocyte subsets, using GSEA through the Bubble GUM software. The human and mouse GeneSets were defined through the same approach based on pre-existing knowledge of equivalency between human and mouse mononuclear phagocytes. A GeneSet specific for B cells was included as a control for the methodology, since the identity of this cell type is clearly established in all species and its homology across species is undisputed. The GeneSets used were named and defined as follows. The transcriptomic fingerprints "B cell," "pDC," "cDC1," "cMo," "ncMo," and "MoDC" consisted in the lists of human/mouse genes showing a high selective expression in the eponym human/mouse cell subset as compared to many other leukocytes

(see Materials and Methods for further details, Data Sheet S2 in Supplementary Material). The transcriptomic fingerprints "cDC vs. Mo/MP" and "Mo/MP vs. cDC" consisted in the lists of human/mouse genes expressed in cDC to higher levels than in Mo/MP, and reciprocally (Data Sheet S2 in Supplementary Material). All possible pairwise comparisons between sheep cell subsets were performed to assess their respective expression of the transcriptomic fingerprints of human and mouse mononuclear phagocyte subsets, using the Bubble GUM software for calculations and graphical output. Results are represented as bubbles, in a color matching that of the cell subset in which the GeneSet was enriched. Stronger and more significant enrichments are represented by bigger and darker bubbles, as illustrated in the legend box of the figure. Specifically, the surface area of bubbles is proportional to the absolute value of the normalized enrichment score (NES). The color intensity of dots is indicative of the false-discovery rate (FDR) statistical value.

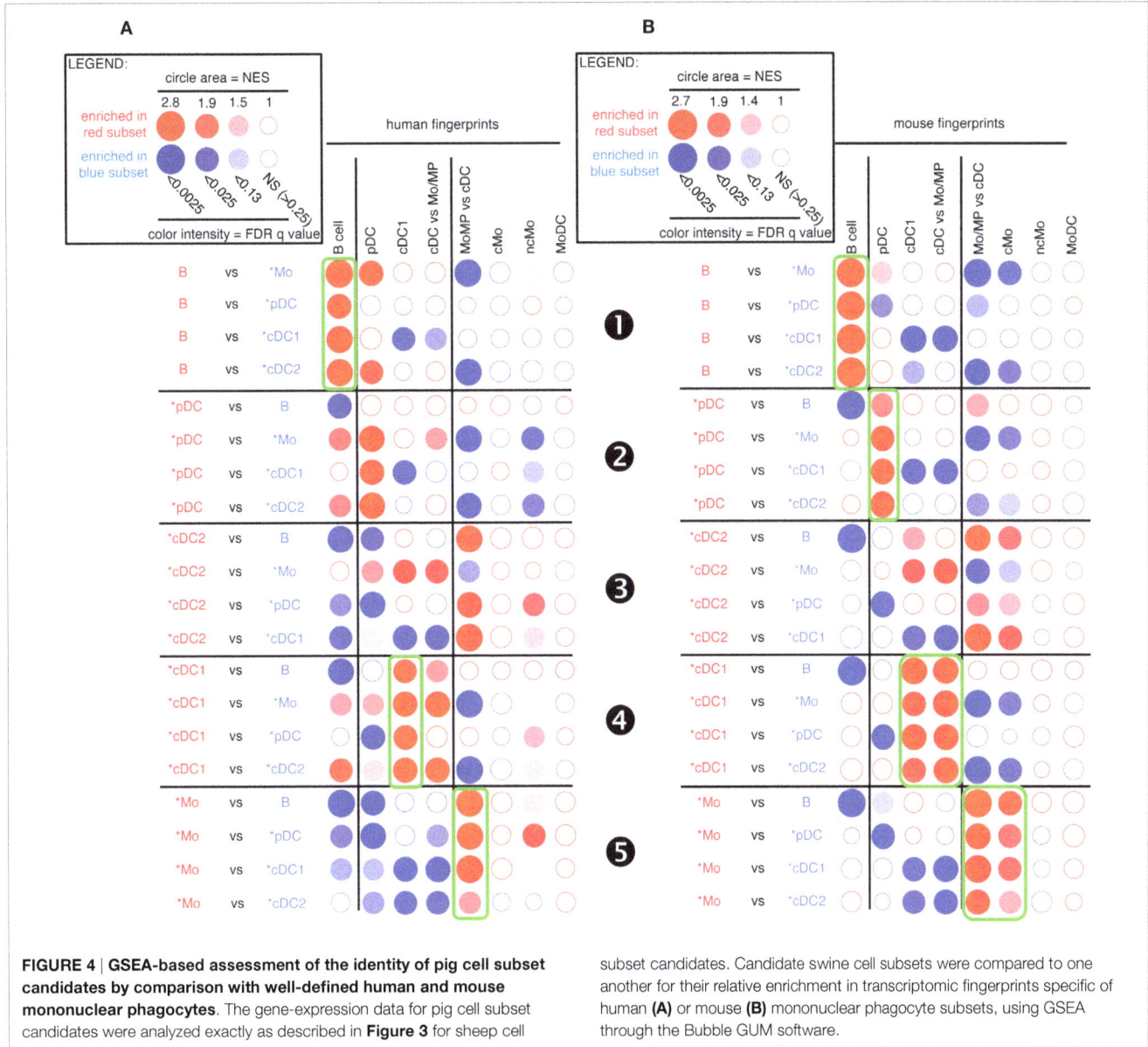

FIGURE 4 | GSEA-based assessment of the identity of pig cell subset candidates by comparison with well-defined human and mouse mononuclear phagocytes. The gene-expression data for pig cell subset candidates were analyzed exactly as described in **Figure 3** for sheep cell subset candidates. Candidate swine cell subsets were compared to one another for their relative enrichment in transcriptomic fingerprints specific of human **(A)** or mouse **(B)** mononuclear phagocyte subsets, using GSEA through the Bubble GUM software.

and the resulting datasets were normalized across species. All B cells from the four mammalian species grouped together in a specific branch of the tree, rather than each with other immune cells of the same species. This finding validates hierarchical clustering as an alternative method for identifying homologous mononuclear phagocytes across species. A closer examination of the dendrogram shows that the different cell types grouped in two major branches. The first one encompassed all the known and candidate cells of the monocytic lineages and pig *cDC2, and split further into two subgroups, one including all the identified or candidate MoDC, and the other one including all the identified or candidate Mo/MP and pig *cDC2. The second branch encompassed all the other cell types known or hypothesized not to belong to the monocytic lineage. This branch further split into two sub-branches, one constituted of the group of B cells and of the group of identified or candidate pDC, and the other

constituted of identified or candidate cDC subsets except pig *cDC2. The common clustering of B and pDC transcriptome can be explained by the shared gene-expression program between B and pDC as mentioned above. Hence, this analysis confirmed the conclusion already drawn from the GSEA analyses, namely the monocytic nature of sheep and pig *Mo/MP and *MoDC, as well as the homology between pig, sheep, mouse, and human *pDC/pDC. Moreover, the hierarchical clustering analysis allowed to better define the nature of sheep and pig *cDC2. Specifically, it confirmed the hypothesis that sheep *cDC2 belong to the cDC family, while, on the contrary to our *a priori* assignment, it shows that pig *cDC2 rather resemble Mo than cDC. However, within the branch of monocytic cells, this analysis grouped Mo/MP by species of origin rather than by cMo vs. ncMo subsets. Similarly, this analysis grouped cDC by species rather than by cDC1 vs. cDC2 subsets.

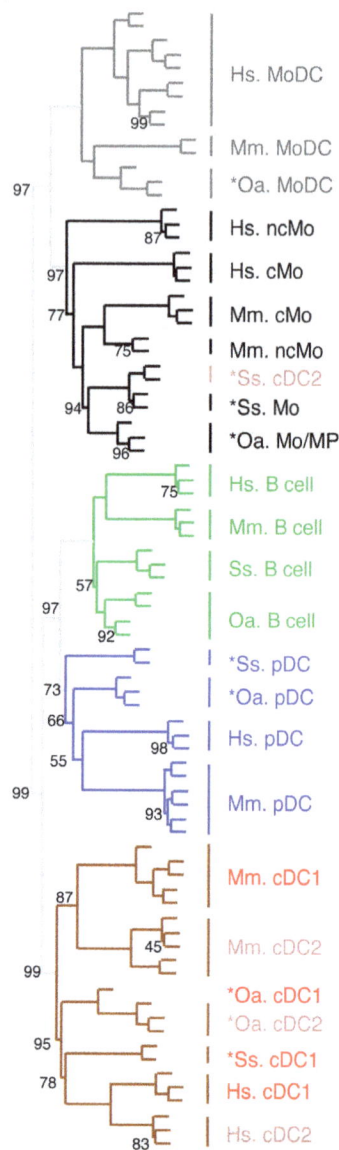

FIGURE 5 | Confirming and completing homology assignment of sheep and pig candidate Mo/MP, pDC, and cDC by unsupervised hierarchical clustering with human and mouse cell types. The datasets of each species were filtered and cross-normalized in order to allow mixing them all together for global analysis of the relationships between sheep, pig, mouse, and human mononuclear phagocyte subsets by using unsupervised hierarchical clustering. In brief, this analysis is focused on 1926 unique orthologous genes (i) for which a functional and specific ProbeSets was present on the microarrays for each species and (ii) which were found to be differentially expressed in each species between at least two subsets of mononuclear phagocytes. For each species and each of these 1926 genes, the expression data was then transformed to a mean = 0 and a variance = 1, in order to cross-normalize expression values to a similar dynamic range between the different datasets. For each cell type, the initials of the scientific name of the species of origin are indicated as a prefix: Hs, human; Mm, mouse; Ss, pig; and Oa, sheep. The robustness of the tree was tested by multiscale bootstrap resampling using Pearson's correlation as distance and average linkage as cluster method, with 1000 iterations at 10 different dataset sizes comprised between 50 and 140% of the complete dataset. An AU (approximately unbiased) p-value (percentage) was calculated and placed on the nodes of the cluster dendrogram. Missing percentages correspond to 100%.

Identification of Similarity Between Subsets of DC and of Mo Across Species Through Hierarchical Clustering Analyses Focused on These Cell Types

The expression patterns of genes outside of the cell types of interest may mask similarity between cDC or Mo subsets, as previously reported (15). Hence, we further evaluated the similarities between subsets of cDC on the one hand, and of Mo/MP on the other hand, by re-analyzing their gene-expression profiles focusing only on the genes that showed significant variation in their expression across DC subsets (**Figure 6**) or Mo/MP (**Figure 7**) in each species. Pig data were not used in the analysis focused on cDC, because, pig *cDC2 belonged to the monocytic branch and not to the DC branch of **Figure 5**. Sheep data were not used in the analysis focused on Mo/MP, because only one subset of sheep Mo/MP had been purified. Remarkably, these focused analyses grouped samples by cell types rather than by species. The cDC-focused hierarchical clustering confirmed the conclusion drawn from GSEA that sheep, mouse, and human cDC1/*cDC1 are homologs, and refined our understanding of the identity of sheep *cDC2 by showing their homology to mouse and human cDC2 (**Figure 6**). The Mo/MP-focused hierarchical clustering allowed to newly identify pig homologs to mouse and human cMo vs. ncMo (**Figure 7**). Pig *cDC2 correspond to ncMo and pig *Mo correspond to cMo. In a complementary phenotypic FACS analysis, we confirmed that likewise human ncMo as compared cMo, pig *cDC2 express higher membrane levels of CD16 and CD163 as compared to pig *Mo (Figure S6 in Supplementary Material).

Altogether, our comparative analyses of the gene-expression profiles of mononuclear phagocyte subsets across mammals

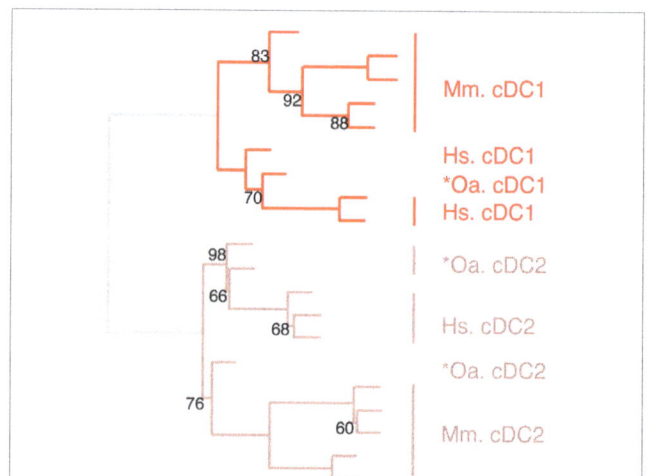

FIGURE 6 | Confirming homology assignment of sheep cDC1 and cDC2 candidates by unsupervised cross-species hierarchical clustering focused on cDC subsets. An unsupervised cross-species hierarchical clustering analysis was performed as described in **Figure 5**, but focused only on cDC subsets. The corresponding filtered dataset included 868 unique orthologous genes found regulated between cDC1 and cDC2 from human (Hs), mouse (Mm), and sheep (Oa). Pig cDC could not be included in this analysis due to the lack of data on proper pig cDC2.

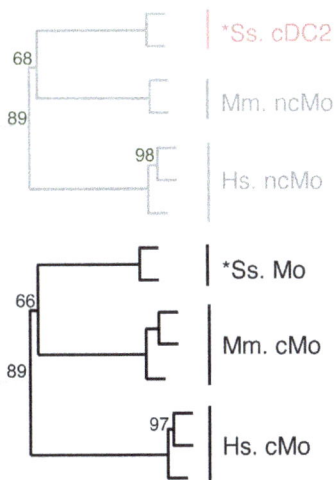

FIGURE 7 | Completing homology assignment of pig cDC2 DC candidate to non-classical Mo subset by unsupervised cross-species hierarchical clustering focused on Mo subsets. An unsupervised cross-species hierarchical clustering analysis was performed as described in **Figure 5**, but focused only on cells from the monocyte branch of the tree obtained in **Figure 5**. The corresponding filtered dataset included 191 unique orthologous genes found regulated between cMo and ncMo from human (Hs), mouse (Mm), and pig (Ss). Sheep data could not be included in this analysis due to lack of data on subsets of sheep monocytes.

indicated that the complex specialization of these cells into distinct subsets is conserved across mammals for both DC and Mo. Subset grouping did not indicate existence of a relationship between transcriptomic proximity of subsets and phylogenetic closeness of species. The conserved organization across distant mammals suggests that the mononuclear phagocyte complexity arose in a common mammalian ancestor and that the different subsets can be considered as homologous subsets across mammals.

Evidences for Homologous cDC and Mo/MP Lineages Across Warm-Blooded Vertebrates

We recently generated the transcriptomic profile of MP, total cDC, and B cells from chicken spleen and found similarities with human and mouse corresponding immune cell subsets by GSEA (31). In order to extend our subset homology analysis to non-mammalian vertebrates, we normalized and processed the transcriptomic data in a hierarchical clustering analysis as described above, using mammalian and chicken Mo/MP, B cells, and cDC subsets (**Figure 8**). There again, a tree consisting of two main branches was obtained, corresponding to a split between Mo/MP and B cells/DC. In the cDC branch, the cDC1 subset clustered together and included the chicken total cDC. The chicken MP grouped with the mammalian Mo/MP. Whereas this analysis is still partial due to limited knowledge and availability on marker sets for sorting immune cell subsets in chicken, it shows that our transcriptomic comparative approach can be used to define subset homology throughout vertebrates. It also further supports that separation of mononuclear phagocytes into Mo/MP

and cDC occurred early during vertebrate evolution and must already have been in place in the common ancestor of reptiles (including birds) and mammals.

Identification of Mononuclear Phagocyte Gene-Expression Signatures Across Mammals

Taking advantage of our multi-species microarray data, we sought to identify core gene-expression signatures that should universally define at the molecular level each of the mononuclear phagocyte subset and that should hold biological relevance based on their selective and conserved expression in homologous subsets throughout mammalian evolution. Absolute signatures ["Min (test) vs. Max (ref)" method, see Materials and methods] encompassed all genes selectively expressed at higher levels in the cell subset of interest (index population) as compared to all the other cell subsets studied (comparator populations), in all species studied. An absolute signature was computed for B cells in order to validate the approach by comparison of the gene list obtained with the advanced knowledge available on the biology of this lymphocyte population. Absolute signatures were also found for pDC, cDC1, and MoDC. Relative signatures encompassed genes selectively expressed to higher levels in one or several cell subsets of interest (index population) as compared to a selection of other cell subsets (comparator populations). The choice of index and comparator populations was largely based on the branching of different cell subsets in hierarchical clustering (**Figure 5**), or on known sharing of specific functions between cell subsets in mouse or human. The conserved absolute and relative gene-expression signatures in mononuclear phagocyte subsets are listed in **Table 1** and Data Sheet S3 in Supplementary Material. In several instances, Ingenuity Pathway Analysis (IPA) mapped a high proportion of the genes to gene interaction networks (**Figure 9** for the DC lineage subsets, **Figure 10** for the monocytic lineage subsets and Figure S7 in Supplementary Material), and revealed predicted upstream regulators (**Figure 11A**) and canonical pathways and functions (**Figure 11B**) that are described thereafter for B cells, DC lineage subsets, and Mo/MP categories. Although certain functions or pathways were enriched in several gene signatures, the genes responsible for the enrichments differed (Data Sheet S4 in Supplementary Material) and pointed out to different, complementary contributions of the distinct cell types to the corresponding functions or pathways.

The conserved B cell signature that we use as our reference subset (**Table 1**) includes a regulatory gene network directed to immunoglobulin production (Figure S7 in Supplementary Material), with PAX5 as an upstream regulator ($p = 10^{-5.8}$) (**Figure 11A**). SOX11 ($p = 10^{-8}$) and FOXO1 ($p = 10^{-7}$) are predicted to be other upstream regulators in the conserved B signature (**Figure 11A**), in agreement with existing knowledge. As expected, this signature is associated to B lymphocyte ontogeny and functions [e.g., "development of B lymphocytes" $p = 10^{-10.4}$, "antibody response" ($p = 10^{-7.5}$), "proliferation of B lymphocytes" ($p = 10^{-10.5}$), and "morphology of B lymphocytes" ($p = 10^{-7.5}$) as well as to the "B cell receptor signaling" pathway ($p = 10^{-7.3}$)] (**Figure 11B**). The B cell signature also pinpoints to genes without any known function in B cells yet, such as the cell cycle gene

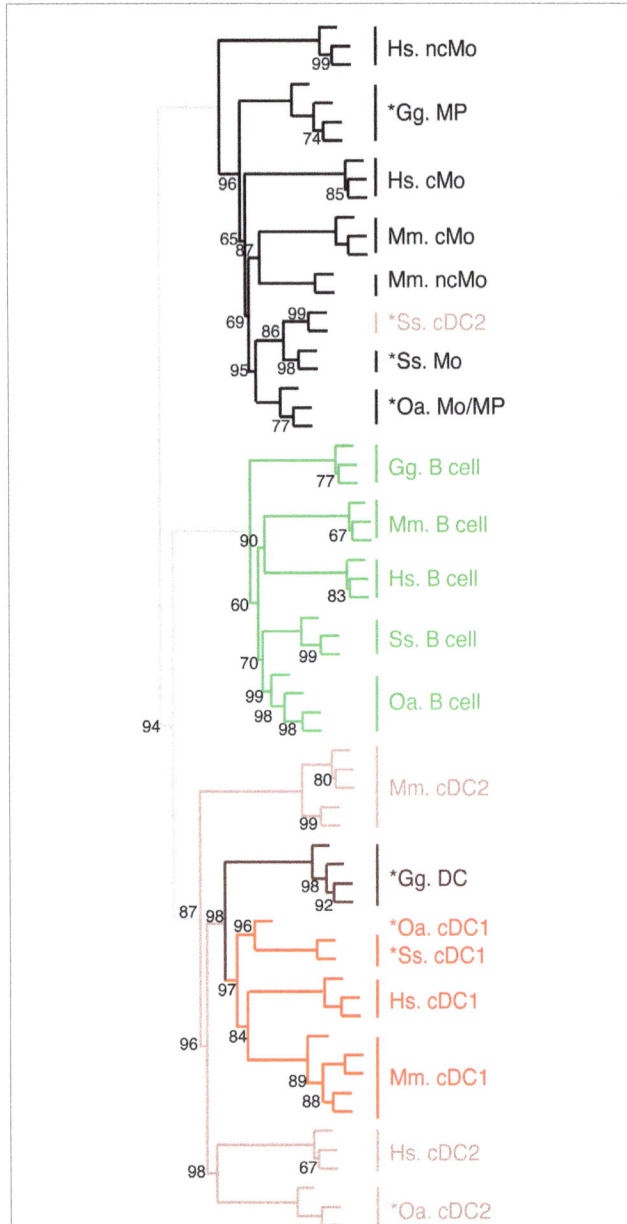

FIGURE 8 | Unsupervised cross-species hierarchical clustering including a chicken dataset demonstrates a conserved organization of vertebrate mononuclear phagocytes in the two main lineages of Mo/MP vs. cDC. An unsupervised cross-species hierarchical clustering analysis was performed as described in **Figure 5**, but including gene-expression data from chicken (Gg prefix for *Gallus gallus*) and focused only on the cell types commonly sorted in all five vertebrate species, i.e., B cells, Mo/MP, and cDC. The corresponding filtered dataset included 388 unique orthologous genes found regulated across cell subsets in each species.

which encodes for a major known regulator of pDC development (56) and other genes whose role is not yet known in this subset, with three of them coding for potential cell surface markers or targeting molecules, i.e., the low density lipoprotein receptor-related protein 8 (*LRP8*), tetraspanin 13 (*TSPAN13*), and a zinc-family transporter protein member (*SLC30A5*) (**Table 1**). These genes, except *SLC30A5*, map to a common network (**Figure 9A**). No functional annotation was found significantly enriched in the pDC absolute signature due to the low number of associated genes. Interestingly, the pDC vs. cDC relative signature includes genes belonging to a regulatory network pointing to IFN $-\alpha/\beta$ production (**Figure 9B**) and retrieves as a major putative upstream regulator X-box binding protein 1 (*XBP1*) ($p = 10^{-15}$) (**Figure 11A**), a transcription factor involved in mouse DC development (57). The pDC vs. cDC relative signature was also enriched for "proliferation of B lymphocytes" ($p = 10^{-4}$), "morphology of B lymphocytes" ($p = 10^{-5}$), and "B cell receptor signaling" pathway ($p = 10^{-2.9}$), similarly to the conserved B cell signature (**Figure 11B**). These observations are consistent with the known usage downstream of mouse and human pDC endocytic receptors of a signaling pathway akin to that of the B cell receptor (58). This known pDC signaling pathway involves the products of *SYK*, *BLNK*, and *PIK3AP1*, three of the six genes responsible for the enrichment of the "B cell receptor signaling" pathway in the conserved pDC vs. cDC gene signature (Data Sheet S4 in Supplementary Material), as well as *CARD11* which contributes to the enrichment for the annotation "proliferation of B lymphocytes" in the pDC vs. cDC signature. This strongly suggests that this signaling pathway is conserved in pDC of all mammalian species. Beside *TCF4* which encodes for a major known regulator of both B and pDC development (52), several other genes associated to B cell biology are found in the pDC vs. cDC relative signature (**Table 1**), namely *CD79B*, *PTPRCAP*, *SEMA4D*, *CTCF*, *IFR1*, and *MEF2C*. This suggests that additional biological processes shared between B cells and pDC remain to be identified.

No absolute signature could be generated for cDC but interesting informations were obtained with relative signatures, i.e., the cDC vs. Mo/MP and cDC vs. pDC. The cDC vs. Mo/MP signature includes *FLT3*, a key gene in mouse DC development (59) as well as many genes of a regulatory network including *BCL11A*, *HLA-DOA*, *HLA-DRA*, *HLA-DMB*, *HLA-DOB*, *CD74*, the axone guidance neuron navigator *NAV1* and the MHC class 2 transcription regulator *RFX5* (**Figure 9C**). In relation to this network, *IL27* ($p = 10^{-4.4}$), *IFNG* ($p = 10^{-2.6}$), and NFkB ($10^{-2.8}$) were retrieved as putative upstream regulators (**Figure 11A**). The cDC vs. Mo/MP signature was enriched for canonical pathways such as "antigen presentation" ($p = 10^{-9.6}$), "DC maturation" ($p = 10^{-4.6}$), and "T helper cell differentiation" ($p = 10^{-6.2}$) (**Figure 11B**). The cDC vs. pDC signature includes a main regulatory network encompassing *PIK3CB*, *ICAM1*, *CLEC7A*, *HLA-DRA*, *IL1B*, and *LGALS3* (**Figure 9D**) and is enriched for "functions of antigen-presenting cells" ($p = 10^{-11.1}$), "inflammatory response" ($p = 10^{-8.9}$), "bacterial infection" ($p = 10^{-7.9}$), "migration of cells" ($p = 10^{-11.4}$), and "clathrin-mediated endocytosis signaling" pathway ($p = 10^{-4.6}$) (**Figure 11B**). TNF ($p = 10^{-9}$), RELA ($10^{-5.8}$), NFKB1 ($10^{-6.1}$),

RAD17 (53) or the *SP140* gene that encodes a nuclear body protein (54) (**Table 1**). Altogether, the results of the functional analysis of the conserved signature of B cells support the biological relevance of the conserved gene signatures generated by our approach.

In the conserved signatures corresponding to the DC lineage, the pDC signature is restricted to few genes including *RUNX2*,

TABLE 1 | Conserved gene signatures for mammalian mononuclear phagocytic cell subsets.

Cell subset gene signatures	genes conserved in 3/3 or 4/4 species[a]	genes conserved in 2/3 or 3/4 species[b]
B cell	*TRAF5; SP140; RAD17; MEF2C; MBD4;* **FCRL1**[c,d,e]; **CD19**	**VPREB3**; *RFX5;* **PAX5**; *BACH2; AFF3; SWAP70; PLEKHA2;* **MS4A1**; *DMXL1;* **CR2**; **CD79B**; *CD22;* **BLK**; *ELL3; STRBP;* **EBF1**
cDC vs Mo/MP	*NAV1; MSI2;* **HLA-DMB**; **FLT3**; *BCL11A*	*RFX5; PLEKHA5;* **HLA-DOA**; *BCAT2; AFF3; FAM149A; APOBEC3H; UVRAG; SPINT2; PDXP;* **HLA-DOB**; **CD74**; *CD5; AP1S3;* **HLA-DRA**
cDC vs pDC	*WDR41;* **WDFY3**; *TPM4;* **TLR2**; *SPI1; SNX14; SNX10; SERPINB1;* **SAMHD1**; *RIN3; REL; RAB32; NHSL1; NCOR2; NAV1;* **MARCKS**; *LYZ; LGALS3; KLF3;* **JAK2**; *ITGA5; IL4I1;* **IL1B**; **IFNGR1**; **IFI30**; **ID2**; **ICAM1**; **HLA-DMB**; *GCA; FGL2; F11R; ETV3; DOCK7; DENND4A; CXCL16; CLEC7A; CHSY1;* **BATF3**; **ATP2B1**; *ARRB1; ARHGAP22;* **ANPEP**; *AIM1; AIF1; AHR; ADAM8*	*YWHAH;* **TPCN1**; **TDRD7**; *SNX21;* **SLC7A10**; *SIPA1L3;* **RGS12**; **MYO1D**; *MRC2; METRNL; MEA1;* **LRRK2**; *LRRC8C; LOXL3;* **HLA-DQB2**; *HAVCR2; FGF17; EHF; DOK1; DGKH; ATXN1; ASB2; ARHGAP26; ACTR3; RNF144B; PLEKHO2; MYOF; LPCAT2; KANK1; FAM114A1; DENND5A; ZNF524; VASP; SULT1A1; SPRED1; SNX8; SH3BP1; SH3BGRL;* **RELB**; *RALB;* **RAC1**; *PTPN12; PLEKHO1;* **PIK3CB**; *PAK1; NR4A1; NAB2; LFNG; JUNB;* **IFNGR2**; *IER2; HFE; FAM49B; EPSTI1; EGR1; EFHD2; DHRS3; CTBP2; COTL1;* **CD74**; *CD63; CBFB; C9ORF72; C1ORF21;* **BCL6**; *BASP1; ANXA5; SR140; PKM2;* **HLA-DRA**; *RGS4; TMSB4X; GMIP;* **MAST2**; *CXCL9; DNAJA4; KIF14; MTUS1; RABGGTA;* **RTN1**; *SYNJ1; TBX3*
DCs vs (Mo/MP & MoDC)	*MSI2; BCL11A*	*RAB34;* **PDCD1LG2**; *CHCHD7; CCL17; CARM1; AUH;* **VEGFA**; *UBA3; TUBA1A; TSKU; TMEM159; SLC48A1; SIGMAR1; RNF181; PTGR1;* **NOS2**; *IKBIP; FAM162A; BHLHE40*
MoDC	*TPI1; NDUFV2;* **FCGR2B**; *CD200R1; ALDOA*	
MoDC vs Mo/MP	*TPI1; SLC2A1; SLAMF1; PRNP; PPA2; POLR1D; PLAU; PALLD; NDUFV2; NARF; MRPL4;* **IL1R2**; **FCGR2B**; *EGLN3; DGKA; CSNK2B; CISH; CD200R1; AVPI1; ALDOA; ADAMTSL4*	*ZNF747; ZNF219; WIBG; VDR; SLC45A4; ROGDI; RASSF7; RAB34; RAB33A; PDE6D;* **PDCD1LG2**; *PBX2; NAGS; KCNK6;* **ICOSLG**; **HRH1**; *GOLGA8B; GOLGA8A; ETHE1; ERCC6; DVL2; DGUOK; CLEC10A; CHN2; CHCHD7;* **CD209**; *CCNG2; CCL17; CARM1; C1ORF122; AUH; ANKRD37; ZEB1;* **VEGFA**; *UBA3; TUBA1A; TSKU; TMEM159; TCTEX1D2; STRA13; SPATA24; SNRNP27; SLC48A1; SIGMAR1; S1PR3; RNF181; RMND1; RAB7A; PTGR1; PIGU; PI4K2A; OST4; NSL1;* **NOS2**; *NAE1; MT1A; MORN4; LMF2; JKAMP; IKBIP; IFT46; HAUS4; GLTPD1; GATC; FAM162A; FAM13A; FAM134A; ESYT1; ERI2; EEPD1; DNLZ; DHRS11; DCTPP1; CENPW; BHLHE40; APOO; AKIP1; CD1B; CGREF1; NOSTRIN; OLFM4;* **GAS6**; *SLC27A3*
(Mo/MP & MoDC) vs DCs	**CEBPB**; *CCDC93;* **C5AR1**	**TLR8**; *FTL; DOK3;* **CD68**
Mo/MP vs cDC	**TLR4**; **SOD2**; *RBMS1;* **LAMP2**; *GLUL; FNDC3B; CYBB;* **CEBPB**; *CCPG1; CCDC93;* **C5AR1**	**TLR8**; *SNX27; RHOQ; OSTM1; KIF1B; FTL; DUSP6; DOK3;* **CTSD**; **CTSB**; **CD68**; *HERC5; IPMK; DPYD*
Mo/MP vs MoDC	*WDR33; VPS13D; UBE2D2; TRA2A; STAG2; SFPQ; NSD1;* **NFKB1**; *NADK; ITPR1; CFLAR; ARFGEF1*	*ZNF407; VPS13C; USP31; SLC16A4; SKAP2; PRKCH; PPFIA1; PIAS2; MDN1; MAP3K5; LRRC8D; CHM; AKAP13; ACTR3; SFRS2IP; RAD51L1; NAT12; MYST3; CDC2L5; ZNF830; ZBED5; TPPP3; TMEM164; TGS1; TBC1D8B; SNRNP35; SMEK1; SLC38A10; SHISA2; RSRC2; REV1; RALGAPB; PWWP2A; PRRC2B; PBRM1; NLRC5; MOGS; MAP7D1; LUC7L3; LIMCH1; KDM4C; ISY1; IP6K1; HNRNPUL2; HNRNPU; HNRNPK; HNRNPH2; HNRNPH1; HNRNPD; HNRNPA1; FOXN3; FAM173B; FAM159B; ERVW-1; CELF2; C9; NUP210L; PDZK1; ALMS1; LAMB1; METTL3; PAIP1*
pDC	**RUNX2**	*LRP8; INPP4A; TSPAN13; SLC30A5; GPM6B*
pDC vs cDC	*UBR2; UBE2H; TMED3;* **TCF4**; *TARBP1;* **SYK**; *STT3B; SPCS3; SNX5; SLC39A7; SIT1; SEMA4D; SEC61A1; SCYL3; SCAMP2; SAP30BP;* **RUNX2**; *RDH11; RASGRP2; RABAC1; PPAPDC1B; PGM3; PARN; PAG1; OGT; NUCB2; MSI2; MEF2C; LMAN2; IQCB1; IFT52; HBS1L; GPAM; GORASP2; FKBP2; FAM3C; EIF2AK3; DERL1; DDOST; DAD1; CYBB; COPA; CDC42SE2; CD4; CD164; BTRC;* **BLNK**; *BCL7A; ATP2A3; ATG5*	*ZXDC; VPS13A; UEVLD; TNRC6B; TMEM63A; TAF9B; TAF1A; SUSD1; STOML1; ST6GALNAC4; SSR2; SRPRB; SPG20; SLC38A6; SLC38A1; SLC25A36; SGCB; SERPINI1; SEC24C; SAP130; RAPGEF2; RALGPS1; RAB28; RAB11FIP2; PTAR1; PIK3AP1; OSTM1; NRP1; MYB; MGAT4A; MCOLN2; MCOLN1; LRP8; KIF13B; KIAA0226;* **IRF7**; *INPP4A; IMPACT; HIVEP1; FKBP8; FANCD2; FAM122B; DMTF1; CSTF1; CREB3L2; COBLL1; CBX4; CANX; ATG4D; ANKRD28; ANKIB1; AGBL3; AFF3; TPRG1L; RNF144A; IFI27L1; FAM65B; ELMOD3; DCAF7; CARS2; ZMYND11; YPEL3; USP24; TUBGCP6; TSPAN13; TRAM1; TOE1; TMEM138; TM9SF1; TCTA; SURF4; STAMBPL1; SSR3; SPCS2; SPATA13; SNX9; SLC7A5; SLC44A2; SLC30A5; SEPP1; SCAND1; SCAMP3; RHOH; RHBDF2; RHBDD1; REXO2; QDPR; PYCR2; PTPRCAP; PRMT7; POLD1; PEX5; NSUN3; MTMR9; LPGAT1; INTS7;* **IFNAR1**; *HM13; GRAP; GANAB; FNDC3A; FASTK; EXOC7; ELOF1; ELMOD2; CTCF; COPE; COMMD6; CNP; CIRBP; CDS2; CD79B;* **CARD11**; *C19ORF10; C16ORF80; C10ORF88; BTD; BET1; ARHGAP12; AHI1; WDR51B; SAPS3; MLF1IP; KIAA1370; CYBASC3; CEP110; CCDC111; ANUBL1; MME; PTPRS; ATF2; GPM6B; MON2; PPM1A; TM7SF3; TMCO1; UGCG; ZDHHC14; ZNF521; TMED10; PAIP1*
cDC2		*FCER1A*
cDC1	**XCR1**; *WDFY4; FNBP1;* **FLT3**; **CADM1**	*SNX22; GCET2*
cDC2 vs (pDC & cDC1)	*TRPS1; STK24; SLC16A3;* **SIRPA**; *SIGLEC8; S100A4; RIN2;* **REL**; **PILRA**; *NFAM1; NCF2; MAFB;* **LRP1**; *ITGAM;* **IL1R2**; **IL1B**; *IGSF6;* **IFI30**; **FHL3**; *EPB41L3; DOCK4; DHRS3; CSF3R;* **CSF1R**; *CLEC4A;* **CD300A**; *C19ORF59; ADRBK2;* **TREM1**	*TNFRSF1B;* **TLR8**; *TICAM2; STK10; SP2; SLFN12; SIGLEC9; SIGLEC7; RNASE2; PHF21A; LST1; LIMD2;* **LILRB2**; **LILRB1**; **LILRA6**; **LILRA3**; *IFITM2; GNGT2; GBP4; FAM111A; EMR1; DPP10; DENND1A;* **DDX58**; *CDKN2B; CD300LF; CD300LB;* **CD209**; *C10ORF11; ADAP1; CLEC6A; DAGLB; WDR45L; SIGLEC5; SFRS5; S100A12; PLEC1; MYST1;* **MX2**; **MS4A8B**; *LRRC33; HSPA6; GK3P; GAPDH; FAM45B; CEBPD;* **CD1E**; *CD1B; FCER1A; KSR1;* **OAS2**; *PTGER3*

(Continued)

TABLE 1 | Continued

Cell subset gene signatures	genes conserved in 3/3 or 4/4 species[a]	genes conserved in 2/3 or 3/4 species[b]
cDC2 vs (pDC & cDC2)	**XCR1**; WDFY4; ST3GAL5; RAB32; PPT1; PPA1; LRRC1; KIAA1598; FNBP1; **FLT3**; CALM1; **CADM1**	SNX22; PPAP2A; PLEKHA5; GRAMD2; DENND1B; CLEC1A; ATXN1; FAM114A1; HEPACAM2; PI4K2A; PLEKHO2; WDR91; TRIO; RALB; PKP4; PDLIM7; G3BP2; **BCL6**; ATPIF1; GCET2; BRWD2; FGD6; MYO9A
ncMo vs cMo[f]	ACAT2; ACE; ACOT9; ADRBK2; ANKRD42; APOA2; ASB2; BDKRB2; BGLAP; C1ORF112; C1ORF56; C20ORF112; CAPZB; CBX4; CD4; CD83; CDH24; CHD5; CSF1R; CYP2R1; DCBLD1; DDB2; DDIT4; DLGAP4; FBP1; GABBR1; GLMN; GNE; GNPNAT1; GPT; GRHPR; HEY1; HN1; IL12RB1; IL17A; IL2RG; KCNMA1; KCTD11; KNDC1; LMX1B; LUZP1; MAFF; MPZL1; MUTYH; MYOD1; NCAPH2; NCOR2; NFKBIA; NPAS2; NUB1; PCK1; PDCD4; PGR; PITPNM1; PLEKHH1; PMF1; PMVK; POLR3H; RAB25; RAD52; RFC5; RHOF; RSAD1; RWDD3; SECISBP2; SERPINA1; SH2D3C; SIRT5; SLC37A1; SMS; ST3GAL1; ST3GAL5; TBC1D8; TCF7L2; TNNC1; U2AF1L4; UNG; WDR76	
cMo vs ncMo[f]	AACS; ABHD5; AGTPBP1; ALDH2; ALOX5AP; ANXA1; AOAH; ARL8B; ATP6V1A; ATP6V1B2; ATP6V1C1; AUH; B4GALT1; C19ORF59; C5ORF15; CCR1; CD164; CD84; CETN2; CLTA; COPB2; CSF3R; CYP27A1; DCLRE1A; DNAJC10; ECE1; EHD4; **EIF2AK2**; EIF2AK3; ENSA; ENTPD7; ERP29; EXOC5; F13A1; F5; FAM102B; FAM63A; FBXL5; FBXO9; FN1; GBE1; GNA12; GNPAT; GSN; GYS1; HMGB2; **IL1R2**; **IL1RN**; ITM2B; KEAP1; LACTB; LCN2; LEO1; LMAN1; LMNB1; **LYZ**; MBD5; MBIP; MGA; MPP1; NHLRC2; NISCH; NKRF; NPC1; NSF; NUCB2; PAM; PARP8; PDE2A; PGD; PLCB1; PNPLA8; PON2; PREPL; PRKAR1A; PRUNE; PSMA1; PSTPIP1; PUM2; PXK; PYGL; RAB27A; RAB3D; RABGAP1L; RARS; RHOT1; RMI1; RNF130; RPGR; RSC1A1; **S100A8**; SCRN3; SCYL1; SDCBP; SEC22C; SELL; SENP5; SERPINB1; SHB; SIGLEC1; SLC16A7; SLC25A44; SLC35B3; SLC39A9; ST8SIA4; TBC1D2; TEX2; TGM1; TM6SF1; TMEM161B; TMEM71; TPCN1; TREML2; TRIP11; TSHZ1; UBE4A; UMPS; USP10; UXS1; VAPB; VNN3; VPS37B; WDTC1; XBP1; ZMYM4	

[a] Genes conserved in 4/4 species, or in 3/3 species for cDC2, cMo, and ncMo since only three species could contribute to the analysis.
[b] Genes conserved in 3/4 species or 2/3 species.
[c] Genes in bold were previously demonstrated to play a significant role in the development or functions of the population of interest.
[d] Underlined genes were annotated as located in "plasma membrane" according to Ingenuity Pathway Analysis.
[e] Genes highlighted in gray have been previously identified as signatures genes for the corresponding mouse and human cell populations in our earlier study (15).
[f] Signature genes of the relative cMo vs. ncMo and of the ncMo vs. cMo signatures were provided only for the 3/3 species selection since the gene lists for the 2/3 species selection encompassed hundreds of genes.

P38 MAPK $(10^{-5.5})$, IFNG $(10^{-5.4})$, and to a lesser extent STAT3 $(10^{-4.7})$ and CSF2 $(10^{-4.3})$ are predicted as putative upstream regulators in this signature (**Figure 11A**). In addition, this cDC vs. pDC signature includes *BATF3*, a gene highly expressed in cDC that is key in cDC1 development in mouse and human (60), as well as *ARHGAP22*, a gene involved in actin cytoskeleton regulation (61), that was initially described as a top gene of the absolute cDC signature common to human and mouse (15). Altogether, the relative gene signatures of cDC emphasize their nature of highly endocytic, motile, and expert antigen-presenting cells throughout species.

The conserved cDC1 signature encompasses genes with known contribution in the biology of this lineage, such as *XCR1*, *FLT3*, and *CADM1* (59), as well as additional genes which biological function in this subset remains enigmatic, such as the germinal center B-cell-expressed transcript 2 protein (*GCET2*),

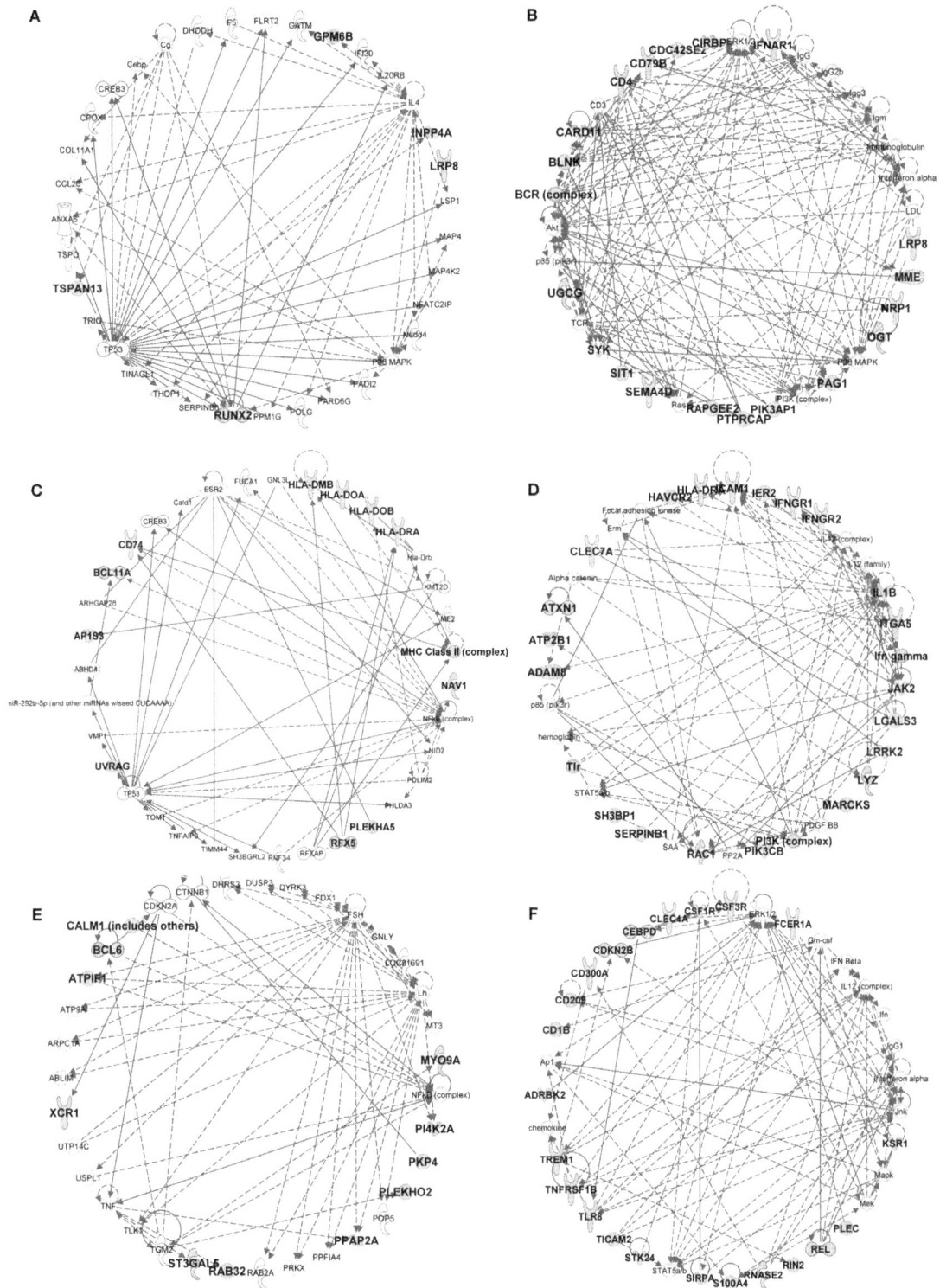

FIGURE 9 | IPA gene interaction networks of the conserved signatures in subsets of the DC lineage. The conserved signatures of subsets of the DC lineage were analyzed in Ingenuity Pathway Analysis which generates networks based on the connectivity of the genes in each signature (in boldface) but also on their connectivity with genes not belonging to the signature (in plain characters). The identified networks are displayed as graphs showing the molecular relationships between genes/gene products. Genes are represented as nodes, and the biological relationship between two nodes is represented as an edge (line). The edges can represent direct (continuous) or indirect (dashed) relationships between nodes. Selected networks generated by IPA and covering parts of conserved cell-specific signatures are displayed: **(A)** pDC signature network, **(B)** pDC vs. cDC signature network, **(C)** cDC vs. Mo/MP signature network, **(D)** cDC vs. pDC signature network, **(E)** cDC1 vs. (pDC and cDC2) signature network, **(F)** cDC2 vs. (pDC and cDC1) signature network.

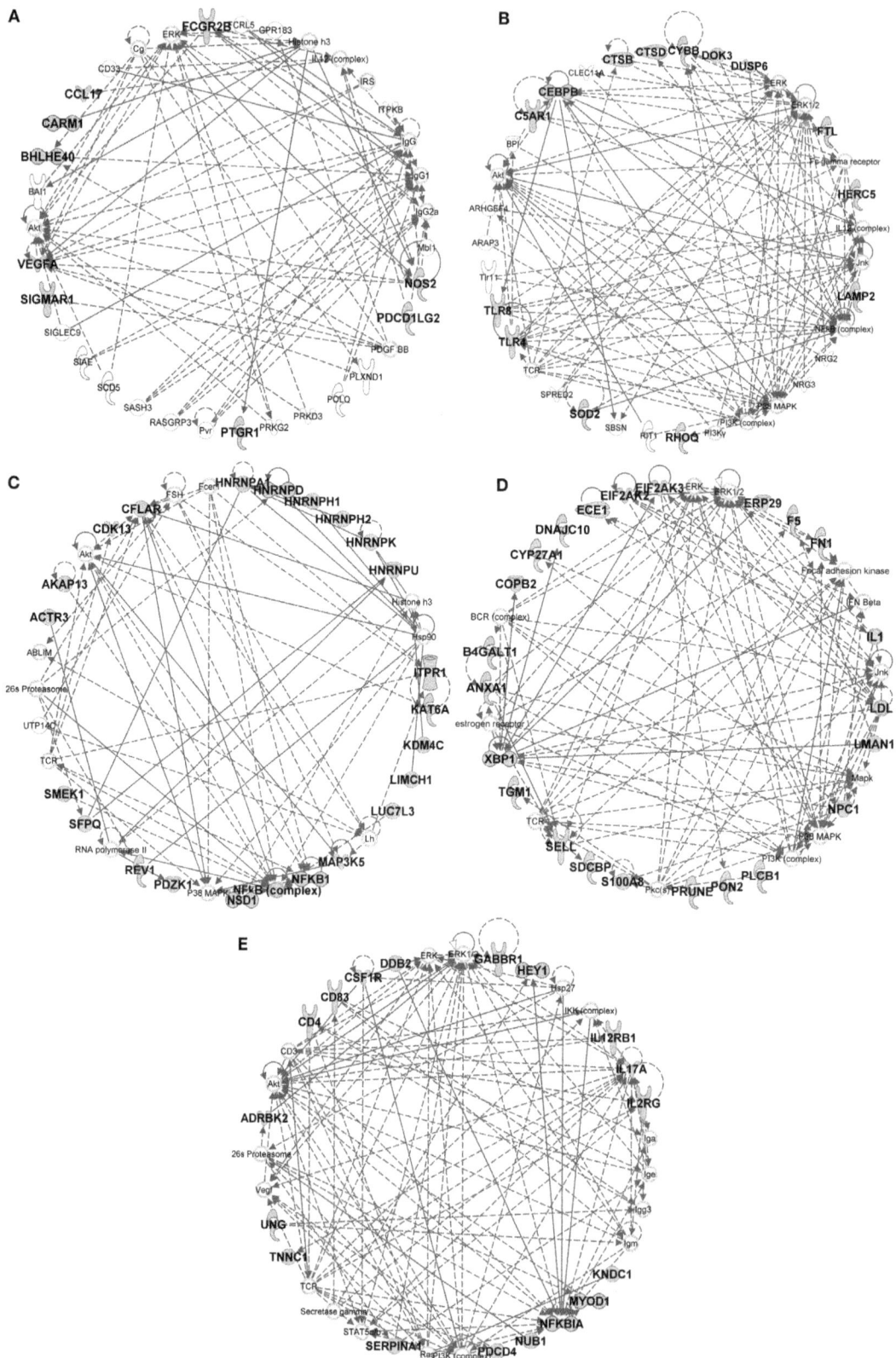

FIGURE 10 | IPA gene interaction networks of the conserved signatures in subsets of the monocytic lineage. The conserved signature of subsets of the monocytic lineage were analyzed in Ingenuity Pathway Analysis as in

Figure 9. The selected networks displayed are: **(A)** MoDC signature network, **(B)** Mo/MP vs. cDC signature network, **(C)** Mo/MP vs. MoDC signature network, **(D)** cMo vs. ncMo signature network, **(E)** ncMo vs. cMo signature network.

Defining mononuclear phagocyte subset homology across several distant warm-blooded vertebrates...

129

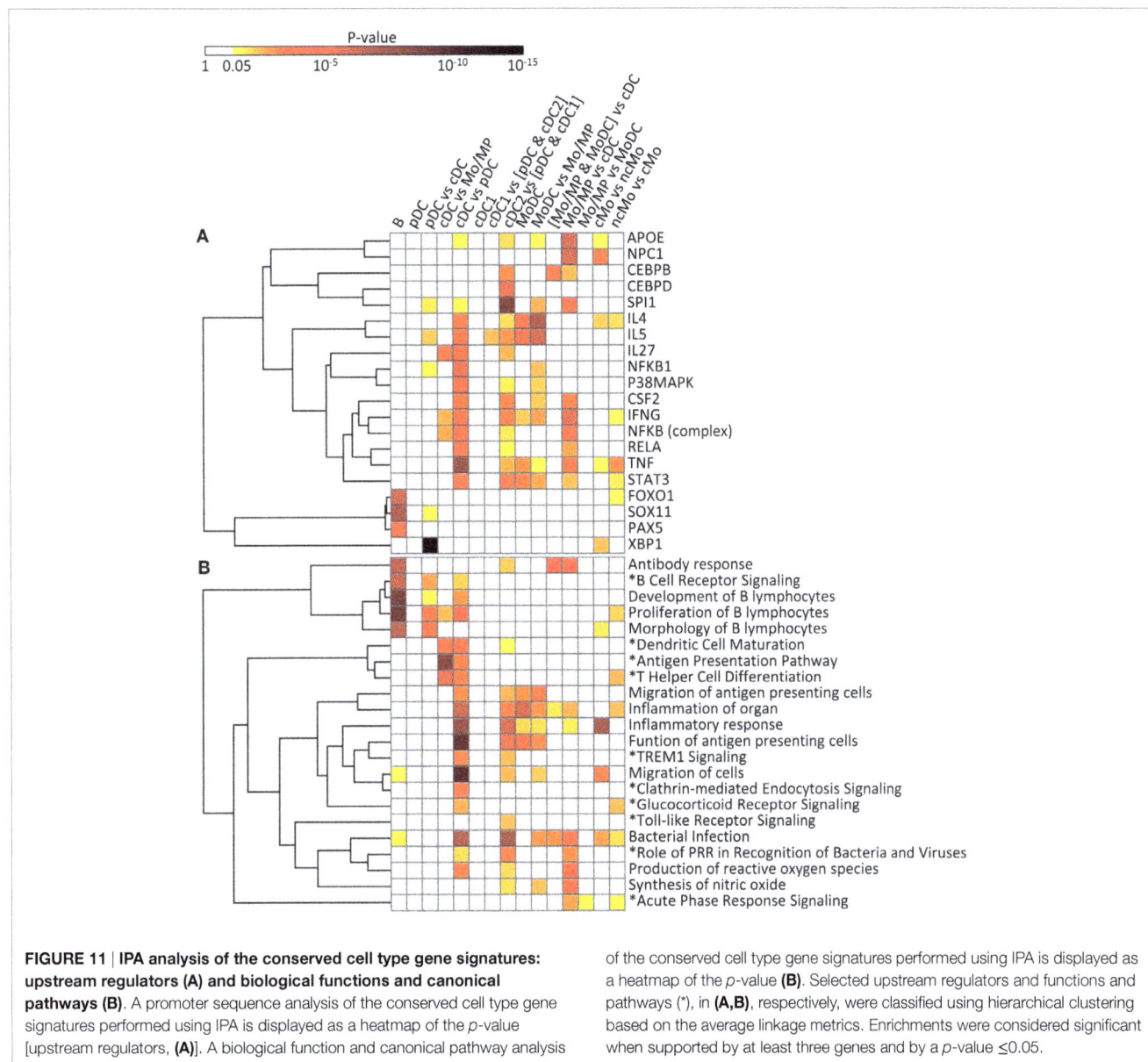

FIGURE 11 | IPA analysis of the conserved cell type gene signatures: upstream regulators (A) and biological functions and canonical pathways (B). A promoter sequence analysis of the conserved cell type gene signatures performed using IPA is displayed as a heatmap of the *p*-value [upstream regulators, **(A)**]. A biological function and canonical pathway analysis of the conserved cell type gene signatures performed using IPA is displayed as a heatmap of the *p*-value **(B)**. Selected upstream regulators and functions and pathways (*), in **(A,B)**, respectively, were classified using hierarchical clustering based on the average linkage metrics. Enrichments were considered significant when supported by at least three genes and by a *p*-value ≤0.05.

the WDFY family member 4 (*WDFY4*) whose polymorphism is associated to autoimmune diseases (62), and two intracellular trafficking proteins, a formin-binding protein (*FNBP1*) (63) and Sorting Nexin-22 (*SNX22*) (64) (**Table 1**). The conserved cDC1 vs. (pDC and cDC2) relative signature provides a longer list of genes belonging to an interaction network that includes *BCL6*, a transcriptional repressor that was recently found involved in the specification of cDC1 (17) as well as *XCR1* and *CALM1* (**Figure 9E**). IPA did not retrieve significant annotations for the cDC1 absolute or relative gene signatures. This emphasizes how little is currently known on the molecular regulation of the functions specific to cDC1s, such as cross-presentation. Future studies investigating in mouse cDC1s the functional role of the genes identified here as being part of the conserved cDC1 signatures will advance our understanding of the functions of these cells and their molecular regulation.

The absolute cDC2 conserved signature was empty. Many genes of the relative cDC2 vs. (pDC and cDC1) signature belong to a network that includes *SIRPα* (*CD172A*), a selective marker of cDC2 within the DC lineage (44), together with *CSFR1*, *TREM1*, *CLEC4A* (also known as *DCIR*), *CD1B*, and *RELB* which is known to control mouse cDC2 differentiation (65) (**Figure 9F**). A second network includes *ITGAM* (*CD11b*), a marker used to identify mouse cDC2, *CLEC6A*, and *IL1B* (Figure 7 in Supplementary Material). SPI1 ($p = 10^{-9.8}$), CEBPD ($10^{-6.2}$), and CEBPB ($10^{-3.6}$) are predicted upstream regulators, as well as CSF2 ($10^{-3.8}$), STAT3 ($10^{-4.4}$), and IFNG (10^{-4}) which were already enriched in the cDC vs. pDC signature (**Figure 11A**). This cDC2 relative signature also includes *IFI30*, also known as *GILT*, a lysosomal thiol reductase important in MHC class II and class I antigen processing (66, 67) (**Table 1**). This relative signature is enriched for "function of antigen-presenting cells" ($p = 10^{-6.2}$), "inflammatory response" ($p = 10^{-6.9}$), and for the pathways "TREM1 Signaling" ($10^{-2.5}$), "Toll-like receptor canonical signaling" ($10^{-2.5}$), and "role of pattern recognition receptors in recognition of bacteria and viruses" ($p = 10^{-3.9}$) (**Figure 11B**). Other genes were uncovered that may be important regulators of the function of cDC2s or which product could be used to identify or target these cells, including the genes coding for plasma membrane proteins such as glycoprotein *CD300A*, the sialic binding lectin *SIGLEC8*, and the paired immunoglobin-like type 2 receptor *PILRA*. This conserved relative signature shows that within the DC lineage throughout species, cDC2 express specific networks of genes related to pathogen sensing, antigen presentation, IL-1β production, and inflammation.

In the conserved signatures corresponding to the monocytic lineage, the absolute MoDC and relative MoDC vs. Mo/MP signatures are enriched for "inflammation of organ" ($p = 10^{-6.9}$ and $p = 10^{-3.3}$), "function of antigen-presenting cells" ($p = 10^{-5.1}$ and $p = 10^{-3.6}$), and "migration of antigen-presenting cells" ($p = 10^{-3.3}$ and $p = 10^{-4.3}$) (**Figure 11B**) and encompasses *NOS2*, *CCL17*, *VEGFA*, and *FCGR2B* that map to a common major network (**Figure 10A**). IL-4 ($p = 10^{-5.7}$) and IL5 ($p = 10^{-5.9}$) are predicted regulators (**Figure 11A**). Among other genes of interest that had not yet been associated to MoDC are the triose phosphate isomerase *TPI1*, the NADH dehydrogenase flavoprotein *NDUFV2*, the aldolase *ALDOA*, and the *CD200R1* gene that encodes for an inhibitory cell surface receptor of MP functions (68) (**Table 1**). The relative MoDC vs. Mo/MP signature encompasses additional genes that participate in "migration of cells" ($p = 10^{-2.3}$, with *S1PR3*, *CCL17*, and *SLC2A1*), "bacterial infection" ($p = 10^{-3.2}$, with *CD1B*, *CD209*, and *FCGR2B*), and "synthesis of nitric oxide" ($10^{-2.5}$, with *PLAU*, *IL1R2*, and *NOS2*) (**Figure 11B** and Data Sheet S4 in Supplementary Material). The conserved MoDC signatures indicate a dominant association of this subset to inflammation, as well as to DC functional properties when compared to Mo/MP across species.

Most of the genes in the Mo/MP vs. cDC conserved signature had been previously identified as overexpressed in murine MP, such as *TLR4*, *CEBPB*, *C5AR1*, and *SOD2* (9, 15) (**Table 1** and **Figure 10B**). A significant proportion of the genes within this signature are related to "inflammation of organ" ($p = 10^{-2.7}$), "production of reactive oxygen species" ($p = 10^{-4.4}$), "synthesis of nitric oxide" ($p = 10^{-3.9}$), "bacterial infection" ($p = 10^{-4.6}$), "role of pattern recognition receptor in recognition of bacteria and viruses" ($p = 10^{-3.3}$), and "acute phase response signaling" ($p = 10^{-2.9}$) (**Figure 11B**). Putative upstream regulators are NPC1 ($p = 10^{-6.8}$), APOE ($p = 10^{-6.8}$), IFNG ($p = 10^{-6.4}$), SPI1 ($p = 10^{-5.2}$), and NFkB ($p = 10^{-5.1}$) (**Figure 11A**). Additional proteins are potential transcriptional regulators of importance in Mo/MP, such as the RNA-binding protein RBMS1 and the cell cycle progression factor CCPG1. The Mo/MP vs. MoDC signature includes a gene network centered on *NFkB* and *MAP3K5* (**Figure 10C**). Overall, the conserved Mo/MP relative signatures support the association of Mo/MP to inflammation and oxidative stress across species.

The conserved comparative signature of cMo vs. ncMo retrieved genes belonging to a network with *IL1*, fibronectin (*FN*), *S100A8*,

and *XBP1* (**Figure 10D**), the latter being proposed as an upstream regulator ($10^{-2.5}$) together with *NPC1* ($p = 10^{-4}$) (**Figure 11A**), and is strongly associated to "inflammatory response" ($p = 10^{-8.4}$) (**Figure 11B**). The reciprocal ncMo vs. cMo signature includes a gene network with *IL17A*, *CSFR1*, *NFKBIA*, and *serpinA1* (**Figure 10E**), and is significantly associated to the "glucocorticoid receptor signaling pathway" ($p = 10^{-2.5}$) and to some extent to the "inflammation of organ" ($p = 10^{-2.6}$) (**Figure 11B**). These relative signatures indicate that cMo have a conserved gene program directed to strong inflammation, whereas ncMo, a poorly understood subset, might be exquisitely regulated by glucocorticoids as suggested in the literature (69, 70).

Altogether, the mononuclear phagocyte system from distantly related mammals is composed of a diversity of subsets that belong to the DC or to the Mo/MP lineage and express discriminating gene signatures involved in distinct regulatory networks and biological functions conserved through mammalian evolution. In most instances, the subset signatures also point to several unexpected genes and upstream regulators that are likely to be important in the subset biology since their selective expression pattern across subsets of mononuclear cells is conserved across species.

Phylogenetic Evidences for the Existence of a Gene Repertoire for Mononuclear Phagocyte Subsets in Birds and Bony Fishes

The existence of orthologous genes of the conserved mononuclear phagocyte subset signatures in reptile/birds, fishes, and agnathans, would indicate that the genetic equipment for mononuclear phagocyte subset diversity is available in vertebrate species distant from mammals. In the case of birds and reptiles, it remains unknown whether they have pDC, cDC1, and cDC2 subsets homologous to mammals. An orthology analysis of selected genes from conserved subset signatures revealed that most genes possess a unique ortholog in birds and reptiles, with conserved synteny with human, for instance *XCR1*, *BATF3*, *RUNX2*, *TSPAN13*, and *CSF1R* (**Table 2**). Furthermore, these same genes also possess one or more orthologs in fish. Multiple orthologs in fish are often due to the whole genome duplication that occurred during the evolution of teleosts, and to further local duplications. Importantly, fish co-orthologs of mononuclear phagocyte subset genes are generally supported by conserved synteny. Genes duplicated in fish may have been subjected to sub-functionalization, as it is the case for many immune genes duplicated in this group of vertebrates; however, some markers have a unique counterpart in fish genomes (like *BATF3*, *RFX5*, and *CIITA*), with copy loss possibly due to detrimental effects of duplication. The case of *MHC class II* is particular: although fish *MHC class II* genes are not always considered as true orthologs of human *MHC class II* genes, their sequences show the hallmarks of *bona fide* class II antigen-presenting receptors and they likely have similar functions. For c-type lectin-like (*CLEC*) molecules, no true orthologs can be identified in fish nor in birds/reptiles, as each branch of vertebrates – even each group of mammals – shows its own set of expanded *CLEC* genes. Altogether these data show that a repertoire of conserved genes for mononuclear phagocyte subsets exists in bony fishes and reptiles, which constitutes a list of candidates for relevant markers.

TABLE 2 | Search for the existence of orthologs in reptile/birds, fishes, and agnathans for selected genes of the conserved mononuclear phagocyte subset signatures.

Cell subset gene signatures	Gene	Reptiles/birds		Fishes		Agnathans (lamprey)	
		Orthologs	Conserved synteny	Orthologs	Conserved synteny	Orthologs	Conserved synteny
cDC vs. Mo/MP	HLA-DR	–	–	–	–	–	–
(MHC-related	HLA-DM	?	–	–	–	–	–
molecules)	HLA-DO	–	–	–	–	–	–
	CD74	+ (1)	Yes	+ (Multiple)	Yes	–	–
	CIITA	+ (1)	Yes	+ (1)	Yes	–	–
cDC vs. Mo/MP	NAV1	+ (1)	Yes	+ (Multiple)	Yes	–	–
	RFX5	+ (1)	Yes	+ (1)	Yes	–	–
	BCL11A	+ (1)	Yes	+ (Multiple)	Yes	–	–
MoMP vs. cDC	CEBPB	+ (1)	Yes	+ (1)	Yes	–	–
	C5AR1	+ (1)[a]	Yes[b]	+ (1)[c]		–	–
	SOD2	+ (1)	Yes	+ (1)	Yes	+ (1)	?
	APOE	–	–	+ (Multiple)	Yes	–	–
	TLR4	+ (1)	Yes	+[d]	Unclear	–	–
cDC1	XCR1	+ (1)	Yes	+ (Multiple)	Yes[e]	–	–
	FLT3	+ (1)	Yes	+ (1)	Yes	–	–
cDC vs. pDC	BATF3	+ (1)	Yes	+ (1)	Yes	–	–
	ARHGAP22	+ (1)	Yes	+ (Multiple)	Yes	+ (1)	?
	CLEC7A	–	–	–[f]	–	–	–
B cells	CD79B	+ (1)	Yes	+ (1)	Loose[g]	–	–
	PAX5	+ (1)	Yes	+ (1)	Yes	(+)[h]	?
	CD19	–	–	–	–	–	–
pDC	RUNX2	+ (1)	Yes	+ (1)[i]	Yes	+ (1)	?
	TSPAN13	+ (1)	Yes	+ (Multiple)	Yes	+ (2)	? (for both)
cDC2 vs. (cDC1	IFI30/GILT	+ (1)	Yes	+ (Multiple)	Yes	+ (1)	?
and pDC)	CSF1R	+ (1)	Yes	+ (Multiple)	Yes	–[j]	–
	SIRPA	?[k]	–	–	–	–	–
	TREM1	–[l]	–	–	–	–	–
	CLEC4A	–	–	–[f]	–	–	–
	CLEC6A	–	–	–[f]	–	–	–
More or less	CLEC9A	–	–	–[f]	–	–	–
cDC1-specific	CLNK	+ (Turkey)	Yes	+ (1)	Yes	+ (1)[m]	?

[a]Birds have one co-ortholog of human C5RA1 and C5RA2; [b]only in the lizard Anolis, not in available bird genomes; [c]fish generally have one co-ortholog of human C5RA1 and C5RA2; [d]only in some species: zebrafish, catfish, and salmonids; [e]see Ref. (36); [f]for all CLEC, no true ortholog, each deep branch of vertebrates has its own set of expanded CLEC; [g]the neighborhood is not conserved but zebrafish CD79B is close to Arhgap27 and Plekhm1 that are on the same human chromosome (chr17) as CD79B but at 20 megabases; CD79B genes often are not annotated in fish genomes. In zebrafish, CD79B is ENSDARG00000088902; [h]a lamprey gene ortholog to PAX5 has been identified and was selectively expressed in lamprey VLRB+ cells which resemble B lymphocytes (55); however, this gene is not identified in the current publicly available assembly of the lamprey genome; [i]duplicated in zebrafish and cavefish; [j]a lamprey gene is a co-ortholog to all vertebrate CSF1R, PDGFR, KIT, FLT3, etc.; [k]in birds species, several genes are co-orthologs of all mammalian SIRPs including SIRPA; [l]bird TREM-like genes are more closely related to TREM2 rather than to TREM1; [m]co-ortholog of CLNK, BLNK, and other related genes.

The presence of BATF3 and XCR1 are hints at possible existence of cDC1 in these species, as BATF3 specifically controls cDC1 development in mice (71) and XCR1 expression is strictly associated to cDC1 in several mammals (34, 36, 40, 41). In contrast, the lamprey does not have identified orthologs for many of the genes selected from the transcriptomic fingerprints of the subsets of mammalian mononuclear phagocytes (**Table 2**). Agnathans, including lampreys and myxines, harbor three adaptive immune cell types, each expressing a specific class of variable lymphocyte receptors, VLRC, VLRA, and VLRB, and showing transcriptomic and functional commonalities with gnathostome γδ T lymphocytes, αβ T lymphocytes, and B lymphocytes, respectively (55, 72). However, it is uncertain whether or not the activation of agnathan lymphocytes requires APCs, and if so, to which extent these cells could resemble gnathostome APCs (72). Contrary to the situation in birds and fishes, our observations do not support the existence in the lamprey of gene sets similar to those defining the transcriptomic fingerprints of the mononuclear phagocytes

of mammals. Although incomplete assembly and annotation of the genome of the lamprey do not allow drawing definitive conclusions, our observations are consistent with the lack in agnathans of MHC functional homologs and of the particular proteasome machinery used by mammalian APCs for antigen processing (72). Altogether, this phylogenetic study shows that the repertoire of key genes characterizing the diversity of the mononuclear phagocytes in mammals were already present in the common ancestor of tetrapods and fishes but might be largely absent in agnathans.

Discussion

Our computational transcriptomic meta-analysis indicates that the complex organization of the mononuclear phagocyte system shows conservation throughout distantly related mammals, a finding that appears to extend to chicken, a non-mammalian vertebrate. In the present work, by using GSEA and hierarchical clustering for unbiased pan-genomic analysis of the molecular identity of immune cell subsets across four vertebrate species, we convincingly established the existence of strong homologies between these cell types across mammals, beyond the already known existence of B cells in all species. Specifically, we could align across mammals cDC1, cDC2, pDC, MoDC, Mo/MP, and cMo vs. ncMo. In addition, we found that many of the genes that we showed to be selectively expressed in distinct mononuclear phagocyte subsets in mammals have existing orthologs in bony fishes while this appears not to be the case in lamprey. Thus, our study suggests that conserved mononuclear phagocyte subsets might exist in all gnathostomes but not in agnathans. However, this hypothesis will require to be tested experimentally, by re-examining the presence of orthologous genes in lamprey upon completion of the genome assembly and its annotation, by identifying and studying candidate mononuclear phagocyte subsets in bony fishes, and by determining whether similar cells exist in sharks, rays, and lamprey. For example, orthologous genes of the conserved mononuclear phagocyte signatures (**Table 2**) could be targeted by the CRISPR/Cas9 technology with a reporter gene marker in order to identify and character-ize mononuclear phagocyte subsets in bony fishes (73), with for certain genes the need to test several putative orthologs in fish due to genome duplication.

The two methodologies that we used to assess subset homolo-gies across species, i.e., hierarchical clustering and GSEA, display complementary functionalities. Hierarchical clustering on filtered, centered, reduced, and aggregated datasets has the advantage of integrating all samples together into a single analy-sis and of providing a global overview of the homologies between cell subsets of various species (15, 17, 18, 74, 75). However, the integration of distinct datasets requires a cross-normalization procedure which consists in a rather profound mathematical transformation of the data. The normalization procedure artificially increases the variance for genes with only small dif-ferences in their initial signal intensities between the different cell types studied. Conversely, it comparatively decreases the variance for genes with high differences in their initial signal intensities between the different cell types studied. To limit the biases that this normalization introduces, it is thus necessary to select only the orthologous genes that vary strongly in their expression across the cell types examined within each species. Another corollary is that this analysis can only be applied to genes that have known orthologs in all species. If one ortholog is missing in only one species, the gene must be removed from the analysis. Hence, this method should be used with caution, only under conditions where dataset normalization does not yield too strong biases in gene-expression profiles. It is also not appropriate when the structures of the different datasets are too different (i.e., the number and potential identities of cell types vary too much across datasets), because the dynamic ranges of gene expression between datasets are not expected to be the same and should therefore not be forced to similar-ity. Even under conditions where the experimental design is favorable to the use of hierarchical clustering, GSEA ensures of the robustness of interpretation. GSEA has been used by us and others to perform cross-species comparisons (5, 19, 29,

42, 76–78). GSEA notably displays advantages and drawbacks distinct from those of hierarchical clustering. First, it is easier to perform GSEA since dedicated ready-to-use stand-alone programs are available which do not require bio-informatics expertise. Second, GSEA is more sensitive, notably to detect overlaps of common functions/gene networks between cell populations or cellular contaminations, as exemplified with sheep *pDC enriched in human and mouse B cell fingerprints. This higher sensitivity is linked to (i) the fact that GSEA can detect coordinate regulation of gene modules (geneset-based approach) and thus does not rely on the strong regulation of few single genes (single gene-based approach), (ii) the fact that GSEA, when applied to multiple species, takes into account all genes that have orthologous counterparts in the considered species and is not restricted only to highly variable genes. Third, GSEA can perform cross-platform comparison without any cross-normalization thus without any supplementary artificial manipulation of the expression data. Finally, it can be performed on multiple datasets, even if their structures are different. However, GSEA presents the limitation of performing pairwise comparisons whose results can be integrated and visu-alized with our Bubble GUM software, but it nevertheless does not provide a global trans-species overview of subset homology. Overall, in order to increase confidence in the interpretation of the results, it is important to combine both approaches and verify that they both lead to consistent conclusions.

Our subset assignment methodology demonstrates similar-ity or proximity between subsets across species but not strict identity. Besides possible intrinsic transcriptomic differences between species, one of the reasons that explain this limitation is the process of subset identification itself, which makes use of different surface markers. Whenever possible, similar marker combinations were used such as CADM1 and CD172 that are known to be conserved markers across human, mouse, and sheep cDC subsets (42). However, mAb anti-CD11c did not exist for the initial gating in pig and the mAbs in the exclusion pool were not the same in pig and sheep. Moreover, existing marker combina-tions are not always specific and can lead to cross-contamination between different cell subsets. Indeed, the GSEA of the sheep *cDC2 revealed that they may have been contaminated by pDC, despite our attempt to avoid this problem through exclusion of CD45RB-expressing cells. It remains possible that pDC express-ing minimal levels of CD45RB were still present in the sorted *cDC2 population, and not in the sheep cDC1 subset. However, since sheep *cDC2 were found in the correct cDC branch of the hierarchical clustering, their contamination by pDC is likely to have been limited. Similarly, it is likely that the sorted sheep *pDC include residual B cells, explaining the enrichment for the human B cell fingerprint at a level above expectation: indeed after exclu-sion of B cells with a pan-B cell marker, sheep *pDC were selected with a mAb directed to CD45RB, which may react with residual B cells that have escaped the pan-B cell exclusion. Yet, sheep *pDC still cluster with other species pDC, separately from B cells. In the case of pig, pDC were selected using markers not expressed by B cells and they displayed an enrichment for B cell fingerprints at a level encountered in GSEA analyses of mouse pDC (Figure 4 in Supplementary Material). Finally, our approach was able to demonstrate that *a priori* assignment of subset identity based on the expression of a few membrane markers could be wrong, like in the case of the pig *cDC2. Moreover, our approach had the power to properly re-assign cell subset identity, demonstrating that pig *cDC2 were actually homologous to mouse and human ncMo. Another laboratory analyzed the transcriptome of similar pig cells sorted as CD14^{low} CD163^{high} cells, but they could not assign them to classical nor to non-classical human Mo, due to differences in bio-informatics approaches in this study (79) and in ours.

Our study will help improving in the near future the toolbox available in each species for rigorous and consistent phenotypic identification of cell subsets, thanks to our identification of novel, conserved, and specific, combinations of surface markers for each cell subset, which should allow generating more appropriate staining reagents. For instance, fluorescently labeled recombinant XCL1 could theoretically be used in any species to rigorously identify and sort cDC1 (38, 41). In addition, cell surface proteins encoded by genes shown here to be selectively expressed in a conserved manner in specific subsets of mononuclear phagocytes represent new candidate markers to refine and homogenize phenotypic identification of these cells across species, such as LRP8, TSPAN13, NRP1, and SLC30A5 for pDC, FCGR2B, and CD200R1 for MoDC, SIGLEC8 and IGSF6 for cDC2, and CSF1R, TLR4, and C5AR1 for Mo/MP (**Table 3**). However, these potential new markers for subset identification need to be validated at the protein level.

The subset-specific signatures that are conserved throughout distant mammals included variable number of genes that were sometimes far lower than the numbers of genes in the human/mouse common signatures. There are several explanations to this finding. There is a contribution of the very high stringency of the "Min (test) vs. Max (ref)" ≥1x method that we used to establish the signatures, since any gene which was not consistently found overexpressed in all the replicates of all the species was excluded. As an example, the gene *DNAJC7*, identified as specific of pDC in our previous work (15) was removed from the human pDC

signature because its "Min (test) vs. Max (ref)" ratio was equal to 0.933, due to a single lower human pDC replicate compared to a single replicate found with a higher signal in human MoDC. There is also a contribution of incomplete mapping of the genome of some of the species studied, leading to an underestimation of the number of orthologous genes that could be queried across all species. For example, *POU2F2*, more highly expressed in human and murine B cells as compared to many other immune cells, has not been mapped yet to the pig genome while it has been mapped to the genome of more distant species such as the spotted gar with a 1-to-1 orthology relationship. Another prominent cause is linked to technical limitations of the microarray approach, such as lack of ProbeSets against certain genes in certain species. This is notably the case for the gene *CLEC9A*, known to be specific of cDC1 but for which no ProbeSet exists in the human Affymetrix HG U133 plus2 gene chip. Sometimes, low signal-to-noise ratio for certain ProbeSets can also be responsible for the loss of putative interesting signature genes, such as *ZNF521* (*Zfp521* in mouse) found to be highly specific of pDC in mouse and human while the pig and sheep orthologous ProbeSet remains at the background level whatever the cell type considered. Recent technological advances now allow performing high throughput RNA sequencing at single cell levels with high sensitivity and processivity, which could solve most of the above issues; indeed, all expressed genes should be detected without any bias and analysis at the single cell level should alleviate any issue of cross-contamination between cell types. Therefore, the generation of gene-expression data for many individual cells of the same type should increase statistical power to define genes co-expressed at the single cell level and defining cell type-specific transcriptomic modules (22). Single cell gene-expression profiling recently allowed the unbiased and *de novo* identification of the different cell types of spleen (80) and central nervous system (81, 82) via the description of their molecular identity, starting from the bulk population of all the cells that could be extracted from the organ, without any prior enrichment procedure, based on the use of potentially confounding phenotypic marker combinations. However, this strategy is

TABLE 3 | Proposition of marker combination for oligo-phenotyping of mononuclear phagocytic cell subsets across species.

Exclusion	Anti-CD3, anti-NK cells, and anti-B cells, if available[a]						
Targeted cell population	pDC	cDC1	cDC2	MoDC	cMo	ncMo	MP
Combination of known markers[b]	FLT3$^+$ SIRPα^{lo} MHC-II^{lo}	FLT3^{hi} SIRPα^{lo} MHC-II$^+$ CD11c$^+$ CADM1^{hi}	FLT3$^+$ SIRPα$^+$ MHC-II$^+$ CD11c$^+$ CADM1^{lo}	FLT3$^-$ SIRPα$^+$ MHC-II$^+$ CD11c$^+$	FLT3$^-$ SIRPα$^+$	FLT3$^-$ SIRPα$^+$	FLT3$^-$ SIRPα$^+$
New additional candidates[c]	LRP8$^+$ TSPAN13^{hi} SLC30A5$^+$ NRP1$^+$	XCR1$^+$	SIGLEC8$^+$ IGSF6$^+$	FCGR2B^{hi} CD200R1$^+$	CSF1R^{int} CCR1$^+$ C19ORF59$^+$	CSF1R^{hi} CD83$^+$	CSF1R$^+$ TLR4^{hi} C5AR1$^+$

[a] *Exclusion with anti-CD3, anti-NK cells, and anti-B cell markers is desirable when appropriate tools are available.*

[b] *A combination of known markers including FLT3, MHC-II, CD11c, SIRPα, and CADM1 allows a first step of identification of subset candidates but is at risk of contamination by sister cell types, or may be incomplete due to non-availability of one of the marker. FLT3 labeling may be performed by using recombinant His-tag FLT3L generated for the relevant species as recently proposed in a review (21).*

[c] *New additional candidate markers for refinement of subset identification are derived from the identification of genes encoding cell surface molecules from the conserved cell subset gene signatures.*

still extremely difficult to apply to species which genome has not yet been completely assembled, as well as to very rare cell types recovered upon prior phenotype-based enrichment. Moreover, to obtain information of sufficient completeness on functionally important genes for which few mRNA are expressed per cell, it is necessary to sequence at a sufficient depth of about one million reads per cell, which today still represents a very high cost when multiplied by the number of individual cells and conditions. Finally, the interpretation of the RNA-seq data on single cells is still largely based on the transcriptomic/molecular identity of cell types that are deduced from microarray analysis of purified cell pools. Hence, our work constitutes a major advancement in the field and is a necessary step before an eventual, later, refinement of the definition of cell subsets and their associated molecular signatures using single cell RNA-seq. The canonical gene-expression signatures that we generated can be used to distinguish and identify cell subsets in other vertebrate species. The cDC1 signature and the cDC2 vs. cDC1 signatures could be evaluated in chicken cDC sorted as single cells to determine whether this population includes only cDC1, as suggested by the trans-vertebrate hierarchical clustering, or a mixture of cDC1 and cDC2.

The conservation of gene signatures and interacting gene networks in homologous cell subsets throughout evolution is likely to bear strong biological meaning. Indeed, many genes of the conserved signatures were already known for their functions in these cells, validating the biological relevance of our signatures. In several instances, the same functional annotations were enriched in distinct subset signatures, but the genes responsible for the enrichments differed. For example, the genes responsible for the enrichment of the pathway "role of pattern recognition receptor in recognition of bacteria and viruses" were *TLR4*, *TLR8*, and *C5AR1* for the Mo/MP vs. cDC signature, *TLR2*, *CLEC7A*, *IL1B*, and *PIK3CB* for the cDC vs. pDC signature, and *TLR8*, *CLEC6A*, *DDX58*, *OAS2*, and *IL1B* for the cDC2 vs. (pDC and cDC1) signature. This analysis shows that cDC and MP express different sets of pattern recognition receptors for detection of viruses and bacteria, and that, within DC, cDC2 are also equipped differently from cDC1 and pDC for sensing of viruses and bacteria. These observations extended to other mammalian species the previous reports that human and mouse cDC2 are preferentially equipped with PRR targeting bacteria or involved in cytosolic sensing of viral infection (83, 84), and that TLR4 is very weakly expressed on pDC and cDC as compared to Mo/MP (83, 85). Similarly, different subset signatures were all enriched for "inflammatory response," "inflammation of organs," and "bacterial infection" but due to different genes. Altogether, this analysis indicates that different mononuclear phagocyte subsets express distinct and specific gene-expression modules which can sometimes contribute in a complementary way to the same general biological process in a conserved manner throughout evolution. Within the conserved gene-expression programs in mononuclear phagocyte subsets, we identified novel candidate genes and putative upstream regulators which likely contribute to the control of the ontogeny or functions of the corresponding cell type. For instance, the *FNBP1* and *SNX22* encoded proteins may be involved

in the specific intracellular trafficking properties promoting antigen cross-presentation by cDC1, *ARHGAP22* and *NAV1* could modulate the organization of the cytoskeleton of cDC to control their mobility or antigen presentation functions, and the transcription regulators *BCL11A* and *MSL2A* may control specific gene networks in cDC. *BLC11A* is known to be key in murine pDC development (50) but it may have a specific role in cDC homeostasis, as inferred from a previous study (86). Our study thus opens the way for deciphering the sets of genes encoding functional cellular modules and their specifying transcription factors in subsets of mononuclear cells, in order to further improve and connect together the molecular and functional definitions of these cell types across species (22, 23).

Conclusion

Our meta-analysis that combines cell sorting and comparative transcriptomic analysis was implemented as a methodology pipeline that could be used by biologists with minimal training in bio-informatics for subsequent extension to other species and to other complex cellular systems. Our study should lead to the identification of homologous mononuclear phagocyte subsets in species other than sheep and pigs, and which are of importance for biomedical investigations, such as bats, rabbits, ferrets, guinea pigs, possibly zebrafishes, and in species of veterinary importance including pets and animals of the food economy. The characterization of mononuclear phagocyte subsets in these species will allow manipulating their immune responses against diseases for the sustainability of our environment.

Author Contributions

ISC and MD directed research and wrote the paper with input from TPVM. TPVM carried out bio-informatics analyses with input from MD. JEY performed most of the cell purification experiments and analyzed data, with input from CU (microarray hybridization, blood processing), SR (pDC isolation), MB (cell sorting), MM (microarray hybridization and analysis), HM (chicken array data), PQ (chicken array data), NB (pig cell phenotyping). PB performed phylogenetic analyses. GF and HS provided key cell types (MoDC) and reagents (unique mAb, non-commercially available). LJ performed array annotations.

Acknowledgments

The authors thank Chantal Kang and Michel Bonneau from the Centre d'Imagerie Interventionnelle for providing pig blood and the Unité Commune d'Expérimentation Animale du Centre de Jouy-en-Josas for providing sheep blood and spleen. We thank Christelle Thibault-Carpentier from the Plate-forme Biopuces (Strasbourg, France) for performing one of the microarray experiments (http://www-microarrays.u-strasbg.fr). This work was supported by institutional funding from Agence Nationale de la Recherche (ANR) PhyloGenDC, ANR-09-BLAN-0073-02. We thank the European

Commission 7th Framework Programme (Initial Training Network FishforPharma, PITNGA- 2011-289209). TPVM was supported through the PhyloGenDC ANR grant and the European Research Council (FP7/2007-2013 Grant Agreement no. 281225 to MD for the Systems Dendritic project).

References

1. Ziegler-Heitbrock L, Ancuta P, Crowe S, Dalod M, Grau V, Hart DN, et al. Nomenclature of monocytes and dendritic cells in blood. *Blood* (2010) 116(16):e74–80. doi:10.1182/blood-2010-02-258558

2. Mildner A, Yona S, Jung S. A close encounter of the third kind: monocyte-derived cells. *Adv Immunol* (2013) 120:69–103. doi:10.1016/B978-0-12-417028-5.00003-X

3. Tamoutounour S, Guilliams M, Montanana Sanchis F, Liu H, Terhorst D, Malosse C, et al. Origins and functional specialization of macrophages and of conventional and monocyte-derived dendritic cells in mouse skin. *Immunity* (2013) 39(5):925–38. doi:10.1016/j.immuni.2013.10.004

4. Bain CC, Bravo-Blas A, Scott CL, Gomez Perdiguero E, Geissmann F, Henri S, et al. Constant replenishment from circulating monocytes maintains the macrophage pool in the intestine of adult mice. *Nat Immunol* (2014) 15(10):929–37. doi:10.1038/ni.2967

5. McGovern N, Schlitzer A, Gunawan M, Jardine L, Shin A, Poyner E, et al. Human dermal CD14(+) cells are a transient population of monocyte-derived macrophages. *Immunity* (2014) 41(3):465–77. doi:10.1016/j.immuni.2014.08.006

6. Molawi K, Wolf Y, Kandalla PK, Favret J, Hagemeyer N, Frenzel K, et al. Progressive replacement of embryo-derived cardiac macrophages with age. *J Exp Med* (2014) 211(11):2151–8. doi:10.1084/jem.20140639

7. Lavin Y, Winter D, Blecher-Gonen R, David E, Keren-Shaul H, Merad M, et al. Tissue-resident macrophage enhancer landscapes are shaped by the local microenvironment. *Cell* (2014) 159(6):1312–26. doi:10.1016/j.cell.2014.11.018

8. Gosselin D, Link VM, Romanoski CE, Fonseca GJ, Eichenfield DZ, Spann NJ, et al. Environment drives selection and function of enhancers controlling tissue-specific macrophage identities. *Cell* (2014) 159(6):1327–40. doi:10.1016/j.cell.2014.11.023

9. Gautier EL, Shay T, Miller J, Greter M, Jakubzick C, Ivanov S, et al. Gene-expression profiles and transcriptional regulatory pathways that underlie the identity and diversity of mouse tissue macrophages. *Nat Immunol* (2012) 13(11):1118–28. doi:10.1038/ni.2419

10. Cheong C, Matos I, Choi JH, Dandamudi DB, Shrestha E, Longhi MP, et al. Microbial stimulation fully differentiates monocytes to DC-SIGN/CD209(+) dendritic cells for immune T cell areas. *Cell* (2010) 143(3):416–29. doi:10.1016/j.cell.2010.09.039

11. Xu Y, Zhan Y, Lew AM, Naik SH, Kershaw MH. Differential development of murine dendritic cells by GM-CSF versus Flt3 ligand has implications for inflammation and trafficking. *J Immunol* (2007) 179(11):7577–84. doi:10.4049/jimmunol.179.11.7577

12. Mayer CT, Ghorbani P, Nandan A, Dudek M, Arnold-Schrauf C, Hesse C, et al. Selective and efficient generation of functional Batf3-dependent CD103+ dendritic cells from mouse bone marrow. *Blood* (2014) 24:3081–91. doi:10.1182/blood-2013-12-545772

13. Greter M, Lelios I, Pelczar P, Hoeffel G, Price J, Leboeuf M, et al. Stroma-derived interleukin-34 controls the development and maintenance of Langerhans cells and the maintenance of microglia. *Immunity* (2012) 37(6):1050–60. doi:10.1016/j.immuni.2012.11.001

14. Schlitzer A, Ginhoux F. Organization of the mouse and human DC network. *Curr Opin Immunol* (2014) 26:90–9. doi:10.1016/j.coi.2013.11.002

15. Robbins SH, Walzer T, Dembele D, Thibault C, Defays A, Bessou G, et al. Novel insights into the relationships between dendritic cell subsets in human and mouse revealed by genome-wide expression profiling. *Genome Biol* (2008) 9(1):R17. doi:10.1186/gb-2008-9-1-r17

16. Haniffa M, Shin A, Bigley V, McGovern N, Teo P, See P, et al. Human tissues contain CD141(hi) cross-presenting dendritic cells with functional homology to mouse CD103(+) nonlymphoid dendritic cells. *Immunity* (2012) 37(1):60–73. doi:10.1016/j.immuni.2012.04.012

17. Watchmaker PB, Lahl K, Lee M, Baumjohann D, Morton J, Kim SJ, et al. Comparative transcriptional and functional profiling defines conserved programs of intestinal DC differentiation in humans and mice. *Nat Immunol* (2014) 15(1):98–108. doi:10.1038/ni.2768

18. Cros J, Cagnard N, Woollard K, Patey N, Zhang SY, Senechal B, et al. Human CD14dim monocytes patrol and sense nucleic acids and viruses via TLR7 and TLR8 receptors. *Immunity* (2010) 33(3):375–86. doi:10.1016/j.immuni.2010.08.012

19. Crozat K, Guiton R, Guilliams M, Henri S, Baranek T, Schwartz-Cornil I, et al. Comparative genomics as a tool to reveal functional equivalences between human and mouse dendritic cell subsets. *Immunol Rev* (2010) 234(1):177–98. doi:10.1111/j.0105-2896.2009.00868.x

20. Ingersoll MA, Spanbroek R, Lottaz C, Gautier EL, Frankenberger M, Hoffmann R, et al. Comparison of gene expression profiles between human and mouse monocyte subsets. *Blood* (2010) 115(3):e10–9. doi:10.1182/blood-2009-07-235028

21. Summerfield A, Auray G, Ricklin M. Comparative dendritic cell biology of veterinary mammals. *Annu Rev Anim Biosci* (2014) 3:533–57. doi:10.1146/annurev-animal-022114-111009

22. Achim K, Arendt D. Structural evolution of cell types by step-wise assembly of cellular modules. *Curr Opin Genet Dev* (2014) 27:102–8. doi:10.1016/j.gde.2014.05.001

23. Arendt D. The evolution of cell types in animals: emerging principles from molecular studies. *Nat Rev Genet* (2008) 9(11):868–82. doi:10.1038/nrg2416

24. Chevallier N, Berthelemy M, Le Rhun D, Laine V, Levy D, Schwartz-Cornil I. Bovine leukemia virus-induced lymphocytosis and increased cell survival mainly involve the CD11b+ B-lymphocyte subset in sheep. *J Virol* (1998) 72(5):4413–20.

25. Fiebach AR, Guzylack-Piriou L, Python S, Summerfield A, Ruggli N. Classical swine fever virus N(pro) limits type I interferon induction in plasmacytoid dendritic cells by interacting with interferon regulatory factor 7. *J Virol* (2011) 85(16):8002–11. doi:10.1128/JVI.00330-11

26. Ruscanu S, Pascale F, Bourge M, Hemati B, Elhmouzi-Younes J, Urien C, et al. The double-stranded RNA bluetongue virus induces type I interferon in plasmacytoid dendritic cells via a MYD88-dependent TLR7/8-independent signaling pathway. *J Virol* (2012) 86(10):5817–28. doi:10.1128/JVI.06716-11

27. Foulon E, Foucras G. Two populations of ovine bone marrow-derived dendritic cells can be generated with recombinant GM-CSF and separated on CD11b expression. *J Immunol Methods* (2008) 339(1):1–10. doi:10.1016/j.jim.2008.07.012

28. Ruscanu S, Jouneau L, Urien C, Bourge M, Lecardonnel J, Moroldo M, et al. Dendritic cell subtypes from lymph nodes and blood show contrasted gene expression programs upon bluetongue virus infection. *J Virol* (2013) 87(16):9333–43. doi:10.1128/JVI.00631-13

29. Marquet F, Vu Manh TP, Maisonnasse P, Elhmouzi-Younes J, Urien C, Bouguyon E, et al. Pig skin includes dendritic cell subsets transcriptomically related to human CD1a and CD14 dendritic cells presenting different migrating behaviors and T cell activation capacities. *J Immunol* (2014) 193(12):5883–93. doi:10.4049/jimmunol.1303150

30. Casel P, Moreews F, Lagarrigue S, Klopp C. sigReannot: an oligo-set re-annotation pipeline based on similarities with the ensemble transcripts and unigene clusters. *BMC Proc* (2009) 3(Suppl 4):S3. doi:10.1186/1753-6561-3-S4-S3

31. Vu Manh TP, Marty H, Sibille P, Le Vern Y, Kaspers B, Dalod M, et al. Existence of conventional dendritic cells in *Gallus gallus* revealed by comparative gene expression profiling. *J Immunol* (2014) 192(10):4510–7. doi:10.4049/jimmunol.1303405

32. Baranek T, Vu Manh TP, Alexandre Y, Maqbool MA, Cabeza JZ, Tomasello E, et al. Differential responses of immune cells to type I interferon contribute to host resistance to viral infection. *Cell Host Microbe* (2012) 12(4):571–84. doi:10.1016/j.chom.2012.09.002

33. Subramanian A, Kuehn H, Gould J, Tamayo P, Mesirov JP. GSEA-P: a desktop application for gene set enrichment analysis. *Bioinformatics* (2007) 23(23):3251–3. doi:10.1093/bioinformatics/btm369

34. Gurka S, Hartung E, Becker M, Kroczek RA. Mouse conventional dendritic cells can be universally classified based on the mutually exclusive expression of XCR1 and SIRPa. *Front Immunol* (2015) 6:35. doi:10.3389/fimmu.2015.00035

35. Dorner BG, Dorner MB, Zhou X, Opitz C, Mora A, Guttler S, et al. Selective expression of the chemokine receptor XCR1 on cross-presenting dendritic cells determines cooperation with CD8+ T cells. *Immunity* (2009) 31(5):823–33. doi:10.1016/j.immuni.2009.08.027

36. Crozat K, Guiton R, Contreras V, Feuillet V, Dutertre CA, Ventre E, et al. The XC chemokine receptor 1 is a conserved selective marker of mammalian cells homologous to mouse CD8{alpha}+ dendritic cells. *J Exp Med* (2010) 207:1283–92. doi:10.1084/jem.20100223

37. Bachem A, Guttler S, Hartung E, Ebstein F, Schaefer M, Tannert A, et al. Superior antigen cross-presentation and XCR1 expression define human CD11c+CD141+ cells as homologues of mouse CD8+ dendritic cells. *J Exp Med* (2010) 207(6):1273–81. doi:10.1084/jem.20100348

38. Crozat K, Tamoutounour S, Vu Manh TP, Fossum E, Luche H, Ardouin L, et al. Cutting edge: expression of XCR1 defines mouse lymphoid-tissue resident and migratory dendritic cells of the CD8{alpha}+ type. *J Immunol* (2011) 187(9):4411–5. doi:10.4049/jimmunol.1101717

39. Bachem A, Hartung E, Guttler S, Mora A, Zhou X, Hegemann A, et al. Expression of XCR1 characterizes the Batf3-dependent lineage of dendritic cells capable of antigen cross-presentation. *Front Immunol* (2012) 3:214. doi:10.3389/fimmu.2012.00214

40. Balan S, Ollion V, Colletti N, Chelbi R, Montanana-Sanchis F, Liu H, et al. Human XCR1+ dendritic cells derived in vitro from CD34+ progenitors closely resemble blood dendritic cells, including their adjuvant responsiveness, contrary to monocyte-derived dendritic cells. *J Immunol* (2014) 193(4):1622–35. doi:10.4049/jimmunol.1401243

41. Dutertre CA, Jourdain JP, Rancez M, Amraoui S, Fossum E, Bogen B, et al. TLR3-responsive, XCR1+, CD141(BDCA-3)+/CD8alpha+-equivalent dendritic cells uncovered in healthy and simian immunodeficiency virus-infected rhesus macaques. *J Immunol* (2014) 192(10):4697–708. doi:10.4049/jimmunol.1302448

42. Contreras V, Urien C, Guiton R, Alexandre Y, Vu Manh TP, Andrieu T, et al. Existence of CD8{alpha}-like dendritic cells with a conserved functional specialization and a common molecular signature in distant mammalian species. *J Immunol* (2010) 185:3313–25. doi:10.4049/jimmunol.1000824

43. Dutertre CA, Wang LF, Ginhoux F. Aligning bona fide dendritic cell populations across species. *Cell Immunol* (2014) 291(1–2):3–10. doi:10.1016/j.cellimm.2014.08.006

44. Guilliams M, Henri S, Tamoutounour S, Ardouin L, Schwartz-Cornil I, Dalod M, et al. From skin dendritic cells to a simplified classification of human and mouse dendritic cell subsets. *Eur J Immunol* (2010) 40(8):2089–94. doi:10.1002/eji.201040498

45. Hajjoubi S, Rival-Gervier S, Hayes H, Floriot S, Eggen A, Piumi F, et al. Ruminants genome no longer contains whey acidic protein gene but only a pseudogene. *Gene* (2006) 370:104–12. doi:10.1016/j.gene.2005.11.025

46. de Jong WW, van Dijk MA, Poux C, Kappe G, van Rheede T, Madsen O. Indels in protein-coding sequences of euarchontoglires constrain the rooting of the eutherian tree. *Mol Phylogenet Evol* (2003) 28(2):328–40. doi:10.1016/S1055-7903(03)00116-7

47. Douady CJ, Douzery EJ. Molecular estimation of eulipotyphlan divergence times and the evolution of "insectivore". *Mol Phylogenet Evol* (2003) 28(2):285–96. doi:10.1016/S1055-7903(03)00119-2

48. Pascale F, Contreras V, Bonneau M, Courbet A, Chilmonczyk S, Bevilacqua C, et al. Plasmacytoid dendritic cells migrate in afferent skin lymph. *J Immunol* (2008) 180(9):5963–72. doi:10.4049/jimmunol.180.9.5963

49. Summerfield A, Guzylack-Piriou L, Schaub A, Carrasco CP, Tache V, Charley B, et al. Porcine peripheral blood dendritic cells and natural interferon-producing cells. *Immunology* (2003) 110(4):440–9. doi:10.1111/j.1365-2567.2003.01755.x

50. Ippolito GC, Dekker JD, Wang YH, Lee BK, Shaffer AL III, Lin J, et al. Dendritic cell fate is determined by BCL11A. *Proc Natl Acad Sci U S A* (2014) 111(11):E998–1006. doi:10.1073/pnas.1319228111

51. Cao T, Ueno H, Glaser C, Fay JW, Palucka AK, Banchereau J. Both Langerhans cells and interstitial DC cross-present melanoma antigens and efficiently activate antigen-specific CTL. *Eur J Immunol* (2007) 37(9):2657–67. doi:10.1002/eji.200636499

52. Cisse B, Caton ML, Lehner M, Maeda T, Scheu S, Locksley R, et al. Transcription factor E2-2 is an essential and specific regulator of plasmacytoid dendritic cell development. *Cell* (2008) 135(1):37–48. doi:10.1016/j.cell.2008.09.016

53. Fredebohm J, Wolf J, Hoheisel JD, Boettcher M. Depletion of RAD17 sensitizes pancreatic cancer cells to gemcitabine. *J Cell Sci* (2013) 126(Pt 15):3380–9. doi:10.1242/jcs.124768

54. Bloch DB, de la Monte SM, Guigaouri P, Filippov A, Bloch KD. Identification and characterization of a leukocyte-specific component of the nuclear body. *J Biol Chem* (1996) 271(46):29198–204. doi:10.1074/jbc.271.46.29198

55. Hirano M, Guo P, McCurley N, Schorpp M, Das S, Boehm T, et al. Evolutionary implications of a third lymphocyte lineage in lampreys. *Nature* (2013) 501(7467):435–8. doi:10.1038/nature12467

56. Sawai CM, Sisirak V, Ghosh HS, Hou EZ, Ceribelli M, Staudt LM, et al. Transcription factor Runx2 controls the development and migration of plasmacytoid dendritic cells. *J Exp Med* (2013) 210(11):2151–9. doi:10.1084/jem.20130443

57. Iwakoshi NN, Pypaert M, Glimcher LH. The transcription factor XBP-1 is essential for the development and survival of dendritic cells. *J Exp Med* (2007) 204(10):2267–75. doi:10.1084/jem.20070525

58. Cao W, Zhang L, Rosen DB, Bover L, Watanabe G, Bao M, et al. BDCA2/Fc epsilon RI gamma complex signals through a novel BCR-like pathway in human plasmacytoid dendritic cells. *PLoS Biol* (2007) 5(10):e248. doi:10.1371/journal.pbio.0050248

59. Haniffa M, Collin M, Ginhoux F. Ontogeny and functional specialization of dendritic cells in human and mouse. *Adv Immunol* (2013) 120:1–49. doi:10.1016/B978-0-12-417028-5.00001-6

60. Murphy TL, Tussiwand R, Murphy KM. Specificity through cooperation: BATF-IRF interactions control immune-regulatory networks. *Nat Rev Immunol* (2013) 13(7):499–509. doi:10.1038/nri3470

61. Mori M, Saito K, Ohta Y. ARHGAP22 localizes at endosomes and regulates actin cytoskeleton. *PLoS One* (2014) 9(6):e100271. doi:10.1371/journal.pone.0100271

62. Yang W, Shen N, Ye DQ, Liu Q, Zhang Y, Qian XX, et al. Genome-wide association study in Asian populations identifies variants in ETS1 and WDFY4 associated with systemic lupus erythematosus. *PLoS Genet* (2010) 6(2):e1000841. doi:10.1371/journal.pgen.1000841

63. Tsujita K, Kondo A, Kurisu S, Hasegawa J, Itoh T, Takenawa T. Antagonistic regulation of F-BAR protein assemblies controls actin polymerization during podosome formation. *J Cell Sci* (2013) 126(Pt 10):2267–78. doi:10.1242/jcs.122515

64. Worby CA, Dixon JE. Sorting out the cellular functions of sorting nexins. *Nat Rev Mol Cell Biol* (2002) 3(12):919–31. doi:10.1038/nrm974

65. Shih VF, Davis-Turak J, Macal M, Huang JQ, Ponomarenko J, Kearns JD, et al. Control of RelB during dendritic cell activation integrates canonical and noncanonical NF-kappaB pathways. *Nat Immunol* (2012) 13(12):1162–70. doi:10.1038/ni.2446

66. Singh R, Cresswell P. Defective cross-presentation of viral antigens in GILT-free mice. *Science* (2010) 328(5984):1394–8. doi:10.1126/science.1189176

67. West LC, Grotzke JE, Cresswell P. MHC class II-restricted presentation of the major house dust mite allergen Der p 1 Is GILT-dependent: implications for allergic asthma. *PLoS One* (2013) 8(1):e51343. doi:10.1371/journal.pone.0051343

68. Mihrshahi R, Barclay AN, Brown MH. Essential roles for Dok2 and RasGAP in CD200 receptor-mediated regulation of human myeloid cells. *J Immunol* (2009) 183(8):4879–86. doi:10.4049/jimmunol.0901531

69. Moniuszko M, Liyanage NP, Doster MN, Parks RW, Grubczak K, Lipinska D, et al. Glucocorticoid treatment at moderate doses of SIV-infected rhesus macaques decreases the frequency of circulating CD14CD16 monocytes but does not alter the tissue virus reservoir. *AIDS Res Hum Retroviruses* (2015) 31(1):115–26. doi:10.1089/AID.2013.0220

70. Fingerle-Rowson G, Angstwurm M, Andreesen R, Ziegler-Heitbrock HW. Selective depletion of CD14+ CD16+ monocytes by glucocorticoid therapy. *Clin Exp Immunol* (1998) 112(3):501–6. doi:10.1046/j.1365-2249.1998.00617.x

71. Hildner K, Edelson BT, Purtha WE, Diamond M, Matsushita H, Kohyama M, et al. Batf3 deficiency reveals a critical role for CD8alpha+ dendritic cells in cytotoxic T cell immunity. *Science* (2008) 322(5904):1097–100. doi:10.1126/science.1164206

72. Kasahara M, Sutoh Y. Two forms of adaptive immunity in vertebrates: similarities and differences. *Adv Immunol* (2014) 122:59–90. doi:10.1016/B978-0-12-800267-4.00002-X

73. Irion U, Krauss J, Nusslein-Volhard C. Precise and efficient genome editing in zebrafish using the CRISPR/Cas9 system. *Development* (2014) 141(24):4827–30. doi:10.1242/dev.115584

74. Ellwood-Yen K, Graeber TG, Wongvipat J, Iruela-Arispe ML, Zhang J, Matusik R, et al. Myc-driven murine prostate cancer shares molecular features with human prostate tumors. *Cancer Cell* (2003) 4(3):223–38. doi:10.1016/S1535-6108(03)00197-1

75. Lee JS, Chu IS, Mikaelyan A, Calvisi DF, Heo J, Reddy JK, et al. Application of comparative functional genomics to identify best-fit mouse models to study human cancer. *Nat Genet* (2004) 36(12):1306–11. doi:10.1038/ng1481

76. Vu Manh TP, Alexandre Y, Baranek T, Crozat K, Dalod M. Plasmacytoid, conventional, and monocyte-derived dendritic cells undergo a profound and convergent genetic reprogramming during their maturation. *Eur J Immunol* (2013) **43**(7):1706–15. doi:10.1002/eji.201243106

77. Lowes MA, Suarez-Farinas M, Krueger JG. Immunology of psoriasis. *Annu Rev Immunol* (2014) **32**:227–55. doi:10.1146/annurev-immunol-032713-120225

78. Artyomov MN, Munk A, Gorvel L, Korenfeld D, Cella M, Tung T, et al. Modular expression analysis reveals functional conservation between human Langerhans cells and mouse cross-priming dendritic cells. *J Exp Med* (2015) **212**(5):743–57. doi:10.1084/jem.20131675

79. Fairbairn L, Kapetanovic R, Beraldi D, Sester DP, Tuggle CK, Archibald AL, et al. Comparative analysis of monocyte subsets in the pig. *J Immunol* (2013) **190**(12):6389–96. doi:10.4049/jimmunol.1300365

80. Jaitin DA, Kenigsberg E, Keren-Shaul H, Elefant N, Paul F, Zaretsky I, et al. Massively parallel single-cell RNA-seq for marker-free decomposition of tissues into cell types. *Science* (2014) **343**(6172):776–9. doi:10.1126/science.1247651

81. Usoskin D, Furlan A, Islam S, Abdo H, Lonnerberg P, Lou D, et al. Unbiased classification of sensory neuron types by large-scale single-cell RNA sequencing. *Nat Neurosci* (2015) **18**(1):145–53. doi:10.1038/nn.3881

82. Zeisel A, Manchado ABM, Codeluppi S, Lönnerberg P, La Manno G, Juréus A, et al. Cell types in the mouse cortex and hippocampus revealed by single-cell RNA-seq. *Science* (2015) **347**(6226):1138–42. doi:10.1126/science.aaa1934

83. Crozat K, Vivier E, Dalod M. Crosstalk between components of the innate immune system: promoting anti-microbial defenses and avoiding immunopathologies. *Immunol Rev* (2009) **227**(1):129–49. doi:10.1111/j.1600-065X.2008.00736.x

84. Luber CA, Cox J, Lauterbach H, Fancke B, Selbach M, Tschopp J, et al. Quantitative proteomics reveals subset-specific viral recognition in dendritic cells. *Immunity* (2010) **32**(2):279–89. doi:10.1016/j.immuni.2010.01.013

85. Miller JC, Brown BD, Shay T, Gautier EL, Jojic V, Cohain A, et al. Deciphering the transcriptional network of the dendritic cell lineage. *Nat Immunol* (2012) **13**(9):888–99. doi:10.1038/ni.2370

86. Wu X, Satpathy AT, Kc W, Liu P, Murphy TL, Murphy KM. Bcl11a controls Flt3 expression in early hematopoietic progenitors and is required for pDC development in vivo. *PLoS One* (2013) **8**(5):e64800. doi:10.1371/journal.pone.0064800

11

Revisiting mouse peritoneal macrophages: heterogeneity, development, and function

Alexandra dos Anjos Cassado[1]*, Maria Regina D'Império Lima[1] and Karina Ramalho Bortoluci[2,3]

[1] Departamento de Imunologia, Instituto de Ciências Biomédicas, Universidade de São Paulo, São Paulo, Brazil, [2] Centro de Terapia Celular e Molecular (CTC-Mol), Universidade Federal de São Paulo, São Paulo, Brazil, [3] Departamento de Ciências Biológicas, Campus Diadema, Universidade Federal de São Paulo, São Paulo, Brazil

Edited by:
Florent Ginhoux,
Singapore Immunology Network,
Singapore

Reviewed by:
Xinjian Chen,
University of Utah, USA
Thomas Marichal,
University of Liège, Belgium

*Correspondence:
Alexandra dos Anjos Cassado,
Departamento de Imunologia,
Instituto de Ciências Biomédicas,
Universidade de São Paulo, Av.
Professor Lineu Prestes, 1730,
Cidade Universitária, São Paulo, SP
05508-900, Brazil
alecassado@hotmail.com

Tissue macrophages play a crucial role in the maintenance of tissue homeostasis and also contribute to inflammatory and reparatory responses during pathogenic infection and tissue injury. The high heterogeneity of these macrophages is consistent with their adaptation to distinct tissue environments and specialization to develop niche-specific functions. Although peritoneal macrophages are one of the best-studied macrophage populations, recently it was demonstrated the co-existence of two subsets in mouse peritoneal cavity (PerC), which exhibit distinct phenotypes, functions, and origins. These macrophage subsets have been classified, according to their morphology, as large peritoneal macrophages (LPMs) and small peritoneal macrophages (SPMs). LPMs, the most abundant subset under steady state conditions, express high levels of F4/80 and low levels of class II molecules of the major histocompatibility complex (MHC). LPMs appear to be originated from embryogenic precursors, and their maintenance in PerC is regulated by expression of specific transcription factors and tissue-derived signals. Conversely, SPMs, a minor subset in unstimulated PerC, have a F4/80^{low}MHC-II^{high} phenotype and are generated from bone-marrow-derived myeloid precursors. In response to infectious or inflammatory stimuli, the cellular composition of PerC is dramatically altered, where LPMs disappear and SPMs become the prevalent population together with their precursor, the inflammatory monocyte. SPMs appear to be the major source of inflammatory mediators in PerC during infection, whereas LPMs contribute for gut-associated lymphoid tissue-independent and retinoic acid-dependent IgA production by peritoneal B-1 cells. In the previous years, considerable efforts have been made to broaden our understanding of LPM and SPM origin, transcriptional regulation, and functional profile. This review addresses these issues, focusing on the impact of tissue-derived signals and external stimulation in the complex dynamics of peritoneal macrophage populations.

Keywords: peritoneal macrophages, peritoneal cavity, LPM, SPM, origin

Introduction

Macrophages are resident cells found in almost all tissues of the body, where they assume specific phenotypes and develop distinct functions. Tissue macrophages are considered as immune sentinels because of their strategic localization and their ability to initiate and modulate immune responses during pathogenic infection or tissue injury and to contribute to the maintenance of tissue homeostasis (1–3). Macrophages were first identified in the late 19th century by Élie Metchnikoff (1845–1916) and designated as large phagocytes (4, 5). Based on their phagocytic activity, macrophages were first classified as cells from the reticuloendothelial system, which also comprised endothelial cells, fibroblasts, spleen and lymphoid reticular cells, Kupffer cells, splenocytes, and monocytes (6). However, because endocytosis performed by endothelial cells is a process that is distinct from phagocytosis, by the late 1960s a new classification system for mononuclear phagocytic cells as cells from "mononuclear phagocytic system" (MPS) was proposed (7). The MPS was defined as a group of phagocytic cells sharing morphological and functional similarities, including pro-monocytes, monocytes, macrophages, dendritic cells (DCs), and their bone marrow (BM) progenitors (7–12). Although the phagocytic cells play similar roles in orchestrating the immune response and maintaining tissue homeostasis (11), they represent cell populations that are extremely heterogeneous (13), and the general classification of mononuclear cells in a unique system is currently under intense discussion (12, 14). In this context, Guilliams et al. suggested a classification of MPS cells based primarily on their ontogeny and secondary on their location, function, and phenotype, promoting a better classification under both steady state and inflammatory conditions (14).

In the last few years, a complex scenario to describe macrophage origins has been developed (15–19), replacing the simplistic view of myeloid precursors giving rise to blood monocytes that, in turn, originate tissue macrophages (20–22). For example, resident macrophages from brain, lung, liver, peritoneum, and spleen are not differentiated from monocytes; instead, they are derived from an embryonic precursor and maintained by self-renewal (23–27). In addition to resident macrophages, infiltrating monocytes are also found in injured tissues, where they can differentiate into inflammatory macrophages or TNF-α- and inducible nitric oxide synthase (iNOS)-producing (Tip)-DCs (28). Currently, it is accepted that inflammatory macrophages and tissue-resident macrophages comprise developmentally and functionally distinct populations (3, 14, 17, 18, 29).

Under steady state conditions, some tissues and serous cavities, including lung, spleen, and the peritoneal cavity (PerC), present distinct resident macrophage subpopulations. In the spleen, at least three macrophage subsets are found: red pulp, metalophilic, and marginal zone macrophages (30). In the PerC, two peritoneal macrophage subsets have been described: large peritoneal macrophage (LPM) and small peritoneal macrophage (SPM) (31). Mouse peritoneal macrophages are among the best-studied macrophage populations in terms of cell biology, development, and inflammatory responses (24, 31–42). Peritoneal macrophages play key roles in the control of infections and inflammatory pathologies (43, 44), as well as in the maintenance of immune response robustness (40). Therefore, this review will discuss recent advances in our understanding of peritoneal macrophage subsets characterization, origin and functions, and the accurate experimental approaches to analyze them.

Identification of Peritoneal Macrophages

Cohn and collaborators introduced the study of peritoneal macrophages (45–48). Indeed, a representative portion of the current knowledge regarding macrophage biology, such as their function, specialization, and development stems from studies performed using peritoneal macrophages as a cellular source. However, the existence of two resident macrophage subsets present in the PerC was described recently (31). These macrophage subsets were designated LPM and SPM according to their size. LPMs and SPMs were initially identified based on their differential expression of F4/80 and CD11b, where LPMs express high levels of F4/80 and CD11b while SPMs show F4/80^{low}CD11b^{low} phenotype (**Table 1**). CD11b is an integrin that, together with CD18, forms the CR3 heterodimer (13, 30, 49), but is not exclusively expressed on macrophages and is found on several others cell types, including polymorphonuclear cells (50, 51), DCs (52), and at low levels on B lymphocytes (53, 54). F4/80, a 160 kD glycoprotein from the epidermal growth factor (EGF)-transmembrane 7 (TM7) family, is expressed by macrophages in several organs, such as the kidney (55), BM (56), epithelium (57), lung (58, 59), lymphoid organs (60), and among others (61, 62), and it is not found on fibroblasts, polymorphonuclear cells, and lymphocytes (63). However, peritoneal eosinophils show low levels of F4/80 (31) and some macrophage subpopulations exhibit low levels or do not express F4/80, such as white pulp and marginal zone splenic macrophages (30). Therefore, F4/80 expression levels distinguish macrophage subpopulations, including those residing in the same tissue, such as subsets found in the spleen and PerC (30, 31, 35). In this sense, the great majority (approximately 90%) of F4/80$^+$CD11b$^+$ cells present in the PerC from several mouse strains, including BALB/c, C57BL/6, 129/S6, FVB/N, SJL/J, and RAG$^{-/-}$, express high levels of these molecules and correspond to the LPM subset, whereas the minor SPM subset expresses low levels of these markers (31).

An accurate evaluation of SPMs and LPMs by flow cytometry and optical microscopy revealed that in addition to the differential expression of CD11b and F4/80, SPMs and LPMs display unique morphologies and phenotypes. LPMs assume the

TABLE 1 | Phenotypic profile of SPMs and LPMs.

Surface molecule	LPMs	SPMs
F4/80	+++	+
CD11b	+++	+
CD11c	+	−
MHC-II	+	++
GR1	+	−
Ly6C	−	−
c-kit	−	−
CD62L	−	++
Dectin-1	+	++
DC-Sign	−	++
TLR4	++	+
CD80	++	+
CD86	+++	+
CD40	++	+
12/15-LOX	+	−
TIM4	+	−

classical morphology described for macrophages after adherence, exhibiting prominent vacuolization and abundant cytoplasm, whereas SPMs display a polarized morphology in culture, presenting dendrites similar to DCs (35). Moreover, the analysis of a complex panel of cell surface molecules (**Table 1**) demonstrated that SPMs express higher levels of MHC-II (IA^b), dectin-1, and DC-sign endocytic receptors than LPMs. Moreover, half of SPM subset expresses high levels CD62L (31, 35, 36). Conversely, LPMs express higher levels of toll like receptor (TLR)-4 and co-stimulatory molecules in comparison to SPMs (31, 35, 36).

Given that PerC is a singular compartment where specialized immune cells reside and interact, including macrophages, B cells, DCs, eosinophils, mast cells, neutrophils, T cells, natural killer (NK), and invariant NKT cells (31, 32, 35, 36, 64), the identification of myeloid cells from PerC based on cell surface molecules is still a complex matter, particularly in terms of distinguishing macrophage subsets from DCs and inflammatory monocytes. The expression of 12/15-lipoxygenase (LOX), Tim4, and Ly6B has also been examined to discriminate heterogeneous macrophage subsets in PerC under steady state conditions and during peritonitis (24, 37, 38, 42). The high expression of 12/15-LOX and Tim4 was observed in peritoneal macrophages, which also express high levels of F4/80 and CD11b, correlating with the phenotype and frequencies observed for LPMs (24, 31, 37, 38, 42). Conversely, 12/15-LOX$^-$ cells and SPM share the same CD11b$^+$F4/80^{low}MHCII^{high} phenotype; however, 12/15-LOX$^-$ cells express high levels of CD11c and co-stimulatory molecules, suggesting that 12/15-LOX$^-$ cells and SPMs are, at least in part, distinct populations (31, 35, 37). Despite similarities in cell morphology and MHC-II expression presented by SPMs and DCs, the possibility that SPMs may be part of the peritoneal DC pool is excluded by the smaller size, the distinct and lack of the CD11b and F4/80 expression presented by DCs and, primarily, by the lower expression of CD11c (HL3 or N418 clones of monoclonal anti-CD11c) on SPMs compared with LPMs or typical peritoneal DCs (31, 35).

Given the cell complexity present in PerC and the importance of the development of efficient strategies to correctly identify macrophage subsets as well as to avoid contamination by other cell populations and misinterpretation of peritoneal macrophage studies, our group has proposed a simple way to identify peritoneal macrophage subsets using a four-color flow cytometry staining panel. From doublet, CD19^{high} and CD11c^{high} discarded selected cell populations; the analysis of F4/80$^+$ cells based on MHCII expression defines three distinct subpopulations, F4/80^{high}IA^{b-neg}, F4/80^{low}IA^{b-high}, and F4/80^{low}IA^{b-neg}, which correspond, respectively, to LPMs, SPMs, and granulocytes (35).

Origin and Development of LPM and SPM

The theories that explain the origin of macrophages have been completely reformulated in the last few years. The differentiation process of monocytes, macrophages, and DCs that occurs in the BM starts with the earliest progenitor, the hematopoietic stem cell (HSC), and follows the common myeloid progenitor (CMP) and the granulocyte and macrophage progenitor (GMP) (16). The clonotypic BM-resident precursor differentiated from

GMP, termed the macrophage-DC precursor (MDP), expresses high levels of the fractalkine receptor CX3CR1, c-kit, and CD115, and gives rise to circulating blood monocytes, some macrophage populations and a common DC precursor (CDP), but does not originate granulocytes (15, 65, 66). The recruitment of monocyte subsets under steady state or inflammatory and pathological conditions depends on particular chemokines and the expression of their counterpart's receptors. The Ly6C$^+$ monocyte subset migrates via a CCR2-dependent pathway, whereas Ly6C$^-$ appears to migrate in response to CX3CR1 signaling (67). Under steady state conditions, extravasated monocytes do not contribute to the pool of resident macrophages in many tissues (3, 15, 16). In inflammatory settings, the Ly6C$^+$ monocyte subset differentiates into inflammatory macrophages and monocyte-derived DCs, such as Tip-DCs (15, 16).

Recent accumulating evidence supports the prenatal origin of tissue-resident macrophages and the idea that they are maintained locally by self-renewal throughout adult life, both in the steady state and after cell turnover, which is predominantly independent of hematopoiesis (17, 18, 23–27, 29, 68, 69). Microglia, Langerhans cells, Kupffer cells, red pulp splenic macrophages, lung, and peritoneal macrophages are originated from embryogenic precursor and proliferative cells maintained by self-renewal (23–27, 69–71). Fetal-liver monocytes or primitive macrophages found in the yolk sac, an extraembryonic tissue, have been related with the origin of tissue-resident macrophages. In this context, recent date using yolk sac macrophages depletion and fate-mapping models demonstrated that yolk sac macrophages, which are generated from early erythro-myeloid progenitors (EMPs), are important for development of macrophages in mid-gestation; however in adulthood, only microglia is maintained by these embryogenic precursor (69). In contrast, fetal monocytes that are derived from late EMPs give rise to tissue-resident macrophages from liver, lung, skin, kidney and spleen (69). The exception to the origin of resident macrophages is intestinal macrophages, which are continuously repopulated by circulating monocytes (72).

Understanding the dynamics of maintenance and recruitment of peritoneal macrophages is of particular interest since these cells are involved in physiological as well as pathological processes, such as peritonitis, tumors, and pancreatitis (40, 43, 44). Early studies demonstrated that peritoneal macrophages are maintained in PerC through self-renewal in the steady state or under inflammatory conditions (73–76). The omentum, a fat tissue that connects the abdominal organs, is also involved in peritoneal macrophage development through the proliferative capacities of omental macrophages (75, 76). The combination of these early observations, which were acquired recently, with the technical advances to correctly identify the peritoneal macrophage subsets has permitted the ontogeny of the peritoneal macrophage subsets to be elucidated (24, 31, 36, 39, 40, 42).

Under steady state conditions, LPMs appear to be maintained by self-renewal and independent of hematopoiesis (26, 36), whereas SPMs are originated from circulating monocytes (31, 36, 40) (**Figure 1**). Dates from Schulz et al. suggest that, in general, F4/80 expression by tissue macrophages correlated with yolk sac (F4/80^{high}) and not hematopoietic (F4/80^{low}) progenitors (25). In the CX3CR1$^{GFP/WT}$ mice, Cain et al. (36) showed the presence of

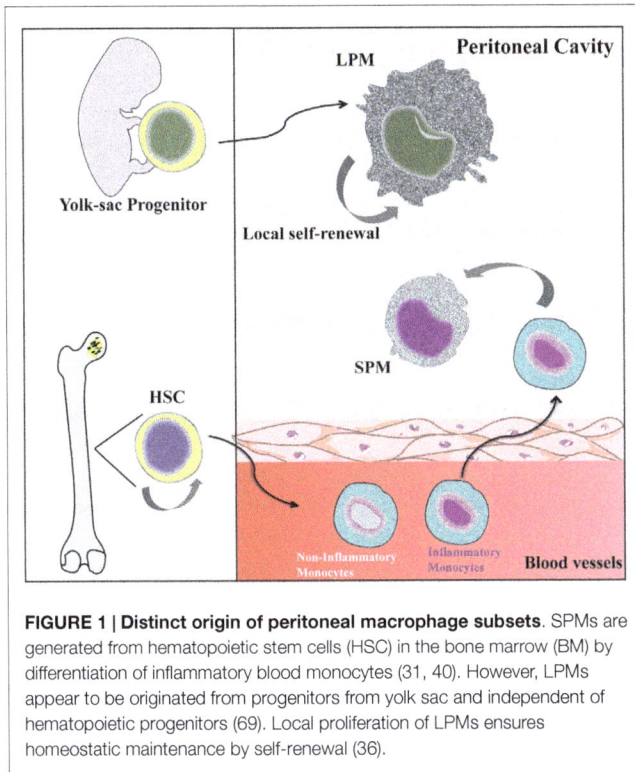

FIGURE 1 | Distinct origin of peritoneal macrophage subsets. SPMs are generated from hematopoietic stem cells (HSC) in the bone marrow (BM) by differentiation of inflammatory blood monocytes (31, 40). However, LPMs appear to be originated from progenitors from yolk sac and independent of hematopoietic progenitors (69). Local proliferation of LPMs ensures homeostatic maintenance by self-renewal (36).

GFP$^+$ cells in DC and SPM pool, but not in the LPM population. Conversely, in the CX$_3$CR1CreRosa26R-FGFP mice, which show the active and past expression of CX3CR1, the presence of GFP$^+$ cells was found within DC, SPM, and LPM populations. These data indicate that SPMs are short-lived cells, whereas LPMs have a more distant ontogenic relationship with a CX3CR1$^+$ progenitor, corroborating the idea that they originate from the yolk sac (36). However, in chimeric C57BL/6 mice reconstituted with C57BL/6-CD45.1 BM, around 80% of SPMs and more than 70% of LPMs are CD45.1-expressing cells, demonstrating that both peritoneal macrophage subsets differentiate from BM precursors after ablation of peritoneal macrophages induced by irradiation (36). Data from our group suggest that PerC recruited Ly6C$^+$ monocytes could give rise to SPMs during inflammatory conditions (31). Confirming that SPMs are generated via the differentiation of inflammatory monocytes recruited to PerC, reduced numbers of SPMs are found in the PerC of CCR2$^{-/-}$ mice (40).

The analysis of Ki67 and phosphorylated histone H3 (pHH3 at a discrete stage of mitosis) staining and the quantification of cell cycle and basal DNA content revealed that the number of proliferating F4/80^{high}CD11b^{high} cells decreases in 12-week-old mice compared with proliferation capacity of this population in newborn mice (15 days to 4 weeks) (24). After 12–16 weeks, the number of F4/80^{high}CD11b^{high} cells in PerC is maintained under a low rate of proliferation, which suggests that the number of F4/80^{high}CD11b^{high} peritoneal cells increases during mouse development until PerC acquires sufficient homeostatic cell numbers (24). Indeed, BrdU-labeled LPM frequencies after a single BrdU pulse were 7 and 15-fold lower than those found in HSC and GMP, respectively. Moreover, the presence of BrdU$^+$ LPMs was detectable 14 days after BrdU pulse, suggesting that they are a

long-lived population, i.e., maintained at low levels of proliferation (36). Conversely, the detection of low numbers of proliferating SPMs at 6–10 days after one pulse of BrdU suggests that these cells have a low proliferation rate under steady state conditions and are short-lived cells (36).

Studies with mice deficient in CCAAT/enhancer binding protein (C/EBP)b also support the notion that LPMs and SPMs represent distinct ontogenies, because in the absence of this transcription factor, PerC did not contain LPMs and exhibited increased numbers of SPMs (36). Interestingly, adoptively transferred SPMs differentiated into LPMs in Cebpb$^{-/-}$ mice. However, in control mice that have normal numbers of LPMs, only a small frequency of transferred SPMs acquired the F4/80^{hi}MHCII^{low}CD93$^+$ phenotype of LPMs. Based on these results, the authors proposed that under physiological conditions, SPMs appear to contribute in only a small way to generate LPMs, but SPMs could be involved in the maintenance of LPMs in situations where this pool has been greatly reduced, such as under inflammatory conditions or following radiation ablation (36). These data are consistent with the findings of Yona et al. (26), which demonstrated the presence of monocyte-derived cells in the LPM compartment 8 weeks after the i.p. injection of thioglycollate. Together with LPMs, a subset of proliferating BM-derived inflammatory macrophage has also been associated with self-renewal mechanisms during the resolution of peritonitis induced by zymosan and thioglycollate (42). Conversely, LPMs do not seem to contribute to the SPM pool, even during inflammation. Our group demonstrated that adoptively transferred CFDA-SE-labeled LPMs 1 h after LPS stimulation retained its phenotype, and no CFDA-SE$^+$ cells were found in the SPM compartment until 2 days after stimulation (31).

In the last year, a great advance in the understanding of the transcriptional control of peritoneal macrophages provided novel insights into this scenario (39, 40). The zinc finger transcription factor GATA-binding protein 6 (GATA6) appears to regulate the majority of peritoneal macrophage-specific genes (PMSGs). Of note, GATA6 is selectively expressed by LPMs (40). Accordingly, the number of LPMs were greatly reduced in peritoneal lavages from GATA6-KO^{mye} and Mac-GATA6 KO mice, which have a GATA6 deficiency in all myeloid cells or only in the macrophage lineages, respectively (39, 40). Interestingly, retinoic acid (RA) is the extracellular factor that regulates GATA-6-specific gene expression in LPMs, because vitamin A depleted (VAD; the RA precursor) mice exhibited a decrease in GATA6 expression and LPM numbers (40). Moreover, the stimulation of peritoneal macrophages from VAD mice with all-trans RA restored the expression of GATA-6 and many PMSGs at levels found in peritoneal macrophages from control mice. In addition to the regulation of gene expression profiling in peritoneal macrophages, GATA-6 appears to be involved in the control of the proliferation, survival, and metabolism of these cells (39, 77). GATA-6-deficient macrophages demonstrate an altered proliferation state during peritonitis (39). Moreover, Lyz2-Cre × GATA6$^{(flox/flox)}$ mice also exhibit reduced numbers of peritoneal macrophages, which could be explained by the perturbation in their metabolism, culminating in the high frequency of cell death found in this compartment (77). Despite great contributions to our understanding in the involvement of GATA-6 in peritoneal macrophage development, metabolism, self-maintenance, and survival, the existence of distinct pathways that could govern the transcriptional regulation of SPMs remains largely unknown.

In addition to transcriptional regulation, signaling factors derived from the microenvironment also play an essential role in promoting the development and phenotype of tissue-resident macrophages. For example, TGF-β1 signaling is required for the development of the microglia population and to regulate a microglia expression program through the Smad tissue factors (78–80). Heme has been shown to induce Spi-c, a transcription factor important for red pulp macrophage development (81, 82). Finally, in PerC, omentum-derived RA promotes the expression of GATA-6 in the LPM subset, determining its localization and functions (40), even if the factors that maintain the SPM pool under steady state conditions still remain to be elucidated.

Dynamics and Function of Peritoneal Macrophage Subsets

Mouse PerC is a compartment where many cell types co-habitat and interact, similar to the secondary lymphoid organs. In addition, PerC is a unique body compartment that contains B-1 cells (83). Under steady state conditions, the peritoneal cells comprise LPMs, SPMs, B-1 cells, conventional B-2 cells, T cells, NK cells, DCs, and granulocytes (mostly eosinophils) (31, 35). B1 cells constitute the majority of the PerC cell population, whereas the SPM and LPM frequencies represent 30–35% of total peritoneal cells (31, 35) (**Figure 2A**). However, after inflammatory or infectious stimuli, there is a dramatic alteration in cell numbers and the frequencies of each of PerC cell subpopulation. With regard to the myeloid compartment, modifications in PerC cell composition include the disappearance of LPMs, increases in SPM frequency and numbers, and a massive recruitment of inflammatory monocytes (24, 31, 35, 36, 40) (**Figure 2B**).

The "macrophage disappearance reaction" (MDR) in PerC has been extensively described during delayed-type hypersensitivity (DTH) and acute inflammatory processes (84). MDR has been associated with cell death, emigration to draining lymph nodes, or adherence of macrophages to structural tissues. LPMs are the unique peritoneal macrophage subset that disappears from PerC, which is attributed not to cell death but rather to their migration to the omentum (31, 40). LPM disappearance in response to inflammatory stimuli is accompanied by an increase in SPM and inflammatory monocyte numbers (24, 31, 35, 36, 40) (**Figure 2B**), and has been correlated with the renewal and improvement of immune conditions of the PerC (35). Adherent peritoneal cells from naive mice, which are composed primarily of LPM, exhibit

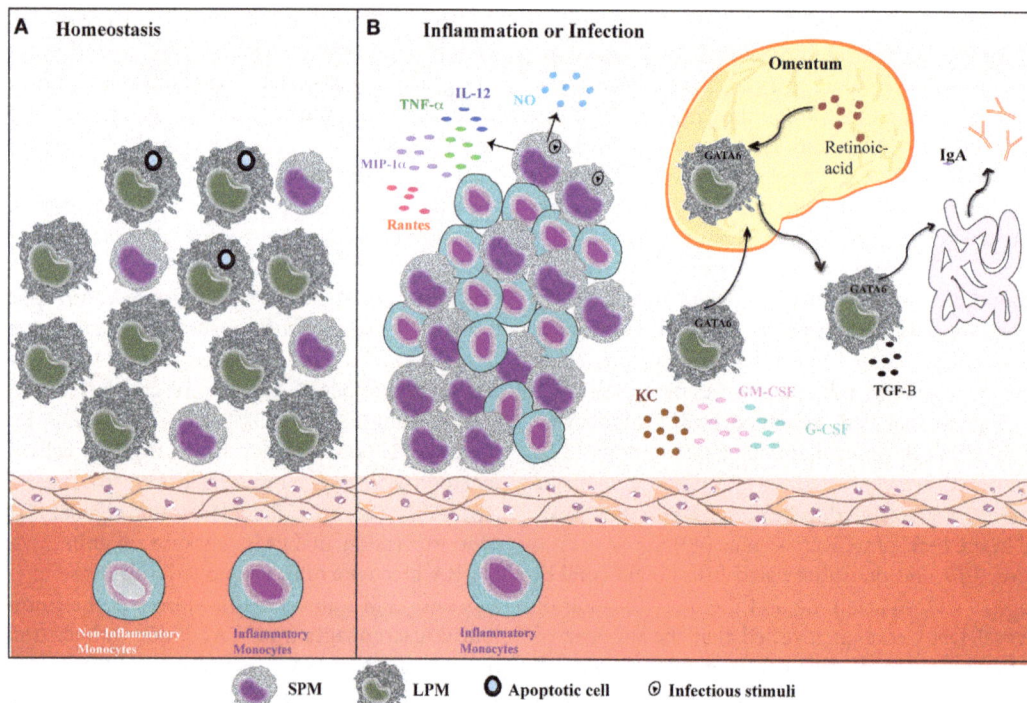

FIGURE 2 | Summary of the dynamic of peritoneal macrophage subsets. (A) Under homeostatic conditions, peritoneal macrophages comprise two subsets LPMs and SPMs (31). LPMs, which are the major peritoneal macrophage population, appear to be responsible for phagocytosis of apoptotic cell and tissue repair (36). (B) At the outset of inflammation, the myeloid compartment is modified in general by disappearance of LPMs, increase of SPMs numbers, and monocytes influx (31, 35, 36, 40). The changes in the myeloid cells from zymosan, *T. cruzi*, and LPS stimulated or thioglicollate-elicited PerC result in the gain of immune state (35, 36). SPMs from zymosan and *T. cruzi* stimulated mice contribute to effector function of PerC through secretion of high levels of NO and presence of IL-12-producing cells (35). In response to LPS *in vivo*, SPMs produce several inflammatory cytokines, such as IL-12, MIP-1α, TNF-α, and RANTES, whereas LPMs produce enhanced amounts of G-CSF, GM-CSF, and KC (36). LPMs, which migrate to omentum by a retinoic acid and GATA-6-dependent way in response to *in vivo* LPS stimulation or vitamin-A deprivation, return to PerC and appear to be correlated with GALT-independent and TGF-β2-dependent IgA production by B-1 cells in the intestine (40).

a high frequency of cells stained for β-galactosamine (β-gal), a senescence marker (85–87). These cells are unable to secrete NO in response to LPS challenge (35). In contrast, adherent peritoneal cells from *Trypanosoma cruzi* or zymosan-stimulated mice in which the main cell population constitutes SPMs and monocytes (F4/80^{low}MHCII^{int}Ly-6C$^+$), respectively, display a significant reduction in the frequency of β-gal-positive cells and secrete high levels of NO in response to LPS (35). The frequency of IL-12-producing cells after *in vitro* LPS plus IFN-γ stimulation was also higher within myelo-monocytic cells from mice exposed to zymosan and *T. cruzi* than the frequencies of IL-12-producing cells found in unstimulated mice (35). In response to *Staphylococcus epidermidis* cell-free (SES) supernatant *in vivo* stimulation, F4/80^{low}CD11b$^+$ cells (consisting of SPMs and DCs) produced enhanced levels of IL-1β, IL-1α, TNF-α, and IL-12 in the presence or absence of subsequent SES treatment (37). In contrast, the supernatants of adherent cells from naïve mice treated with SES were found to contain high levels of MCP-1, MCP-1α, MIP-1β, and G-CSF (37). It is important to note that 4 days after thioglycollate injection, peritoneal cells, an extensively studied cell population (88–91), also consist primarily of SPMs and inflammatory monocytes (31, 40). The increase in SPM numbers and the influx of inflammatory monocytes that will give rise to SPMs greatly contribute to the improvement of the capacity of PerC to deal with inflammatory stimuli. Indeed, although neither SPMs nor LPMs produce significant levels of pro- or anti-inflammatory cytokines under steady state conditions (35–37), SPMs appear to develop a pro-inflammatory profile in response to *in vitro* stimuli. SPMs produced high levels of TNF-α, MIP-1α, and RANTES in response to LPS, whereas LPMs were the unique population that produced abundant levels of G-CSF, GM-CSF, and KC in response to the same stimulus (36) (**Figure 2B**).

The NO secretion and pro-inflammatory cytokine production are the most important functions of activated macrophages by inflammatory stimulation and assigns the M1 profile (13, 34, 92–97). The functional profile of peritoneal macrophages was previous studied by our group and others (33, 34). Peritoneal macrophages from Th1-prone mouse strains (C57BL/6 and B10.A) are easily activated to produce NO in response to rIFN-γ or LPS, characterizing the M1 profile. In contrast, macrophages from Th2-prone mouse strains (BALB/c and DBA/2) exhibit a weak NO response as a consequence of high levels of spontaneously secreted TGF-β1 (34). Moreover, the cells from C57BL/6 IL-12p40-deficient mice have a bias toward the M2 profile, indicating that IL-12 is required for M1 polarization of peritoneal macrophages (33). Although LPMs from naïve mice can produce NO after *in vitro* LPS stimulation, SPMs produce higher levels of NO than LPMs following *in vivo* LPS stimulation. The NO secretion by LPMs was also detected by flow cytometry in *Escherichia coli* inoculated mice (31), whereas nitrite was not produced *in vitro* by LPS-stimulated adherent peritoneal cells from control mice, which is composed mainly by LPMs (35). In addition, adherent cells obtained 48 h after *T. cruzi* infection, which are mostly composed by SPMs, were the unique source of NO without *in vitro* subsequent challenge with LPS (35). In resume, the SPM and LPM subsets cannot be accommodated in the M1/M2 framework considering the NO secretion. However, considering phagocytic

assays, SPMs appear to develop an efficient profile to control infections as M1 macrophages, whereas LPMs assume a role in the maintenance of PerC physiological conditions as M2 or alternative macrophages. Despite the preserved phagocytic ability of LPMs, higher numbers of zymosan and *E. coli* were found inside of SPMs at early time points after i.p. injection (31, 35). Conversely, at 1 h after challenge, LPMs appear to present a higher phagocytic index of apoptotic thymocytes in comparison to SPMs (36) (**Figure 2A**).

In addition, it was recently demonstrated that LPMs have a unique ability to induce gut-associated lymphoid tissue (GALT)-independent IgA production by peritoneal B-1 cells (40) (**Figure 2B**). RA and TGF-β2 are the most critical factors to induce IgA class switching, and the production of TGF-β2 is regulated by the *Tgfb2* and *Ltbp1* genes, which are expressed by LPMs in a GATA-6-dependent manner. This process is regulated by the abundant presence of RA in the omentum, which is responsible for the induction of GATA-6 expression in LPMs that migrates to this tissue. The dynamic of LPM migration between the PerC and the omentum after the stimulation of PerC is correlated with their disappearance and the return to basal numbers of LPMs later after stimulation with LPS, zymosan, and thioglycollate (24, 31, 35, 36, 39, 40). This observation suggests that LPMs can return to PerC to resolve an infectious or inflammatory process. Therefore, the presence of two specialized macrophage subsets in PerC is crucial to maintain the health of this compartment under different situations.

Concluding Remarks

Peritoneal macrophages represent one of the most studied macrophage populations. However, the existence of two phenotypically and functionally distinct subsets, LPMs and SPMs, residing in the PerC was recognized recently (31). In the last year, great advances in our understanding of the transcriptional regulation of peritoneal macrophages have brought novel insights into the identification of LPMs and SPMs (39, 40). GATA-6, an LPM-restricted transcription factor, regulates many PMSGs, including those related to the maintenance of LPMs in PerC (40) and those that determine their function (40), metabolism, proliferation, and cell survival (39, 77). Under steady state conditions, LPMs appear to originate independently from hematopoietic precursors and retained the ability to proliferate *in situ*, maintaining physiological numbers (26, 36). Conversely, SPMs appear to originate from circulating monocytes (31, 36, 40), and their numbers increase remarkably under inflammatory conditions. Of note, SPMs together with their precursor, the inflammatory monocyte population, are the major myeloid populations present in elicited PerC, and are an excellent resource to study the biology of inflammatory macrophages. SPMs and LPMs exhibit specialized functions in the PerC, where SPMs present a pro-inflammatory functional profile, and LPMs appear to have a role in the maintenance of PerC physiological conditions. Moreover, the particular interactions between macrophage subsets and other peritoneal cell populations appear to play crucial roles in PerC immune state. Although the consequences of the crosstalk between SPMs and peritoneal T and B lymphocytes remain to be clarified, LPMs are required for GALT-independent and RA-dependent IgA production by peritoneal B-1 cells (40). Finally, the elucidation of the influence of soluble factors and the microbiota on the maintenance of LPM/SPM ratios in PerC, and the role of these subsets in the systemic immune response are the future challenges for this field.

References

1. Taylor PR, Gordon S. Monocyte heterogeneity and innate immunity. *Immunity* (2003) **19**(1):2–4. doi:10.1016/S1074-7613(03)00178-X

2. Gordon S. The macrophage: past, present and future. *Eur J Immunol* (2007) **37**(Suppl 1):S9–17. doi:10.1002/eji.200737638

3. Wynn TA, Chawla A, Pollard JW. Macrophage biology in development, homeostasis and disease. *Nature* (2013) **496**(7446):445–55. doi:10.1038/nature12034

4. Metchnikoff E. *Leçons sur la Pathologie Comparée de l'inflammation faites à l'Institut Pasteur en 1891.* Paris: Masson (1892).

5. Van M. [Elie Metchnikoff, 1845-1916]. *Voeding* (1964) **25**:351–6.

6. Aschoff L. Das reticuloendotheliale system. *Ergeb Inn Med Kinderheilkd* (1924) **26**:1–117.

7. van Furth R, Cohn ZA, Hirsch JG, Humphrey JH, Spector WG, Langevoort HL. The mononuclear phagocyte system: a new classification of macrophages, monocytes, and their precursor cells. *Bull World Health Organ* (1972) **46**(6):845–52.

8. van Furth R. Current view of the mononuclear phagocyte system. *Haematol Blood Transfus* (1981) **27**:3–10.

9. van Furth R. The mononuclear phagocyte system. *Verh Dtsch Ges Pathol* (1980) **64**:1–11.

10. Hume DA, Ross IL, Himes SR, Sasmono RT, Wells CA, Ravasi T. The mononuclear phagocyte system revisited. *J Leukoc Biol* (2002) **72**(4):621–7.

11. Hume DA. The mononuclear phagocyte system. *Curr Opin Immunol* (2006) **18**(1):49–53. doi:10.1016/j.coi.2005.11.008

12. Geissmann F, Gordon S, Hume DA, Mowat AM, Randolph GJ. Unravelling mononuclear phagocyte heterogeneity. *Nat Rev Immunol* (2010) **10**(6):453–60. doi:10.1038/nri2784

13. Gordon S, Taylor PR. Monocyte and macrophage heterogeneity. *Nat Rev Immunol* (2005) **5**(12):953–64. doi:10.1038/nri1733

14. Guilliams M, Ginhoux F, Jakubzick C, Naik SH, Onai N, Schraml BU, et al. Dendritic cells, monocytes and macrophages: a unified nomenclature based on ontogeny. *Nat Rev Immunol* (2014) **14**(8):571–8. doi:10.1038/nri3712

15. Auffray C, Sieweke MH, Geissmann F. Blood monocytes: development, heterogeneity, and relationship with dendritic cells. *Annu Rev Immunol* (2009) **27**:669–92. doi:10.1146/annurev.immunol.021908.132557

16. Geissmann F, Manz MG, Jung S, Sieweke MH, Merad M, Ley K. Development of monocytes, macrophages, and dendritic cells. *Science* (2010) **327**(5966):656–61. doi:10.1126/science.1178331

17. Davies LC, Jenkins SJ, Allen JE, Taylor PR. Tissue-resident macrophages. *Nat Immunol* (2013) **14**:986–95. doi:10.1038/ni.2705

18. Ginhoux F, Jung S. Monocytes and macrophages: developmental pathways and tissue homeostasis. *Nat Rev Immunol* (2014) **14**(6):392–404. doi:10.1038/nri3671

19. Ginhoux F, Merad M. Ontogeny and homeostasis of Langerhans cells. *Immunol Cell Biol* (2010) **88**(4):387–92. doi:10.1038/icb.2010.38

20. van Furth R. [The origin of mononuclear phagocytes]. *Ned Tijdschr Geneeskd* (1967) **111**(48):2208.

21. van Furth R, Cohn ZA. The origin and kinetics of mononuclear phagocytes. *J Exp Med* (1968) **128**(3):415–35. doi:10.1084/jem.128.3.415

22. van Furth R. Origin and kinetics of monocytes and macrophages. *Semin Hematol* (1970) **7**(2):125–41.

23. Ginhoux F, Greter M, Lebeouf M, Nandi S, See P, Gokhan S, et al. Fate mapping analysis reveals that adult microglia derive from primitive macrophages. *Science* (2010) **330**(6005):841–5. doi:10.1126/science.1194637

24. Davies LC, Rosas M, Smith PJ, Fraser DJ, Jones SA, Taylor PR. A quantifiable proliferative burst of tissue macrophages restores homeostatic macrophage populations after acute inflammation. *Eur J Immunol* (2011) **41**(8):2155–64. doi:10.1002/eji.201141817

25. Schulz C, Gomez Perdiguero E, Chorro L, Szabo-Rogers H, Cagnard N, Kierdorf K, et al. A lineage of myeloid cells independent of Myb and hematopoietic stem cells. *Science* (2012) **336**(6077):86–90. doi:10.1126/science.1219179

26. Yona S, Kim KW, Wolf Y, Mildner A, Varol D, Breker M, et al. Fate mapping reveals origins and dynamics of monocytes and tissue macrophages under homeostasis. *Immunity* (2013) **38**(1):79–91. doi:10.1016/j.immuni.2012.12.001

27. Hashimoto D, Chow A, Noizat C, Teo P, Beasley MB, Leboeuf M, et al. Tissue-resident macrophages self-maintain locally throughout adult life with minimal contribution from circulating monocytes. *Immunity* (2013) **38**(4):792–804. doi:10.1016/j.immuni.2013.04.004

28. Auffray C, Fogg DK, Narni-Mancinelli E, Senechal B, Trouillet C, Saederup N, et al. CX3CR1+ CD115+ CD135+ common macrophage/DC precursors and the role of CX3CR1 in their response to inflammation. *J Exp Med* (2009) **206**(3):595–606. doi:10.1084/jem.20081385

29. Sieweke MH, Allen JE. Beyond stem cells: self-renewal of differentiated macrophages. *Science* (2013) **342**(6161):1242974. doi:10.1126/science.1242974

30. Taylor PR, Martinez-Pomares L, Stacey M, Lin HH, Brown GD, Gordon S. Macrophage receptors and immune recognition. *Annu Rev Immunol* (2005) **23**:901–44. doi:10.1146/annurev.immunol.23.021704.115816

31. Ghosn EE, Cassado AA, Govoni GR, Fukuhara T, Yang Y, Monack DM, et al. Two physically, functionally, and developmentally distinct peritoneal macrophage subsets. *Proc Natl Acad Sci U S A* (2010) **107**(6):2568–73. doi:10.1073/pnas.0915000107

32. Schleicher U, Hesse A, Bogdan C. Minute numbers of contaminant CD8+ T cells or CD11b+CD11c+ NK cells are the source of IFN-gamma in IL-12/IL-18-stimulated mouse macrophage populations. *Blood* (2005) **105**(3):1319–28. doi:10.1182/blood-2004-05-1749

33. Bastos KR, Alvarez JM, Marinho CR, Rizzo LV, Lima MR. Macrophages from IL-12p40-deficient mice have a bias toward the M2 activation profile. *J Leukoc Biol* (2002) **71**(2):271–8.

34. Mills CD, Kincaid K, Alt JM, Heilman MJ, Hill AM. M-1/M-2 macrophages and the Th1/Th2 paradigm. *J Immunol* (2000) **164**(12):6166–73. doi:10.4049/jimmunol.164.12.6166

35. Cassado Ados A, de Albuquerque JA, Sardinha LR, Buzzo Cde L, Faustino L, Nascimento R, et al. Cellular renewal and improvement of local cell effector activity in peritoneal cavity in response to infectious stimuli. *PLoS One* (2011) **6**(7):e22141. doi:10.1371/journal.pone.0022141

36. Cain DW, O'Koren EG, Kan MJ, Womble M, Sempowski GD, Hopper K, et al. Identification of a tissue-specific, C/EBPbeta-dependent pathway of differentiation for murine peritoneal macrophages. *J Immunol* (2013) **191**(9):4665–75. doi:10.4049/jimmunol.1300581

37. Dioszeghy V, Rosas M, Maskrey BH, Colmont C, Topley N, Chaitidis P, et al. 12/15-Lipoxygenase regulates the inflammatory response to bacterial products in vivo. *J Immunol* (2008) **181**(9):6514–24. doi:10.4049/jimmunol.181.9.6514

38. Rosas M, Thomas B, Stacey M, Gordon S, Taylor PR. The myeloid 7/4-antigen defines recently generated inflammatory macrophages and is synonymous with Ly-6B. *J Leukoc Biol* (2010) **88**(1):169–80. doi:10.1189/jlb.0809548

39. Rosas M, Davies LC, Giles PJ, Liao CT, Kharfan B, Stone TC, et al. The transcription factor Gata6 links tissue macrophage phenotype and proliferative renewal. *Science* (2014) **344**(6184):645–8. doi:10.1126/science.1251414

40. Okabe Y, Medzhitov R. Tissue-specific signals control reversible program of localization and functional polarization of macrophages. *Cell* (2014) **157**(4):832–44. doi:10.1016/j.cell.2014.04.016

41. Wang C, Yu X, Cao Q, Wang Y, Zheng G, Tan TK, et al. Characterization of murine macrophages from bone marrow, spleen and peritoneum. *BMC Immunol* (2013) **14**:6. doi:10.1186/1471-2172-14-6

42. Davies LC, Rosas M, Jenkins SJ, Liao CT, Scurr MJ, Brombacher F, et al. Distinct bone marrow-derived and tissue-resident macrophage lineages proliferate at key stages during inflammation. *Nat Commun* (2013) **4**:1886. doi:10.1038/ncomms2877

Acknowledgments

This work was supported by Fundação de Amparo à Pesquisa do Estado de São Paulo (FAPESP – Brazil) Proc 2013/16010-5 and 2013/07140-2, Brazilian Research Council (CNPq-Brazil), CAPES, and INCTV.

43. Dahdah A, Gautier G, Attout T, Fiore F, Lebourdais E, Msallam R, et al. Mast cells aggravate sepsis by inhibiting peritoneal macrophage phagocytosis. *J Clin Invest* (2014) **124**(10):4577–89. doi:10.1172/JCI75212

44. Machado MCC, Coelho AMM. *Role of Peritoneal Macrophages on Local And Systemic Inflammatory Response in Acute Pancreatitis*. São Paulo: InTech (2012). doi:10.5772/25639

45. Cohn ZA. Determinants of infection in the peritoneal cavity. I. Response to and fate of *Staphylococcus aureus* and *Staphylococcus albus* in the mouse. *Yale J Biol Med* (1962) **35**:12–28.

46. Cohn ZA. Determinants of infection in the peritoneal cavity. II. Factors influencing the fate of *Staphylococcus aureus* in the mouse. *Yale J Biol Med* (1962) **35**:29–47.

47. Cohn ZA. Determinants of infection in the peritoneal cavity. III. The action of selected inhibitors on the fate of *Staphylococcus aureus* in the mouse. *Yale J Biol Med* (1962) **35**:48–61.

48. Steinman RM, Moberg CL. Zanvil Alexander Cohn 1926-1993. *J Exp Med* (1994) **179**(1):1–30. doi:10.1084/jem.179.1.1

49. Taylor PR, Brown GD, Geldhof AB, Martinez-Pomares L, Gordon S. Pattern recognition receptors and differentiation antigens define murine myeloid cell heterogeneity ex vivo. *Eur J Immunol* (2003) **33**(8):2090–7. doi:10.1002/eji.200324003

50. Hickstein DD, Ozols J, Williams SA, Baenziger JU, Locksley RM, Roth GJ. Isolation and characterization of the receptor on human neutrophils that mediates cellular adherence. *J Biol Chem* (1987) **262**(12):5576–80.

51. Petty HR, Todd RF III. Receptor-receptor interactions of complement receptor type 3 in neutrophil membranes. *J Leukoc Biol* (1993) **54**(5):492–4.

52. Shortman K, Liu YJ. Mouse and human dendritic cell subtypes. *Nat Rev Immunol* (2002) **2**(3):151–61. doi:10.1038/nri746

53. Kantor AB, Stall AM, Adams S, Herzenberg LA. Differential development of progenitor activity for three B-cell lineages. *Proc Natl Acad Sci U S A* (1992) **89**(8):3320–4. doi:10.1073/pnas.89.8.3320

54. Ghosn EE, Yang Y, Tung J, Herzenberg LA. CD11b expression distinguishes sequential stages of peritoneal B-1 development. *Proc Natl Acad Sci U S A* (2008) **105**(13):5195–200. doi:10.1073/pnas.0712350105

55. Hume DA, Gordon S. Mononuclear phagocyte system of the mouse defined by immunohistochemical localization of antigen F4/80. Identification of resident macrophages in renal medullary and cortical interstitium and the juxtaglomerular complex. *J Exp Med* (1983) **157**(5):1704–9. doi:10.1084/jem.157.5.1704

56. Hume DA, Loutit JF, Gordon S. The mononuclear phagocyte system of the mouse defined by immunohistochemical localization of antigen F4/80: macrophages of bone and associated connective tissue. *J Cell Sci* (1984) **66**:189–94.

57. Hume DA, Perry VH, Gordon S. The mononuclear phagocyte system of the mouse defined by immunohistochemical localisation of antigen F4/80: macrophages associated with epithelia. *Anat Rec* (1984) **210**(3):503–12. doi:10.1002/ar.1092100311

58. Bedoret D, Wallemacq H, Marichal T, Desmet C, Quesada Calvo F, Henry E, et al. Lung interstitial macrophages alter dendritic cell functions to prevent airway allergy in mice. *J Clin Invest* (2009) **119**(12):3723–38. doi:10.1172/JCI39717

59. Guilliams M, De Kleer I, Henri S, Post S, Vanhoutte L, De Prijck S, et al. Alveolar macrophages develop from fetal monocytes that differentiate into long-lived cells in the first week of life via GM-CSF. *J Exp Med* (2013) **210**(10):1977–92. doi:10.1084/jem.20131199

60. Hume DA, Robinson AP, MacPherson GG, Gordon S. The mononuclear phagocyte system of the mouse defined by immunohistochemical localization of antigen F4/80. Relationship between macrophages, Langerhans cells, reticular cells, and dendritic cells in lymphoid and hematopoietic organs. *J Exp Med* (1983) **158**(5):1522–36. doi:10.1084/jem.158.5.1522

61. Hume DA, Perry VH, Gordon S. Immunohistochemical localization of a macrophage-specific antigen in developing mouse retina: phagocytosis of dying neurons and differentiation of microglial cells to form a regular array in the plexiform layers. *J Cell Biol* (1983) **97**(1):253–7. doi:10.1083/jcb.97.1.253

62. Hume DA, Halpin D, Charlton H, Gordon S. The mononuclear phagocyte system of the mouse defined by immunohistochemical localization of antigen F4/80: macrophages of endocrine organs. *Proc Natl Acad Sci U S A* (1984) **81**(13):4174–7. doi:10.1073/pnas.81.13.4174

63. Austyn JM, Gordon S. F4/80, a monoclonal antibody directed specifically against the mouse macrophage. *Eur J Immunol* (1981) **11**(10):805–15. doi:10.1002/eji.1830111013

64. Ghosn EE, Yang Y, Tung J, Herzenberg LA, Herzenberg LA. CD11b expression distinguishes sequential stages of peritoneal B-1 development. *Proc Natl Acad Sci U S A* (2008) **105**(13):5195–200. doi:10.1073/pnas.0712350105

65. Fogg DK, Sibon C, Miled C, Jung S, Aucouturier P, Littman DR, et al. A clonogenic bone marrow progenitor specific for macrophages and dendritic cells. *Science* (2006) **311**(5757):83–7. doi:10.1126/science.1117729

66. Landsman L, Varol C, Jung S. Distinct differentiation potential of blood monocyte subsets in the lung. *J Immunol* (2007) **178**(4):2000–7. doi:10.4049/jimmunol.178.4.2000

67. Geissmann F, Jung S, Littman DR. Blood monocytes consist of two principal subsets with distinct migratory properties. *Immunity* (2003) **19**(1):71–82. doi:10.1016/S1074-7613(03)00174-2

68. Chorro L, Geissmann F. Development and homeostasis of 'resident' myeloid cells: the case of the Langerhans cell. *Trends Immunol* (2010) **31**(12):438–45. doi:10.1016/j.it.2010.09.003

69. Hoeffel G, Chen J, Lavin Y, Low D, Almeida FF, See P, et al. C-myb(+) erythro-myeloid progenitor-derived fetal monocytes give rise to adult tissue-resident macrophages. *Immunity* (2015) **42**(4):665–78. doi:10.1016/j.immuni.2015.03.011

70. Ajami B, Bennett JL, Krieger C, Tetzlaff W, Rossi FM. Local self-renewal can sustain CNS microglia maintenance and function throughout adult life. *Nat Neurosci* (2007) **10**(12):1538–43. doi:10.1038/nn2014

71. Chorro L, Sarde A, Li M, Woollard KJ, Chambon P, Malissen B, et al. Langerhans cell (LC) proliferation mediates neonatal development, homeostasis, and inflammation-associated expansion of the epidermal LC network. *J Exp Med* (2009) **206**(13):3089–100. doi:10.1084/jem.20091586

72. Zigmond E, Jung S. Intestinal macrophages: well educated exceptions from the rule. *Trends Immunol* (2013) **34**(4):162–8. doi:10.1016/j.it.2013.02.001

73. Parwaresch MR, Wacker HH. Origin and kinetics of resident tissue macrophages. Parabiosis studies with radiolabelled leucocytes. *Cell Tissue Kinet* (1984) **17**(1):25–39.

74. Melnicoff MJ, Horan PK, Breslin EW, Morahan PS. Maintenance of peritoneal macrophages in the steady state. *J Leukoc Biol* (1988) **44**(5):367–75.

75. Daems WT, de Bakker JM. Do resident macrophages proliferate? *Immunobiology* (1982) **161**(3–4):204–11. doi:10.1016/S0171-2985(82)80075-2

76. Wijffels JF, Hendrickx RJ, Steenbergen JJ, Eestermans IL, Beelen RH. Milky spots in the mouse omentum may play an important role in the origin of peritoneal macrophages. *Res Immunol* (1992) **143**(4):401–9. doi:10.1016/S0923-2494(05)80072-0

77. Gautier EL, Ivanov S, Williams JW, Huang SC, Marcelin G, Fairfax K, et al. Gata6 regulates aspartoacylase expression in resident peritoneal macrophages and controls their survival. *J Exp Med* (2014) **211**(8):1525–31. doi:10.1084/jem.20140570

78. Makwana M, Jones LL, Cuthill D, Heuer H, Bohatschek M, Hristova M, et al. Endogenous transforming growth factor beta 1 suppresses inflammation and promotes survival in adult CNS. *J Neurosci* (2007) **27**(42):11201–13. doi:10.1523/JNEUROSCI.2255-07.2007

79. Abutbul S, Shapiro J, Szaingurten-Solodkin I, Levy N, Carmy Y, Baron R, et al. TGF-beta signaling through SMAD2/3 induces the quiescent microglial phenotype within the CNS environment. *Glia* (2012) **60**(7):1160–71. doi:10.1002/glia.22343

80. Butovsky O, Jedrychowski MP, Moore CS, Cialic R, Lanser AJ, Gabriely G, et al. Identification of a unique TGF-beta-dependent molecular and functional signature in microglia. *Nat Neurosci* (2014) **17**(1):131–43. doi:10.1038/nn.3599

81. Kohyama M, Ise W, Edelson BT, Wilker PR, Hildner K, Mejia C, et al. Role for Spi-C in the development of red pulp macrophages and splenic iron homeostasis. *Nature* (2009) **457**(7227):318–21. doi:10.1038/nature07472

82. Haldar M, Kohyama M, So AY, Kc W, Wu X, Briseno CG, et al. Heme-mediated SPI-C induction promotes monocyte differentiation into iron-recycling macrophages. *Cell* (2014) **156**(6):1223–34. doi:10.1016/j.cell.2014.01.069

83. Baumgarth N. The double life of a B-1 cell: self-reactivity selects for protective effector functions. *Nat Rev Immunol* (2011) **11**(1):34–46. doi:10.1038/nri2901

84. Barth MW, Hendrzak JA, Melnicoff MJ, Morahan PS. Review of the macrophage disappearance reaction. *J Leukoc Biol* (1995) **57**(3):361–7.

85. Lloberas J, Celada A. Effect of aging on macrophage function. *Exp Gerontol* (2002) **37**(12):1325–31. doi:10.1016/S0531-5565(02)00125-0

86. Herrero C, Sebastian C, Marques L, Comalada M, Xaus J, Valledor AF, et al. Immunosenescence of macrophages: reduced MHC class II gene expression. *Exp Gerontol* (2002) **37**(2–3):389–94. doi:10.1016/S0531-5565(01)00205-4

87. Dimri GP, Lee X, Basile G, Acosta M, Scott G, Roskelley C, et al. A biomarker that identifies senescent human cells in culture and in aging skin in vivo. *Proc Natl Acad Sci U S A* (1995) **92**(20):9363–7. doi:10.1073/pnas.92.20.9363

88. Takahashi M, Galligan C, Tessarollo L, Yoshimura T. Monocyte chemoattractant protein-1 (MCP-1), not MCP-3, is the primary chemokine required for monocyte recruitment in mouse peritonitis induced with thioglycollate or zymosan A. *J Immunol* (2009) **183**(5):3463–71. doi:10.4049/jimmunol.0802812

89. Cohn ZA. Activation of mononuclear phagocytes: fact, fancy, and future. *J Immunol* (1978) **121**(3):813–6.

90. Leijh PC, van Zwet TL, ter Kuile MN, van Furth R. Effect of thioglycolate on phagocytic and microbicidal activities of peritoneal macrophages. *Infect Immun* (1984) **46**(2):448–52.

91. Lagasse E, Weissman IL. Flow cytometric identification of murine neutrophils and monocytes. *J Immunol Methods* (1996) **197**(1–2):139–50. doi:10.1016/0022-1759(96)00138-X

92. Mantovani A, Sozzani S, Locati M, Allavena P, Sica A. Macrophage polarization: tumor-associated macrophages as a paradigm for polarized M2 mononuclear phagocytes. *Trends Immunol* (2002) **23**(11):549–55. doi:10.1016/S1471-4906(02)02302-5

93. Gordon S. Alternative activation of macrophages. *Nat Rev Immunol* (2003) **3**(1):23–35. doi:10.1038/nri978

94. Mantovani A, Sica A, Sozzani S, Allavena P, Vecchi A, Locati M. The chemokine system in diverse forms of macrophage activation and polarization. *Trends Immunol* (2004) **25**(12):677–86. doi:10.1016/j.it.2004.09.015

95. Martinez FO, Helming L, Gordon S. Alternative activation of macrophages: an immunologic functional perspective. *Annu Rev Immunol* (2009) **27**:451–83. doi:10.1146/annurev.immunol.021908.132532

96. Mosser DM. The many faces of macrophage activation. *J Leukoc Biol* (2003) **73**(2):209–12. doi:10.1189/jlb.0602325

97. Mosser DM, Edwards JP. Exploring the full spectrum of macrophage activation. *Nat Rev Immunol* (2008) **8**(12):958–69. doi:10.1038/nri2448

12

Fate mapping of dendritic cells

Mateusz Pawel Poltorak and Barbara Ursula Schraml *

Institute for Medical Microbiology, Immunology and Hygiene, Technische Universität München, Munich, Germany

Edited by:
Shalin Naik,
Walter & Eliza Hall Institute, Australia

Reviewed by:
Sam Basta,
Queen's University, Canada
Theresa T. Lu,
Weill Cornell Medical Center, USA

*Correspondence:
Barbara Ursula Schraml,
Institute for Medical Microbiology,
Immunology and Hygiene, Technische
Universität München, Trogerstraße
30, Munich 81675, Germany
barbara.schraml@tum.de

Dendritic cells (DCs) are a heterogeneous group of mononuclear phagocytes with versatile roles in immunity. They are classified predominantly based on phenotypic and functional properties, namely their stellate morphology, expression of the integrin CD11c, and major histocompatibility class II molecules, as well as their superior capacity to migrate to secondary lymphoid organs and stimulate naïve T cells. However, these attributes are not exclusive to DCs and often change within inflammatory or infectious environments. This led to debates over cell identification and questioned even the mere existence of DCs as distinct leukocyte lineage. Here, we review experimental approaches taken to fate map DCs and discuss how these have shaped our understanding of DC ontogeny and lineage affiliation. Considering the ontogenetic properties of DCs will help to overcome the inherent shortcomings of purely phenotypic- and function-based approaches to cell definition and will yield a more robust way of DC classification.

Keywords: dendritic cell, ontogeny, fate mapping, lineage tracing, mononuclear phagocyte

Introduction

Dendritic cells (DCs) were originally identified in mouse spleen for their unique stellate morphology, their ability to adhere to certain glass surfaces and their superior capacity to activate naïve T lymphocytes that distinguished them from macrophages (MØs) (1–3). Mostly for historical reasons, DCs are considered part of the mononuclear phagocyte (MP) system, which groups all highly phagocytic cells derived from monocytes or their precursors based on the premise that tissue MØs arise from monocytes (4–9). This presumed relatedness of DCs, monocytes, and MØs coupled to the lack of reliable ways to distinguish MP subtypes has caused continuous debates over accurate cell-type identification and has led some to question whether DCs in fact constitute an independent cell lineage (6, 7, 10–14). However, today we have conclusive evidence demonstrating that DCs, monocytes, and MØs have distinct cellular origin and we further distinguish plasmacytoid DCs (pDCs) from two subsets of so-called conventional or classical DCs (cDCs) based on unique developmental requirements (7, 15–19). Nonetheless, DCs remain defined based on phenotypic and functional properties that often overlap with those of monocytes or MØs (19), although some have suggested a shift in paradigm toward a nomenclature that takes cell ontogeny into account (6, 7, 10).

Dendritic cells are generally identified by their high expression of major histocompatibility complex class II molecules (MHCII) and of the integrin CD11c, as well as their superior capacity to migrate from non-lymphoid to lymphoid organs and stimulate naïve T cells (3, 20–22). However, these characteristics are not absolute and can change in situations of inflammation or infection, thus complicating cell identification (6, 7, 23, 24). For instance, CD11c, considered the hallmark surface marker of DCs, is also found on B, T, and NK cells as well as some monocytes, MØs, and eosinophils (25–32). Dendritic protrusions have also been observed in some MØs and T cells (33–35). Further, surface markers, such as F4/80, CD14, or CD64 (Fc-gamma receptor 1), generally associated with monocytes or MØs can be found on DCs (36–38). One might argue that the most defining feature of DCs is their ability to activate T cells, however such definition discounts the fact that DCs potently regulate innate immune responses independent of their ability to migrate to lymphoid organs or stimulate T cells (39–44). Conversely, non-DCs can carry antigen to lymph nodes and activate naïve T cells in some instances (45–47).

Therefore, morphological and functional properties, as well as the expression of surface markers are insufficient to clearly distinguish DCs from monocytes and MØs, raising the necessity to find a more robust way of cell identification. Recent studies in mouse and human indicate that DCs, MØs, and monocytes have unique ontogenetic properties and thus can be considered distinct cell lineages (36, 48–54). Here, we review approaches that have been employed to track and define the progeny of DC precursors *in vivo* and discuss how such "fate mapping" approaches have improved our understanding of DC heterogeneity and ontogeny. These studies lay the foundation for moving toward cell ontogeny as a major lineage-determining criterion, which will allow for a more reliable and precise classification of DCs and DC subsets.

DC Development

Dendritic cells are short-lived and their maintenance relies on constant replenishment from bone marrow progenitors that originate from hematopoietic stem cells (HSCs) (19, 55). In the classic model of DC development monocytes and DCs arise from bipotent progenitors, so-called MØ and DC progenitors (MDPs) (**Figure 1**) (56). MDPs further give rise to common DC progenitors (CDPs) restricted to the generation of pDCs and cDCs (**Figure 1**) (57, 58). pDCs terminally differentiate in the bone

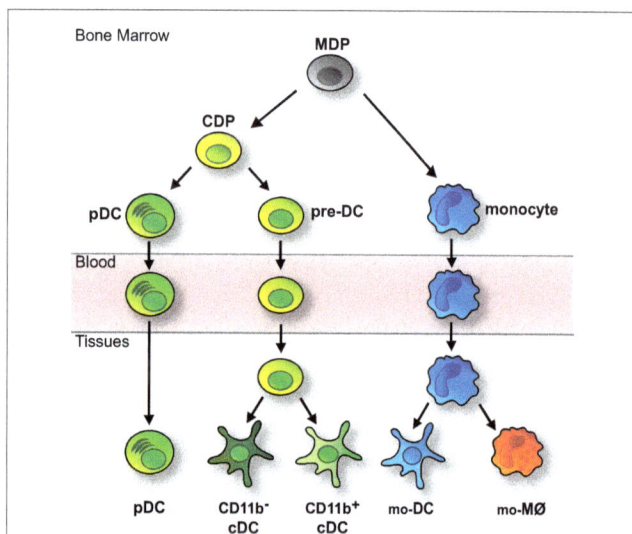

FIGURE 1 | Classic model of DC development. DCs and monocytes are ancestrally related and arise from bi-potential MDPs residing in the bone marrow. MDPs further differentiate into monocytes and CDPs, which are restricted to the generation of various types of DCs. CDPs give rise to pDCs, which fully develop in the bone marrow, and pre-DCs, which migrate through the blood to tissues, where they fully differentiate into CD11b⁻ (including CD8α⁺ cDCs in lymphoid tissue and migratory CD103⁺ cDCs in non-lymphoid tissue) and CD11b⁺ cDCs. Monocytes complete their development in the bone marrow and reach peripheral tissues via the bloodstream. There they further differentiate into monocyte-derived DCs (mo-DCs) or MØs (mo-MØs) in response to environmental cues.

marrow, thus exit the bone marrow as fully developed cells and reach peripheral organs via the blood stream (**Figure 1**) (15, 59). In contrast, cDCs arise from another developmental intermediate termed pre-DC, which exits the bone marrow and migrates through the blood to seed lymphoid and non-lymphoid tissues (60, 61). There, pre-DCs terminally differentiate into cDCs, including the main CD11b⁻ and CD11b⁺ subtypes (**Figure 1**) (60–63). In lymphoid tissues these are CD8α⁺CD11b⁻ and CD11b⁺ resident cDCs, whereas in non-lymphoid tissues they comprise CD103⁺CD11b⁻ and CD11b⁺ migratory cDCs (3, 60–63). Like pDCs, monocytes complete their development in the bone marrow but in tissues they differentiate into cells with DC- or MØ-like features (**Figure 1**) (23, 24, 64, 65). This plasticity is remarkably prominent in inflammatory or infectious environments, when monocyte-derived cells with qualities of DCs have been referred to as TNF-α/iNOS-producing DCs (Tip-DCs), monocyte-derived DCs (mo-DCs), and/or inflammatory DCs (23, 24, 64, 65).

Although most of our knowledge concerning DC development is derived from mouse studies, developmental parallels have been observed in other species (66–73). Especially the identification of putative equivalent DC progenitor populations in human holds promise for future research (72, 73). Yet, some uncertainties remain. Common lymphoid progenitors (CLPs) can give rise to DC descendants upon adoptive transfer (74), although it is now thought that DCs originate predominantly from myeloid progenitors (75, 76). Nonetheless, some pDCs, but not cDCs, show evidence of VDJ gene rearrangements, potentially indicating lymphoid lineage heritage (15, 59, 77). However, it remains unclear whether evidence of *Rag* gene expression history necessarily means that pDCs have dual lymphoid and myeloid origin. Contrary to the dogma that monocytes and DCs share a common immediate ancestor, recent data suggest that lineage divergence of HSC-derived myeloid cells occurs much earlier than previously predicted and that monocytes and DCs might arise independent of a bi-potential developmental intermediate (49, 78, 79). Elucidating such unresolved aspects pertaining to DC ontogeny may solve uncertainties in determining lineage affiliation, which, in turn, will aid to further decipher the unique functions of DCs in immunity.

Fate Mapping

Understanding cell development requires models with which the relationship of a precursor cell and its progeny can be defined *in vivo*. Such "fate mapping" can be achieved in various ways and relies on the selective labeling of the cell(s) of interest so that consequently the development of the marked cell can be followed in its natural environment (80). Tracing progenitors *in vivo* also offers the possibility to determine the fate of populations when lineage affiliation is most heavily debated, namely following experimental manipulation to generate conditions of inflammation or infection. While most fate mapping strategies follow the progeny of bulk cell populations, recently developed techniques have enabled the tracing of single cells, thus providing valuable information regarding their developmental potential at the clonal level (80, 81). In all fate mapping experiments, it is important to consider that their interpretation is dependent on the use of select, faithful and stable markers (82).

Precursor Transfers

The transfer of purified and pre-marked precursor cells into congenic recipients is the most accessible form of fate mapping as a variety of labeling options can be used to distinguish between donor and host cells (**Figure 2A**) (80). As a result, precursor transfers are commonly used to study cell development and lineage relationships and remain a standard protocol for defining the stemness of progenitor cells (80). Such experiments rely on the ability to purify sufficient precursors that, after cell isolation, retain the capacity to home to the appropriate anatomical niche and expand sufficiently into detectable progeny. To circumvent such limitations transfer studies are often combined with protocols to induce leukopenia, such as irradiation, in order to increase the niche available for cell engraftment (**Figure 2A**) (80). However, these manipulations can alter developmental signals, which, in turn, might impact on the interpretation of results (18, 54, 83). To best mimic the endogenous cellular environment, progenitors have been returned directly to their organs of origin, for instance by intra-bone injection (84).

The DC progenitors MDP, CDP, and pre-DC were in part defined by assessing their developmental potential after adoptive transfer into mice (56–58, 60, 61, 84–86). In such experiments, MDPs give rise to DCs and monocytes, whereas CDPs and pre-DCs are restricted to the generation of DCs but do not generate monocytes or other leukocyte lineages (56–58, 61, 84–86). In combination with experiments assessing the differentiation potential of single progenitors *in vitro* (56–58), these studies have significantly shaped our view of DC development (**Figure 1**). Surprisingly, the existence of MDP as a bi-potential intermediate for DCs and monocytes has recently been questioned when single CX$_3$CR1$^+$ MDPs were unable to generate both DCs and monocytes upon differentiation *in vitro* (78). The authors further found that adoptively transferred CX$_3$CR1$^+$ MDPs, not only gave rise to DCs and monocytes but also neutrophils (78). However, such multi-potency of MDPs was not observed in earlier studies (52, 56, 61, 85, 86) and is not evident in genetic CX$_3$CR1 fate mapping experiments (50). It is possible that these discrepancies may be explained by experimental variation such as differences in cell isolation, the timing of analysis or variances in the niche available for cell engraftment following irradiation (18, 54, 83). In light of these results it is noteworthy, however, that upon adoptive transfer MDPs exhibit pDC potential only in some studies (52, 86) but not others (56, 85), whereas the presumed downstream CDPs produce both pDCs and cDCs (57, 58). Taken together these experiments raise some doubt about the existence of a MDP as a key developmental intermediate for monocytes, cDCs, and pDCs. However, resolving this matter will require the use of better models to trace single cells *in vivo* as experiments relying on the isolation and analysis of bulk progenitor populations are inherently prone to disparities in gating strategy or cell purity.

In DC ontogeny, these issues are augmented because MDP and CDP exhibit substantial phenotypic overlap: both lack lineage-defining markers, are characterized by expression of CX$_3$CR1, CD115 (M-CSFR, Csf1r) as well as CD135 (FMS-like tyrosine

kinase 3, FLT3) and, until recently, CDP could only be distinguished from MDP by lower expression of the receptor tyrosine kinase CD117 (c-kit) (56–58, 61, 86). We have recently found that the C-type lectin receptor DNGR-1 (Clec9a) marks cells resembling CDPs (36). Surprisingly, upon adoptive transfer, DNGR-1$^+$CD115$^+$ progenitors exhibit cDC-restricted differentiation potential and do not generate pDCs (36), suggesting that DNGR-1 marks cDC-restricted progenitors. These data are in line with a recent study demonstrating a strong bias for CD115$^+$ CDPs to generate cDCs, whereas pDCs arise predominantly from CD115 negative cells (79). Therefore, cDCs and pDCs appear to have distinct developmental intermediates that can be distinguished by expression of CD115 (79) and DNGR-1 (36). Since CD115$^+$ CDPs presumably express DNGR-1 (36), it is unclear why some CD115$^+$ CDPs show combined cDC and pDC potential in clonal assays (57, 58, 79). It is possible that antibody-mediated triggering of DNGR-1 or growth factor receptors, such as CD115, during cell isolation skews DC differentiation toward a particular DC sub-lineage in an unforeseeable manner. The developmental potential of progenitors may also be influenced by the specific culture conditions used (78) or DCs could exhibit a degree of developmental plasticity (87). Nonetheless, the existence of a putative intermediate monocyte-restricted progenitor downstream of MDP (common monocyte progenitor, cMoP) (52) alongside the aforementioned pDC- and cDC-restricted progenitors supports a model in which monocytes, cDCs, and pDCs develop independently. The genuine point of lineage divergence, however, remains to be determined.

Questions regarding the lineage affiliation of DCs have been muddled significantly by the developmental plasticity of monocytes (6, 24). The phenotypic transformation of monocytes into DC-like cells is most prominent in inflamed environments (8, 19, 23, 24). It can also be mimicked *in vitro* by culturing monocytes in the presence of GM-CSF (granulocyte-macrophage colony-stimulating factor) ± IL-4 (Interleukin-4) (88, 89). However, *in vivo* the inflammation-induced differentiation of monocytes into cells with attributes of DCs appears GM-CSF-independent (90), highlighting that the developmental requirements underlying this phenotypic conversion *in vitro* might differ from those involved *in vivo*. In the absence of experimentally induced infection or inflammation, adoptively transferred monocytes readily acquire CD11c and MHCII expression as well as functional features of DCs in non-lymphoid tissues (91–95). This phenotypic conversion is also observed after adoptive transfer into unirradiated hosts, which most closely mimics steady-state conditions (63). In contrast, transferred monocytes do not generate DCs in lymphoid organs, even if the niche for engraftment is opened by depletion of CD11c$^+$ cells (84). Importantly, in non-lymphoid tissues monocytes exclusively generate CD11b$^+$, but not CD103$^+$CD11b$^-$ cells, which is in contrast to CDPs and pre-DCs that generate CD11b$^+$ as well as CD103$^+$CD11b$^-$ cDCs (63, 91–95). Therefore, CD11c$^+$MHCII$^+$CD11b$^+$ cells in non-lymphoid tissues appear to constitute a population of mixed cellular origin that can arise from monocytic progenitors as well as pre-DCs. Adoptive transfer experiments do not allow to determine the relative contribution of each progenitor to this population,

A   Precursor transfer

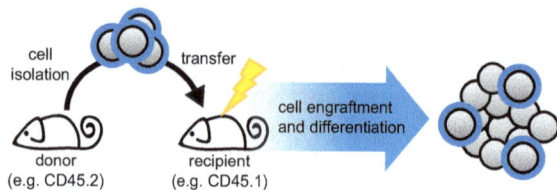

B   Reporter genes

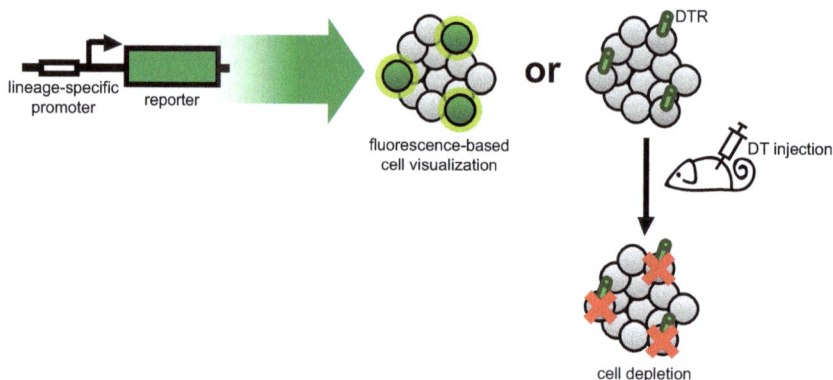

C   Cellular barcoding

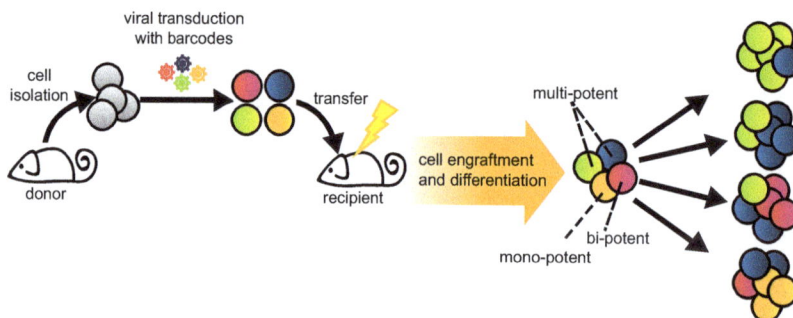

D   Genetic lineage tracing

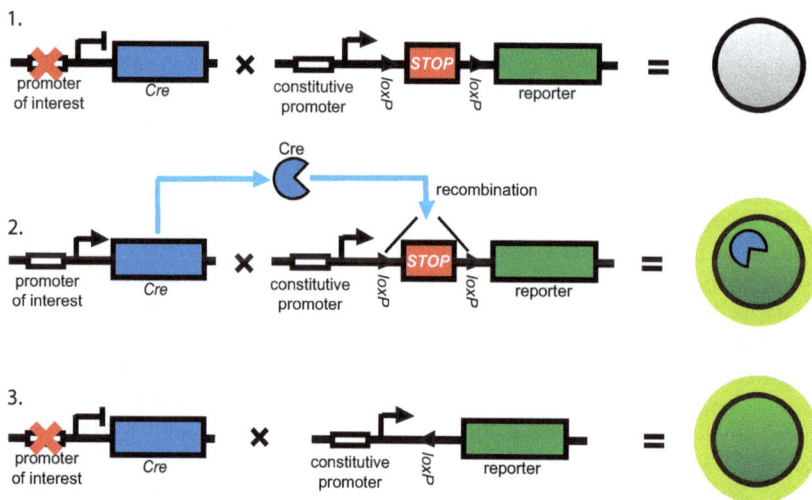

FIGURE 2 | Strategies to fate map DCs.

(Continued)

FIGURE 2 | Continued

(A) Progenitors are adoptively transferred to assess their differentiation in the physiological context. Graft-derived cells are distinguished from host cells based on pre-defined labels, for instance congenic markers. This method is often combined with strategies to increase the niche available for cell engraftment, such as irradiation. **(B)** In transgenic approaches, lineage-restricted promoters can be used to drive a reporter gene. Target cell populations can be visualized by the expression of fluorescent proteins or can be depleted. In the latter case, cell-restricted expression of DTR allows for conditional cell ablation following DT injection. **(C)** Progenitors are transduced *in vitro* with semi-random DNA sequences (barcodes) by retro- or lentiviral vectors and subsequently

transferred into irradiated congenic recipients. After differentiation, cell progeny are analyzed for their barcode repertoire using deep sequencing or microarray. The representation of a given barcode in multiple cell populations indicates multi-potency of the transferred cell. **(D)** Expression of Cre recombinase is driven by a lineage or cell-specific promoter. Additionally, a reporter gene, usually a fluorescent protein, is placed under control of a constitutive promoter. Expression of the reporter is blocked by inserting a *loxP*-flanked *STOP* cassette (1). Cre expression leads to excision of the *STOP* cassette resulting in expression of the reporter gene (2). Since the promoter-driving reporter gene expression is constitutively active, the target cell is irreversibly labeled irrespective of continuous Cre expression (3).

although surrogate markers such as CD64 or Mar-1 can serve to distinguish monocyte-derived cells from bona fide pre-DC-derived cDCs (46, 93, 94).

Notably, in irradiated hosts transferred monocytes can also generate $CD11c^+MHCII^+$ cells of the epidermis, which resemble Langerhans cells (LCs) (96–99). LCs exhibit many phenotypic and functional features of DCs, such as the capacity to migrate to lymphoid organs and stimulate naïve T cells, and have long been considered a prototypical DC population (96–99). However, we now realize that the majority of LCs is established before birth and maintained under steady-state conditions by self-renewal from local progenitors (96, 97, 99–102). These properties thus ontogenetically separate LCs from bone marrow-derived DCs or monocyte-derived cells. Moreover, monocytes may not necessarily adopt features of DCs or MØs upon entry into tissues, as a recent study indicates that monocyte can also exist in tissues without further differentiation (45). When considering this immense plasticity it will be crucial to elucidate the environmental cues that shape the diverse fates of monocytes to further dissect the full functional spectrum of monocytes and monocyte-derived cells.

Lineage Restricted Reporters

When the availability of isolatable progenitor cells is limiting and when populations are ontogenetically heterogeneous or might be influenced by alterations in their surroundings, determining lineage affiliation requires models to trace cells directly in their natural environment. One way to achieve this is by engineering models in which lineage-restricted promoters or genetic elements drive the expression of reporter genes (**Figure 2B**) (80, 82, 103). It is important to bear in mind that such experiments assume that the expression of the selected marker is restricted to the cell lineage in question and therefore, the choice of stable and specific markers is essential (80, 82, 103). Additionally, the genetic elements used to drive expression of the reporter must faithfully mimic endogenous gene expression (80, 82, 103).

Genetic elements of the *Itgax* gene, which encodes CD11c, have extensively been used to generate reporters to study DCs (82, 103). As such, transgenic mice in which the CD11c promoter drives the expression of fluorescent proteins (**Figure 2B**) have been key to visualizing the distribution and cellular interactions of DCs in a variety of tissues, including lymphoid organs, heart, lung, and skin (103–107). But fate mapping can also be achieved by cell deletion. Transgenic expression of primate diphtheria toxin receptor (DTR) renders murine cells susceptible to diphtheria toxin (DT)-induced cell death and, thus, enables inducible target

cell depletion (**Figure 2B**) (82, 108). In this sense mice in which DTR expression is controlled by the elements of the CD11c promoter have been widely used to characterize the *in vivo* functions of DCs (28, 109–112). In part through analyzing such reporter mice, however, it has become evident that CD11c expression is not entirely restricted to DCs. It is also expressed on alveolar MØs, $Ly6C^{low}$ as well as activated monocytes, plasmablasts, NK cells, and some T cells (25–29, 113). In addition, CD11c-driven fate reporter expression varies depending on the specific promoter elements used for transgenesis. CD11c.DTR mice, which were generated by conventional transgenesis using a 5.5-kb promoter element of the *Itgax* gene (109, 114), efficiently deplete most CD11c-expressing cDCs, LCs, alveolar, splenic marginal zone, and metallophilic MØs, as well as plasmablasts and T cells (27, 109, 115). However, DT-induced cell depletion in these mice is incomplete and spares certain cell types that transcribe their endogenous *Itgax* allele, including pDCs and NK cells (82, 115). Additionally, prolonged cell depletion using CD11c.DTR mice requires the use of bone marrow chimeras, possibly because of aberrant DTR expression on non-immune cells (82, 108, 112). Notably, this is not the case in CD11c.DOG and CD11c.LuciDTR mice, which were generated using bacterial artificial chromosome (BAC) transgenesis to place DTR under control of the extended regulatory region of the *Itgax* gene and in which DTR expression seems to more faithfully represent endogenous CD11c expression (28, 112, 115, 116). In all models, the occurrence of systemic neutrophilia and monocytosis following $CD11c^+$ cell depletion (28, 115, 117) adds another layer of complexity to deciphering the cellular function and lineage affiliation of DCs.

The realization that CD11c is not restricted to DCs in all instances nurtured the search for more specific lineage-defining markers. Two groups simultaneously identified the transcription factor Zbtb46 (zDC, Btbd4) as ideal candidate to distinguish cDCs, as it is expressed in pre-DCs and cDCs but not in pDCs or their precursors (37, 38). Consistently, $CD8\alpha^+$ and $CD11b^+$ cDCs in lymphoid organs as well as $CD103^+$ cDCs in non-lymphoid organs uniformly express Zbtb46 as assessed in Zbtb46-GFP (37) and Zbtb46-DTR (38) reporter mice generated by site-directed mutagenesis. In contrast, $CD11c^+MHCII^+CD11b^+$ cells in non-lymphoid organs, including lung, small intestine, and kidney, exhibit partial Zbtb46 expression (37, 38) indicating that they represent a heterogeneous population. This is consistent with reports demonstrating that these cells are of mixed monocyte and pre-DC origin (63, 91, 92, 95). Subsequently, Zbtb46 reporter mice have been used to help establish lineage relationships in a variety of tissues including heart, pancreas, tumors, and thymus (118–121). The fact that Zbtb46 expression is also found in human DCs suggests that it may also help to identify DCs across species (48, 122).

Nevertheless, the use of Zbtb46 as lineage-defining marker requires a note of caution. Zbtb46 expression is downregulated after DC stimulation and it is found in some non-immune cells (37, 123). Despite its prominent expression in the cDC lineage, Zbtb46 appears largely dispensable for cDC development (37, 123). Instead, it may reinforce DC-specific transcriptional programs (37) and/or suppress DC activation (123). Interestingly, monocytes activated in the presence of GM-CSF ± IL-4 uniformly induce Zbtb46 expression, whereas monocyte-derived Tip-DCs that are generated following infection with *Listeria monocytogenes* do not (37). This raises the possibility that Zbtb46 may control DC-like features of monocyte-derived cells in some inflammatory situations and it will be interesting to determine if Zbtb46 controls transcriptional programs in monocytes. These data also highlight that despite its selective expression on cDC progenitors and their descendants, Zbtb46 is not necessarily an indicator of cell ontogeny.

Identifying Common Developmental Requirements

Establishing that the development and/or delineation of a cell type depends on a certain transcription or growth factor constitutes a powerful way of fate mapping that has extensively been applied to MPs (42, 51, 63, 124–141). We can now clearly delineate DCs into distinct subpopulations based on the transcriptional programs that govern their development. pDCs are distinguished from two subsets of cDCs by their dependence on E2-2 (67, 142). The differentiation of pre-DCs into CD8α^+ cDCs in lymphoid organs and CD103$^+$CD11b$^-$ cDCs in non-lymphoid tissues is controlled by a set of transcription factors, including Irf8, Nfil-3, Id2, and Batf3 (124–128). Therefore, CD8α^+ cDCs and CD103$^+$ cDCs represent a developmentally related lineage of cDCs (6, 7). Notably, these cells also exhibit a degree of functional relatedness that is, for instance, exemplified by their superior capacity to activate CD8$^+$ T cells (124, 143–145). In contrast, the development of CD11b$^+$ cDCs from pre-DCs is controlled by distinct transcription factors, including RelB, RbpJ, PU.1, and Irf4 (42, 129–136). Notably, expression of CD24 separates pre-DCs into cells that preferentially generate either CD8α^+ or CD11b$^+$ cDCs in spleen (60) suggesting a stepwise differentiation of pre-DCs into cDCs. It will be interesting to determine whether such heterogeneity of pre-DCs also exists in the bone marrow. Notably, the extent of transcription factor dependence is linked to the genetic background of the particular mouse strain analyzed (146–148), indicating that transcriptional requirements are not always absolute or redundant factors exist (148). Consistently, CD8α^+ DCs can develop in the absence of Batf3, Id2, and Nfil-3 (149). The local microenvironment may also contribute to shaping the diversity of the DC compartment, as in some tissues, such as the spleen and intestinal system, CD11b$^+$ cDCs can be divided into ontogenetically and functionally distinct subpopulations (36, 42, 91, 95, 131). Importantly, some of the transcription factors controlling DC differentiation in mice have also been implicated in the development of human DCs (67, 69, 71) and putative equivalent

DC subpopulations exist in rat, chicken, sheep, and pig (150–153), highlighting that DC populations are conserved across species.

While several growth factors have been linked to DC differentiation, the development of all DC subsets is strongly dependent on FLT3 ligand (FLT3L) and downstream signaling events (7, 18, 154). FLT3L administration potently expands pDCs and cDCs in mice and humans (72, 73, 85, 155–157). *In vitro*, FLT3L promotes the differentiation of bone marrow progenitors from mice, humans, and pigs into functional subsets of DCs (66, 158, 159). Mice lacking FLT3L display a severe deficiency in DCs, which is also apparent, although to a lesser extent, in mice lacking its receptor CD135 or mice treated with CD135 inhibitors (63, 137, 160, 161). In contrast, FLT3L appears largely dispensable for monocyte and MØ development (137) and, therefore, FLT3L dependency is often used delineate DCs *in vivo* (18, 65, 162). The interpretation of fate mapping using mice deficient in CD135 or its ligand is however complicated by the fact that these animals also exhibit abnormalities in other hematopoietic lineages, including B, T, and NK cells (137, 163) and show evidence of systemic neutrophilia and monocytosis, as has been reported in other DC-deficient models (112, 117).

Despite the prominent expression of CD135 on DC progenitors it remains to be clarified exactly at what stage of cellular differentiation FLT3L impacts on DC development. Consistent with a role for FLT3L early in development, a reduction of bone marrow CDPs in FLT3L deficient animals has been reported but ranges from a mere twofold decrease (164) to near complete absence (78). In contrast, the numbers of MDPs and splenic pre-DCs appear largely unaffected by CD135 deficiency (85). The observation that pre-DC frequencies in non-lymphoid organs of FLT3L-deficient mice are reduced (63) and that transfer of DCs into a FLT3L-deficient environment decreases their homeostatic proliferation (85) indicates a role for FLT3L in the peripheral expansion of DCs rather than their differentiation. This interpretation would equally be consistent with the observation that DCs that develop in the absence of FLT3L are functional (137). In light of this finding it will be interesting to determine, to what extent FLT3L impacts on the development and functional regulation of other MPs. Addition of FLT3L to purified human monocytes cultured with GM-CSF ± IL-4 increases their T cell stimulatory capacity (165), although it is not clear whether this is also the case for murine monocytes. Culture of murine bone marrow with GM-CSF and IL-4 presumably mimics monocyte differentiation under the same conditions (166). When FLT3 signaling is inhibited in such bulk cultures the T cell stimulatory capacity of the output cells is reduced (161). Therefore, these data raise the possibility that FLT3L might influence monocyte differentiation into cells with functional properties of DCs also in the murine system, although a direct causality remains to be demonstrated. Further, comparative gene expression profiling revealed that upon migration to lymph nodes LCs induce CD135 expression (167), indicating that they might be capable of responding to FLT3L. Therefore, it is conceivable that FLT3L may control certain functional aspects generally associated with DCs, such as antigen presentation, in ontogenetically distinct MP subtypes, which will be interesting to formally address in the context of FLT3L or CD135 deficiency.

Dendritic cell progenitors also express CD115, the receptor for MØ colony-stimulating factor (M-CSF) (56–58, 61, 86). However, compared to the dominant role of FLT3L in DC differentiation, M-CSF-deficiency only mildly impacts on DC development (168). M-CSF deficient osteopetrotic (op/op) mice exhibit a two- to threefold reduction in splenic cDCs and pDCs, respectively, but the remaining DCs are capable of stimulating a mixed lymphocyte reaction and induce costimulatory molecules upon activation, thus appear functional (168). In contrast, M-CSF is strongly required for monocyte and MØ development (141, 169). Therefore, the observation that mice lacking CD115 exhibit reduced frequencies of CD11c$^+$MHCII$^+$CD11b$^+$ cells in non-lymphoid organs (63, 91) likely reflects the ontogenetic heterogeneity of this population (63, 91–95). Consistently, M-CSF is also required for the generation of monocyte-derived cells with features of DCs during inflammation (90). Nonetheless, M-CSF may play a role in DC development. It can promote DC differentiation *in vitro* and *in vivo* even in the absence of FLT3L, although DCs generated by M-CSF alone phenotypically and functionally differ from those induced by FLT3L (170). M-CSF-induced DC poeisis is also more efficient in FLT3L-sufficient conditions (170). *In vivo*, antibody-mediated blockade of M-CSF in pregnant mice reduces pre-DC extravasation, translating into a reduction of CD11b$^+$ DCs in the pregnant uterus (171). Whether M-CSF affects pre-DC migration also in other tissues and whether it acts in a cell intrinsic manner or by promoting the production of chemotactic factors by other cells remains to be determined (171).

In purified monocytes, GM-CSF induces phenotypic and functional attributes of DCs (88, 89, 172). Similarly, purified CD115$^+$ MDPs respond to GM-CSF by differentiating into CD11c$^+$MHCII$^+$ DCs (85) and GM-CSF deficiency leads to a slight reduction of bone marrow MDPs and CDPs (164). However, GM-CSF is dispensable for the differentiation of lymphoid tissue DCs (85, 173) and, therefore, it seemed likely that GM-CSF would selectively regulate the differentiation of monocytes into cells resembling DCs (23). This speculation also lead to the hypothesis that monocytes cultured in the presence of GM-CSF represent the counterpart of mo-DCs generated under conditions of inflammation/infection *in vivo* (23). Surprisingly, GM-CSF does not appear to control monocyte differentiation *in vivo* (90) and thus, GM-CSF elicited monocyte-derived cells are unlikely to be fully equivalent to inflammatory monocyte-derived cells. Rather, GM-CSF influences the homeostasis of cDCs in a variety, but not all, non-lymphoid tissues, most likely by promoting cell survival (90). Importantly, GM-CSF deficiency leads to a greater reduction of CD103$^+$ cDCs than of CD11b$^+$ cDCs (90). However, the extent of cDC reduction in the absence of GM-CSF apparently relates to the markers used for cell identification (90, 147, 164). This is most likely because GM-CSF regulates certain phenotypic as well as functional features of DCs, such as CD103 expression (174) or their ability to cross-present antigen (90, 174, 175). Therefore, the above-mentioned growth factors not only influence lineage decisions but also impact on the functional regulation of DCs, monocytes, and MØs. Elucidating the exact roles of FLT3L, GM-CSF, and M-CSF in each cell type will help to decipher the functional heterogeneity of MPs.

Cellular Barcoding

The biggest challenge for fate mapping is to trace the developmental plasticity of individual cells. This can now be achieved using "cellular barcoding," in which progenitors are tagged *in vitro* with semi-random, non-coding DNA sequences by transduction using retro- or lentiviral vectors (**Figure 2C**) (81). Therefore, the barcodes are heritable and by choosing conditions of low transduction efficiency one can ensure that each cell receives only a single barcode. Subsequently, barcode-labeled progenitors are adoptively transferred in numbers low enough to minimize the chance that two identically barcoded cells are transferred into the same recipient (**Figure 2C**). After differentiation *in vivo*, cell progeny are analyzed for their barcode repertoire using deep sequencing or custom microarray. Since each barcode represents an individual progenitor, the presence of the same barcode in more than one cell type indicates that they were generated from a single precursor (multi-potent or bi-potent, **Figure 2C**). On the other hand, if a barcode is only found in one cell type, the progenitor generated only a single cell lineage (mono-potent, **Figure 2C**) (81).

During maturation, HSCs are thought to progressively lose their self-renewal ability and become increasingly limited in their differentiation potential, ultimately giving rise to lineage-restricted progenitors (55, 176). Lymphoid primed multi-potent progenitors (LMPPs) are developmental intermediates downstream of HSCs that can give rise to various, but not all, cell lineages and are thus considered multi-potent (55, 176). Surprisingly, in barcoding experiments only a minority (3%) of single LMPPs exhibits true multi-potency, defined as the ability to generate all of the following cell lineages: B cells, DCs, and myeloid cells (monocytes and neutrophils) (49). Rather, single LMPPs differ drastically in terms of their cellular output: 10% of the progenitors contribute primarily to B cells, 10% primarily to myeloid cells but about 50% of transferred LMPPs produce predominantly DCs (49). The remaining fraction of progenitors exhibits bi-potentiality to generate combinations of the examined cell lineages (49). Therefore, LMPPs are multi-potent when analyzed as a population, however single cells exhibit unexpected lineage bias that is imprinted early in development. Why the majority of LMPPs is DC-committed (49), even though DCs constitute a minority lineage compared to B cells, remains to be clarified, although it is possible that some progenitors proliferate better than others or have certain competitive advantages. A major lineage divergence toward DCs seems to occur before or at the LMPP stage, as most HSCs analyzed by the same method are multi-potent, although even HSCs exhibit a degree of lineage bias (49, 177). Since CDPs might arise directly from LMPPs without additional developmental intermediates (79), these data infer that DCs diverge as a developmental lineage distinct from other myeloid cells early on (49).

This, again, questions the existence of a bi-potential MDP as central intermediate in the development of DCs and monocytes. Yet, it is noteworthy that even though DC-biased LMPPs are fivefold more frequent than bi-potent myeloid/DC LMPPs, mono-potent and bi-potent progenitors contribute equally to the final DC pool (49). Therefore, bi-potent progenitors seem to play a significant part in generating DCs, potentially because they have a proliferative advantage. Resolving these issues will require further refinement of the technique at hand. The differentiation potential of progenitors may be influenced by cell isolation, processing or *in vitro* manipulation (80) and virus-mediated transformation might skew cell fate in an unforeseeable manner, as evidenced by the fact that barcoded LMPPs cannot generate T cells (49, 81). This also means that barcoding does not yet uncover the full potential of single progenitors. The early lineage bias of HSCs and LMPPs suggests that cell development may follow a model of graded commitment rather than proceeding in a truly stepwise manner (178). It will be interesting to determine, to what extent this process is regulated by epigenetic modification and how inflammatory processes might impact on lineage divergence. Future studies will benefit from the development of models allowing for *in vivo* barcoding of single cells but the labor-intensive quantification and analysis of barcoding experiments makes it difficult to follow populations in real time.

Genetic Lineage Tracing

Dynamic mapping of populations of distinct origin *in vivo* can be achieved using genetic lineage tracing based on *Cre-loxP* technology (**Figure 2D**) (80, 179). It relies on inducible reporter genes that are placed under the control of constitutively active promoters, such as the *Rosa26* locus. The reporter is most commonly a fluorescent protein that is preceded by a *loxP*-flanked *STOP* cassette and, therefore, its expression is induced only after Cre recombinase (Cre) mediated excision of the stop codon (**Figure 2D**). Since this form of labeling is genetic it is also heritable, meaning that any cell expressing Cre will pass on the label to all progeny, irrespective of continuous recombinase expression (**Figure 2D**). Since the promoter driving the reporter gene is constitutively active, labeling is irreversible and not affected by fluctuations in gene expression (**Figure 2D**) (80).

By crossing mice expressing Cre under the control of the *Clec9a* locus to *Rosa26-STOP-flox*-enhanced-yellow fluorescent protein (YFP) reporter mice (180), we have recently generated the first genetic model to trace the progeny of DNGR-1$^+$ CDPs and pre-DCs (36). In these mice, YFP expression is restricted to DCs but is not found in monocytes or MØs even in inflammatory conditions, as tested after intestinal inflammation or infection with *L. monocytogenes* (36). Nonetheless, certain limitations need to be taken into account. DNGR-1 is also expressed on CD8α$^+$/CD103$^+$ cDCs and to a lower extent on pDCs (36, 71, 181, 182) and, therefore, in these populations labeling is not a strict indicator of cell ontogeny. Further, labeling of CDP and pre-DC progeny in mice heterozygous for Cre is incomplete, possibly due to a delay in Cre protein synthesis and DNA recombination in rapidly cycling progenitors (36). Consistently, penetrance of the YFP label is increased in mice homozygous for Cre (36). The efficiency of lineage tracing experiments in such cases or when Cre expression is low may be improved by using alternate reporter constructs in which the *loxP* sites are positioned closer together, thus facilitating recombination (183).

Genetic lineage tracing does not require prior knowledge of which markers are expressed by the output cells and, thus, enables unbiased monitoring of cell ontogeny. Therefore, we were able to identify CDP-derived cells in cell populations previously thought to constitute monocytes/MØs based on the expression of surface markers, such as CD64 (36). CD64$^+$ CDP-derived cells do not express *Clec9a* message and are especially frequent in kidneys, although the presence of few YFP$^+$ cells in the CD64$^+$ component of lung and small intestine indicates that atypical CDP-derived cells also exist in other tissues (36). CD64$^+$ kidney DCs resemble yolk sac-derived F4/80^{hi} tissue-resident MØs, appear to lack Zbtb46 expression (37) and their affiliation as DCs or MØs has been debated (184). We, therefore, used adoptive transfer as additional method to confirm cell ontogeny. Surprisingly, neither purified DNGR-1$^+$ CDPs nor total bone marrow generated F4/80^{hi}CD64$^+$ CDP progeny in kidneys 1 week after adoptive transfer into irradiated recipients (36). Since kidney DCs reportedly have a slow turnover (185), it is possible that CDPs had insufficient opportunity to reach their renal niche and expand during short-term transfer experiments. Consistent with this notion, F4/80^{hi}CD64$^+$ kidney leukocytes were efficiently generated from bone marrow progenitors in long-term reconstitution experiments (36). Therefore, our data strongly support a CDP origin of CD64$^+$ kidney leukocytes, despite their phenotypic resemblance to monocytes or MØs (36). These data exemplify the power of lineage tracing in following cell ontogeny in an unbiased way, although it is possible that DNGR-1 is expressed on yet unidentified developmental intermediates.

Addressing this possibility might require tamoxifen-inducible Cre constructs that can be used to pulse label progenitor populations (80). In the future, combinatorial approaches, such as "split-Cre" fragments controlled by two different promoters (186) or an intersection where Cre and the inducible reporter are driven by two cell-specific promoters (187, 188) may be of benefit to generate improved models to lineage trace DCs. The identification of CDP-derived cells with attributes of monocytes/MØs exemplifies the insufficiency of phenotypic properties, such as surface markers, as means of accurate cell identification of MPs. It also raises the question why cells of distinct ontogeny but overlapping phenotype exist in the same tissue. Further elucidation of the specific functions of MPs in immunity will benefit from lineage tracing approaches that result in target cell deletion through the use of inducible DTR or DT subunit modules (82, 112, 189, 190).

Conclusion

The studies discussed above have significantly advanced our understanding of DC ontogeny but have also uncovered some uncertainties (**Figure 3**). While the bone marrow origin of DCs and monocytes is undisputed, the exact developmental intermediates and branching points between HSCs and DC progenitors remain to be clarified. Current data indicate that lineage imprinting toward DCs and monocytes may occur as early as LMPPs, potentially through epigenetic modification (**Figure 3**). This realization constitutes a major conceptual shift as it puts in question the existence of a bi-potential MDP and the resulting relatedness of DCs and monocytes. A definitive resolution of this question requires increasingly refined methods to genetically trace single progenitors or select DC and monocyte lineages. Nonetheless, it is clear that cDCs, pDCs, and monocytes can be separated based on their descendance from committed

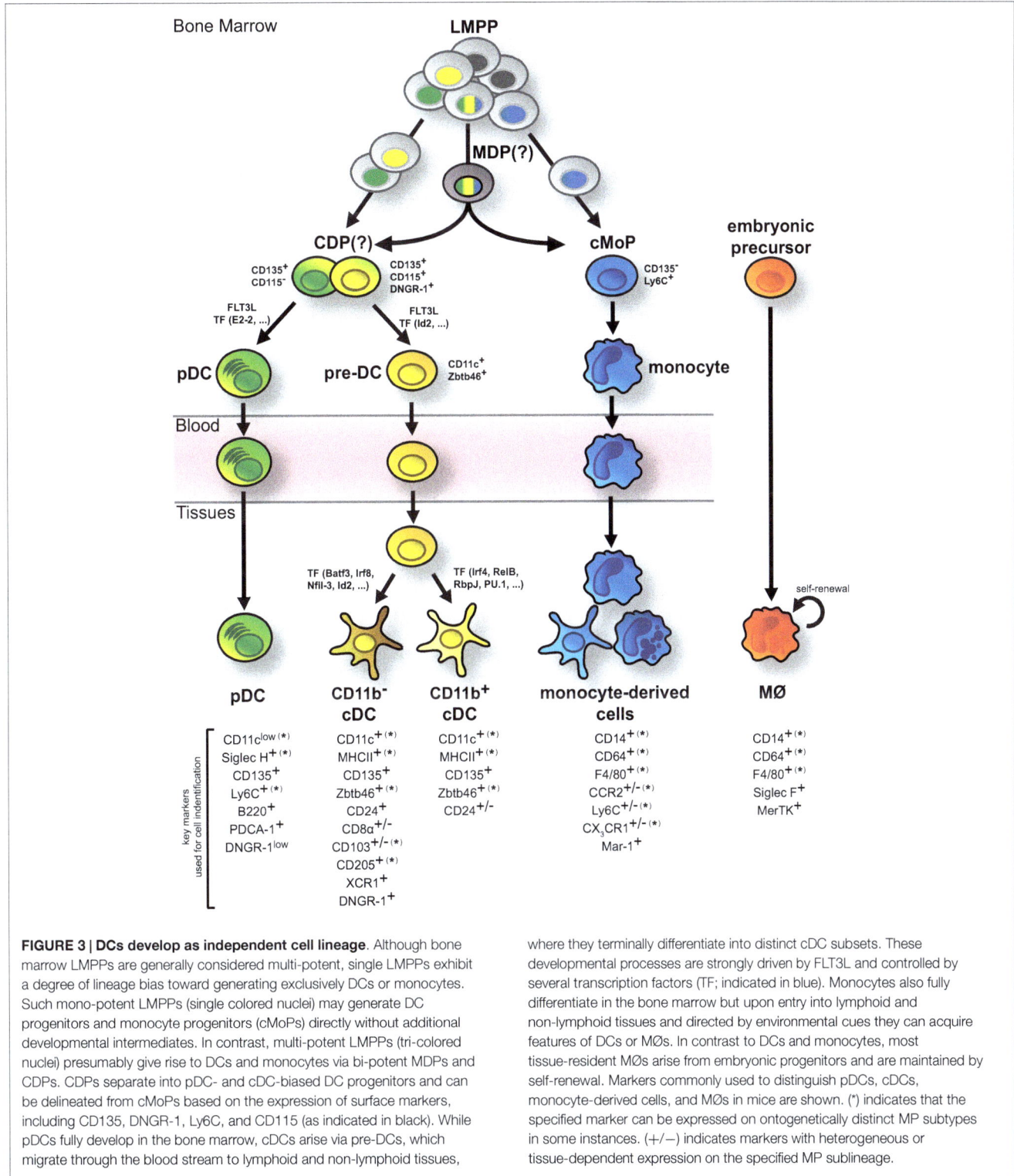

FIGURE 3 | DCs develop as independent cell lineage. Although bone marrow LMPPs are generally considered multi-potent, single LMPPs exhibit a degree of lineage bias toward generating exclusively DCs or monocytes. Such mono-potent LMPPs (single colored nuclei) may generate DC progenitors and monocyte progenitors (cMoPs) directly without additional developmental intermediates. In contrast, multi-potent LMPPs (tri-colored nuclei) presumably give rise to DCs and monocytes via bi-potent MDPs and CDPs. CDPs separate into pDC- and cDC-biased DC progenitors and can be delineated from cMoPs based on the expression of surface markers, including CD135, DNGR-1, Ly6C, and CD115 (as indicated in black). While pDCs fully develop in the bone marrow, cDCs arise via pre-DCs, which migrate through the blood stream to lymphoid and non-lymphoid tissues, where they terminally differentiate into distinct cDC subsets. These developmental processes are strongly driven by FLT3L and controlled by several transcription factors (TF; indicated in blue). Monocytes also fully differentiate in the bone marrow but upon entry into lymphoid and non-lymphoid tissues and directed by environmental cues they can acquire features of DCs or MØs. In contrast to DCs and monocytes, most tissue-resident MØs arise from embryonic progenitors and are maintained by self-renewal. Markers commonly used to distinguish pDCs, cDCs, monocyte-derived cells, and MØs in mice are shown. (*) indicates that the specified marker can be expressed on ontogenetically distinct MP subtypes in some instances. (+/−) indicates markers with heterogeneous or tissue-dependent expression on the specified MP sublineage.

developmental intermediates (**Figure 3**). Their differentiation is further driven by unique factors indicating that their developmental paths are distinct (**Figure 3**). In stark contrast to pDCs, cDCs, and monocytes, most tissue MØs arise from embryonic progenitors and are predominantly maintained by self-renewal into adulthood (**Figure 3**).

Taken together, these data unequivocally establish that DCs, monocytes, and MØs develop as unique cellular entities and although one could argue that most of this knowledge is derived from mouse studies, developmental parallels have been observed in other species (66–73). Despite these advances, we are at a loss for a universal definition of DCs that is readily accessible to experts within and outside the field of MP biology. In light of this recognition, it has been suggested to revise the current nomenclature of MPs into a system that takes cell ontogeny into account when defining subpopulations (6). Such system would greatly aid our understanding of phagocyte biology as it remains uncertain to what extent the cellular origin of DCs, monocytes, and MØs determines the unique functionality of these cells in immunity and/or tissue homeostasis. While global profiling has revealed a role for the local tissue microenvironment in shaping the transcriptional landscape of DCs, monocytes, and MØs from different organs, certain gene signatures and transcriptional features are set by ontogeny (167, 191–193). Therefore, the full functional diversity of DCs, monocytes, and MØs is likely shaped by both nature (ontogeny) and nurture

(the environment). Since ontogeny is immutable it provides a more robust common denominator for cell definition that enables deciphering cellular functions without assuming preconceived functional or phenotypic relationships. DC classification based on cell ancestry is a work in progress but its implementation will ultimately yield a more robust and transparent way of cell definition.

Acknowledgments

BS and MP are funded by the German Research Foundation (Emmy Noether Grant: Schr 1444/1-1). We thank members of the Schraml lab and Clarissa Prazeres da Costa for critical reading of the manuscript.

References

1. Steinman RM, Cohn ZA. Identification of a novel cell type in peripheral lymphoid organs of mice. I. Morphology, quantitation, tissue distribution. *J Exp Med* (1973) 137(5):1142–62. doi:10.1084/jem.137.5.1142

2. Steinman RM, Witmer MD. Lymphoid dendritic cells are potent stimulators of the primary mixed leukocyte reaction in mice. *Proc Natl Acad Sci U S A* (1978) 75(10):5132–6. doi:10.1073/pnas.75.10.5132

3. Steinman RM, Idoyaga J. Features of the dendritic cell lineage. *Immunol Rev* (2010) 234(1):5–17. doi:10.1111/j.0105-2896.2009.00888.x

4. van Furth R, Cohn ZA. The origin and kinetics of mononuclear phagocytes. *J Exp Med* (1968) 128(3):415–35. doi:10.1084/jem.128.3.415

5. van Furth R, Cohn ZA, Hirsch JG, Humphrey JH, Spector WG, Langevoort HL. The mononuclear phagocyte system: a new classification of macrophages, monocytes, and their precursor cells. *Bull World Health Organ* (1972) 46(6):845–52.

6. Guilliams M, Ginhoux F, Jakubzick C, Naik SH, Onai N, Schraml BU, et al. Dendritic cells, monocytes and macrophages: a unified nomenclature based on ontogeny. *Nat Rev Immunol* (2014) 14(8):571–8. doi:10.1038/nri3712

7. Hashimoto D, Miller J, Merad M. Dendritic cell and macrophage heterogeneity in vivo. *Immunity* (2011) 35(3):323–35. doi:10.1016/j.immuni.2011.09.007

8. Gordon S, Taylor PR. Monocyte and macrophage heterogeneity. *Nat Rev Immunol* (2005) 5(12):953–64. doi:10.1038/nri1733

9. van Furth R. Identification of mononuclear phagocytes: overview and definitions. In: Adams DO, Edelson PJ, Koren H, editors. *Methods for Studying Mononuclear Phagocytes*. New York: Academic Press (1980). p. 243–52.

10. Schraml BU, Reis e Sousa C. Defining dendritic cells. *Curr Opin Immunol* (2015) 32:13–20. doi:10.1016/j.coi.2014.11.001

11. Geissmann F, Gordon S, Hume DA, Mowat AM, Randolph GJ. Unravelling mononuclear phagocyte heterogeneity. *Nat Rev Immunol* (2010) 10(6):453–60. doi:10.1038/nri2784

12. Hume DA. Macrophages as APC and the dendritic cell myth. *J Immunol* (2008) 181(9):5829–35. doi:10.4049/jimmunol.181.9.5829

13. Hume DA. The mononuclear phagocyte system. *Curr Opin Immunol* (2006) 18(1):49–53. doi:10.1016/j.coi.2005.11.008

14. Hume DA, Ross IL, Himes SR, Sasmono RT, Wells CA, Ravasi T. The mononuclear phagocyte system revisited. *J Leukoc Biol* (2002) 72(4):621–7.

15. Reizis B. Regulation of plasmacytoid dendritic cell development. *Curr Opin Immunol* (2010) 22(2):206–11. doi:10.1016/j.coi.2010.01.005

16. Heath WR, Carbone FR. Dendritic cell subsets in primary and secondary T cell responses at body surfaces. *Nat Immunol* (2009) 10(12):1237–44. doi:10.1038/ni.1822

17. Briseño CG, Murphy TL, Murphy KM. Complementary diversification of dendritic cells and innate lymphoid cells. *Curr Opin Immunol* (2014) 29C:69–78. doi:10.1016/j.coi.2014.04.006

18. Merad M, Sathe P, Helft J, Miller J, Mortha A. The dendritic cell lineage: ontogeny and function of dendritic cells and their subsets in the steady state and the inflamed setting. *Annu Rev Immunol* (2013) 31:563–604. doi:10.1146/annurev-immunol-020711-074950

19. Geissmann F, Manz MG, Jung S, Sieweke MH, Merad M, Ley K. Development of monocytes, macrophages, and dendritic cells. *Science* (2010) 327(5966):656–61. doi:10.1126/science.1178331

20. Nussenzweig MC, Steinman RM, Witmer MD, Gutchinov B. A monoclonal antibody specific for mouse dendritic cells. *Proc Natl Acad Sci U S A* (1982) 79(1):161–5. doi:10.1073/pnas.79.1.161

21. Steinman RM, Kaplan G, Witmer MD, Cohn ZA. Identification of a novel cell type in peripheral lymphoid organs of mice. V. Purification of spleen dendritic cells, new surface markers, and maintenance in vitro. *J Exp Med* (1979) 149(1):1–16. doi:10.1084/jem.149.1.1

22. Metlay JP, Witmer-Pack MD, Agger R, Crowley MT, Lawless D, Steinman RM. The distinct leukocyte integrins of mouse spleen dendritic cells as identified with new hamster monoclonal antibodies. *J Exp Med* (1990) 171(5):1753–71. doi:10.1084/jem.171.5.1753

23. Shortman K, Naik SH. Steady-state and inflammatory dendritic-cell development. *Nat Rev Immunol* (2007) 7(1):19–30. doi:10.1038/nri1996

24. Mildner A, Yona S, Jung S. A close encounter of the third kind: monocyte-derived cells. *Adv Immunol* (2013) 120:69–103. doi:10.1016/B978-0-12-417028-5.00003-X

25. Carlens J, Wahl B, Ballmaier M, Bulfone-Paus S, Forster R, Pabst O. Common gamma-chain-dependent signals confer selective survival of eosinophils in the murine small intestine. *J Immunol* (2009) 183(9):5600–7. doi:10.4049/jimmunol.0801581

26. Vermaelen K, Pauwels R. Accurate and simple discrimination of mouse pulmonary dendritic cell and macrophage populations by flow cytometry: methodology and new insights. *Cytometry A* (2004) 61(2):170–7. doi:10.1002/cyto.a.20064

27. Probst HC, Tschannen K, Odermatt B, Schwendener R, Zinkernagel RM, Van Den Broek M. Histological analysis of CD11c-DTR/GFP mice after in vivo depletion of dendritic cells. *Clin Exp Immunol* (2005) 141(3):398–404. doi:10.1111/j.1365-2249.2005.02868.x

28. Hochweller K, Striegler J, Hämmerling GJ, Garbi N. A novel CD11c.DTR transgenic mouse for depletion of dendritic cells reveals their requirement for homeostatic proliferation of natural killer cells. *Eur J Immunol* (2008) 38(10):2776–83. doi:10.1002/eji.200838659

29. Huleatt JW, Lefrançois L. Antigen-driven induction of CD11c on intestinal intraepithelial lymphocytes and CD8+ T cells in vivo. *J Immunol* (1995) 154(11):5684–93.

30. Rubtsov AV, Rubtsova K, Fischer A, Meehan RT, Gillis JZ, Kappler JW, et al. Toll-like receptor 7 (TLR7)-driven accumulation of a novel CD11c(+) B-cell population is important for the development of autoimmunity. *Blood* (2011) 118(5):1305–15. doi:10.1182/blood-2011-01-331462

31. Drutman SB, Kendall JC, Trombetta ES. Inflammatory spleen monocytes can upregulate CD11c expression without converting into dendritic cells. *J Immunol* (2012) 188(8):3603–10. doi:10.4049/jimmunol.1102741

32. Hebel K, Griewank K, Inamine A, Chang H-D, Müller-Hilke B, Fillatreau S, et al. Plasma cell differentiation in T-independent type 2 immune responses is independent of CD11c(high) dendritic cells. *Eur J Immunol* (2006) 36(11):2912–9. doi:10.1002/eji.200636356

33. Kim K-W, Vallon-Eberhard A, Zigmond E, Farache J, Shezen E, Shakhar G, et al. In vivo structure/function and expression analysis of the CX3C chemokine fractalkine. *Blood* (2011) **118**(22):e156–67. doi:10.1182/blood-2011-04-348946

34. Farache J, Koren I, Milo I, Gurevich I, Kim KW, Zigmond E, et al. Luminal bacteria recruit CD103+ dendritic cells into the intestinal epithelium to sample bacterial antigens for presentation. *Immunity* (2013) **38**(3):581–95. doi:10.1016/j.immuni.2013.01.009

35. Bergstresser PR, Sullivan S, Streilein JW, Tigelaar RE. Origin and function of Thy-1+ dendritic epidermal cells in mice. *J Invest Dermatol* (1985) **85**(1 Suppl):85s–90s. doi:10.1111/1523-1747.ep12275516

36. Schraml BU, van Blijswijk J, Zelenay S, Whitney PG, Filby A, Acton SE, et al. Genetic tracing via DNGR-1 expression history defines dendritic cells as a hematopoietic lineage. *Cell* (2013) **154**(4):843–58. doi:10.1016/j.cell.2013.07.014

37. Satpathy AT, Kc W, Albring JC, Edelson BT, Kretzer NM, Bhattacharya D, et al. Zbtb46 expression distinguishes classical dendritic cells and their committed progenitors from other immune lineages. *J Exp Med* (2012) **209**(6):1135–52. doi:10.1084/jem.20120030

38. Meredith MM, Liu K, Darrasse-Jèze G, Kamphorst AO, Schreiber HA, Guermonprez P, et al. Expression of the zinc finger transcription factor zDC (Zbtb46, Btbd4) defines the classical dendritic cell lineage. *J Exp Med* (2012) **209**(6):1153–65. doi:10.1084/jem.20112675

39. Mashayekhi M, Sandau MM, Dunay IR, Frickel EM, Khan A, Goldszmid RS, et al. CD8α(+) dendritic cells are the critical source of interleukin-12 that controls acute infection by *Toxoplasma gondii* tachyzoites. *Immunity* (2011) **35**(2):249–59. doi:10.1016/j.immuni.2011.08.008

40. Reis e Sousa C, Hieny S, Scharton-Kersten T, Jankovic D, Charest H, Germain RN, et al. In vivo microbial stimulation induces rapid CD40 ligand-independent production of interleukin 12 by dendritic cells and their redistribution to T cell areas. *J Exp Med* (1997) **186**(11):1819–29. doi:10.1084/jem.186.11.1819

41. Whitney PG, Bar E, Osorio F, Rogers NC, Schraml BU, Deddouche S, et al. Syk signaling in dendritic cells orchestrates innate resistance to systemic fungal infection. *PLoS Pathog* (2014) **10**(7):e1004276. doi:10.1371/journal.ppat.1004276

42. Satpathy AT, Briseño CG, Lee JS, Ng D, Manieri NA, Kc W, et al. Notch2-dependent classical dendritic cells orchestrate intestinal immunity to attaching-and-effacing bacterial pathogens. *Nat Immunol* (2013) **14**(9):937–48. doi:10.1038/ni.2679

43. Kinnebrew MA, Buffie CG, Diehl GE, Zenewicz LA, Leiner I, Hohl TM, et al. Interleukin 23 production by intestinal CD103(+)CD11b(+) dendritic cells in response to bacterial flagellin enhances mucosal innate immune defense. *Immunity* (2012) **36**(2):276–87. doi:10.1016/j.immuni.2011.12.011

44. Arora P, Baena A, Yu KOA, Saini NK, Kharkwal SS, Goldberg MF, et al. A single subset of dendritic cells controls the cytokine bias of natural killer T cell responses to diverse glycolipid antigens. *Immunity* (2014) **40**(1):105–16. doi:10.1016/j.immuni.2013.12.004

45. Jakubzick C, Gautier EL, Gibbings SL, Sojka DK, Schlitzer A, Johnson TE, et al. Minimal differentiation of classical monocytes as they survey steady-state tissues and transport antigen to lymph nodes. *Immunity* (2013) **39**(3):599–610. doi:10.1016/j.immuni.2013.08.007

46. Plantinga M, Guilliams M, Vanheerswynghels M, Deswarte K, Branco-Madeira F, Toussaint W, et al. Conventional and monocyte-derived CD11b(+) dendritic cells initiate and maintain T helper 2 cell-mediated immunity to house dust mite allergen. *Immunity* (2013) **38**(2):322–35. doi:10.1016/j.immuni.2012.10.016

47. Pozzi LA, Maciaszek JW, Rock KL. Both dendritic cells and macrophages can stimulate naive CD8 T cells in vivo to proliferate, develop effector function, and differentiate into memory cells. *J Immunol* (2005) **175**(4):2071–81. doi:10.4049/jimmunol.175.4.2071

48. McGovern N, Schlitzer A, Gunawan M, Jardine L, Shin A, Poyner E, et al. Human dermal CD14+ cells are a transient population of monocyte-derived macrophages. *Immunity* (2014) **41**(3):465–77. doi:10.1016/j.immuni.2014.08.006

49. Naik SH, Perié L, Swart E, Gerlach C, van Rooij N, de Boer RJ, et al. Diverse and heritable lineage imprinting of early haematopoietic progenitors. *Nature* (2013) **496**(7444):229–32. doi:10.1038/nature12013

50. Yona S, Kim K-W, Wolf Y, Mildner A, Varol D, Breker M, et al. Fate mapping reveals origins and dynamics of monocytes and tissue macrophages under homeostasis. *Immunity* (2013) **38**(1):79–91. doi:10.1016/j.immuni.2012.12.001

51. Schulz C, Gomez Perdiguero E, Chorro L, Szabo-Rogers H, Cagnard N, Kierdorf K, et al. A lineage of myeloid cells independent of Myb and hematopoietic stem cells. *Science* (2012) **336**(6077):86–90. doi:10.1126/science.1219179

52. Hettinger J, Richards DM, Hansson J, Barra MM, Joschko A-C, Krijgsveld J, et al. Origin of monocytes and macrophages in a committed progenitor. *Nat Immunol* (2013) **14**(8):821–30. doi:10.1038/ni.2638

53. Haniffa M, Ginhoux F, Wang X-N, Bigley V, Abel M, Dimmick I, et al. Differential rates of replacement of human dermal dendritic cells and macrophages during hematopoietic stem cell transplantation. *J Exp Med* (2009) **206**(2):371–85. doi:10.1084/jem.20081633

54. Hashimoto D, Chow A, Noizat C, Teo P, Beasley MB, Leboeuf M, et al. Tissue-resident macrophages self-maintain locally throughout adult life with minimal contribution from circulating monocytes. *Immunity* (2013) **38**(4):792–804. doi:10.1016/j.immuni.2013.04.004

55. Iwasaki H, Akashi K. Myeloid lineage commitment from the hematopoietic stem cell. *Immunity* (2007) **26**(6):726–40. doi:10.1016/j.immuni.2007.06.004

56. Fogg DK, Sibon C, Miled C, Jung S, Aucouturier P, Littman DR, et al. A clonogenic bone marrow progenitor specific for macrophages and dendritic cells. *Science* (2006) **311**(5757):83–7. doi:10.1126/science.1117729

57. Naik SH, Sathe P, Park H-Y, Metcalf D, Proietto AI, Dakic A, et al. Development of plasmacytoid and conventional dendritic cell subtypes from single precursor cells derived in vitro and in vivo. *Nat Immunol* (2007) **8**(11):1217–26. doi:10.1038/ni1522

58. Onai N, Obata-Onai A, Schmid MA, Ohteki T, Jarrossay D, Manz MG. Identification of clonogenic common Flt3+M-CSFR+ plasmacytoid and conventional dendritic cell progenitors in mouse bone marrow. *Nat Immunol* (2007) **8**(11):1207–16. doi:10.1038/ni1518

59. Shortman K, Sathe P, Vremec D, Naik S, O'Keeffe M. Plasmacytoid dendritic cell development. *Adv Immunol* (2013) **120**:105–26. doi:10.1016/B978-0-12-417028-5.00004-1

60. Naik SH, Metcalf D, Van Nieuwenhuijze A, Wicks I, Wu L, O'Keeffe M, et al. Intrasplenic steady-state dendritic cell precursors that are distinct from monocytes. *Nat Immunol* (2006) **7**(6):663–71. doi:10.1038/ni1340

61. Liu K, Victora GD, Schwickert TA, Guermonprez P, Meredith MM, Yao K, et al. In vivo analysis of dendritic cell development and homeostasis. *Science* (2009) **324**(5925):392–7. doi:10.1126/science.1170540

62. Liu K, Waskow C, Liu X, Yao K, Hoh J, Nussenzweig M. Origin of dendritic cells in peripheral lymphoid organs of mice. *Nat Immunol* (2007) **8**(6):578–83. doi:10.1038/ni1462

63. Ginhoux F, Liu K, Helft J, Bogunovic M, Greter M, Hashimoto D, et al. The origin and development of nonlymphoid tissue CD103+ DCs. *J Exp Med* (2009) **206**(13):3115–30. doi:10.1084/jem.20091756

64. Serbina NV, Salazar-Mather TP, Biron CA, Kuziel WA, Pamer EG. TNF/iNOS-producing dendritic cells mediate innate immune defense against bacterial infection. *Immunity* (2003) **19**(1):59–70. doi:10.1016/S1074-7613(03)00171-7

65. Haniffa M, Collin M, Ginhoux F. Ontogeny and functional specialization of dendritic cells in human and mouse. *Adv Immunol* (2013) **120**:1–49. doi:10.1016/B978-0-12-417028-5.00001-6

66. Guzylack-Piriou L, Alves MP, McCullough KC, Summerfield A. Porcine Flt3 ligand and its receptor: generation of dendritic cells and identification of a new marker for porcine dendritic cells. *Dev Comp Immunol* (2010) **34**(4):455–64. doi:10.1016/j.dci.2009.12.006

67. Cisse B, Caton ML, Lehner M, Maeda T, Scheu S, Locksley R, et al. Transcription factor E2-2 is an essential and specific regulator of plasmacytoid dendritic cell development. *Cell* (2008) **135**(1):37–48. doi:10.1016/j.cell.2008.09.016

68. Maraskovsky E, Daro E, Roux E, Teepe M, Maliszewski CR, Hoek J, et al. In vivo generation of human dendritic cell subsets by Flt3 ligand. *Blood* (2000) **96**(3):878–84.

69. Hambleton S, Salem S, Bustamante J, Bigley V, Boisson-Dupuis S, Azevedo J, et al. IRF8 mutations and human dendritic-cell immunodeficiency. *N Engl J Med* (2011) **365**(2):127–38. doi:10.1056/NEJMoa1100066

70. Harada S, Kimura T, Fujiki H, Nakagawa H, Ueda Y, Itoh T, et al. Flt3 ligand promotes myeloid dendritic cell differentiation of human hematopoietic progenitor cells: possible application for cancer immunotherapy. *Int J Oncol* (2007) **30**(6):1461–8. doi:10.3892/ijo.30.6.1461

71. Poulin LF, Reyal Y, Uronen-Hansson H, Schraml BU, Sancho D, Murphy KM, et al. DNGR-1 is a specific and universal marker of mouse and human Batf3-dependent dendritic cells in lymphoid and nonlymphoid tissues. *Blood* (2012) 119(25):6052–62. doi:10.1182/blood-2012-01-406967

72. Breton G, Lee J, Zhou YJ, Schreiber JJ, Keler T, Puhr S, et al. Circulating precursors of human CD1c+ and CD141+ dendritic cells. *J Exp Med* (2015) 212(3):401–13. doi:10.1084/jem.20141441

73. Lee J, Breton G, Oliveira TYK, Zhou YJ, Aljoufi A, Puhr S, et al. Restricted dendritic cell and monocyte progenitors in human cord blood and bone marrow. *J Exp Med* (2015) 212(3):385–99. doi:10.1084/jem.20141442

74. Traver D, Akashi K, Manz M, Merad M, Miyamoto T, Engleman EG, et al. Development of CD8alpha-positive dendritic cells from a common myeloid progenitor. *Science* (2000) 290(5499):2152–4. doi:10.1126/science.290.5499.2152

75. Schlenner SM, Madan V, Busch K, Tietz A, Läufle C, Costa C, et al. Fate mapping reveals separate origins of T cells and myeloid lineages in the thymus. *Immunity* (2010) 32(3):426–36. doi:10.1016/j.immuni.2010.03.005

76. Luche H, Ardouin L, Teo P, See P, Henri S, Merad M, et al. The earliest intrathymic precursors of CD8α(+) thymic dendritic cells correspond to myeloid-type double-negative 1c cells. *Eur J Immunol* (2011) 41(8):2165–75. doi:10.1002/eji.201141728

77. Sathe P, Vremec D, Wu L, Corcoran L, Shortman K. Convergent differentiation: myeloid and lymphoid pathways to murine plasmacytoid dendritic cells. *Blood* (2013) 121(1):11–9. doi:10.1182/blood-2012-02-413336

78. Sathe P, Metcalf D, Vremec D, Naik SH, Langdon WY, Huntington ND, et al. Lymphoid tissue and plasmacytoid dendritic cells and macrophages do not share a common macrophage-dendritic cell-restricted progenitor. *Immunity* (2014) 41(1):104–15. doi:10.1016/j.immuni.2014.05.020

79. Onai N, Kurabayashi K, Hosoi-Amaike M, Toyama-Sorimachi N, Matsushima K, Inaba K, et al. A clonogenic progenitor with prominent plasmacytoid dendritic cell developmental potential. *Immunity* (2013) 38(5):943–57. doi:10.1016/j.immuni.2013.04.006

80. Kretzschmar K, Watt FM. Lineage tracing. *Cell* (2012) 148(1–2):33–45. doi:10.1016/j.cell.2012.01.002

81. Naik SH, Schumacher TN, Perié L. Cellular barcoding: a technical appraisal. *Exp Hematol* (2014) 42(8):598–608. doi:10.1016/j.exphem.2014.05.003

82. Bar-On L, Jung S. Defining dendritic cells by conditional and constitutive cell ablation. *Immunol Rev* (2010) 234(1):76–89. doi:10.1111/j.0105-2896.2009.00875.x

83. Plett PA, Frankovitz SM, Orschell-Traycoff CM. In vivo trafficking, cell cycle activity, and engraftment potential of phenotypically defined primitive hematopoietic cells after transplantation into irradiated or nonirradiated recipients. *Blood* (2002) 100(10):3545–52. doi:10.1182/blood.V100.10.3545

84. Varol C, Landsman L, Fogg DK, Greenshtein L, Gildor B, Margalit R, et al. Monocytes give rise to mucosal, but not splenic, conventional dendritic cells. *J Exp Med* (2007) 204(1):171–80. doi:10.1084/jem.20061011

85. Waskow C, Liu K, Darrasse-Jèze G, Guermonprez P, Ginhoux F, Merad M, et al. The receptor tyrosine kinase Flt3 is required for dendritic cell development in peripheral lymphoid tissues. *Nat Immunol* (2008) 9(6):676–83. doi:10.1038/ni.1615

86. Auffray C, Fogg DK, Narni-Mancinelli E, Senechal B, Trouillet C, Saederup N, et al. CX3CR1+ CD115+ CD135+ common macrophage/DC precursors and the role of CX3CR1 in their response to inflammation. *J Exp Med* (2009) 206(3):595–606. doi:10.1084/jem.20081385

87. Schlitzer A, Loschko J, Mair K, Vogelmann R, Henkel L, Einwächter H, et al. Identification of CCR9- murine plasmacytoid DC precursors with plasticity to differentiate into conventional DCs. *Blood* (2011) 117(24):6562–70. doi:10.1182/blood-2010-12-326678

88. Sallusto F, Lanzavecchia A. Efficient presentation of soluble antigen by cultured human dendritic cells is maintained by granulocyte/macrophage colony-stimulating factor plus interleukin 4 and downregulated by tumor necrosis factor alpha. *J Exp Med* (1994) 179(4):1109–18. doi:10.1084/jem.179.4.1109

89. Romani N, Gruner S, Brang D, Kämpgen E, Lenz A, Trockenbacher B, et al. Proliferating dendritic cell progenitors in human blood. *J Exp Med* (1994) 180(1):83–93. doi:10.1084/jem.180.1.83

90. Greter M, Helft J, Chow A, Hashimoto D, Mortha A, Agudo-Cantero J, et al. GM-CSF controls nonlymphoid tissue dendritic cell homeostasis but is dispensable for the differentiation of inflammatory dendritic cells. *Immunity* (2012) 36(6):1031–46. doi:10.1016/j.immuni.2012.03.027

91. Bogunovic M, Ginhoux F, Helft J, Shang L, Hashimoto D, Greter M, et al. Origin of the lamina propria dendritic cell network. *Immunity* (2009) 31(3):513–25. doi:10.1016/j.immuni.2009.08.010

92. Varol C, Vallon-Eberhard A, Elinav E, Aychek T, Shapira Y, Luche H, et al. Intestinal lamina propria dendritic cell subsets have different origin and functions. *Immunity* (2009) 31(3):502–12. doi:10.1016/j.immuni.2009.06.025

93. Tamoutounour S, Henri S, Lelouard H, de Bovis B, de Haar C, van der Woude CJ, et al. CD64 distinguishes macrophages from dendritic cells in the gut and reveals the Th1-inducing role of mesenteric lymph node macrophages during colitis. *Eur J Immunol* (2012) 42(12):3150–66. doi:10.1002/eji.201242847

94. Langlet C, Tamoutounour S, Henri S, Luche H, Ardouin L, Grégoire C, et al. CD64 expression distinguishes monocyte-derived and conventional dendritic cells and reveals their distinct role during intramuscular immunization. *J Immunol* (2012) 188(4):1751–60. doi:10.4049/jimmunol.1102744

95. Scott CL, Bain CC, Wright PB, Sichien D, Kotarsky K, Persson EK, et al. CCR2(+)CD103(-) intestinal dendritic cells develop from DC-committed precursors and induce interleukin-17 production by T cells. *Mucosal Immunol* (2014) 8(2):327–39. doi:10.1038/mi.2014.70

96. Merad M, Manz MG, Karsunky H, Wagers A, Peters W, Charo I, et al. Langerhans cells renew in the skin throughout life under steady-state conditions. *Nat Immunol* (2002) 3(12):1135–41. doi:10.1038/ni852

97. Ginhoux F, Tacke F, Angeli V, Bogunovic M, Loubeau M, Dai X-M, et al. Langerhans cells arise from monocytes in vivo. *Nat Immunol* (2006) 7(3):265–73. doi:10.1038/ni1307

98. Merad M, Ginhoux F, Collin M. Origin, homeostasis and function of Langerhans cells and other langerin-expressing dendritic cells. *Nat Rev Immunol* (2008) 8(12):935–47. doi:10.1038/nri2455

99. Romani N, Clausen BE, Stoitzner P. Langerhans cells and more: langerin-expressing dendritic cell subsets in the skin. *Immunol Rev* (2010) 234(1):120–41. doi:10.1111/j.0105-2896.2009.00886.x

100. Ginhoux F, Merad M. Ontogeny and homeostasis of Langerhans cells. *Immunol Cell Biol* (2010) 88(4):387–92. doi:10.1038/icb.2010.38

101. Hoeffel G, Wang Y, Greter M, See P, Teo P, Malleret B, et al. Adult Langerhans cells derive predominantly from embryonic fetal liver monocytes with a minor contribution of yolk sac-derived macrophages. *J Exp Med* (2012) 209(6):1167–81. doi:10.1084/jem.20120340

102. Chorro L, Sarde A, Li M, Woollard KJ, Chambon P, Malissen B, et al. Langerhans cell (LC) proliferation mediates neonatal development, homeostasis, and inflammation-associated expansion of the epidermal LC network. *J Exp Med* (2009) 206(13):3089–100. doi:10.1084/jem.20091586

103. Hume DA. Applications of myeloid-specific promoters in transgenic mice support in vivo imaging and functional genomics but do not support the concept of distinct macrophage and dendritic cell lineages or roles in immunity. *J Leukoc Biol* (2011) 89(4):525–38. doi:10.1189/jlb.0810472

104. Lindquist RL, Shakhar G, Dudziak D, Wardemann H, Eisenreich T, Dustin ML, et al. Visualizing dendritic cell networks in vivo. *Nat Immunol* (2004) 5(12):1243–50. doi:10.1038/ni1139

105. Thornton EE, Looney MR, Bose O, Sen D, Sheppard D, Locksley R, et al. Spatiotemporally separated antigen uptake by alveolar dendritic cells and airway presentation to T cells in the lung. *J Exp Med* (2012) 209(6):1183–99. doi:10.1084/jem.20112667

106. Choi J-H, Do Y, Cheong C, Koh H, Boscardin SB, Oh Y-S, et al. Identification of antigen-presenting dendritic cells in mouse aorta and cardiac valves. *J Exp Med* (2009) 206(3):497–505. doi:10.1084/jem.20082129

107. Khanna KM, Blair DA, Vella AT, McSorley SJ, Datta SK, Lefrançois L. T cell and APC dynamics in situ control the outcome of vaccination. *J Immunol* (2010) 185(1):239–52. doi:10.4049/jimmunol.0901047

108. Bennett CL, Clausen BE. DC ablation in mice: promises, pitfalls, and challenges. *Trends Immunol* (2007) 28(12):525–31. doi:10.1016/j.it.2007.08.011

109. Jung S, Unutmaz D, Wong P, Sano G-I, de los Santos K, Sparwasser T, et al. In vivo depletion of CD11c+ dendritic cells abrogates priming of CD8+ T cells by exogenous cell-associated antigens. *Immunity* (2002) 17(2):211–20. doi:10.1016/S1074-7613(02)00365-5

110. Tittel AP, Heuser C, Ohliger C, Knolle PA, Engel DR, Kurts C. Kidney dendritic cells induce innate immunity against bacterial pyelonephritis. *J Am Soc Nephrol* (2011) 22(8):1435–44. doi:10.1681/ASN.2010101072

111. Bar-On L, Birnberg T, Kim K-W, Jung S. Dendritic cell-restricted CD80/86 deficiency results in peripheral regulatory T-cell reduction but is not associated

with lymphocyte hyperactivation. *Eur J Immunol* (2011) **41**(2):291–8. doi:10. 1002/eji.201041169

112. van Blijswijk J, Schraml BU, Reis e Sousa C. Advantages and limitations of mouse models to deplete dendritic cells. *Eur J Immunol* (2013) **43**(1):22–6. doi:10.1002/eji.201243022

113. Racine R, Chatterjee M, Winslow GM. CD11c expression identifies a population of extrafollicular antigen-specific splenic plasmablasts responsible for CD4 T-independent antibody responses during intracellular bacterial infection. *J Immunol* (2008) **181**(2):1375–85. doi:10.4049/jimmunol.181.2. 1375

114. Brocker T, Riedinger M, Karjalainen K. Driving gene expression specifically in dendritic cells. *Adv Exp Med Biol* (1997) **417**:55–7. doi:10.1007/ 978-1-4757-9966-8_9

115. Tittel AP, Heuser C, Ohliger C, Llanto C, Yona S, Hämmerling GJ, et al. Functionally relevant neutrophilia in CD11c diphtheria toxin receptor transgenic mice. *Nat Methods* (2012) **9**(4):385–90. doi:10.1038/nmeth.1905

116. Bar-On L, Jung S. Defining in vivo dendritic cell functions using CD11c-DTR transgenic mice. *Methods Mol Biol* (2010) **595**:429–42. doi:10.1007/ 978-1-60761-421-0_28

117. Jiao J, Dragomir A-C, Kocabayoglu P, Rahman AH, Chow A, Hashimoto D, et al. Central role of conventional dendritic cells in regulation of bone marrow release and survival of neutrophils. *J Immunol* (2014) **192**(7):3374–82. doi:10.4049/jimmunol.1300237

118. Gardner JM, Metzger TC, McMahon EJ, Au-Yeung BB, Krawisz AK, Lu W, et al. Extrathymic aire-expressing cells are a distinct bone marrow-derived population that induce functional inactivation of CD4+ T Cells. *Immunity* (2013) **39**(3):560–72. doi:10.1016/j.immuni.2013.08.005

119. Epelman S, Lavine KJ, Beaudin AE, Sojka DK, Carrero JA, Calderon B, et al. Embryonic and adult-derived resident cardiac macrophages are maintained through distinct mechanisms at steady state and during inflammation. *Immunity* (2014) **40**(1):91–104. doi:10.1016/j.immuni.2013.11. 019

120. Broz ML, Binnewies M, Boldajipour B, Nelson AE, Pollack JL, Erle DJ, et al. Dissecting the tumor myeloid compartment reveals rare activating antigen-presenting cells critical for T cell immunity. *Cancer Cell* (2014) **26**(5):638–52. doi:10.1016/j.ccell.2014.09.007

121. Ferris ST, Carrero JA, Mohan JF, Calderon B, Murphy KM, Unanue ER. A minor subset of Batf3-dependent antigen-presenting cells in islets of langerhans is essential for the development of autoimmune diabetes. *Immunity* (2014) **41**(4):657–69. doi:10.1016/j.immuni.2014.09.012

122. Segura E, Touzot M, Bohineust A, Cappuccio A, Chiocchia G, Hosmalin A, et al. Human inflammatory dendritic cells induce Th17 cell differentiation. *Immunity* (2013) **38**(2):336–48. doi:10.1016/j.immuni.2012.10.018

123. Meredith MM, Liu K, Kamphorst AO, Idoyaga J, Yamane A, Guermonprez P, et al. Zinc finger transcription factor zDC is a negative regulator required to prevent activation of classical dendritic cells in the steady state. *J Exp Med* (2012) **209**(9):1583–93. doi:10.1084/jem.20121003

124. Hildner K, Edelson BT, Purtha WE, Diamond M, Matsushita H, Kohyama M, et al. Batf3 deficiency reveals a critical role for CD8alpha+ dendritic cells in cytotoxic T cell immunity. *Science* (2008) **322**(5904):1097–100. doi:10.1126/ science.1164206

125. Edelson BT, Kc W, Juang R, Kohyama M, Benoit LA, Klekotka PA, et al. Peripheral CD103+ dendritic cells form a unified subset developmentally related to CD8alpha+ conventional dendritic cells. *J Exp Med* (2010) **207**(4):823–36. doi:10.1084/jem.20091627

126. Schiavoni G, Mattei F, Sestili P, Borghi P, Venditti M, Morse HC, et al. ICSBP is essential for the development of mouse type I interferon-producing cells and for the generation and activation of CD8alpha(+) dendritic cells. *J Exp Med* (2002) **196**(11):1415–25. doi:10.1084/jem.20021263

127. Hacker C, Kirsch RD, Ju X-S, Hieronymus T, Gust TC, Kuhl C, et al. Transcriptional profiling identifies Id2 function in dendritic cell development. *Nat Immunol* (2003) **4**(4):380–6. doi:10.1038/ni903

128. Kashiwada M, Pham NL, Pewe LL, Harty JT, Rothman PB. NFIL3/E4BP4 is a key transcription factor for CD8alpha(+) dendritic cell development. *Blood* (2011) **117**(23):6193–7. doi:10.1182/blood-2010-07-295873

129. Suzuki S, Honma K, Matsuyama T, Suzuki K, Toriyama K, Akitoyo I, et al. Critical roles of interferon regulatory factor 4 in CD11bhighCD8alpha- dendritic cell development. *Proc Natl Acad Sci U S A* (2004) **101**(24):8981–6. doi:10.1073/pnas.0402139101

130. Caton ML, Smith-Raska MR, Reizis B. Notch-RBP-J signaling controls the homeostasis of CD8- dendritic cells in the spleen. *J Exp Med* (2007) **204**(7):1653–64. doi:10.1084/jem.20062648

131. Lewis KL, Caton ML, Bogunovic M, Greter M, Grajkowska LT, Ng D, et al. Notch2 receptor signaling controls functional differentiation of dendritic cells in the spleen and intestine. *Immunity* (2011) **35**(5):780–91. doi:10.1016/j. immuni.2011.08.013

132. Schlitzer A, McGovern N, Teo P, Zelante T, Atarashi K, Low D, et al. IRF4 transcription factor-dependent CD11b+ dendritic cells in human and mouse control mucosal IL-17 cytokine responses. *Immunity* (2013) **38**(5):970–83. doi:10.1016/j.immuni.2013.04.011

133. Persson EK, Uronen-Hansson H, Semmrich M, Rivollier A, Hägerbrand K, Marsal J, et al. IRF4 transcription-factor-dependent CD103(+)CD11b(+) dendritic cells drive mucosal T helper 17 cell differentiation. *Immunity* (2013) **38**(5):958–69. doi:10.1016/j.immuni.2013.03.009

134. Tamura T, Tailor P, Yamaoka K, Kong HJ, Tsujimura H, O'shea JJ, et al. IFN regulatory factor-4 and -8 govern dendritic cell subset development and their functional diversity. *J Immunol* (2005) **174**(5):2573–81. doi:10.4049/ jimmunol.174.5.2573

135. Wu L, D'Amico A, Winkel KD, Suter M, Lo D, Shortman K. RelB is essential for the development of myeloid-related CD8alpha- dendritic cells but not of lymphoid-related CD8alpha+ dendritic cells. *Immunity* (1998) **9**(6):839–47. doi:10.1016/S1074-7613(00)80649-4

136. Guerriero A, Langmuir PB, Spain LM, Scott EW. PU.1 is required for myeloid-derived but not lymphoid-derived dendritic cells. *Blood* (2000) **95**(3):879–85.

137. McKenna HJ, Stocking KL, Miller RE, Brasel K, De Smedt T, Maraskovsky E, et al. Mice lacking flt3 ligand have deficient hematopoiesis affecting hematopoietic progenitor cells, dendritic cells, and natural killer cells. *Blood* (2000) **95**(11):3489–97.

138. Kohyama M, Ise W, Edelson BT, Wilker PR, Hildner K, Mejia C, et al. Role for Spi-C in the development of red pulp macrophages and splenic iron homeostasis. *Nature* (2009) **457**(7227):318–21. doi:10.1038/nature07472

139. Wang Y, Szretter KJ, Vermi W, Gilfillan S, Rossini C, Cella M, et al. IL-34 is a tissue-restricted ligand of CSF1R required for the development of Langerhans cells and microglia. *Nat Immunol* (2012) **13**(8):753–60. doi:10.1038/ni.2360

140. Greter M, Lelios I, Pelczar P, Hoeffel G, Price J, Leboeuf M, et al. Stroma-derived interleukin-34 controls the development and maintenance of langerhans cells and the maintenance of microglia. *Immunity* (2012) **37**(6):1050–60. doi:10.1016/j.immuni.2012.11.001

141. Wiktor-Jedrzejczak WW, Ahmed A, Szczylik C, Skelly RR. Hematological characterization of congenital osteopetrosis in op/op mouse. Possible mechanism for abnormal macrophage differentiation. *J Exp Med* (1982) **156**(5):1516–27. doi:10.1084/jem.156.5.1516

142. Ghosh HS, Cisse B, Bunin A, Lewis KL, Reizis B. Continuous expression of the transcription factor e2-2 maintains the cell fate of mature plasmacytoid dendritic cells. *Immunity* (2010) **33**(6):905–16. doi:10.1016/j.immuni.2010.11. 023

143. Helft J, Manicassamy B, Guermonprez P, Hashimoto D, Silvin A, Agudo J, et al. Cross-presenting CD103+ dendritic cells are protected from influenza virus infection. *J Clin Invest* (2012) **122**(11):4037–47. doi:10.1172/JCI60659

144. Desch AN, Randolph GJ, Murphy K, Gautier EL, Kedl RM, Lahoud MH, et al. CD103+ pulmonary dendritic cells preferentially acquire and present apoptotic cell-associated antigen. *J Exp Med* (2011) **208**(9):1789–97. doi:10. 1084/jem.20110538

145. Bedoui S, Whitney PG, Waithman J, Eidsmo L, Wakim L, Caminschi I, et al. Cross-presentation of viral and self antigens by skin-derived CD103+ dendritic cells. *Nat Immunol* (2009) **10**(5):488–95. doi:10.1038/ni.1724

146. Murphy KM. Transcriptional control of dendritic cell development. *Adv Immunol* (2013) **120**:239–67. doi:10.1016/B978-0-12-417028-5.00009-0

147. Edelson BT, Bradstreet TR, Kc W, Hildner K, Herzog JW, Sim J, et al. Batf3-dependent CD11b(low/-) peripheral dendritic cells are GM-CSF-independent and are not required for Th cell priming after subcutaneous immunization. *PLoS One* (2011) **6**(10):e25660. doi:10.1371/journal.pone.0025660

148. Tussiwand R, Lee W-L, Murphy TL, Mashayekhi M, Wumesh KC, Albring JC, et al. Compensatory dendritic cell development mediated by BATF-IRF interactions. *Nature* (2012) **490**(7421):502–7. doi:10.1038/nature11531

149. Seillet C, Jackson JT, Markey KA, Brady HJM, Hill GR, MacDonald KPA, et al. CD8α+ DCs can be induced in the absence of transcription factors Id2, Nfil3, and Batf3. *Blood* (2013) **121**(9):1574–83. doi:10.1182/blood-2012-07-445650

150. Contreras V, Urien C, Guiton R, Alexandre Y, Vu Manh T-P, Andrieu T, et al. Existence of CD8α-like dendritic cells with a conserved functional specialization and a common molecular signature in distant mammalian species. *J Immunol* (2010) **185**(6):3313–25. doi:10.4049/jimmunol.1000824

151. Marquet F, Bonneau M, Pascale F, Urien C, Kang C, Schwartz-Cornil I, et al. Characterization of dendritic cells subpopulations in skin and afferent lymph in the swine model. *PLoS One* (2011) **6**(1):e16320. doi:10.1371/journal.pone.0016320

152. Vu Manh TP, Marty H, Sibille P, Le Vern Y, Kaspers B, Dalod M, et al. Existence of conventional dendritic cells in Gallus gallus revealed by comparative gene expression profiling. *J Immunol* (2014) **192**(10):4510–7. doi:10.4049/jimmunol.1303405

153. Hubert FX, Voisine C, Louvet C, Heslan M, Josien R. Rat plasmacytoid dendritic cells are an abundant subset of MHC class II+ CD4+CD11b-OX62- and type I IFN-producing cells that exhibit selective expression of toll-like receptors 7 and 9 and strong responsiveness to CpG. *J Immunol* (2004) **172**(12):7485–94. doi:10.4049/jimmunol.172.12.7485

154. Sathaliyawala T, O'Gorman WE, Greter M, Bogunovic M, Konjufca V, Hou ZE, et al. Mammalian target of rapamycin controls dendritic cell development downstream of Flt3 ligand signaling. *Immunity* (2010) **33**(4):597–606. doi:10.1016/j.immuni.2010.09.012

155. Brasel K, McKenna HJ, Morrissey PJ, Charrier K, Morris AE, Lee CC, et al. Hematologic effects of flt3 ligand in vivo in mice. *Blood* (1996) **88**(6):2004–12.

156. Maraskovsky E, Brasel K, Teepe M, Roux ER, Lyman SD, Shortman K, et al. Dramatic increase in the numbers of functionally mature dendritic cells in Flt3 ligand-treated mice: multiple dendritic cell subpopulations identified. *J Exp Med* (1996) **184**(5):1953–62. doi:10.1084/jem.184.5.1953

157. Pulendran B, Banchereau J, Burkeholder S, Kraus E, Guinet E, Chalouni C, et al. Flt3-ligand and granulocyte colony-stimulating factor mobilize distinct human dendritic cell subsets in vivo. *J Immunol* (2000) **165**(1):566–72. doi:10.4049/jimmunol.165.1.566

158. Naik SH, Proietto AI, Wilson NS, Dakic A, Schnorrer P, Fuchsberger M, et al. Cutting edge: generation of splenic CD8+ and CD8- dendritic cell equivalents in Fms-like tyrosine kinase 3 ligand bone marrow cultures. *J Immunol* (2005) **174**(11):6592–7. doi:10.4049/jimmunol.174.11.6592

159. Poulin LF, Salio M, Griessinger E, Anjos-Afonso F, Craciun L, Chen J-L, et al. Characterization of human DNGR-1+ BDCA3+ leukocytes as putative equivalents of mouse CD8alpha+ dendritic cells. *J Exp Med* (2010) **207**(6):1261–71. doi:10.1084/jem.20092618

160. Tussiwand R, Onai N, Mazzucchelli L, Manz MG. Inhibition of natural type I IFN-producing and dendritic cell development by a small molecule receptor tyrosine kinase inhibitor with Flt3 affinity. *J Immunol* (2005) **175**(6):3674–80. doi:10.4049/jimmunol.175.6.3674

161. Whartenby KA, Calabresi PA, McCadden E, Nguyen B, Kardian D, Wang T, et al. Inhibition of FLT3 signaling targets DCs to ameliorate autoimmune disease. *Proc Natl Acad Sci U S A* (2005) **102**(46):16741–6. doi:10.1073/pnas.0506088102

162. Jenkins SJ, Hume DA. Homeostasis in the mononuclear phagocyte system. *Trends Immunol* (2014) **35**(8):358–67. doi:10.1016/j.it.2014.06.006

163. Mackarehtschian K, Hardin JD, Moore KA, Boast S, Goff SP, Lemischka IR. Targeted disruption of the flk2/flt3 gene leads to deficiencies in primitive hematopoietic progenitors. *Immunity* (1995) **3**(1):147–61. doi:10.1016/1074-7613(95)90167-1

164. Kingston D, Schmid MA, Onai N, Obata-Onai A, Baumjohann D, Manz MG. The concerted action of GM-CSF and Flt3-ligand on in vivo dendritic cell homeostasis. *Blood* (2009) **114**(4):835–43. doi:10.1182/blood-2009-02-206318

165. Kim S-W, Choi S-M, Choo YS, Kim I-K, Song B-W, Kim H-S. Flt3 ligand induces monocyte proliferation and enhances the function of monocyte-derived dendritic cells in vitro. *J Cell Physiol* (2014). doi:10.1002/jcp.24824

166. Xu Y, Zhan Y, Lew AM, Naik SH, Kershaw MH. Differential development of murine dendritic cells by GM-CSF versus Flt3 ligand has implications for inflammation and trafficking. *J Immunol* (2007) **179**(11):7577–84. doi:10.4049/jimmunol.179.11.7577

167. Miller JC, Brown BD, Shay T, Gautier EL, Jojic V, Cohain A, et al. Deciphering the transcriptional network of the dendritic cell lineage. *Nat Immunol* (2012) **13**(9):888–99. doi:10.1038/ni.2370

168. MacDonald KPA, Rowe V, Bofinger HM, Thomas R, Sasmono T, Hume DA, et al. The colony-stimulating factor 1 receptor is expressed on dendritic cells during differentiation and regulates their expansion. *J Immunol* (2005) **175**(3):1399–405. doi:10.4049/jimmunol.175.3.1399

169. Dai XM, Ryan GR, Hapel AJ, Dominguez MG, Russell RG, Kapp S, et al. Targeted disruption of the mouse colony-stimulating factor 1 receptor gene results in osteopetrosis, mononuclear phagocyte deficiency, increased primitive progenitor cell frequencies, and reproductive defects. *Blood* (2002) **99**(1):111–20. doi:10.1182/blood.V99.1.111

170. Fancke B, Suter M, Hochrein H, O'Keeffe M. M-CSF: a novel plasmacytoid and conventional dendritic cell poietin. *Blood* (2008) **111**(1):150–9. doi:10.1182/blood-2007-05-089292

171. Tagliani E, Shi C, Nancy P, Tay C-S, Pamer EG, Erlebacher A. Coordinate regulation of tissue macrophage and dendritic cell population dynamics by CSF-1. *J Exp Med* (2011) **208**(9):1901–16. doi:10.1084/jem.20110866

172. Schreurs MW, Eggert AA, de Boer AJ, Figdor CG, Adema GJ. Generation and functional characterization of mouse monocyte-derived dendritic cells. *Eur J Immunol* (1999) **29**(9):2835–41. doi:10.1002/(SICI)1521-4141(199909)29:09<2835::AID-IMMU2835>3.3.CO;2-H

173. Vremec D, Lieschke GJ, Dunn AR, Robb L, Metcalf D, Shortman K. The influence of granulocyte/macrophage colony-stimulating factor on dendritic cell levels in mouse lymphoid organs. *Eur J Immunol* (1997) **27**(1):40–4. doi:10.1002/eji.1830270107

174. Zhan Y, Carrington EM, Van Nieuwenhuijze A, Bedoui S, Seah S, Xu Y, et al. GM-CSF increases cross-presentation and CD103 expression by mouse CD8⁺ spleen dendritic cells. *Eur J Immunol* (2011) **41**(9):2585–95. doi:10.1002/eji.201141540

175. Sathe P, Pooley J, Vremec D, Mintern J, Jin J-O, Wu L, et al. The acquisition of antigen cross-presentation function by newly formed dendritic cells. *J Immunol* (2011) **186**(9):5184–92. doi:10.4049/jimmunol.1002683

176. Luc S, Buza-Vidas N, Jacobsen SE. Biological and molecular evidence for existence of lymphoid-primed multipotent progenitors. *Ann N Y Acad Sci* (2007) **1106**:89–94. doi:10.1196/annals.1392.023

177. Verovskaya E, Broekhuis MJ, Zwart E, Ritsema M, van Os R, de Haan G, et al. Heterogeneity of young and aged murine hematopoietic stem cells revealed by quantitative clonal analysis using cellular barcoding. *Blood* (2013) **122**(4):523–32. doi:10.1182/blood-2013-01-481135

178. Paul F, Amit I. Plasticity in the transcriptional and epigenetic circuits regulating dendritic cell lineage specification and function. *Curr Opin Immunol* (2014) **30C**:1–8. doi:10.1016/j.coi.2014.04.004

179. Vorhagen S, Jackow J, Mohor SG, Tanghe G, Tanrikulu L, Skazik-Vogt C, et al. Lineage tracing mediated by cre-recombinase activity. *J Invest Dermatol* (2015) **135**(1):e28. doi:10.1038/jid.2014.472

180. Srinivas S, Watanabe T, Lin CS, William CM, Tanabe Y, Jessell TM, et al. Cre reporter strains produced by targeted insertion of EYFP and ECFP into the ROSA26 locus. *BMC Dev Biol* (2001) **1**:4. doi:10.1186/1471-213X-1-4

181. Caminschi I, Proietto AI, Ahmet F, Kitsoulis S, Shin Teh J, Lo JCY, et al. The dendritic cell subtype-restricted C-type lectin Clec9A is a target for vaccine enhancement. *Blood* (2008) **112**(8):3264–73. doi:10.1182/blood-2008-05-155176

182. Sancho D, Mourão-Sá D, Joffre OP, Schulz O, Rogers NC, Pennington DJ, et al. Tumor therapy in mice via antigen targeting to a novel, DC-restricted C-type lectin. *J Clin Invest* (2008) **118**(6):2098–110. doi:10.1172/JCI34584

183. Liu J, Willet SG, Bankaitis ED, Xu Y, Wright CVE, Gu G. Non-parallel recombination limits cre-loxP-based reporters as precise indicators of conditional genetic manipulation. *Genesis* (2013) **51**(6):436–42. doi:10.1002/dvg.22384

184. Nelson PJ, Rees AJ, Griffin MD, Hughes J, Kurts C, Duffield J. The renal mononuclear phagocytic system. *J Am Soc Nephrol* (2012) **23**(2):194–203. doi:10.1681/ASN.2011070680

185. Dong X, Swaminathan S, Bachman LA, Croatt AJ, Nath KA, Griffin MD. Antigen presentation by dendritic cells in renal lymph nodes is linked to systemic and local injury to the kidney. *Kidney Int* (2005) **68**(3):1096–108. doi:10.1111/j.1523-1755.2005.00502.x

186. Hirrlinger J, Requardt RP, Winkler U, Wilhelm F, Schulze C, Hirrlinger PG. Split-CreERT2: temporal control of DNA recombination mediated by split-Cre protein fragment complementation. *PLoS One* (2009) **4**(12):e8354. doi:10.1371/journal.pone.0008354

187. Schreiber HA, Loschko J, Karssemeijer RA, Escolano A, Meredith MM, Mucida D, et al. Intestinal monocytes and macrophages are required for T cell polarization in response to *Citrobacter rodentium*. *J Exp Med* (2013) **210**(10):2025–39. doi:10.1084/jem.20130903

188. Diehl GE, Longman RS, Zhang J-X, Breart B, Galan C, Cuesta A, et al. Microbiota restricts trafficking of bacteria to mesenteric lymph nodes by CX(3)CR1(hi) cells. *Nature* (2013) **494**(7435):116–20. doi:10.1038/nature11809

189. Ohnmacht C, Pullner A, King SBS, Drexler I, Meier S, Brocker T, et al. Constitutive ablation of dendritic cells breaks self-tolerance of CD4 T cells and results in spontaneous fatal autoimmunity. *J Exp Med* (2009) **206**(3):549–59. doi:10.1084/jem.20082394

190. Voehringer D, Liang H-E, Locksley RM. Homeostasis and effector function of lymphopenia-induced "memory-like" T cells in constitutively T cell-depleted mice. *J Immunol* (2008) **180**(7):4742–53. doi:10.4049/jimmunol.180.7.4742

191. Gautier EL, Shay T, Miller J, Greter M, Jakubzick C, Ivanov S, et al. Gene-expression profiles and transcriptional regulatory pathways that underlie the identity and diversity of mouse tissue macrophages. *Nat Immunol* (2012) **13**(11):1118–28. doi:10.1038/ni.2419

192. Lavin Y, Winter D, Blecher-Gonen R, David E, Keren-Shaul H, Merad M, et al. Tissue-resident macrophage enhancer landscapes are shaped by the local microenvironment. *Cell* (2014) **159**(6):1312–26. doi:10.1016/j.cell.2014.11.018

193. Gosselin D, Link VM, Romanoski CE, Fonseca GJ, Eichenfield DZ, Spann NJ, et al. Environment drives selection and function of enhancers controlling tissue-specific macrophage identities. *Cell* (2014) **159**(6):1327–40. doi:10.1016/j.cell.2014.11.023

13

Functions of Vγ4 T cells and dendritic epidermal T cells on skin wound healing

Yashu Li[1], Jun Wu[1,2,3], Gaoxing Luo[1,2]* and Weifeng He[1,2]*

[1] State Key Laboratory of Trauma, Burn and Combined Injury, Institute of Burn Research, Southwest Hospital, Third Military Medical University (Army Medical University), Chongqing, China, [2] Chongqing Key Laboratory for Disease Proteomics, Chongqing, China, [3] Department of Burns, The First Affiliated Hospital, Sun Yat-Sen University, Guangzhou, China

Edited by:
Kenth Gustafsson,
University College London,
United Kingdom

Reviewed by:
Tomasz Zal,
University of Texas MD Anderson
Cancer Center, United States
Xing Chang,
Shanghai Institutes for Biological
Sciences (CAS), China

*Correspondence:
Gaoxing Luo
logxw@yahoo.com;
Weifeng He
whe761211@hotmail.com

Wound healing is a complex and dynamic process that progresses through the distinct phases of hemostasis, inflammation, proliferation, and remodeling. Both inflammation and re-epithelialization, in which skin γδ T cells are heavily involved, are required for efficient skin wound healing. Dendritic epidermal T cells (DETCs), which reside in murine epidermis, are activated to secrete epidermal cell growth factors, such as IGF-1 and KGF-1/2, to promote re-epithelialization after skin injury. Epidermal IL-15 is not only required for DETC homeostasis in the intact epidermis but it also facilitates the activation and IGF-1 production of DETC after skin injury. Further, the epidermal expression of IL-15 and IGF-1 constitutes a feedback regulatory loop to promote wound repair. Dermis-resident Vγ4 T cells infiltrate into the epidermis at the wound edges through the CCR6-CCL20 pathway after skin injury and provide a major source of IL-17A, which enhances the production of IL-1β and IL-23 in the epidermis to form a positive feedback loop for the initiation and amplification of local inflammation at the early stages of wound healing. IL-1β and IL-23 suppress the production of IGF-1 by DETCs and, therefore, impede wound healing. A functional loop may exist among Vγ4 T cells, epidermal cells, and DETCs to regulate wound repair.

Keywords: wound healing, dendritic epidermal T cell, Vgamma 4 T cell, IL-17A, IGF-1, re-epithelialization

SKIN γδ T CELLS ARE HEAVILY INVOLVED IN THE WOUND HEALING PROCESS

Wound healing is a complicated repair process to recover the integrity of skin. This process is orchestrated by four overlapping phases, which are clotting, inflammation, re-epithelialization, and remolding (1). Murine γδ T cells as important components of skin immunity engage in inflammation and re-epithelialization in wound repair (2–4). Several subsets of γδ T cells with distinct functions exist in skin tissue: dendritic epidermal T cells (DETCs), which uniformly express an invariant Vγ5Vδ1 TCR (according to Heilig and Tonegawa's nomenclature) and exclusively reside in the murine epidermis (>90%), primarily provide IGF-1 and KGF-1/2 in the epidermis to enhance re-epithelialization and thereby promote skin wound repair. Vγ4 T cells, a dominant subset of murine peripheral and dermal γδ T cells (approximately 50%), provide an early major source of IL-17A to initiate and amplify local inflammation after skin damage (5, 6). Interestingly, although inflammation is required for efficient skin wound healing, excessive inflammation has a negative impact on skin wound repair. In line with this notion, IL-17A, a potent pro-inflammation factor, exhibits dual roles in skin wound closure (7–9). It has been reported that dermal Vγ4 T cells infiltrate the epidermis and interact with epidermal cells to form an IL-17A-IL-1/23 positive feedback loop for amplifying local inflammation. Li et al. recently revealed that Vγ4 T cells suppress IGF-1 production by DETCs through the IL-17A-IL-1/23 loop and thus delay skin wound healing (10). This suggests that a potential functional link exists between Vγ4 T cells, epidermal cells, and DETCs in the wound-healing process. This review focuses on the functional diversity of skin-resident γδ T cell subsets in wound repair.

THE DEVELOPMENT OF Vγ5 T CELLS IN THE THYMUS

Vγ5 T cells are the first generated γδ T cells on embryonic day (ED) 13 in the early fetal thymus, but they are no longer produced after ED 18 (11, 12). Their γ and δ chains are identically rearranged to Vγ5-Jγ1Cγ1 and Vδ1-Dδ2-Jδ2Cδ, respectively, with invariant canonical junctional sequences (13). IL-7 signaling is required for the rearrangement of TCRγ but not TCRαβ (14). IL-7 and IL-7 receptors are responsible for the recombination of the TCRγ locus by regulating locus accessibility to the V(D)J recombinase (15). Positive selection is necessary for the maturation of Vγ5 T cells in the fetal thymus, which depends on the engagement of TCR and some ligands expressed by thymic stromal cells (16). Skint-1 is essential for the positive selection of Vγ5 T cells in the murine fetal thymus (17). CD122 (the β chain of IL-2/IL-15 receptor, IL-2/IL-15Rβ) and the skin-homing receptors are induced on Vγ5 T cells after positive selection in the fetal thymus, and are crucial for Vγ5 T cells to migrate into the epidermis (16). Vγ5 T cells gain a "memory-like" pre-activation phenotype of CD44+CD122+CD25− before exiting the thymus (11).

THE MIGRATION AND RESIDENCY OF Vγ5 T CELLS IN THE EPIDERMIS

During ED 15.5–16.5, Vγ5 T cells egress from the thymus and move to the epidermal layer of skin (18). The expression of CCR6 is reduced, whereas the expression of sphingosine-1-phosphate receptor 1 (S1P1) is increased on mature Vγ5 T cells, both of which allow mature Vγ5 T cells to exit but retrain immature cells in the thymus (16). Furthermore, the expressions of skin-homing molecules CCR10, CCR4, E, and P selectin ligands, and integrin αE are also markedly increased on the surface of Vγ5 T cells to help them migrate and reside in the epidermis (16, 19, 20). CCR9, CCR7, and CD62L have low expression on Vγ5 T cells, indicating that Vγ5 T cells are not able to migrate into secondary lymphoid organs (21).

Vγ5 T cells exclusively reside in the murine epidermis and comprise over 90% of murine epidermal T lymphocytes (22). Since the characteristic feature of epidermal Vγ5 T cells is their highly dynamic dendritic morphology, they are named DETCs. DETCs are anchored in the upper epidermis, and most dendrites are immobilized and apical toward keratinocyte tight junctions, while the remaining dendrites are projected to the basal

epidermis and extend and contract in a highly mobile state (23). Keratinocytes predominantly express E-cadherin and DETCs express E-cadherin receptor integrin αEβ7, especially at the ends of apical dendrites, which assist in anchoring the dendrites of DETCs in the epidermis (23).

HOMEOSTASIS OF DETCs IN THE EPIDERMIS

Dendritic epidermal T cells slowly expand under a steady state to maintain skin homeostasis (23). Vγ5 TCR and Skint-1 signaling are required for DETC homeostasis in the epidermis (17). Several secreted factors also contribute to the maintenance of DETCs. For example, CD122 expressed on DETCs is essential for their proliferation and survival in both fetal thymus and skin (24). IL-15Rα (CD215) is highly expressed on the surface of DETCs (11). IL-15 helps the survival and proliferation of DETCs upon TCR engagement (25). Furthermore, IL-15 can interact with CD122 of DETCs to maintain their localization and homeostasis in the epidermis (25). Importantly, DETCs secrete a small amount of IGF-1 in a steady state to sustain survival and prevent apoptosis of keratinocytes to maintain epidermis homeostasis (4). Liu et al. and Bai et al. recently reported that DETC-derived IGF-1 is positively correlated with keratinocyte-derived IL-15, which is partially controlled by the mTOR signaling pathway (26, 27). In addition, epidermal IGF-1 and IL-15 cooperate to promote the homeostasis of DETCs in diabetic animals (28). CD122 and CD69 are regarded as activation markers on DETCs (23). Vγ5 TCR signals preserve the expression of CD69 and CD122, which help DETCs stay in a state of pre-activation (23).

γδ T CELLS IN THE HUMAN EPIDERMIS

Murine DETCs lack an exact counterpart in humans. γδ T cells have a TCR that expresses the Vδ1 chain reside in both the epidermis and dermis of human skin (29). γδ T cells that exist in peripheral blood express the Vδ2 TCR (30). Cutaneous leukocyte antigen (CLA), which is the ligand for E-selectin and a skin-homing marker, is also expressed on epidermal- and dermal-resident Vδ1 T cells and αβ T cells, while CLA expression on Vδ2 T cells from the blood is low (29). No significant or distinct differences in CLA expression on Vδ1 T cells or αβ T cells exist between the epidermis and dermis. Vδ1 TCR comprise about 10–20% of T cells in the human epidermis and dermis, respectively, while the ratio of Vδ1 T/αβ T cells is less in the epidermis than in the dermis. Epidermal-resident Vδ1 T cells and αβ T cells are activated after acute injury and produce IGF-1 to promote wound repair. However, both Vδ1 T cells and αβ T cells separated from chronic non-healing wounds do not secrete IGF-1, indicating that their function is impaired in chronic wounds compared with T cells isolated from acute wounds (29). Moreover, human blood Vδ2 T cells can be recruited to the skin inflamed with psoriasis. These Vδ2T cells are CLA- and CCR6-positive, secrete proinflammatory cytokines, such as IL-17A, TNF-α, and IFN-γ, and produce psoriasis chemokines, such as IL-8, CCL3, CCL4, and CCL5 (31).

CLA$^+$Vδ2 T cells not only upregulate the production of IGF-1, but also activate keratinocytes dependent on TNF-α and IFN-γ (31). Skint-1 is absent in humans, which may partially explain why the development of TCR chains is different between humans and mice (17).

THE ACTIVATION OF DETCs UPON SKIN INJURY

The activation of DETCs around wound edges is necessary for their proliferation and the secretion of epidermal growth factor during wound healing (3, 4, 32). Once keratinocytes get stressed or damaged, the morphology of DETCs changes from dendritic to round at the wound edge 1 h later (3). TCR complexes locate at the apical dendrite ends under a steady state, but migrate to the basal epidermis upon wounding (23). Functional changes follow the morphological changes, and the expression of IGF-1 is markedly increased in DETCs (4). Differing from αβ T cells, γδ T cells are directly activated by TCR signaling in a non-major histocompatibility complex (MHC)-restricted manner (33). TCRs on DETCs sense some unidentified ligands expressed on damaged keratinocytes after skin injury (34). Apart from TCR signaling, other co-stimulatory molecules [such as NKG2D, junctional adhesion molecule-like protein (JAML), and 2B4] and T cell growth factors have been recently demonstrated to contribute to the activation of DETCs (35–38).

NKG2D

NKG2D is a C-type lectin-like stimulatory receptor, expressed on activated CD8$^+$T cells, macrophages, NK1.1$^+$T cells, and DETCs (39). NKG2D has two alternative splicing isoforms: NKG2D-S (short) and NKG2D-L (long) (40). DETCs constitutively express NKG2D-S, NKG2D-L, and cell surface protein NKG2D (40). NKG2D signals act as co-stimulatory signals for CD8$^+$T cells, or directly trigger cytotoxicity and cytokine production in activated murine NK cells (39). Without TCR engagement, NKG2D signals are sufficient to trigger cytotoxicity and IFN-γ production in DETCs (40). NKG2D ligands, which belong to MHC-class-I related proteins, are expressed under stress conditions, such as infection, tumorigenesis, and tissue damage (41). Retinoic acid early inducible-1 α-ε, mouse UL16-binding protein-like transcript 1, and histocompatibility a-c (H60 a-c) are known NKG2D ligands in mice, and among them H60c has been detected in skin (42). H60c protein is inductively expressed on keratinocytes at wound margins (43). The interaction between NKG2D and H60c is necessary for KGF secretion by DETCs in wound repair (43).

JUNCTIONAL ADHESION MOLECULE-LIKE PROTEIN (JAML)

Junctional adhesion molecule-like protein is expressed on the surface of DETCs (44). Coxsackie and adenovirus receptor (CAR) is induced on damaged keratinocytes and acts as a functional ligand for JAML (44). The JAML–CAR interaction is necessary for the activation of DETCs and cytokine production of TNF-α and KGF-1 (44). Blocking the interaction between JAML and CAR impedes wound healing, suggesting that JAML–CAR provides a costimulatory signal for the activation of DETCs during wound repair (44).

2B4

2B4, a 66-kD glycoprotein, is expressed on NK and T cells and kills tumor targets by non-MHC-restricted mechanisms (45). 2B4 is also detected on DETCs and helps DETCs to mediate cytotoxic killing against skin-derived tumors (46). IL-2 upregulates 2B4 expression and enhances cytotoxic ability of DETCs (46). Whether 2B4 participates in wound healing has not been clarified.

TOLL-LIKE RECEPTOR (TLR) 4

Toll-like receptor 4 is the primary signaling receptor for lipopolysaccharide (LPS) (47). MD2 assists TLR4 for intracellular distribution and accelerates TLR4 for LPS recognition (48). In the steady state, the expression of TLR4-MD2 is lacking on the surface of DETCs, while during cutaneous inflammation, TLR4-MD2 expression is improved on DETCs when they migrate from the epidermis (47). The roles of TLR4-MD2 in wound healing are not known.

IL-15

IL-2 and IL-15 participate in the survival and activation of γδ T cells (25, 49, 50). Both interact with α, β, or γ$_c$ chains of the receptor complexes (50, 51). Although IL-15 is similar to IL-2 in its biological properties and three-dimensional configuration, IL-15 is more important than IL-2 for the survival and proliferation of DETCs upon TCR engagement, because IL-15 and not IL-2 is expressed in the epidermis under a steady state (25). In addition, mature fetal Vγ5 thymocytes and DETCs express the IL-15Rβ chain (CD122) but not IL-2Rα (CD25) (11). Compared to wild-type controls, the number of mature Vγ5 T cells is reduced in the fetal thymus and DETCs are absent in IL-15$^{-/-}$ mice, while the number of mature Vγ5 T cells is normal in the fetal thymus and DETCs survive in the adult skin of IL-2$^{-/-}$ mice (25). Therefore, IL-15 rather than IL-2 seems to be necessary for DETC homeostasis in the skin (25, 35). Moreover, activated DETCs secrete IL-15 but fail to produce IL-2, indicating that IL-15 is more important than IL-2 for the efficient activation of DETCs at the early stages of wound healing. Liu et al. and Wang et al. have demonstrated that IL-15 rescues the insufficient activation of DETCs and increases IGF-1 production by DETCs, and IGF-1 in turn induces keratinocytes to secrete IL-15 in diabetic mice (26, 28). Their work indicates that IL-15 and IGF-1 are positively correlated in the wounded epidermis to promote re-epithelialization (**Figure 1**). Furthermore, the regulation of the IGF-1-IL-15 loop partially depends on the mTOR pathway (26–28).

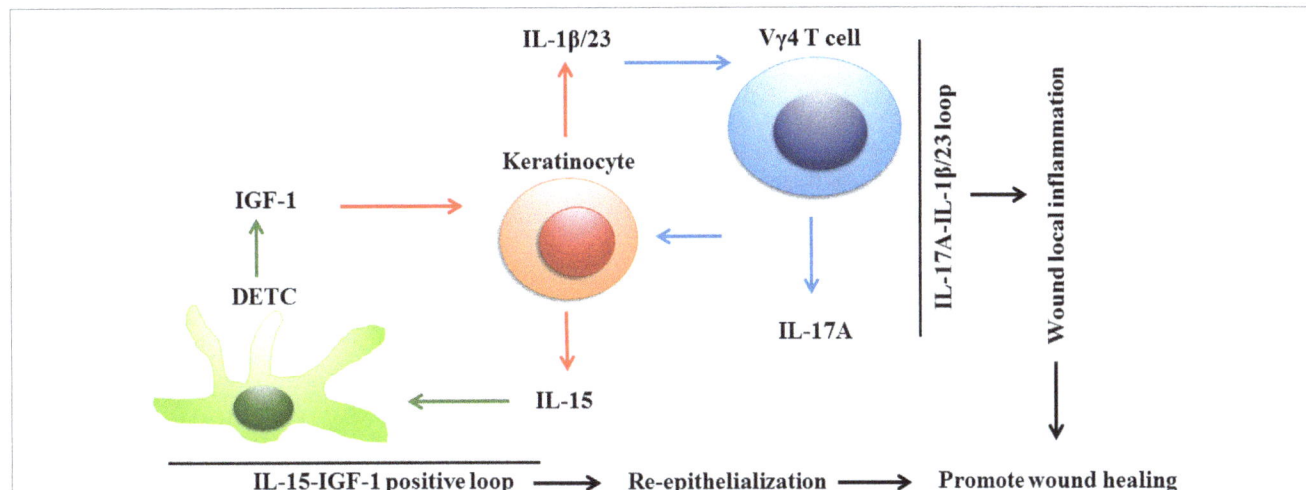

FIGURE 1 | Dendritic epidermal T cells (DETCs), keratinocytes, and Vγ4 T cells constitute two correlated loops to improve wound repair in diabetic mice. Upon skin injury in diabetic mice, keratinocyte-derived IL-15 increases IGF-1 production by DETCs, which in turn enhances keratinocytes to secrete IL-15. A positive correlation between IL-15 and IGF-1 is formed in wounded epidermis, and thereby amplifies IGF-1 production in epidermis for promoting re-epithelialization and wound healing. Meanwhile, keratinocytes could also interact with Vγ4 T cells, which infiltrate in epidermis to form an IL-17A-IL-1β/IL-23 feedback loop to augment local inflammation for efficient skin wound healing. In the wounds of diabetic mice, DETC-mediated IL-15-IGF-1 correlation and Vγ4 T cell-mediated IL-17A-IL-1β/IL-23 loop are coordinated to improve the defects of diabetic wound healing through enhancing re-epithelialization and local inflammation, respectively.

ACTIVATED DETCs PROMOTE RE-EPITHELIALIZATION BY PRODUCING IGF-1 AND KGF-1/2

IGF-1 is primarily produced in the liver, but it is also derived from DETCs in the epidermis. DETCs constitutively generate IGF-1, and keratinocytes express IGF-1R under normal conditions (4). IGF-1 combined with IGF-1R can trigger phosphoinositide 3-kinase and mitogen-activated protein kinase pathways to protect keratinocytes from apoptosis and differentiation (4, 52, 53). Beyond secreting a small amount of IGF-1 in the steady state, DETCs also express IGF-1R to maintain survival in the epidermis *via* an autocrine pathway (4). Phosphorylated IGF-1R is increased at wound margins 24 h after injury, and upregulated IGF-1 protects keratinocytes from apoptosis in damaged areas to assist re-epithelialization (4).

Dendritic epidermal T cells do not secrete KGFs (KGF-1 and KGF-2) in homeostasis conditions, but rapidly produce KGFs upon wounding (3). Keratinocytes constitutively express KGF receptor FGFR2-IIIb, and thus KGFs derived from DETCs can bind FGFR2-IIIb receptor to induce the proliferation and migration of keratinocytes during the re-epithelial phase of wound healing (3, 54). FGFR2-IIIb is not expressed on DETCs, showing that KGFs do not reversely regulate the effector functions of DETCs under stressed conditions (3). DETCs can also secrete TGF-β to aid tissue repair; release GM-CSF XCL1, CCL3, CCL4, CCL5, and hyaluronan to recruit leukocytes to wound sites; and produce IL-17, IFN-γ, and TNF-α to facilitate inflammation (55, 56).

THE DEVELOPMENT OF Vγ4 T CELLS IN THE THYMUS

Vγ4 TCR is rearranged in the late fetal thymus from ED 17 until birth and afterward (57, 58). Vγ4 T cells develop into two main subsets: IL-17A+Vγ4 T cells with the phenotype of CCR6+CD27−, and IFN-γ+Vγ4 T cells with CCR6−CD27+ (59). Certain embryonic thymus conditions are required for γδ T cells to acquire the capacity to produce IL-17A. IL-7 is necessary for the development of γδ T17 cells in the thymus, which can promote the accessibility of the TCR γ locus to V(D)J recombinase and regulate the differentiation of γδ T cells preferentially toward the CD27−IL-17A+ subset (15, 60). CCR6+CD27−γδ T17 cells express the subunit of IL-17A/F receptor IL-17RC, which is not detected on CCR6−CD27+γδ T cells (61). In the absence of IL-17A, CCR6+CD27−γδ T17 cells become overabundant in the thymus and secondary lymphoid organs, indicating that the development and homeostasis of γδ T17 cells is restricted by IL-17A in a negative feedback loop (61). Moreover, transcription factor Sox13 is required for the maturation of IL-17A+Vγ4 T cells in the neonatal thymus, and its mutation is able to protect mice from psoriasis-like dermatitis (62).

Vγ4 T CELLS ARE THE DOMINANT SUBSET OF MURINE DERMAL γδ T CELLS

When exiting the thymus, Vγ4 T cells have obtained stem cell-like properties of self-renewal and are radiation resistant (63). Vγ4 T cells are localized to the secondary lymphoid organs as the dominant subset of murine peripheral γδ T cells, and they are also distributed in the dermal layer of murine skin (63). Vγ4 T cells comprise nearly 50% of dermal γδ T cells, though Vγ1, Vγ5, Vγ6, and Vγ7 T cells also exist in the dermis (64). Vγ4 T cells, as the major γδ T cells in the dermis, are capable of secreting IL-17A and IFN-γ, which play distinctive roles in autoimmune diseases, graft rejection, antiviral immunity, and antitumor responses (6, 10, 33, 65).

Vγ4 T CELLS PROVIDE THE MAJOR SOURCE OF IL-17A AT THE EARLY STAGE OF SKIN INFLAMMATION

Vγ4 T cells have been reported to participate in autoimmune diseases and skin graft rejection at the early stages by producing IL-17A (10, 33, 62, 66). IFN-γ-positive Vγ4 T cells play a protective role in antitumor immunity, but they do not contribute in skin transplantation and wound healing (10, 33, 67). Which cytokine Vγ4 T cells secrete may depend on local circumstances. As it is well-known that Th17 cells are a major source of IL-17A in the adaptive immune response, Vγ4 T cells act as an innate source of IL-17A before Th17 cells play their roles (68). Vγ4 T cells have some features in common with Th17 cells, such as IL-23 receptor, CCR6, and RORγ (68). However, Vγ4 T cells have gained the potent ability to produce IL-17A and express dectin-1 and TLRs when they egress from the thymus and, therefore, they can directly interact with pathogens and secrete IL-17A as the first line of defense against bacterial pathogens (61, 68). Vγ4 T cells also produce IL-17A to induce psoriasis-like skin inflammation, and IL-17A-positive T cells expand promptly in draining lymph nodes when exposed to the inflammatory agent imiquimod (64, 69). Furthermore, we have reported recently that Vγ4 T cells provide a major source of IL-17A in the epidermis at the early stages of wounding. Approximately half of the epidermal IL-17A-positive cells are Vγ4 T cells after skin injury (67). IL-17A production in the epidermis is dramatically decreased after the depletion of Vγ4 T cells in wild-type mice, but it is significantly enhanced in *Tcrδ*^{−/−} animals by the addition of freshly isolated Vγ4 T cells onto wound beds (67). In addition, Vγ4 T cells also migrate to noninflamed skin and peripheral lymph nodes, and respond faster and stronger to a second imiquimod challenge (69). Expanded Vγ4 T cells in lymph nodes can infiltrate back into inflammatory skin *via* S1P1 with similar migratory mechanisms as conventional αβ T cells (70). Of note, we purchased anti-Vγ4 TCR (UC3-10A6) antibody (Ab) from BioXcell to deplete Vγ4 T cells according to previous research (71, 72). However, *in vivo* treatment with both GL3 and UC7-13D5 antibodies against TCR, as identified by Koenecke et al., caused TCR internalization instead of γδ T cell depletion (73). Therefore, we cannot exclude the possibility that Ab treatment cannot eliminate γδ T cells, but instead decreases TCR complexes on the cellular surface.

Vγ4 T CELL-DERIVED IL-17A AND EPIDERMAL IL-1β/IL-23 FORM A POSITIVE FEEDBACK LOOP TO AMPLIFY LOCAL INFLAMMATION AFTER SKIN INJURY

The IL-1β/IL-23-IL-17A axis is critical for the initiation and amplification of inflammatory responses (5, 6, 69, 74, 75). IL-17A has been demonstrated to act upstream to enhance epidermal IL-1β/IL-23 production in a skin graft transplantation model (10). Furthermore, IL-1β and IL-23 production in the epidermis of wound edges is weakened by a deficiency or blockage of IL-17A but enhanced by the addition of rIL-17A (67). Depletion of Vγ4 T cells reduces epidermal IL-1β/IL-23 production, but supplementing wild-type rather than *Il-17a*^{−/−} Vγ4 T cells onto wound beds promotes epidermal IL-1β/IL-23 production in *Tcrδ*^{−/−} mice (67). Therefore, we regard that IL-17A secreted by Vγ4 T cells and IL-1β/IL-23 derived from epidermal cells may form a positive feedback loop in the epidermis around wounds to amplify local inflammation after skin injury. In addition, IL-17A-producing γδ T cells express high levels of CCR6 on their surface and are recruited by CCL20 to inflammatory sites (76). CCL20 neutralization dramatically decreases the infiltration of Vγ4 T cells into the epidermis around wounds and reduces epidermal IL-17A production. Together, IL-1β/IL-23 and CCL20 have the ability to amplify IL-17A production by Vγ4 T cells and thereby exacerbate local inflammation in the epidermis after skin injury (**Figure 2**). However, the capability of Vγ4 T cells to produce IL-17A can be repressed by B and T lymphocyte attenuator, which decreases the accumulation of Vγ4 T cells in imiquimod-induced inflammatory skin and draining lymph nodes (77). Whether IL-17A production by Vγ4 T cells is negatively regulated by BLTA in wound healing needs further investigation.

IL-17A IS REQUIRED FOR EFFICIENT SKIN WOUND HEALING, BUT EXCESSIVE IL-17A RETARDS WOUND REPAIR

IL-17A is an important pro-inflammatory cytokine that plays a critical role in the initiation and amplification of inflammation responses. IL-17A is required for efficient skin wound healing. *Il-17a*^{−/−} mice exhibit defects in wound repair, which can be restored by the addition of rIL-17A and IL-17A-producing DETCs (9). Moreover, IL-17A production is reduced in skin around wounds of diabetic mice, and IL-17A-positive Vγ4 T cells transferred to the wound bed can improve wound healing (78). In addition, DETCs provide a source of epidermal IL-17A after skin injury, which accelerates wound healing by inducing epidermal keratinocytes to express the host-defense molecules β-defensin 3 and RegIIIγ (9). However, Rodero et al. reported a contradictory role for IL-17A in skin wound repair and found that the application of an IL-17A-neutralizing Ab onto the wound bed significantly promoted wound healing (8). To reconcile these conflicting roles of IL-17A in skin wound healing, Li et al. blocked IL-17A with an overdose of neutralizing Ab (200 μg/wound) in wound margins, which led to defective skin wound closure, indicating that IL-17A is essential for efficient wound healing. However, the addition of a moderate dose of anti-IL-17A neutralizing Ab (20 μg/wound) significantly improved skin wound repair. In addition, a high dose of rIL-17A (200 ng/wound) rather than low or medium doses (2 or 20 ng/wound) injected into the wound bed prominently delayed skin wound healing, suggesting that excessive IL-17A has a negative impact on skin wound repair (67). These facts strongly suggest that IL-17A plays dual roles: moderate IL-17A

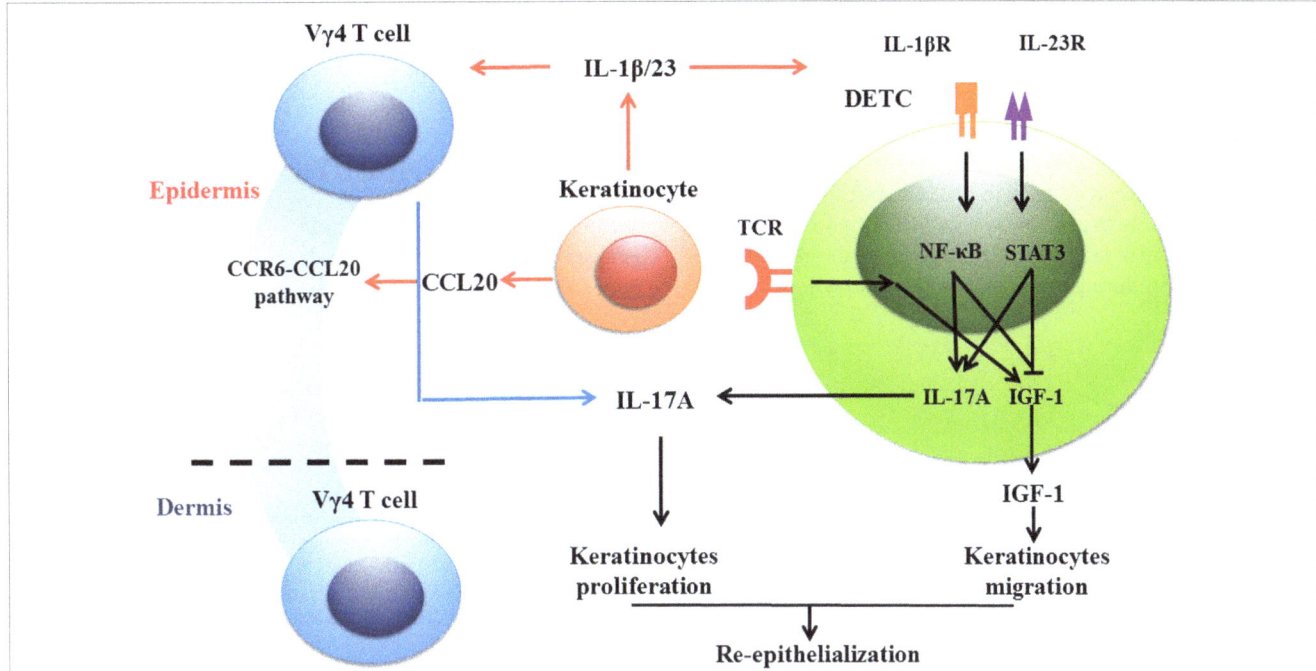

FIGURE 2 | Vγ4 T cells inhibit the production of IGF-1 in dendritic epidermal T cells (DETCs) *via* IL-17A-IL-1β/IL-23 loop and thereby delay wound healing in normal mice. Once keratinocytes get stressed or damaged, DETCs are activated quickly through TCRs by sensing some as-yet-unknown antigens expressed by damaged keratinocytes in a non-major histocompatibility complex restricted manner. Dermal Vγ4 T cells are attracted to the epidermis around wounds by CCL20, which is increased in wounded epidermis, and provide an early source of IL-17A to delay skin wound healing. IL-17 secreted by Vγ4 T cells inhibits IGF-1 production by DETCs in an indirective way where IL-1β and IL-23 produced by keratinocytes act as the bridge between them. IL-1β and IL-23 are key cytokines to suppress IGF-1 production by DETCs. Besides, Vγ4 T cell-derived IL-17A facilitates keratinocytes to secrete IL-1β and IL-23, and thus forms a positive feedback with keratinocyte-derived IL-1β and IL-23. It is worthy to note that IL-1β is more effective than IL-23 in our investigation. NF-κB pathway plays a crucial role in IL-1β-suppressed IGF-1 expression in DETCs. In conclusion, for wounds in normal mice, Vγ4 T cell-mediated IL-17A-IL-1β/IL-23 loop has a negative impact on IGF-1 production by DETCs and thereby delays skin wound closure.

is required for efficient skin wound healing, but excessive IL-17A dampens skin wound closure. We consider that these dual roles do not coexist at the same time, but rather depend on the concentration of IL-17A under the circumstances. Mild amounts of IL-17A at the wound edge re-establish the antimicrobial skin barrier after skin injury by inducing epidermal keratinocytes to express antimicrobial peptides and proteins, but superfluous IL-17A induces the IL-1β/IL-23-IL-17A loop to amplify local inflammation, thus inhibiting wound repair.

Vγ4 T CELLS INHIBIT IGF-1 PRODUCTION OF DETCs TO DELAY SKIN WOUND CLOSURE THROUGH IL-17A

Vγ4 T cells secrete IL-17A to delay wound repair, but IL-17A fails to directly affect the pro-healing function of DETCs, as IGF-1 expression by DETCs is not able to be directly reduced by IL-17A (67). However, epidermal IL-1β and IL-23 are key factors for the suppression of IGF-1 in DETCs. Taken together with the positive loop of IL-17A and IL-1β/IL-23, it is very likely that IL-1β and IL-23 act as the bridge between Vγ4 T cells and DETCs. Furthermore, IL-1β and IL-23 notably promote the phosphorylation of NF-κB and STAT3 and facilitate their translocation from

the cytoplasm to the nucleus in DETCs. However, IL-1β shows more significant effects on DETCs than IL-23, which only exhibits synergic inhibition with IL-1β (67). Therefore, we consider that IL-17A produced by Vγ4 T cells indirectly impedes DETCs to secrete IGF-1 and delays skin wound closure with mediators IL-1β and IL-23 (**Figure 2**).

A POTENTIAL FUNCTIONAL LINK BETWEEN Vγ4 T CELLS, EPIDERMAL CELLS, AND DETCs IN SKIN WOUND HEALING

Upon skin injury, epidermal cells interact with DETCs to form an IL-15-IGF-1 loop to amplify IGF-1 production in the epidermis for re-epithelialization (26, 28). Meanwhile, epidermal cells can also interact with epidermis-infiltrating Vγ4 T cells to form an IL-17A-IL-1β/IL-23 loop to augment local inflammation (78). In diabetic wounds, the DETC-mediated IL-15-IGF-1 loop and Vγ4 T cell-mediated IL-17A-IL-1β/IL-23 loop improve the defects of diabetic wound healing by enhancing re-epithelialization and local inflammation, respectively (**Figure 1**). However, in normal wounds, the Vγ4 T cell-mediated IL-17A-IL-1β/IL-23 loop has a negative impact on IGF-1 production by DETCs and thereby delays skin wound closure (**Figure 2**). This indicates that a balance exists between Vγ4 T cell-derived IL-17A and DETC-derived IGF-1 for optimal skin wound healing.

Some interesting issues need to be further investigated in the near future: the precise underlying mechanisms of IL-1β and IL-23 inhibition of IGF-1 production in DETCs, and the influence of co-stimulatory molecules on the two loops during wound healing. Whether epidermal stem cells are involved in the regulation of IL-17A and IGF-1 during re-epithelialization, and how IL-17A and

IGF-1 play roles in the homeostasis, migration, proliferation, and differentiation of epidermal stem cells still remains unknown.

AUTHOR CONTRIBUTIONS

Writing the original draft: YL and WH. Writing reviews and editing: WH, GL, and JW. Funding acquisition: WH and GL. Supervision: GL and WH.

REFERENCES

1. Ben Amar M, Wu M. Re-epithelialization: advancing epithelium frontier during wound healing. *J R Soc Interface* (2014) 11:20131038. doi:10.1098/rsif.2013.1038

2. Boismenu R, Havran WL. Modulation of epithelial cell growth by intraepithelial gamma delta T cells. *Science* (1994) 266:1253–5. doi:10.1126/science.7973709

3. Jameson J, Ugarte K, Chen N, Yachi P, Fuchs E, Boismenu R, et al. A role for skin gammadelta T cells in wound repair. *Science* (2002) 296:747–9. doi:10.1126/science.1069639

4. Sharp LL, Jameson JM, Cauvi G, Havran WL. Dendritic epidermal T cells regulate skin homeostasis through local production of insulin-like growth factor 1. *Nat Immunol* (2005) 6:73–9. doi:10.1038/ni1152

5. Nielsen MM, Lovato P, MacLeod AS, Witherden DA, Skov L, Dyring-Andersen B, et al. IL-1beta-dependent activation of dendritic epidermal T cells in contact hypersensitivity. *J Immunol* (2014) 192:2975–83. doi:10.4049/jimmunol.1301689

6. Yoshiki R, Kabashima K, Honda T, Nakamizo S, Sawada Y, Sugita K, et al. IL-23 from Langerhans cells is required for the development of imiquimod-induced psoriasis-like dermatitis by induction of IL-17A-producing gammadelta T cells. *J Invest Dermatol* (2014) 134(7):1912–21. doi:10.1038/jid.2014.98

7. Takagi N, Kawakami K, Kanno E, Tanno H, Takeda A, Ishii K, et al. IL-17A promotes neutrophilic inflammation and disturbs acute wound healing in skin. *Exp Dermatol* (2017) 26:137–44. doi:10.1111/exd.13115

8. Rodero MP, Hodgson SS, Hollier B, Combadiere C, Khosrotehrani K. Reduced Il17a expression distinguishes a Ly6c(lo)MHCII(hi) macrophage population promoting wound healing. *J Invest Dermatol* (2013) 133:783–92. doi:10.1038/jid.2012.368

9. MacLeod AS, Hemmers S, Garijo O, Chabod M, Mowen K, Witherden DA, et al. Dendritic epidermal T cells regulate skin antimicrobial barrier function. *J Clin Invest* (2013) 123:4364–74. doi:10.1172/JCI70064

10. Li Y, Huang Z, Yan R, Liu M, Bai Y, Liang G, et al. Vgamma4 gammadelta T cells provide an early source of IL-17A and accelerate skin graft rejection. *J Invest Dermatol* (2017) 137:2513–22. doi:10.1016/j.jid.2017.03.043

11. Van Beneden K, De Creus A, Stevenaert F, Debacker V, Plum J, Leclercq G. Expression of inhibitory receptors Ly49E and CD94/NKG2 on fetal thymic and adult epidermal TCR V gamma 3 lymphocytes. *J Immunol* (2002) 168:3295–302. doi:10.4049/jimmunol.168.7.3295

12. Havran WL, Allison JP. Origin of Thy-1+ dendritic epidermal cells of adult mice from fetal thymic precursors. *Nature* (1990) 344:68–70. doi:10.1038/344068a0

13. Xiong N, Raulet DH. Development and selection of gammadelta T cells. *Immunol Rev* (2007) 215:15–31. doi:10.1111/j.1600-065X.2006.00478.x

14. Kang J, Volkmann A, Raulet DH. Evidence that gammadelta versus alphabeta T cell fate determination is initiated independently of T cell receptor signaling. *J Exp Med* (2001) 193:689–98. doi:10.1084/jem.193.6.689

15. Schlissel MS, Durum SD, Muegge K. The interleukin 7 receptor is required for T cell receptor gamma locus accessibility to the V(D)J recombinase. *J Exp Med* (2000) 191:1045–50. doi:10.1084/jem.191.6.1045

16. Xiong N, Kang C, Raulet DH. Positive selection of dendritic epidermal gammadelta T cell precursors in the fetal thymus determines expression of skin-homing receptors. *Immunity* (2004) 21:121–31. doi:10.1016/j.immuni.2004.06.008

17. Boyden LM, Lewis JM, Barbee SD, Bas A, Girardi M, Hayday AC, et al. Skint1, the prototype of a newly identified immunoglobulin superfamily gene cluster, positively selects epidermal gammadelta T cells. *Nat Genet* (2008) 40:656–62. doi:10.1038/ng.108

18. Ikuta K, Kina T, MacNeil I, Uchida N, Peault B, Chien YH, et al. A developmental switch in thymic lymphocyte maturation potential occurs at the level of hematopoietic stem cells. *Cell* (1990) 62:863–74. doi:10.1016/0092-8674(90)90262-D

19. Jiang X, Campbell JJ, Kupper TS. Embryonic trafficking of gammadelta T cells to skin is dependent on E/P selectin ligands and CCR4. *Proc Natl Acad Sci U S A* (2010) 107:7443–8. doi:10.1073/pnas.0912943107

20. Schon MP, Schon M, Parker CM, Williams IR. Dendritic epidermal T cells (DETC) are diminished in integrin alphaE(CD103)-deficient mice. *J Invest Dermatol* (2002) 119:190–3. doi:10.1046/j.1523-1747.2002.17973.x

21. Jin Y, Xia M, Saylor CM, Narayan K, Kang J, Wiest DL, et al. Cutting edge: intrinsic programming of thymic gammadeltaT cells for specific peripheral tissue localization. *J Immunol* (2010) 185:7156–60. doi:10.4049/jimmunol.1002781

22. Asarnow DM, Kuziel WA, Bonyhadi M, Tigelaar RE, Tucker PW, Allison JP. Limited diversity of gamma delta antigen receptor genes of Thy-1+ dendritic epidermal cells. *Cell* (1988) 55:837–47. doi:10.1016/0092-8674(88)90139-0

23. Chodaczek G, Papanna V, Zal MA, Zal T. Body-barrier surveillance by epidermal gammadelta TCRs. *Nat Immunol* (2012) 13:272–82. doi:10.1038/ni0612-621d

24. Ye SK, Maki K, Lee HC, Ito A, Kawai K, Suzuki H, et al. Differential roles of cytokine receptors in the development of epidermal gamma delta T cells. *J Immunol* (2001) 167:1929–34. doi:10.4049/jimmunol.167.4.1929

25. De Creus A, Van Beneden K, Stevenaert F, Debacker V, Plum J, Leclercq G. Developmental and functional defects of thymic and epidermal V gamma 3 cells in IL-15-deficient and IFN regulatory factor-1-deficient mice. *J Immunol* (2002) 168:6486–93. doi:10.4049/jimmunol.168.12.6486

26. Liu Z, Liang G, Gui L, Li Y, Liu M, Bai Y, et al. Weakened IL-15 production and impaired mTOR activation alter dendritic epidermal T cell homeostasis in diabetic mice. *Sci Rep* (2017) 7:6028. doi:10.1038/s41598-017-05950-5

27. Bai Y, Xu R, Zhang X, Zhang X, Hu X, Li Y, et al. Differential role of rapamycin in epidermis-induced IL-15-IGF-1 secretion via activation of Akt/mTORC2. *Cell Physiol Biochem* (2017) 42:1755–68. doi:10.1159/000479443

28. Wang Y, Bai Y, Li Y, Liang G, Jiang Y, Liu Z, et al. IL-15 enhances activation and IGF-1 production of dendritic epidermal T cells to promote wound healing in diabetic mice. *Front Immunol* (2017) 8:1557. doi:10.3389/fimmu.2017.01557

29. Toulon A, Breton L, Taylor KR, Tenenhaus M, Bhavsar D, Lanigan C, et al. A role for human skin-resident T cells in wound healing. *J Exp Med* (2009) 206:743–50. doi:10.1084/jem.20081787

30. Tyler CJ, Doherty DG, Moser B, Eberl M. Human Vgamma9/Vdelta2 T cells: innate adaptors of the immune system. *Cell Immunol* (2015) 296:10–21. doi:10.1016/j.cellimm.2015.01.008

31. Laggner U, Di Meglio P, Perera GK, Hundhausen C, Lacy KE, Ali N, et al. Identification of a novel proinflammatory human skin-homing Vgamma9Vdelta2 T cell subset with a potential role in psoriasis. *J Immunol* (2011) 187:2783–93. doi:10.4049/jimmunol.1100804

32. Macleod AS, Havran WL. Functions of skin-resident gammadelta T cells. *Cell Mol Life Sci* (2011) 68:2399–408. doi:10.1007/s00018-011-0702-x

33. He W, Hao J, Dong S, Gao Y, Tao J, Chi H, et al. Naturally activated V gamma 4 gamma delta T cells play a protective role in tumor immunity through

expression of eomesodermin. *J Immunol* (2010) 185:126–33. doi:10.4049/jimmunol.0903767

34. Komori HK, Witherden DA, Kelly R, Sendaydiego K, Jameson JM, Teyton L, et al. Cutting edge: dendritic epidermal gammadelta T cell ligands are rapidly and locally expressed by keratinocytes following cutaneous wounding. *J Immunol* (2012) 188:2972–6. doi:10.4049/jimmunol.1100887

35. Baccala R, Witherden D, Gonzalez-Quintial R, Dummer W, Surh CD, Havran WL, et al. Gamma delta T cell homeostasis is controlled by IL-7 and IL-15 together with subset-specific factors. *J Immunol* (2005) 174:4606–12. doi:10.4049/jimmunol.174.8.4606

36. Bauer S, Groh V, Wu J, Steinle A, Phillips JH, Lanier LL, et al. Activation of NK cells and T cells by NKG2D, a receptor for stress-inducible MICA. *Science* (1999) 285:727–9. doi:10.1126/science.285.5428.727

37. Keyes BE, Liu S, Asare A, Naik S, Levorse J, Polak L, et al. Impaired epidermal to dendritic T cell signaling slows wound repair in aged skin. *Cell* (2016) 167:1323–38.e14. doi:10.1016/j.cell.2016.10.052

38. Whang MI, Guerra N, Raulet DH. Costimulation of dendritic epidermal gammadelta T cells by a new NKG2D ligand expressed specifically in the skin. *J Immunol* (2009) 182:4557–64. doi:10.4049/jimmunol.0802439

39. Jamieson AM, Diefenbach A, McMahon CW, Xiong N, Carlyle JR, Raulet DH. The role of the NKG2D immunoreceptor in immune cell activation and natural killing. *Immunity* (2002) 17:19–29. doi:10.1016/S1074-7613(02)00333-3

40. Nitahara A, Shimura H, Ito A, Tomiyama K, Ito M, Kawai K. NKG2D ligation without T cell receptor engagement triggers both cytotoxicity and cytokine production in dendritic epidermal T cells. *J Invest Dermatol* (2006) 126:1052–8. doi:10.1038/sj.jid.5700112

41. Eagle RA, Trowsdale J. Promiscuity and the single receptor: NKG2D. *Nat Rev Immunol* (2007) 7:737–44. doi:10.1038/nri2144

42. Takada A, Yoshida S, Kajikawa M, Miyatake Y, Tomaru U, Sakai M, et al. Two novel NKG2D ligands of the mouse H60 family with differential expression patterns and binding affinities to NKG2D. *J Immunol* (2008) 180:1678–85. doi:10.4049/jimmunol.180.3.1678

43. Yoshida S, Mohamed RH, Kajikawa M, Koizumi J, Tanaka M, Fugo K, et al. Involvement of an NKG2D ligand H60c in epidermal dendritic T cell-mediated wound repair. *J Immunol* (2012) 188:3972–9. doi:10.4049/jimmunol.1102886

44. Witherden DA, Verdino P, Rieder SE, Garijo O, Mills RE, Teyton L, et al. The junctional adhesion molecule JAML is a costimulatory receptor for epithelial gammadelta T cell activation. *Science* (2010) 329:1205–10. doi:10.1126/science.1192698

45. Garni-Wagner BA, Purohit A, Mathew PA, Bennett M, Kumar V. A novel function-associated molecule related to non-MHC-restricted cytotoxicity mediated by activated natural killer cells and T cells. *J Immunol* (1993) 151:60–70.

46. Schuhmachers G, Ariizumi K, Mathew PA, Bennett M, Kumar V, Takashima A. 2B4, a new member of the immunoglobulin gene superfamily, is expressed on murine dendritic epidermal T cells and plays a functional role in their killing of skin tumors. *J Invest Dermatol* (1995) 105:592–6. doi:10.1111/1523-1747.ep12323533

47. Shimura H, Nitahara A, Ito A, Tomiyama K, Ito M, Kawai K. Up-regulation of cell surface Toll-like receptor 4-MD2 expression on dendritic epidermal T cells after the emigration from epidermis during cutaneous inflammation. *J Dermatol Sci* (2005) 37:101–10. doi:10.1016/j.jdermsci.2004.11.006

48. Nagai Y, Akashi S, Nagafuku M, Ogata M, Iwakura Y, Akira S, et al. Essential role of MD-2 in LPS responsiveness and TLR4 distribution. *Nat Immunol* (2002) 3:667–72. doi:10.1038/ni809

49. Tschachler E, Steiner G, Yamada H, Elbe A, Wolff K, Stingl G. Dendritic epidermal T cells: activation requirements and phenotypic characterization of proliferating cells. *J Invest Dermatol* (1989) 92:763–8. doi:10.1111/1523-1747.ep12722546

50. Edelbaum D, Mohamadzadeh M, Bergstresser PR, Sugamura K, Takashima A. Interleukin (IL)-15 promotes the growth of murine epidermal gamma delta T cells by a mechanism involving the beta- and gamma c-chains of the IL-2 receptor. *J Invest Dermatol* (1995) 105:837–43. doi:10.1111/1523-1747.ep12326260

51. Giri JG, Ahdieh M, Eisenman J, Shanebeck K, Grabstein K, Kumaki S, et al. Utilization of the beta and gamma chains of the IL-2 receptor by the novel cytokine IL-15. *EMBO J* (1994) 13:2822–30.

52. Edmondson SR, Thumiger SP, Werther GA, Wraight CJ. Epidermal homeostasis: the role of the growth hormone and insulin-like growth factor systems. *Endocr Rev* (2003) 24:737–64. doi:10.1210/er.2002-0021

53. Sadagurski M, Yakar S, Weingarten G, Holzenberger M, Rhodes CJ, Breitkreutz D, et al. Insulin-like growth factor 1 receptor signaling regulates skin development and inhibits skin keratinocyte differentiation. *Mol Cell Biol* (2006) 26:2675–87. doi:10.1128/MCB.26.7.2675-2687.2006

54. Johnson DE, Williams LT. Structural and functional diversity in the FGF receptor multigene family. *Adv Cancer Res* (1993) 60:1–41. doi:10.1016/S0065-230X(08)60821-0

55. Jameson JM, Cauvi G, Sharp LL, Witherden DA, Havran WL. Gammadelta T cell-induced hyaluronan production by epithelial cells regulates inflammation. *J Exp Med* (2005) 201:1269–79. doi:10.1084/jem.20042057

56. Havran WL, Jameson JM. Epidermal T cells and wound healing. *J Immunol* (2010) 184:5423–8. doi:10.4049/jimmunol.0902733

57. Heilig JS, Tonegawa S. Diversity of murine gamma genes and expression in fetal and adult T lymphocytes. *Nature* (1986) 322:836–40. doi:10.1038/322836a0

58. Takagaki Y, Nakanishi N, Ishida I, Kanagawa O, Tonegawa S. T cell receptor-gamma and -delta genes preferentially utilized by adult thymocytes for the surface expression. *J Immunol* (1989) 142:2112–21.

59. Ribot JC, deBarros A, Pang DJ, Neves JF, Peperzak V, Roberts SJ, et al. CD27 is a thymic determinant of the balance between interferon-gamma- and interleukin 17-producing gammadelta T cell subsets. *Nat Immunol* (2009) 10:427–36. doi:10.1038/ni.1717

60. Michel ML, Pang DJ, Haque SF, Potocnik AJ, Pennington DJ, Hayday AC. Interleukin 7 (IL-7) selectively promotes mouse and human IL-17-producing gammadelta cells. *Proc Natl Acad Sci U S A* (2012) 109:17549–54. doi:10.1073/pnas.1204327109

61. Haas JD, Ravens S, Duber S, Sandrock I, Oberdorfer L, Kashani E, et al. Development of interleukin-17-producing gammadelta T cells is restricted to a functional embryonic wave. *Immunity* (2012) 37:48–59. doi:10.1016/j.immuni.2012.06.003

62. Gray EE, Ramirez-Valle F, Xu Y, Wu S, Wu Z, Karjalainen KE, et al. Deficiency in IL-17-committed Vgamma4(+) gammadelta T cells in a spontaneous Sox13-mutant CD45.1(+) congenic mouse substrain provides protection from dermatitis. *Nat Immunol* (2013) 14:584–92. doi:10.1038/ni.2585

63. Pang DJ, Neves JF, Sumaria N, Pennington DJ. Understanding the complexity of gammadelta T-cell subsets in mouse and human. *Immunology* (2012) 136:283–90. doi:10.1111/j.1365-2567.2012.03582.x

64. Cai Y, Shen X, Ding C, Qi C, Li K, Li X, et al. Pivotal role of dermal IL-17-producing gammadelta T cells in skin inflammation. *Immunity* (2011) 35:596–610. doi:10.1016/j.immuni.2011.08.001

65. Costa MF, de Negreiros CB, Bornstein VU, Valente RH, Mengel J, Henriques M, et al. Murine IL-17+ Vgamma4 T lymphocytes accumulate in the lungs and play a protective role during severe sepsis. *BMC Immunol* (2015) 16:36. doi:10.1186/s12865-015-0098-8

66. Blink SE, Caldis MW, Goings GE, Harp CT, Malissen B, Prinz I, et al. gammadelta T cell subsets play opposing roles in regulating experimental autoimmune encephalomyelitis. *Cell Immunol* (2014) 290:39–51. doi:10.1016/j.cellimm.2014.04.013

67. Li Y, Wang Y, Zhou L, Liu M, Liang G, Yan R, et al. Vgamma4 T cells inhibit the pro-healing functions of dendritic epidermal T cells to delay skin wound closure through IL-17A. *Front Immunol* (2018) 9:240. doi:10.3389/fimmu.2018.00240

68. Martin B, Hirota K, Cua DJ, Stockinger B, Veldhoen M. Interleukin-17-producing gammadelta T cells selectively expand in response to pathogen products and environmental signals. *Immunity* (2009) 31:321–30. doi:10.1016/j.immuni.2009.06.020

69. Ramirez-Valle F, Gray EE, Cyster JG. Inflammation induces dermal Vgamma4+ gammadeltaT17 memory-like cells that travel to distant skin and accelerate secondary IL-17-driven responses. *Proc Natl Acad Sci U S A* (2015) 112:8046–51. doi:10.1073/pnas.1508990112

70. Maeda Y, Seki N, Kataoka H, Takemoto K, Utsumi H, Fukunari A, et al. IL-17-producing Vgamma4+ gammadelta T cells require sphingosine 1-phosphate receptor 1 for their egress from the lymph nodes under homeostatic and inflammatory conditions. *J Immunol* (2015) 195:1408–16. doi:10.4049/jimmunol.1500599

71. Hartwig T, Pantelyushin S, Croxford AL, Kulig P, Becher B. Dermal IL-17-producing gammadelta T cells establish long-lived memory in the skin. *Eur J Immunol* (2015) 45:3022–33. doi:10.1002/eji.201545883

72. Suryawanshi A, Veiga-Parga T, Rajasagi NK, Reddy PB, Sehrawat S, Sharma S, et al. Role of IL-17 and Th17 cells in herpes simplex virus-induced

corneal immunopathology. *J Immunol* (2011) 187:1919–30. doi:10.4049/jimmunol.1100736

73. Koenecke C, Chennupati V, Schmitz S, Malissen B, Forster R, Prinz I. In vivo application of mAb directed against the gammadelta TCR does not deplete but generates "invisible" gammadelta T cells. *Eur J Immunol* (2009) 39:372–9. doi:10.1002/eji.200838741

74. Sutton CE, Lalor SJ, Sweeney CM, Brereton CF, Lavelle EC, Mills KH. Interleukin-1 and IL-23 induce innate IL-17 production from gammadelta T cells, amplifying Th17 responses and autoimmunity. *Immunity* (2009) 31:331–41. doi:10.1016/j.immuni.2009.08.001

75. van der Fits L, Mourits S, Voerman JS, Kant M, Boon L, Laman JD, et al. Imiquimod-induced psoriasis-like skin inflammation in mice is mediated via the IL-23/IL-17 axis. *J Immunol* (2009) 182:5836–45. doi:10.4049/jimmunol.0802999

76. Mabuchi T, Singh TP, Takekoshi T, Jia GF, Wu X, Kao MC, et al. CCR6 is required for epidermal trafficking of gammadelta-T cells in an IL-23-induced model of psoriasiform dermatitis. *J Invest Dermatol* (2013) 133:164–71. doi:10.1038/jid.2012.260

77. Bekiaris V, Sedy JR, Macauley MG, Rhode-Kurnow A, Ware CF. The inhibitory receptor BTLA controls gammadelta T cell homeostasis and inflammatory responses. *Immunity* (2013) 39:1082–94. doi:10.1016/j.immuni.2013.10.017

78. Liu Z, Xu Y, Zhang X, Liang G, Chen L, Xie J, et al. Defects in dermal Vgamma4 gamma delta T cells result in delayed wound healing in diabetic mice. *Am J Transl Res* (2016) 8:2667–80.

14

Blood monocytes and their subsets: established features and open questions

*Loems Ziegler-Heitbrock** *

Helmholtz Zentrum München, Asklepios Fachkliniken München-Gauting, Gauting, Germany

Edited by:
Florent Ginhoux,
Singapore Immunology Network,
Singapore

Reviewed by:
Anne Hosmalin,
Cochin Institute, France
Muzlifah Aisha Haniffa,
Newcastle University, UK
Claudia Jakubzick,
National Jewish Health, USA

*Correspondence:
Loems Ziegler-Heitbrock,
Helmholtz Zentrum München,
Asklepios Fachkliniken
München-Gauting, Enzianstr.3,
82211 Herrsching, Germany
lzh@monocyte.eu

In contrast to the past reliance on morphology, the identification and enumeration of blood monocytes are nowadays done with monoclonal antibodies and flow cytometry and this allows for subdivision into classical, intermediate, and non-classical monocytes. Using specific cell surface markers, dendritic cells in blood can be segregated from these monocytes. While in the past, changes in monocyte numbers as determined in standard hematology counters have not had any relevant clinical impact, the subset analysis now has uncovered informative changes that may be used in management of disease.

Keywords: monocyte subsets, nomenclature, classical monocytes, intermediate monocytes, non-classical monocytes

The Definition of Monocytes

The term monocyte is used for blood cells of a lineage called monocytes/macrophages or mononuclear phagocytes. These blood monocytes are bone marrow-derived leukocytes that are functionally characterized by the ability to phagocytose, to produce cytokines, and to present antigen. In early studies, they had been identified based on glass adherence and morphology (1). Also, cytochemistry for specific enzymes like monocyte-specific esterase (2, 3) has been employed, while the standard approach in clinical hematology relies on physical properties of these cells including light scatter.

In bone marrow, the monocytes derive from myelo-monocytic stem cells, which give rise to more direct precursors like monoblasts and pro-monocytes. These cells earlier were identified based on morphology (4) such that the monoblast was an ill-defined cell type. More recently in the mouse model, a Ly6C+ CD115+ CD117+ monoblast-type cell, termed common monocyte progenitor (cMoP), was identified in bone marrow and spleen and this cell is able to proliferate and give rise to the different monocyte subsets (5). A cMoP monoblast type of cell remains to be identified for man and other species.

The number of circulating blood monocytes in man can strongly increase within minutes by stress or exercise followed by a rapid return to baseline levels. These recruited cells are thought to come from what is called the marginal pool (6). This compartment describes areas of reduced blood velocity close to the endothelium of venules and here cells can loosely adhere and can be mobilized in a catecholamine-dependent fashion (7). These marginal pool monocytes can have an adhesion molecule pattern distinct from monocytes found in blood at rest.

In addition, CD11b^{high} (CD90, B220, CD49b, NK1.1, Ly-6G, F4/80, I-Ab, CD11c)^{low} cells are mobilized from the spleen after severe injury (8). These cells have monocyte morphology and their transcriptome matches with that of blood monocytes. Furthermore, CD11b+ Ly6C^{hi} monocytes can be mobilized from bone marrow to blood in infectious disease models (9), and adoptively transferred monocytes were shown to return to the bone marrow (10) in the mouse. What remains to be determined is whether the spleen and bone marrow compartments also contribute to the pool of monocytes that can be mobilized by stress and exercise.

When under homeostatic or inflammatory conditions, the monocytes have migrated into tissue; then by definition, these cells are called macrophages. Cells newly emigrated into the lung have been termed monocytes in some studies [e.g., Ref. (11)]. Since monocytes, once they have arrived in tissue, will start to transform into larger cells and rapidly lose their monocyte characteristics, others have called these recently emigrated cells "small macrophages" (12).

More detailed studies in the mouse have demonstrated tissue cells with characteristics close to blood monocytes (13, 14). However, these cells in the lymph node show a gene expression pattern that distinguishes them from the blood cells (14) and in the skin they show increased expression of lysozyme and CD68, markers typical of mature macrophages (13). Therefore, more data are required in the mouse model and obviously also in man before a consensus can be reached whether we use the term tissue monocyte or whether we continue to call these cells macrophages. Until these issues have been resolved, the term monocyte should be restricted to cells in the blood compartment and the bone marrow and spleen reservoirs that can replenish the blood monocyte pool.

Definition of Blood Monocytes Based on Cells Surface Markers

As explained above, monocytes initially had been identified by function and morphology and these criteria have been misleading especially when disease processes altered these features. Therefore, attempts have been made to define unequivocal criteria for monocytes. Here, monoclonal antibodies against cell surface molecules have been proposed. In man, CD14 has been used as a marker (15), and in the mouse, CD115 is often employed (16). CD115 identifies the M-CSF receptor and has the main drawback that in the mouse, it is downregulated on blood monocytes with inflammation (17). Also, the question is whether such markers are sufficiently specific and do not react with other cell types like dendritic cells (DCs). In fact, part of the CD1c+ blood DCs in man can express low-level CD14 (18) and also human B cells have been reported to express some CD14 (19). Therefore, monocytes can be identified with markers like CD14 and CD115, but this should be supported by additional markers and by functional studies. Interestingly, when searching for macrophage-specific transcripts in the mouse, CD64 and MerTK have emerged (20). While CD64 is absent from non-classical monocytes in man, MerTK is a molecule that might prove informative for blood monocytes in different species. In addition, staining for CD16, which is used for monocyte subset definition (see below), will at the same time help to exclude DCs in human blood.

Dissection of Monocytes from Dendritic Cells

Dendritic cells were first described by Steinman and Cohn as stellate cells isolated from mouse spleen (21). Over the years, there have been debates as to whether these cells are a distinct lineage or part of the mononuclear phagocyte system. A common precursor for monocytes and DCs was described in the mouse (22), but the existence of this cell was later disputed (23) suggesting that DCs

and monocytes may diverge at an earlier multi-potent progenitor stage (24).

However, the demonstration that monocytes can be used to generate DCs *in vitro* by adding GM-CSF and IL-4 suggested a close relationship between monocytes and DCs (25). Later, transcriptome analysis demonstrated that such monocyte-derived DCs rather resemble macrophages than DCs from lymphoid tissue (26). Therefore, these *in vitro* generated monocyte-derived cells are potent antigen-presenting cells, but they do not represent *bona fide* DCs; they rather belong to the monocyte/macrophage lineage. Still not resolved is the question whether in tissue the monocyte-derived cells with high levels of class II expression and with high antigen-presenting capacity should be termed monocyte-derived DC (13, 27, 28) or activated macrophages.

In addition to DCs in tissue, cells with DC properties have been described in blood based on the expression of CD68, CD1c, or CD141 (29, 30). Transcriptome analysis has demonstrated that these cells and the monocytes belong to different clusters (26, 31). These data suggest that blood DCs can be segregated from monocytes and macrophages as a separate lineage.

The data also demonstrate the power of transcriptomic analysis in defining and dissecting leukocyte populations like monocytes and DCs. Ontogeny can help in such a definition, but in men, adoptive transfer is limited to strategies like transfer of bone marrow stem cells, and informative mutations are rare. Also, the ontogeny approach needs to be used with caution since a defined progenitor cell can give rise to clearly distinct cell populations. An informative example is the megakaryocyte–erythrocyte progenitor (MEP) cell, which gives rise to either megakaryocytes and their platelet progeny or to erythroblasts and their red blood cell progeny (32, 33). Megakaryocytes and erythroblasts have a distinct transcriptome (34), and they are involved in distinct functions, i.e., in blood clotting and oxygen transport, respectively. Therefore, although having a common ontogeny, these cells belong to clearly separate lineages. This example illustrates that ontogeny can provide a framework, but a comprehensive analyses like transcriptomics and the analysis of cell function are required for dissecting cell types and for developing a nomenclature. Therefore, in order to assign a novel leukocyte population in blood or tissue to either monocytes or DCs, a straight-forward approach is to analyze the transcriptome (and other omics like the proteome, lipidome, glycome, or metabolome) of these cells in comparison to typical monocytes and DCs and to then ask whether the novel cell type co-clusters with either prototypic monocytes or DCs (26).

Monocyte Subpopulations

Evidence for monocyte subpopulations has come from experiments using differential flotation in counter-current elutriation (35) and from differential binding to antibody-coated red blood cells, which has defined populations with different functions (36). With the use of monoclonal antibodies and flow cytometry, tools have become available to clearly define, enumerate, and isolate monocyte subsets based on the differential expression of CD14 and CD16 cell-surface markers (37).

In 2010, an international consortium under the auspices of the IUIS and the WHO has proposed a nomenclature for monocyte

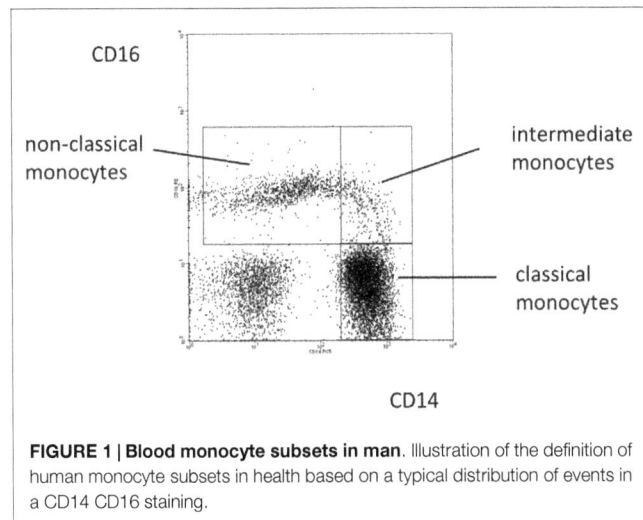

FIGURE 1 | Blood monocyte subsets in man. Illustration of the definition of human monocyte subsets in health based on a typical distribution of events in a CD14 CD16 staining.

subpopulations (38). The proposal defined the major population of CD14^{high} cells found in human blood as classical monocytes and the minor population of cells with low CD14 and high CD16 as non-classical monocytes. A population in between these two subsets was termed intermediate monocytes (see **Figure 1**).

While an unequivocal approach to defining the intermediate monocytes has not been developed, as yet (39), a host of studies on intermediate monocytes has been published since the 2010 proposal. In fact, a search for the term "intermediate monocyte" under Google Scholar has revealed more than 100 studies on these cells since 2010. These reports have described an expansion of intermediate monocytes in various inflammatory diseases and these cells have been shown to be of prognostic relevance in cardiovascular disease (40). The use of additional markers for delineation of intermediate monocytes has been suggested (41) and it remains to be shown whether markers, such as CCR2 or slan, will improve the definition of these cells.

The same nomenclature as proposed for man can be used in other species [reviewed in Ref. (42)]. The respective cells can be very similar to men as seen for non-human primates (43, 44). In species like the mouse, the classical and non-classical monocyte subsets can be identified as well, but different markers like CD115, Ly6C, and CD43 are used (16, 45). Also in species like rat, pig, cow, and horse, classical and non-classical monocytes can be defined and even intermediate monocytes have been described in some animals (42). It is predicted that the nomenclature of monocyte subsets will be applicable to all mammalian species.

In human blood, a population of slan-positive cells has been described as DCs, but phenotypic analysis has shown that these cells are CD14-low and CD16-high (46), functional studies demonstrated a high capacity to produce TNF (47), and clinical studies showed that these cells are depleted by glucocorticoid treatment (48). These features are identical to what has been reported as characteristics of non-classical CD14+CD16++ monocytes (37, 49, 50). Also, the increased absolute numbers of slan-positive monocytes and of non-classical monocytes show a clear correlation in HIV-infected patients (51), and part of the non-classical monocytes has been shown to be slan-positive (52–54). Collectively, these findings suggest that the slan-positive cells belong to the non-classical monocytes.

There may be additional monocyte subsets including Fcepsilon-RI-positive cells (55), which were found with a median of 2.5% among CD14-positive blood monocytes in a pediatric cohort (56) and these cells may be involved in IgE clearance (57). Also, proliferating monocytes have been described (58) as well as precursors for fibrocytes (59) and osteoclasts (60). For all of these cell types, further characterization is awaited.

Clinical Implications of Monocyte Numbers

Monocyte numbers as defined in the hematology lab using light scatter properties have not contributed much to diagnosis and monitoring of disease, but with the definition of monocyte subsets by flow cytometry, informative patterns have emerged. For example, severe infection will increase the number of non-classical and intermediate monocytes (61–63). Here, it remains to be analyzed whether such an increase can predict prognosis, as has been suggested (64). Furthermore, therapy with glucocorticoids leads to a decrease of non-classical monocytes, which appears to be due to a selective induction of apoptosis in the non-classical monocytes while classical monocytes even increase in number under glucocorticoids (50, 65). Also, blockade of the M-CSF pathway can lead to depletion of non-classical monocytes (66–68). A likely explanation is that M-CSF signaling via the CD115 M-CSF receptor is required for the classical monocytes to mature into non-classical monocytes. Again still to be determined is whether such a drug-induced depletion can be used to predict therapeutic response in inflammatory diseases. Still unresolved is the mechanism of depletion of non-classical monocytes in three siblings within one family (69). Here, more families with this type of defect need to be analyzed in order to identify the gene and the mechanisms involved. Finally, the absolute count of intermediate monocytes was shown to predict cardiovascular events (70, 71). Hence, analysis of monocyte subsets by flow cytometry now provides clinically useful parameters in various settings. What remains to be established in this context is an unequivocal dissection of the non-classical and the intermediate monocytes.

References

1. van Furth R, Cohn ZA. The origin and kinetics of mononuclear phagocytes. *J Exp Med* (1968) **128**:415–35. doi:10.1084/jem.128.3.415
2. Tucker SB, Pierre RV, Jordon RE. Rapid identification of monocytes in a mixed mononuclear cell preparation. *J Immunol Methods* (1977) **14**:267–9. doi:10.1016/0022-1759(77)90137-5
3. Uphoff CC, Drexler HG. Biology of monocyte-specific esterase. *Leuk Lymphoma* (2000) **39**:257–70. doi:10.3109/10428190009065825
4. Goud TJ, Schotte C, van Furth R. Identification and characterization of the monoblast in mononuclear phagocyte colonies grown in vitro. *J Exp Med* (1975) **142**:1180–99. doi:10.1084/jem.142.5.1180
5. Hettinger J, Richards DM, Hansson J, Barra MM, Joschko AC, Krijgsveld J, et al. Origin of monocytes and macrophages in a committed progenitor. *Nat Immunol* (2013) **14**:821–30. doi:10.1038/ni.2638
6. Klonz A, Wonigeit K, Pabst R, Westermann J. The marginal blood pool of the rat contains not only granulocytes, but also lymphocytes, NK-cells and monocytes: a second intravascular compartment, its cellular composition, adhesion

molecule expression and interaction with the peripheral blood pool. *Scand J Immunol* (1996) **44**:461–9. doi:10.1046/j.1365-3083.1996.d01-334.x

7. Steppich B, Dayyani F, Gruber R, Lorenz R, Mack M, Ziegler-Heitbrock HWL. Selective mobilization of CD14(+)CD16(+) monocytes by exercise. *Am J Physiol Cell Physiol* (2000) **279**:C578–86.

8. Swirski FK, Nahrendorf M, Etzrodt M, Wildgruber M, Cortez-Retamozo V, Panizzi P, et al. Identification of splenic reservoir monocytes and their deployment to inflammatory sites. *Science* (2009) **325**:612–6. doi:10.1126/science.1175202

9. Shi C, Jia T, Mendez-Ferrer S, Hohl TM, Serbina NV, Lipuma L, et al. Bone marrow mesenchymal stem and progenitor cells induce monocyte emigration in response to circulating toll-like receptor ligands. *Immunity* (2011) **34**:590–601. doi:10.1016/j.immuni.2011.02.016

10. Varol C, Landsman L, Fogg DK, Greenshtein L, Gildor B, Margalit R, et al. Monocytes give rise to mucosal, but not splenic, conventional dendritic cells. *J Exp Med* (2007) **204**:171–80. doi:10.1084/jem.20061011

11. Alexis N, Soukup J, Ghio A, Becker S. Sputum phagocytes from healthy individuals are functional and activated: a flow cytometric comparison with cells in bronchoalveolar lavage and peripheral blood. *Clin Immunol* (2000) **97**:21–32. doi:10.1006/clim.2000.4911

12. Frankenberger M, Eder C, Hofer TP, Heimbeck I, Skokann K, Kassner G, et al. Chemokine expression by small sputum macrophages in COPD. *Mol Med* (2011) **17**:762–70. doi:10.2119/molmed.2010.00202

13. Tamoutounour S, Guilliams M, Montanana Sanchis F, Liu H, Terhorst D, Malosse C, et al. Origins and functional specialization of macrophages and of conventional and monocyte-derived dendritic cells in mouse skin. *Immunity* (2013) **39**:925–38. doi:10.1016/j.immuni.2013.10.004

14. Jakubzick C, Gautier EL, Gibbings SL, Sojka DK, Schlitzer A, Johnson TE, et al. Minimal differentiation of classical monocytes as they survey steady-state tissues and transport antigen to lymph nodes. *Immunity* (2013) **39**:599–610. doi:10.1016/j.immuni.2013.08.007

15. Ziegler-Heitbrock HWL, Ulevitch RJ. CD14: cell surface receptor and differentiation marker. *Immunol Today* (1993) **14**:121–5. doi:10.1016/0167-5699(93)90212-4

16. Sunderkotter C, Nikolic T, Dillon MJ, Van Rooijen N, Stehling M, Drevets DA, et al. Subpopulations of mouse blood monocytes differ in maturation stage and inflammatory response. *J Immunol* (2004) **172**:4410–7. doi:10.4049/jimmunol.172.7.4410

17. Drevets DA, Schawang JE, Mandava VK, Dillon MJ, Leenen PJ. Severe *Listeria monocytogenes* infection induces development of monocytes with distinct phenotypic and functional features. *J Immunol* (2010) **185**:2432–41. doi:10.4049/jimmunol.1000486

18. Schwarz H, Schmittner M, Duschl A, Horejs-Hoeck J. Residual endotoxin contaminations in recombinant proteins are sufficient to activate human CD1c+ dendritic cells. *PLoS One* (2014) **9**:e113840. doi:10.1371/journal.pone.0113840

19. Ziegler-Heitbrock HWL, Pechumer H, Petersmann I, Durieux JJ, Vita N, Labeta MO, et al. CD14 is expressed and functional in human B cells. *Eur J Immunol* (1994) **24**:1937–40. doi:10.1002/eji.1830240835

20. Gautier EL, Shay T, Miller J, Greter M, Jakubzick C, Ivanov S, et al. Gene-expression profiles and transcriptional regulatory pathways that underlie the identity and diversity of mouse tissue macrophages. *Nat Immunol* (2012) **13**:1118–28. doi:10.1038/ni.2419

21. Steinman RM, Cohn ZA. Identification of a novel cell type in peripheral lymphoid organs of mice. I. Morphology, quantitation, tissue distribution. *J Exp Med* (1973) **137**:1142–62. doi:10.1084/jem.137.5.1142

22. Fogg DK, Sibon C, Miled C, Jung S, Aucouturier P, Littman DR, et al. A clonogenic bone marrow progenitor specific for macrophages and dendritic cells. *Science* (2006) **311**:83–7. doi:10.1126/science.1117729

23. Sathe P, Metcalf D, Vremec D, Naik SH, Langdon WY, Huntington ND, et al. Lymphoid tissue and plasmacytoid dendritic cells and macrophages do not share a common macrophage-dendritic cell-restricted progenitor. *Immunity* (2014) **41**:104–15. doi:10.1016/j.immuni.2014.05.020

24. Onai N, Ohteki T. Bipotent or oligopotent? A macrophage and DC progenitor revisited. *Immunity* (2014) **41**:5–7. doi:10.1016/j.immuni.2014.07.004

25. Sallusto F, Lanzavecchia A. Efficient presentation of soluble antigen by cultured human dendritic cells is maintained by granulocyte/macrophage colony-stimulating factor plus interleukin 4 and downregulated by tumor necrosis factor alpha. *J Exp Med* (1994) **179**:1109–18. doi:10.1084/jem.179.4.1109

26. Robbins SH, Walzer T, Dembele D, Thibault C, Defays A, Bessou G, et al. Novel insights into the relationships between dendritic cell subsets in human and mouse revealed by genome-wide expression profiling. *Genome Biol* (2008) **9**:R17. doi:10.1186/gb-2008-9-1-r17

27. Randolph GJ, Inaba K, Robbiani DF, Steinman RM, Muller WA. Differentiation of phagocytic monocytes into lymph node dendritic cells in vivo. *Immunity* (1999) **11**:753–61. doi:10.1016/S1074-7613(00)80149-1

28. Landsman L, Varol C, Jung S. Distinct differentiation potential of blood monocyte subsets in the lung. *J Immunol* (2007) **178**:2000–7. doi:10.4049/jimmunol.178.4.2000

29. Strobl H, Scheinecker C, Riedl E, Csmarits B, Bello-Fernandez C, Pickl WF, et al. Identification of CD68+lin- peripheral blood cells with dendritic precursor characteristics. *J Immunol* (1998) **161**:740–8.

30. Dzionek A, Fuchs A, Schmidt P, Cremer S, Zysk M, Miltenyi S, et al. BDCA-2, BDCA-3, and BDCA-4: three markers for distinct subsets of dendritic cells in human peripheral blood. *J Immunol* (2000) **165**:6037–46. doi:10.4049/jimmunol.165.11.6037

31. Frankenberger M, Hofer TP, Marei A, Dayyani F, Schewe S, Strasser C, et al. Transcript profiling of CD16-positive monocytes reveals a unique molecular fingerprint. *Eur J Immunol* (2012) **42**:957–74. doi:10.1002/eji.201141907

32. Debili N, Coulombel L, Croisille L, Katz A, Guichard J, Breton-Gorius J, et al. Characterization of a bipotent erythro-megakaryocytic progenitor in human bone marrow. *Blood* (1996) **88**:1284–96.

33. Klimchenko O, Mori M, Distefano A, Langlois T, Larbret F, Lecluse Y, et al. A common bipotent progenitor generates the erythroid and megakaryocyte lineages in embryonic stem cell-derived primitive hematopoiesis. *Blood* (2009) **114**:1506–17. doi:10.1182/blood-2008-09-178863

34. Macaulay IC, Tijssen MR, Thijssen-Timmer DC, Gusnanto A, Steward M, Burns P, et al. Comparative gene expression profiling of in vitro differentiated megakaryocytes and erythroblasts identifies novel activatory and inhibitory platelet membrane proteins. *Blood* (2007) **109**:3260–9. doi:10.1182/blood-2006-07-036269

35. Norris DA, Morris RM, Sanderson RJ, Kohler PF. Isolation of functional subsets of human peripheral blood monocytes. *J Immunol* (1979) **123**:166–72.

36. Zembala M, Uracz W, Ruggiero I, Mytar B, Pryjma J. Isolation and functional characteristics of FcR+ and FcR- human monocyte subsets. *J Immunol* (1984) **133**:1293–9.

37. Passlick B, Flieger D, Ziegler-Heitbrock HWL. Identification and characterization of a novel monocyte subpopulation in human peripheral blood. *Blood* (1989) **74**:2527–34.

38. Ziegler-Heitbrock L, Ancuta P, Crowe S, Dalod M, Grau V, Hart DN, et al. Nomenclature of monocytes and dendritic cells in blood. *Blood* (2010) **116**:e74–80. doi:10.1182/blood-2010-02-258558

39. Ziegler-Heitbrock L, Hofer TP. Toward a refined definition of monocyte subsets. *Front Immunol* (2013) **4**:23. doi:10.3389/fimmu.2013.00023

40. Rogacev KS, Cremers B, Zawada AM, Seiler S, Binder N, Ege P, et al. CD14++CD16+ monocytes independently predict cardiovascular events: a cohort study of 951 patients referred for elective coronary angiography. *J Am Coll Cardiol* (2012) **60**:1512–20. doi:10.1016/j.jacc.2012.07.019

41. Wong KL, Yeap WH, Tai JJ, Ong SM, Dang TM, Wong SC. The three human monocyte subsets: implications for health and disease. *Immunol Res* (2012) **53**:41–57. doi:10.1007/s12026-012-8297-3

42. Ziegler-Heitbrock L. Monocyte subsets in man and other species. *Cell Immunol* (2014) **289**:135–9. doi:10.1016/j.cellimm.2014.03.019

43. Kim WK, Sun Y, Do H, Autissier P, Halpern EF, Piatak M Jr, et al. Monocyte heterogeneity underlying phenotypic changes in monocytes according to SIV disease stage. *J Leukoc Biol* (2010) **87**:557–67. doi:10.1189/jlb.0209082

44. Kwissa M, Nakaya HI, Oluoch H, Pulendran B. Distinct TLR adjuvants differentially stimulate systemic and local innate immune responses in nonhuman primates. *Blood* (2012) **119**:2044–55. doi:10.1182/blood-2011-10-388579

45. Ingersoll MA, Spanbroek R, Lottaz C, Gautier EL, Frankenberger M, Hoffmann R, et al. Comparison of gene expression profiles between human and mouse monocyte subsets. *Blood* (2010) **115**:e10–9. doi:10.1182/blood-2009-07-235028

46. de Baey A, Mende I, Riethmueller G, Baeuerle PA. Phenotype and function of human dendritic cells derived from M-DC8(+) monocytes. *Eur J Immunol* (2001) **31**:1646–55. doi:10.1002/1521-4141(200106)31:6<1646::AID-IMMU1646>3.0.CO;2-X

47. Schakel K, Kannagi R, Kniep B, Goto Y, Mitsuoka C, Zwirner J, et al. 6-Sulfo LacNAc, a novel carbohydrate modification of PSGL-1, defines an inflammatory type of human dendritic cells. *Immunity* (2002) **17**:289–301. doi:10.1016/S1074-7613(02)00393-X

48. Thomas K, Dietze K, Wehner R, Metz I, Tumani H, Schultheiss T, et al. Accumulation and therapeutic modulation of 6-sulfo LacNAc(+) dendritic cells in multiple sclerosis. *Neurol Neuroimmunol Neuroinflamm* (2014) **1**:e33. doi:10.1212/NXI.0000000000000033

49. Belge KU, Dayyani F, Horelt A, Siedlar M, Frankenberger M, Frankenberger B, et al. The proinflammatory CD14+CD16+DR++ monocytes are a major source of TNF. *J Immunol* (2002) **168**:3536–42. doi:10.4049/jimmunol.168.7.3536

50. Fingerle-Rowson G, Angstwurm M, Andreesen R, Ziegler-Heitbrock HWL. Selective depletion of CD14+ CD16+ monocytes by glucocorticoid therapy. *Clin Exp Immunol* (1998) **112**:501–6. doi:10.1046/j.1365-2249.1998.00617.x

51. Dutertre CA, Amraoui S, DeRosa A, Jourdain JP, Vimeux L, Goguet M, et al. Pivotal role of M-DC8(+) monocytes from viremic HIV-infected patients in TNFalpha overproduction in response to microbial products. *Blood* (2012) **120**:2259–68. doi:10.1182/blood-2012-03-418681

52. Siedlar M, Frankenberger M, Ziegler-Heitbrock LH, Belge KU. The M-DC8-positive leukocytes are a subpopulation of the CD14+ CD16+ monocytes. *Immunobiology* (2000) **202**:11–7. doi:10.1016/S0171-2985(00)80047-9

53. Cros J, Cagnard N, Woollard K, Patey N, Zhang SY, Senechal B, et al. Human CD14dim monocytes patrol and sense nucleic acids and viruses via TLR7 and TLR8 receptors. *Immunity* (2010) **33**:375–86. doi:10.1016/j.immuni.2010.08.012

54. Wong KL, Tai JJ, Wong WC, Han H, Sem X, Yeap WH, et al. Gene expression profiling reveals the defining features of the classical, intermediate, and nonclassical human monocyte subsets. *Blood* (2011) **118**:e16–31. doi:10.1182/blood-2010-12-326355

55. Maurer D, Fiebiger E, Reininger B, Wolff-Winiski B, Jouvin MH, Kilgus O, et al. Expression of functional high affinity immunoglobulin E receptors (Fc epsilon RI) on monocytes of atopic individuals. *J Exp Med* (1994) **179**:745–50. doi:10.1084/jem.179.2.745

56. Dehlink E, Baker AH, Yen E, Nurko S, Fiebiger E. Relationships between levels of serum IgE, cell-bound IgE, and IgE-receptors on peripheral blood cells in a pediatric population. *PLoS One* (2010) **5**:e12204. doi:10.1371/journal.pone.0012204

57. Greer AM, Wu N, Putnam AL, Woodruff PG, Wolters P, Kinet JP, et al. Serum IgE clearance is facilitated by human FcepsilonRI internalization. *J Clin Invest* (2014) **124**:1187–98. doi:10.1172/JCI68964

58. Clanchy FI, Holloway AC, Lari R, Cameron PU, Hamilton JA. Detection and properties of the human proliferative monocyte subpopulation. *J Leukoc Biol* (2006) **79**:757–66. doi:10.1189/jlb.0905522

59. Bucala R, Spiegel LA, Chesney J, Hogan M, Cerami A. Circulating fibrocytes define a new leukocyte subpopulation that mediates tissue repair. *Mol Med* (1994) **1**:71–81.

60. Komano Y, Nanki T, Hayashida K, Taniguchi K, Miyasaka N. Identification of a human peripheral blood monocyte subset that differentiates into osteoclasts. *Arthritis Res Ther* (2006) **8**:R152. doi:10.1186/ar2046

61. Fingerle G, Pforte A, Passlick B, Blumenstein M, Strobel M, Ziegler-Heitbrock HWL. The novel subset of CD14+/CD16+ blood monocytes is expanded in sepsis patients. *Blood* (1993) **82**:3170–6.

62. Horelt A, Belge KU, Steppich B, Prinz J, Ziegler-Heitbrock L. The CD14+CD16+ monocytes in erysipelas are expanded and show reduced cytokine production. *Eur J Immunol* (2002) **32**:1319–27. doi:10.1002/1521-4141(200205)32:5<1319::AID-IMMU1319>3.0.CO;2-2

63. Shalova IN, Kajiji T, Lim JY, Gomez-Pina V, Fernandez-Ruiz I, Arnalich F, et al. CD16 regulates TRIF-dependent TLR4 response in human monocytes and their subsets. *J Immunol* (2012) **188**:3584–93. doi:10.4049/jimmunol.1100244

64. Fingerle-Rowson G, Auers J, Kreuzer E, Fraunberger P, Blumenstein M, Ziegler-Heitbrock LH. Expansion of CD14+CD16+ monocytes in critically ill cardiac surgery patients. *Inflammation* (1998) **22**:367–79. doi:10.1023/A:1022316815196

65. Dayyani F, Belge KU, Frankenberger M, Mack M, Berki T, Ziegler-Heitbrock L. Mechanism of glucocorticoid-induced depletion of human CD14+CD16+ monocytes. *J Leukoc Biol* (2003) **74**:33–9. doi:10.1189/jlb.1202612

66. Korkosz M, Bukowska-Strakova K, Sadis S, Grodzicki T, Siedlar M. Monoclonal antibodies against macrophage colony-stimulating factor diminish the number of circulating intermediate and nonclassical (CD14(++)CD16(+)/CD14(+)CD16(++)) monocytes in rheumatoid arthritis patient. *Blood* (2012) **119**:5329–30. doi:10.1182/blood-2012-02-412551

67. MacDonald KP, Palmer JS, Cronau S, Seppanen E, Olver S, Raffelt NC, et al. An antibody against the colony-stimulating factor 1 receptor depletes the resident subset of monocytes and tissue- and tumor-associated macrophages but does not inhibit inflammation. *Blood* (2010) **116**:3955–63. doi:10.1182/blood-2010-02-266296

68. Lenzo JC, Turner AL, Cook AD, Vlahos R, Anderson GP, Reynolds EC, et al. Control of macrophage lineage populations by CSF-1 receptor and GM-CSF in homeostasis and inflammation. *Immunol Cell Biol* (2012) **90**:429–40. doi:10.1038/icb.2011.58

69. Frankenberger M, Ekici AB, Angstwurm MW, Hoffmann H, Hofer TP, Heimbeck I, et al. A defect of CD16-positive monocytes can occur without disease. *Immunobiology* (2013) **218**:169–74. doi:10.1016/j.imbio.2012.02.013

70. Heine GH, Ortiz A, Massy ZA, Lindholm B, Wiecek A, Martinez-Castelao A, et al. Transplant, monocyte subpopulations and cardiovascular risk in chronic kidney disease. *Nat Rev Nephrol* (2012) **8**:362–9. doi:10.1038/nrneph.2012.41

71. Rogacev KS, Zawada AM, Emrich I, Seiler S, Bohm M, Fliser D, et al. Lower Apo A-I and lower HDL-C levels are associated with higher intermediate CD14++CD16+ monocyte counts that predict cardiovascular events in chronic kidney disease. *Arterioscler Thromb Vasc Biol* (2014) **34**:2120–7. doi:10.1161/ATVBAHA.114.304172

15

Microglia versus myeloid cell nomenclature during brain inflammation

Melanie Greter*, Iva Lelios and Andrew Lewis Croxford

Institute of Experimental Immunology, University of Zurich, Zurich, Switzerland

Edited by:
Martin Guilliams,
Ghent University – VIB, Belgium

Reviewed by:
Steffen Jung,
Weizmann Institute of Science, Israel
Marco Prinz,
University of Freiburg, Germany

*Correspondence:
Melanie Greter,
Institute of Experimental Immunology,
University of Zurich,
Winterthurerstrasse 190, Zurich 8057,
Switzerland
greter@immunology.uzh.ch

As immune sentinels of the central nervous system (CNS), microglia not only respond rapidly to pathological conditions but also contribute to homeostasis in the healthy brain. In contrast to other populations of the myeloid lineage, adult microglia derive from primitive myeloid precursors that arise in the yolk sac early during embryonic development, after which they self-maintain locally and independently of blood-borne myeloid precursors. Under neuro-inflammatory conditions such as experimental autoimmune encephalomyelitis, circulating monocytes invade the CNS parenchyma where they further differentiate into macrophages or inflammatory dendritic cells. Often it is difficult to delineate resident microglia from infiltrating myeloid cells using currently known markers. Here, we will discuss the current means to reliably distinguish between these populations, and which recent advances have helped to make clear definitions between phenotypically similar, yet functionally diverse myeloid cell types.

Keywords: microglia, macrophage, monocyte, dendritic cell, CNS inflammation

Introduction

Most tissues are populated by incredibly diverse and abundant myeloid cells. By contrast, the central nervous system (CNS) harbors comparatively few myeloid cell subsets. This is likely due to the immune privilege and relative isolation enjoyed by the CNS compared to other non-lymphoid tissues such as the gut or the lung, which are continually confronted with foreign entities. In the steady state, the CNS houses several populations of myeloid cells with distinct localizations including perivascular, choroid plexus, and meningeal macrophages/dendritic cells (DCs) and microglia, which are the most abundant (1). Microglia are considered the resident macrophages of the brain given that they are the only myeloid cells present in the CNS parenchyma. Microglia perform both homeostatic and immune-related functions and constitute about 5–20% of all cells in the CNS (2). They use their "ramified" morphology to act as immune sentinels, extending specialized processes, and sampling the local environment for foreign bodies (3, 4). Numerous recent reports have unmasked additional functions for microglia other than being simply the brain's intrinsic immune system. For example, microglia are also critical for neuronal development, adult neurogenesis, learning-dependent synapse formation, and brain homeostasis (5–7). Microglia are classified as tissue resident macrophages but are clearly ontogenically distinct from other members of the mononuclear phagocyte system (MPS), which includes DCs, monocytes, and macrophages. Microglia originate from primitive macrophages that derive from erythro-myeloid precursors in the yolk sac (8–10). These primitive yolk sac macrophages colonize the developing brain in mice as early as embryonic day 9.5 (8). Throughout adult life microglia remain of embryonic origin in the healthy CNS and maintain themselves locally without any detectable contribution from circulating myeloid progenitors including monocytes. Yolk sac macrophages and microglia precursors in the developing brain express high levels of the fractalkine receptor (CX3CR1) and are positive for the integrin alpha M (Itgam, also know as CD11b; macrophage-1 antigen, Mac-1), F4/80, and the macrophage-colony stimulating factor receptor 1 (Csf-1R, CD115) similar to adult microglia as described below (8). Compared to adult microglia, however, microglia precursors are CD45^{hi}. The development of microglia is dependent on Csf-1R (CD115), the transcription factors PU.1 and Irf8 but is independent of Myb, which is crucial for the development of hematopoietic stem cells (HSCs) (8, 9, 11, 12). In contrast to microglia, recent adoptive transfer and fate-mapping studies revealed that other macrophage populations are either embryonically derived from definitive hematopoiesis (e.g., alveolar or heart macrophages) or are constantly replaced by circulating monocytes (e.g., dermal or gut macrophages) (10, 13–18). Aside from the unique ontogeny of microglia within the MPS, a clear classification of microglia compared to other tissue macrophages in terms of phenotype and function has been difficult. Only recently, transcriptome and epigenetic analysis identified genes uniquely expressed and regulated by microglia but not by other macrophage populations (19–24). These studies might be useful to classify and distinguish microglia from other myeloid cells.

Microglia Markers in Steady State

In steady state conditions, microglia express surface markers typically present on many other tissue macrophages and/or monocytes such as CD11b, F4/80, Fc-gamma receptor 1 (CD64), and CD115 (Csf-1R), ionized calcium-binding adapter molecule 1 (Iba-1) and

proto-oncogene tyrosine-protein kinase MER (MerTK) (**Figure 1**) (19). In contrast to microglia, which are γ-irradiation resistant, perivascular myeloid cells are replaced by bone marrow (BM)-derived precursors after total-body irradiation and BM transplantation (25–28). However, the exact ontogeny of (non-microglia) myeloid cells associated with the CNS and whether they are also able to maintain themselves locally is, to date, not known (29). These perivascular cells are equipped to present antigen (varying levels of MHCII and CD11c). Whether they represent a homogeneous distinct population or a heterogeneous population of macrophages and/or DCs is not entirely resolved. In the past, a cell expressing F4/80 was deemed to be a macrophage, whereas a cell expressing CD11c was considered a DC. It is clear now that subsets of DCs can also express F4/80 and certain macrophage populations express CD11c. Upon Flt3L treatment, a CD11c$^+$MHCII$^+$ population in the meninges and choroid plexus expanded, which is indicative of the DC lineage whose development is dependent on Flt3L signaling (27, 28). In addition, a limited number of CD11c$^+$ myeloid cells were also described to be in a juxtavascular location in the CNS parenchyma (30). These cells might, however, represent *bona fide* microglia expressing CD11c in certain regions of the brain. Further studies are required to dissect the ontogeny and characterize these elusive myeloid cells associated with the CNS in the steady state.

As common to many other macrophage populations including microglia, most of the CNS-associated myeloid cells also express CD11b, CD115, Iba-1, and F4/80 (31). Therefore, apart from their location, the only available means to unequivocally distinguish microglia from other CNS-resident myeloid populations (CNS-associated macrophages/DCs) and circulating monocytes

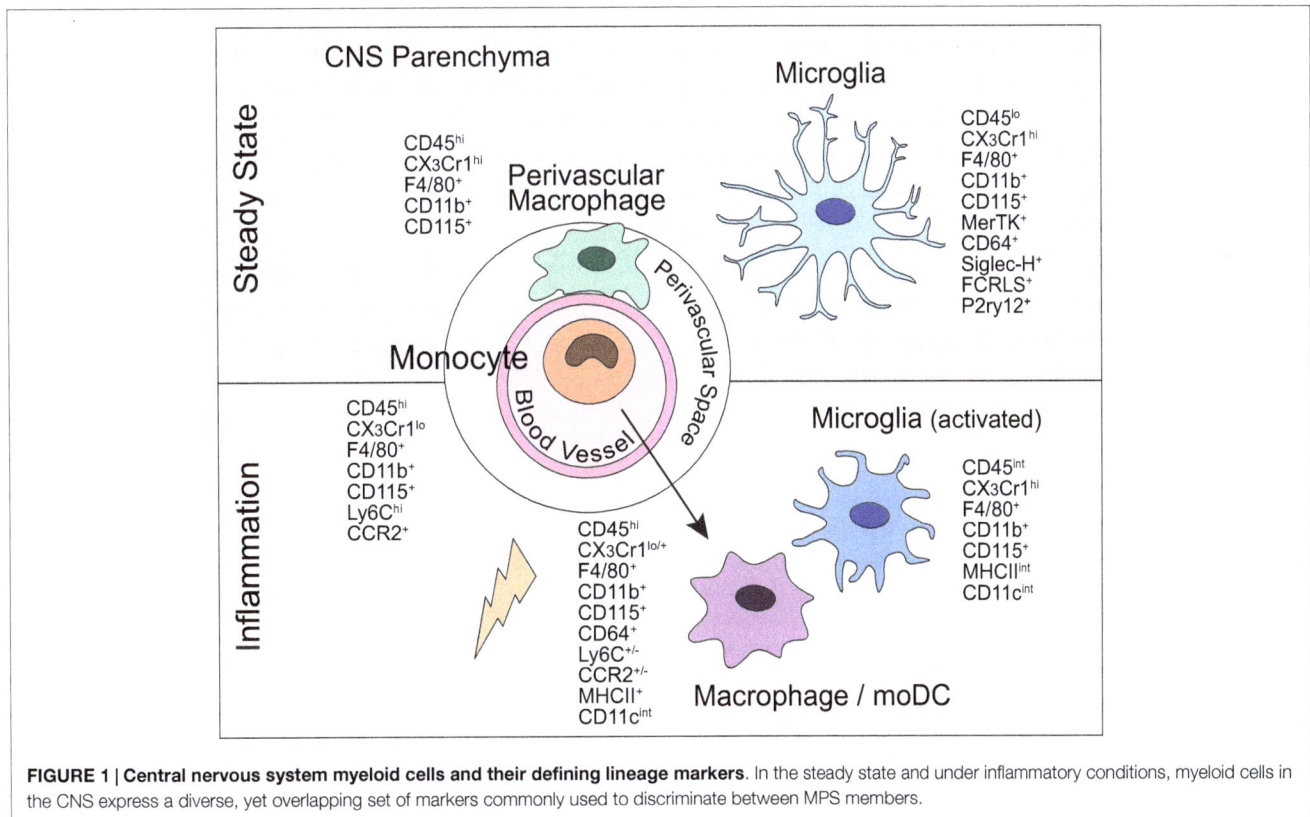

FIGURE 1 | **Central nervous system myeloid cells and their defining lineage markers**. In the steady state and under inflammatory conditions, myeloid cells in the CNS express a diverse, yet overlapping set of markers commonly used to discriminate between MPS members.

is by the reduced expression of the common leukocyte antigen CD45, which is readily detectable by flow cytometry. Adult microglia, unlike most other tissue macrophages, constitutively express high levels of the fractalkine receptor CX3CR1 (32). A major advance in the microglia field has been the generation of *Cx3Cr1^{creER}* mice (32–34). This tamoxifen-inducible Cre-recombinase under the CX3CR1 promoter allows for microglia-specific gene targeting. Despite the fractalkine receptor being expressed by monocytes and myeloid precursors in the BM, microglia remain a self-contained population in the CNS and therefore remain targeted long after ceasing of tamoxifen administration, returning the short-lived, circulating myeloid cells to their wild-type origin.

Only recently, gene expression studies have identified surface markers and transcription factors specifically expressed by steady state microglia but not by other macrophage populations or monocytes. These include, for example, sialic acid-binding immunoglobulin-type lectin H (Siglec-H), Fc receptor-like S (Fcrls), and purinergic receptor P2Y G-protein coupled 12 (P2ry12) (20, 21). Furthermore, microglia seem to be the only hematopoietic cell population that specifically expresses Sal-like 1 (Sall1), a transcription factor that plays a crucial role in kidney development (35). While previous studies have reported expression of Sall1 only by stromal cells, in the adult CNS this factor is expressed exclusively by microglia. These aforementioned gene expression studies have compared the transcriptome of microglia to either macrophages derived from the spleen, the lung, the peritoneum, or to monocytes. Whether these microglia core signature markers are also expressed by CNS-associated macrophages/DCs remains to be shown.

Microglia Markers in Inflammation

In contrast to the healthy brain, during neuro-inflammation the picture becomes far more complicated. A hallmark of microglia is their rapid activation after a CNS insult, resulting in their migration toward injury, proliferation, and their change in morphology. They take on a more "amoeboid" shape with shorter and thicker processes, display increased immunoreactivity for Iba-1 and upregulate CD45. Experimental autoimmune encephalomyelitis (EAE), which is a mouse model for multiple sclerosis (MS), is characterized by infiltration of T cells, monocytes, and neutrophils. Monocytes and their progeny [macrophages/monocyte-derived DCs (moDCs)] are undoubtedly the prevailing cell type in the lesions (see below). However, activated microglia are also clearly detected in the vicinity of the inflammatory lesions. The downregulation of Ly6C by monocytes upon their differentiation adds complexity to the separation of these two distinct cell types based on the commonly used cell surface markers. Additionally, molecules involved in antigen presentation and T cell stimulation, which are barely detectable in steady state microglia, are expressed to some level by microglia already at disease onset and retain expression throughout disease progression. These markers include major histocompatibility complex class II (MHCII), CD11c (also known as integrin alpha X, Itgax), CD80 (B7-1), CD86 (B7-2), and CD40 (36–38). Under these conditions, it is considerably more difficult to distinguish these activated

microglia from inflammatory monocyte-derived cells. Similar changes in microglia surface markers have been observed in mouse models of neurodegenerative disease such as Alzheimer's disease (AD), Parkinson's disease, and amyotrophic lateral sclerosis (ALS) (20, 31, 39).

A recent study, however, has used differentially expressed chemokine receptors on the surface of microglia and monocytes to distinguish those two myeloid populations and study their function during neuro-inflammation. Microglia were identified by their high expression of CX3CR1 whereas infiltrating monocytes, which subsequently differentiate into macrophages/DCs, were defined by their high expression of C–C chemokine receptor type 2 (CCR2), a receptor mediating monocyte recruitment to sites of inflammation. Gene expression profiles from macrophages versus "embryonically derived" microglia at different stages of EAE show that despite some similarities between these inflammatory cell types, microglia exhibit a distinct molecular signature (40). This genetic distinction reflects a different function of resident microglia and infiltrating monocytes under pathological circumstances. While monocyte-derived macrophages seemed to be the effector cell type causing CNS damage, microglia might have a regulatory function and could play a role in tissue repair and homeostasis (40). Another report also showed that monocytes recruited to the CNS in EAE do not acquire microglia-signature genes (21). These studies will unquestionably help attributing unique functions to microglia and CNS-invading myeloid cells in different pathological conditions in the brain.

Whether microglia-specific surface markers and transcription factors alter their expression between steady state and inflammation remains unclear. The microglia-specific ATP receptor P2ry12 was downregulated under inflammatory conditions such as lipopolysaccharide (LPS) systemic injection or in SOD1 mouse-model of ALS (41, 42). On the other hand, P2ry12 and Fcrls continue to be expressed by microglia in EAE but are not expressed by infiltrating monocytes (21). Therefore, further studies need to be undertaken in order to better characterize these "new" phenotypic microglia markers under pathological conditions.

Monocyte-Derived Microglia

Even though microglia homeostasis is maintained through local self-renewal, under certain conditions circulating precursors can give rise to microglia-like cells. For example, early studies using BM-chimeras showed that up to 10–20% of microglia were reconstituted by donor-derived cells 6-12 months after total-body irradiation and BM transplantation (43). However, this engraftment of BM-derived microglia can only be seen upon blood–brain-barrier (BBB) disruption (e.g. irradiation) and becomes minimal in models where the BBB is unperturbed (e.g. protection of the head during irradiation and parabiosis) (44, 45). This clearly indicates that under steady state conditions, monocytes or BM-derived myeloid progenitors do not infiltrate the CNS parenchyma and thus do not give rise to adult microglia. Similarly, as described above in experimental models of neuro-inflammation, monocytes infiltrate the CNS and differentiate into effector cells resembling phenotypically activated microglia. Despite these similarities, monocyte-derived cells do not persist in the CNS after inflammation has been resolved and thus do not contribute to long-lived microglia (46).

Finally, local administration of ganciclovir to transgenic mice expressing the thymidine kinase of herpes simplex virus under the CD11b promoter (CD11b-HSVTK) leads to a rapid depletion of microglia (47, 48). Subsequently, BM-derived cells enter the CNS and differentiate into long-lived microglia-like cells. Notably, while these cells form a network filling the niche for embryonically derived microglia, they do not obtain a complete microglia phenotype. Monocyte-derived "microglia" in this model show a less-ramified morphology and a higher expression of CD45 compared to yolk sac-derived microglia resembling more activated microglia (48). It has not yet been investigated whether these monocyte-derived "microglia" functionally resemble embryonically derived microglia or whether they acquire microglia-signature genes (Siglec-H, Fcrls, P2ry12) as described above.

Recent studies used a Csf-1R inhibitor to deplete microglia. Upon treatment stop, microglia were repopulated within 1 week (49, 50). These studies showed that new "microglia" were derived from CNS-resident nestin-positive precursors and resembled embryonically derived microglia in response to an inflammatory stimulus. Animals with newly repopulated "microglia" did not display any impairment in behavior, cognition, or motor function compared to control animals (50).

Monocytes and Monocyte-Derived Cells in CNS Inflammation

Brain inflammation or "encephalitis" invariably results in a reshaping of the myeloid cell populations inhabiting the CNS. An inflammatory response brought on by either infection or autoimmune manifestations results in a rapid increase in blood-derived cellularity to this otherwise dormant site. Despite EAE being fully dependent on T helper cells (51), the vast majority of the inflammatory infiltrate seen in EAE is of myeloid derivation. Two types of monocytes exist including the classical monocytes ($Ly6C^{hi}CCR2^{+}CX_3CR1^{lo}$) and the non-classical monocytes ($Ly6C^{lo}CCR2^{-}CX_3CR1^{hi}$). Here, we will only discuss $Ly6C^{hi}$ monocytes given that during neuro-inflammation, this is the subset recruited to the brain. Engraftment of phagocytes derived from circulating $CCR2^{+}$ monocytes has also been shown in an AD mouse model (39). $Ly6C^{hi}$ monocytes egress from the BM and cross the BBB in a CCR2-dependent manner (52, 53) followed by their differentiation into macrophages/moDCs and upregulation of a set of cell surface markers (e.g., MHCII, CD11c) expressed on a wide variety of MPS members. Likewise, microglia progressively alter their phenotype to resemble more classically activated macrophages during CNS inflammation, infection, and neuronal or myelin damage (54).

$Ly6C^{+}$ monocytes were shown to migrate into the CNS prior to disease onset and precede the development of paralysis and subsequent clinical manifestations of EAE, when "DC-like" cells are found in abundance in the inflamed tissue (55, 56). This corresponds well with a previous report showing $CD205^{+}$ myeloid cells accumulating in the meninges, choroid plexus, and subpial space of the spinal cord and in perivascular cuffs in demyelinating lesions during acute disease (57). $CD11b^{+}$ DCs within the inflamed CNS were demonstrated to be critical for the propagation of EAE (27, 58, 59). Further phenotypical characterization would be required to demarcate their lineage whether they resemble moDCs or are more similar to classical DCs. Indeed, monocyte-derived antigen presenting cells (APCs) have been shown to be required for optimal priming of T cells in models of infection (60). Current evidence suggests that phenotypically similar macrophages in the CNS can not only contribute to the generation of inflammatory lesions and perform a pathogenic role in the demyelination process but also contribute to regenerative repair mechanisms to resolve inflammation (61, 62). These studies emphasize that distinct functions are attributed to the different subsets of myeloid cells in the course of a CNS inflammation. As such, a complete understanding of cell types based on surface phenotype alone would be of great benefit both in preclinical models of CNS inflammation and also in human patients.

Even with the knowledge we now possess on myeloid cell diversity, it is still commonplace in the literature using animal models of CNS inflammation to use a simplistic $CD45^{hi}CD11b^{hi}$ gating strategy to separate CNS infiltrating, blood-derived myeloid cells from CNS-resident, embryonically derived microglia ($CD45^{low}$) (63). Efforts to sort cells using a broad $CD45^{hi}CD11b^{hi}$ surface phenotype from within the inflamed CNS will inevitably result in analysis of multiple cell types, lacking any of the desired specificity. Indeed, without the removal of $Ly6G^{hi}$ cells during sorting, a mixed population is inevitable and expression profiles subsequently attributed to moDCs are either confused with, or heavily influenced by, an abundant neutrophil contamination. Even if the effort is taken to remove neutrophils, moDCs at various stages of development will be incorporated. This distinction is increasingly important given that both neutrophils and moDCs have been shown to mediate BBB permeability and demyelination, and that different pathogenic mechanisms are likely active in the two populations during the same inflammation (40, 64).

We know that at least four clearly distinct cell types share this rather non-specific $CD45^{hi}CD11b^{hi}$ surface phenotype in an inflamed CNS, namely neutrophils ($CD11b^{+}Ly6G^{+}$), monocytes ($CD11b^{+}Ly6C^{hi}CX_3CR1^{low}$), and their progeny such as moDCs and/or activated macrophages (**Figure 1**). The latter two cell types represent most likely the same population with just different names assigned by different studies. $Ly6C^{hi}$ monocytes that have migrated into the CNS can further be subdivided into numerous differentiation stages characterized by the upregulation of CD11c and MHCII, with the concomitant downregulation of Ly6C and CCR2. Upon differentiation and upregulation of MHCII, monocytes are then called moDCs/activated macrophages. Thus, moDCs in the CNS are characterized by the expression of $CD11b^{+}F4/80^{+}MHCII^{+}CD11c^{int}Ly6C^{+/-}$. These moDCs/activated macrophages also express CD64 and likely also MerTK, which both are universally expressed by tissue macrophages including microglia (19). Interestingly, it has been shown that monocytes recruited to the CNS during EAE do not express the newly identified microglia markers Fcrls and P2ry12 highlighting again the diverse ontogeny of these cell types and suggesting that the microglia-signature genes are indeed specific to microglia rather than location (CNS) specific (21). The FcεRIα (MAR-1) has been suggested to represent a moDC marker. Whether moDCs in the inflamed CNS express MAR-1 has so far not been analyzed (65). Perhaps a more functional distinction should be drawn on the level of relevance for the inflammatory process to persist. CNS-infiltrating myeloid cells with DC-like morphology express MHCII, CD40, and CD86, all of which have critical roles in multiple inflammatory models (66). The CD86/CD28 interaction between T cells and APCs is of critical importance for T cell activation. Furthermore, the CD40/CD40L interaction induces a maturation pathway within the inflamed CNS, resulting in further costimulatory capabilities and proinflammatory cytokine expression (67, 68). The levels of CD40 on monocyte-derived cells in the inflamed CNS are variable but generally not as high as on classical DCs.

After activation, inflammatory macrophages can not only express a wide range of inflammatory cytokines but also oxygen-based chemically reactive molecules involved in host defense. The route an activated macrophage takes depends largely on the T cell and/or NK cell-derived cytokines present during their activation. For example, activation in the presence of LPS and IFN-γ leads to a "classical" activation (often called "M1"), resulting in secretion of high levels of TNF-α, iNOS, IL-1, IL-6, and IL-12. Conversely, activation of the same cells in the presence of IL-4 and IL-10 will result in rapid upregulation of IL-10, production of Arginase-1, and upregulation of the mannose receptor CD206, generating a macrophage capable of suppressing T cell activity (called "M2") (69). This intracellular divergence in phenotype illustrates that an apparently similar cell expressing F4/80, CD64, CD11b, MHCII, and CD11c on its surface may, in fact, differ greatly in its function. Indeed, markers identifying both M1 and

M2 macrophage populations have been shown synergistically in CNS biopsies obtained from MS patients. CD40, CD64, CD86, and CD32, mannose receptor and CD163 were co-expressed in the large majority of foamy macrophages found in lesional CNS (70). Therefore, surface characterization of inflammatory macrophages would appear insufficient and may mask different macrophage populations in direct opposition to each other, depending on the type of inflammation taking place. Generally in the steady state, tissue macrophages display an "M2-like" phenotype and are critical for tissue homeostasis. Interestingly, in a model of spinal cord injury, it was shown that M2 macrophages (CD11b[+]F4/80[+]CX$_3$CR1[hi]Ly6C[lo]) are beneficial and promote recovery (62).

Conclusion

Under steady state conditions, site specific and phenotypic characteristics exist to distinguish between microglia and other CNS-associated macrophages. As with almost all innate and adaptive immune cell types, consensus with respect to nomenclature in CNS-resident versus CNS-infiltrating myeloid cells has not been effectively reached under inflammatory conditions. The advent of microarray technology and next generation sequencing will serve to provide more useful ways to distinguish between these two apparently similar, yet ever more functionally diverse cell types. An ever-increasing variety of previously unappreciated, and non-immune homeostatic functions performed by macrophages are now beginning to emerge, making a more detailed separation of these cell types highly desirable (71). Ultimately, better characterization and dissection of the various myeloid cells in an inflamed brain will help deciphering the specialized functions of the different members of the MPS in pathological conditions.

References

1. Katsumoto A, Lu H, Miranda AS, Ransohoff RM. Ontogeny and functions of central nervous system macrophages. *J Immunol* (2014) 193:2615–21. doi:10.4049/jimmunol.1400716
2. Lawson LJ, Perry VH, Dri P, Gordon S. Heterogeneity in the distribution and morphology of microglia in the normal adult mouse brain. *Neuroscience* (1990) 39:151–70. doi:10.1016/0306-4522(90)90229-W
3. Davalos D, Grutzendler J, Yang G, Kim JV, Zuo Y, Jung S, et al. ATP mediates rapid microglial response to local brain injury in vivo. *Nat Neurosci* (2005) 8:752–8. doi:10.1038/nn1472
4. Nimmerjahn A, Kirchhoff F, Helmchen F. Resting microglial cells are highly dynamic surveillants of brain parenchyma in vivo. *Science* (2005) 308:1314–8. doi:10.1126/science.1110647
5. Paolicelli RC, Bolasco G, Pagani F, Maggi L, Scianni M, Panzanelli P, et al. Synaptic pruning by microglia is necessary for normal brain development. *Science* (2011) 333:1456–8. doi:10.1126/science.1202529
6. Schafer DP, Lehrman EK, Kautzman AG, Koyama R, Mardinly AR, Yamasaki R, et al. Microglia sculpt postnatal neural circuits in an activity and complement-dependent manner. *Neuron* (2012) 74:691–705. doi:10.1016/j.neuron.2012.03.026
7. Zhan Y, Paolicelli RC, Sforazzini F, Weinhard L, Bolasco G, Pagani F, et al. Deficient neuron-microglia signaling results in impaired functional brain connectivity and social behavior. *Nat Neurosci* (2014) 17:400–6. doi:10.1038/nn.3641
8. Ginhoux F, Greter M, Leboeuf M, Nandi S, See P, Gokhan S, et al. Fate mapping analysis reveals that adult microglia derive from primitive macrophages. *Science* (2010) 330:841–5. doi:10.1126/science.1194637
9. Kierdorf K, Erny D, Goldmann T, Sander V, Schulz C, Perdiguero EG, et al. Microglia emerge from erythromyeloid precursors via Pu.1- and Irf8-dependent pathways. *Nat Neurosci* (2013) 16:273–80. doi:10.1038/nn.3318
10. Perdiguero EG, Klapproth K, Schulz C, Busch K, Azzoni E, Crozet L, et al. Tissue-resident macrophages originate from yolk-sac-derived erythro-myeloid progenitors. *Nature* (2014) 518(7540):547–51. doi:10.1038/nature13989
11. Schulz C, Gomez Perdiguero E, Chorro L, Szabo-Rogers H, Cagnard N, Kierdorf K, et al. A lineage of myeloid cells independent of Myb and hematopoietic stem cells. *Science* (2012) 336:86–90. doi:10.1126/science.1219179
12. Shiau CE, Kaufman Z, Meireles AM, Talbot WS. Differential requirement for irf8 in formation of embryonic and adult macrophages in zebrafish. *PLoS One* (2015) 10:e0117513. doi:10.1371/journal.pone.0117513
13. Varol C, Landsman L, Fogg DK, Greenshtein L, Gildor B, Margalit R, et al. Monocytes give rise to mucosal, but not splenic, conventional dendritic cells. *J Exp Med* (2007) 204:171–80. doi:10.1084/jem.20061011
14. Bogunovic M, Ginhoux F, Helft J, Shang L, Hashimoto D, Greter M, et al. Origin of the lamina propria dendritic cell network. *Immunity* (2009) 31:513–25. doi:10.1016/j.immuni.2009.08.010
15. Guilliams M, De Kleer I, Henri S, Post S, Vanhoutte L, De Prijck S, et al. Alveolar macrophages develop from fetal monocytes that differentiate into long-lived cells in the first week of life via GM-CSF. *J Exp Med* (2013) 210:1977–92. doi:10.1084/jem.20131199
16. Bain CC, Bravo-Blas A, Scott CL, Gomez Perdiguero E, Geissmann F, Henri S, et al. Constant replenishment from circulating monocytes maintains the macrophage pool in the intestine of adult mice. *Nat Immunol* (2014) 15:929–37. doi:10.1038/ni.2967
17. Epelman S, Lavine KJ, Beaudin AE, Sojka DK, Carrero JA, Calderon B, et al. Embryonic and adult-derived resident cardiac macrophages are maintained

through distinct mechanisms at steady state and during inflammation. *Immunity* (2014) **40**:91–104. doi:10.1016/j.immuni.2013.11.019

18. Hoeffel G, Chen J, Lavin Y, Low D, Almeida FF, See P, et al. C-myb(+) erythro-myeloid progenitor-derived fetal monocytes give rise to adult tissue-resident macrophages. *Immunity* (2015) **42**:665–78. doi:10.1016/j.immuni.2015.03.011

19. Gautier EL, Shay T, Miller J, Greter M, Jakubzick C, Ivanov S, et al. Gene-expression profiles and transcriptional regulatory pathways that underlie the identity and diversity of mouse tissue macrophages. *Nat Immunol* (2012) **13**:1118–28. doi:10.1038/ni.2419

20. Chiu IM, Morimoto ET, Goodarzi H, Liao JT, O'Keeffe S, Phatnani HP, et al. A neurodegeneration-specific gene-expression signature of acutely isolated microglia from an amyotrophic lateral sclerosis mouse model. *Cell Rep* (2013) **4**:385–401. doi:10.1016/j.celrep.2013.06.018

21. Butovsky O, Jedrychowski MP, Moore CS, Cialic R, Lanser AJ, Gabriely G, et al. Identification of a unique TGF-beta-dependent molecular and functional signature in microglia. *Nat Neurosci* (2014) **17**:131–43. doi:10.1038/nn.3599

22. Gosselin D, Link VM, Romanoski CE, Fonseca GJ, Eichenfield DZ, Spann NJ, et al. Environment drives selection and function of enhancers controlling tissue-specific macrophage identities. *Cell* (2014) **159**:1327–40. doi:10.1016/j.cell.2014.11.023

23. Lavin Y, Winter D, Blecher-Gonen R, David E, Keren-Shaul H, Merad M, et al. Tissue-resident macrophage enhancer landscapes are shaped by the local microenvironment. *Cell* (2014) **159**:1312–26. doi:10.1016/j.cell.2014.11.018

24. Zeisel A, Munoz-Manchado AB, Codeluppi S, Lonnerberg P, La Manno G, Jureus A, et al. Brain structure. Cell types in the mouse cortex and hippocampus revealed by single-cell RNA-seq. *Science* (2015) **347**:1138–42. doi:10.1126/science.aaa1934

25. Hickey WF, Kimura H. Perivascular microglial cells of the CNS are bone marrow-derived and present antigen in vivo. *Science* (1988) **239**:290–2. doi:10.1126/science.3276004

26. Lassmann H, Hickey WF. Radiation bone marrow chimeras as a tool to study microglia turnover in normal brain and inflammation. *Clin Neuropathol* (1993) **12**:284–5.

27. Greter M, Heppner FL, Lemos MP, Odermatt BM, Goebels N, Laufer T, et al. Dendritic cells permit immune invasion of the CNS in an animal model of multiple sclerosis. *Nat Med* (2005) **11**:328–34. doi:10.1038/nm1197

28. Anandasabapathy N, Victora GD, Meredith M, Feder R, Dong B, Kluger C, et al. Flt3L controls the development of radiosensitive dendritic cells in the meninges and choroid plexus of the steady-state mouse brain. *J Exp Med* (2011) **208**:1695–705. doi:10.1084/jem.20102657

29. Hashimoto D, Chow A, Noizat C, Teo P, Beasley MB, Leboeuf M, et al. Tissue-resident macrophages self-maintain locally throughout adult life with minimal contribution from circulating monocytes. *Immunity* (2013) **38**:792–804. doi:10.1016/j.immuni.2013.04.004

30. Prodinger C, Bunse J, Kruger M, Schiefenhovel F, Brandt C, Laman JD, et al. CD11c-expressing cells reside in the juxtavascular parenchyma and extend processes into the glia limitans of the mouse nervous system. *Acta Neuropathol* (2011) **121**:445–58. doi:10.1007/s00401-010-0774-y

31. Prinz M, Priller J, Sisodia SS, Ransohoff RM. Heterogeneity of CNS myeloid cells and their roles in neurodegeneration. *Nat Neurosci* (2011) **14**:1227–35. doi:10.1038/nn.2923

32. Yona S, Kim KW, Wolf Y, Mildner A, Varol D, Breker M, et al. Fate mapping reveals origins and dynamics of monocytes and tissue macrophages under homeostasis. *Immunity* (2013) **38**:79–91. doi:10.1016/j.immuni.2012.12.001

33. Goldmann T, Wieghofer P, Muller PF, Wolf Y, Varol D, Yona S, et al. A new type of microglia gene targeting shows TAK1 to be pivotal in CNS autoimmune inflammation. *Nat Neurosci* (2013) **16**:1618–26. doi:10.1038/nn.3531

34. Parkhurst CN, Yang G, Ninan I, Savas JN, Yates JR III, Lafaille JJ, et al. Microglia promote learning-dependent synapse formation through brain-derived neurotrophic factor. *Cell* (2013) **155**:1596–609. doi:10.1016/j.cell.2013.11.030

35. Nishinakamura R, Matsumoto Y, Nakao K, Nakamura K, Sato A, Copeland NG, et al. Murine homolog of SALL1 is essential for ureteric bud invasion in kidney development. *Development* (2001) **128**:3105–15.

36. Juedes AE, Ruddle NH. Resident and infiltrating central nervous system APCs regulate the emergence and resolution of experimental autoimmune encephalomyelitis. *J Immunol* (2001) **166**:5168–75. doi:10.4049/jimmunol.166.8.5168

37. Ponomarev ED, Shriver LP, Maresz K, Dittel BN. Microglial cell activation and proliferation precedes the onset of CNS autoimmunity. *J Neurosci Res* (2005) **81**:374–89. doi:10.1002/jnr.20488

38. Almolda B, Gonzalez B, Castellano B. Antigen presentation in EAE: role of microglia, macrophages and dendritic cells. *Front Biosci* (2011) **16**:1157–71. doi:10.2741/3781

39. Mildner A, Schlevogt B, Kierdorf K, Bottcher C, Erny D, Kummer MP, et al. Distinct and non-redundant roles of microglia and myeloid subsets in mouse models of Alzheimer's disease. *J Neurosci* (2011) **31**:11159–71. doi:10.1523/JNEUROSCI.6209-10.2011

40. Yamasaki R, Lu H, Butovsky O, Ohno N, Rietsch AM, Cialic R, et al. Differential roles of microglia and monocytes in the inflamed central nervous system. *J Exp Med* (2014) **211**:1533–49. doi:10.1084/jem.20132477

41. Haynes SE, Hollopeter G, Yang G, Kurpius D, Dailey ME, Gan WB, et al. The P2Y12 receptor regulates microglial activation by extracellular nucleotides. *Nat Neurosci* (2006) **9**:1512–9. doi:10.1038/nn1805

42. Butovsky O, Jedrychowski MP, Cialic R, Krasemann S, Murugaiyan G, Fanek Z, et al. Targeting miR-155 restores abnormal microglia and attenuates disease in SOD1 mice. *Ann Neurol* (2015) **77**:75–99. doi:10.1002/ana.24304

43. Kennedy DW, Abkowitz JL. Kinetics of central nervous system microglial and macrophage engraftment: analysis using a transgenic bone marrow transplantation model. *Blood* (1997) **90**(3):986–93.

44. Ajami B, Bennett JL, Krieger C, Tetzlaff W, Rossi FM. Local self-renewal can sustain CNS microglia maintenance and function throughout adult life. *Nat Neurosci* (2007) **10**:1538–43. doi:10.1038/nn2014

45. Mildner A, Schmidt H, Nitsche M, Merkler D, Hanisch UK, Mack M, et al. Microglia in the adult brain arise from Ly-6ChiCCR2+ monocytes only under defined host conditions. *Nat Neurosci* (2007) **10**:1544–53. doi:10.1038/nn2015

46. Ajami B, Bennett JL, Krieger C, McNagny KM, Rossi FM. Infiltrating monocytes trigger EAE progression, but do not contribute to the resident microglia pool. *Nat Neurosci* (2011) **14**:1142–9. doi:10.1038/nn.2887

47. Heppner FL, Greter M, Marino D, Falsig J, Raivich G, Hovelmeyer N, et al. Experimental autoimmune encephalomyelitis repressed by microglial paralysis. *Nat Med* (2005) **11**:146–52. doi:10.1038/nm0405-455

48. Varvel NH, Grathwohl SA, Baumann F, Liebig C, Bosch A, Brawek B, et al. Microglial repopulation model reveals a robust homeostatic process for replacing CNS myeloid cells. *Proc Natl Acad Sci U S A* (2012) **109**:18150–5. doi:10.1073/pnas.1210150109

49. Elmore MR, Najafi AR, Koike MA, Dagher NN, Spangenberg EE, Rice RA, et al. Colony-stimulating factor 1 receptor signaling is necessary for microglia viability, unmasking a microglia progenitor cell in the adult brain. *Neuron* (2014) **82**:380–97. doi:10.1016/j.neuron.2014.02.040

50. Elmore MR, Lee RJ, West BL, Green KN. Characterizing newly repopulated microglia in the adult mouse: impacts on animal behavior, cell morphology, and neuroinflammation. *PLoS One* (2015) **10**:e0122912. doi:10.1371/journal.pone.0122912

51. Schreiner B, Heppner FL, Becher B. Modeling multiple sclerosis in laboratory animals. *Semin Immunopathol* (2009) **31**:479–95. doi:10.1007/s00281-009-0181-4

52. Fife BT, Huffnagle GB, Kuziel WA, Karpus WJ. CC chemokine receptor 2 is critical for induction of experimental autoimmune encephalomyelitis. *J Exp Med* (2000) **192**:899–905. doi:10.1084/jem.192.6.899

53. Gaupp S, Pitt D, Kuziel WA, Cannella B, Raine CS. Experimental autoimmune encephalomyelitis (EAE) in CCR2(-/-) mice: susceptibility in multiple strains. *Am J Pathol* (2003) **162**:139–50. doi:10.1016/S0002-9440(10)63805-9

54. Kreutzberg GW. Microglia: a sensor for pathological events in the CNS. *Trends Neurosci* (1996) **19**:312–8. doi:10.1016/0166-2236(96)10049-7

55. King IL, Dickendesher TL, Segal BM. Circulating Ly-6C+ myeloid precursors migrate to the CNS and play a pathogenic role during autoimmune demyelinating disease. *Blood* (2009) **113**:3190–7. doi:10.1182/blood-2008-07-168575

56. Mildner A, Mack M, Schmidt H, Bruck W, Djukic M, Zabel MD, et al. CCR2+Ly-6Chi monocytes are crucial for the effector phase of autoimmunity in the central nervous system. *Brain* (2009) **132**:2487–500. doi:10.1093/brain/awp144

57. Serafini B, Columba-Cabezas S, Di Rosa F, Aloisi F. Intracerebral recruitment and maturation of dendritic cells in the onset and progression of experimental autoimmune encephalomyelitis. *Am J Pathol* (2000) **157**:1991–2002. doi:10.1016/S0002-9440(10)64838-9

58. McMahon EJ, Bailey SL, Castenada CV, Waldner H, Miller SD. Epitope spreading initiates in the CNS in two mouse models of multiple sclerosis. *Nat Med* (2005) **11**:335–9. doi:10.1038/nm1202

59. Bailey SL, Schreiner B, McMahon EJ, Miller SD. CNS myeloid DCs presenting endogenous myelin peptides 'preferentially' polarize CD4+ T(H)-17 cells in relapsing EAE. *Nat Immunol* (2007) **8**:172–80. doi:10.1038/ni1430

60. Schreiber HA, Loschko J, Karssemeijer RA, Escolano A, Meredith MM, Mucida D, et al. Intestinal monocytes and macrophages are required for T cell polarization in response to *Citrobacter rodentium*. *J Exp Med* (2013) **210**:2025–39. doi:10.1084/jem.20130903

61. Kigerl KA, Gensel JC, Ankeny DP, Alexander JK, Donnelly DJ, Popovich PG. Identification of two distinct macrophage subsets with divergent effects causing either neurotoxicity or regeneration in the injured mouse spinal cord. *J Neurosci* (2009) **29**:13435–44. doi:10.1523/JNEUROSCI.3257-09.2009

62. Shechter R, Miller O, Yovel G, Rosenzweig N, London A, Ruckh J, et al. Recruitment of beneficial M2 macrophages to injured spinal cord is orchestrated by remote brain choroid plexus. *Immunity* (2013) **38**:555–69. doi:10.1016/j.immuni.2013.02.012

63. Vainchtein ID, Vinet J, Brouwer N, Brendecke S, Biagini G, Biber K, et al. In acute experimental autoimmune encephalomyelitis, infiltrating macrophages are immune activated, whereas microglia remain immune suppressed. *Glia* (2014) **62**:1724–35. doi:10.1002/glia.22711

64. Aube B, Levesque SA, Pare A, Chamma E, Kebir H, Gorina R, et al. Neutrophils mediate blood-spinal cord barrier disruption in demyelinating neuroinflammatory diseases. *J Immunol* (2014) **193**:2438–54. doi:10.4049/jimmunol.1400401

65. Plantinga M, Guilliams M, Vanheerswynghels M, Deswarte K, Branco-Madeira F, Toussaint W, et al. Conventional and monocyte-derived CD11b(+) dendritic cells initiate and maintain T helper 2 cell-mediated immunity to house dust mite allergen. *Immunity* (2013) **38**:322–35. doi:10.1016/j.immuni.2012.10.016

66. Banchereau J, Steinman RM. Dendritic cells and the control of immunity. *Nature* (1998) **392**:245–52. doi:10.1038/32588

67. Stout RD, Suttles J, Xu J, Grewal IS, Flavell RA. Impaired T cell-mediated macrophage activation in CD40 ligand-deficient mice. *J Immunol* (1996) **156**:8–11.

68. Becher B, Durell BG, Miga AV, Hickey WF, Noelle RJ. The clinical course of experimental autoimmune encephalomyelitis and inflammation is controlled by the expression of CD40 within the central nervous system. *J Exp Med* (2001) **193**:967–74. doi:10.1084/jem.193.8.967

69. Mosser DM, Edwards JP. Exploring the full spectrum of macrophage activation. *Nat Rev Immunol* (2008) **8**:958–69. doi:10.1038/nri2448

70. Vogel DY, Vereyken EJ, Glim JE, Heijnen PD, Moeton M, Van Der Valk P, et al. Macrophages in inflammatory multiple sclerosis lesions have an intermediate activation status. *J Neuroinflammation* (2013) **10**:35. doi:10.1186/1742-2094-10-35

71. Muller PA, Koscso B, Rajani GM, Stevanovic K, Berres ML, Hashimoto D, et al. Crosstalk between muscularis macrophages and enteric neurons regulates gastrointestinal motility. *Cell* (2014) **158**:300–13. doi:10.1016/j.cell.2014.04.050

16

Role of dendritic cells in natural immune control of HIV-1 infection

Enrique Martin-Gayo [1]* and Xu G. Yu [2]*

[1] *Hospital Universitario de la Princesa, Universidad Autónoma de Madrid, Madrid, Spain,* [2] *Ragon Institute of MGH, MIT, and Harvard, Massachusetts General Hospital, Harvard Medical School, Boston, MA, United States*

Edited by:
Paul Urquhart Cameron,
The University of Melbourne, Australia

Reviewed by:
Philippe Benaroch,
Centre National de la Recherche
Scientifique (CNRS), France
Laura Fantuzzi
Istituto Superiore di Sanità (ISS), Italy

*Correspondence:
Enrique Martin-Gayo
enrique.martin@uam.es
Xu G. Yu
xyu@mgh.harvard.edu

Dendritic cells (DCs) are professional antigen-presenting cells that link innate and adaptive immunity and are critical for the induction of protective immune responses against pathogens. Proportions of these cells are markedly decreased in the blood of untreated HIV-1-infected individuals, suggesting they might be intrinsically involved in HIV-1 pathogenesis. However, despite several decades of active research, the precise role and contribution of these cells to protective or detrimental host responses against HIV-1 are still remarkably unclear. Recent studies have shown that DCs possess a fine-tuned machinery to recognize HIV-1 replication products through a variety of innate pathogen sensing mechanisms, which may be instrumental for generating both cellular and humoral protective immune responses in persons who naturally control HIV-1 replication. Yet, dysregulated and abnormal activation of DCs might also contribute to sustained inflammation and immune activation accelerating disease progression during chronic progressive infection. Emerging data also suggest that DCs can influence the induction of potent broadly-neutralizing antibodies, and may, for this reason, have to be considered as important components of future HIV-1 vaccination strategies. Apart from their involvement in antiviral host immunity, at least a subgroup of DCs seem intrinsically susceptible to HIV-1 infection and may serve as a viral target cell population. Indeed recent studies suggest that specific DC subpopulations residing in the genital mucosa are preferentially infected by HIV-1 and play an active role in sexual transmission; therefore, DCs may contribute to viral dissemination and possible persistence of the viral reservoirs through either direct or indirect mechanisms. Here, we analyze the distinct and partially opposing roles of DCs during HIV-1 disease pathogenesis, with a focus on implications of DC biology natural immune control and HIV cure research efforts.

Keywords: dendritic cell, HIV-1 controller, IFN, Tfh, bNAb, vaccine

INTRODUCTION

Dendritic cells (DCs) represent a heterogeneous family of immune cells that link innate and adaptive immunity. The main function of these innate cells is to capture, process, and present antigens to adaptive immune cells and mediate their polarization into effector cells (1). DCs can be subdivided in two main subtypes: plasmacytoid (pDC) and myeloid (mDC) DCs, which specialize in the recognition of different pathogen associated molecular patterns (PAMPs) due to the unique distribution of Pattern Recognition Receptors (PRR), such as toll-like receptors, C-type lectins and intracellular nucleic acid sensors (2–4). As a result, mDCs and pDCs can efficiently induce $CD4^+$ and $CD8^+$ T cell responses against different types of pathogens. In addition, both mDCs and pDCs are also capable of interacting with Natural Killer (NK) cells, which are particularly relevant during viral infections (5). Therefore, the contribution of different DC subtypes to immune responses against microbial infections seems to be highly complex and be influenced by context- and pathogen-dependent factors.

During HIV-1 infection, several effector components of the innate and adaptive immune system are involved in the host antiviral response, and although these immune responses seem unable to prevent the establishment of the infection, they can influence HIV-1 disease progression. Effective immune control of HIV-1 infection occurs in rare population of HIV-1 infected individuals who are able to spontaneously control HIV-1 replication in the absence of antiretroviral therapy, and to maintain undetectable levels of viral replication as measured by commercial PCR assays. In these individuals, long-lived polyfunctional HIV-1-specific CD8+ T cells have been identified as the main biological correlate of spontaneous immune control of HIV-1 (6–8). However, the contribution of DCs to durable immune control of HIV-1 is still a relatively unexplored area and a matter of active debate. During the last years, new relevant data about DC biology in the context of HIV-1 infection have become available, specifically with regards to DC susceptibility to infection, to DC-mediated immune regulation and to direct host-pathogen interactions between DC and HIV-1. In this review, we have focused on consolidating the most recent advances on DC biology in the context of HIV-1 immunopathology, and on providing a detailed evaluation of the role of DC in HIV-1 immune control.

ANATOMICAL LOCALIZATION AND ACTIVATION OF DCs DURING HIV/SIV INFECTION

DCs are physiologically distributed in mucosal and lymphoid tissues where they capture antigens and present them to T cells, but a small proportion of mDCs and pDCs are also circulating in the blood. mDCs can be identified as lineage marker negative cells that display high surface levels of CD11c and HLA-DR (9) while pDCs are CD11c−HLA-DR+ cells characterized by surface expression of the C-type lectin BDCA2, high levels of the alpha chain of the receptor for interleukin-3 (CD123) and the immunoglobulin superfamily receptor immunoglobulin-like transcript 7 (ILT7) (10). Upon HIV-1 infection, the anatomical distribution of DCs is dramatically altered and lower proportions of pDCs and mDCs are present in the blood of infected untreated individuals (11–13). The extent of the depletion of circulating mDC is correlated with rapid disease progression during HIV-1 and SIV infections (14, 15). Interestingly, proportions of circulating pDCs are more profoundly reduced in HIV-1 progressors in contrast to controllers (16); although the exact mechanisms responsible for these differences remain unknown. Despite these discrepancies, circulating pDCs from both controllers and progressors are characterized by upregulated expression of the gut homing integrin α4β7, suggesting selective trafficking to mucosal intestinal tissue where the majority of

HIV-1-infected cells reside (17). Similarly, higher levels of activation in gut resident mDCs and pDCs seem to be associated with changes in gut microbiota and immune homeostasis (18). In addition to migration to the gut, preferential recruitment of pDCs to the lymph nodes also occurs in HIV-1-infected subjects (19).

Besides the changes in anatomical distribution, circulating and tissue-resident DCs display an activated phenotype defined by upregulation of costimulatory molecules in infected individuals (11–13). In fact, higher levels of activation in blood DCs seem to correlate with plasma viremia in progressors (20). In contrast, less pronounced phenotypical signs of immune activation, combined with increased functionality have been described in mDCs from the blood of HIV controllers (21). Interestingly, highly activated mDCs residing in the lymph nodes from HIV-1+ patients seem to co-express inhibitory costimulatory molecules such as PD-L1 and are still capable of responding to TLR stimulation, in contrast to cells from peripheral blood (19).

A hallmark of circulating pDCs from the blood of HIV-1+ individuals is the expression of high basal levels of type I interferon (IFN) and IFN-stimulated genes, likely reflecting abnormal immune activation (22). Interestingly, this higher baseline activation of IFN-dependent immune activity seems to make pDCs from progressors refractory to antigenic stimulation (23, 24), and paradoxically reduces their ability to secrete appropriate levels of IFN-α upon PRR stimulation. In contrast, pDCs from controllers maintain IFN-α secretion levels that are comparable to those of healthy individuals. Consistent with these findings, microscopy-based studies indicated differences in the trafficking of intracellular TNF-related apoptosis-inducing ligand (TRAIL) in pDCs from controllers and healthy donors compared to progressors. TRAIL is a molecule known to induce apoptosis of CD4+ T cells through a mechanism regulated by the alarmin High Mobility Group Box 1 (HMGB1) (23, 25). While TRAIL seems to be recycled from the membrane of pDCs in controllers after exposure to HIV-1, pDCs from viremic patients appear to constitutively express TRAIL on the membrane, which may contribute to unspecific induction of cell death in CD4+ T cells and accelerate cell loss and immunodeficiency (26) (**Figure 1**). Overall, these data indicate that pDCs from controllers maintain a functional profile that is similar to healthy persons. In contrast, pDCs from progressors exhibit a hyperactivated state characterized by constitutive TRAIL up-regulation, higher basal levels of IFN-dependent immune responses, and a reduced ability to produce IFN-α in responses to antigen exposure, most likely as a result of generalized immune activation that makes cells refractory to microbial stimulation (**Figure 1**). The normal functional profile of pDCs in controllers therefore could be a consequence, rather than a cause of viral immune control. Notably, the initiation of antiretroviral therapy does not revert the decline in pDC frequency and function observed during progressive infection, suggesting an irreversible defect in pDC physiology in progressors after prolonged exposure to high viremia (27). Interestingly, less pathogenic HIV-2 strains induce lower levels of type I IFN expression in pDCs compared to HIV-1, suggesting that lower levels of pDC activation could be associated with immune control of the infection (28). In addition to pDCs,

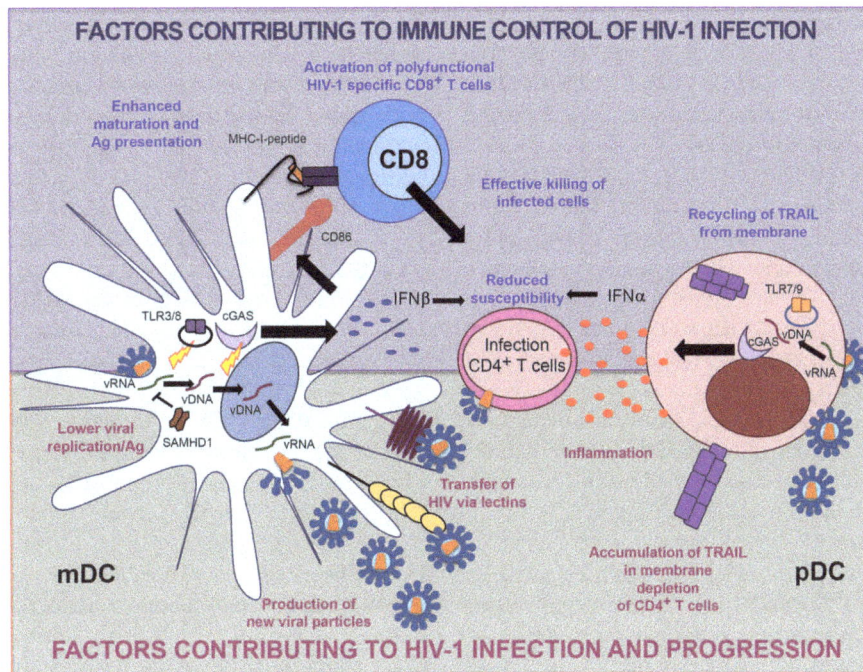

FIGURE 1 | Schematic representation of factors in human mDC and pDC contributing to immune control vs. progression of HIV-1 infection.

recent works in the SIV model have suggested that additional cell types might be responsible for abnormal activation of type I IFN responses at later stages of progressive infection (29). Together, cumulative information from recent studies suggests that DC distribution and function might be critically altered during HIV-1 infection and that preservation of physiological DC distribution and function is associated with immune control of the infection.

DCs AS VEHICLES FOR HIV-1 TRANSMISSION AND DISSEMINATION

During the last few years, several studies have shown that DCs have the ability to transfer HIV-1 particles to target CD4+ T cells and facilitate their infection, in a process known as trans-infection (30). This phenomenon starts with the transference of HIV-1 virions to pockets in the membrane of DCs, where they accumulate and are subsequently actively transferred to T cells through virological synapses (31). The stability of such transmission events depends on the expression of adhesion molecules, such as Intercellular Adhesion Molecule (ICAM) (32) and the actin assembly machinery (33). In order to transfer viruses to CD4+ T cells, DCs require the capture of HIV-1 particles through the lectin Dendritic Cell Intercellular Adhesion Molecule-3-Grabbing Non-integrin (DC-SIGN) (34, 35), and the Sialic acid-binding Immunoglobulin-type Lectin 1 (SIGLEC-1) receptor (36). Recent reports have shown that the ability of DCs to facilitate HIV-1 trans-infection is acquired upon activation with inflammatory molecules associated with poor HIV-1 prognosis, such as IFNα and LPS (37). These

stimuli have been shown to induce SIGLEC-1 expression and therefore, enhance the capture and transfer of viral particles. Notably, circulating mDCs but not pDCs facilitate trans-infection of HIV-1 (32). Therefore, increased basal levels of immune activation and high-level viremia might contribute to disease progression through facilitation of viral trans-infection by mDCs. Consistent with a role of mDCs in supporting HIV-1 trans-infection in lymphoid tissues, it was shown that depletion of lymph node-resident mDCs in tissue-suspension cultures reduced the efficiency of HIV-1 infection of CD4+ T cells (38). However, mutations of SIGLEC-1, which naturally occur in small proportions of individuals, did not seem to provide protection from HIV-1 infection or attenuation of HIV-1 disease progression (39), suggesting that classical, SIGLEC-1-independent HIV-1 dissemination within in the host remains the predominant mechanisms fueling viral infection *in vivo*.

DC SUSCEPTIBILITY TO HIV-1 INFECTION AND HOST RESTRICTION FACTORS

Most DCs express the coreceptor CD4 (40), and therefore are in principle susceptible to infection with HIV-1. However, DCs seem to represent a more hostile and restrictive environment for HIV-1 than CD4+ T cells, for reasons that are not completely clear. While initial studies suggested that monocyte derived DCs (MDDCs) are highly resistant to infection with HIV-1 (41), primary mDCs are able to support some levels of HIV-1 replication, at least *in vitro* (42–44).

The main restriction factor that limits HIV-1 replication in MDDCs and mDCs seems to be the cytoplasmic protein SAM domain and HD domain-containing protein 1 (SAMHD1), which is highly expressed in myeloid cells and is able to block HIV-1 replication at the retro-transcription level by depleting endogenous intracellular pools of dNTPs (45), and by directly degrading viral RNA (46). While it is clear that SAMHD1 is a key factor limiting replication of HIV-1 in MDDCs (47) and inhibiting further spread of virions to T cells (48), recent studies demonstrated that MDDCs can actually support productive infection with HIV-1 to a certain degree, despite high levels of expression of this restriction factor (49). The functional ability of SAMHD1 to restrict HIV-1 replication is regulated by phosphorylation mediated by host kinases from the cyclin-dependent kinase family (50). Interestingly, the functionally active, de-phosphorylated form of SAMHD1 is preferentially found in primary DCs isolated *ex vivo* from human blood, which potentially could contribute to a higher resistance to infection (51). However, it is unclear whether restriction of HIV-1 by SAMHD1 in mDCs might truly benefit the host, since restriction of HIV-1 replication via SAMHD1 may impair cytoplasmic viral immune recognition in mDCs and impair their ability to prime HIV-1-specific T cells. On the other hand, interactions of mDCs with T cells induce downregulation of SAMHD1 expression (52), allowing human primary mDCs to be more permissive to infection (44). Importantly, recent data indicate that primary CD1c$^+$ and CD141$^+$ mDC subtypes might differ in their susceptibility to HIV-1 infection. In this regard, expression of the endosomal protein RAB15 prevents fusion of viral particles in CD141$^+$ mDCs and induces a higher level of cell-intrinsic resistance to infection with HIV-1 and HIV-2 compared to CD1c$^+$ mDCs (53). Further proof for the susceptibility of primary mDCs to HIV-1 infection was provided by a recent study identifying a distinct population of CD1a$^+$ mDCs residing in the vaginal mucosa, which supported CCR5-tropic but not CXCR4-tropic HIV-1 replication, in contrast to vaginal Langerhans cells (LC). These data suggest that these vaginal mDCs might play an active role in the selection of transmitted viral variants during heterosexual HIV-1 acquisition (54).

In the context of immune control of HIV-1 infection, recent studies suggest that monocytes and mDCs from HIV-1 controllers restrict early HIV-1 replication steps, specifically at the level of viral integration (44, 55, 56) while restriction of viral reverse transcription is less obvious, possibly due to lower induction of SAMHD1 expression in HIV-1 controllers upon exposure to HIV-1. This specific replicative pattern of HIV-1 may enable enhanced cytoplasmic sensing of accumulated HIV-1 reverse transcripts, which represent the primary substrate for innate immune recognition, and facilitate antigen processing and presentation (44, 56). Interestingly, although SAMHD1 is thought to be an interferon inducible gene, DCs and CD4$^+$ T cells fail to induce its expression in the presence of type I IFNs (57). Therefore, higher permissiveness of mDCs from controllers to viral reverse transcription may represent a key element for supporting cytoplasmic detection of HIV-1 and for inducing potent cell-intrinsic responses that lead to the effective activation of HIV-1-specific T cells (**Figure 1**).

Although SAMHD1 is recognized as a critical host factor limiting HIV-1 replication in myeloid cells, alternative SAMHD1-independent restriction mechanisms might also be playing a role in effective immunological control of HIV-1 replication. Among them, recognition of the HIV-1 capside by cyclophilin A (41, 58) and TRIM5 α (59, 60) or endogenous levels of β-catenin (52), could be actively contributing to block HIV-1 replication in myeloid cells. In addition, some studies suggest that HIV-1 could trigger TLR activation in DCs (61). Indeed, activation of MDDCs through TLR4 and TLR3 resulted in inhibition of HIV-1 replication steps in DC, while simultaneously increasing their ability to prime HIV-specific CD8$^+$ T cells (62). Therefore, TLR-dependent activation of DC could play a relevant role for inducing highly-functional cellular immune responses against HIV-1. Supporting this idea, polymorphisms in the TLR3 gene confer resistance to HIV-1 infection (63). In fact, it was recently suggested that TLR activation could be playing an active role in the detection of HIV-1 by primary CD141$^+$ mDCs (53). Therefore, more studies are required to investigate the mutual interplay between viral restriction in DCs and immune control of HIV-1, and to determine the contribution of myeloid cells to persisting viral reservoirs during suppressive antiretroviral therapy.

INNATE IMMUNE RESPONSES TO HIV-1 IN DCs

DCs are, in principle, capable of inducing secretion of type I IFNs upon recognition of viral nucleic acids, which subsequently leads to transcription of interferon stimulated genes (ISGs) and the upregulation of class II HLA and costimulatory molecules. As a result of such cell-intrinsic, IFN-dependent immune responses, mature DCs become more restrictive for viral replication, while the expression of molecules involved in antigen presentation and co-stimulation is increased. Whether mDCs can induce secretion of type I IFNs in response to HIV-1 is still highly controversial. In MDDCs, HIV-1 seems to be able to induce expression of several IFN-related genes in the absence of actual production of IFN α/β due to the selective activation of IRF-1 mediated signaling instead of inducing phosphorylation of IRF3, which is known to be required for induction of type I IFNs (64). However, the intracellular DNA sensor cGAS is expressed by myeloid cells (65) and is able of producing cGAMP second messengers upon recognition of HIV-1 DNA (66), leading to the activation of the sensor STING and the signal transducer TBK-1, which promote IFN β production (67, 68). Thus, primary DCs are, in principle, able to sense and induce type I IFN upon exposure to cytoplasmic HIV-1 DNA. In fact, activation of cGAS seems to be required for the transcription of IFN β by primary mDCs and MDDCs in the context of HIV-1 and other viral infections (44, 69, 70). This DNA-dependent mechanism of viral sensing leading to type I IFN responses might be more active in human CD1c$^+$ mDCs compared to CD141$^+$ mDCs (53). Interestingly, cGAS triggers TLR9-independent activation of primary pDCs in response to intracellular DNA (71, 72). However, current phenotypic markers for pDC identify a heterogenous cell population that, in addition to *bona fide* pDC, contains pre-DC precursors of mDCs, which could also be differentially contributing to the observed responses to HIV-1 (73). Such heterogeneity could be the result of different pre-pDC and/or pDCs originated from either lymphoid or myeloid precursors with different functional properties (74–76). Therefore, more studies are required to elucidate the contribution of TLR-independent sensing of HIV-1 in pDCs. Finally, activation of the cGAS pathway by HIV-1 might involve interactions with additional host factors such as the newly identified NONO protein, which apparently is able to bind cGAS and the HIV-1 capsid and facilitate innate sensing of HIV-1 DNA in dendritic cells (77).

Importantly, preserved or enhanced induction of IFN responses has been described in both primary pDCs (23, 24) and mDCs (44) from HIV-1 controllers exposed to HIV-1. A recent single-cell RNAseq study has identified a highly functional population of $CD64^{Hi}CD86^{Hi}$ PD-L1^{Hi} mDCs characterized by a strong type I IFN signature that is induced more efficiently in HIV-1 controllers than in progressors or healthy individuals in response to HIV-1 (78). The induction of such highly functional mDCs depended on the activation of TBK-1, which acts downstream of several intracellular sensing pathways including cGAS and TLR-3. Therefore, enhanced innate recognition of HIV-1 by both pDCs and mDCs might be a contributing factor to develop effective HIV-1 specific immunity in these individuals (**Figure 1**). However, HIV-1 might have evolved to minimize such mechanisms of viral DNA recognition, since HIV-1 Vif and Vpr are capable of inactivating TBK-1 which is downstream of the cGAS-STING pathway (79). Therefore, additional mechanisms such as alterations in the activation threshold of intracellular sensing pathways might be playing a role in DCs from controllers. In addition to sensing of viral DNA by cGAS, viral immune recognition in DCs could be connected with the RIG-I pathway, which may also contribute to activation of DCs in response to HIV-1 (80). In fact, communication and collaboration between RIG-I and DNA sensing pathways has been reported to amplify innate immune responses against intracellular viral DNA (81, 82). Although no genetic alterations in genes encoding for innate immune sensors for HIV-1 have been found in GWAS studies including large HIV-1-infected populations, a more targeted analysis of innate immune genes may in the future allow to identify immunogenetic polymorphisms in the innate immune system that facilitate innate immune sensing and natural viral control in specific subgroups of HIV-1 controllers.

ANTIGEN PRESENTING CELL FUNCTION OF DCs AND ADAPTIVE IMMUNITY AGAINST HIV-1

Given associations between the polyfunctionality of T cell responses and natural progression of HIV-1 infection (8), several studies have focused on the function of DCs as professional antigen presenting cells (APC) and how these cells are involved in the priming of adaptive immune cells. As mentioned before, both mDCs and pDCs can respond and mature to a certain degree in response to HIV-1, but may become exhausted and hyporesponsive during chronic progressive infection, which might impact their antigen-presenting cell function (83). In the case of pDCs, infection with HIV-1 seems to turn these cells more tolerogenic, and increase their potential to drive polarization of CD4$^+$ T cells into immunosuppressive T regulatory cells (84). On the other hand, although pDCs can activate CD8$^+$ T cells through cross-presentation (85), no studies have yet analyzed the impact of pDCs on the priming of HIV-1-specific cytotoxic CD8$^+$ T cell responses. In contrast, while circulating mDCs from healthy individuals are functionally incapable of efficiently priming T cells *in vitro* after exposure to HIV-1 (86), effective antigen presenting functions of mDCs from HIV-1 elite controllers is associated with enhanced abilities to prime HIV-specific CD8$^+$ T cells in these patients (44, 78). In addition to mDCs, recent *in vitro* studies have shown that MDDCs can acquire HIV-1 antigens from Langerhans cells, become activated and induce cross-presentation to CD8$^+$ T cells (87), suggesting that these cells may also be potentially able to prime protective HIV-1-specific cytotoxic CD8$^+$ T cells. In addition, independent studies have shown that MDDCs can mediate cross-presentation of immuno-dominant HIV-1 peptides and activate HIV-1-specific CD8$^+$ T cells (88). However, MDDC are not a physiological DC subset present and in fact more closely resemble inflammatory DCs (89). Nevertheless, a recent evaluation suggested that primary CD141$^+$ mDCs, obtain HIV-1 antigens from infected CD1c$^+$ mDCs for cross-presentation to CD8$^+$ T cells (53). Interestingly, DCs infected with HIV-1 can also present endogenous viral peptides and mediate activation of HIV-1-specific CD4$^+$ T cells (90). Therefore, mDCs and MDDCs might be involved in the priming of effective HIV-1-specific T cell responses observed in controllers.

Although highly functional HIV-1-specific CD8$^+$ T cell responses were identified as the main correlate of antiviral immune defense (91), the discovery of broadly neutralizing antibodies (bNAbs) against multiple strains of HIV-1 (92, 93), has led to a great interest in understanding their potential contribution to spontaneous immunological control of HIV-1. Recent works have indeed identified a subpopulation of HIV-1 viremic controllers who develop bNAbs in the absence of high levels of viremia or immune activation (94). In previous studies in viremic HIV-1-infected progressors, the induction of bNAbs was associated with the presence of CXCR5$^+$ PD-1$^+$ T follicular helper cells (Tfh) in the blood (95). Although Tfh cells facilitate B cell maturation and immunoglobulin class switching in lymphoid tissue (96), peripheral CXCR5$^+$ PD-1$^+$ CD4$^+$ T lymphocytes have been proposed to act as peripheral counterparts of Tfh cells (pTfh) (95, 97) and could serve as a peripheral biomarker of high germinal center Tfh cell activity. Therefore, the priming of Tfh cells by mDC might be important in HIV-1 controllers capable of inducing antibodies with broader neutralizing activity. Supporting this possibility, mDCs from controller neutralizers are more efficient in priming CD4$^+$ T cells into long lived PD-1^{Lo} Tfh precursors, which can differentiate into functional PD-1^{Hi} Tfh effector cells upon antigenic stimulation (98). Importantly, higher frequencies of PD-1^{Lo} Tfh precursors in the blood are correlated with higher breadth of Ab neutralization in this subset of controllers. Compatible with an indirect role of DCs for influencing humoral immunity through polarization of Tfh, mDCs from controller neutralizers are characterized by high levels of CD40, a molecule previously involved in Tfh cell differentiation (99), and display distinct transcriptional patterns that differ from those present in CD64^{Hi} PD-L1^{Hi} mDCs from elite controllers with high CD8$^+$ T cell priming potential (98). Therefore, these findings might suggest a range of functional specializations of mDCs from controllers that may contribute to immune viral control through different immune mechanisms. To which degree individual DC subpopulations influence other components of the innate and adaptive immune system and contribute to control of HIV-1 is still an open question that requires further study.

CONCLUSIONS

In this review, we have summarized recent advances in understanding DC biology in the context of HIV-1 immune control. While studies have revealed multiple mechanisms by which DCs might contribute to controlling HIV-1 (**Figure 1**), future studies will be necessary to evaluate the complexity of individual DC subsets in promoting beneficial versus detrimental effects during HIV-1 infection. Similarly, our knowledge about the intrinsic ability of pDCs and mDCs to sense and respond to HIV-1 has greatly improved over the last few years, but translating this insight into improved and more specific adjuvants for future preventive and therapeutic HIV-1 vaccines represents a considerable challenge. The development to new humanized animal models that recapitulate human DC biology will likely be critical to identify effective vaccination strategies based on DCs. Together, DCs are emerging as critical players of effective immune responses in HIV-1 and a closer understanding of these cells might contribute to the development of novel effective vaccines or immunotherapies.

AUTHOR CONTRIBUTIONS

EM-G and XY conceived and wrote the manuscript.

REFERENCES

1. Shortman K, Liu YJ. Mouse and human dendritic cell subtypes. *Nat Rev Immunol.* (2002) 2:151–61. doi: 10.1038/nri746
2. Jarrossay D, Napolitani G, Colonna M, Sallusto F, Lanzavecchia A. Specialization and complementarity in microbial molecule recognition by human myeloid and plasmacytoid dendritic cells. *Eur J Immunol.* (2001) 31:3388–93. doi: 10.1002/1521-4141(200111)31:11>3388::AIDIMMU3388<3.0.CO;2-Q
3. Szabo A, Rajnavolgyi E. Collaboration of Toll-like and RIG-I-like receptors in human dendritic cells: tRIGgering antiviral innate immune responses. *Am J Clini Exp Immunol.* (2013) 2:195–207.
4. Unterholzner L. The interferon response to intracellular DNA: why so many receptors? *Immunobiology.* (2013) 218:1312–21. doi: 10.1016/j.imbio.2013.07.007
5. Marcenaro E, Carlomagno S, Pesce S, Moretta A, Sivori S. NK/DC crosstalk in anti-viral response. *Adv Exp Med Biol.* (2012) 946:295–308. doi: 10.1007/978-1-4614-0106-3_17
6. Blankson JN. Effector mechanisms in HIV-1 infected elite controllers: highly active immune responses? *Antiviral Research.* (2010) 85:295–302. doi: 10.1016/j.antiviral.2009.08.007
7. Hersperger AR, Martin JN, Shin LY, Sheth PM, Kovacs CM, Cosma GL, et al. Increased HIV-specific CD8+ T-cell cytotoxic potential in HIV elite controllers is associated with T-bet expression. *Blood.* (2011) 117:3799–808. doi: 10.1182/blood-2010-12-322727
8. Saez-Cirion A, Lacabaratz C, Lambotte O, Versmisse P, Urrutia A, Boufassa F, et al. HIV controllers exhibit potent CD8 T cell capacity to suppress HIV infection *ex vivo* and peculiar cytotoxic T lymphocyte activation phenotype. *Proc Natl Acad Sci USA.* (2007) 104:6776–81. doi: 10.1073/pnas.0611244104
9. Kadowaki N, Ho S, Antonenko S, Malefyt RW, Kastelein RA, Bazan F, et al. Subsets of human dendritic cell precursors express different toll-like receptors and respond to different microbial antigens. *J Exp Med.* (2001) 194:863–9. doi: 10.1084/jem.194.6.863
10. Tavano B, Galao RP, Graham DR, Neil SJ, Aquino VN, Fuchs D, et al. Ig-like transcript 7, but not bone marrow stromal cell antigen 2 (also known as HM1.24, tetherin, or CD317), modulates plasmacytoid dendritic cell function in primary human blood leukocytes. *J Immunol.* (2013) 190:2622–30. doi: 10.4049/jimmunol.1202391
11. Barron MA, Blyveis N, Palmer BE, MaWhinney S, Wilson CC. Influence of plasma viremia on defects in number and immunophenotype of blood dendritic cell subsets in human immunodeficiency virus 1-infected individuals. *J Infect Dis.* (2003) 187:26–37. doi: 10.1086/345957
12. Dillon SM, Robertson KB, Pan SC, Mawhinney S, Meditz AL, Folkvord JM, et al. Plasmacytoid and myeloid dendritic cells with a partial activation phenotype accumulate in lymphoid tissue during asymptomatic chronic HIV-1 infection. *J Acqu Immune Defic Syndromes.* (2008) 48:1–12. doi: 10.1097/QAI.0b013e3181664b60
13. Sabado RL, O'Brien M, Subedi A, Qin L, Hu N, Taylor E, et al. Evidence of dysregulation of dendritic cells in primary HIV infection. *Blood.* (2010) 116:3839–52. doi: 10.1182/blood-2010-03-273763
14. Diao Y, Geng W, Fan X, Cui H, Sun H, Jiang Y, et al. Low CD1c + myeloid dendritic cell counts correlated with a high risk of rapid disease progression during early HIV-1 infection. *BMC Infect Dis.* (2015) 15:342. doi: 10.1186/s12879-015-1092-8
15. Malleret B, Karlsson I, Maneglier B, Brochard P, Delache B, Andrieu T, et al. Effect of SIVmac infection on plasmacytoid and CD1c+ myeloid dendritic cells in cynomolgus macaques. *Immunology.* (2008) 124:223–33. doi: 10.1111/j.1365-2567.2007.02758.x
16. Donaghy H, Pozniak A, Gazzard B, Qazi N, Gilmour J, Gotch F, et al. Loss of blood CD11c(+) myeloid and CD11c(-) plasmacytoid dendritic cells in patients with HIV-1 infection correlates with HIV-1 RNA virus load. *Blood.* (2001) 98:2574–6. doi: 10.1182/blood.V98.8.2574
17. Lehmann C, Jung N, Forster K, Koch N, Leifeld L, Fischer J, et al. Longitudinal analysis of distribution and function of plasmacytoid dendritic cells in peripheral blood and gut mucosa of HIV infected patients. *J Infect Dis.* (2014) 209:940–9. doi: 10.1093/infdis/jit612
18. Cunningham CR, Champhekar A, Tullius MV, Dillon BJ, Zhen A, de la Fuente JR, et al. Type I and type II interferon coordinately regulate suppressive dendritic cell fate and function during viral persistence. *PLoS Pathogens.* (2016) 12:e1005356. doi: 10.1371/journal.ppat.1005356
19. Carranza P, Del Rio Estrada PM, Diaz Rivera D, Ablanedo-Terrazas Y, Reyes-Teran G. Lymph nodes from HIV-infected individuals harbor mature

dendritic cells and increased numbers of PD-L1+ conventional dendritic cells. *Human Immunol.* (2016) 77:584–93. doi: 10.1016/j.humimm.2016.05.019

20. Huang J, Burke P, Yang Y, Seiss K, Beamon J, Cung T, et al. Soluble HLA-G inhibits myeloid dendritic cell function in HIV-1 infection by interacting with leukocyte immunoglobulin-like receptor B2. *J Virol.* (2010) 84:10784–91. doi: 10.1128/JVI.01292-10

21. Huang J, Burke PS, Cung TD, Pereyra F, Toth I, Walker BD, et al. Leukocyte immunoglobulin-like receptors maintain unique antigen-presenting properties of circulating myeloid dendritic cells in HIV-1-infected elite controllers. *J Virol.* (2010) 84:9463–71. doi: 10.1128/JVI.01009-10

22. Lehmann C, Harper JM, Taubert D, Hartmann P, Fatkenheuer G, Jung N, et al. Increased interferon alpha expression in circulating plasmacytoid dendritic cells of HIV-1-infected patients. *J Acquired Immune Defic Syndr.* (2008) 48:522–30. doi: 10.1097/QAI.0b013e31817f97cf

23. Barblu L, Machmach K, Gras C, Delfraissy JF, Boufassa F, Leal M, et al. Plasmacytoid dendritic cells (pDCs) from HIV controllers produce interferon-alpha and differentiate into functional killer pDCs under HIV activation. *J Infect Dis.* (2012) 206:790–801. doi: 10.1093/infdis/jis384

24. Machmach K, Leal M, Gras C, Viciana P, Genebat M, Franco E, et al. Plasmacytoid dendritic cells reduce HIV production in elite controllers. *J Virol.* (2012) 86:4245–52. doi: 10.1128/JVI.07114-11

25. Saidi H, Bras M, Formaglio P, Melki MT, Charbit B, Herbeuval JP, et al. HMGB1 Is Involved in IFN-alpha production and TRAIL expression by HIV-1-exposed plasmacytoid dendritic cells: impact of the crosstalk with NK cells. *PLoS Pathog.* (2016) 12:e1005407. doi: 10.1371/journal.ppat.1005407

26. Herbeuval JP, Grivel JC, Boasso A, Hardy AW, Chougnet C, Dolan MJ, et al. CD4+ T-cell death induced by infectious and noninfectious HIV-1: role of type 1 interferon-dependent, TRAIL/DR5-mediated apoptosis. *Blood.* (2005) 106:3524–31. doi: 10.1182/blood-2005-03-1243

27. Lichtner M, Rossi R, Rizza MC, Mengoni F, Sauzullo I, Massetti AP, et al. Plasmacytoid dendritic cells count in antiretroviral-treated patients is predictive of HIV load control independent of CD4+ T-cell count. *Curr HIV Res.* (2008) 6:19–27. doi: 10.2174/157016208783571937

28. Royle CM, Graham DR, Sharma S, Fuchs D, Boasso A. HIV-1 and HIV-2 differentially mature plasmacytoid dendritic cells into IFN-producing cells or APCs. *J Immunol.* (2014) 193:3538–48. doi: 10.4049/jimmunol.1400860

29. Kader M, Smith AP, Guiducci C, Wonderlich ER, Normolle D, Watkins SC, et al. Blocking TLR7- and TLR9-mediated IFN-alpha production by plasmacytoid dendritic cells does not diminish immune activation in early SIV infection. *PLoS Pathog.* (2013) 9:e1003530. doi: 10.1371/journal.ppat.1003530

30. Kijewski SD, Gummuluru S. A mechanistic overview of dendritic cell-mediated HIV-1 trans infection: the story so far. *Future Virol.* (2015) 10:257–69. doi: 10.2217/fvl.15.2

31. Dale BM, Alvarez RA, Chen BK. Mechanisms of enhanced HIV spread through T-cell virological synapses. *Immunol Rev.* (2013) 251:113–24. doi: 10.1111/imr.12022

32. Groot F, van Capel TM, Kapsenberg ML, Berkhout B, de Jong EC. Opposing roles of blood myeloid and plasmacytoid dendritic cells in HIV-1 infection of T cells: transmission facilitation versus replication inhibition. *Blood.* (2006) 108:1957–64. doi: 10.1182/blood-2006-03-010918

33. Menager MM, Littman DR. Actin dynamics regulates dendritic cell-mediated transfer of HIV-1 to T cells. *Cell.* (2016) 164:695–709. doi: 10.1016/j.cell.2015.12.036

34. Geijtenbeek TB, Kwon DS, Torensma R, van Vliet SJ, van Duijnhoven GC, Middel J, et al. DC-SIGN, a dendritic cell-specific HIV-1-binding protein that enhances trans-infection of T cells. *Cell.* (2000) 100:587–97. doi: 10.1016/S0092-8674(00)80694-7

35. Kwon DS, Gregorio G, Bitton N, Hendrickson WA, Littman DR. DC-SIGN-mediated internalization of HIV is required for trans-enhancement of T cell infection. *Immunity.* (2002) 16:135–44. doi: 10.1016/S1074-7613(02)00259-5

36. Izquierdo-Useros N, Lorizate M, McLaren PJ, Telenti A, Krausslich HG, Martinez-Picado J. HIV-1 capture and transmission by dendritic cells: the role of viral glycolipids and the cellular receptor Siglec-1. *PLoS Pathog.* (2014) 10:e1004146. doi: 10.1371/journal.ppat.1004146

37. Pino M, Erkizia I, Benet S, Erikson E, Fernandez-Figueras MT, Guerrero D, et al. HIV-1 immune activation induces Siglec-1 expression and enhances viral trans-infection in blood and tissue myeloid cells. *Retrovirology.* (2015) 12:37. doi: 10.1186/s12977-015-0160-x

38. Reyes-Rodriguez AL, Reuter MA, McDonald D. Dendritic cells enhance HIV infection of memory CD4(+) T cells in human lymphoid tissues. *AIDS Res Human Retroviruses.* (2016) 32:203–10. doi: 10.1089/aid.2015.0235

39. Martinez-Picado J, McLaren PJ, Erkizia I, Martin MP, Benet S, Rotger M, et al. Identification of Siglec-1 null individuals infected with HIV-1. *Nat Comm.* (2016) 7:12412. doi: 10.1038/ncomms12412

40. Patterson S, Rae A, Hockey N, Gilmour J, Gotch F. Plasmacytoid dendritic cells are highly susceptible to human immunodeficiency virus type 1 infection and release infectious virus. *J Virol.* (2001) 75:6710–3. doi: 10.1128/JVI.75.14.6710-6713.2001

41. Manel N, Hogstad B, Wang Y, Levy DE, Unutmaz D, Littman DR. A cryptic sensor for HIV-1 activates antiviral innate immunity in dendritic cells. *Nature.* (2010) 467:214–7. doi: 10.1038/nature09337

42. Smed-Sorensen A, Lore K, Vasudevan J, Louder MK, Andersson J, Mascola JR, et al. Differential susceptibility to human immunodeficiency virus type 1 infection of myeloid and plasmacytoid dendritic cells. *J Virol.* (2005) 79:8861–9. doi: 10.1128/JVI.79.14.8861-8869.2005

43. Cameron PU, Handley AJ, Baylis DC, Solomon AE, Bernard N, Purcell DF, et al. Preferential infection of dendritic cells during human immunodeficiency virus type 1 infection of blood leukocytes. *J Virol.* (2007) 81:2297–306. doi: 10.1128/JVI.01795-06

44. Martin-Gayo E, Buzon MJ, Ouyang Z, Hickman T, Cronin J, Pimenova D, et al. Potent cell-intrinsic immune responses in dendritic cells facilitate HIV-1-specific T cell immunity in HIV-1 elite controllers. *PLoS Pathog.* (2015) 11:e1004930. doi: 10.1371/journal.ppat.1004930

45. Lahouassa H, Daddacha W, Hofmann H, Ayinde D, Logue EC, Dragin L, et al. SAMHD1 restricts the replication of human immunodeficiency virus type 1 by depleting the intracellular pool of deoxynucleoside triphosphates. *Nat Immunol.* (2012) 13:223–8. doi: 10.1038/ni.2236

46. Ryoo J, Choi J, Oh C, Kim S, Seo M, Kim SY, et al. The ribonuclease activity of SAMHD1 is required for HIV-1 restriction. *Nat Med.* (2014) 20:936–41. doi: 10.1038/nm.3626

47. Hrecka K, Hao C, Gierszewska M, Swanson SK, Kesik-Brodacka M, Srivastava S, et al. Vpx relieves inhibition of HIV-1 infection of macrophages mediated by the SAMHD1 protein. *Nature.* (2011) 474:658–61. doi: 10.1038/nature10195

48. Puigdomenech I, Casartelli N, Porrot F, Schwartz O. SAMHD1 restricts HIV-1 cell-to-cell transmission and limits immune detection in monocyte-derived dendritic cells. *J Virol.* (2013) 87:2846–56. doi: 10.1128/JVI.02514-12

49. Hertoghs N, van der Aar AM, Setiawan LC, Kootstra NA, Gringhuis SI, Geijtenbeek TB. SAMHD1 degradation enhances active suppression of dendritic cell maturation by HIV-1. *J Immunol.* (2015) 194:4431–7. doi: 10.4049/jimmunol.1403016

50. Cribier A, Descours B, Valadao AL, Laguette N, Benkirane M. Phosphorylation of SAMHD1 by cyclin A2/CDK1 regulates its restriction activity toward HIV-1. *Cell Rep.* (2013) 3:1036–43. doi: 10.1016/j.celrep.2013.03.017

51. Bloch N, O'Brien M, Norton TD, Polsky SB, Bhardwaj N, Landau NR. HIV type 1 infection of plasmacytoid and myeloid dendritic cells is restricted by high levels of SAMHD1 and cannot be counteracted by Vpx. *AIDS Res Human Retroviruses.* (2014) 30:195–203. doi: 10.1089/aid.2013.0119

52. Aljawai Y, Richards MH, Seaton MS, Narasipura SD, Al-Harthi L. beta-Catenin/TCF-4 signaling regulates susceptibility of macrophages and resistance of monocytes to HIV-1 productive infection. *Curr HIV Res.* (2014) 12:164–73. doi: 10.2174/1570162X12666140526122249

53. Silvin A, Yu CI, Lahaye X, Imperatore F, Brault JB, Cardinaud S, et al. Constitutive resistance to viral infection in human CD141(+) dendritic cells. *Sci Immunol.* (2017) 2:eaai8071. doi: 10.1126/sciimmunol.aai8071

54. Pena-Cruz V, Agosto LM, Akiyama H, Olson A, Moreau Y, Larrieux JR, et al. HIV-1 replicates and persists in vaginal epithelial dendritic cells. *J Clini Invest.* (2018) 128:3439–44. doi: 10.1172/JCI98943

55. Saez-Cirion A, Hamimi C, Bergamaschi A, David A, Versmisse P, Melard A, et al. Restriction of HIV-1 replication in macrophages and CD4+ T cells from HIV controllers. *Blood.* (2011) 118:955–64. doi: 10.1182/blood-2010-12-327106

56. Hamimi C, David A, Versmisse P, Weiss L, Bruel T, Zucman D, et al. Dendritic cells from HIV controllers have low susceptibility to HIV-1 infection *in vitro* but high capacity to capture HIV-1 particles. *PLoS ONE.* (2016) 11:e0160251. doi: 10.1371/journal.pone.0160251

57. St. Gelais C, de Silva S, Amie SM, Coleman CM, Hoy H, Hollenbaugh JA, et al. SAMHD1 restricts HIV-1 infection in dendritic cells (DCs) by dNTP depletion, but its expression in DCs and primary CD4+ T-lymphocytes cannot be upregulated by interferons. *Retrovirology*. (2012) 9:105. doi: 10.1186/1742-4690-9-105

58. Lahaye X, Satoh T, Gentili M, Cerboni S, Silvin A, Conrad C, et al. Nuclear envelope protein SUN2 promotes cyclophilin-a-dependent steps of HIV replication. *Cell Rep*. (2016) 15:879–92. doi: 10.1016/j.celrep.2016.03.074

59. Pertel T, Hausmann S, Morger D, Zuger S, Guerra J, Lascano J, et al. TRIM5 is an innate immune sensor for the retrovirus capsid lattice. *Nature*. (2011) 472:361–5. doi: 10.1038/nature09976

60. Portilho DM, Fernandez J, Ringeard M, Machado AK, Boulay A, Mayer M, et al. Endogenous TRIM5alpha function is regulated by SUMOylation and nuclear sequestration for efficient innate sensing in dendritic cells. *Cell Rep*. (2016) 14:355–69. doi: 10.1016/j.celrep.2015.12.039

61. Ben Haij N, Planes R, Leghmari K, Serrero M, Delobel P, Izopet J, et al. HIV-1 tat protein induces production of proinflammatory cytokines by human dendritic cells and monocytes/macrophages through engagement of TLR4-MD2-CD14 complex and activation of NF-kappaB pathway. *PLoS ONE*. (2015) 10:e0129425. doi: 10.1371/journal.pone.0129425

62. Cardinaud S, Urrutia A, Rouers A, Coulon PG, Kervevan J, Richetta C, et al. Triggering of TLR-3,−4, NOD2, and DC-SIGN reduces viral replication and increases T-cell activation capacity of HIV-infected human dendritic cells. *Eur J Immunol*. (2017) 47:818–29. doi: 10.1002/eji.201646603

63. Sironi M, Biasin M, Cagliani R, Forni D, De Luca M, Saulle I, et al. A common polymorphism in TLR3 confers natural resistance to HIV-1 infection. *J Immunol*. (2012) 188:818–23. doi: 10.4049/jimmunol.1102179

64. Harman AN, Lai J, Turville S, Samarajiwa S, Gray L, Marsden V, et al. HIV infection of dendritic cells subverts the IFN induction pathway via IRF-1 and inhibits type 1 IFN production. *Blood*. (2011) 118:298–308. doi: 10.1182/blood-2010-07-297721

65. Sun L, Wu J, Du F, Chen X, Chen ZJ. Cyclic GMP-AMP synthase is a cytosolic DNA sensor that activates the type I interferon pathway. *Science*. (2013) 339:786–91. doi: 10.1126/science.1232458

66. Gao D, Wu J, Wu YT, Du F, Aroh C, Yan N, et al. Cyclic GMP-AMP synthase is an innate immune sensor of HIV and other retroviruses. *Science*. (2013) 341:903–6. doi: 10.1126/science.1240933

67. Zhang H, Tang K, Zhang Y, Ma R, Ma J, Li Y, et al. Cell-free tumor microparticle vaccines stimulate dendritic cells via cGAS/STING signaling. *Cancer Immunol Res*. (2015) 3:196–205. doi: 10.1158/2326-6066.CIR-14-0177

68. Shu C, Li X, Li P. The mechanism of double-stranded DNA sensing through the cGAS-STING pathway. *Cytokine Growth Factor Rev*. (2014) 25:641–8. doi: 10.1016/j.cytogfr.2014.06.006

69. Lahaye X, Satoh T, Gentili M, Cerboni S, Conrad C, Hurbain I, et al. The capsids of HIV-1 and HIV-2 determine immune detection of the viral cDNA by the innate sensor cGAS in dendritic cells. *Immunity*. (2013) 39:1132–42. doi: 10.1016/j.immuni.2013.11.002

70. Yoh SM, Schneider M, Seifried J, Soonthornvacharin S, Akleh RE, Olivieri KC, et al. PQBP1 is a proximal sensor of the cGAS-dependent innate response to HIV-1. *Cell*. (2015) 161:1293–305. doi: 10.1016/j.cell.2015.04.050

71. Bode C, Fox M, Tewary P, Steinhagen F, Ellerkmann RK, Klinman D, et al. Human plasmacytoid dentritic cells elicit a Type I Interferon response by sensing DNA via the cGAS-STING signaling pathway. *Eur J Immunol*. (2016) 46:1615–21. doi: 10.1002/eji.201546113

72. Paijo J, Doring M, Spanier J, Grabski E, Nooruzzaman M, Schmidt T, et al. cGAS senses human cytomegalovirus and induces type I interferon responses in human monocyte-derived cells. *PLoS Pathog*. (2016) 12:e1005546. doi: 10.1371/journal.ppat.1005546

73. See P, Dutertre CA, Chen J, Gunther P, McGovern N, Irac SE, et al. Mapping the human DC lineage through the integration of high-dimensional techniques. *Science*. (2017) 356:eaag3009. doi: 10.1126/science.aag3009

74. Rodrigues PF, Alberti-Servera L, Eremin A, Grajales-Reyes GE, Ivanek R, Tussiwand R. Distinct progenitor lineages contribute to the heterogeneity of plasmacytoid dendritic cells. *Nat Immunol*. (2018) 19:711–22. doi: 10.1038/s41590-018-0136-9

75. Herman JS, Sagar, Grun D. FateID infers cell fate bias in multipotent progenitors from single-cell RNA-seq data. *Nat Methods*. (2018) 15:379–86. doi: 10.1038/nmeth.4662

76. Martin-Gayo E, Gonzalez-Garcia S, Garcia-Leon MJ, Murcia-Ceballos A, Alcain J, Garcia-Peydro M, et al. Spatially restricted JAG1-Notch signaling in human thymus provides suitable DC developmental niches. *J Exp Med*. (2017) 214:3361–79. doi: 10.1084/jem.20161564

77. Lahaye X, Gentili M, Silvin A, Conrad C, Picard L, Jouve M, et al. NONO Detects the nuclear HIV capsid to promote cgas-mediated innate immune activation. *Cell*. (2018) 175:488–501.e22. doi: 10.1016/j.cell.2018.08.062

78. Martin-Gayo E, Cole MB, Kolb KE, Ouyang Z, Cronin J, Kazer SW, et al. A reproducibility-based computational framework identifies an inducible, enhanced antiviral state in dendritic cells from HIV-1 elite controllers. *Genome Biol*. (2018) 19:10. doi: 10.1186/s13059-017-1385-x

79. Harman AN, Nasr N, Feetham A, Galoyan A, Alshehri AA, Rambukwelle D, et al. HIV blocks interferon induction in human dendritic cells and macrophages by dysregulation of TBK1. *J Virol*. (2015) 89:6575–84. doi: 10.1128/JVI.00889-15

80. Berg RK, Melchjorsen J, Rintahaka J, Diget E, Soby S, Horan KA, et al. Genomic HIV RNA induces innate immune responses through RIG-I-dependent sensing of secondary-structured RNA. *PLoS ONE*. (2012) 7:e29291. doi: 10.1371/journal.pone.0029291

81. Ablasser A, Bauernfeind F, Hartmann G, Latz E, Fitzgerald KA, Hornung V. RIG-I-dependent sensing of poly(dA:dT) through the induction of an RNA polymerase III-transcribed RNA intermediate. *Nat Immunol*. (2009) 10:1065–72. doi: 10.1038/ni.1779

82. Ahlers LR, Bastos RG, Hiroyasu A, Goodman AG. Invertebrate iridescent virus 6, a DNA virus, stimulates a mammalian innate immune response through RIG-I-like receptors. *PLoS ONE*. (2016) 11:e0166088. doi: 10.1371/journal.pone.0166088

83. Kaushik S, Teque F, Patel M, Fujimura SH, Schmidt B, Levy JA. Plasmacytoid dendritic cell number and responses to Toll-like receptor 7 and 9 agonists vary in HIV Type 1-infected individuals in relation to clinical state. *AIDS Res Human Retroviruses*. (2013) 29:501–10. doi: 10.1089/aid.2012.0200

84. Manches O, Fernandez MV, Plumas J, Chaperot L, Bhardwaj N. Activation of the noncanonical NF-kappaB pathway by HIV controls a dendritic cell immunoregulatory phenotype. *Proc Natl Acad Sci USA*. (2012) 109:14122–7. doi: 10.1073/pnas.1204032109

85. Tel J, Schreibelt G, Sittig SP, Mathan TS, Buschow SI, Cruz LJ, et al. Human plasmacytoid dendritic cells efficiently cross-present exogenous Ags to CD8+ T cells despite lower Ag uptake than myeloid dendritic cell subsets. *Blood*. (2013) 121:459–67. doi: 10.1182/blood-2012-06-435644

86. Granelli-Piperno A, Shimeliovich I, Pack M, Trumpfheller C, Steinman RM. HIV-1 selectively infects a subset of nonmaturing BDCA1-positive dendritic cells in human blood. *J Immunol*. (2006) 176:991–8. doi: 10.4049/jimmunol.176.2.991

87. van den Berg LM, Cardinaud S, van der Aar AM, Sprokholt JK, de Jong MA, Zijlstra-Willems EM, et al. Langerhans cell-dendritic cell cross-talk via langerin and hyaluronic acid mediates antigen transfer and cross-presentation of HIV-1. *J Immunol*. (2015) 195:1763–73. doi: 10.4049/jimmunol.1402356

88. Dinter J, Duong E, Lai NY, Berberich MJ, Kourjian G, Bracho-Sanchez E, et al. Variable processing and cross-presentation of HIV by dendritic cells and macrophages shapes CTL immunodominance and immune escape. *PLoS Pathog*. (2015) 11:e1004725. doi: 10.1371/journal.ppat.1004725

89. Segura E, Touzot M, Bohineust A, Cappuccio A, Chiocchia G, Hosmalin A, et al. Human inflammatory dendritic cells induce Th17 cell differentiation. *Immunity*. (2013) 38:336–48. doi: 10.1016/j.immuni.2012.10.018

90. Coulon PG, Richetta C, Rouers A, Blanchet FP, Urrutia A, Guerbois M, et al. HIV-infected dendritic cells present endogenous MHC class II-restricted antigens to HIV-specific CD4+ T cells. *J Immunol*. (2016) 197:517–32. doi: 10.4049/jimmunol.1600286

91. Walker BD, Yu XG. Unravelling the mechanisms of durable control of HIV-1. *Nat Rev Immunol*. (2013) 13:487–98. doi: 10.1038/nri3478

92. Kwong PD, Mascola JR. Human antibodies that neutralize HIV-1: identification, structures, and B cell ontogenies. *Immunity*. (2012) 37:412–25. doi: 10.1016/j.immuni.2012.08.012

93. Burton DR, Mascola JR. Antibody responses to envelope glycoproteins in HIV-1 infection. *Nat Immunol*. (2015) 16:571–6. doi: 10.1038/ni.3158

94. Doria-Rose NA, Klein RM, Daniels MG, O'Dell S, Nason M, Lapedes A, et al. Breadth of human immunodeficiency virus-specific neutralizing activity in

sera: clustering analysis and association with clinical variables. *J Virol.* (2010) 84:1631–6. doi: 10.1128/JVI.01482-09

95. Locci M, Havenar-Daughton C, Landais E, Wu J, Kroenke MA, Arlehamn CL, et al. Human circulating PD-1+CXCR3-CXCR5+ memory Tfh cells are highly functional and correlate with broadly neutralizing HIV antibody responses. *Immunity.* (2013) 39:758–69. doi: 10.1016/j.immuni. 2013.08.031

96. Breitfeld D, Ohl L, Kremmer E, Ellwart J, Sallusto F, Lipp M, et al. Follicular B helper T cells express CXC chemokine receptor 5, localize to B cell follicles, and support immunoglobulin production. *J Exp Med.* (2000) 192:1545–52. doi: 10.1084/jem.192.11.1545

97. Morita R, Schmitt N, Bentebibel SE, Ranganathan R, Bourdery L, Zurawski G, et al. Human blood CXCR5(+)CD4(+) T cells are counterparts of T follicular cells and contain specific subsets that differentially support antibody secretion. *Immunity.* (2011) 34:108–21. doi: 10.1016/j.immuni.2011.01.009

98. Martin-Gayo E, Cronin J, Hickman T, Ouyang Z, Lindqvist M, Kolb KE, et al. Circulating CXCR5+CXCR3+PD-1lo Tfh-like cells in HIV-1 controllers with neutralizing antibody breadth. *JCI Insight.* (2017) 2:e89574. doi: 10.1172/jci.insight.89574

99. Ma CS, Suryani S, Avery DT, Chan A, Nanan R, Santner-Nanan B, et al. Early commitment of naive human CD4(+) T cells to the T follicular helper (T(FH)) cell lineage is induced by IL-12. *Immunol Cell Biol.* (2009) 87:590–600. doi: 10.1038/icb.2009.64

17

The debate about dendritic cells and macrophages in the kidney

Catherine Gottschalk and Christian Kurts *

Institute of Experimental Immunology, Rheinische Friedrich-Wilhelms-Universität Bonn, Bonn, Germany

Edited by:
Martin Guilliams,
Ghent University – VIB, Belgium

Reviewed by:
Barbara Ursula Schraml,
Technische Universität München,
Germany
Molly Ingersoll,
Institut Pasteur, France

*Correspondence:
Christian Kurts,
Institute of Experimental Immunology,
Rheinische Friedrich-Wilhelms-
Universität Bonn, Sigmund-Freud-Str.
25, Bonn D-53127, Germany
ckurts@uni-bonn.de

The mononuclear phagocyte system includes macrophages and dendritic cells (DCs), which are usually classified by morphology, phenotypical characteristics, and function. In the last decades, large research communities have gathered substantial knowledge on the roles of these cells in immune homeostasis and anti-infectious defense. However, these communities developed to a degree independent from each other, so that the nomenclature and functions of the numerous DC and macrophage subsets overlap, resulting in the present intense debate about the correct nomenclature. This controversy has also reached the field of experimental nephrology. At present, no mutually accepted way to distinguish renal DC and macrophages is available, so that many important roles in acute and chronic kidney disease have been ascribed to both DCs and macrophages. In this perspective article, we discuss the causes and consequences of the overlapping DC–macrophage classification systems, functional roles of DCs and macrophages, and the transferability of recent findings from other disciplines to the renal mononuclear phagocyte system from the nephrologist's point of view.

Keywords: dendritic cells, macrophages, kidney, flow-cytometry, glomerulonephritis

Introduction

The current intense debate regarding the classification and nomenclature of dendritic cells (DCs) and macrophages has reached also the field of experimental nephrology. Numerous kidney diseases are immune mediated, such as the different forms of glomerulonephritis, and research over the last years has described important, yet overlapping roles of both cells types.

Macrophages and DCs are often considered distinct cell types based on their morphology and function. Macrophages were defined as large vacuolar cells that are highly phagocytic and modulate immune responses by production of immune mediators (1, 2), whereas DCs were characterized as stellate migratory cells that act as sentinels in non-lymphoid tissues and enter lymphoid tissues upon antigen encounter, present antigen and subsequently activate naïve T lymphocytes (3–5). Following these original descriptions, two research areas developed that more or less independently studied macrophages and DCs. This artificial separation has contributed to the emergence of different names for similar or the same cell types, thereby adding to the current confusion about their identity and function. In particular, advances in multi-color flow cytometry and gene-analyses enabled researchers to define many DC and macrophage subsets by the expression of a variety of surface molecules (6). As cell surface markers are easy to determine, they are widely used to classify mononuclear phagocytes, although they are rather unspecific and their expression patterns in the murine and human systems differ substantially.

Also in the kidney, surface markers and functional parameters have been used to propose several classification systems of mononuclear phagocytes. However, these systems overlap, comparable to the situation in other non-lymphoid organs, resulting in great uncertainty among experimental nephrologists regarding the correct terminology. Here, we discuss the present state of our knowledge on renal mononuclear phagocytes in health and disease and problems resulting from the current nomenclature debate from a nephrologist point of view.

The Network of Renal DCs and Macrophages

The kidney parenchyma consists of the outer renal cortex and the inner renal medulla. Numerous individual functional units, the nephrons, span both compartments. The cortex contains glomeruli and proximal tubuli of the nephrons, which generate the primary urine. The medulla contains the loop of Henle, which generates a high osmolarity that is required for water reabsorption from the primary urine. The distal tubules end in collecting ducts through which the concentrated urine is transported into the renal pelvis and on through the ureters into the bladder. The space between the tubules is known as tubulointerstitium and contains blood vessels, fibroblasts, and numerous cells of the hematopoietic system that had been classified by pathologists as constituents of the reticuloendothelial system [reviewed in Ref. (7)].

Early immunological studies had classified the tubulointerstitial mononuclear cells as macrophages due to their F4/80 expression (8). During the early 1990s, several groups reported that these tubulointerstitial cells morphologically resembled DCs in humans and rodents (9–12), whereas cells with the typical morphology of macrophages were described to reside mainly in the kidney capsule, the intravasal lumina, and the pelvic wall of healthy kidneys (13). The use of CX$_3$CR1-reporter mice and live cell imaging illustrated the intricate tubulointerstitial network of dendritiform processes that these cells use to constantly probe the environment, suggestive of DCs in action (14–16). The nomenclature debate intensified when it became clear that the vast majority of renal mononuclear phagocytes possess the phenotype CD11c$^+$ CD11b$^+$ F4/80$^+$ CX$_3$CR1$^+$ (17), which allows classification of both macrophages and DCs.

Notably, CX$_3$CR1 exhibits relative organ specificity for renal mononuclear phagocytes: these cells were >50% reduced in the kidneys, but not in other organs (except the intestine) of CX$_3$CR1-deficient mice. This may be explained by the comparatively high renal expression of its ligand CX$_3$CL1 (18). Notably, those CX$_3$CR1$^+$ phagocytes that co-express CD11c and exert DC functionality were reduced even by more than 75% (18–20). This may result from an effect of CX$_3$CR1 on CD11c expression, but this has yet to be shown. Interestingly, CX$_3$CR1 regulated the numbers of the CD11c$^+$ and the CD11c$^-$ renal mononuclear phagocytes by different mechanisms: it promoted homeostatic and inflammatory recruitment of the former, whereas it prevented *in situ* proliferation of the latter under inflammatory conditions (20). Assuming that CD11c distinguishes renal DCs and macrophages, this difference would be consistent with recent reports that the number of tissue macrophages is regulated by local proliferation (21), whereas DC numbers are usually thought to be regulated by immigration and emigration (22).

The kidney also contains a minor subset of CD103$^+$ DCs, which constitute <5% of all renal CD11c$^+$ phagocytes and lack expression of CX$_3$CR1, CD11b, and F4/80 (23), whose function currently is unclear. There are neither CD11b$^+$ CD103$^+$ DCs nor plasmacytoid DCs in the healthy kidney (24).

Functionality and Phenotype of Renal Mononuclear Phagocytes

Researchers from both, the DC and the macrophage fields, have investigated kidney mononuclear phagocytes defined by cell surface markers in homeostasis and models of renal disease. Many important roles were shown in models of acute renal injury and in chronic immune-mediated kidney disease (**Table 1**), such as cytokine production or T cell-crosstalk in response to tissue injury or infection (17, 25–33). However, none of these functions is generally accepted to be exclusive for DCs or macrophages. Moreover, many nephrologists trained by the DC and macrophage communities still use CD11c and F4/80 to identify DCs and macrophages, respectively (see **Table 1**), even though 70–90% of renal mononuclear phagocytes co-express these two markers (17), implying that they studied cellular subsets that largely overlap. Also, the tools used for loss-of-function studies cannot clearly discriminate between DCs and macrophages: CD11c–DTR mice are used to deplete kidney DCs, CD11b–DTR mice for depleting kidney macrophages but the expression of CD11c and CD11b on kidney mononuclear phagocytes is too heterogeneous for this black-and-white thinking (34). Clodronate liposomes are used for both purposes (35–38). All kidney mononuclear phagocytes are phagocytic (34) which might render them sensitive to clodronate liposomes.

The consequence of this overlap is well illustrated by two recent studies examining how CX$_3$CR1 affects renal disease: both studies agreed that mononuclear phagocytes are substantially reduced in the kidneys of CX$_3$CR1-deficient mice. However, one of them noted a higher susceptibility to renal candidiasis and attributed this to the loss of renal macrophages (19), while the other documented protection against glomerulonephritis and assigned this to the loss of renal DCs (18). A possible explanation for this different classifications is that glomerulonephritis is driven mostly by phagocytes in the kidney cortex, in which glomeruli are located, whereas anti-infectious activity seem to be primarily due to phagocytes in the medulla, through which pathogens enter the kidney (18). Medullary phagocytes express significantly less CD11c than those in the cortex, which may bias their classification as DCs. The causes for these phenotypical and functional differences between medullary and cortical mononuclear phagocytes are unknown, but may result from differences in osmolarity, pH, and oxygen tension between these compartments, to which the mononuclear phagocytes may adapt. This would be in line with the current view that the tissue microenvironment dictates the organ-specific plasticity of macrophages (39, 40), and thus, perhaps also of renal mononuclear phagocytes.

Re-Defining Kidney Mononuclear Phagocyte Nomenclature

The current definitions of renal DCs and macrophages are not mutually exclusive, so that renal mononuclear phagocytes may fulfill the definitions of both cell types simultaneously. This creates

TABLE 1 | Summary of the functions of mononuclear phagocyte subsets in renal diseases, which have been attributed to either renal DC or macrophages, based on marker expression and/or disease attenuation or aggravation after cell depletion.

Disease	Function and associated cell type	Classification of associated cell types
Acute renal injury	**Pro-inflammatory**	**Pro-inflammatory**
Ischemia/re-perfusion	I. TNFa secretion DC (26, 63, 64) Macrophages (63, 64) II. Th activation Macrophages (63)	DC CD45+, CD11c+, MHCII+, CD11b+, CD16+, F4/80+, CD68+, CD4−, CD8−, CD205−, 33D1−, CD169− CD204− (26) Macrophages Sensitive to liposomal dichloromethylene bisphosphonate (clodronate liposome) treatment, F4/80+ (63) Sensitive to clodronate liposome treatment (64)
	Anti-inflammatory I. Tissue regeneration DC (33) Macrophages (67) II. Suppression of TNFa, IL-6, CXCL2, CCL2 production by IRF4 upregulation DC (65) III. Prevention of renal failure DC (66)	**Anti-inflammatory** DC Sensitive to clodronate liposome treatment, CD45+, MHCII+,CD11c+, F4/80+ (33) Sensitive to clodronate liposome treatment, CD45+, MHCII+, CD11c+ (65) Sensitive to clodronate liposome treatment, CD11b+ (66) Macrophages Sensitive to clodronate liposome treatment, F4/80+ (67)
Unilateral ureter obstruction (UUO)	**Pro-inflammatory** I. Antigen presentation to CD4+ T cells DC (27) II. Accumulation of Th17 cells DC (28) III. TNFa, TGFb production DC (28, 68) Macrophages (68) IV. Tubular apoptosis DC (68) Macrophages (68) V. Renal fibrosis DC (68, 69) Macrophages (68, 69)	**Pro-inflammatory** DC CD11c+, T cell stimulatory, phagocytotic (27) CD45+, CD11c+, F4/80−, Ly6C− or CD45+, CD11c+, F4/80−, Ly6C−, sensitive to clodronate liposome treatment (28) CD45+ CD11c+, F4/80+ (sensitive to clodronate liposomes) or F4/80− (not sensitive to clodronate liposomes) (68) Macrophages CD45+ F4/80+, CD11c−, sensitive to clodronate liposomes (68) CD45+, CD11b+, Csfr1R-GFP+, CD11c−; depletion in CD11b–DTR mice (69)
Adriamycin nephropathy, cisplatin nephropathy, crystal nephropathy	**Pro-inflammatory** I. Aggravation of kidney injury in adriamycin-induced nephropathy Macrophages (25) II. IL-1b secretion after inflammasome activation DC (29)	**Pro-inflammatory** DC In vitro studies with bone marrow derived DC; renal CD45+, CD11c+ cells; sensitive to clodronate liposome depletion and diphtheria toxin in CD11c–DTRg mice (29) Macrophages CD45+, MHCII+, CD11c+, F4/80+, CD68+, CD204+, CD206+, CD103−; morphology, phagocytic capacity, ontogeny (25)
	Anti-inflammatory I. Protective against cisplatin nephropathy, induction of IL-10 DC (70)	**Anti-inflammatory** DC CD45+, MHCII+, CD11c+, CD11b+, F4/80+; morphology of GFP+ cells in CD11c–DTRtg mice (70)
Chronic renal disease	**Accumulating**	**Accumulating**
Glomerulonephritides	I. Population changes during nephrotoxic nephritis DC (17)	DC CD11c+, CD11b+, F4/80+; morphology, lysosomal content, phagocytic activity, microbicidal effector functions, expression of T cell costimulatory molecules, T cell activation (17)
	Pro-inflammatory I. Crescent formation Macrophages (71) II. T cell infiltration and activation DC (32, 72) III. Chemokine expression DC (73)	**Pro-inflammatory** DC MHCII+, CD11c+, F4/80− (72) MHCII+, CD11c+ CD11b+, CD8−, B220−; depletion in CD11c–DTR mice; antigen presentation and T cell activation function (32) Chemokine expression by CD11b+ CD11c+ DC was analyzed in lymphoid organs (73) Macrophages Sensitive to diphtheria toxin in CD11b-DTR mice, CD68+ (71)

(Continued)

TABLE 1 | Continued

Disease	Function and associated cell type	Classification of associated cell types
	Anti-inflammatory	**Anti-inflammatory**
	I. Induction of IL-10 secretion by CD4 T cells	DC
	DC (31)	Morphology; MHCII+, CD11c+, CD11b+, sensitive to diphtheria toxin in CD11c-DTR mice (31)
	II. Recruitment of regulatory CXCR6+ iNKT cells DC (74)	CD45+, CD11c+, depletion in CD11c-luciDTR mice (74)
Infection	**Anti-infectious**	**Anti-infectious**
	I. Bacterial clearance	DC
	DC (18, 30)	MHCII+, CD45+, CD11c+, CD11b+, F4/80+, CX₃CR1+ CD103⁻; depletion in
	II. Candida protection	CD11c–DTR mice (18, 30)
	Macrophages (19)	Enrichment by Flt3L administration, sorted by CD11c purification (75)
	III. Response to infectious stimuli, chemokine	Macrophages
	secretion, migration	MHCII+, F4/80+, CD11b+, CD11cˡᵒ; morphology (19)
	DC (75)	

confusion, especially among those nephrologists that are more interested in disease relevance than in semantics. A recent proposal for a unified nomenclature has been based on cellular ontogeny: it proposes an initial division of mononuclear phagocytes into macrophages, monocytes and monocyte-derived cells and DCs (so-called "level 1 nomenclature") (41). This classification was based on the following facts: (1) most adult macrophages in tissues are successors of an embryonic precursor and maintained through self-renewal (42–46), (2) a common monocyte progenitor (cMoP) exists, which gives rise to monocytes (47), and (3) conventional DC (cDC) and plasmacytoid DC but not monocytes or macrophages arise from a common DC precursor (CDP) (48, 49). Thus, tissue-resident macrophages were classified by their origin from embryonic (yolk sac and fetal monocytes)-derived erythro-myeloid progenitors (46, 50) and DC were classified as cells arising from hematopoietic stem cell-derived precursors, identified by genetic tracing via DNGR1 (CLEC9A) (51), which are distinct from monocyte/macrophage precursors. Finally, monocyte-derived cells differentiate from cMoP that can exert macrophage- or DC-like functions and express markers associated with either (41). This classification does not resolve the question whether monocyte-derived macrophages and monocyte-derived DCs are ontogenically distinct or whether one cell type displays high plasticity in different microenvironments. To include cell function, location, and morphology, the authors suggested to add a "level 2" nomenclature to the level 1 classification (41).

While this nomenclature proposal might bring order into the ever increasing numbers DC and macrophage subsets, one major concern remains: without fate mapping tools, the origin of a phagocyte in a given tissue is usually not apparent, so that surrogate markers need to be used. Several markers for distinguishing phagocytes derived from different precursors are currently being discussed, but as we shall see below, they fail to discriminate renal DCs and macrophages.

One of these markers, CD64, alone or in combination with CCR2 or MerTK, has been reported to identify monocyte-derived macrophages and to be able to discriminate DC from non-DC in the intestine, the muscle and spleen (52–55), and the skin (56).

DNGR1, when combined with genetic fate mapping technology, was shown to mark CDP and pre-DC (51), whereas Csf1r can be used for fate mapping of yolk sac derived (myb independent) tissue macrophages (46). In the kidney, most mononuclear phagocytes express CD64, low levels of CD11b and high levels of F4/80, which is not the case in other organs. However, 30% of CD64+ cells co-expressed the DNGR1-fate mapper, indicating that CD64 expression, despite the evidence for specificity in other organs, does not differentiate CDP-derived from monocyte-derived cells in the kidney (51). Similarly, another fate mapping study that used Myb and PU.1 dependency for defining CD11bʰⁱ monocytes or macrophages and F4/80ᵇʳⁱᵍʰᵗ tissue macrophages derived by adult or embryonic hematopoiesis, respectively, found a dual origin in kidney macrophages as well (45). These findings highlight the difficulties when basing cellular classification solely on ontogeny when ontogeny is based on surrogate markers. Furthermore, transferring ontogeny-based nomenclature to human mononuclear phagocytes in tissue might prove impracticable.

A classification approach based on transcriptome analysis reported that CD11c+ MHC II+ cells in the kidney expressed a set of core DC markers characteristic of DCs in non-lymphoid tissues, that is absent from macrophages, including Zbtb64, Flt3, and CCR7 (57). These "core DC markers" had been defined by analyzing cDCs except the CD11b+ non-lymphoid tissue–DC, because of the great heterogeneity of CD11b+ cells. However, these constitute the vast majority of kidney mononuclear phagocytes.

Another classification approach is based on mononuclear phagocyte functionality. However, observed functions generally represent a snapshot of a cell within a specific context and time frame. Demonstrating that a phagocyte performs a given function under certain conditions at a certain time-point does not imply that this is a general feature of this cell. Furthermore, there is no clear demarcation between exclusive DC and macrophage functions. For example, macrophages phagocytose and degrade material. However, under certain conditions DCs do that too, albeit less efficiently [reviewed in Ref. (58)]. On the other hand, DCs classically activate naïve T cells, but macrophages can do that too, albeit less efficiently (59, 60). Furthermore, the ability to stimulate T cells is difficult to determine on a single cell basis. A recent study differentiated renal mononuclear phagocytes into five phenotypically and functionally distinct populations (34). In that study, mononuclear phagocyte populations were differentiated by CD11c, CD11b, F4/80, CD103, CD14, CD16, and CD64 expression in juvenile and adult mice of different strains. Functional analyses and fate mapping studies were used for further characterization. In line with the complexity of kidney mononuclear phagocyte subsets observed by others and us (17, 45, 51), the study revealed that all subsets expressed CD68 that is usually used to identify macrophages and that all subsets were phagocytic but showed differences in their antigen presentation capacity. Fate mapping experiments identified one population with a dual origin, two populations that were closely related to monocytes, whereas the remaining two were not. Notably, the largest population not only showed the phenotypical and functional characteristics of reparative macrophages (M2) but also had significant antigen presentation function and most likely emigrated from the kidney under inflammatory conditions. Additionally, this population differed significantly between mouse strains, which might explain immunological differences between those strains. The authors concluded that functions are more related to context than separate lineage and suggested their marker combination as an unbiased approach to identify kidney mononuclear phagocyte populations (34). These findings are consistent with recent concepts that macrophage fine differentiation is shaped by the tissue microenvironment (39, 40).

Concluding Remarks

As a consequence of the separate development of the DC and macrophage research communities, the functional and phenotypic definitions of these cell types overlap substantially. Thus, scientists from both communities often study the same cells, perhaps unaware of, or ignoring progress and concepts in the other field. The false assumption that classifying a mononuclear phagocyte as a macrophage implies that it is not a DC, and vice versa, hampers communication between researchers from both fields. Some studies have focused on arguing about subsets and semantics (61), perhaps hoping to "claim territory" for their own communities. This may result in highly citable or controversial publications, but it does not advance our understanding of mononuclear phagocytes, neither in the kidney nor elsewhere.

An overlapping classification system, such as the existing one, is certainly not desirable. An improvement is needed. It is unrealistic to assume that either the DC or the macrophage community

will accept the nomenclature of the other field. Drawing a line that segregates mononuclear phagocytes into DCs or macrophages will unlikely be acceptable to both fields. Furthermore, there are currently no unambiguous discriminatory parameters; for any new parameters introduced, exceptions are reported quickly, such as for CD64 and DNGR1-fate tracking in the kidney. Still, an improved classification system is needed. How can we reach a consensus?

First, the purpose of the revised classification system needs to be defined. Clinicians are interested in cellular entities that are useful for diagnostic or therapeutic purposes and translational immunologists often study the functions of cellular subsets. Basic immunologists may favor ontogeny, which is biologically the cleanest and most logical approach. However, mononuclear phagocytes adapt their gene enhancer landscape according to the tissue of residence independently of the precursor they originated from (39), an ontogeny-based nomenclature may lead to different cell types with similar functionality, or to cells of the same name with different functionality depending on the organ they reside in. Moreover, the origin of a mononuclear phagocyte in a given tissue is not obviously apparent, because unique discriminatory parameters are missing. Thus, ontogeny, although theoretically logical, will be difficult to use for routine research. At the end of the day, a classification system needs to be convenient and feasible, or it will not be used.

The late Ralph Steinman remarked "The DC is a functional state" (personal communication). Indeed, at the age of single cell transcriptomics, it becomes clear that several transcriptional programs may run simultaneously in individual mononuclear phagocytes, and confer a spectrum of functionalities that are more or less consistent with the current concepts of a DC, of a macrophage, or both. Current technical advances will undoubtedly allow distinguishing far more functional states of mononuclear phagocytes. In the field of renal immunology, experts coming from the DC and macrophage communities have jointly suggested avoiding the DC–macrophage controversy altogether by referring to mononuclear phagocytes (preferentially using a "catchier" name for these cells), with different degrees of DC- or macrophage-, or other functionalities (62). It remains to be seen whether basic immunologists and scientists studying mononuclear phagocytes in other organs feel that this is useful or not.

Acknowledgments

We thank Natalio Garbi and Wolfgang Kastenmüller for helpful discussions. This work was supported by the Deutsche Forschungsgemeinschaft (grants KFO228 and SFBTR57). CK is a member of the Excellence-Cluster ImmunoSensation in Bonn.

References

1. Mantovani B, Rabinovitch M, Nussenzweig V. Phagocytosis of immune complexes by macrophages. Different roles of the macrophage receptor sites for complement (C3) and for immunoglobulin (IgG). *J Exp Med* (1972) **135**(4):780–92. doi:10.1084/jem.135.4.780

2. MacMicking J, Xie QW, Nathan C. Nitric oxide and macrophage function. *Annu Rev Immunol* (1997) **15**:323–50. doi:10.1146/annurev.immunol.15.1.323

3. Steinman RM, Cohn ZA. Identification of a novel cell type in peripheral lymphoid organs of mice. I. Morphology, quantitation, tissue distribution. *J Exp Med* (1973) **137**(5):1142–62. doi:10.1084/jem.137.5.1142

4. Inaba K, Metlay JP, Crowley MT, Steinman RM. Dendritic cells pulsed with protein antigens in vitro can prime antigen-specific, MHC-restricted T cells in situ. *J Exp Med* (1990) **172**(2):631–40. doi:10.1084/jem.172.2.631

5. Steinman RM. The dendritic cell system and its role in immunogenicity. *Annu Rev Immunol* (1991) **9**:271–96. doi:10.1146/annurev.iy.09.040191.001415

6. Hashimoto D, Miller J, Merad M. Dendritic cell and macrophage heterogeneity in vivo. *Immunity* (2011) **35**(3):323–35. doi:10.1016/j.immuni.2011.09.007

7. Saba TM. Physiology and physiopathology of the reticuloendothelial system. *Arch Intern Med* (1970) **126**(6):1031–52. doi:10.1001/archinte.126.6.1031

8. Hume DA, Gordon S. Mononuclear phagocyte system of the mouse defined by immunohistochemical localization of antigen F4/80. Identification of resident macrophages in renal medullary and cortical interstitium and the juxtaglomerular complex. *J Exp Med* (1983) **157**(5):1704–9. doi:10.1084/jem.157.5.1704

9. Markovic-Lipkovski J, Muller CA, Risler T, Bohle A, Muller GA. Association of glomerular and interstitial mononuclear leukocytes with different forms of glomerulonephritis. *Nephrol Dial Transplant* (1990) **5**(1):10–7. doi:10.1093/ndt/5.1.10

10. Cuzic S, Ritz E, Waldherr R. Dendritic cells in glomerulonephritis. *Virchows Arch B Cell Pathol Incl Mol Pathol* (1992) **62**(6):357–63. doi:10.1007/BF02899704

11. Austyn JM, Hankins DF, Larsen CP, Morris PJ, Rao AS, Roake JA. Isolation and characterization of dendritic cells from mouse heart and kidney. *J Immunol* (1994) **152**(5):2401–10.

12. Kaissling B, Le Hir M. Characterization and distribution of interstitial cell types in the renal cortex of rats. *Kidney Int* (1994) **45**(3):709–20. doi:10.1038/ki.1994.95

13. Kaissling B, Hegyi I, Loffing J, Le Hir M. Morphology of interstitial cells in the healthy kidney. *Anat Embryol (Berl)* (1996) **193**(4):303–18. doi:10.1007/BF00186688

14. Soos TJ, Sims TN, Barisoni L, Lin K, Littman DR, Dustin ML, et al. CX3CR1+ interstitial dendritic cells form a contiguous network throughout the entire kidney. *Kidney Int* (2006) **70**(3):591–6. doi:10.1038/sj.ki.5001567

15. Snelgrove SL, Kausman JY, Lo C, Lo C, Ooi JD, Coates PT, et al. Renal dendritic cells adopt a pro-inflammatory phenotype in obstructive uropathy to activate T cells but do not directly contribute to fibrosis. *Am J Pathol* (2012) **180**(1):91–103. doi:10.1016/j.ajpath.2011.09.039

16. Devi S, Li A, Westhorpe CL, Lo CY, Abeynaike LD, Snelgrove SL, et al. Multiphoton imaging reveals a new leukocyte recruitment paradigm in the glomerulus. *Nat Med* (2013) **19**(1):107–12. doi:10.1038/nm.3024

17. Kruger T, Benke D, Eitner F, Lang A, Wirtz M, Hamilton-Williams EE, et al. Identification and functional characterization of dendritic cells in the healthy murine kidney and in experimental glomerulonephritis. *J Am Soc Nephrol* (2004) **15**(3):613–21. doi:10.1097/01.ASN.0000114553.36258.91

18. Hochheiser K, Heuser C, Krause TA, Teteris S, Ilias A, Weisheit C, et al. Exclusive CX3CR1 dependence of kidney DCs impacts glomerulonephritis progression. *J Clin Invest* (2013) **123**(10):4242–54. doi:10.1172/JCI70143

19. Lionakis MS, Swamydas M, Fischer BG, Plantinga TS, Johnson MD, Jaeger M, et al. CX3CR1-dependent renal macrophage survival promotes *Candida* control and host survival. *J Clin Invest* (2013) **123**(12):5035–51. doi:10.1172/JCI71307

20. Engel DR, Krause TA, Snelgrove SL, Thiebes S, Hickey MJ, Boor P, et al. CX3CR1 reduces kidney fibrosis by inhibiting local proliferation of profibrotic macrophages. *J Immunol* (2015) **194**(4):1628–38. doi:10.4049/jimmunol.1402149

21. Jenkins SJ, Ruckerl D, Cook PC, Jones LH, Finkelman FD, van Rooijen N, et al. Local macrophage proliferation, rather than recruitment from the blood, is a signature of TH2 inflammation. *Science* (2011) **332**(6035):1284–8. doi:10.1126/science.1204351

22. Liu K, Nussenzweig MC. Origin and development of dendritic cells. *Immunol Rev* (2010) **234**(1):45–54. doi:10.1111/j.0105-2896.2009.00879.x

23. Ginhoux F, Liu K, Helft J, Bogunovic M, Greter M, Hashimoto D, et al. The origin and development of nonlymphoid tissue CD103+ DCs. *J Exp Med* (2009) **206**(13):3115–30. doi:10.1084/jem.20091756

24. Kurts C, Panzer U, Anders HJ, Rees AJ. The immune system and kidney disease: basic concepts and clinical implications. *Nat Rev Immunol* (2013) **13**(10):738–53. doi:10.1038/nri3523

25. Cao Q, Wang Y, Wang XM, Lu J, Lee VW, Ye Q, et al. Renal F4/80+ CD11c+ mononuclear phagocytes display phenotypic and functional characteristics of macrophages in health and in adriamycin nephropathy. *J Am Soc Nephrol* (2015) **26**(2):349–63. doi:10.1681/ASN.2013121336

26. Dong X, Swaminathan S, Bachman LA, Croatt AJ, Nath KA, Griffin MD. Resident dendritic cells are the predominant TNF-secreting cell in early renal ischemia-reperfusion injury. *Kidney Int* (2007) **71**(7):619–28. doi:10.1038/sj.ki.5002132

27. Dong X, Swaminathan S, Bachman LA, Croatt AJ, Nath KA, Griffin MD. Antigen presentation by dendritic cells in renal lymph nodes is linked to systemic and local injury to the kidney. *Kidney Int* (2005) **68**(3):1096–108. doi:10.1111/j.1523-1755.2005.00502.x

28. Dong X, Bachman LA, Miller MN, Nath KA, Griffin MD. Dendritic cells facilitate accumulation of IL-17 T cells in the kidney following acute renal obstruction. *Kidney Int* (2008) **74**(10):1294–309. doi:10.1038/ki.2008.394

29. Mulay SR, Kulkarni OP, Rupanagudi KV, Migliorini A, Darisipudi MN, Vilaysane A, et al. Calcium oxalate crystals induce renal inflammation by NLRP3-mediated IL-1beta secretion. *J Clin Invest* (2013) **123**(1):236–46. doi:10.1172/JCI63679

30. Tittel AP, Heuser C, Ohliger C, Knolle PA, Engel DR, Kurts C. Kidney dendritic cells induce innate immunity against bacterial pyelonephritis. *J Am Soc Nephrol* (2011) **22**(8):1435–41. doi:10.1681/ASN.2010101072

31. Scholz J, Lukacs-Kornek V, Engel DR, Specht S, Kiss E, Eitner F, et al. Renal dendritic cells stimulate IL-10 production and attenuate nephrotoxic nephritis. *J Am Soc Nephrol* (2008) **19**(3):527–37. doi:10.1681/ASN.2007060684

32. Heymann F, Meyer-Schwesinger C, Hamilton-Williams EE, Hammerich L, Panzer U, Kaden S, et al. Kidney dendritic cell activation is required for progression of renal disease in a mouse model of glomerular injury. *J Clin Invest* (2009) **119**(5):1286–97. doi:10.1172/JCI38399

33. Kim MG, Boo CS, Ko YS, Lee HY, Cho WY, Kim HK, et al. Depletion of kidney CD11c+ F4/80+ cells impairs the recovery process in ischaemia/reperfusion-induced acute kidney injury. *Nephrol Dial Transplant* (2010) **25**(9):2908–21. doi:10.1093/ndt/gfq183

34. Kawakami T, Lichtnekert J, Thompson LJ, Karna P, Bouabe H, Hohl TM, et al. Resident renal mononuclear phagocytes comprise five discrete populations with distinct phenotypes and functions. *J Immunol* (2013) **191**(6):3358–72. doi:10.4049/jimmunol.1300342

35. Griffin MD. Mononuclear phagocyte depletion strategies in models of acute kidney disease: what are they trying to tell us? *Kidney Int* (2012) **82**(8):835–7. doi:10.1038/ki.2012.164

36. Tittel AP, Heuser C, Ohliger C, Llanto C, Yona S, Hammerling GJ, et al. Functionally relevant neutrophilia in CD11c diphtheria toxin receptor transgenic mice. *Nat Methods* (2012) **9**(4):385–90. doi:10.1038/nmeth.1905

37. Weisser SB, van Rooijen N, Sly LM. Depletion and reconstitution of macrophages in mice. *J Vis Exp* (2012) (66):4105. doi:10.3791/4105

38. Jung S, Unutmaz D, Wong P, Sano G, De los Santos K, Sparwasser T, et al. In vivo depletion of CD11c+ dendritic cells abrogates priming of CD8+ T cells by exogenous cell-associated antigens. *Immunity* (2002) **17**(2):211–20. doi:10.1016/S1074-7613(02)00365-5

39. Lavin Y, Winter D, Blecher-Gonen R, David E, Keren-Shaul H, Merad M, et al. Tissue-resident macrophage enhancer landscapes are shaped by the local microenvironment. *Cell* (2014) **159**(6):1312–26. doi:10.1016/j.cell.2014.11.018

40. Gosselin D, Link VM, Romanoski CE, Fonseca GJ, Eichenfield DZ, Spann NJ, et al. Environment drives selection and function of enhancers controlling tissue-specific macrophage identities. *Cell* (2014) **159**(6):1327–40. doi:10.1016/j.cell.2014.11.023

41. Guilliams M, Ginhoux F, Jakubzick C, Naik SH, Onai N, Schraml BU, et al. Dendritic cells, monocytes and macrophages: a unified nomenclature based on ontogeny. *Nat Rev Immunol* (2014) **14**(8):571–8. doi:10.1038/nri3712

42. Ajami B, Bennett JL, Krieger C, McNagny KM, Rossi FM. Infiltrating monocytes trigger EAE progression, but do not contribute to the resident microglia pool. *Nat Neurosci* (2011) **14**(9):1142–9. doi:10.1038/nn.2887

43. Ginhoux F, Greter M, Leboeuf M, Nandi S, See P, Gokhan S, et al. Fate mapping analysis reveals that adult microglia derive from primitive macrophages. *Science* (2010) **330**(6005):841–5. doi:10.1126/science.1194637

44. Mildner A, Schmidt H, Nitsche M, Merkler D, Hanisch UK, Mack M, et al. Microglia in the adult brain arise from *Ly-6ChiCCR2+* monocytes only under defined host conditions. *Nat Neurosci* (2007) **10**(12):1544–53. doi:10.1038/nn2015

45. Schulz C, Gomez Perdiguero E, Chorro L, Szabo-Rogers H, Cagnard N, Kierdorf K, et al. A lineage of myeloid cells independent of Myb and hematopoietic stem cells. *Science* (2012) **336**(6077):86–90. doi:10.1126/science.1219179

46. Gomez Perdiguero E, Klapproth K, Schulz C, Busch K, Azzoni E, Crozet L, et al. Tissue-resident macrophages originate from yolk-sac-derived erythro-myeloid progenitors. *Nature* (2015) **518**(7540):547–51. doi:10.1038/nature13989

47. Hettinger J, Richards DM, Hansson J, Barra MM, Joschko AC, Krijgsveld J, et al. Origin of monocytes and macrophages in a committed progenitor. *Nat Immunol* (2013) **14**(8):821–30. doi:10.1038/ni.2638

48. Naik SH, Sathe P, Park HY, Metcalf D, Proietto AI, Dakic A, et al. Development of plasmacytoid and conventional dendritic cell subtypes from single precursor cells derived in vitro and in vivo. *Nat Immunol* (2007) **8**(11):1217–26. doi:10.1038/ni1522

49. Onai N, Obata-Onai A, Schmid MA, Ohteki T, Jarrossay D, Manz MG. Identification of clonogenic common Flt3+M-CSFR+ plasmacytoid and conventional dendritic cell progenitors in mouse bone marrow. *Nat Immunol* (2007) **8**(11):1207–16. doi:10.1038/ni1518

50. Ginhoux F, Jung S. Monocytes and macrophages: developmental pathways and tissue homeostasis. *Nat Rev Immunol* (2014) **14**(6):392–404. doi:10.1038/nri3671

51. Schraml BU, van Blijswijk J, Zelenay S, Whitney PG, Filby A, Acton SE, et al. Genetic tracing via DNGR-1 expression history defines dendritic cells as a hematopoietic lineage. *Cell* (2013) **154**(4):843–58. doi:10.1016/j.cell.2013.07.014

52. Plantinga M, Guilliams M, Vanheerswynghels M, Deswarte K, Branco-Madeira F, Toussaint W, et al. Conventional and monocyte-derived CD11b(+) dendritic cells initiate and maintain T helper 2 cell-mediated immunity to house dust mite allergen. *Immunity* (2013) **38**(2):322–35. doi:10.1016/j.immuni.2012.10.016

53. Tamoutounour S, Henri S, Lelouard H, de Bovis B, de Haar C, van der Woude CJ, et al. CD64 distinguishes macrophages from dendritic cells in the gut and reveals the Th1-inducing role of mesenteric lymph node macrophages during colitis. *Eur J Immunol* (2012) **42**(12):3150–66. doi:10.1002/eji.201242847

54. Langlet C, Tamoutounour S, Henri S, Luche H, Ardouin L, Gregoire C, et al. CD64 expression distinguishes monocyte-derived and conventional dendritic cells and reveals their distinct role during intramuscular immunization. *J Immunol* (2012) **188**(4):1751–60. doi:10.4049/jimmunol.1102744

55. Gautier EL, Shay T, Miller J, Greter M, Jakubzick C, Ivanov S, et al. Gene-expression profiles and transcriptional regulatory pathways that underlie the identity and diversity of mouse tissue macrophages. *Nat Immunol* (2012) **13**(11):1118–28. doi:10.1038/ni.2419

56. Tamoutounour S, Guilliams M, Montanana Sanchis F, Liu H, Terhorst D, Malosse C, et al. Origins and functional specialization of macrophages and of conventional and monocyte-derived dendritic cells in mouse skin. *Immunity* (2013) **39**(5):925–38. doi:10.1016/j.immuni.2013.10.004

57. Miller JC, Brown BD, Shay T, Gautier EL, Jojic V, Cohain A, et al. Deciphering the transcriptional network of the dendritic cell lineage. *Nat Immunol* (2012) **13**(9):888–99. doi:10.1038/ni.2370

58. Savina A, Amigorena S. Phagocytosis and antigen presentation in dendritic cells. *Immunol Rev* (2007) **219**:143–56. doi:10.1111/j.1600-065X.2007.00552.x

59. Desmedt M, Rottiers P, Dooms H, Fiers W, Grooten J. Macrophages induce cellular immunity by activating Th1 cell responses and suppressing Th2 cell responses. *J Immunol* (1998) **160**(11):5300–8.

60. Moser M. Regulation of Th1/Th2 development by antigen-presenting cells in vivo. *Immunobiology* (2001) **204**(5):551–7. doi:10.1078/0171-2985-00092

61. Hume DA. Macrophages as APC and the dendritic cell myth. *J Immunol* (2008) **181**(9):5829–35. doi:10.4049/jimmunol.181.9.5829

62. Nelson PJ, Rees AJ, Griffin MD, Hughes J, Kurts C, Duffield J. The renal mononuclear phagocytic system. *J Am Soc Nephrol* (2012) **23**(2):194–203. doi:10.1681/ASN.2011070680

63. Day YJ, Huang L, Ye H, Linden J, Okusa MD. Renal ischemia-reperfusion injury and adenosine 2A receptor-mediated tissue protection: role of macrophages. *Am J Physiol Renal Physiol* (2005) **288**(4):F722–31. doi:10.1152/ajprenal.00378.2004

64. Jo SK, Sung SA, Cho WY, Go KJ, Kim HK. Macrophages contribute to the initiation of ischaemic acute renal failure in rats. *Nephrol Dial Transplant* (2006) **21**(5):1231–9. doi:10.1093/ndt/gfk047

65. Lassen S, Lech M, Rommele C, Mittruecker HW, Mak TW, Anders HJ. Ischemia reperfusion induces IFN regulatory factor 4 in renal dendritic cells, which suppresses postischemic inflammation and prevents acute renal failure. *J Immunol* (2010) **185**(3):1976–83. doi:10.4049/jimmunol.0904207

66. Lech M, Avila-Ferrufino A, Allam R, Segerer S, Khandoga A, Krombach F, et al. Resident dendritic cells prevent postischemic acute renal failure by help of single Ig IL-1 receptor-related protein. *J Immunol* (2009) **183**(6):4109–18. doi:10.4049/jimmunol.0900118

67. Sola A, Weigert A, Jung M, Vinuesa E, Brecht K, Weis N, et al. Sphingosine-1-phosphate signalling induces the production of Lcn-2 by macrophages to promote kidney regeneration. *J Pathol* (2011) **225**(4):597–608. doi:10.1002/path.2982

68. Kitamoto K, Machida Y, Uchida J, Izumi Y, Shiota M, Nakao T, et al. Effects of liposome clodronate on renal leukocyte populations and renal fibrosis in murine obstructive nephropathy. *J Pharmacol Sci* (2009) **111**(3):285–92. doi:10.1254/jphs.09227FP

69. Lin SL, Castano AP, Nowlin BT, Lupher ML Jr, Duffield JS. Bone marrow Ly6Chigh monocytes are selectively recruited to injured kidney and differentiate into functionally distinct populations. *J Immunol* (2009) **183**(10):6733–43. doi:10.4049/jimmunol.0901473

70. Tadagavadi RK, Reeves WB. Renal dendritic cells ameliorate nephrotoxic acute kidney injury. *J Am Soc Nephrol* (2010) **21**(1):53–63. doi:10.1681/ASN.2009040407

71. Duffield JS, Tipping PG, Kipari T, Cailhier JF, Clay S, Lang R, et al. Conditional ablation of macrophages halts progression of crescentic glomerulonephritis. *Am J Pathol* (2005) **167**(5):1207–19. doi:10.1016/S0002-9440(10)61209-6

72. Bagavant H, Deshmukh US, Wang H, Ly T, Fu SM. Role for nephritogenic T cells in lupus glomerulonephritis: progression to renal failure is accompanied by T cell activation and expansion in regional lymph nodes. *J Immunol* (2006) **177**(11):8258–65. doi:10.4049/jimmunol.177.11.8258

73. Ishikawa S, Sato T, Abe M, Nagai S, Onai N, Yoneyama H, et al. Aberrant high expression of B lymphocyte chemokine (BLC/CXCL13) by C11b+CD11c+ dendritic cells in murine lupus and preferential chemotaxis of B1 cells towards BLC. *J Exp Med* (2001) **193**(12):1393–402. doi:10.1084/jem.193.12.1393

74. Riedel JH, Paust HJ, Turner JE, Tittel AP, Krebs C, Disteldorf E, et al. Immature renal dendritic cells recruit regulatory CXCR6(+) invariant natural killer T cells to attenuate crescentic GN. *J Am Soc Nephrol* (2012) **23**(12):1987–2000. doi:10.1681/ASN.2012040394

75. Coates PT, Colvin BL, Ranganathan A, Duncan FJ, Lan YY, Shufesky WJ, et al. CCR and CC chemokine expression in relation to Flt3 ligand-induced renal dendritic cell mobilization. *Kidney Int* (2004) **66**(5):1907–17. doi:10.1111/j.1523-1755.2004.00965.x

18

Investigating evolutionary conservation of dendritic cell subset identity and functions

Thien-Phong Vu Manh[1,2,3], Nicolas Bertho[4], Anne Hosmalin[5,6,7,8], Isabelle Schwartz-Cornil[4] and Marc Dalod[1,2,3]*

[1] UM2, Centre d'Immunologie de Marseille-Luminy (CIML), Aix-Marseille University, Marseille, France, [2] U1104, Institut National de la Santé et de la Recherche Médicale (INSERM), Marseille, France, [3] UMR7280, Centre National de la Recherche Scientifique (CNRS), Marseille, France, [4] Virologie et Immunologie Moléculaires UR892, Institut National de la Recherche Agronomique, Jouy-en-Josas, France, [5] INSERM U1016, Institut Cochin, Paris, France, [6] CNRS UMR8104, Paris, France, [7] Université Paris Descartes, Paris, France, [8] Assistance Publique-Hôpitaux de Paris (AP-HP), Hôpital Cochin, Paris, France

Edited by:
*Florent Ginhoux,
Singapore Immunology Network,
Singapore*

Reviewed by:
*Richard A. Kroczek,
Robert Koch-Institute, Germany
Muzlifah Aisha Haniffa,
Newcastle University, UK*

*Correspondence:
*Marc Dalod,
Centre d'Immunologie de
Marseille-Luminy (CIML), Parc
Scientifique et Technologique de
Luminy, Case 906, Marseille Cedex
09 F-13288, France
dalod@ciml.univ-mrs.fr*

Dendritic cells (DCs) were initially defined as mononuclear phagocytes with a dendritic morphology and an exquisite efficiency for naïve T-cell activation. DC encompass several subsets initially identified by their expression of specific cell surface molecules and later shown to excel in distinct functions and to develop under the instruction of different transcription factors or cytokines. Very few cell surface molecules are expressed in a specific manner on any immune cell type. Hence, to identify cell types, the sole use of a small number of cell surface markers in classical flow cytometry can be deceiving. Moreover, the markers currently used to define mononuclear phagocyte subsets vary depending on the tissue and animal species studied and even between laboratories. This has led to confusion in the definition of DC subset identity and in their attribution of specific functions. There is a strong need to identify a rigorous and consensus way to define mononuclear phagocyte subsets, with precise guidelines potentially applicable throughout tissues and species. We will discuss the advantages, drawbacks, and complementarities of different methodologies: cell surface phenotyping, ontogeny, functional characterization, and molecular profiling. We will advocate that gene expression profiling is a very rigorous, largely unbiased and accessible method to define the identity of mononuclear phagocyte subsets, which strengthens and refines surface phenotyping. It is uniquely powerful to yield new, experimentally testable, hypotheses on the ontogeny or functions of mononuclear phagocyte subsets, their molecular regulation, and their evolutionary conservation. We propose defining cell populations based on a combination of cell surface phenotyping, expression analysis of hallmark genes, and robust functional assays, in order to reach a consensus and integrate faster the huge but scattered knowledge accumulated by different laboratories on different cell types, organs, and species.

Keywords: mononuclear phagocytes, comparative genomics, human, non-human primates, mouse, pig, sheep, chicken

Introduction

The immune system includes a large variety of myeloid and lymphoid cell types which develop through distinct ontogenic pathways, express specific phenotypes, and exert specialized functions.

The mononuclear phagocytes form a complex group of myeloid cells that encompass three major cell types, i.e., monocytes, macrophages, and dendritic cells (DC), together with their proximal progenitors. These three cell types contribute to maintain host integrity by shaping the innate and adaptive immune defense, a generic function related to their common phagocytic properties and their capacity to present antigen to T cells. These functions are also shared by other types of professional antigen-presenting cells (APCs), in particular B lymphocytes. However, different types of APCs are primarily devoted to distinct functions (**Figure 1**). B cells produce antibodies. Monocytes patrol the organism for the detection of pathogens and dominantly display inflammatory and oxidative stress response. Macrophages mainly perform microbicidal, scavenging, and tissue trophic/maintenance functions. DC are uniquely efficient for antigen-specific activation of naïve T lymphocytes, a process called T-cell priming. Indeed, DC were initially defined by their dendritic morphology and their exquisite capacity for T-cell priming. DC include two main cell types, the plasmacytoid DC (pDC) that are expert in type I interferon synthesis upon viral stimulation and the conventional DC (cDC) that are specialized in antigen capture, processing, and presentation for T-cell priming. Two cDC subsets can be distinguished based on a further segregation of functions. $XCR1^+$ cDC1 are particularly efficient in $CD8^+$ T-cell activation and cross-presentation, at least in mice. $XCR1^-$ cDC2 are most efficient for T helper cell priming, in particular polarization toward Th2 or Th17, and for the promotion of humoral immunity. Importantly, an additional layer of complexity is generated by the plasticity of the different mononuclear cell types, which display modified phenotypes and functions contingent to the anatomical microenvironment where they reside or when exposed to pathogens or inflammation. For instance, monocytes adopt a dendritic morphology and antigen-presentation functions in inflammatory settings (1–3) as well as when located in the dermis (4–6), leading to their designation as monocyte-derived DC (MoDC). Langerhans cells, long considered to be DC due to their morphology and antigen-presentation function, are now known as a type of tissue macrophages (7–13). More generally, the gene expression programs, phenotypes, and functional properties of macrophages are strongly influenced by their tissue of residence. Finally, not only $XCR1^+$ cDC but also other DC subsets including pDC and $XCR1^-$ cDC can also efficiently cross-present antigens to $CD8^+$

T cells when appropriately stimulated (14–22). Thus, the plasticity of the mononuclear phagocyte responses superimposes onto the segregation of phenotypes and functions attributed to subsets (**Figure 2**), which can lead to confusion in the definition of the different cell types if only based on functional assays. Hence, morphologic, phenotypic, and functional criteria are not sufficient to rigorously define mononuclear phagocyte subsets, and to properly discriminate what are distinct cell types as opposed to different developmental or activation states of a given cell type. Complementary or robust alternative criteria are needed to rigorously define the identity of the mononuclear phagocyte subsets.

Mononuclear phagocyte subsets were recently shown to develop from distinct progenitors and/or under the instruction of different transcription factors or cytokines. cDC and pDC derive from a dedicated bone marrow precursor, the common DC progenitor, with a differentiation potential strictly restricted to this hematopoietic lineage. pDC and cDC homeostasis exquisitely depends on the growth factor FLT3-L. pDC development strictly depends on the transcription factors TCF4 (E2-2) and SPIB both in mouse and human, $XCR1^+$ cDC development on the master transcription factor IRF8 at least in mice, and $XCR1^-$ cDC development on IRF4. Macrophages derive from a monocytic precursor, either of embryonic origin as in the case of Langerhans cells and microglia, or at least in part from circulating blood monocytes as in the case of gut macrophages. Egress of classical monocytes from the bone marrow into the blood strictly depends on the chemokine receptor CCR2. As a consequence, in competitive mixed bone marrow reconstitution experiments in mice, all cell types derived from circulating blood monocytes are primarily reconstituted from wild-type cells and not from CCR2-deficient cells. Hence, it has been proposed that the study of their developmental pathway, in other words ontogeny, was the best way to classify mononuclear phagocyte cell types, at least in the mouse model where the knowledge in DC subset properties is also the most advanced. Indeed, in this model, genetically modified animals unambiguously permit to track the development of cell types and to dissect their phenotypes and functions, in different contexts *in vivo*. However, the identity and functions of the different mononuclear phagocyte subsets need to be established outside of the mouse model, in animal species where ontogenic studies cannot be easily conducted, in order to accelerate translation of our advanced knowledge on the functioning of the mouse immune

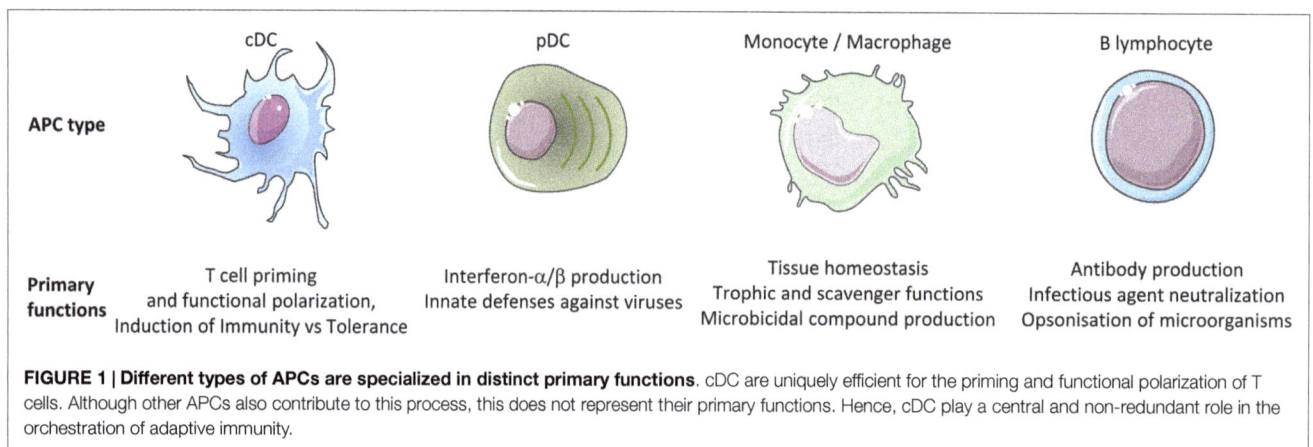

	cDC	pDC	Monocyte / Macrophage	B lymphocyte
APC type				
Primary functions	T cell priming and functional polarization, Induction of Immunity vs Tolerance	Interferon-α/β production Innate defenses against viruses	Tissue homeostasis Trophic and scavenger functions Microbicidal compound production	Antibody production Infectious agent neutralization Opsonisation of microorganisms

FIGURE 1 | Different types of APCs are specialized in distinct primary functions. cDC are uniquely efficient for the priming and functional polarization of T cells. Although other APCs also contribute to this process, this does not represent their primary functions. Hence, cDC play a central and non-redundant role in the orchestration of adaptive immunity.

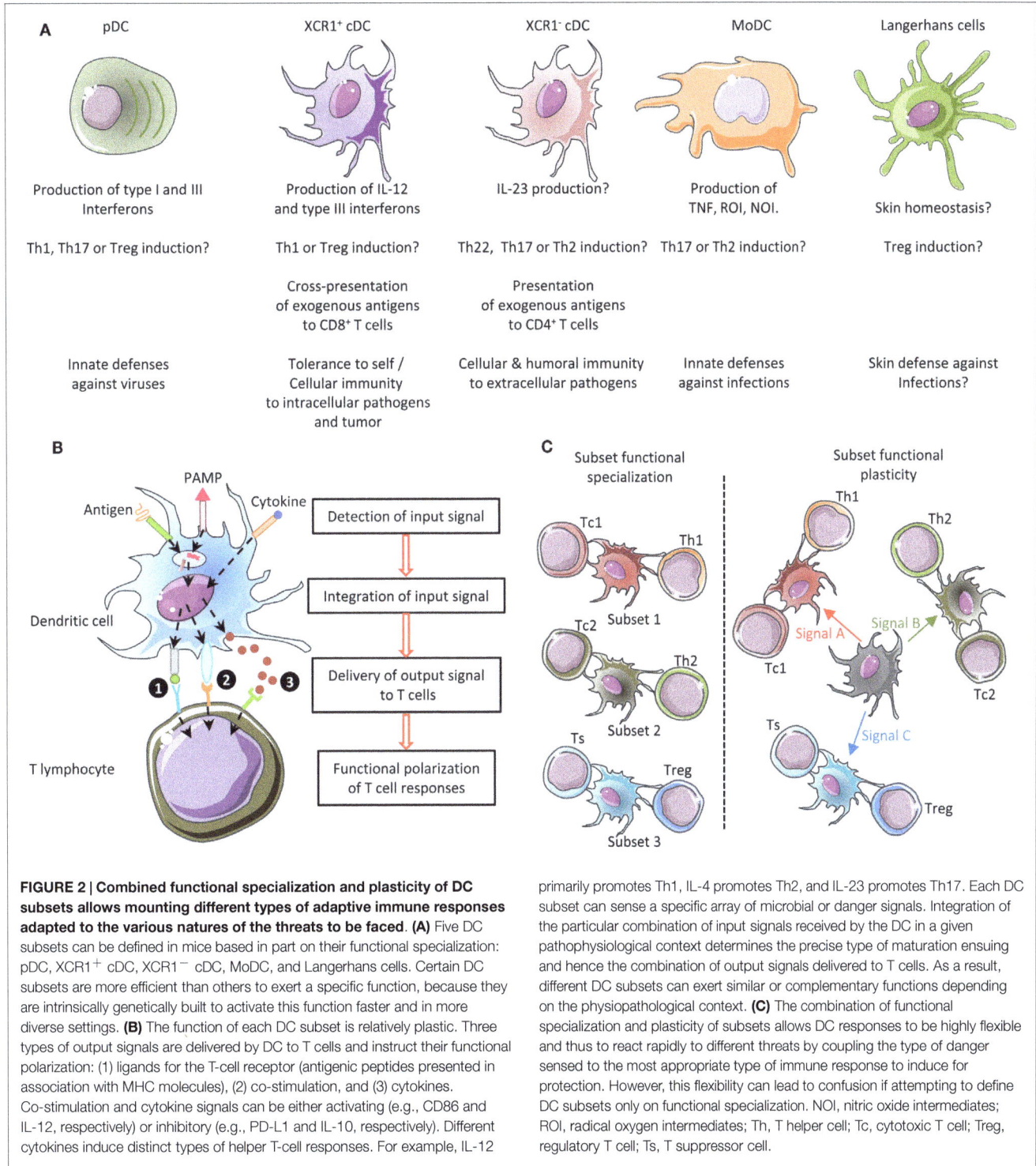

FIGURE 2 | Combined functional specialization and plasticity of DC subsets allows mounting different types of adaptive immune responses adapted to the various natures of the threats to be faced. (A) Five DC subsets can be defined in mice based in part on their functional specialization: pDC, XCR1$^+$ cDC, XCR1$^-$ cDC, MoDC, and Langerhans cells. Certain DC subsets are more efficient than others to exert a specific function, because they are intrinsically genetically built to activate this function faster and in more diverse settings. **(B)** The function of each DC subset is relatively plastic. Three types of output signals are delivered by DC to T cells and instruct their functional polarization: (1) ligands for the T-cell receptor (antigenic peptides presented in association with MHC molecules), (2) co-stimulation, and (3) cytokines. Co-stimulation and cytokine signals can be either activating (e.g., CD86 and IL-12, respectively) or inhibitory (e.g., PD-L1 and IL-10, respectively). Different cytokines induce distinct types of helper T-cell responses. For example, IL-12 primarily promotes Th1, IL-4 promotes Th2, and IL-23 promotes Th17. Each DC subset can sense a specific array of microbial or danger signals. Integration of the particular combination of input signals received by the DC in a given pathophysiological context determines the precise type of maturation ensuing and hence the combination of output signals delivered to T cells. As a result, different DC subsets can exert similar or complementary functions depending on the physiopathological context. **(C)** The combination of functional specialization and plasticity of subsets allows DC responses to be highly flexible and thus to react rapidly to different threats by coupling the type of danger sensed to the most appropriate type of immune response to induce for protection. However, this flexibility can lead to confusion if attempting to define DC subsets only on functional specialization. NOI, nitric oxide intermediates; ROI, radical oxygen intermediates; Th, T helper cell; Tc, cytotoxic T cell; Treg, regulatory T cell; Ts, T suppressor cell.

system toward clinical and/or economical applications to sustain global human health. Very promising vaccine and immunomodulatory strategies have been developed in mouse models based on DC subset targeting (23–35). The translation of these strategies to human and other species has not yet reached the expected success, likely due to insufficient knowledge in the identity and function of homologous DC subsets across species. This knowledge is needed in biomedical model species, primarily in non-human primates, and also in alternative models such as pigs that share physiological and anatomical similarities with humans – for instance skin and lung structural properties – and that present sensitivity to human pathogens of great importance for public health such as influenza. In addition, this knowledge is needed for companion and sport animals, and for animals of the agro-economy, such as ruminants, pigs, poultry, and fishes, with the goal to improve vaccination strategies against pathogens responsible for major economic losses, to decrease antibiotic use and to ameliorate animal welfare. These species, as well as wild animals, are also targets or reservoirs for major zoonotic pathogens whose control could thus benefit from new vaccine strategies targeting DC subsets in these animal species. This raises the question how to best define DC subset identity and functions in a way that can be extrapolated from mouse to human and other species, for clinical applications as well as for a better understanding of the evolution of the immune system.

Different Methodologies to Define the Identity of Immune Cell Types, with Their Advantages and Drawbacks

Several methodologies have been proposed to define cell types. They include cell surface phenotyping and morphology, ontogeny, functional characterization, molecular profiling at population level, and molecular profiling at single cell level. We will discuss the specific drawbacks and advantages of each of these approaches (**Table 1**).

Cell Surface Phenotyping and Morphology

Cell surface phenotyping generally is a mandatory first step for all other proposed methodologies aiming at defining DC subsets. It may be skipped only for particular experiments of molecular profiling at single cell level and perhaps for functional tests based on validated protocols for specific depletion of the targeted cell subset *in vivo*. Indeed, phenotypic characterization/identification of DC subsets is necessary either to purify them for morphological analysis, functional assays, or molecular profiling, or to compare their characteristics in tissues or bulk cell suspensions (expression of lineage reporters in cell fate mapping experiments, anatomical location, maturation status, cytokine production, interactions with T cells...). Phenotypic characterization through cell surface phenotyping by flow cytometry is the method of DC subset identification the easiest to perform and the most frequently used. No single cell surface marker has been found to be sufficient for identification of a given DC subset, except for XCR1 expression on mouse and human XCR1$^+$ cDC (18, 36–42) and maybe BDCA2 or LILRA4 expression on human pDC (43–46). Thus, to rigorously identify any given DC subset in any species with a limited risk of contamination by another cell type, most of the time complex combinations of multiple markers are required, often including the use of exclusion marker to ensure lack of contamination of the cell population targeted by other cell types sharing with it many positive markers. For example, the CD8α$^+$ subset of mouse pDC can heavily contaminate mouse lymphoid organ-resident XCR1$^+$ cDC when defined phenotypically as Lineage$^-$ CD11c$^+$CD8α$^+$ (47–49). This problem can be solved by exclusion of SiglecH$^+$ or CCR9$^+$ cells or by using XCR1 as a positive marker. Similarly, other cells including MoDC or activated CD1c (BDCA1)$^+$ XCR1$^-$ cDC can heavily contaminate human XCR1$^+$ cDC when defined phenotypically as Lineage$^-$ HLA-DR$^+$CD141 (BDCA3)$^+$ (41, 50, 51). This problem can be solved by using CADM1 or XCR1 as additional positive markers (41, 52). Rigorous phenotypic identification of XCR1$^-$ cDC (mouse CD11b$^+$ cDC and human CD1c$^+$ cDC) can be much more challenging,

since these cells can be difficult to discriminate from MoDC, in particular under inflammation settings (53, 54). Identification of DC based on oligoparameter phenotyping is even more at risk of inaccuracy in other species, due to the limited panel of available antibodies directed to surface markers and to the poor knowledge in surface marker expression selectivity in non-DC cell types. However, major advances have recently been made to refine strategies for DC subset identification by cell surface phenotyping, in part based on novel knowledge gained through ontogeny and molecular profiling studies as will be discussed below. Hence, protocols for DC subset identification by cell surface phenotyping might soon become standardized, at least in mouse and human. This would allow better comparison of data across laboratories and limit the risk of use of inappropriate protocols leading to improper data interpretation. Special attention should be given to enzymatic dissociation that can strongly modify cell surface marker detection. Ideally, universal phenotyping protocols could be designed, allowing to considerably simplify the current nomenclatures for DC subsets by using the same name and similar marker combinations to identify homologous cell types irrespective of their tissues and species of origin (55–57). Moreover, the markers used to define and name DC subsets could be chosen based on their relevance to the biology of these cells, contrary to the current situation where the markers used were discovered fortuitously/empirically and may not be linked to the biology of the eponymous cells, as is the case for CD8α and CD141 for mouse and human XCR1$^+$ cDC, respectively. However, when identifying a potentially new subset of DC or studying in a novel context a potentially known DC subset, a number of precautions need to be taken for data interpretation, including confirmation of conclusions by complementary methods such as ontogeny, functional, or molecular profiling studies.

Ontogeny

Ontogeny studies in mice, in particular studies on the dependence of DC subset development on transcription factors, have been instrumental in identifying the homologies between lymphoid tissue-resident CD8α$^+$ cDC and the CD103$^+$CD11b$^-$ cDC present in non-lymphoid tissues and migrating into the draining lymph nodes once activated (58). These studies, together with gene expression profiling analyses (9, 40), ultimately allowed grouping mouse CD8α$^+$ cDC and CD103$^+$CD11b$^-$ cDC together under the umbrella of the XCR1$^+$ cDC subset (38, 40, 59, 60). The recent discrimination of mouse CD11b$^+$ cDC from MoDC has also been largely based on the analysis of the role of specific chemokine or growth factor receptors on cell type development *in vivo*, namely CCR2 dependence as a characteristic of monocytic origin and FLT3 dependence as a proof of cDC identity (2, 3, 6, 61). In addition, mouse CD11b$^+$ cDC development was shown to selectively depend on the IRF4 transcription factor (62, 63). Moreover, the establishment of the concept that mouse *bona fide* DC constitute a separate hematopoietic lineage, and the discrimination between mouse CD11b$^+$ cDC and MoDC, were confirmed using mutant animals allowing to track natural precursor–progeny relationships *in vivo* through irreversible fluorescent tagging of all daughter cells of a given type of hematopoietic progenitor, based on Cre-mediated conditional activation

TABLE 1 | Different methodologies to define DC subsets with their advantages and drawbacks[a].

	Methodology				
	Cell surface phenotyping	Ontogeny	Functional characterization	Molecular profiling	
				At the population level	At the single cell level
Dependency on cell surface phenotyping	Not applicable	Yes but **methodology allows assessing the risk of cell type cross-contamination**	Yes, risk of bias. Data quality heavily depends on rigor of the cell surface phenotyping procedure used to identify cell types	Yes, risk of bias. Data quality heavily depends on rigor of the cell surface phenotyping procedure used to identify cell types. ***A posteriori* analyses can allow rigorously assessing the risk of cell type cross-contamination**	**No** ***Ab initio* identification of cell types without use of prior knowledge on their identity**
Experimental feasibility	**Good**	Difficult for most species except mouse	Depends on the species studied and the functions tested	**Good** Needs comparison with sister cell types and potential contaminants	Challenging both for data generation and data analysis. Commercial solutions exist for data generation but are expensive. Needs to balance cost and sequencing depth. Data analysis still in a large part dependent upon knowledge from molecular profiling at the population level
Protocol standardization	**Achievable soon** but currently limited. Currently used markers defined fortuitously/empirically, generally unrelated to cell biology, and different between tissues, species, and laboratories	Difficult	Difficult. The most subject to variations. Multiplicity of protocols depending on the functions tested, the tissues used and the species studied including its genetics, and even on the laboratories	**Good** **Routine technology for data generation** **Democratization of bioinformatics analyses**	Should happen upon technology maturation and democratization
Frequency of use	**Most frequent**	Mostly by specialists	**Frequent** Depending on the species studied and the functions tested	**Increasing frequency**	Very rare but high potential
Advancement of knowledge	The less informative	**Generally dichotomic information allowing relatively easy classification. Relevant to cell biology**	**Yes** **The most relevant for clinical and veterinary applications**	**Yes** **Generation of novel hypotheses on the ontogeny or functions of cells and their molecular regulation. Identification of conserved and biologically relevant cell surface markers. Identification of candidate molecular targets to manipulate cell functions**	**Yes** **Same advantages as molecular profiling at the population level. In addition, i) unbiased identification of cell types and associated transcriptomic signatures, ii) strong potential for identification of new cell types, iii) evaluation of intra-cell type heterogeneity, and iv) rigorous identification of cellular modules constituted of genes co-expressed in single cells and contributing to the same biological function**

[a]Advantages are indicated in bold font and drawbacks in plain font.

of a floxed reporter gene under the control of the constitutive Rosa26 promoter, an experimental strategy-coined fate mapping (64). Based on the important contribution of ontogenic studies for rigorous delineation of the identity of mouse DC subsets and of their lineage relationships, it has been proposed to use ontogeny as a primary methodology for the classification of mononuclear cell subsets in all species (57). Recent methodological progress has now made rigorous ontogenic studies applicable to human DC subsets, by using surrogate models of DC development from human CD34$^+$ hematopoietic progenitors, either *in vitro* (41, 65, 66) or *in vivo* in alymphoid mice (66–68). Such approaches have allowed demonstrating remarkable similarities in the ontogeny of mouse and human DC subsets. For example, knock-down experiments performed by transducing human CD34$^+$ hematopoietic progenitors with shRNA-expressing lentiviral vectors allowed to show that human pDC development critically depends on the transcription factor SPIB including *in vivo* in humanized mice (67), and that human XCR1$^+$ cDC development depends on the transcription factor BATF3 *in vitro* but not *in vivo* in humanized mice (68). Moreover, the pathway for the development of human pDC, XCR1$^+$ cDC, and XCR1$^-$ cDC was very recently demonstrated to be similar to that described for mouse DC subsets, with the identification of the human homologs to the mouse common DC progenitor and pre-cDC (66, 69). The role of candidate genes susceptible to affect DC development can even be assessed *in vivo* in humans in the rare cases where patients have been identified with primary immune deficiencies resulting from natural mutations in such genes (70). Strategies are being developed to actively search for human primary immunodeficiencies affecting DC development as experiments of nature allowing deciphering the molecular mechanisms regulating this biological process (71). However, ontogenic studies will often not be applicable in human for rigorous assessment of the identity of DC subsets, for example when studying a potentially known DC subset in a novel physiopathological context, including characterization of the DC subsets present in steady-state non-lymphoid tissues (50) or infiltrating tumors and their draining lymph nodes (72, 73) or isolated from infected/inflamed tissues. In addition, rigorous ontogenic studies will be very difficult to perform in many species, because (i) precursor/progeny relationships remain very difficult to evaluate *in vivo* through cell fate mapping or cell transfer experiments, (ii) *in vivo* analysis of cell subset development dependence on growth factors or transcription factors cannot be reasonably done due to operational and/or financial reasons, and (iii) *in vitro* models of *bona fide* DC development are currently lacking (74). Hence, the use of other methodologies will be necessary to prove DC subset identity in these various conditions.

Functional Characterization

Ideally, cell types should be defined based on the array of functions they can exert, because this definition links identity to function and is hence the most relevant to understand the functioning of the immune system and to harness the biology of DC subsets for improving health care of humans and of other species. In addition, cell type definitions based on their functional specialization could be the most universal across tissues and species. However, functional assays are often the hardest to perform experimentally and

can be the most subject to variations depending on assays and experimental conditions. This is especially the case for assays aiming at comparing the ability of different DC subsets to activate T cells. If one aims at precisely comparing the cell-intrinsic ability of different DC subsets to process and present antigens, a number of potentially confounding factors must be taken into account to design the experiment in order to reduce the risk of inappropriate interpretation of results. Adequate steps must be taken to preserve the viability of DC subsets and control for it. This implies adding to each isolated DC subset the appropriate cytokines or growth factors necessary for their survival, for example GM-CSF for cDC and IL-3 for human pDC. For instance, sorted XCR1$^+$ cDC show a lower *ex vivo* survival as compared to XCR1$^-$ cDC in mice and sheep (75, 76). Sorting of DC subset by positive selections may affect DC subset responses due to antibody-mediated receptor stimulation (43, 77–79). This also implies including a positive control consisting in DC subsets pulsed with optimal epitopic peptides, to assess on antigen-specific T-cell priming by DC the impact of other factors than DC subset-intrinsic differences in antigen processing and presentation, not only differences in DC subset viability but also in delivery of co-stimulation or cytokine signals. In this regard, for a fair comparison between DC subsets, they should each be matured by stimulation with an appropriate adjuvant. PolyI:C is much more efficient than LPS for the activation of human XCR1$^+$ DC while it is the reverse for the activation of human MoDC. TLR7 or TLR9 ligands, but not TLR3 or TLR8 ligands, are potent activators of human pDC. Another layer of complexity is due to fundamental differences in the design of experiments in different species. While the gold standard for antigen processing and presentation assays in mice is the measurement of the activation of TcR-transgenic naïve T cells, this is not possible in other species where various surrogate readouts are used including antigen-specific re-activation of antigen-experienced T-cell clones or polyclonal T-cell lines or even proliferation of allogeneic T cells. It is known that significant differences exist in mice in the signals required for naïve T-cell priming, antigen-experienced T-cell re-activation, or allogeneic T-cell proliferation induction. Therefore, the same exact function is not fairly tested in different species. Furthermore, in species outside mice and humans, the use of epitopic peptide control requires to have accurate MHC typing and knowledge of the corresponding optimal peptides, which are generally unavailable. In addition, for accessibility reasons, the DC subsets used generally derived from different anatomic compartments depending on the species. For example, spleen DC subsets are often used in mice, blood, or tonsil DC in humans and lymph DC in sheep, which can further confound rigorous interpretation of the results when differences are observed between species. Finally, while inbred mice with defined sanitary status are generally used to limit the variability of the responses between individuals, this is not the case for other species including humans where the considerable heterogeneity in the genotypes, environments, and immune histories of individuals contribute to the strong variability of their responses (80). Hence, even for mouse experiments, there is a strong need for standardization of functional assays assessing the ability of DC subsets to process and present antigens and to functionally polarize T cells. Moreover, when attempting to compare DC subset functional specialization across two species, efforts should be made to use comparable experimental designs in both species. Thus, while functional characterization is highly desirable when identifying a potentially new subset of DC or studying in a novel context a potentially known DC subset, the identity of DC subsets must first be studied through alternative approaches measuring cell type-specific parameters that are less strongly influenced by the tissue microenvironment and the genetic or immune history of populations, and for which experimental protocols are relatively well standardized.

Molecular Profiling at the Population Level

As the ontogeny and functions of cell types are instructed by specific gene expression modules, cell type identity can be defined by its molecular fingerprinting, including through gene expression profiling (81, 82). Homologous cell types between species can be defined as "those cells that evolved from the same precursor cell type in the last common ancestor" (82). This implies that homologous cell types must exhibit closer molecular fingerprints and gene expression programs than non-homologous cell types. Thus, it should be possible to decipher the identity of immune cell types of virtually all vertebrate species, by establishing their gene signatures and comparing them to the transcriptomic finger-prints of the well-characterized immune cell types of the mouse referent species. This is indeed an approach we pioneered to compare mouse spleen and human blood DC subsets (39) and later extended to comparison with sheep lymph cDC subsets (76), mouse DC subsets across tissues (40), as well as chicken spleen and pig skin mononuclear phagocyte subsets (83, 84). This approach allowed us to rigorously demonstrate for the first time to the best of our knowledge that human $CD1c^+$ cDC and $CD141^+$ cDC were homologous to mouse $CD11b^+$ cDC and $CD8\alpha^+$ cDC, respectively (39, 85). This was later confirmed by us and others based on phenotypic, functional, and ontogeny studies (18, 37, 50, 65, 86). In addition, this approach permitted to show that cDC split into $XCR1^+$ and $XCR1^-$ subsets in migrating skin lymph DC in sheep, a species belonging to the Laurasiatherians, which is a mammalian order distant from the mouse and human Euarchontoglires (76). This approach also provided the first compelling evidence for existence of *bona fide* cDC and macrophages in chicken, showing that diversification in mononuclear phagocyte cell types appeared in a common ancestor to mammals and reptiles (83). Comparative transcriptomics also led to recognize CADM1 and SIRPα as surface molecules whose conserved expression throughout distant species can be used as a first phenotyping step to identify $XCR1^+$ and $XCR1^-$ cDC subsets in any mammal (76). Notably, CADM1 is a highly conserved molecule, presenting about 90% identity across mammalian orthologs, thus allowing using commercial anti-human CADM1 antibodies for cellular staining in distant species (76, 84). We found the *Xcr1* gene among genes specifically expressed in mouse spleen $CD8\alpha^+$ DC when compared to a number of other immune cell types [see Supplementary Material "Additional file 5; gb-2008-9-1-r17-s5.xls" from Robbins et al. (39), specifically in the "CD8a_DC_gene_signature" established from our microarray data and confirmed from our own re-analysis of the microarray dataset independently generated by Dudziak et al. (87)]. Specific expression of the Xcr1 protein on mouse lymphoid tissue-resident

$CD8\alpha^+$ DC and its functions were first unveiled in the pioneering report from the group of Kroczek (36), who showed that $CD8^+$ T-cell cross-priming depends on their ability to secrete the Xcr1 ligand Xcl1 in experimental models where either the OVA coupled to an anti-CD205 Ab or OVA-expressing allogeneic pre-B cells are administrated *in vivo*. Xcr1 expression on $CD8\alpha^+$ DCs was also found to be critical for the optimal induction of $CD8^+$ T-cell responses upon *Listeria monocytogenes* infection (18). Importantly, comparative transcriptomics revealed XCR1 as a specific and universal marker for $XCR1^+$ cDC across tissues and species. This was initially shown in human, mice, and sheep (18, 37, 76) and subsequently in non-human primates and pigs (18, 37, 38, 40, 52, 59, 60). Altogether, these studies were critical for the current proposal of cDC subset classification into $XCR1^+$ and $XCR1^-$ cDC (38, 40). Many other recent studies have demonstrated the power of gene expression profiling to determine with a high degree of certainty the identity of mononuclear phagocyte subsets in a tissue where they had not been rigorously studied before or to identify homologous subsets of mononuclear phagocytes across species (5, 6, 8, 9, 50, 88–90). Importantly, standardized protocols for generation and analysis of gene expression data are routinely performed in many laboratories, platforms, or commercial companies in many countries. The corresponding costs have strongly decreased over the last decade and continue to go down. Hence, gene expression profiling at the population level is a very robust and reproducible methodology that is feasible in virtually all species where tools are available or can be developed to phenotypically identify and purify candidate cell subsets. However, potentially confounding factors must be taken into account to design experiments in order to reduce the risk of inappropriate interpretation of results (**Figure 3**). First and foremost, great care and rigor must be exerted in designing the experimental sampling protocol for cell subset purification, inasmuch as minor contamination by another cell type can dramatically impact the gene expression profile obtained. Hence, it is critical to carefully design the marker combination used to purify the different cell populations to be studied, and to control cell purity prior to the generation of the gene expression data. Second, to allow proper analysis of the gene expression profiles of the targeted cell type, appropriate cell type controls must be included, encompassing sister cell types as well as cell types that could be potential contaminants due to their expression of several of the markers used for positive selection of the targeted cell type. These controls are critical to allow assessing the risk of contamination by another cell type (49).

Molecular Profiling at the Single Cell Level

Recent technological advances now allow performing high throughput RNA sequencing at single cell levels with high sensitivity and processivity. Transcriptomic analyses at the single cell level could solve most of the issues raised in the previous section for molecular profiling at the population level. Indeed, because it alleviates the necessity to purify cells on imperfect and potentially confounding phenotypic marker combinations, analysis at the single cell level should allow unbiased identification of potentially all cell types and their associated transcriptomic signatures. It also solves the issue of cross-contamination between cell types, since the identity of each single cell is established *a*

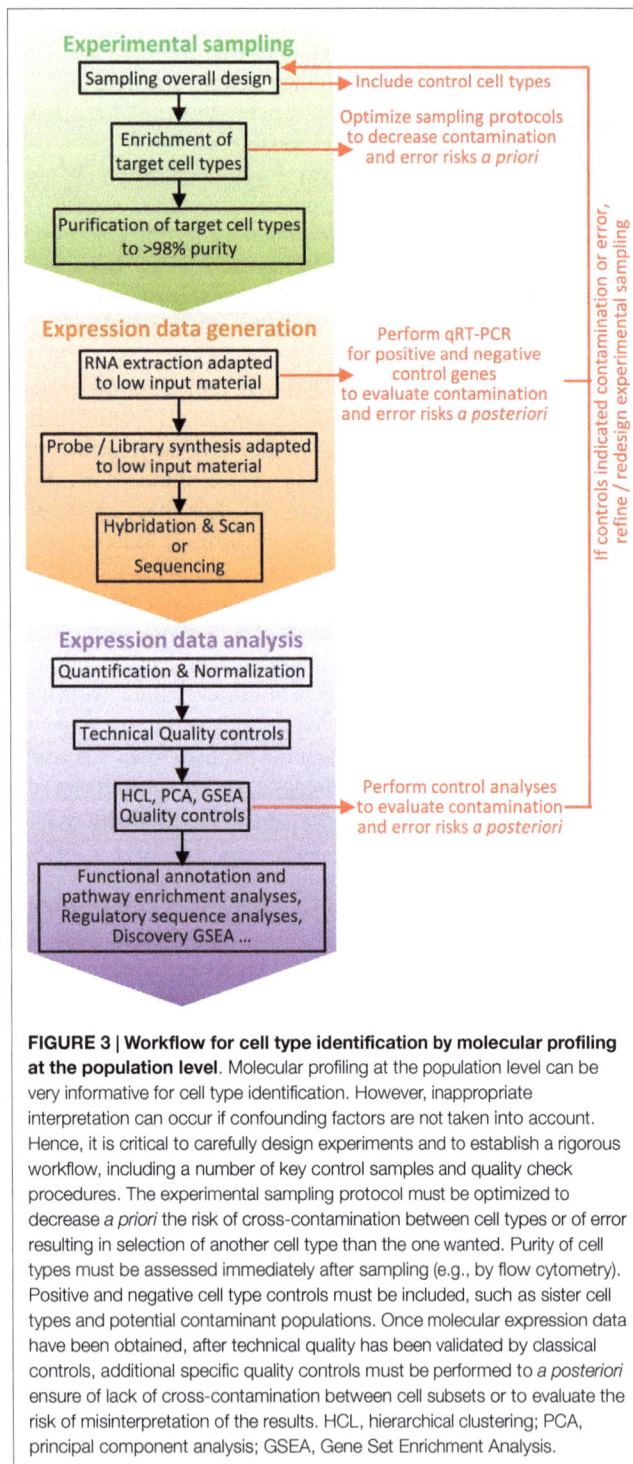

FIGURE 3 | Workflow for cell type identification by molecular profiling at the population level. Molecular profiling at the population level can be very informative for cell type identification. However, inappropriate interpretation can occur if confounding factors are not taken into account. Hence, it is critical to carefully design experiments and to establish a rigorous workflow, including a number of key control samples and quality check procedures. The experimental sampling protocol must be optimized to decrease a priori the risk of cross-contamination between cell types or of error resulting in selection of another cell type than the one wanted. Purity of cell types must be assessed immediately after sampling (e.g., by flow cytometry). Positive and negative cell type controls must be included, such as sister cell types and potential contaminant populations. Once molecular expression data have been obtained, after technical quality has been validated by classical controls, additional specific quality controls must be performed to a posteriori ensure of lack of cross-contamination between cell subsets or to evaluate the risk of misinterpretation of the results. HCL, hierarchical clustering; PCA, principal component analysis; GSEA, Gene Set Enrichment Analysis.

of their molecular identity, starting from the bulk population of all the cells that could be extracted from the organ without any prior enrichment procedure. However, molecular profiling at the single cell level cannot be used without prior phenotype-based enrichment for very rare cell types, and it is difficult to apply to species in which genome has not yet been completely assembled. To obtain complete information, including on functionally important genes for which few mRNA are expressed per cell, it is necessary to sequence at a sufficient depth of about one million reads per cell, which today still represents a very high cost when multiplied by the number of individual cells and conditions. This is all the more the case since, likewise for molecular profiling at population level, correct interpretation of the data requires that sister cell types as well as cell types that could be potential contaminants are included in the experimental design. Moreover, the technology for single cell RNA sequencing is not yet democratized, since it is challenging both for sample preparation and for data analysis. For standardization of high quality sample preparation, commercial solutions exist but are very expensive. For data analysis, there is no consensus yet on how the data should be mathematically modeled for adequate removal of background signal and for discrimination of false negative signal due to sampling bias in the pool of the cell mRNA as opposed to true lack of gene expression. In addition, the interpretation of the RNA-seq data on single cells is still largely based on the transcriptomic/molecular identity of cell types that are deduced from microarray analysis of purified cell pools (91). Hence, molecular profiling at the population level currently represents a more sustainable strategy for most laboratories.

Recent Advances Brought by Comparative Transcriptomics at the Population Level for Defining the Identity and Functions of Mononuclear Phagocyte Subsets and Their Molecular Regulation

In this section, we will review major advances brought forward by comparative transcriptomics at the population level for defining the identity and functions of mononuclear phagocyte subsets and their molecular regulation.

Gene expression profiling of cell types with apparent ambiguous phenotype or functions allowed to rigorously establish their identity, which could be achieved properly strictly contingent to their comparison with all candidate sister cell subsets as well as more distantly related cell types. Hence, we and others showed that human blood Lineage$^-$CD16$^+$ cells are non-classical monocytes (39, 88) and not DC as was sometimes claimed (94–96). Similarly, analysis of human skin CD14$^+$ cell expression of the transcriptomic fingerprints independently established for cDC, monocytes, and macrophages provided critical evidence that these cells are monocyte-derived macrophages (5) while they were previously designated as DC (4). Transcriptomic analyses were also instrumental to demonstrate the homology of this human dermal cell type with the murine CD11b$^+$Ly6C$^-$CD64^{lo-hi} (6) and pig CD163$^+$ (84) skin subsets. We were also able to show that cell populations claimed to correspond to novel cell types actually corresponded to a distinct differentiation or activation state of an already known cell type, for example establishing that the

posteriori based on the analysis of its gene expression program. In addition, the generation of gene expression data for many individual cells of the same type should increase statistical power to define genes co-expressed at the single cell level and to define cell type-specific transcriptomic modules (81). As a proof-of-principle, single cell gene expression profiling recently allowed the unbiased and de novo identification of the different cell types of spleen (91) and central nervous system (92, 93) via the description so-called interferon killer DC correspond to a particular activation state of NK cells (39). Furthermore, we showed that, upon many types of *in vivo* or *in vitro* stimulation, human and murine pDC and cDC undergo a remodeling of their gene expression program related to their plasticity, including induction of NFκB and IFN target genes, but still keep the canonical gene expression associated to their subset identity (41, 97). In particular, contrary to what other researchers hypothesized (98), gene expression profiling showed that activated pDC are not undergoing a cell fate conversion into a novel type of cDC (97).

Gene expression profiling also allowed aligning subsets of mononuclear cells across tissues (6, 8, 9, 40, 55, 99), establishing cell type homologies across species (5, 39, 50, 76, 83–85, 88, 89, 100), and rigorously examining the proximity of *in vitro*-derived subsets of mononuclear cells with those naturally existing *in vivo* (39, 41, 66, 101). These studies allowed significantly advancing the ontogeny and functional characterization of mononuclear phagocyte subsets based on the novel hypotheses that can be inferred from the analysis of the gene expression programs of the cells and from their comparison with other well-characterized cell types.

The study of the functional specialization of human DC subsets was strongly boosted by the demonstration of their transcriptomic homologies with mouse DC subsets (39, 85) which was recognized as a major breakthrough in the field (37, 53, 102–104) and acknowledged to have been impossible to draw from studies based on a limited set of molecular markers (105). In particular, this led to test whether human XCR1$^+$ cDC could be more efficient for cross-presentation than other human DC subsets. Even though the extent to which human XCR1$^+$ cDC are more efficient for cross-presentation than other human DC subsets is debated, the results from the functional studies performed independently by many teams concurrently demonstrate that these cells excel at cross-presentation of cell-associated antigens (18, 19, 37, 41, 86, 106) and of particulate antigens delivered through FcγR, through late endosomal targeting (21, 107) or upon polyI:C stimulation (18, 41, 86, 108). In addition, in sheep, the skin lymph migrating XCR1$^+$ cDC spontaneously displayed a higher efficiency of

soluble antigen-presentation to specific CD8$^+$ T cells, as compared to XCR1$^-$ cDC (76).

Based on the demonstration of the striking transcriptomic similarities between mouse and human subsets of mononuclear cells, and on knowledge on the ontogeny of these cells in the mouse (109, 110), we proposed that, similar to their mouse counterparts, human pDC and cDC constitute a specific family of cells within the hematopoietic tree, should derive from a common progenitor with a DC-restricted differentiation potential, and could be derived *in vitro* from human CD34$^+$ progenitor cells in part under the instruction of the FLT3-L growth factor (39, 85), all of which was later confirmed experimentally (41, 65, 66, 69, 111, 112).

Very importantly, comparative genomics of immune cell subsets yielded conserved transcriptomic fingerprints for each of these cell types (39), a novel knowledge which considerably accelerated the deciphering of the molecular mechanisms regulating the development and functions of leukocytes as reviewed in **Table 2** (18, 36, 59, 100, 113–127). Finally, this approach uniquely allowed identifying conserved and biologically relevant cell surface markers for each subset of mononuclear cells which could enable considerably simplifying the nomenclature for DC subsets by using the same name and similar marker combinations to identify homologous cell types irrespective of their tissues and species of origin (55–57).

Conclusion and Perspectives

While it might be the case in the future for single cell RNA-seq, currently no single method is sufficient to allow the best possible classification of DC. Hence, ideally, all available methods (cell surface phenotyping, gene expression profiling, functional analyses, and ontogeny) should be combined together to define DC subset identity. However, such a combination of approaches cannot be used to define cell subsets in many instances due to technical, financial, or ethical limitations. Taking these limitations into consideration, the data reviewed here show that comparative transcriptomics at the population level is currently the most robust

TABLE 2 | Genes which selective expression pattern in immune cell types was uncovered through comparative genomics and which functions in these cells were deciphered later.

Transcriptomic signature[a]	Gene symbol (alias)	Function
pDC	PACSIN1	Necessary for pDC production of type I interferons upon TLR7/9 stimulation (115)
	RUNX2	Necessary for terminal differentiation of pDC in, and their egress from, bone marrow (114)
	TCF4 (E2-2)	Master transcription factor instructing pDC development and functions (113)
	BCL11A	Necessary for pDC development (116, 117)
cDC	ZBTB46 (BTBD4)	Transcription factor that appears to be a specific marker of the cDC and endothelial lineages and which limits spontaneous cDC maturation (118, 119, 128)
	BATF3 (9130211I03Rik)	Transcription factor which can be critical for development of XCR1$^+$ cDC depending on the context (121)
cDC above pDC	BCL6	Promotes the development of XCR1$^+$ cDC (99, 120)
XCR1$^+$ cDC above XCR1$^-$ cDC and pDC	TLR3	TLR3 triggering induces a very strong activation of mouse and human XCR1$^+$ cDC including a uniquely high production of IFN-β and type III IFN (41, 100, 129, 130)
	RAB11A	Functionally promotes cross-presentation by storing MHC class I in a unique endosomal recycling compartment (122)
Mouse XCR1$^+$ cDC	XCR1	Likely promotes efficient interactions between XCR1$^+$ cDC and NK cells or CD8$^+$ T cells (18, 36)
Pan-T cells	THEMIS (E430004N04Rik)	Sets the signal threshold for positive and negative selection of developing T cells in the thymus (124–127)
	BCL11B	Regulates critical aspects of the development, functions, and homeostasis of T cells (123)

[a]*Transcriptomic signatures conserved between mouse and human unless specified otherwise, first reported in Robbins et al. (39), and encompassing the genes listed in this table.*

and feasible way to define the identity of cell types. Indeed, because the ontogeny and functions of cell types are instructed by specific gene expression modules, cell type identity can be defined in a universal and unbiased way by its molecular fingerprinting, including through gene expression profiling (81). However, due to its dependency on pre-selection of cell populations based on their expression patterns of a few cell surface molecules, gene expression profiling at the cell population level is imperfect and may require iterative steps of refined cell type isolation and gene expression profiling as illustrated in **Figure 3**. Hence, it is all the more important that each step of the procedure is performed and rigorously quality controlled according to the best standards in the field.

Cell purity is fundamental. It is important to design a sampling method specific for each study, through identification of the most robust criteria available in the current state of the art for purification of the target cell type based on phenotypic, morphologic, or anatomical characteristics. Cell enrichment is necessary for rare cell types among bulk populations. It relies on the depletion of other populations (MACS or EasySep™ for instance). The marker combination for negative selection must not unwillingly remove a population of interest. For instance, some antibody cocktails for human DC enrichment use anti-CD16 monoclonal antibodies, so as to deplete NK cells, but this should be proscribed for the study of non-classical, CD16$^+$ monocytes. Positive selection by magnetic or flow cytometry sorting is most often required after cell enrichment. Antibody labeling must be clear-cut, with separate peaks and/or selection of the events with the highest labeling and the lowest potential contamination by other populations. This selection implies the use of marker combinations specific for the population of interest, since specific markers are rarely available. XCR1 is a rare instance of a conserved marker so far only expressed on a discrete DC population. To the best of our knowledge, reliable commercial reagent are available for XCR1 staining only for mouse and rat, but XCR1 staining can also be achieved with fluorescently labeled recombinant XCL1 (40, 41, 52), a strategy that is amenable to many species in which XCL1 sequence is known. CLEC4C alias BDCA2 and LILRA4 alias ILT7 are specific markers for human pDC, but their engagement induces inhibitory signals which for instance reduce pDC production of type I interferons after stimulation (43, 77–79, 131). Although selectively expressed at high levels on human pDC in the blood or lymphoid organs under steady-state conditions, NRP1 alias BDCA4 can be induced on activated cDC and is also expressed on other cell types including neurons, endothelial cells, and tumor cells (132, 133). CD123 is a good marker to help identifying pDC in non-human primates, but it also labels mastocytes which are present in the blood or in lymphoid organs (134). Cell purity must be controlled in each experiment, by flow cytometry re-analysis just after sorting, and as one of the first step of transcriptomic analysis by examining the expression of negative and positive control genes (expression of genes that should be expressed only on other populations including potential contaminants, and expression of genes characteristic for the population of interest including but not restricted to genes coding for the molecules used for positive selection) (**Figure 3**).

The quality and quantity of mRNA must be adequate, even when cell numbers are low. RNA extraction kits adapted to low

cell number samples may be required. mRNA quality must be controlled by electrophoresis. A linear amplification protocol must be used, that has been validated for yielding results from low input RNA showing a strong correlation with the results obtained with higher RNA input and a classical amplification procedure.

For bioinformatics analyses, the dataset must include sister cell types as well as the cell types the most likely to contaminate the cell type of interest, or at least be compatible for integrative analysis with a reference dataset including these control populations. Several independent methods for data analysis should be used, to ensure robustness of interpretation. Beyond relative classification of the cell types of the dataset by classical approaches computing the overall distance between their gene expression programs as performed by hierarchical clustering or principal component analysis, the identity of cell types can also be reliably inferred from the analysis of their relative expression of robust cell type-specific gene signatures established from re-analysis of public gene chip databases and/or from published articles.

Novel advances are being brought through molecular profiling of subsets of mononuclear cells. In addition to steady-state conditions, populations can be analyzed after stimulation to identify the specific activation pathways elicited in pure cell populations or upon interaction between different cell types (41, 97, 135–137). In addition to unbiased analysis of the cellular composition of different organs (91, 93), transcriptomic profiling at the single cell level will allow studying heterogeneity in gene expression within one cell type with the hope to link it to functional heterogeneity (138) and eventually with the former history/epigenetic imprinting of each cell. Comparative transcriptomic studies allowed us and others to identify in humans, non-human primates, pig, sheep, and chicken cDC subsets homologous to those well described in mice (5, 18, 39, 50, 52, 62, 76, 83–85). These studies suggest that similar cDC subsets already existed in the last common ancestor of birds and mammals. Conserved gene modules appear during evolution to elicit new functions (81, 82). For instance, regarding T helper lineage diversification during evolution, contrary to bony fishes, the elephant shark, a cartilaginous fish, has been reported to lack genes encoding for critical transcription factors or cytokines instructing the development or involved in the functions of Th2, Th17, and Treg cells, such as RORC and FOXP3, IL-4, IL-21, IL-23, and IL-2 (139). This suggests that the genes required for the development of the different T helper lineages might have appeared progressively as modules during evolution starting in bony fishes and with late development of the Treg and Th17 lineages (81). Comparative genomics of mononuclear phagocyte subsets and single cell gene expression profiling will critically help identifying novel gene modules and their associated immune functions. In pDC, evolutionarily conserved co-expression of *TCF4, RUNX2, TLR7, TLR9, UNC93B1, MYD88, IRAK4, IRF7*, and *PACSIN1* might represent part of a gene module instructing the functional specialization of this cell type in high level production of type I interferon in response to sensing of oligonucleotide sequences of viral or autologous origin. In XCR1$^+$ cDC, evolutionarily conserved co-expression of *CLEC9A, SYK, RAB11A, RAB7B, SEPT3, SNX22, TLR3, CADM1*, and *XCR1* might represent part of a gene module instructing the functional specialization of this cell type in CD8$^+$ T-cell activation and specifically in cross-presentation of cell-associated antigens.

In any case, the discovery of the sets of genes that are tightly co-expressed in DC subsets across various tissues and species, not only at the population level but also at the single cell level, should allow identifying the gene modules instructing DC subset functions. Characterization of the members of these gene modules which role in DC is unknown yet should strongly contribute to increase our knowledge on DC subset functional specialization and their molecular regulation. Of note, not all of these gene modules might harbor the same differential pattern of expression between DC subsets in different animal species. Some functions have gained or lost expression in specific cell subsets in some species which should correlate with similar changes in the expression patterns of the corresponding gene modules. For instance, IL-12 is produced both by pDC and cDC in mice, but only by cDC in humans, while antigen cross-presentation appears to be more strongly associated with XCR1$^+$ cDC in mice than in humans (18, 19, 22). Isolation and comparison of mononuclear phagocyte subsets from homologous organs in different species may help understand how the anatomical compartmentalization of these cells is established and affects their functions, including local interaction with specific cell types and chemokines. Dating when during evolution pDC as well as classical and non-classical monocyte subsets appeared, and in which anatomical compartments they reside in the species the most distant to humans and mice, may give novel insights into the core functions of these populations.

In vivo manipulation of DC can promote and orient immune responses based on the intrinsic functional properties of the DC subset targeted and can be advantageously used for prophylactic vaccination or immunotherapy against cancer or infections. This strategy can benefit from the knowledge gained from the expression profiling of DC subsets and their alignment across species. Notably, based on their homology with mouse XCR1$^+$ cDC, human XCR1$^+$ cDC can be considered as a promising target when cross-presentation is desirable, in particular for fighting cancer or infections by intracellular pathogens (23, 24, 29, 72, 73, 140–143). Moreover, because it is specifically expressed in XCR1$^+$ cDC in a conserved manner in evolution, and it has been successfully used for *in vivo* delivery of antigens specifically to XCR1$^+$ cDC to vaccinate mice (23, 24), XCR1 can be considered for a universal DC targeting strategy in potentially all vertebrate species. Interestingly, the targeting of XCR1 can be achieved with targeting units composed of recombinant XCL1 fused to protective antigens in the form of vaccibodies (24), a strategy that is amenable to many species in which the XCL1 sequence is known. Although more broadly expressed in the DC lineage at least in mice, CLEC9A is also an interesting target since it directly promotes cross-presentation of the material it binds, probably by delivering it into appropriate endosomes (144, 145), and because it is selectively expressed to high levels on XCR1$^+$ cDC in humans, sheep, and mice (25, 32, 76, 146) although it may not be the case in some other species such as pig. Arguments in favor or against the targeting of XCR1$^+$ cDC in the clinic are summarized in **Table 3**. The identification of XCR1$^+$ cDC in companion and sport animals, and in animals of the agro-economy, such as ruminants, pigs, poultry, and fishes, will allow designing better vaccines to protect

TABLE 3 | The PROs and CONs for *in vivo* targeting of XCR1$^+$ cDC[a].

	PROs	CONs
Cross-presentation efficiency	Higher for blood and skin XCR1$^+$ cDC, especially for cell-associated antigens	Disputed for XCR1$^+$ cDC from secondary lymphoid organs (19, 22) depending on intracellular compartment of antigen delivery (21)
Anatomical localization	Present in lymphoid and non-lymphoid tissues, enabling subcutaneous, intradermal, or oral vaccination	Low efficiency of human XCR1$^+$ cDC for induction of mucosa-homing CD8$^+$ T cells (151)?
Frequency	Few cells can mediate important functions *in vivo*. Quality matters more than quantity	Very few numbers of XCR1$^+$ cDC in most tissues
Specificity of targeting	Very specific expression of XCR1 as opposed to the broader expression of CD141, DEC205, and CLEC9A. Precise targeting and better pharmacodynamics	Too specific, limiting biological effect to just one DC subset, may not induce strong enough or broad enough immune responses
Responsiveness to adjuvants	Very good responsiveness to Poly:C. Poly:C is a very potent adjuvant for the induction of strong, polyfunctional CD8$^+$ T-cell responses which might result in part from TLR3 triggering in XCR1$^+$ cDC	Poly:C may primarily work by activating other targets, i.e., non-immune cells expressing TLR3 or cells activated through MDA5
Proof of concept achieved in mice	XCR1$^+$ cDC are critical for anti-tumoral responses in mice (72, 121, 152, 153). XCR1 targeting works in mice (23, 24). XCR1 bio-equivalency in human, macaques, mouse, pig, and sheep, same gene expression pattern and biological function. Hence, higher probability of translation to human of mechanistic studies in animals	Many previous failures of mouse to human translation
In vitro model	Ability to generate *in vitro* and manipulate *bona fide* human XCR1$^+$ cDC from CD34$^+$ cord blood progenitors (41, 65, 66, 69, 111, 112)	
Cytokine production	XCR1$^+$ cDC can produce IL-12 but maybe optimal conditions to induce this function remain to be identified (50, 65, 66, 143). Mouse and human XCR1$^+$ cDC are high producers of beta and type III interferons upon Poly:C stimulation (41, 100, 129, 130)	Human XCR1$^+$ cDC are very poor producers of IL-12 (70, 108)
Clinical data	Gene expression profiling of human tumors suggest that infiltration by XCR1$^+$ cDC but not other myeloid cells is of good prognosis both in mice and humans (72)	Formal measurements of XCR1$^+$ cDC infiltration in human tumors and of its beneficial role for disease control remain to be established

[a]*More details and bibliographical references can be found in the main text of this review.*

TABLE 4 | Practical guidelines for consistent definition of DC subsets across mouse and human tissues with potential applicability to other mammals.

Characterization	XCR1⁻ cDC2		XCR1⁺ cDC1		pDC	
	High or positive	Negative or low	High or positive	Negative or low	High or positive	Negative or low
Conserved phenotype	CD11c^{high} MHC-II^{high} FLT3$^+$ SIRPα$^+$	CD3$^-$ CD19$^-$ CD14$^{-/low}$ CD206$^{-/low}$ CD123$^-$	CD11c$^{low-to-high}$ MHC-II^{high} FLT3$^+$ XCR1$^+$ CADM1$^+$	CD3$^-$ CD19$^-$ CD14$^{-/low}$ CD206$^{-/low}$ CD123$^-$	MHC-II^{int} FLT3$^+$	CD3$^-$ CD19$^-$ CD14$^{-/low}$ CD206$^{-/low}$ CD19$^-$
Critical species-specific phenotypic markers					Mouse: Siglec-H or Ccr9 Human: CD123 and CLEC4C (BDCA2) or ILT7 (LILRA4)	
Hallmark genes (18, 37, 39)	*FLT3* *TLR8* *ZBTB46* **IRF4**^{a}	*XCR1* *RAB7B* *GCET2* *TLR4* *IRF8* *TCF4* *RUNX2* *SPIB*	*FLT3* *XCR1* *CADM1* *TLR3* *RAB7B* *GCET2* *ZBTB46* **IRF8** **BATF3**	*TLR4* *TLR7* *IRF4* *TCF4* *RUNX2* *SPIB*	*FLT3* *TLR7* *TLR9* *PACSIN1* **IRF8** **TCF4** **RUNX2** **SPIB** **BCL11A**	*XCR1* *CADM1* *TLR3* *TLR8* *RAB7B* *GCET2* *ZBTB46* *BATF3* *CADM1*
Hallmark cytokine production	IL-23 production? (62)		Type III interferon production upon TLR3 triggering (41, 100, 129, 130)		Production of type I and III interferons in response to TLR7/9 triggering	
Hallmark antigen-presentation functions	High efficiency for CD4$^+$ T-cell activation		High efficiency for CD8$^+$ T-cell activation, in particular through cross-presentation of cell-associated antigens			

^{a}*Master transcription factors critical for cell subset development are indicated in bold font.*

them against infections in order to ameliorate animal welfare and to prevent pandemics causing severe economic losses. It will also contribute to a global public health strategy because some of these animal species as well as wild animals are targets or reservoirs for major zoonotic pathogens. The identification of XCR1$^+$ cDC in rhesus macaques and in pigs opens the way to preclinical vaccination studies in these species which are close to humans. Vaccibodies based on XCL1 dimers coupled to influenza or SIV proteins are planned to be used for vaccination of pigs or rhesus macaques, respectively, and induction of immune responses and protection against infection. pDC targeting could also be considered as an interesting alternative for vaccination against viruses or tumors (20, 147, 148), or for the induction of cross-tolerance to treat autoimmune diseases or food allergies (149, 150).

A synthetic list of phenotypic, transcriptomic, and functional hallmarks which have already allowed conserved identification of different DC and monocyte subsets in humans and mice is presented in **Table 4**. The present Special Issue and future workshop on DC nomenclature will help reach a consensus panel for practical definition of the populations, in order to integrate faster the huge, but scattered knowledge accumulated by different laboratories in different cell types, species, and organs.

Acknowledgments

The studies performed in the laboratories are supported by funding from INSERM, CNRS, INRA, a CNRS-AP-HP collaboration, Aix-Marseille University, Université Paris Descartes, Université Paris Diderot, ANRS, the Agence Nationale de la Recherche (ANR) (PhyloGenDC, ANR-09-BLAN-0073-02 and DCskin-VacFlu, ANR-11-ISV3-0001), the "Integrative Biology of Emerging Infectious Diseases" Labex (ANR-10-LABX-62-IBEID), the DCBIOL Labex (ANR-11-LABEX-0043, grant ANR-10-IDEX-0001-02 PSL*), and the A*MIDEX project (ANR-11-IDEX-0001-02) funded by the French Government's "Investissements d'Avenir" program managed by the ANR, as well as the European Community's Seventh Framework Programme FP7/2007–2013 (European Research Council Starting Grant Agreement number 281225 for MD including salary support to TPVM). We thank past and present members of our laboratories and Dr. Rémi Cheynier for their contribution to studies on DC and for discussions. Schematic representations of cells are adapted from Servier Medical Art Powerpoint image bank (smart.servier.fr/servier-medical-art). We apologize for not quoting certain studies because of space limitations.

References

1. Cheong C, Matos I, Choi JH, Dandamudi DB, Shrestha E, Longhi MP, et al. Microbial stimulation fully differentiates monocytes to DC-SIGN/CD209(+) dendritic cells for immune T cell areas. *Cell* (2010) 143(3):416–29. doi:10.1016/j.cell.2010.09.039

2. Plantinga M, Guilliams M, Vanheerswynghels M, Deswarte K, Branco-Madeira F, Toussaint W, et al. Conventional and monocyte-derived CD11b(+) dendritic cells initiate and maintain T helper 2 cell-mediated immunity to house dust mite allergen. *Immunity* (2013) 38(2):322–35. doi:10.1016/j.immuni.2012.10.016

3. Langlet C, Tamoutounour S, Henri S, Luche H, Ardouin L, Gregoire C, et al. CD64 expression distinguishes monocyte-derived and conventional dendritic cells and reveals their distinct role during intramuscular immunization. *J Immunol* (2012) 188(4):1751–60. doi:10.4049/jimmunol.1102744

4. Klechevsky E, Morita R, Liu M, Cao Y, Coquery S, Thompson-Snipes L, et al. Functional specializations of human epidermal Langerhans cells and CD14+ dermal dendritic cells. *Immunity* (2008) **29**(3):497–510. doi:10.1016/j.immuni.2008.07.013

5. McGovern N, Schlitzer A, Gunawan M, Jardine L, Shin A, Poyner E, et al. Human dermal CD14(+) cells are a transient population of monocyte-derived macrophages. *Immunity* (2014) **41**(3):465–77. doi:10.1016/j.immuni.2014.08.006

6. Tamoutounour S, Guilliams M, Montanana Sanchis F, Liu H, Terhorst D, Malosse C, et al. Origins and functional specialization of macrophages and of conventional and monocyte-derived dendritic cells in mouse skin. *Immunity* (2013) **39**(5):925–38. doi:10.1016/j.immuni.2013.10.004

7. Hoeffel G, Wang Y, Greter M, See P, Teo P, Malleret B, et al. Adult Langer-hans cells derive predominantly from embryonic fetal liver monocytes with a minor contribution of yolk sac-derived macrophages. *J Exp Med* (2012) **209**(6):1167–81. doi:10.1084/jem.20120340

8. Gautier EL, Shay T, Miller J, Greter M, Jakubzick C, Ivanov S, et al. Gene-expression profiles and transcriptional regulatory pathways that underlie the identity and diversity of mouse tissue macrophages. *Nat Immunol* (2012) **13**(11):1118–28. doi:10.1038/ni.2419

9. Miller JC, Brown BD, Shay T, Gautier EL, Jojic V, Cohain A, et al. Deciphering the transcriptional network of the dendritic cell lineage. *Nat Immunol* (2012) **13**(9):888–99. doi:10.1038/ni.2370

10. Greter M, Lelios I, Pelczar P, Hoeffel G, Price J, Leboeuf M, et al. Stroma-derived interleukin-34 controls the development and maintenance of Langer-hans cells and the maintenance of microglia. *Immunity* (2012) **37**(6):1050–60. doi:10.1016/j.immuni.2012.11.001

11. Perdiguero EG, Klapproth K, Schulz C, Busch K, Azzoni E, Crozet L, et al. Tissue-resident macrophages originate from yolk-sac-derived erythro-myeloid progenitors. *Nature* (2014) **518**(7540):547–51. doi:10.1038/nature13989

12. Schulz C, Gomez Perdiguero E, Chorro L, Szabo-Rogers H, Cagnard N, Kierdorf K, et al. A lineage of myeloid cells independent of Myb and hematopoietic stem cells. *Science* (2012) **336**(6077):86–90. doi:10.1126/science.1219179

13. Wang Y, Szretter KJ, Vermi W, Gilfillan S, Rossini C, Cella M, et al. IL-34 is a tissue-restricted ligand of CSF1R required for the development of Langerhans cells and microglia. *Nat Immunol* (2012) **13**(8):753–60. doi:10.1038/ni.2360

14. Dadaglio G, Fayolle C, Zhang X, Ryffel B, Oberkampf M, Felix T, et al. Antigen targeting to CD11b+ dendritic cells in association with TLR4/TRIF signaling promotes strong CD8+ T cell responses. *J Immunol* (2014) **193**(4):1787–98. doi:10.4049/jimmunol.1302974

15. Desch AN, Gibbings SL, Clambey ET, Janssen WJ, Slansky JE, Kedl RM, et al. Dendritic cell subsets require cis-activation for cytotoxic CD8 T-cell induction. *Nat Commun* (2014) **5**:4674. doi:10.1038/ncomms5674

16. Mouries J, Moron G, Schlecht G, Escriou N, Dadaglio G, Leclerc C. Plasmacytoid dendritic cells efficiently cross-prime naive T cells in vivo after TLR activation. *Blood* (2008) **112**(9):3713–22. doi:10.1182/blood-2008-03-146290

17. Ballesteros-Tato A, Leon B, Lund FE, Randall TD. Temporal changes in dendritic cell subsets, cross-priming and costimulation via CD70 control CD8(+) T cell responses to influenza. *Nat Immunol* (2010) **11**(3):216–24. doi:10.1038/ni.1838

18. Crozat K, Guiton R, Contreras V, Feuillet V, Dutertre CA, Ventre E, et al. The XC chemokine receptor 1 is a conserved selective marker of mammalian cells homologous to mouse CD8alpha+ dendritic cells. *J Exp Med* (2010) **207**(6):1283–92. doi:10.1084/jem.20100223

19. Segura E, Durand M, Amigorena S. Similar antigen cross-presentation capacity and phagocytic functions in all freshly isolated human lymphoid organ-resident dendritic cells. *J Exp Med* (2013) **210**(5):1035–47. doi:10.1084/jem.20121103

20. Hoeffel G, Ripoche AC, Matheoud D, Nascimbeni M, Escriou N, Lebon P, et al. Antigen crosspresentation by human plasmacytoid dendritic cells. *Immunity* (2007) **27**(3):481–92. doi:10.1016/j.immuni.2007.07.021

21. Cohn L, Chatterjee B, Esselborn F, Smed-Sorensen A, Nakamura N, Chalouni C, et al. Antigen delivery to early endosomes eliminates the superiority of human blood BDCA3+ dendritic cells at cross presentation. *J Exp Med* (2013) **210**(5):1049–63. doi:10.1084/jem.20121251

22. Mittag D, Proietto AI, Loudovaris T, Mannering SI, Vremec D, Shortman K, et al. Human dendritic cell subsets from spleen and blood are similar in phenotype and function but modified by donor health status. *J Immunol* (2011) **186**(11):6207–17. doi:10.4049/jimmunol.1002632

23. Hartung E, Becker M, Bachem A, Reeg N, Jakel A, Hutloff A, et al. Induction of potent CD8 T cell cytotoxicity by specific targeting of antigen to cross-presenting dendritic cells in vivo via murine or human XCR1. *J Immunol* (2015) **194**(3):1069–79. doi:10.4049/jimmunol.1401903

24. Fossum E, Grodeland G, Terhorst D, Tveita AA, Vikse E, Mjaaland S, et al. Vaccine molecules targeting Xcr1 on cross-presenting DCs induce protective CD8(+) T-cell responses against influenza virus. *Eur J Immunol* (2015) **45**(2):624–35. doi:10.1002/eji.201445080

25. Caminschi I, Proietto AI, Ahmet F, Kitsoulis S, Shin Teh J, Lo JC, et al. The dendritic cell subtype-restricted C-type lectin Clec9A is a target for vaccine enhancement. *Blood* (2008) **112**(8):3264–73. doi:10.1182/blood-2008-05-155176

26. Caminschi I, Vremec D, Ahmet F, Lahoud MH, Villadangos JA, Murphy KM, et al. Antibody responses initiated by Clec9A-bearing dendritic cells in normal and Batf3(-/-) mice. *Mol Immunol* (2012) **50**(1–2):9–17. doi:10.1016/j.molimm.2011.11.008

27. Idoyaga J, Lubkin A, Fiorese C, Lahoud MH, Caminschi I, Huang Y, et al. Comparable T helper 1 (Th1) and CD8 T-cell immunity by targeting HIV gag p24 to CD8 dendritic cells within antibodies to Langerin, DEC205, and Clec9A. *Proc Natl Acad Sci U S A* (2011) **108**(6):2384–9. doi:10.1073/pnas.1019547108

28. Lahoud MH, Ahmet F, Kitsoulis S, Wan SS, Vremec D, Lee CN, et al. Targeting antigen to mouse dendritic cells via Clec9A induces potent CD4 T cell responses biased toward a follicular helper phenotype. *J Immunol* (2011) **187**(2):842–50. doi:10.4049/jimmunol.1101176

29. Li J, Ahmet F, Sullivan LC, Brooks AG, Kent SJ, De Rose R, et al. Antibodies targeting Clec9A promote strong humoral immunity without adjuvant in mice and non-human primates. *Eur J Immunol* (2014) **45**(3):854–64. doi:10.1002/eji.201445127

30. Park HY, Light A, Lahoud MH, Caminschi I, Tarlinton DM, Shortman K. Evolution of B cell responses to Clec9A-targeted antigen. *J Immunol* (2013) **191**(10):4919–25. doi:10.4049/jimmunol.1301947

31. Joffre OP, Sancho D, Zelenay S, Keller AM, Reis e Sousa C. Efficient and versatile manipulation of the peripheral CD4+ T-cell compartment by antigen targeting to DNGR-1/CLEC9A. *Eur J Immunol* (2010) **40**(5):1255–65. doi:10.1002/eji.201040419

32. Sancho D, Mourao-Sa D, Joffre OP, Schulz O, Rogers NC, Pennington DJ, et al. Tumor therapy in mice via antigen targeting to a novel, DC-restricted C-type lectin. *J Clin Invest* (2008) **118**(6):2098–110. doi:10.1172/JCI34584

33. Bonifaz LC, Bonnyay DP, Charalambous A, Darguste DI, Fujii S, Soares H, et al. In vivo targeting of antigens to maturing dendritic cells via the DEC-205 receptor improves T cell vaccination. *J Exp Med* (2004) **199**(6):815–24. doi:10.1084/jem.20032220

34. Nchinda G, Kuroiwa J, Oks M, Trumpfheller C, Park CG, Huang Y, et al. The efficacy of DNA vaccination is enhanced in mice by targeting the encoded protein to dendritic cells. *J Clin Invest* (2008) **118**(4):1427–36. doi:10.1172/JCI34224

35. Do Y, Didierlaurent AM, Ryu S, Koh H, Park CG, Park S, et al. Induction of pulmonary mucosal immune responses with a protein vaccine targeted to the DEC-205/CD205 receptor. *Vaccine* (2012) **30**(45):6359–67. doi:10.1016/j.vaccine.2012.08.051

36. Dorner BG, Dorner MB, Zhou X, Opitz C, Mora A, Guttler S, et al. Selective expression of the chemokine receptor XCR1 on cross-presenting dendritic cells determines cooperation with CD8+ T cells. *Immunity* (2009) **31**(5):823–33. doi:10.1016/j.immuni.2009.08.027

37. Bachem A, Guttler S, Hartung E, Ebstein F, Schaefer M, Tannert A, et al. Superior antigen cross-presentation and XCR1 expression define human CD11c+CD141+ cells as homologues of mouse CD8+ dendritic cells. *J Exp Med* (2010) **207**(6):1273–81. doi:10.1084/jem.20100348

38. Gurka S, Hartung E, Becker M, Kroczek RA. Mouse conventional dendritic cells can be universally classified based on the mutually exclusive expression of XCR1 and SIRPα. *Front Immunol* (2015) **6**:35. doi:10.3389/fimmu.2015.00035

39. Robbins SH, Walzer T, Dembele D, Thibault C, Defays A, Bessou G, et al. Novel insights into the relationships between dendritic cell subsets in human and mouse revealed by genome-wide expression profiling. *Genome Biol* (2008) **9**(1):R17. doi:10.1186/gb-2008-9-1-r17

40. Crozat K, Tamoutounour S, Vu Manh TP, Fossum E, Luche H, Ardouin L, et al. Cutting edge: expression of XCR1 defines mouse lymphoid-tissue resident and migratory dendritic cells of the CD8alpha+ type. *J Immunol* (2011) 187(9):4411–5. doi:10.4049/jimmunol.1101717

41. Balan S, Ollion V, Colletti N, Chelbi R, Montanana-Sanchis F, Liu H, et al. Human XCR1+ dendritic cells derived in vitro from CD34+ progenitors closely resemble blood dendritic cells, including their adjuvant responsiveness, contrary to monocyte-derived dendritic cells. *J Immunol* (2014) 193(4):1622–35. doi:10.4049/jimmunol.1401243

42. Yamazaki C, Miyamoto R, Hoshino K, Fukuda Y, Sasaki I, Saito M, et al. Conservation of a chemokine system, XCR1 and its ligand, XCL1, between human and mice. *Biochem Biophys Res Commun* (2010) 397(4):756–61. doi:10.1016/j.bbrc.2010.06.029

43. Dzionek A, Sohma Y, Nagafune J, Cella M, Colonna M, Facchetti F, et al. BDCA-2, a novel plasmacytoid dendritic cell-specific type II C-type lectin, mediates antigen capture and is a potent inhibitor of interferon alpha/beta induction. *J Exp Med* (2001) 194(12):1823–34. doi:10.1084/jem.194.12.1823

44. Rissoan MC, Duhen T, Bridon JM, Bendriss-Vermare N, Peronne C, de Saint Vis B, et al. Subtractive hybridization reveals the expression of immunoglobulin-like transcript 7, Eph-B1, granzyme B, and 3 novel transcripts in human plasmacytoid dendritic cells. *Blood* (2002) 100(9):3295–303. doi:10.1182/blood-2002-02-0638

45. Cao W, Rosen DB, Ito T, Bover L, Bao M, Watanabe G, et al. Plasmacytoid dendritic cell-specific receptor ILT7-Fc epsilonRI gamma inhibits toll-like receptor-induced interferon production. *J Exp Med* (2006) 203(6):1399–405. doi:10.1084/jem.20052454

46. Dzionek A, Fuchs A, Schmidt P, Cremer S, Zysk M, Miltenyi S, et al. BDCA-2, BDCA-3, and BDCA-4: three markers for distinct subsets of dendritic cells in human peripheral blood. *J Immunol* (2000) 165(11):6037–46. doi:10.4049/jimmunol.165.11.6037

47. Dalod M, Salazar-Mather TP, Malmgaard L, Lewis C, Asselin-Paturel C, Briere F, et al. Interferon alpha/beta and interleukin 12 responses to viral infections: pathways regulating dendritic cell cytokine expression in vivo. *J Exp Med* (2002) 195(4):517–28. doi:10.1084/jem.20011672

48. Bar-On L, Birnberg T, Lewis KL, Edelson BT, Bruder D, Hildner K, et al. CX3CR1+ CD8alpha+ dendritic cells are a steady-state population related to plasmacytoid dendritic cells. *Proc Natl Acad Sci U S A* (2010) 107(33):14745–50. doi:10.1073/pnas.1001562107

49. Vu Manh T-P, Dalod M. Characterization of dendritic cell subsets through gene expression analysis. *Methods Mol Biol* (2015) (in press).

50. Haniffa M, Shin A, Bigley V, McGovern N, Teo P, See P, et al. Human tissues contain CD141hi cross-presenting dendritic cells with functional homology to mouse CD103+ nonlymphoid dendritic cells. *Immunity* (2012) 37(1):60–73. doi:10.1016/j.immuni.2012.04.012

51. Chu CC, Ali N, Karagiannis P, Di Meglio P, Skowera A, Napolitano L, et al. Resident CD141 (BDCA3)+ dendritic cells in human skin produce IL-10 and induce regulatory T cells that suppress skin inflammation. *J Exp Med* (2012) 209(5):935–45. doi:10.1084/jem.20112583

52. Dutertre CA, Jourdain JP, Rancez M, Amraoui S, Fossum E, Bogen B, et al. TLR3-responsive, XCR1+, CD141(BDCA-3)+/CD8alpha+-equivalent dendritic cells uncovered in healthy and simian immunodeficiency virus-infected rhesus macaques. *J Immunol* (2014) 192(10):4697–708. doi:10.4049/jimmunol.1302448

53. Dutertre CA, Wang LF, Ginhoux F. Aligning bona fide dendritic cell populations across species. *Cell Immunol* (2014) 291(1–2):3–10. doi:10.1016/j.cellimm.2014.08.006

54. Mildner A, Jung S. Development and function of dendritic cell subsets. *Immunity* (2014) 40(5):642–56. doi:10.1016/j.immuni.2014.04.016

55. Guilliams M, Henri S, Tamoutounour S, Ardouin L, Schwartz-Cornil I, Dalod M, et al. From skin dendritic cells to a simplified classification of human and mouse dendritic cell subsets. *Eur J Immunol* (2010) 40(8):2089–94. doi:10.1002/eji.201040498

56. Ziegler-Heitbrock L, Ancuta P, Crowe S, Dalod M, Grau V, Hart DN, et al. Nomenclature of monocytes and dendritic cells in blood. *Blood* (2010) 116(16):e74–80. doi:10.1182/blood-2010-02-258558

57. Guilliams M, Ginhoux F, Jakubzick C, Naik SH, Onai N, Schraml BU, et al. Dendritic cells, monocytes and macrophages: a unified nomenclature based on ontogeny. *Nat Rev Immunol* (2014) 14(8):571–8. doi:10.1038/nri3712

58. Merad M, Sathe P, Helft J, Miller J, Mortha A. The dendritic cell lineage: ontogeny and function of dendritic cells and their subsets in the steady state and the inflamed setting. *Annu Rev Immunol* (2013) 31:563–604. doi:10.1146/annurev-immunol-020711-074950

59. Bachem A, Hartung E, Guttler S, Mora A, Zhou X, Hegemann A, et al. Expression of XCR1 characterizes the Batf3-dependent lineage of dendritic cells capable of antigen cross-presentation. *Front Immunol* (2012) 3:214. doi:10.3389/fimmu.2012.00214

60. Becker M, Guttler S, Bachem A, Hartung E, Mora A, Jakel A, et al. Ontogenic, phenotypic, and functional characterization of XCR1(+) dendritic cells leads to a consistent classification of intestinal dendritic cells based on the expression of XCR1 and SIRPalpha. *Front Immunol* (2014) 5:326. doi:10.3389/fimmu.2014.00326

61. Tamoutounour S, Henri S, Lelouard H, de Bovis B, de Haar C, van der Woude CJ, et al. CD64 distinguishes macrophages from dendritic cells in the gut and reveals the Th1-inducing role of mesenteric lymph node macrophages during colitis. *Eur J Immunol* (2012) 42(12):3150–66. doi:10.1002/eji.201242847

62. Schlitzer A, McGovern N, Teo P, Zelante T, Atarashi K, Low D, et al. IRF4 transcription factor-dependent CD11b+ dendritic cells in human and mouse control mucosal IL-17 cytokine responses. *Immunity* (2013) 38(5):970–83. doi:10.1016/j.immuni.2013.04.011

63. Persson EK, Uronen-Hansson H, Semmrich M, Rivollier A, Hagerbrand K, Marsal J, et al. IRF4 transcription-factor-dependent CD103(+)CD11b(+) dendritic cells drive mucosal T helper 17 cell differentiation. *Immunity* (2013) 38(5):958–69. doi:10.1016/j.immuni.2013.03.009

64. Schraml BU, van Blijswijk J, Zelenay S, Whitney PG, Filby A, Acton SE, et al. Genetic tracing via DNGR-1 expression history defines dendritic cells as a hematopoietic lineage. *Cell* (2013) 154(4):843–58. doi:10.1016/j.cell.2013.07.014

65. Poulin LF, Salio M, Griessinger E, Anjos-Afonso F, Craciun L, Chen JL, et al. Characterization of human DNGR-1+ BDCA3+ leukocytes as putative equivalents of mouse CD8alpha+ dendritic cells. *J Exp Med* (2010) 207(6):1261–71. doi:10.1084/jem.20092618

66. Lee J, Breton G, Oliveira TY, Zhou YJ, Aljoufi A, Puhr S, et al. Restricted dendritic cell and monocyte progenitors in human cord blood and bone marrow. *J Exp Med* (2015) 212(3):385–99. doi:10.1084/jem.20141442

67. Schotte R, Nagasawa M, Weijer K, Spits H, Blom B. The ETS transcription factor Spi-B is required for human plasmacytoid dendritic cell development. *J Exp Med* (2004) 200(11):1503–9. doi:10.1084/jem.20041231

68. Poulin LF, Reyal Y, Uronen-Hansson H, Schraml BU, Sancho D, Murphy KM, et al. DNGR-1 is a specific and universal marker of mouse and human Batf3-dependent dendritic cells in lymphoid and nonlymphoid tissues. *Blood* (2012) 119(25):6052–62. doi:10.1182/blood-2012-01-406967

69. Breton G, Lee J, Zhou YJ, Schreiber JJ, Keler T, Puhr S, et al. Circulating precursors of human CD1c+ and CD141+ dendritic cells. *J Exp Med* (2015) 212(3):401–13. doi:10.1084/jem.20141441

70. Hambleton S, Salem S, Bustamante J, Bigley V, Boisson-Dupuis S, Azevedo J, et al. IRF8 mutations and human dendritic-cell immunodeficiency. *N Engl J Med* (2011) 365(2):127–38. doi:10.1056/NEJMoa1100066

71. Jardine L, Barge D, Ames-Draycott A, Pagan S, Cookson S, Spickett G, et al. Rapid detection of dendritic cell and monocyte disorders using CD4 as a lineage marker of the human peripheral blood antigen-presenting cell compartment. *Front Immunol* (2013) 4:495. doi:10.3389/fimmu.2013.00495

72. Broz ML, Binnewies M, Boldajipour B, Nelson AE, Pollack JL, Erle DJ, et al. Dissecting the tumor myeloid compartment reveals rare activating antigen-presenting cells critical for T cell immunity. *Cancer Cell* (2014) 26(5):638–52. doi:10.1016/j.ccell.2014.09.007

73. Woo SR, Corrales L, Gajewski TF. Innate immune recognition of cancer. *Annu Rev Immunol* (2015) 33:445–74. doi:10.1146/annurev-immunol-032414-112043

74. Summerfield A, Auray G, Ricklin M. Comparative dendritic cell biology of veterinary mammals. *Annu Rev Anim Biosci* (2015) 3:533–57. doi:10.1146/annurev-animal-022114-111009

75. Vremec D, Hansen J, Strasser A, Acha-Orbea H, Zhan Y, O'Keeffe M, et al. Maintaining dendritic cell viability in culture. *Mol Immunol* (2015) 63(2):264–7. doi:10.1016/j.molimm.2014.07.011

76. Contreras V, Urien C, Guiton R, Alexandre Y, Vu Manh TP, Andrieu T, et al. Existence of CD8alpha-like dendritic cells with a conserved functional

specialization and a common molecular signature in distant mammalian species. *J Immunol* (2010) **185**(6):3313–25. doi:10.4049/jimmunol.1000824

77. Fanning SL, George TC, Feng D, Feldman SB, Megjugorac NJ, Izaguirre AG, et al. Receptor cross-linking on human plasmacytoid dendritic cells leads to the regulation of IFN-alpha production. *J Immunol* (2006) **177**(9):5829–39. doi:10.4049/jimmunol.177.9.5829

78. Jahn PS, Zanker KS, Schmitz J, Dzionek A. BDCA-2 signaling inhibits TLR-9-agonist-induced plasmacytoid dendritic cell activation and antigen presentation. *Cell Immunol* (2010) **265**(1):15–22. doi:10.1016/j.cellimm.2010.06.005

79. Tavano B, Boasso A. Effect of immunoglobin-like transcript 7 cross-linking on plasmacytoid dendritic cells differentiation into antigen-presenting cells. *PLoS One* (2014) **9**(2):e89414. doi:10.1371/journal.pone.0089414

80. Duffy D, Rouilly V, Libri V, Hasan M, Beitz B, David M, et al. Functional analysis via standardized whole-blood stimulation systems defines the boundaries of a healthy immune response to complex stimuli. *Immunity* (2014) **40**(3):436–50. doi:10.1016/j.immuni.2014.03.002

81. Achim K, Arendt D. Structural evolution of cell types by step-wise assembly of cellular modules. *Curr Opin Genet Dev* (2014) **27**:102–8. doi:10.1016/j.gde.2014.05.001

82. Arendt D. The evolution of cell types in animals: emerging principles from molecular studies. *Nat Rev Genet* (2008) **9**(11):868–82. doi:10.1038/nrg2416

83. Vu Manh TP, Marty H, Sibille P, Le Vern Y, Kaspers B, Dalod M, et al. Existence of conventional dendritic cells in *Gallus gallus* revealed by comparative gene expression profiling. *J Immunol* (2014) **192**(10):4510–7. doi:10.4049/jimmunol.1303405

84. Marquet F, Vu Manh TP, Maisonnasse P, Elhmouzi-Younes J, Urien C, Bouguyon E, et al. Pig skin includes dendritic cell subsets transcriptomically related to human CD1a and CD14 dendritic cells presenting different migrating behaviors and T cell activation capacities. *J Immunol* (2014) **193**(12):5883–93. doi:10.4049/jimmunol.1303150

85. Crozat K, Guiton R, Guilliams M, Henri S, Baranek T, Schwartz-Cornil I, et al. Comparative genomics as a tool to reveal functional equivalences between human and mouse dendritic cell subsets. *Immunol Rev* (2010) **234**(1):177–98. doi:10.1111/j.0105-2896.2009.00868.x

86. Jongbloed SL, Kassianos AJ, McDonald KJ, Clark GJ, Ju X, Angel CE, et al. Human CD141+ (BDCA-3)+ dendritic cells (DCs) represent a unique myeloid DC subset that cross-presents necrotic cell antigens. *J Exp Med* (2010) **207**(6):1247–60. doi:10.1084/jem.20092140

87. Dudziak D, Kamphorst AO, Heidkamp GF, Buchholz VR, Trumpfheller C, Yamazaki S, et al. Differential antigen processing by dendritic cell subsets in vivo. *Science* (2007) **315**(5808):107–11. doi:10.1126/science.1136080

88. Cros J, Cagnard N, Woollard K, Patey N, Zhang SY, Senechal B, et al. Human CD14dim monocytes patrol and sense nucleic acids and viruses via TLR7 and TLR8 receptors. *Immunity* (2010) **33**(3):375–86. doi:10.1016/j.immuni.2010.08.012

89. Ingersoll MA, Spanbroek R, Lottaz C, Gautier EL, Frankenberger M, Hoffmann R, et al. Comparison of gene expression profiles between human and mouse monocyte subsets. *Blood* (2010) **115**(3):e10–9. doi:10.1182/blood-2009-07-235028

90. Segura E, Touzot M, Bohineust A, Cappuccio A, Chiocchia G, Hosmalin A, et al. Human inflammatory dendritic cells induce Th17 cell differentiation. *Immunity* (2013) **38**(2):336–48. doi:10.1016/j.immuni.2012.10.018

91. Jaitin DA, Kenigsberg E, Keren-Shaul H, Elefant N, Paul F, Zaretsky I, et al. Massively parallel single-cell RNA-seq for marker-free decomposition of tissues into cell types. *Science* (2014) **343**(6172):776–9. doi:10.1126/science.1247651

92. Usoskin D, Furlan A, Islam S, Abdo H, Lonnerberg P, Lou D, et al. Unbiased classification of sensory neuron types by large-scale single-cell RNA sequencing. *Nat Neurosci* (2015) **18**(1):145–53. doi:10.1038/nn.3881

93. Zeisel A, Manchado ABM, Codeluppi S, Lönnerberg P, La Manno G, Juréus A, et al. Cell types in the mouse cortex and hippocampus revealed by single-cell RNA-seq. *Science* (2015) **347**(6226):1138–42. doi:10.1126/science.aaa1934

94. Schakel K, Kannagi R, Kniep B, Goto Y, Mitsuoka C, Zwirner J, et al. 6-Sulfo LacNAc, a novel carbohydrate modification of PSGL-1, defines an inflammatory type of human dendritic cells. *Immunity* (2002) **17**(3):289–301. doi:10.1016/S1074-7613(02)00393-X

95. Lindstedt M, Lundberg K, Borrebaeck CA. Gene family clustering identifies functionally associated subsets of human in vivo blood and tonsillar dendritic cells. *J Immunol* (2005) **175**(8):4839–46. doi:10.4049/jimmunol.175.8.4839

96. Dobel T, Kunze A, Babatz J, Trankner K, Ludwig A, Schmitz M, et al. FcgammaRIII (CD16) equips immature 6-sulfo LacNAc-expressing dendritic cells (slanDCs) with a unique capacity to handle IgG-complexed antigens. *Blood* (2013) **121**(18):3609–18. doi:10.1182/blood-2012-08-447045

97. Vu Manh TP, Alexandre Y, Baranek T, Crozat K, Dalod M. Plasmacytoid, conventional, and monocyte-derived dendritic cells undergo a profound and convergent genetic reprogramming during their maturation. *Eur J Immunol* (2013) **43**(7):1706–15. doi:10.1002/eji.201243106

98. Reizis B, Bunin A, Ghosh HS, Lewis KL, Sisirak V. Plasmacytoid dendritic cells: recent progress and open questions. *Annu Rev Immunol* (2011) **29**:163–83. doi:10.1146/annurev-immunol-031210-101345

99. Watchmaker PB, Lahl K, Lee M, Baumjohann D, Morton J, Kim SJ, et al. Comparative transcriptional and functional profiling defines conserved programs of intestinal DC differentiation in humans and mice. *Nat Immunol* (2014) **15**(1):98–108. doi:10.1038/ni.2768

100. Lauterbach H, Bathke B, Gilles S, Traidl-Hoffmann C, Luber CA, Fejer G, et al. Mouse CD8alpha+ DCs and human BDCA3+ DCs are major producers of IFN-lambda in response to poly IC. *J Exp Med* (2010) **207**(12):2703–17. doi:10.1084/jem.20092720

101. Lundberg K, Albrekt AS, Nelissen I, Santegoets S, de Gruijl TD, Gibbs S, et al. Transcriptional profiling of human dendritic cell populations and models – unique profiles of in vitro dendritic cells and implications on functionality and applicability. *PLoS One* (2013) **8**(1):e52875. doi:10.1371/journal.pone.0052875

102. Reizis B. Regulation of plasmacytoid dendritic cell development. *Curr Opin Immunol* (2010) **22**(2):206–11. doi:10.1016/j.coi.2010.01.005

103. Villadangos JA, Shortman K. Found in translation: the human equivalent of mouse CD8+ dendritic cells. *J Exp Med* (2010) **207**(6):1131–4. doi:10.1084/jem.20100985

104. Naik SH. Demystifying the development of dendritic cell subtypes, a little. *Immunol Cell Biol* (2008) **86**(5):439–52. doi:10.1038/icb.2008.28

105. Soumelis V, Pattarini L, Michea P, Cappuccio A. Systems approaches to unravel innate immune cell diversity, environmental plasticity and functional specialization. *Curr Opin Immunol* (2015) **32C**:42–7. doi:10.1016/j.coi.2014.12.007

106. Deauvieau F, Ollion V, Doffin AC, Achard C, Fonteneau JF, Verronese E, et al. Human natural killer cells promote cross-presentation of tumor cell-derived antigens by dendritic cells. *Int J Cancer* (2015) **136**(5):1085–94. doi:10.1002/ijc.29087

107. Flinsenberg TW, Compeer EB, Koning D, Klein M, Amelung FJ, van Baarle D, et al. Fcgamma receptor antigen targeting potentiates cross-presentation by human blood and lymphoid tissue BDCA-3+ dendritic cells. *Blood* (2012) **120**(26):5163–72. doi:10.1182/blood-2012-06-434498

108. Nizzoli G, Krietsch J, Weick A, Steinfelder S, Facciotti F, Gruarin P, et al. Human CD1c+ dendritic cells secrete high levels of IL-12 and potently prime cytotoxic T-cell responses. *Blood* (2013) **122**(6):932–42. doi:10.1182/blood-2013-04-495424

109. Onai N, Obata-Onai A, Schmid MA, Ohteki T, Jarrossay D, Manz MG. Identification of clonogenic common Flt3+M-CSFR+ plasmacytoid and conventional dendritic cell progenitors in mouse bone marrow. *Nat Immunol* (2007) **8**(11):1207–16. doi:10.1038/ni1518

110. Naik SH, Sathe P, Park HY, Metcalf D, Proietto AI, Dakic A, et al. Development of plasmacytoid and conventional dendritic cell subtypes from single precursor cells derived in vitro and in vivo. *Nat Immunol* (2007) **8**(11):1217–26. doi:10.1038/ni1522

111. Thordardottir S, Hangalapura BN, Hutten T, Cossu M, Spanholtz J, Schaap N, et al. The aryl hydrocarbon receptor antagonist StemRegenin 1 promotes human plasmacytoid and myeloid dendritic cell development from CD34+ hematopoietic progenitor cells. *Stem Cells Dev* (2014) **23**(9):955–67. doi:10.1089/scd.2013.0521

112. Proietto AI, Mittag D, Roberts AW, Sprigg N, Wu L. The equivalents of human blood and spleen dendritic cell subtypes can be generated in vitro from human CD34(+) stem cells in the presence of fms-like tyrosine kinase 3 ligand and thrombopoietin. *Cell Mol Immunol* (2012) **9**(6):446–54. doi:10.1038/cmi.2012.48

113. Cisse B, Caton ML, Lehner M, Maeda T, Scheu S, Locksley R, et al. Transcription factor E2-2 is an essential and specific regulator of plasmacytoid dendritic cell development. *Cell* (2008) 135(1):37–48. doi:10.1016/j.cell.2008.09.016

114. Sawai CM, Sisirak V, Ghosh HS, Hou EZ, Ceribelli M, Staudt LM, et al. Transcription factor Runx2 controls the development and migration of plasmacytoid dendritic cells. *J Exp Med* (2013) 210(11):2151–9. doi:10.1084/jem.20130443

115. Esashi E, Bao M, Wang YH, Cao W, Liu YJ. PACSIN1 regulates the TLR7/9-mediated type I interferon response in plasmacytoid dendritic cells. *Eur J Immunol* (2012) 42(3):573–9. doi:10.1002/eji.201142045

116. Ippolito GC, Dekker JD, Wang YH, Lee BK, Shaffer AL III, Lin J, et al. Dendritic cell fate is determined by BCL11A. *Proc Natl Acad Sci U S A* (2014) 111(11):E998–1006. doi:10.1073/pnas.1319228111

117. Wu X, Satpathy AT, Kc W, Liu P, Murphy TL, Murphy KM. Bcl11a controls Flt3 expression in early hematopoietic progenitors and is required for pDC development in vivo. *PLoS One* (2013) 8(5):e64800. doi:10.1371/journal.pone.0064800

118. Meredith MM, Liu K, Darrasse-Jeze G, Kamphorst AO, Schreiber HA, Guermonprez P, et al. Expression of the zinc finger transcription factor zDC (Zbtb46, Btbd4) defines the classical dendritic cell lineage. *J Exp Med* (2012) 209(6):1153–65. doi:10.1084/jem.20112675

119. Satpathy AT, Kc W, Albring JC, Edelson BT, Kretzer NM, Bhattacharya D, et al. Zbtb46 expression distinguishes classical dendritic cells and their committed progenitors from other immune lineages. *J Exp Med* (2012) 209(6):1135–52. doi:10.1084/jem.20120030

120. Ohtsuka H, Sakamoto A, Pan J, Inage S, Horigome S, Ichii H, et al. Bcl6 is required for the development of mouse CD4+ and CD8alpha+ dendritic cells. *J Immunol* (2011) 186(1):255–63. doi:10.4049/jimmunol.0903714

121. Hildner K, Edelson BT, Purtha WE, Diamond M, Matsushita H, Kohyama M, et al. Batf3 deficiency reveals a critical role for CD8alpha+ dendritic cells in cytotoxic T cell immunity. *Science* (2008) 322(5904):1097–100. doi:10.1126/science.1164206

122. Nair-Gupta P, Baccarini A, Tung N, Seyffer F, Florey O, Huang Y, et al. TLR signals induce phagosomal MHC-I delivery from the endosomal recycling compartment to allow cross-presentation. *Cell* (2014) 158(3):506–21. doi:10.1016/j.cell.2014.04.054

123. Avram D, Califano D. The multifaceted roles of Bcl11b in thymic and peripheral T cells: impact on immune diseases. *J Immunol* (2014) 193(5):2059–65. doi:10.4049/jimmunol.1400930

124. Fu G, Vallee S, Rybakin V, McGuire MV, Ampudia J, Brockmeyer C, et al. Themis controls thymocyte selection through regulation of T cell antigen receptor-mediated signaling. *Nat Immunol* (2009) 10(8):848–56. doi:10.1038/ni.1766

125. Johnson AL, Aravind L, Shulzhenko N, Morgun A, Choi SY, Crockford TL, et al. Themis is a member of a new metazoan gene family and is required for the completion of thymocyte positive selection. *Nat Immunol* (2009) 10(8):831–9. doi:10.1038/ni.1769

126. Kakugawa K, Yasuda T, Miura I, Kobayashi A, Fukiage H, Satoh R, et al. A novel gene essential for the development of single positive thymocytes. *Mol Cell Biol* (2009) 29(18):5128–35. doi:10.1128/MCB.00793-09

127. Lesourne R, Uehara S, Lee J, Song KD, Li L, Pinkhasov J, et al. Themis, a T cell-specific protein important for late thymocyte development. *Nat Immunol* (2009) 10(8):840–7. doi:10.1038/ni.1768

128. Meredith MM, Liu K, Kamphorst AO, Idoyaga J, Yamane A, Guermonprez P, et al. Zinc finger transcription factor zDC is a negative regulator required to prevent activation of classical dendritic cells in the steady state. *J Exp Med* (2012) 209(9):1583–93. doi:10.1084/jem.20121003

129. Yoshio S, Kanto T, Kuroda S, Matsubara T, Higashitani K, Kakita N, et al. Human blood dendritic cell antigen 3 (BDCA3)(+) dendritic cells are a potent producer of interferon-lambda in response to hepatitis C virus. *Hepatology* (2013) 57(5):1705–15. doi:10.1002/hep.26182

130. Zhang S, Kodys K, Li K, Szabo G. Human type 2 myeloid dendritic cells produce interferon-lambda and amplify interferon-alpha in response to hepatitis C virus infection. *Gastroenterology* (2013) 144(2):414–25e7. doi:10.1053/j.gastro.2012.10.034

131. Cao W, Bover L, Cho M, Wen X, Hanabuchi S, Bao M, et al. Regulation of TLR7/9 responses in plasmacytoid dendritic cells by BST2 and ILT7 receptor interaction. *J Exp Med* (2009) 206(7):1603–14. doi:10.1084/jem.20090547

132. Kolodkin AL, Levengood DV, Rowe EG, Tai YT, Giger RJ, Ginty DD. Neuropilin is a semaphorin III receptor. *Cell* (1997) 90(4):753–62. doi:10.1016/S0092-8674(00)80535-8

133. Soker S, Takashima S, Miao HQ, Neufeld G, Klagsbrun M. Neuropilin-1 is expressed by endothelial and tumor cells as an isoform-specific receptor for vascular endothelial growth factor. *Cell* (1998) 92(6):735–45. doi:10.1016/S0092-8674(00)81402-6

134. Nascimbeni M, Perie L, Chorro L, Diocou S, Kreitmann L, Louis S, et al. Plasmacytoid dendritic cells accumulate in spleens from chronically HIV-infected patients but barely participate in interferon-alpha expression. *Blood* (2009) 113(24):6112–9. doi:10.1182/blood-2008-07-170803

135. Baranek T, Manh TP, Alexandre Y, Maqbool MA, Cabeza JZ, Tomasello E, et al. Differential responses of immune cells to type I interferon contribute to host resistance to viral infection. *Cell Host Microbe* (2012) 12(4):571–84. doi:10.1016/j.chom.2012.09.002

136. Ruscanu S, Jouneau L, Urien C, Bourge M, Lecardonnel J, Moroldo M, et al. Dendritic cell subtypes from lymph nodes and blood show contrasted gene expression programs upon Bluetongue virus infection. *J Virol* (2013) 87(16):9333–43. doi:10.1128/JVI.00631-13

137. Banchereau R, Baldwin N, Cepika AM, Athale S, Xue Y, Yu CI, et al. Transcriptional specialization of human dendritic cell subsets in response to microbial vaccines. *Nat Commun* (2014) 5:5283. doi:10.1038/ncomms6283

138. Shalek AK, Satija R, Adiconis X, Gertner RS, Gaublomme JT, Raychowdhury R, et al. Single-cell transcriptomics reveals bimodality in expression and splicing in immune cells. *Nature* (2013) 498(7453):236–40. doi:10.1038/nature12172

139. Venkatesh B, Lee AP, Ravi V, Maurya AK, Lian MM, Swann JB, et al. Elephant shark genome provides unique insights into gnathostome evolution. *Nature* (2014) 505(7482):174–9. doi:10.1038/nature12826

140. van der Aa E, van Montfoort N, Woltman AM. BDCA3CLEC9A human dendritic cell function and development. *Semin Cell Dev Biol* (2014) (in press). doi:10.1016/j.semcdb.2014.05.016

141. Gallois A, Bhardwaj N. A needle in the 'cancer vaccine' haystack. *Nat Med* (2010) 16(8):854–6. doi:10.1038/nm0810-854

142. Tullett KM, Lahoud MH, Radford KJ. Harnessing human cross-presenting CLEC9A(+)XCR1(+) dendritic cells for immunotherapy. *Front Immunol* (2014) 5:239. doi:10.3389/fimmu.2014.00239

143. Meixlsperger S, Leung CS, Ramer PC, Pack M, Vanoaica LD, Breton G, et al. CD141+ dendritic cells produce prominent amounts of IFN-alpha after dsRNA recognition and can be targeted via DEC-205 in humanized mice. *Blood* (2013) 121(25):5034–44. doi:10.1182/blood-2012-12-473413

144. Sancho D, Joffre OP, Keller AM, Rogers NC, Martinez D, Hernanz-Falcon P, et al. Identification of a dendritic cell receptor that couples sensing of necrosis to immunity. *Nature* (2009) 458(7240):899–903. doi:10.1038/nature07750

145. Zelenay S, Keller AM, Whitney PG, Schraml BU, Deddouche S, Rogers NC, et al. The dendritic cell receptor DNGR-1 controls endocytic handling of necrotic cell antigens to favor cross-priming of CTLs in virus-infected mice. *J Clin Invest* (2012) 122(5):1615–27. doi:10.1172/JCI60644

146. Huysamen C, Willment JA, Dennehy KM, Brown GD. CLEC9A is a novel activation C-type lectin-like receptor expressed on BDCA3+ dendritic cells and a subset of monocytes. *J Biol Chem* (2008) 283(24):16693–701. doi:10.1074/jbc.M709923200

147. Tel J, Aarntzen EH, Baba T, Schreibelt G, Schulte BM, Benitez-Ribas D, et al. Natural human plasmacytoid dendritic cells induce antigen-specific T-cell responses in melanoma patients. *Cancer Res* (2013) 73(3):1063–75. doi:10.1158/0008-5472.CAN-12-2583

148. Tel J, Sittig SP, Blom RA, Cruz LJ, Schreibelt G, Figdor CG, et al. Targeting uptake receptors on human plasmacytoid dendritic cells triggers antigen cross-presentation and robust type I IFN secretion. *J Immunol* (2013) 191(10):5005–12. doi:10.4049/jimmunol.1300787

149. Guery L, Hugues S. Tolerogenic and activatory plasmacytoid dendritic cells in autoimmunity. *Front Immunol* (2013) 4:59. doi:10.3389/fimmu.2013.00059

150. Goubier A, Dubois B, Gheit H, Joubert G, Villard-Truc F, Asselin-Paturel C, et al. Plasmacytoid dendritic cells mediate oral tolerance. *Immunity* (2008) 29(3):464–75. doi:10.1016/j.immuni.2008.06.017

151. Yu CI, Becker C, Wang Y, Marches F, Helft J, Leboeuf M, et al. Human CD1c+ dendritic cells drive the differentiation of CD103+ CD8+ mucosal

effector T cells via the cytokine TGF-beta. *Immunity* (2013) **38**(4):818–30. doi:10.1016/j.immuni.2013.03.004

152. Fuertes MB, Kacha AK, Kline J, Woo SR, Kranz DM, Murphy KM, et al. Host type I IFN signals are required for antitumor CD8+ T cell responses through CD8{alpha}+ dendritic cells. *J Exp Med* (2011) **208**(10):2005–16. doi:10.1084/jem.20101159

153. Diamond MS, Kinder M, Matsushita H, Mashayekhi M, Dunn GP, Archambault JM, et al. Type I interferon is selectively required by dendritic cells for immune rejection of tumors. *J Exp Med* (2011) **208**(10):1989–2003. doi:10.1084/jem.20101158

19

Human macrophages clear the biovar microtus strain of *Yersinia pestis* more efficiently than murine macrophages

Qingwen Zhang [1†], Youquan Xin [1†], Haihong Zhao [1], Rongjiao Liu [2], Xiaoqing Xu [1], Yanfeng Yan [2], Zhipeng Kong [1], Tong Wang [2], Zhizhen Qi [1], Qi Zhang [1], Yang You [2], Yajun Song [2], Yujun Cui [2], Ruifu Yang [2*], Xuefei Zhang [1*] and Zongmin Du [2*]

[1] Qinghai Institute for Endemic Disease Prevention and Control, Xining, China, [2] State Key Laboratory of Pathogen and Biosecurity, Beijing Institute of Microbiology and Epidemiology, Beijing, China

Edited by:
Hasan Zaki,
University of Texas Southwestern
Medical Center, United States

Reviewed by:
Roger Derek Pechous,
University of Arkansas for Medical
Sciences, United States
Yuan He,
Wayne State School of Medicine,
United States

*Correspondence:
Zongmin Du
zmduams@163.com
Ruifu Yang
ruifuyang@gmail.com
Xuefei Zhang
1730326847@qq.com

[†] These authors have contributed
equally to this work

Yersinia pestis is the etiological agent of the notorious plague that has claimed millions of deaths in history. Of the four known *Y. pestis* biovars (Antiqua, Medievalis, Orientalis, and Microtus), Microtus strains are unique for being highly virulent in mice but avirulent in humans. Here, human peripheral lymphocytes were infected with the fully virulent 141 strain or the Microtus strain 201, and their transcriptomes were determined and compared. The most notable finding was that robust responses in the pathways for cytokine-cytokine receptor interaction, chemokine signaling pathway, Toll-like receptor signaling and Jak-STAT signaling were induced at 2 h post infection (hpi) in the 201- but not the 141-infected lymphocytes, suggesting that human lymphocytes might be able to constrain infections caused by strain 201 but not 141. Consistent with the transcriptome results, much higher IFN-γ and IL-1β were present in the supernatants from the 201-infected lymphocytes, while inflammatory inhibitory IL-10 levels were higher in the 141-infected lymphocytes. The expressions of CSTD and SLC11A1, both of which are functional components of the lysosome, increased in the 201-infected human macrophage-like U937 cells. Further assessment of the survival rate of the 201 bacilli in the U937 cells and murine macrophage RAW 264.7 cells revealed no viable bacteria in the U937 cells at 32 hpi.; however, about 5–10% of the bacteria were still alive in the RAW264.7 cells. Our results indicate that human macrophages can clear the intracellular *Y. pestis* 201 bacilli more efficiently than murine macrophages, probably by interfering with critical host immune responses, and this could partially account for the host-specific pathogenicity of *Y. pestis* Microtus strains.

Keywords: *Yersinia pestis*, host-specific pathogenicity, biovar microtus, human lymphocyte, transcriptomes

INTRODUCTIONS

Yersina pestis, the causative agent of plague, is responsible for three historical pandemics including the notorious Black Death in medieval Europe (Perry and Fetherston, 1997; Butler, 2009). This lethal pathogen manifests itself as three main clinical forms: bubonic plague, pneumonic plague and septicemia plague (Perry and Fetherston, 1997). Pneumonic plague, the most serious form

Human macrophages clear the biovar microtus strain of Yersinia pestis more efficiently than murine...

217

of the disease, leads to the certain death of victims if not properly treated in time. *Y. pestis* is a highly contagious pathogen that can spread quickly among closely contacted individuals by airborne droplets, thereby posing the potential threat of disease outbreaks in populations under suitable circumstances, which could include bioterrorism (Inglesby et al., 2000). Different *Y. pestis* strains are classified into four biovars (Antiqua, Mediaevalis, Orientalis and Microtus) according to their phenotypic properties (Zhou et al., 2004). Since the announcement of whole genome sequences of *Y. pestis* CO92 (biovar Orientalis) and KIM (biovars Mediaevalis) (Parkhill et al., 2001; Deng et al., 2002), the whole genome sequence data for more than 30 different *Y. pestis* strains, including each of the four biovars, have been released so far (https://www.ncbi.nlm.nih.gov/). Biovar Microtus was recently proposed to be a novel *Y. pestis* biovar because strains isolated from Microtus-related plague foci in China are lethal to microtus and other small rodents but avirulent to larger mammals and human (Fan et al., 1995), and unlike strains belonging to the three classical biovars, those strains can reduce nitrate, ferment rhamnose and melibiose, but not arabinose (Zhou et al., 2004). Biovar Microtus strains are distributed in the *Microtus brandti* plague focus of the Xilin Gol Grassland and the *M. fuscus* plague focus of the Qinghai-Tibet Plateau in China. However, since the first isolation of *Y. pestis* strains in the 1970s from the Microtus-related foci in China, no cases of human plague have been reported to be linked to these strains although serious enzootic plague epidemics have occurred every few years in the same areas. Human volunteers were subcutaneously inoculated with 1.5×10^7 colony forming units (CFU) of the *Y. pestis* strains isolated from Microtus-related foci and the results confirmed for the first time that these strains are avirulent in humans (Fan et al., 1995).

It has been established that *Y. pestis* has evolved recently from *Y. pseudotuberculosis* and only limited genetic diversity has been found among the different *Y. pestis* strains (Achtman et al., 1999, 2004). The genome composition of the different *Y. pestis* biovars are very similar, although there is evidence of frequent events of gene rearrangement, large gene fragments deletion, gene loss, and gene inactivation among those genomes (Song et al., 2004). However, despite the monumental advances in comparative genomics and proteomics that have taken place over recent years, we are no further in elucidating the mechanisms underlying the host-specific pathogenicity of the Microtus strains (Song et al., 2004; Zhou et al., 2012). Therefore, in the present study, human peripheral lymphocytes were infected with the fully virulent biovar Antiqua strain 141 or the human avirulent Microtus strain 201, and their transcriptomes were compared. We observed significantly different responses in, for example, the cytokine-cytokine receptor interaction pathway, chemokine signaling pathway, Toll-like receptor (TLR) signaling pathway, and lysosome pathways between the 201- and 141- infected lymphocytes, suggesting that the ability of human lymphocytes to restrict infection caused by strain 201 or 141 might be very different. IFN-γ and IL-1β, which are recognized to benefit the host defenses against plague, were present at significantly higher levels in the supernatant from the 201-infected lymphocytes, while levels of the inhibitory IL-10 inflammatory cytokines were higher in 141-infected lymphocytes. Further assessment

of the survival of 201 bacilli in human macrophage-like U937 cells and murine macrophage RAW 264.7 cells revealed that human macrophages clear intracellular *Y. pestis* 201 bacilli more efficiently than mice macrophages do. Our results showed that human macrophages exhibit much higher bactericidal activities than mouse macrophages and this finding might partially account for the host-specific pathogenicity of *Y. pestis* Microtus strains.

RESULTS

Overview of the Comparative Transcriptome of Human Peripheral Lymphocytes Infected With Different *Y. pestis* Biovars Strains

Human peripheral lymphocytes were infected with the human-avirulent Microtus 201 strain or the fully virulent Antiqua strain 141 each at a multiplicity of infection (MOI) of 10. The infected lymphocytes harvested at 2, 4, and 8 h post-infection (hpi) were subjected to the total RNA isolation and RNA sequencing (RNA-seq) analysis. Clean reads from the RNA-seq analysis were mapped to the annotated human genome (NCBI36/hg18). Each library comprised around 4.5×10^7 clean reads, 62.83% of which were on average mapped uniquely to the human reference genome (**Table 1**). Gene expression was calculated using the fragments per kilobase of gene per million fragments mapped (FPKM) method (Mortazavi et al., 2008), and the genes whose expression levels were altered by over 2-fold in expression level with false discovery rates (FDRs) ≤ 0.001 were defined as being the differentially expressed genes (Benjamini and Yekutieli, 2005).

First, the transcriptomes of the peripheral lymphocytes infected with strains 201 or 141 were compared with that of the uninfected lymphocytes, respectively. Differentially expressed genes were identified according to the aforementioned criteria and the numbers of genes at each time point post-infection were calculated. More than 260 genes were differentially expressed at 2 hpi between the 201-infected and uninfected lymphocytes, and the number increased greatly as the infection period proceeded and exceeded 1,600 at 8 hpi (**Figure 1A**). In sharp contrast, the transcriptome of the 141-infected peripheral lymphocytes differed very little from that of the uninfected cells at 2 hpi, and fewer than 20 genes were found to be differentially expressed, although the numbers of differentially expressed genes at 8 hpi in the 141- and 201-infected cells were high and comparable (**Figure 1B**).

Further comparative analysis showed once again that the most remarkable gene expression changes between the 201- and 141-infected peripheral lymphocytes occurred at the earliest time point that was analyzed (**Figure 1C**). The differentially expressed genes were relatively fewer at 8 hpi although they were quite abundant in the 201-infected (**Figure 1A**) and the 141-infected (**Figure 1B**) lymphocytes in comparison with the uninfected lymphocytes control. These data demonstrate that the initial contact by the human peripheral lymphocytes with the two *Y. pestis* strains triggered two obviously distinct responses, one

TABLE 1 | RNA-seq library descriptions.

ID	Sample name	Clean reads[a]	Unique match (percent)[b]	Total unmapped reads (percent)[c]	Expressed gene
1	141-2h	44641336	29202733 (65.42%)	13240983 (29.69%)	15462
2	141-4h	45138004	29179029 (64.64%)	13934344 (30.87%)	15477
3	141-8h	44789942	28803453 (64.31%)	14196597 (31.70%)	15568
4	201-2h	44201980	28907751 (65.40%)	13441594 (30.41%)	15547
5	201-4h	44294928	27656449 (62.44%)	13662809 (30.85%)	15506
6	201-8h	44851572	27616280 (61.57%)	14148314 (31.54%)	15570
7	Normal	44701980	28449706 (63.64%)	13266250 (29.68%)	15512

[a]Clean reads represent the number of high quality clean reads.
[b]Unique matches denotes the reads that mapped to unique positions in the reference human genome.
[c]Total unmapped reads denotes the number of reads that couldn't be mapped to the reference genome.
The percentages of the unique matched reads and the unmapped reads were calculated by dividing the values of the unique or unmapped reads by those of the clean reads of clean reads × 100 (values shown in brackets).

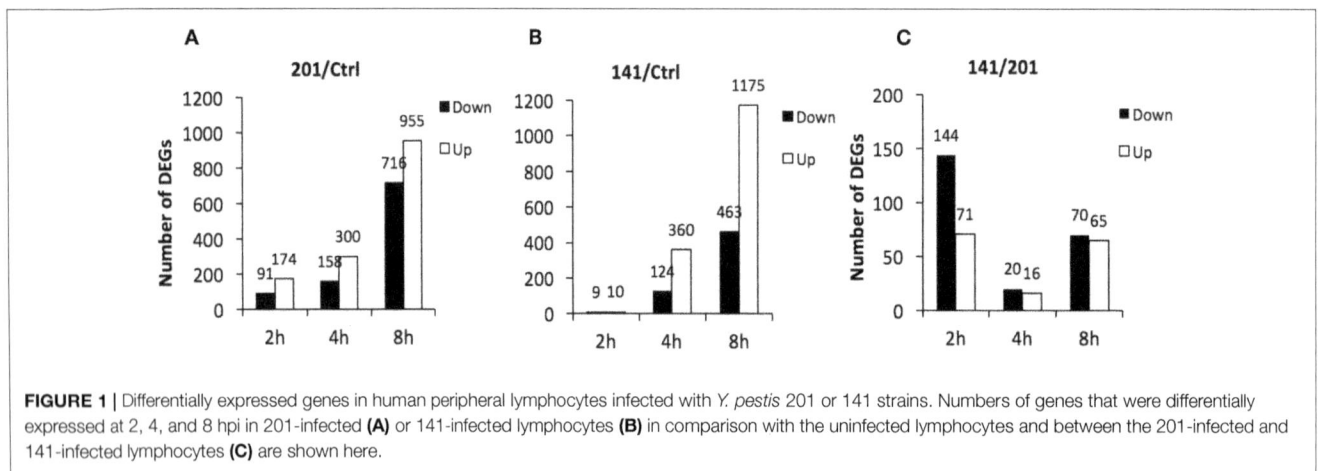

FIGURE 1 | Differentially expressed genes in human peripheral lymphocytes infected with Y. pestis 201 or 141 strains. Numbers of genes that were differentially expressed at 2, 4, and 8 hpi in 201-infected **(A)** or 141-infected lymphocytes **(B)** in comparison with the uninfected lymphocytes and between the 201-infected and 141-infected lymphocytes **(C)** are shown here.

of which could possibly facilitate efficient recognition of 201 bacilli and boost strong immune responses, while the other failed to do so in the case of infection with *Y. pestis* strain 141. It appeared that the 141-infected lymphocytes were strongly inhibited in their ability to deploy an inflammatory immune response because of a vital virulence mechanism present in strain 141, but absent in strain 201, which might originate from some as yet uncharacterized difference(s) between the two strains.

To validate the RNA-seq data, primer pairs for 23 genes were designed for quantitative RT-PCR analysis (**Supplementary Table 1**), among which 12 genes were up-regulated, 7 genes down-regulated and 3 unchanged in the 201-infected lymphocytes in comparison with the uninfected control according to the RNA-seq results. qRT-PCR was performed using the same RNA samples used in RNA-seq library construction as templates, 16s RNA was the reference gene for data normalization, and the log2 ratios of the target gene concentrations in the 201-infected lymphocytes to those of the uninfected control were calculated. The correlation efficiencies (R^2) between the qRT-PCR and RNA-seq analysis results were over 0.90 for the RNA samples collected at 2 and 8 hpi, and 0.83 for the samples collected at 4 hpi (**Supplementary Figure 1** and **Table 2**). These results indicate that the RNA-seq data described in this study are highly reliable.

Multiple Immune Response Pathways Differ in Their Responses to Infection With Strains 201 or 141

To characterize the host transcriptomic response to the human avirulent Microtus strain 201, genes differentially expressed in 201- or 141-infected human peripheral lymphocytes, compared with the uninfected cells, were subjected to the pathway enrichment analysis using the DAVID 6.7 bioinformatics tool (http://david.abcc.ncifcrf.gov) (Huang da et al., 2009). Consistent with the observation that the genes differentially expressed between 201- and 141-infected lymphocytes peaked at 2 hpi, several pathways critical for inducing the intense immune responses required for invading pathogen clearance were significantly enriched in the 201- but not the 141-infected lymphocytes ($p < 0.05$, Fisher's exact test followed by the Bonferroni multiple testing correction) at this time point. These pathways included the cytokine-cytokine receptor interaction pathway, the chemokine signaling pathway, the TLR signaling pathway, and the Jak-STAT signaling pathway, among others (**Table 2**). **Figure 2** shows the heat maps for some of the differentially enriched pathway in details. Strikingly, at the later infection time points (4 and 8 hpi), the pathway enrichment analysis revealed a significant difference between the

Human macrophages clear the biovar microtus strain of Yersinia pestis more efficiently than murine...

219

TABLE 2 | Pathway enrichment analysis of the differentially expressed genes in 201- or 141-infeced peripheral lymphocytes.

Pathway term	201-infected			141-infected		
	2h	4h	8h	2h	4h	8h
hsa04060:Cytokine-cytokine receptor interaction	37	29	70		34	76
hsa04620:Toll-like receptor signaling pathway	13					
hsa04062:Chemokine signaling pathway	16		40		16	45
hsa04630:Jak-STAT signaling pathway	12	16			19	
hsa04640:Hematopoietic cell lineage	9		28			28
hsa04142:Lysosome			30			
hsa04666:Fc gamma R-mediated phagocytosis					11	
hsa04512:ECM-receptor interaction			22		10	
hsa04621:NOD-like receptor signaling pathway			18			21
hsa04672:Intestinal immune network for IgA production			15			16
hsa05330:Allograft rejection			12			
hsa05310:Asthma						13

Pathway enrichment analysis was performed using DAVID 6.7 bioinformatics tool (http://david.abcc.ncifcrf.gov) (Huang da et al., 2009). Only pathways that were significantly (p < 0.05, Fisher's exact test followed by the Bonferroni multiple testing correction) enriched at any time points in the 201- or 141- infected lymphocytes are shown here. Numbers in the table represents the differentially expressed gene numbers in a specific pathway at the indicated time point.

201- and 141-infected lymphocytes in the lysosome pathway, the Fcγ receptor-mediated phagocytosis pathway and the extracellular-matrix (ECM) receptor pathway, suggesting that the phagocytosis and lysosome maturation process could be differentially modulated in the lymphocytes infected with the two different strains.

Pathogens invading the mammalian host are recognized by pathogen recognition receptors (PRRs) via sensing of the conserved ligands on microorganisms (Takeuchi and Akira, 2010), and TLRs represent the major PRRs responsible for bacterial pathogen recognition (Krishnan et al., 2007). The signaling pathways triggered by TLRs exhibit distinct characteristics in several key molecules (**Supplementary Figure 2A**). CD14 protein binds to lipopolysaccharide (LPS), which after transfer to the TLR4/MD-2 complex triggers the downstream signaling required for the production of pro-inflammatory cytokines and type I interferon (Leeuwenberg et al., 1994; Ishihara et al., 2004). TLR adaptor molecule 2 (TICAM2) is an adaptor that transduces signals upon ligand binding to TLR4. Interferon regulatory factor 5 (IRF5) is a transcriptional factor involved in activation of interferon and the immune system. TLR8 recognizes single-stranded RNA from both viruses and bacteria (Sarvestani et al., 2012). The expression levels of *CD14, TLR8, TICAM2,* and *IRF5* were much higher in the 201-infected lymphocytes than in their 141-infected counterparts at 2 hpi, a finding further confirmed by the qRT-PCR analysis (**Supplementary Figure 2B**), although their expression levels after this time point gradually became comparable in the 201- and 141-infected lymphocytes.

As a facultative intracellular pathogen, *Y. pestis* is readily taken up by phagocytes but the pathogen becomes phagocytosis-resistant shortly after its initial intracellular survival and replication (Pujol et al., 2009; Connor et al., 2015). Thus, in systemic infection of the mammalian hosts, the survival of *Y. pestis* in host macrophages at the early stage of infection

are critical for the later disease progress. Our results revealed that the differentially expressed genes were significantly enriched in the lysosome pathway in 201-infected lymphocytes, and 30 genes were significantly changed at 8 hpi (**Table 2**) but not in the 141-infected cells. Although we saw no significant enrichment in this pathway at 2 hpi, several key genes (e.g., *ATP6V0A1, CYSZ, CTSD* and *SLC11A1*) involved in lysosome formation and maturation were expressed at higher levels in the 201-infected lymphocytes at 2 hpi than in the 141-infected cells. *ATP6V0A1* encodes a vacuolar ATPase that mediates acidification of intracellular organelles, and vacuolar acidification is necessary for zymogen activation in the lysosome (Saw et al., 2011). CTSD protein exhibits pepsin-like activity, whereas CTSZ is a lysosome cysteine protease that has been shown to have lysosomal degradative capacity (del Cerro-Vadillo et al., 2006; Liu et al., 2014; Tranchemontagne et al., 2016). SLC11A1 functions as a divalent transition metal transporter involved in host resistance to infection with *Mycobacterium tuberculosis* and *M. leprae*, and also is associated with some inflammatory diseases (Govoni et al., 1999). qRT-PCR analysis of the aforementioned genes expressed in the 201- and 141-infected lymphocytes (**Supplementary Figure 3**) confirmed the RNA-seq results that the genes were expressed at higher levels in the 201-infected lymphocytes than in their 141-infected counterparts, suggesting that the lysosome pathway response differed between the infection with strains 201 and 141.

Cytokine Secretion in Human Lymphocytes Infected With *Y. pestis* Strain 201 Differ Significantly From Those Infected With Strain 141

Next, we sought to compare cytokines and chemokines secretion in lymphocytes infected with the two different strains. Culture supernatants from the infected lymphocytes were collected and

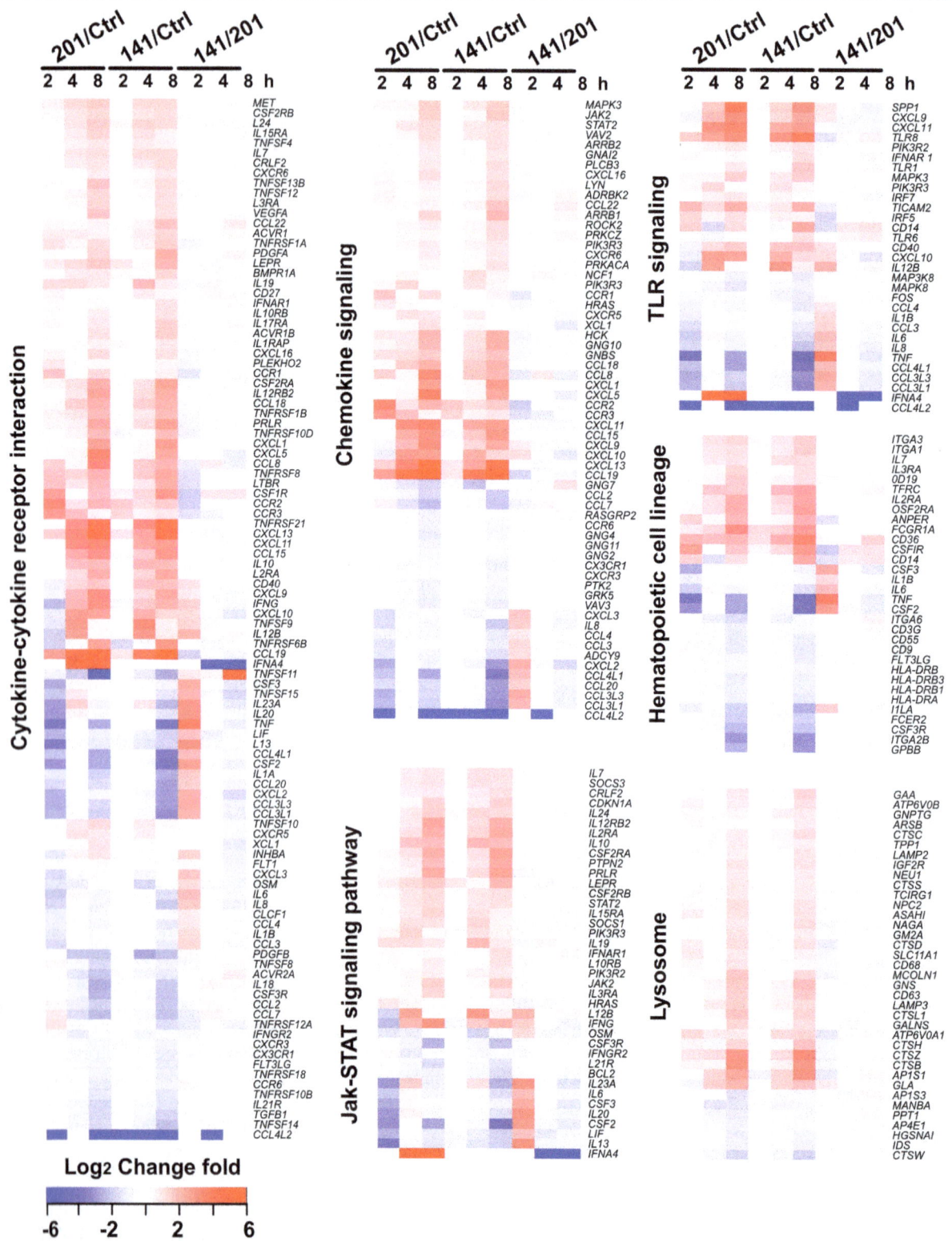

FIGURE 2 | Gene expression patterns of the pathways that showing the significant differences between the *Y. pestis* 201- and 141-infected human peripheral lymphocytes. Some of the critical cellular pathways that were significantly enriched in the genes differentially expressed at 2, 4, or 8 hpi in 201- and 141-infected lymphocytes are shown ($p < 0.05$, Fisher's exact test followed by the Benjamini multiple testing correction). Changes of the differentially expressed genes are presented in different colors as shown in the color key.

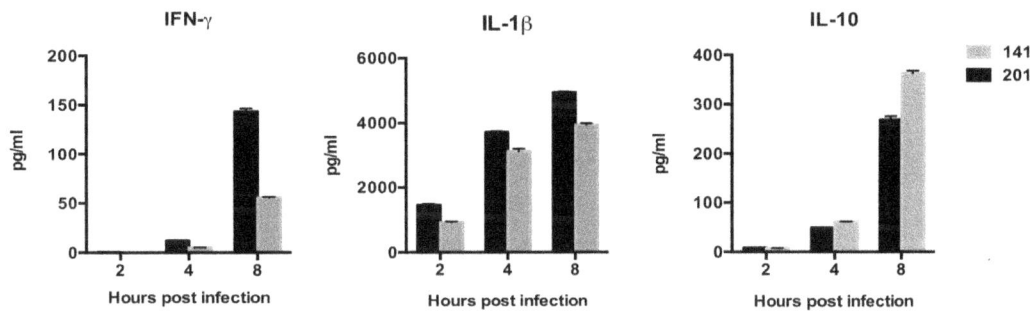

FIGURE 3 | Human peripheral lymphocytes infected with the *Y. pestis* 201 strain secreted more abundant IFN-γ and IL-1β, and less immunosuppressive IL-10 than those infected with the 141 strain. Culture supernatants from the infected lymphocytes were collected at 2, 4, and 8 hpi and analyzed for the different cytokines using BD^{TM} CBA Kits. Significantly higher IFN-γ and IL-1β levels were present in the supernatant from the 201-infected lymphocytes, while inflammatory inhibitory IL-10 was higher in the 141-infected lymphocytes at 8 hpi. Figures were drawn using Graphpad Prism 5.0, and the statistical significance of the differences in the cytokine levels between the different groups of samples were analyzed by two-way ANOVA analysis followed by Bonferroni post-tests.

analyzed for IL-8, IL-1β, IL-6, IL-10, TNF-α, IL-12p, Interferon-γ (IFN-γ), IL-17A using BD^{TM} CBA Kits. Only the cytokines and chemokines that showed statistically significant differences ($p < 0.05$) were taken into account in the following analysis (**Figure 3**). The immunological measurements results showed that the 201-infected lymphocytes secreted much more IFN-γ and IL-1β than those infected with strain 141. IFN-γ is critical for protection against plague and treatment with IFN-γ greatly enhances the bactericidal activity of macrophages against the *Y. pestis* bacterium (Pujol et al., 2005; Parent et al., 2006). *Y. pestis* is known to be able to suppress the production of inflammatory cytokines IL-1β to promote its infection process (Ratner et al., 2016). Production of IFN-γ and IL-1β, both of which paly essential roles in host immune responses, appear to be significantly inhibited in the presence of strain 141 in comparison with the human avirulent strain 201. By contrast, secretion of the immunosuppressive cytokine Interleukin-10 (IL-10) was significantly higher in the 141-infected lymphocytes, unlike their 201-infected counterparts, which accords with previous reports that IL-10-deficient mice are resistant to *Y. pestis* (Turner et al., 2009).

Expression Analysis for Three Essential Proteins in Functional Lysosome in U937 and RAW264.7 Cells Infected With the 201 Strain

The immediate up-regulation of the proteins involved in lysosome activation in 201-infected human lymphocytes promoted us to determine their expression levels after infection with *Y. pestis* strain 201. Human macrophage-like U937 cells primed with phorbol myristate acetate (PMA) and murine macrophage RAW264.7 cells were infected with *Y. pestis* strain 201, and the infected cells collected at 2, 4, 8 hpi were lysed for the immunoblotting detection of ATP6V0A1, SLC11A1, CTSD and CTSZ. Primed U937 cells possess major macrophage characteristics and can therefore resemble lysosome activation quite well in the infected host phagocytes (Koren et al., 1979; Grabenstein et al., 2006). With the exception of ATP6V0A1,

whose lack of detection was probably related to its low abundance, all of the other three proteins were successfully detected (**Figure 4**). The levels of CTSD seemed to be slightly increased in the U937 cells after infection with the 201 strain, and the levels of CTSZ showed no significant alterations during the infection. Because experiments involving human virulent *Y. pestis* strains must be performed at biosafety level 3 laboratory, we were unable to obtain the expression levels of these proteins in 141-infected lymphocytes. However, interestingly, when the RAW264.7 cells were infected with *Y. pestis* strain 201, both the basal expression of CTSD and those after infection were quite low compared with that of the U937 cells, indicating that very low expression of this protease occurred under these conditions. The expression levels of SLC11A1 in both U937 and RAW264.7 cells were significantly increased after infection, and the induction of SLC11A1 in RAW264.7 seemed to be even stronger. The different expression patterns of CTSD and SLC11A1 in U937 and RAW264.7 hinted us that the lysosome activity toward *Y. pestis* strain 201 could be distinctive, which is likely relevant to the host-specific pathogenicity of Microtus strains of *Y. pestis*. However, no significant alteration in CTSZ expression was seen in both U937 and RAW264.7 cells although its' transcript level showed > 10-fold abundance at 8 hpi in the human lymphocytes, probably related to some post-transcriptional regulatory mechanisms.

The Survival Rate of *Y. pestis* Strain 201 Is Significantly Lower in U937 Cells Than in RAW264.7 Cells

Because U937 and RAW264.7 cells seemed to respond to 201 and 141 strains differentially in respect of lysosome activation, we wanted to further characterize the interactions of *Y. pestis* strain 201 with these cell lines. U937 and RAW264.7 cells were infected with the bacteria and their survival percentages were measured using a gentamycin protection assay followed by plating cell lysates on agar plates. The U937 cells were stimulated with PMA (100 ng/mL) for 48 h before they were infected. Both U937 and RAW264.7 cells were infected with the *Y. pestis* 201 strain at

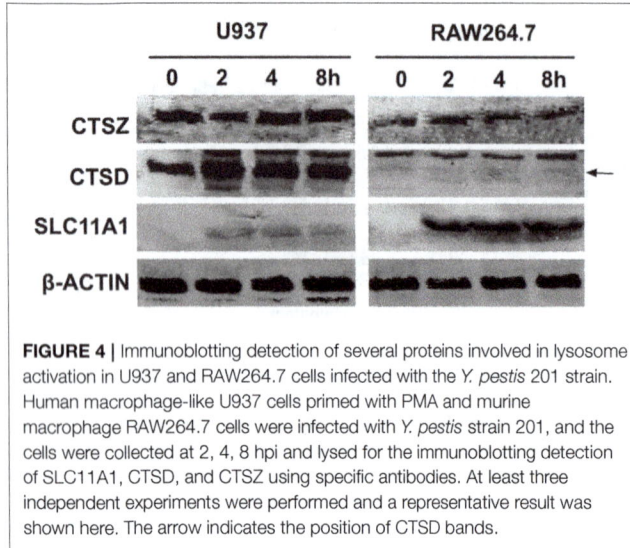

FIGURE 4 | Immunoblotting detection of several proteins involved in lysosome activation in U937 and RAW264.7 cells infected with the *Y. pestis* 201 strain. Human macrophage-like U937 cells primed with PMA and murine macrophage RAW264.7 cells were infected with *Y. pestis* strain 201, and the cells were collected at 2, 4, 8 hpi and lysed for the immunoblotting detection of SLC11A1, CTSD, and CTSZ using specific antibodies. At least three independent experiments were performed and a representative result was shown here. The arrow indicates the position of CTSD bands.

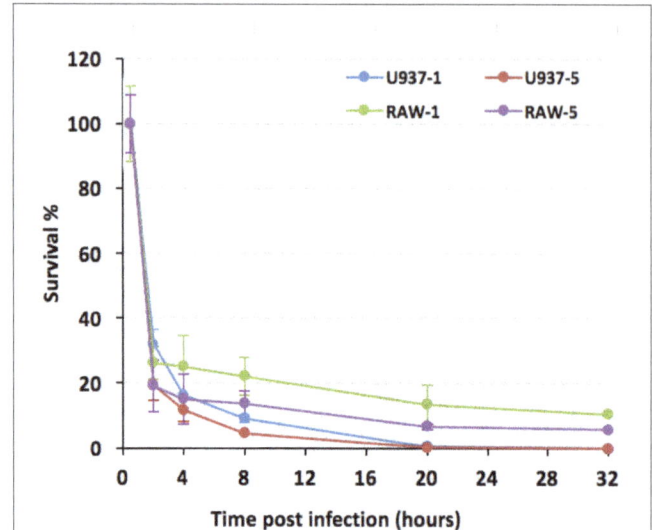

FIGURE 5 | *Y. pestis* strain 201 survival rates were significantly lower in U937 than in RAW264.7 cells. PMA primed U937 and RAW264.7 cells were infected with *Y. pestis* 201 strain at an MOI of 1 or 5, and the living bacteria inside the cells at 2, 4, 8, 20, and 32 hpi were counted by plating the cell lysates onto the agar plates. Experiments were performed in triplicates for three independent times and the similar results were obtained. The survival percentages of the bacteria, as based on a representative result, are shown in average and standard deviation.

a MOI of 1 or 5, and the infections were allowed to proceed for 32 h. The number of live bacteria in U937 and RAW264.7 cells at 0.5 h hpi were designated as the initial value and the survival percent at 2, 4, 8, 20, and 32 hpi were calculated by dividing the corresponding live bacterial cell number to that at 0.5 hpi. The percentage survival of *Y. pestis* bacilli decreased drastically to about 25–30% in both types of cells at 2 hpi at a MOI of 1. This percentage was sustained in RAW264.7 cells at a relatively steady level, dropping only 4% from 2 to 8 hpi; however, the bacteria inside the U937 cells were cleared much faster, as shown by the prominent decrease from 32 to 9% during this time period. At 32 hpi, no living 201 bacteria were detected in the U937 cells, whereas 10.6% of the bacteria were still alive in the RAW264.7 cells (**Figure 5**). When the cells were infected with *Y. pestis* 201 at an MOI of 5, similar results were found, where the percentage survival of the 201 bacteria in the RAW264.7 cells was significantly higher than that in the U937 cells, indicating that human macrophages can clear the *Y. pestis* Microtus strain more efficiently than the murine macrophages.

MATERIALS AND METHODS

Cell Cultures, Bacterial Strains and Infections

Peripheral lymphocytes were isolated from the whole blood collected from 8 healthy donors using Ficoll-Paque Plus (GE Healthcare, USA). The purified lymphocytes were maintained in RPMI-1640 medium supplemented with 10% fetal bovine serum (FBS) and 2 mmol/L of L-glutamine at 37°C in a 5% CO_2 incubator. *Y. pestis* 141 strain is a biovar Antiqua strain, highly virulent in mice and human. *Y. pestis* strain 201 belongs to the biovar Microtus and is highly virulent in mice but avirulent in humans. The glycerol-preserved strains were inoculated into 5 mL of brain heart infusion (BHI) medium and allowed to grow at 26°C for 24 h. The strains recovered were subjected to three consecutive passages and the third passage cultures were grown until they approached an optical

density the (OD) $_{600nm}$ value equal to 1. Bacterial cultures were centrifuged and resuspened in RPMI-1640. The purified human peripheral lymphocytes were infected with the bacterial suspensions of 141 or 201 strains at an MOI of 10, and the infected cells were centrifuged briefly to promote the adhesion of the bacteria to the lymphocytes before incubation at 37°C in a 5% CO_2 incubator. Gentamycin at 100 µg/ml was added after 0.5 h of infection to kill the extracellular bacteria in order to inhibit the proliferation of the bacteria during infection. At the 0.5, 2, 4, and 8 hpi, the culture supernatants from the 201- or 141-infected lymphocytes were collected and centrifuged at 4.000 rpm for 5 min to remove the floating bacteria and cell debris, and the cleared supernatants were stored in small aliquots for the cytokine measurements. The infected lymphocytes were collected and subjected to total RNA isolation.

RNA Isolation, Sequencing and Data Analysis

Total RNAs were purified from the lymphocytes infected with strains 201 or 141 using a TRIzol reagent (Invitrogen, Carlsbad, CA) and the isolated RNAs were treated with RNAse free DNase I (Thermo Fisher Scientific, Waltham, MA). The RNA integrity number, rRNA ratio (28S/18S), and the total RNA concentration in the samples were determined according to the previous descriptions (Du et al., 2014; Yang et al., 2017). Sequencing libraries were prepared using the Illumina Truseq RNA sample preparation kit according to the manufacturer's protocol. Briefly, the first-strand cDNA was generated by First

Human macrophages clear the biovar microtus strain of Yersinia pestis more efficiently than murine...

223

Strand Master Mix and SuperScript II reverse transcription using random primers and the second-strand cDNA was synthesized using DNA polymerase I and RNase H. The fragmented cDNA molecules were end-repaired, purified, and then ligated with PloyA. The sequencing was performed using Illumina HiSeq 2000 (Illumina Inc., USA). The original image data is converted into "raw reads" via base calling. To obtain clean reads of acceptable quality, raw data filtering and quality control were performed to remove the adaptors-containing reads, unknown bases amounting to more than 10%, or low quality bases. The Burrows-Wheeler alignment tool was used to map the clean reads to the human reference genome (NCBI36/hg18) (Li and Durbin, 2009), and Bowtie was used to map to the reference genes (Langmead et al., 2009). Genes and isoform expression levels were quantified by the RSEM software package. We used the FPKM method to calculate the gene expression levels (Mortazavi et al., 2008). The FDR was computed by dividing the number of falsely discovered genes at a given p-value by the number of statistically significant differentially expressed genes comparing the sample to the control at the same p-value (Benjamini and Yekutieli, 2005). The criteria used for differentially expressed genes (DEG) were FDR $\leq$ 0.001 and log2 (FPKM in the test sample/FPKM in the reference) $\geq$ 1. All the data discussed here have been deposited to GEO at NCBI with the accession number of GSE121084.

Quantitative Reverse Transcription PCR (qRT-PCR) Analysis

To confirm gene expression levels by qRT-PCR, RNA samples that were subjected to the RNA-sequencing were used as templets and the qRT-PCR analysis were performed based on SYBR Green I fluorescence using Roche Light Cycler 480. RNA samples from the uninfected, 201- or 141-infected lymphocytes collected at the 2, 4 and 8 hpi were used as templates with primer pairs to amplify the corresponding target genes (**Supplementary Table 2**). Correlations between the sequencing data and the qRT PCR results were calculated using linear regression methodology.

Cytokine Analysis

The cytokines present in the culture supernatants from human lymphocytes infected with 201 or 141 strains for 2, 4, and 8 h were detected using the BD™ Cytometric Bead Array (CBA) Human Inflammatory Cytokine Kit and Human Th1/Th2/Th17 Kit (BD Biosciences, San Jose, CA). The cytokines concentrations were measured by a Becton-Dickinson FACS Caliber flow cytometer. Cytokine levels were processed using Graphpad Prism 5.0, and the statistical significance of the differences in the cytokine levels between the different groups of samples were analyzed by two-way ANOVA analysis followed by Bonferroni post-tests.

Survival Capabilities of *Y. pestis* Strain 201 and 141 in U937 Cell and RAW264.7 Cells

U937 and RAW264.7 cells were maintained in RPMI 1640 medium containing 10% FBS and 2 mmol/L of L-glutamine at

37°C in a 5% CO_2 incubator. U937 cells were primed with PMA (100 ng/mL) for 48 hours before infection. U937 and RAW264.7 cells were seeded onto 24-well plates at a concentration of 4 $\times$ 10^5/mL the day before infection. *Y. pestis* strains 201 and 141 were grown in BHI until they reached OD_{600nm} =1, and each were collected separately by centrifugation and resuspendend in RPMI 1640. The U937 cells or RAW264.7 cells were then infected with strains 201 or 141 at an MOI of 1 or 5, and gentamycin at 100 μg/ml was added to the cells to kill the extracellular bacteria after 0.5 h of infection. At 0.5, 2, 4, 8, 12, 20, 32 hpi, the culture medium from each well was decanted, the cells were thoroughly washed in phosphate-saline buffered saline (PBS), and the infected cells were lysed by addition of sterile H_2O containing 0.1% Triton X-100 for 15 min at room temperature to release the engulfed bacteria. The living bacteria were counted by plating the diluted cell lysate onto to the agar plates in triplicate. This experiment was performed independently three times as biological triplicates and a representative result is shown as the mean $\pm$ SD ($n = 3$).

Immunoblotting Detection of Proteins Involved in Lysosome Pathway in U937 Cell and RAW264.7 Cells

U937 cells were PMA-primed before they were infected with the *Y. pestis* strain as described above. U937 cells and RAW264.7 cells were maintained, seeded and infected with *Y. pestis* strains 201 and 141 at an MOI = 10, according to the method described in the survival capability assays. Gentamycin was added to the cultures to inhibit the over proliferation of the bacteria at 0.5 hpi. U937 and RAW264.7 cells were harvested at 2, 4, and 8 hpi, washed with ice-cold PBS, and lysed with lysis buffer (50 mM Tris-HCl [pH 7.4], 150 mM NaCl, 1% TritonX-100, 1% sodium deoxycholate, 2 mM sodium pyrophosphate, 1 mM EDTA) supplemented with a protease inhibitor complete cocktail (Roche, Basel, Switzerland). Sodium dodecyl sulfate (SDS) loading buffer was added into the cell lysate, which was boiled and then separated by 12% SDS-polyacrylamide gel electrophoresis (SDS-PAGE), followed by transfer onto a polyvinylidene fluoride membrane (GE Healthcare, Piscataway, NJ, USA). Specific proteins were detected using antibodies against Actin, CTSZ, CTSD and SLC11A1(Santa Cruz Biotechnology, Dallas, TX).

DISCUSSION

Yersinia pestis has evolved from *Y. pseudotuberculosis*, a foodborne enteropathogenic pathogen responsible for self-limiting intestinal infections (Morelli et al., 2010). Three pathogenic *Yersinia* bacteria shared a common 70-Kb virulence plasmid that encodes the type three secretion system (T3SS), which can deliver several *Yersinia* virulence effectors called *Yersinia* outer membrane proteins (Yops) into the eukaryotic cell cytosol to paralyze host defenses (Cornelis, 2002). Acquiring pPCP1 and pMT1 plasmids has endowed *Y. pestis* with new virulent factors such as the F1 capsular antigen, the Pla plasminogen activator, and the murine toxin, all of which are critical for this bacterium to be a flea-transmitted lethal pathogen (Prentice and Rahalison, 2007). Massive gene losses have also played very important roles in the evolution of *Y. pestis* (Chain et al., 2004). Indeed, several newly acquired features contributing to the high pathogenicity and flea-bite transmissibility of *Y. pestis* have resulted from reductive evolution. For instance, the Inv and YadA adhesins which are needed for enteropathogenic yersiniae colonization the Peyer's patches and mesenteric lymph nodes, are both pseudogenes in *Y. pestis* (Marra and Isberg, 1997; Parkhill et al., 2001; Heise and Dersch, 2006). Furthermore, with several biosynthesis gene for LPS being inactivated in *Y. pestis*, LPS in this bacterium lacks the O antigen and switches from a hexa-acylated to tetra-acylated form that is poorly recognized by TLR4 (Kawahara et al., 2002; Rebeil et al., 2004). *Y. pestis* Microtus strains are thought to lie on the evolutionarily oldest branch of the *Y. pestis*, which are capable of transmission via fleabites and are also highly virulent in mice but are unable to infect larger mammals such as rabbits, guinea pigs and humans (Fan et al., 1995; Song et al., 2004). Comparative genomic analysis of *Y. pestis* from distinct biovars has found little evidence in their different genome compositions to accounts for the host-specificity displayed by Microtus. Here, we sought to probe into this mystery in the context of the host response and further investigate the interactions between the Microtus strain and human or murine macrophages. Our results indicate that human peripheral lymphocytes respond differently to infections with the *Y. pestis* Microtus strain and the strain virulent for humans after their initial encounter, especially in several innate immunity pathways including the lysosome and TLR signaling pathways (**Figure 2**). We assumed that the PAMPs that are present on the bacterial surface or the degraded small molecules arising from 201 and 141 bacteria are differentially recognized by human peripheral lymphocytes, leading to the distinct innate immune responses in these cells. Our additional *in vitro* study confirmed that the human macrophages were competent at clearing the biovar Microtus strains of *Y. pestis*, unlike the mouse macrophages where this bacterium was able to bacteria to survive for a much longer time (**Figure 5**). This result is consistent with the transcriptomic results that gene expressions in the lysosome pathway were significantly enriched in the 201-infected lymphocytes and two functionally important proteins, CTSD and SLC11A1 that are involved in lysosome activation, showed enhanced expression. It has long been established that the initial interaction between the host macrophages and *Y. pestis* bacilli is critical to systemic infection progression, and our findings hint that human and murine macrophages exhibit distinct recognition or clearing abilities when infected with the Microtus strains. This finding is probably related to some uncharacterized features that differs between Microtus and the other biovars strains, or slight but very critical differences in some receptors between the human and mouse hosts. Further investigations are clearly needed in this area, especially for the discovery of the molecular mechanisms underlying the host-specific pathogenesis properties of the *Y. pestis* biovar Microtus strains.

AUTHOR CONTRIBUTIONS

ZD, RY, and XZ designed and cooperated the study. QWZ, YX, HZ, and ZQ isolated peripheral lymphocytes and performed the cell infection experiments. RL, QZ, and QWZ performed the survival rates analysis of the *Y. pestis* strains and immunoblotting analysis. XX performed qRT-PCR analysis. ZD, YFY, and ZK analyzed the data. YS, YC, TW, and YY provided technical assistance. ZD and QWZ wrote the manuscript.

ACKNOWLEDGMENTS

This work was supported by the National Natural Science Foundation of China (grant no. 31430006,31470242). We thank Sandra Cheesman, Ph.D., from Liwen Bianji, Edanz Group China (www.liwenbianji.cn/ac), for editing the English text of a draft of this manuscript.

SUPPLEMENTARY MATERIAL

Supplementary Figure 1 | Correlations between the expression levels of 23 genes measured by RNA-seq and qRT-PCR were analyzed using the linear regression method.

Supplementary Figure 2 | Several important molecules in the TLR signaling pathway were differentially expressed in the 201- and 141-infected human lymphocytes. Changes in the expression levels of CD14, IRF5, TLR8, and TICAM2 in 201- and 141-infected lymphocytes were plotted against the time point from infection initiation according to the RNA-seq analysis results **(A)**. Expression changes for the same group of genes were plotted according to the qRT-PCR results **(B)**.

Supplementary Figure 3 | Several important molecules in the lysosome pathway were differentially expressed in the 201- and 141-infected human lymphocytes. Changes in the expression levels of ATP6V0A1, CTSD, CTSK and SLC11A1 in the 201- and 141-infected lymphocytes were plotted against the time point from infection initation according to the RNA-seq analysis results **(A)**. Expression changes for the same group of genes were plotted according to the qRT-PCR results **(B)**.

Supplementary Table 1 | Primers used for qRT-PCR.

Supplementary Table 2 | Correlation coefficients between RNA-seq analysis and q-RT PCR results were analyzed using the samples from the lymphocytes infected with strain 201 and the normal cell control.

REFERENCES

Achtman, M., Morelli, G., Zhu, P., Wirth, T., Diehl, I., Kusecek, B., et al. (2004). Microevolution and history of the plague bacillus, *Yersinia pestis. Proc. Natl. Acad. Sci. U.S.A.* 101, 17837–17842. doi: 10.1073/pnas.0408026101

Achtman, M., Zurth, K., Morelli, G., Torrea, G., Guiyoule, A., and Carniel, E. (1999). *Yersinia pestis*, the cause of plague, is a recently emerged clone of *Yersinia pseudotuberculosis. Proc. Natl. Acad. Sci. U.S.A.* 96, 14043–14048. doi: 10.1073/pnas.96. 24.14043

Benjamini, Y., and Yekutieli, D. (2005). Quantitative trait Loci analysis using the false discovery rate. *Genetics* 171, 783–790. doi: 10.1534/genetics.104. 036699

Butler, T. (2009). Plague into the 21st century. *Clin. Infect. Dis.* 49, 736–742. doi: 10.1086/604718

Chain, P. S., Carniel, E., Larimer, F. W., Lamerdin, J., Stoutland, P. O., Regala, W. M., et al. (2004). Insights into the evolution of *Yersinia pestis* through whole-genome comparison with *Yersinia pseudotuberculosis*. *Proc. Natl. Acad. Sci. U.S.A.* 101, 13826–13831. doi: 10.1073/pnas.0404012101

Connor, M. G., Pulsifer, A. R., Price, C. T., Abu Kwaik, Y., and Lawrenz, M. B. (2015). *Yersinia pestis* requires host Rab1b for survival in macrophages. *PLoS Pathog* 11:e1005241. doi: 10.1371/journal.ppat.1005241

Cornelis, G. R. (2002). Yersinia type III secretion: send in the effectors. *J. Cell Biol.* 158, 401–408. doi: 10.1083/jcb.200205077

del Cerro-Vadillo, E., Madrazo-Toca, F., Carrasco-Marín, E., Fernandez-Prieto, L., Beck, C., Leyva-Cobián, F., et al. (2006). Cutting edge: a novel nonoxidative phagosomal mechanism exerted by cathepsin-D controls *Listeria monocytogenes* intracellular growth. *J. Immunol.* 176, 1321–1325. doi: 10.4049/jimmunol.176.3.1321

Deng, W., Burland, V., Plunkett, G., Boutin, A., Mayhew, G. F., Liss, P., et al. (2002). Genome sequence of *Yersinia pestis* KIM. *J. Bacteriol.* 184, 4601–4611. doi: 10.1128/JB.184.16.4601-4611.2002

Du, Z., Yang, H., Tan, Y., Tian, G., Zhang, Q., Cui, Y., et al. (2014). Transcriptomic response to *Yersinia pestis*: RIG-I like receptor signaling response is detrimental to the host against plague. *J. Genet. Genomics* 41, 379–396. doi: 10.1016/j.jgg.2014.05.006

Fan, Z., Luo, Y., Wang, S., Jin, L., Zhou, X., Liu, J., et al. (1995). Microtus brandti plague in the Xilin Gol Grassland was inoffensive to human. *Chin. J. Control Endemic Dis.* 10, 56–57.

Govoni, G., Canonne-Hergaux, F., Pfeifer, C. G., Marcus, S. L., Mills, S. D., Hackam, D. J., et al. (1999). Functional expression of Nramp1 *in vitro* in the murine macrophage line RAW264.7. *Infect. Immun.* 67, 2225–2232.

Grabenstein, J. P., Fukuto, H. S., Palmer, L. E., and Bliska, J. B. (2006). Characterization of phagosome trafficking and identification of PhoP-regulated genes important for survival of *Yersinia pestis* in macrophages. *Infect. Immun.* 74, 3727–3741. doi: 10.1128/IAI.00255-06

Heise, T., and Dersch, P. (2006). Identification of a domain in *Yersinia* virulence factor YadA that is crucial for extracellular matrix-specific cell adhesion and uptake. *Proc. Natl. Acad. Sci. U.S.A.* 103, 3375–3380. doi: 10.1073/pnas.0507749103

Huang da, W., Sherman, B. T., and Lempicki, R. A. (2009). Systematic and integrative analysis of large gene lists using DAVID bioinformatics resources. *Nat. Protoc.* 4, 44–57. doi: 10.1038/nprot.2008.211

Inglesby, T. V., Dennis, D. T., Henderson, D. A., Bartlett, J. G., Ascher, M. S., Eitzen, E., et al. (2000). Plague as a biological weapon: medical and public health management. Working Group on Civilian Biodefense. *JAMA* 283, 2281–2290. doi: 10.1001/jama.283.17.2281

Ishihara, S., Rumi, M. A., Kadowaki, Y., Ortega-Cava, C. F., Yuki, T., Yoshino, N., et al. (2004). Essential role of MD-2 in TLR4-dependent signaling during *Helicobacter pylori*-associated gastritis. *J. Immunol.* 173, 1406–1416. doi: 10.4049/jimmunol.173.2.1406

Kawahara, K., Tsukano, H., Watanabe, H., Lindner, B., and Matsuura, M. (2002). Modification of the structure and activity of lipid A in *Yersinia pestis* lipopolysaccharide by growth temperature. *Infect. Immun.* 70, 4092–4098. doi: 10.1128/IAI.70.8.4092-4098.2002

Koren, H. S., Anderson, S. J., and Larrick, J. W. (1979). *In vitro* activation of a human macrophage-like cell line. *Nature* 279, 328–331. doi: 10.1038/279 328a0

Krishnan, J., Selvarajoo, K., Tsuchiya, M., Lee, G., and Choi, S. (2007). Toll-like receptor signal transduction. *Exp. Mol. Med.* 39, 421–438. doi: 10.1038/emm.2007.47

Langmead, B., Trapnell, C., Pop, M., and Salzberg, S. L. (2009). Ultrafast and memory-efficient alignment of short DNA sequences to the human genome. *Genome Biol.* 10:R25. doi: 10.1186/gb-2009-10-3-r25

Leeuwenberg, J. F., Dentener, M. A., and Buurman, W. A. (1994). Lipopolysaccharide LPS-mediated soluble TNF receptor release and TNF receptor expression by monocytes. Role of CD14, LPS binding protein, and bactericidal/permeability-increasing protein. *J. Immunol.* 152, 5070–5076.

Li, H., and Durbin, R. (2009). Fast and accurate short read alignment with Burrows-Wheeler transform. *Bioinformatics* 25, 1754–1760. doi: 10.1093/bioinformatics/btp324

Liu, B., Fang, M., Hu, Y., Huang, B., Li, N., Chang, C., et al. (2014). Hepatitis B virus X protein inhibits autophagic degradation by impairing lysosomal maturation. *Autophagy* 10, 416–430. doi: 10.4161/auto.27286

Marra, A., and Isberg, R. (1997). Invasin-dependent and invasin-independent pathways for translocation of *Yersinia pseudotuberculosis* across the Peyer's patch intestinal epithelium. *Infect. Immun.* 65, 3412–3421.

Morelli, G., Song, Y., Mazzoni, C. J., Eppinger, M., Roumagnac, P., Wagner, D. M., et al. (2010). *Yersinia pestis* genome sequencing identifies patterns of global phylogenetic diversity. *Nat. Genet.* 42, 1140–1143. doi: 10.1038/ ng.705

Mortazavi, A., Williams, B. A., McCue, K., Schaeffer, L., and Wold, B. (2008). Mapping and quantifying mammalian transcriptomes by RNA-Seq. *Nat. Methods* 5, 621–628. doi: 10.1038/nmeth.1226

Parent, M. A., Wilhelm, L. B., Kummer, L. W., Szaba, F. M., Mullarky, I. K., and Smiley, S. T. (2006). Gamma interferon, tumor necrosis factor alpha, and nitric oxide synthase 2, key elements of cellular immunity, perform critical protective functions during humoral defense against lethal pulmonary *Yersinia pestis* infection. *Infect. Immun.* 74, 3381–3386. doi: 10.1128/IAI.00 185-06

Parkhill, J., Wren, B. W., Thomson, N. R., Titball, R. W., Holden, M. T., Prentice, M. B., et al. (2001). Genome sequence of *Yersinia pestis*, the causative agent of plague. *Nature* 413, 523–527. doi: 10.1038/35097083

Perry, R. D., and Fetherston, J. D. (1997). *Yersinia pestis*–etiologic agent of plague. *Clin. Microbiol. Rev.* 10, 35–66. doi: 10.1128/CMR.10.1.35

Prentice, M. B., and Rahalison, L. (2007). Plague. *Lancet* 369, 1196–1207. doi: 10.1016/S0140-6736(07)60566-2

Pujol, C., Grabenstein, J. P., Perry, R. D., and Bliska, J. B. (2005). Replication of *Yersinia pestis* in interferon gamma-activated macrophages requires ripA, a gene encoded in the pigmentation locus. *Proc. Natl. Acad. Sci. U.S.A.* 102, 12909–12914. doi: 10.1073/pnas.0502849102

Pujol, C., Klein, K. A., Romanov, G. A., Palmer, L. E., Cirota, C., Zhao, Z., et al. (2009). *Yersinia pestis* can reside in autophagosomes and avoid xenophagy in murine macrophages by preventing vacuole acidification. *Infect. Immun.* 77, 2251–2261. doi: 10.1128/IAI.00068-09

Ratner, D., Orning, M. P., Starheim, K. K., Marty-Roix, R., Proulx, M. K., Goguen, J. D., et al. (2016). Manipulation of Interleukin-1beta and Interleukin-18 production by *Yersinia pestis* effectors YopJ and YopM and redundant impact on virulence. *J. Biol. Chem.* 291, 9894–9905. doi: 10.1074/jbc.M115.697698

Rebeil, R., Ernst, R. K., Gowen, B. B., Miller, S. I., and Hinnebusch, B. J. (2004). Variation in lipid A structure in the pathogenic *yersiniae*. *Mol. Microbiol.* 52, 1363–1373. doi: 10.1111/j.1365-2958.2004. 04059.x

Sarvestani, S. T., Williams, B. R., and Gantier, M. P. (2012). Human Toll-like receptor 8 can be cool too: implications for foreign RNA sensing. *J. Interferon Cytokine Res.* 32, 350–361. doi: 10.1089/jir.2012. 0014

Saw, N. M., Kang, S. Y., Parsaud, L., Han, G. A., Jiang, T., Grzegorczyk, K., et al. (2011). Vacuolar H$^{(+)}$-ATPase subunits Voa1 and Voa2 cooperatively regulate secretory vesicle acidification, transmitter uptake, and storage. *Mol. Biol. Cell.* 22, 3394–3409. doi: 10.1091/mbc.e11-02-0155

Song, Y., Tong, Z., Wang, J., Wang, L., Guo, Z., Han, Y., et al. (2004). Complete genome sequence of *Yersinia pestis* strain 91001, an isolate avirulent to humans. *DNA Res.* 11, 179–197. doi: 10.1093/dnares/11.3.179

Takeuchi, O., and Akira, S. (2010). Pattern recognition receptors and inflammation. *Cell* 140, 805–820. doi: 10.1016/j.cell.2010.01.022

Tranchemontagne, Z. R., Camire, R. B., O'Donnell, V. J., Baugh, J., and Burkholder, K. M. (2016). *Staphylococcus aureus* strain USA300 perturbs acquisition of lysosomal enzymes and requires phagosomal acidification for survival inside macrophages. *Infect. Immun.* 84, 241–253. doi: 10.1128/IAI.00704-15

Turner, J. K., Xu, J. L., and Tapping, R. I. (2009). Substrains of 129 mice are resistant to *Yersinia pestis* KIM5: implications for interleukin-10-deficient mice. *Infect. Immun.* 77, 367–373. doi: 10.1128/IAI.01057-08

Yang, H., Wang, T., Tian, G., Zhang, Q., Wu, X., Xin, Y., et al. (2017). Host transcriptomic responses to pneumonic plague reveal that *Yersinia pestis* inhibits both the initial adaptive and innate immune responses in mice. *Int. J. Med. Microbiol.* 307, 64–74. doi: 10.1016/j.ijmm.2016.11.002

Zhou, D., Tong, Z., Song, Y., Han, Y., Pei, D., Pang, X., et al. (2004). Genetics of metabolic variations between *Yersinia pestis* biovars and the proposal of a new biovar, microtus. *J. Bacteriol.* 186, 5147–5152. doi: 10.1128/JB.186.15.5147-5152. 2004

Zhou, L., Ying, W., Han, Y., Chen, M., Yan, Y., Li, L., et al. (2012). A proteome reference map and virulence factors analysis of *Yersinia pestis* 91001. *J. Proteomics* 75, 894–907. doi: 10.1016/j.jprot.2011. 10.004

20

Ontogeny of synovial macrophages and the roles of synovial macrophages from different origins in arthritis

Jiajie Tu [1†], Wenming Hong [1,2†], Yawei Guo [1], Pengying Zhang [1], Yilong Fang [1], Xinming Wang [1,2], Xiaoyun Chen [1], Shanshan Lu [1] and Wei Wei [1*]

[1] Key Laboratory of Anti-Inflammatory and Immune Medicine, Ministry of Education, Anhui Collaborative Innovation Center of Anti-Inflammatory and Immune Medicine, Institute of Clinical Pharmacology, Anhui Medical University, Hefei, China,
[2] Department of Neurosurgery, First Affiliated Hospital of Anhui Medical University, Hefei, China

Edited by:
Manfred B. Lutz,
University of Wuerzburg, Germany

Reviewed by:
Kiwook Kim,
Washington University School of
Medicine in St. Louis, United States
Luigi Racioppi,
University of Naples Federico II, Italy

*Correspondence:
Wei Wei
wwei@ahmu.edu.cn

[†] These authors have contributed
equally to this work

The ontogeny of macrophages in most organ/tissues in human body has been proven. Due to the limited number and inaccessibility of synovial macrophages (SM), the origin of SM has not been fully illuminated. The objective of this study was designed to investigate the ontogeny of SM and to evaluate the role of SM from different origins in arthritis. Two origins of SM, embryonic SM (ESM) and bone marrow SM (BMSM) were identified in Cx3cr1-EGFP mice, CCR2$^{-/-}$ mice and bone marrow (BM) chimera model by using a stringent sorting strategy. The cellular features, including dynamic total cell number, *in situ* proliferation, phagocytosis and expressions of pro-inflammatory and anti-inflammatory genes, of ESM and BMSM were compared. In addition, ESM and BMSM showed different expression patterns in Rheumatoid Arthritis (RA) patients' synovium and during the developmental process of collagen-induced arthritis (CIA) mice. Taken together, these results demonstrated that the SM at least has two origins, ESM and BMSM. The different cellular property and dynamic expression patterns in RA patients/CIA mice highlight the notion that ESM and BMSM might play different role in arthritis.

Keywords: SM, ontogeny, ESM, BMSM, CIA, RA

INTRODUCTION

In the immune system of human body, macrophages are the first line of defense against exogenous impairment. In traditional opinion, macrophages are differentiated from circulating monocytes. However, a series of recent publications have been shown that the origins of macrophages in different tissues/organs are not exclusively derived from circulating monocytes. Yolk sac and fetal liver are two main origins of macrophages at embryonic stage (1–8). Although macrophages in synovium (synovial macrophages, SM) were identified long time ago (9), the specific ontogeny of SM is still not systematically investigated. In addition, the specific function of SM from different origins has not been clarified in RA patients or animal model of arthritis.

Here we used combinational methods of immunostaining, flow cytometry, and chimera model to explore the SM ontogeny from prenatal, perinatal, neonatal until adulthood stage in mice. Embryonic SM (ESM) was present in mice joint synovium at prenatal stage, bone marrow-derived SM (BMSM) appeared around perinatal stage and these two macrophage populations mixed after birth. The cellular features of ESM and BMSM are quite different, and the expression patterns of

ESM and BMSM show a dynamic change during the developmental process of mouse arthritis model, which is further validated by using synovium from RA patients. These results highlighted the notion that the SM that from different origins may play different role in RA, presenting potential pathogenically mechanism and therapeutically targets of RA.

MATERIALS AND METHODS

Mice

This study was approved by experimental animal ethics committee of Anhui Medical University (No. LLSC20160121) (**Supplementary Information**). C57BL/6 mice (Jackson lab code:000664), Cx3cr1+/GFP mice (Jackson lab code:005582), B6.SJL-Ptprc Pepc /BoyJ mice (CD45.1, Jackson lab code:002014), and CCR2$^{-/-}$ mice (Jackson lab code:004999) were purchased from Jackson lab (US) maintained under pathogen–free conditions at the Animal Experimental Center at the Anhui Medical University, China. All mice were used between E12.5 and 9 weeks, unless stated otherwise. All experiments were carried out according to agreement with protocols approved by the Anhui Medical University Institutional Animal Care and Use Committee. To remove the mice synovium, sacrifice mice by cervical dislocation, remove the knee joint by scissors, and tweezer without breaking femur (to exclude the contamination from bone marrow), wash in PBS (1 min), 75% ethanol (30s), and PBS (1 min). Then expose synovium by cutting off ligament and remove the synovium by tweezer. Removed mice synovium were washed and miced in 1,640 medium (Life Technologies) with penicillin/streptomycin. Dissected mice synovium were digested in 1,640 media containing 1 mg/ml of collagenase type 4 (Sigma) and 0.1 mg/ml of deoxyribonuclease I (Sigma), incubation for 1 h at 37°C (200 rpm). After filtration by using sterile nylonmesh (70 μm, Falcon), the single cells from mice synovium were centrifuged and the cell pellet resuspended in fresh medium for the following sorting: CD45$^+$Ly6G$^-$CD11c$^-$CD169$^-$Ly6C$^-$F4/80$^+$CD11b$^+$. Whole experiment carried out under a sterile condition.

Generation of Bone Marrow Chimeras

8-week-old B6.SJL-Ptprc Pepc /BoyJ mice (CD45.1$^+$) mice were lethally irradiated (5 Gy) and then were reconstituted immediately by intravenous infusion of 5×10^6 BM cells from C57BL/6 mice (CD45.2$^+$). Mice were maintained for 8 weeks before blood samples and synovium tissues were used for FACS detection of chimeras.

Patients

This part was approved by the biomedical ethics committee of Anhui Medical University (No. 20160094) (**Supplementary Information**). Informed consent was obtained from all patients and/or their legal guardians. 20 joint synovium biopsies from osteoarthritis (OA) and RA patients were used in this experiment, respectively. The patients' information was included in the **Supplementary Information**. The method used for the isolation of SM from synovial tissue was modified from a method previously described (10). Synovium was replaced from joints of patients with OA and RA. Replaced hyperplastic synovium were washed and miced in 1,640 medium (Life Technologies) with penicillin/streptomycin. Dissected synovial tissue were digested in 1,640 media containing 1 mg/ml of collagenase type 4 (Sigma) and 0.1 mg/ml of deoxyribonuclease I (Sigma), incubation for 2 h at 37°C (200 rpm). Tissues were then vortexed and resuspended in fresh media. After filtration by using sterile nylonmesh (70 μm, Falcon), the single cells from RA patients' symovium were centrifuged and the cell pellet resuspended in fresh media for the following sorting: CD45$^+$CD15$^-$CD1c$^-$CD14$^+$EMR1(F4/80)$^+$CD11b$^+$. Whole experiment carried out under a sterile condition.

Collagen-Induced Arthritis (CIA) Model

CIA Induction and Treatment Type II collagen was dissolved in 0.1 Macetic acid and emulsified with an equal volume of complete Freund's adjuvant to produce a final concentration of 2 mg/ml before incubating overnight at 4°C. DBA/1 mice (Model Animal Research Center of Nanjing University, China) were injected twice with 0.1 ml of this emulsion at the base of the tail. The day of the first immunization was defined as day 0, and the second injection was administered into the back on day 21. Mice were divided into normal group and CIA model group ($n = 8$ per group). The normal and CIA mice were given an equal volume of vehicle.

H&E Staining, Immunofluorescence and Immunohistrochemistry

The H&E staining, immunofluorescence (IF) and immunohistochemistry (IHC) were carried out according to standard protocol from previous studies (11). All antibodies for IF and IHC experiments are listed in **Supplementary Information**.

Flow Cytometry

The flow cytomery was performed according to standard protocol from previous studies (11) by using FACSArial II (BD). Antibodies for flow cytometry are listed in **Supplementary Information**.

Proliferation

For the detection of Ki67 expression, 1×10^5 SM were fixed and stained for 60 min at 20°C with anti-Ki67 (Miltenyi), then analyzed by flow cytometer.

Phagocytosis

Isolated ESM and BMSM (1×10^5 cells) were cultured with Dextran-FITC according to the manufacturer's guidelines (Life Technology) and were analyzed by flow cytometer.

Real Time-Quantitative Polymerase Chain Reaction (RT-qPCR)

ESM and BMSM were isolated by flow cytometry from the synovium, and then total RNA was extracted from SM with Trizol reagent (Life technology). RNA as reverse-transcribed to cDNA with the PrimeScriptTM RT reagent kit (Takara), and gene expression was assayed by quantitative RT-PCR with

SYBR qPCR master mix (Life technology) and the 7,500 qPCR system (Applied Biosystems). cDNA samples were assayed in triplicate and expression was normalized by using endogenous control GAPDH. All primer for qRT-PCR were listed in the **Supplementary Information**.

Statistical Analysis

Groups were compared with Student's *t*-test or, for multiple-group comparisons, one-way analysis of variance followed by a Bonferroni post-test with Prism Software (GraphPad Software).

RESULTS

The Ontogeny of Synovial Macrophages (SM)

A series of previous reports suggested that the macrophages markers F4/80 and CD11b could be used to distinguish embryonic-resident and bone marrow-derived macrophages (1–5, 7). Therefore, we assessed whether expression of F4/80 and CD11b could be used to identify specific macrophage groups in the synovium by using a strict gating strategy CD45$^+$Ly6G$^-$CD11c$^-$CD169$^-$F4/80$^+$CD11b$^+$ (**Supplementary Figure 1**) (1). It's known that mice joint synovium formed around E12.5 (12, 13) (**Figure 1A**) We firstly tried to identify the F4/80$^+$ and CD11b$^+$ SM in mice joint at E12.5. Cx3cr1$^+$/GFP mice were also used for SM identification, in which one allele of the gene encoding the chemokine receptor (CX3CR1) is replaced by green fluorescent protein (GFP). Fluorescence results demonstrated that Cx3cr1$^+$/GFP cells localized around joint at E12.5 (**Figure 1B**). In addition, resident macrophage marker F4/80 were used to validate that embryonic-resident SM (ESM) appears as early as E12.5 by using IHC and FACS assays (**Figures 1C,D**). However, because joint synovium is such a small piece of tissue that only includes several layers of cells, the morphology of synovium is too vague to distinguish from other issues at E12.5. Therefore, we couldn't exclude the possibility that these Cx3cr1$^+$/GFP cells or F4/80$^+$ ESM localize in other issues around joint while not synovium. With developmental process, the morphology of synovium is quite clear at E15.5 and synovium developed as an intact and specific tissue at P7 (**Supplementary Figure 2**). The number of F4/80$^+$ ESM gradually increased with embryonic development (**Figure 1E**). However, before E18.5, CD11b$^+$ SM couldn't be detected.

Around the perinatal period (From E20.5 to P7), after excluding CD11c$^+$ dendritic cells (DCs), Siglec-F$^+$ eosinophils, and Ly6G$^+$ neutrophils, a distinct population of F4/80$^-$CD11b$^+$ cells were found at E20.5 (**Figure 2A**). Most CD11b$^+$F4/80$^-$ BMSM were Ly6C$^+$ (**Figure 2B**), validating that BMSM is indeed derived from monocytes. To distinguish BMSM from monocytes, two pan-macrophage markers, MerTK and CD68, were also evaluated here (**Figure 2B** and **Supplementary Figure 3**). The results showed that there is a transition of Ly6C$^+$ population to MerTK$^+$ BMSM (**Figure 2B**). The mice bone marrow initially formed around E19 (14), suggesting that the bone marrow-derived synovial macrophages (BMSM) already implant in mice synovium at E20.5. In synovium of both neonatal mice and

adult mice, we validated the ESM and BMSM on the expression of the resident-macrophage marker F4/80 and bone marrow-derived macrophage marker CD11b (**Figure 2C**). In general, F4/80$^+$CD11b$^-$ ESM gradually increased from neonatal to adult stage and F4/80$^-$CD11b$^+$ BMSM showed the opposite expression pattern. Interestingly, with the development, there is another mixed population (F4/80$^+$CD11b$^+$) appears at neonatal stage and progressively increased up to adult (**Figure 2C**), implying that the distinction between ESM (F4/80$^+$CD11b$^-$) and BMSM (F4/80$^-$CD11b$^+$) is becoming more indistinct with the mice development after birth. The numbers of total SM from different origins were calculated accordingly (**Supplementary Figure 4**).

The F4/80$^+$CD11b$^+$ SM are the dominant populations in adult mice. To elucidate the origin of this population, we firstly identified the subgroups of F4/80$^+$CD11b$^+$ SM by using monocyte marker Ly6C and Major histocompatibility complex class II (MHC II, a marker of activated macrophages) in adult wild-type mice. There are three subgroups of F4/80$^+$CD11b$^+$ SM:Ly6C$^-$, Ly6C$^+$MHCII$^-$, and Ly6C$^+$MHCII$^+$ (**Figure 2D**). The Ly6C$^+$MHCII$^-$ subgroup showed intermediate expression of CX3CR1^{int}, which is an established status of circulating monocytes. Ly6C$^-$ subgroup expressed high level of CX3CR1^{hi}, which often refers to mature resident macrophages. The expression of CX3CR1 in Ly6C$^+$MHCII$^+$ subgroup is between Ly6C$^+$MHCII$^-$ and Ly6C$^-$ subgroups (**Figure 2E**).

CCR2$^{-/-}$ mice (Ly6C$^+$ monocyte is defective) is used to evaluate the role of classic circulating monocytes to Ly6C$^+$ SM. Compared to adult wild-type mice, the Ly6C$^+$ SM dramatically decreased in CCR2$^{-/-}$ mice (**Figure 2D**). However, the percentage of Ly6C$^-$ SM from CCR2$^{-/-}$ mice is similar to wild-type mice (**Figure 2D**), suggesting that CCR2 deficiency didn't affect resident ESM.

Characterization of SM From Different Origins

To identify ESM and BMSM, we firstly validated them as macrophages by detecting macrophage-specific markers CD64, CD14, and CX3CR1 (**Figure 3A**). To further verify the macrophage character, both ESM and BMSM could engulf FITC-labeled Dextran, although ESM showed higher phagocytic activity (**Figure 3B**). In addition, the expression of some classic pro-inflammatory genes IL-1β and tumor-necrosis factor (TNF)-α and anti-inflammatory genes IL-4 and IL-10 were detected in ESM and BMSM. ESM showed higher expression of IL-4 and IL-10, while BMSM have larger amount of IL-1β and TNF-α (**Figure 3C**). Therefore, ESM and BMSM demonstrated different phenotypic and functional properties.

We next tested the *in situ* proliferation of ESM and BMSM at prenatal and postnatal stages. In E16.5 mice, around 65% of mature ESM had high Ki67 (a marker that exclusively expressed in proliferating cells), and the Ki67$^+$ ESM gradually decreased from prenatal to postnatal stages (**Figure 4A**). PCNA staining further validated that SM proliferates *in situ* at postnatal stage (**Figure 4B**). By 8 weeks of age, nearly no Ki67$^+$ ESM could be detected (**Figure 4A**). However, the Ki67$^+$ BMSM couldn't

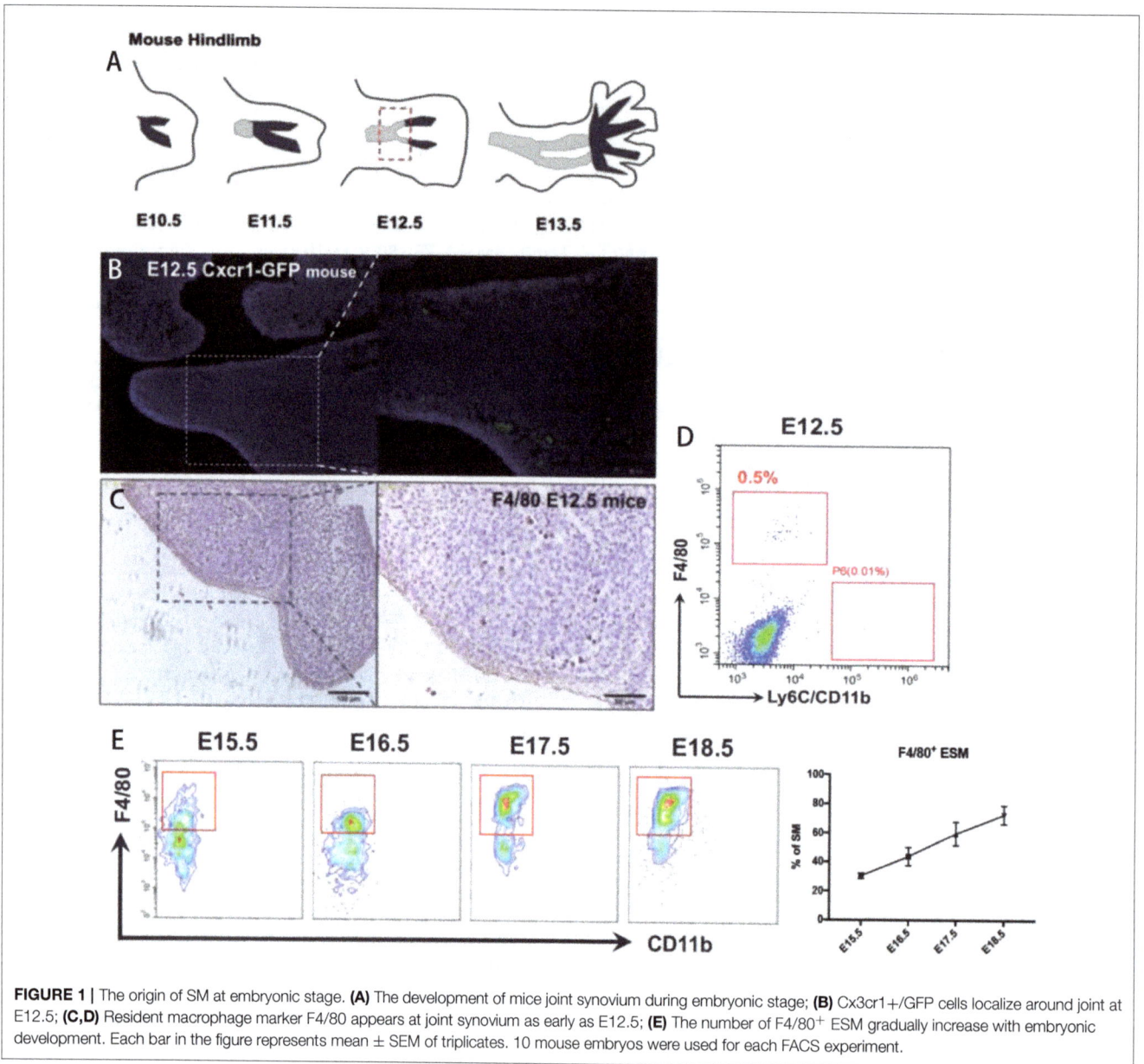

FIGURE 1 | The origin of SM at embryonic stage. **(A)** The development of mice joint synovium during embryonic stage; **(B)** Cx3cr1+/GFP cells localize around joint at E12.5; **(C,D)** Resident macrophage marker F4/80 appears at joint synovium as early as E12.5; **(E)** The number of F4/80+ ESM gradually increase with embryonic development. Each bar in the figure represents mean ± SEM of triplicates. 10 mouse embryos were used for each FACS experiment.

be detected neither during prenatal nor postnatal stages (data not shown). Thus, ESM significantly proliferated *in situ* from embryonic stage to neonatal stage, but gradually diminished after birth.

Previous results demonstrated that the SM was derived from both embryonic and bone marrow-derived macrophages. To further reveal the contribution of bone marrow (BM) cells to ESM and BMSM, bone marrow chimera model was established to further elucidate the contribution of ESM and BMSM in total SM. CD45.1 host mice were irradiated and then CD45.2 BM cells was isolated and transplanted by intravenous injection at caudal vein. After 2 months, joint synovium from chimera mice were isolated and analyzed by FACS. For total SM, over 30% cells are from donor. Almost all ESM were recipient origin (**Figure 4C**), suggesting that ESM was radio-resistant

and didn't need a contribution from BM-derived cells. As expected, BMSM was mostly replaced by circulating cells after transplantation (**Figure 4C**).

The Role of SM From Different Origins in Arthritis

Some macrophage populations show pro-inflammatory cellular feature under arthritis condition. SM has been proven that play an essential role in Rheumatoid arthritis (RA). However, there is little study to investigate specific role of ESM and BMSM during pathogenesis of RA. Here, a Collagen-induced arthritis (CIA) mice model and replaced synovium from RA patients were used to evaluate the role of ESM and BMSM in arthritis. Firstly, synovium was isolated from joints of CIA mice. Compared to normal control, H&E staining and Masson

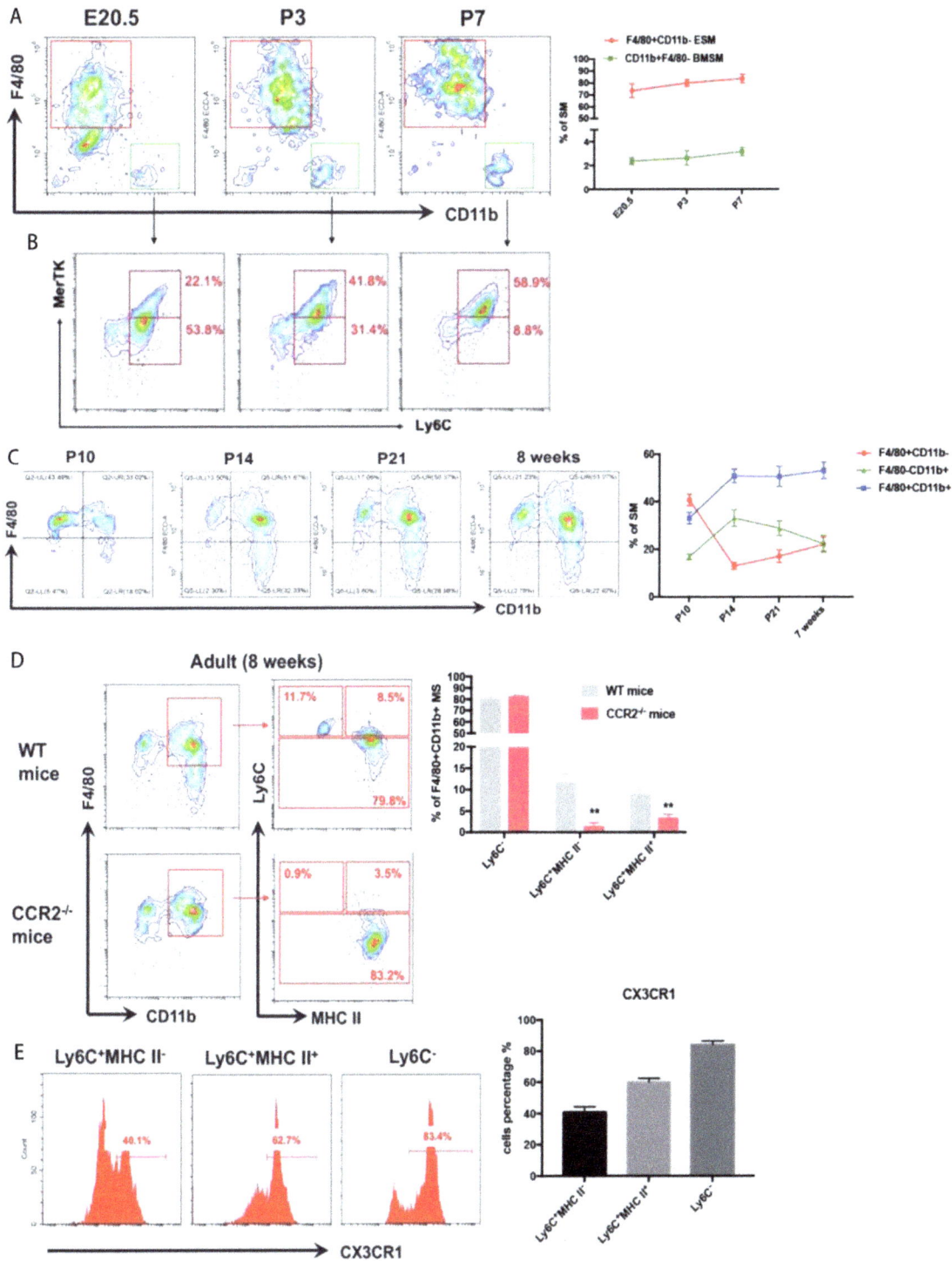

FIGURE 2 | The ontology of SM at perinatal to adult stage. **(A)** F4/80⁻CD11b⁺ BMSM appear in joint synovium at E20.5; **(B)** The expression of Ly6C and MerTK in F4/80⁻CD11b⁺ BMSM; **(C)** ESM gradually increase from neonatal to adult stage and BMSM show the opposite expression pattern. F4/80⁺CD11b⁺ SM appear at neonatal stage and progressively increase up to adult; **(D)** To further divide F4/80⁺CD11b⁺ SM by Ly6C and MHC II in WT mice and CCR2-/- mice. **(E)** The CX3CR1 expression in three subgroups of F4/80⁺CD11b⁺ SM. Each bar in the figure represents mean ± SEM of triplicates. For embryonic stage, 8–10 mouse embryos were used for each FACs experiment. For neonatal stage, 6 cubs were used for each FACS experiment. **P < 0.01.

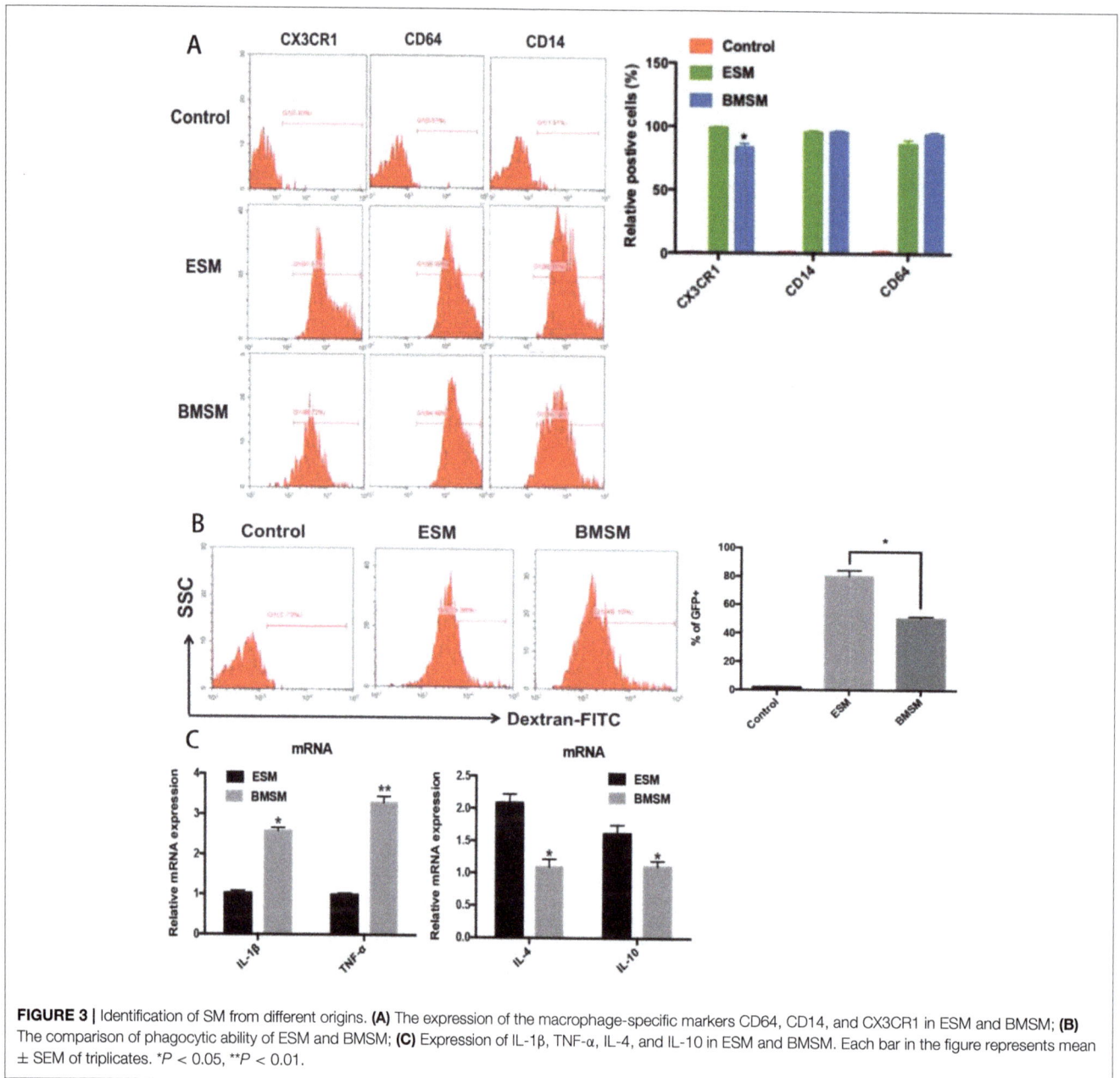

FIGURE 3 | Identification of SM from different origins. **(A)** The expression of the macrophage-specific markers CD64, CD14, and CX3CR1 in ESM and BMSM; **(B)** The comparison of phagocytic ability of ESM and BMSM; **(C)** Expression of IL-1β, TNF-α, IL-4, and IL-10 in ESM and BMSM. Each bar in the figure represents mean ± SEM of triplicates. *$P < 0.05$, **$P < 0.01$.

staining showed that the hyperplasia and hyper-infiltration of immune cells were observed in joint synovium that isolated from CIA mice (**Figures 5A,B**). Next, we detected the dynamic expression pattern of SM in the developmental process of CIA model (**Figure 5C**). ESM (F4/80$^+$CD11b$^-$) gradually decreased from initial developmental stage to peak stage, and then increased at relief stage. On contrary, BMSM (F4/80$^-$CD11b$^+$) showed an opposite expression pattern. We further analyzed the M1(pro-inflammatory) and M2(anti-inflammatory) percentage of ESM and BMSM at different developmental stage of CIA model (**Figure 5D**). FACS results demonstrated that the polarization of ESM skewed to M2 while BMSM showed more M1 phenotype during the CIA development. Taken together,

these data suggested that ESM is M2-like population that was generally repressed in CIA model; while the number of BMSM population significantly increased in CIA model and showed M1-like pro-inflammatory phenotype. To further prove this point in human, synovium from RA patients joint was used. SM from osteoarthritis (OA) patients were used as control, since the inflammatory features of OA are much less than that of RA (**Figure 6A** and **Supplementary Figure 5**). Immunofluorescence results suggested that there are much more CD11b$^+$ cells infiltration in RA synovium than that of OA synovium (**Figure 6A**). In RA synovium, CD11b$^+$ cells localized within and around vessel tube and EMR1(the human homology of mouse F4/80)$^+$ cells appeared outside vessel (**Figure 6B**),

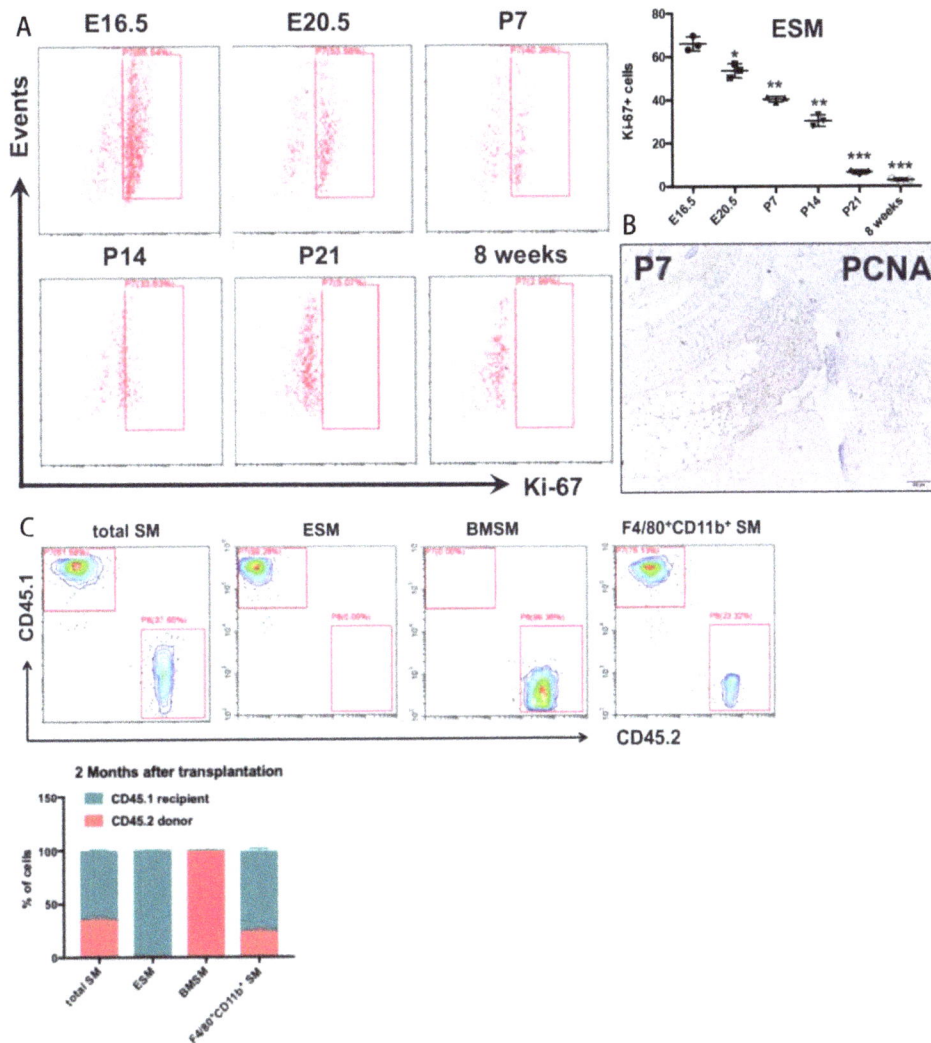

FIGURE 4 | The proliferative ability of SM. **(A)** *in situ* proliferation of the ESM at prenatal and postnatal stages; **(B)** PCNA staining in joint synovium *in situ* at postnatal stage; **(C)** A chimera approach by joining congenic wild-type CD45.1$^+$ and CD45.2$^+$ mice to assess the contribution of non-host cells in synovium. Each bar in the figure represents mean ± SEM of triplicates. $^*P < 0.05$, $^{**}P < 0.01$, $^{***}P < 0.001$.

validating that BMSM recruitment was from circulation and significantly increased in RA synovium. On contrary, EMR1$^+$ cells didn't localize near blood vessel, suggesting that ESM wasn't circulation-orient (**Figure 6B**). EMR1$^+$CD11b$^+$ RA SM was also isolated from RA and OA synovium by using a similar sorting strategy in mice (**Figure 6C**). RA SM showed much more BMSM population (EMR1^{low}CD11b^{high}) and less ESM population (EMR1^{high}CD11b^{low}), mimicking the results from CIA mice (**Figure 6D**). Similar to CIA mice, BMSM population (EMR1^{low}CD11b^{high}) and ESM population (EMR1^{high}CD11b^{low}) from RA synovium demonstrated M1 and M2 polarization, respectively (**Figure 6E**). In summary, these results suggested that BMSM is more like a group of pro-inflammatory SM, which significantly increased in both CIA mice and RA patients, while ESM is functional as an anti-inflammatory role during the same conditions.

DISCUSSION

Recently, the traditional idea that tissue macrophages are exclusively derived from circulating monocytes has been challenged. A series of paper suggested that tissue macrophages stem from both embryonic precursors except circulating monocytes (5, 13, 15). In the current study, we found that SM also has two origins: ESM seeded at joint symnovium at prenatal stage and the classic BMSM also infiltrated in synovium at perinatal stage. This two distinct SM populations gradually mixed together with development after birth. This transition is a potential key step for physiological homeostasis of joint synovium. In addition, we found that ESM are radio-resistant in chimera model. This is a surprising finding because macrophages in most peripheral tissue are, at least partially, radio-sensitive, such as bone marrow macrophages (16), colonic

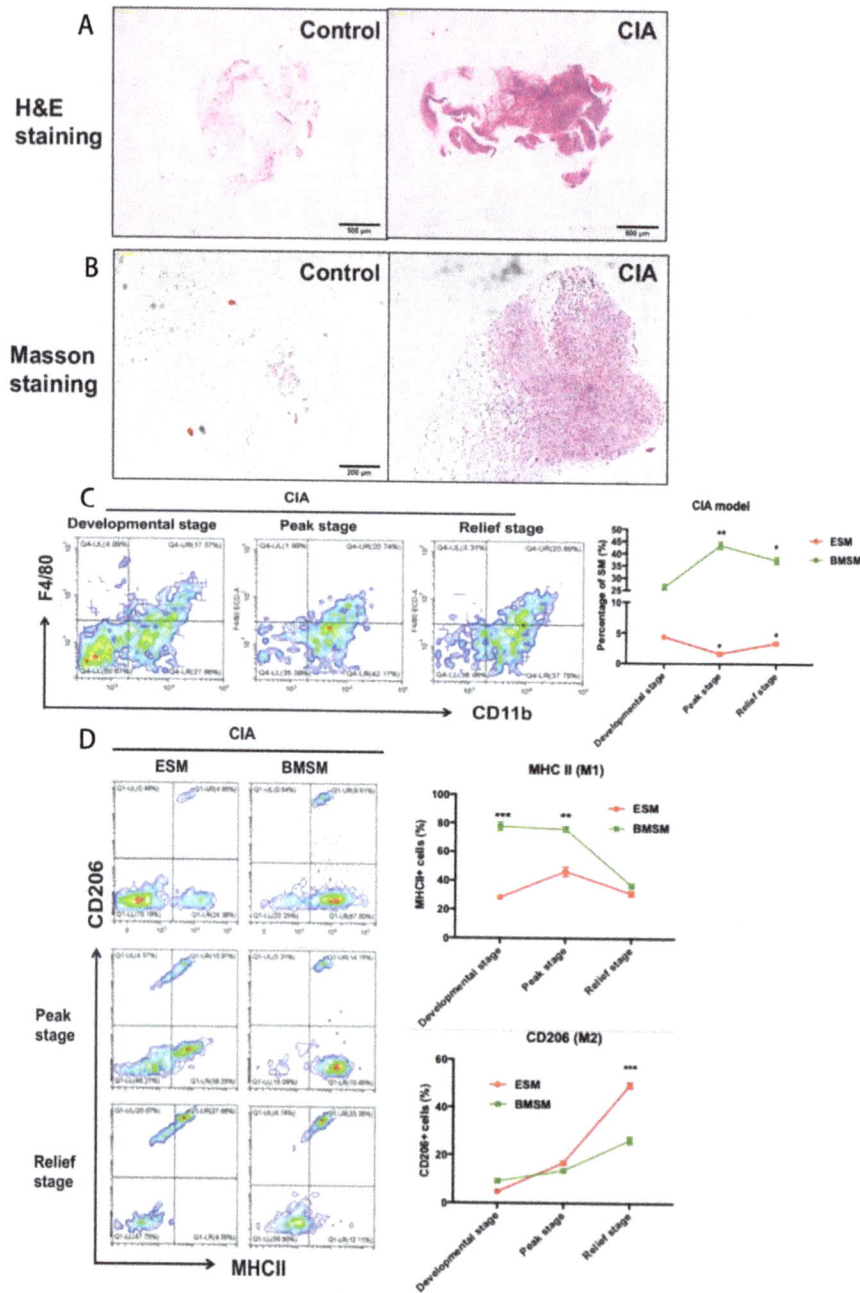

FIGURE 5 | The role of SM from different origins in CIA mice. **(A,B)** H&E staining and Masson staining of synovium and joint of control and CIA mice; **(C)** The dynamic expression pattern of SM in the developmental process of CIA model; **(D)** The expressionfo M1(pro-inflammatory)/M2(anti-inflammatory) markers of ESM and BMSM at different developmental stage of CIA model. Each bar in the figure represents mean ± SEM of triplicates. 5 mice were used for each FACS experiment. *$P < 0.05$, **$P < 0.01$, ***$P < 0.001$.

macrophages (1) and langerhans cells (7). This feature shows the unique character of SM. Therefore, the difference between SM and other resident peripheral tissues should be compared in the future.

The cellular features of these two SM populations are also quite different, including proliferation, phagocytosis and expression of pro-inflammatory and anti-inflammatory genes. Of interest, ESM showed considerable proliferative ability during

the prenatal and neonatal periods (17, 18), suggesting that the cellular phenotype might change with the development. To further elucidate this, embryonic pulse-chase fate mapping system will be used to compare ESM in prenatal, neonatal and adult stages.

Due to the therapeutic potential of changing macrophage phenotype in RA, the polarization of macrophages plays an essential role in many pathological conditions. ESM and

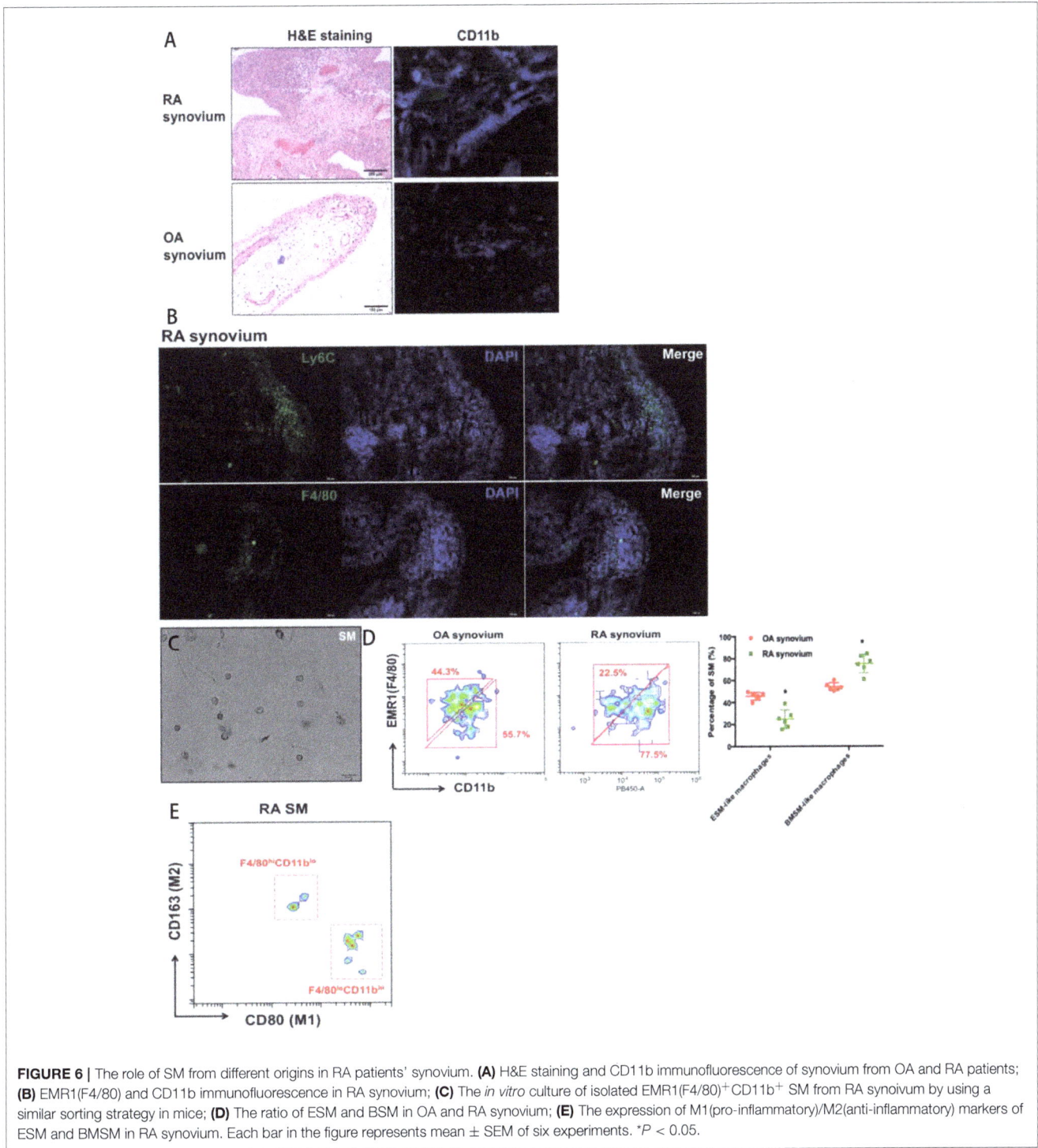

FIGURE 6 | The role of SM from different origins in RA patients' synovium. **(A)** H&E staining and CD11b immunofluorescence of synovium from OA and RA patients; **(B)** EMR1(F4/80) and CD11b immunofluorescence in RA synovium; **(C)** The in vitro culture of isolated EMR1(F4/80)⁺CD11b⁺ SM from RA synoivum by using a similar sorting strategy in mice; **(D)** The ratio of ESM and BSM in OA and RA synovium; **(E)** The expression of M1(pro-inflammatory)/M2(anti-inflammatory) markers of ESM and BMSM in RA synovium. Each bar in the figure represents mean ± SEM of six experiments. *$P < 0.05$.

BMSM from CIA mice show specific M1 and M2 polarization trends, implying that the complexity of SM is also high. Interestingly, compared to CIA mice, we got a similar result of SM polarization from synovium from RA patients, suggesting that altering macrophage polarization in autoimmune inflammation is indeed a potential treatment that definitely warrants further investigation.

ETHICS STATEMENT

This study was carried out in accordance with the recommendations of No. 20160094, biomedical ethics committee of Anhui Medical University with written informed consent from all subjects. All subjects gave written informed consent in accordance with the Declaration of Helsinki. The protocol was approved by the biomedical ethics committee of Anhui Medical University.

This study was carried out in accordance with the recommendations of No. LLSC20160121, experimental animals ethics committee of Anhui Medical University. The protocol was approved by the experimental animals ethics committee of Anhui Medical University.

AUTHOR CONTRIBUTIONS

JT, YG, WH, PZ, XW, YF, XC, and SL performed the experiments and drafted the manuscript. WW revised the manuscript.

FUNDING

This study was supported by the National Natural Science Foundation of China (81673444; 81330081), Natural Science Foundation of Anhui Province for young scholars (1708085QH200) and Grants for Scientific Research of BSKY from Anhui Medical University (4501041101).

SUPPLEMENTARY MATERIAL

Supplementary Figure 1 | The preparation and gating process for ESM and BMSM sorting.

Supplementary Figure 2 | The morphology of mouse synovium at E15.5 and P7 stage.

Supplementary Figure 3 | The expression of pan-macrophage CD68 in F4/80-CD11b+ BMSM during perinatal stage.

Supplementary Figure 4 | The expression pattern of ESM, BMSM, and F4/80$^+$CD11b$^+$ SM at postnatal stage and the total numbers of SM during mice development.

Supplementary Figure 5 | Safranin-O staining of OA and RA synovium.

REFERENCES

1. Bain CC, Bravo-Blas A, Scott CL, Gomez Perdiguero E, Geissmann F, Henri S, et al. Constant replenishment from circulating monocytes maintains the macrophage pool in the intestine of adult mice. *Nat Immunol.* (2014) 15:929–37. doi: 10.1038/ni.2967
2. Ginhoux F, Greter M, Leboeuf M, Nandi S, See P, Gokhan S, et al. Fate mapping analysis reveals that adult microglia derive from primitive macrophages. *Science.* (2010) 330:841–5. doi: 10.1126/science.1194637
3. Yona S, Kim KW, Wolf Y, Mildner A, Varol D, Breker M, et al. Fate mapping reveals origins and dynamics of monocytes and tissue macrophages under homeostasis. *Immunity.* (2013) 38:79–91. doi: 10.1016/j.immuni.2012.12.001
4. Jakubzick C, Gautier EL, Gibbings SL, Sojka DK, Schlitzer A, Johnson TE, et al. Minimal differentiation of classical monocytes as they survey steady-state tissues and transport antigen to lymph nodes. *Immunity.* (2013) 39:599–610. doi: 10.1016/j.immuni.2013.08.007
5. Hashimoto D, Chow A, Noizat C, Teo P, Beasley MB, Leboeuf M, et al. Tissue-resident macrophages self-maintain locally throughout adult life with minimal contribution from circulating monocytes. *Immunity.* (2013) 38:792–804. doi: 10.1016/j.immuni.2013.04.004
6. Ginhoux F, Guilliams M. Tissue-resident macrophage ontogeny and homeostasis. *Cell.* (2016) 44:439–49. doi: 10.1016/j.immuni.2016.02.024
7. Hoeffel G, Wang Y, Greter M, See P, Teo P, Malleret B, et al. Adult Langerhans cells derive predominantly from embryonic fetal liver monocytes with a minor contribution of yolk sac-derived macrophages. *J Exp Med.* (2012) 209:1167–81. doi: 10.1084/jem.20120340
8. Bain CC, Hawley CA, Garner H, Scott CL, Schridde A, Steers NJ, et al. Long-lived self-renewing bone marrow-derived macrophages displace embryo-derived cells to inhabit adult serous cavities. *Nat Commun.* (2016) 7:1–14. doi: 10.1038/ncomms11852
9. Athanasou NA. Synovial macrophages. *Ann Rheum Dis.* (1995) 54:392–4. doi: 10.1136/ard.54.5.392
10. Mandelin AM, Homan PJ, Shaffer AM, Cuda CM, Dominguez ST, Bacalao E, et al. Transcriptional profiling of synovial macrophages using minimally invasive ultrasound-guided synovial biopsies in rheumatoid arthritis. *Arthr Rheumatol.* (2018) 70:841–54. doi: 10.1002/art.40453
11. Tu J, Ng SH, Shui Luk AC, Liao J, Jiang X, Feng B, et al. MicroRNA-29b/Tet1 regulatory axis epigenetically modulates mesendoderm differentiation in mouse embryonic stem cells. *Nucleic Acids Res.* (2015) 43:7805–22. doi: 10.1093/nar/gkv653
12. Pazin DE, Gamer LW, Cox KA, Rosen V. Molecular profiling of synovial joints: Use of microarray analysis to identify factors that direct the development of the knee and elbow. *Dev Dyn.* (2012) 241:1816–26. doi: 10.1002/dvdy.23861
13. Yamagami T, Molotkov A, Zhou CJ. Canonical Wnt signaling activity during synovial joint development. *J Mol Histol.* (2009) 40:311–6. doi: 10.1007/s10735-009-9242-1
14. Baron MH, Isern J, Fraser ST. The embryonic origins of erythropoiesis in mammals. *Blood.* (2012) 119:4828–37. doi: 10.1182/blood-2012-01-153486
15. Schulz C, Perdiguero EG, Chorro L, Szabo-rogers H, Cagnard N, Kierdorf K, et al. A lineage of myeloid cells independent of Myb and hematopoietic stem cells. *Science.* (2012) 336:2–7. doi: 10.1126/science.1219179
16. Gordon S, Plüddemann A. Tissue macrophages : heterogeneity and functions. *BMC Biol.* (2017) 15:1–18. doi: 10.1186/s12915-017-0392-4
17. Davies LC, Rosas M, Smith PJ, Fraser DJ, Jones SA, Taylor PR. A quantifiable proliferative burst of tissue macrophages restores homeostatic macrophage populations after acute inflammation. *Eur J Immunol.* (2011) 41:2155–64. doi: 10.1002/eji.201141817
18. Chorro L, Sarde A, Li M, Woollard KJ, Chambon P, Malissen B, et al. Langerhans cell. (LC) proliferation mediates neonatal development, homeostasis, and inflammation-associated expansion of the epidermal LC network. *J Exp Med.* (2009) 206:3089–100. doi: 10.1084/jem. 20091586

Permissions

All chapters in this book were first published in DCMNC, by Frontiers; hereby published with permission under the Creative Commons Attribution License or equivalent. Every chapter published in this book has been scrutinized by our experts. Their significance has been extensively debated. The topics covered herein carry significant findings which will fuel the growth of the discipline. They may even be implemented as practical applications or may be referred to as a beginning point for another development.

The contributors of this book come from diverse backgrounds, making this book a truly international effort. This book will bring forth new frontiers with its revolutionizing research information and detailed analysis of the nascent developments around the world.

We would like to thank all the contributing authors for lending their expertise to make the book truly unique. They have played a crucial role in the development of this book. Without their invaluable contributions this book wouldn't have been possible. They have made vital efforts to compile up to date information on the varied aspects of this subject to make this book a valuable addition to the collection of many professionals and students.

This book was conceptualized with the vision of imparting up-to-date information and advanced data in this field. To ensure the same, a matchless editorial board was set up. Every individual on the board went through rigorous rounds of assessment to prove their worth. After which they invested a large part of their time researching and compiling the most relevant data for our readers.

The editorial board has been involved in producing this book since its inception. They have spent rigorous hours researching and exploring the diverse topics which have resulted in the successful publishing of this book. They have passed on their knowledge of decades through this book. To expedite this challenging task, the publisher supported the team at every step. A small team of assistant editors was also appointed to further simplify the editing procedure and attain best results for the readers.

Apart from the editorial board, the designing team has also invested a significant amount of their time in understanding the subject and creating the most relevant covers. They scrutinized every image to scout for the most suitable representation of the subject and create an appropriate cover for the book.

The publishing team has been an ardent support to the editorial, designing and production team. Their endless efforts to recruit the best for this project, has resulted in the accomplishment of this book. They are a veteran in the field of academics and their pool of knowledge is as vast as their experience in printing. Their expertise and guidance has proved useful at every step. Their uncompromising quality standards have made this book an exceptional effort. Their encouragement from time to time has been an inspiration for everyone.

The publisher and the editorial board hope that this book will prove to be a valuable piece of knowledge for researchers, students, practitioners and scholars across the globe.

List of Contributors

Alba Martín-Moreno
Sección de Inmunología, Laboratorio InmunoBiología Molecular, Hospital General Universitario Gregorio Marañón (HGUGM), Madrid, Spain
Instituto Investigación Sanitaria Gregorio Marañón (IiSGM), Madrid, Spain

Mª Angeles Muñoz-Fernández
Sección de Inmunología, Laboratorio InmunoBiología Molecular, Hospital General Universitario Gregorio Marañón (HGUGM), Madrid, Spain
Instituto Investigación Sanitaria Gregorio Marañón (IiSGM), Madrid, Spain, Spanish HIV-HGM BioBank, Madrid, Spain
Networking Research Center on Bioengineering, Biomaterials and Nanomedicine (CIBER BBN), Madrid, Spain

David Vremec
The Walter and Eliza Hall Institute, Melbourne, VIC, Australia

Ken Shortman
The Walter and Eliza Hall Institute, Melbourne, VIC, Australia
Department of Medical Biology, The University of Melbourne, Melbourne, VIC, Australia Burnet Institute, Melbourne, VIC, Australia

Muzlifah Haniffa
Human Dendritic Cell Laboratory, Institute of Cellular Medicine, Newcastle University, Newcastle upon Tyne, UK

Gary Reynolds
Human Dendritic Cell Laboratory, Institute of Cellular Medicine, Newcastle University, Newcastle upon Tyne, UK
Musculoskeletal Research Group, Institute of Cellular Medicine, Newcastle University, Newcastle upon Tyne, UK

Mor Gross, Tomer-Meir Salame and Steffen Jung
Department of Immunology, Weizmann Institute of Science, Rehovot, Israel
Biological Services, Weizmann Institute of Science, Rehovot, Israel

Bayan Sudan and Mark A. Wacker
Infectious Diseases Division, Department of Internal Medicine, University of Iowa, Iowa City, IA, USA

Mary E. Wilson and Joel W. Graff
Infectious Diseases Division, Department of Internal Medicine, University of Iowa, Iowa City, IA, USA
Iowa City VA Medical Center, Iowa City, IA, USA

Guillaume Hoeffel and Florent Ginhoux
Singapore Immunology Network (SIgN), Agency for Science, Technology and Research (A*STAR), Singapore, Singapore

Simon Yona
University College London, London, UK

Siamon Gordon
Sir William Dunn School of Pathology, The University of Oxford, Oxford, UK

Martin Guilliams and Lianne van de Laar
Laboratory of Immunoregulation, VIB Inflammation Research Center, Ghent University, Ghent, Belgium
Department of Respiratory Medicine, University Hospital Ghent, Ghent, Belgium

Roxane Tussiwand
Department of Biomedicine, University of Basel, Basel, Switzerland

Emmanuel L. Gautier
INSERM UMR_S 1166, Sorbonne Universités, UPMC Univ Paris 06, Pitié-Salpêtrière Hospital, Paris, France

Thien-Phong Vu Manh and Marc Dalod
UM2, Centre d'Immunologie de Marseille-Luminy, Aix Marseille Université, Marseille, France
U1104, INSERM, Marseille, France
UMR7280, CNRS, Marseille, France

Jamila Elhmouzi-Younes, Céline Urien, Suzana Ruscanu, Luc Jouneau, Nicolas Bertho, Pierre Boudinot and Isabelle Schwartz-Cornil
UR892, Virologie et Immunologie Moléculaires, INRA, Domaine de Vilvert, Jouy-en-Josas, France

Mickaël Bourge
IFR87 La Plante et son Environnement, IMAGIF CNRS, Gif-sur-Yvette, France

Marco Moroldo
CRB GADIE, Génétique Animale et Biologie Intégrative, INRA, Domaine de Vilvert, Jouy-en-Josas, France

Gilles Foucras
UMR1225, Université de Toulouse, INPT, ENVT, Toulouse, France
UMR1225, Interactions Hôtes-Agents Pathogènes, INRA, Toulouse, France

Henri Salmon, Hélène Marty and Pascale Quéré
UMR1282, Infectiologie et Santé Publique, INRA, Nouzilly, France
UMR1282, Université François Rabelais de Tours, Tours, France

Alexandra dos Anjos Cassado and Maria Regina D'Império Lima
Departamento de Imunologia, Instituto de Ciências Biomédicas, Universidade de São Paulo, São Paulo, Brazil

Karina Ramalho Bortoluci
Centro de Terapia Celular e Molecular (CTC-Mol), Universidade Federal de São Paulo, São Paulo, Brazil
Departamento de Ciências Biológicas, Campus Diadema, Universidade Federal de São Paulo, São Paulo, Brazil

Mateusz Pawel Poltorak and Barbara Ursula Schraml
Institute for Medical Microbiology, Immunology and Hygiene, Technische Universität München, Munich, Germany

Yashu Li
State Key Laboratory of Trauma, Burn and Combined Injury, Institute of Burn Research, Southwest Hospital, Third Military Medical University (Army Medical University), Chongqing, China

Gaoxing Luo and Weifeng He
State Key Laboratory of Trauma, Burn and Combined Injury, Institute of Burn Research, Southwest Hospital, Third Military Medical University (Army Medical University), Chongqing, China
Chongqing Key Laboratory for Disease Proteomics, Chongqing, China

Jun Wu
State Key Laboratory of Trauma, Burn and Combined Injury, Institute of Burn Research, Southwest Hospital, Third Military Medical University (Army Medical University), Chongqing, China
Chongqing Key Laboratory for Disease Proteomics, Chongqing, China
Department of Burns, The First Affiliated Hospital, Sun Yat-Sen University, Guangzhou, China

Loems Ziegler-Heitbrock
Helmholtz Zentrum München, Asklepios Fachkliniken München-Gauting, Gauting, Germany

Melanie Greter, Iva Lelios and Andrew Lewis Croxford
Institute of Experimental Immunology, University of Zurich, Zurich, Switzerland

Enrique Martin-Gayo
Hospital Universitario de la Princesa, Universidad Autónoma de Madrid, Madrid, Spain

Xu G. Yu
Ragon Institute of MGH, MIT, and Harvard, Massachusetts General Hospital, Harvard Medical School, Boston, MA, United States

Catherine Gottschalk and Christian Kurts
Institute of Experimental Immunology, Rheinische Friedrich-Wilhelms-Universität Bonn, Bonn, Germany

Thien-Phong Vu Manh and Marc Dalod
UM2, Centre d'Immunologie de Marseille-Luminy (CIML), Aix-Marseille University, Marseille, France
U1104, Institut National de la Santé et de la Recherche Médicale (INSERM), Marseille, France
UMR7280, Centre National de la Recherche Scientifique (CNRS), Marseille, France

Nicolas Bertho and Isabelle Schwartz-Cornil
Virologie et Immunologie Moléculaires UR892, Institut National de la Recherche Agronomique, Jouy-en-Josas, France

Anne Hosmalin
INSERM U1016, Institut Cochin, Paris, France
CNRS UMR8104, Paris, France
Université Paris Descartes, Paris, France
Assistance Publique-Hôpitaux de Paris (AP-HP), Hôpital Cochin, Paris, France

Qingwen Zhang, Youquan Xin, Haihong Zhao, Xiaoqing Xu, Zhipeng Kong, Zhizhen Qi, Qi Zhang and Xuefei Zhang
Qinghai Institute for Endemic Disease Prevention and Control, Xining, China

Rongjiao Liu, Yanfeng Yan, Tong Wang, Yang You, Yajun Song, Yujun Cui, Ruifu Yang and Zongmin Du
State Key Laboratory of Pathogen and Biosecurity, Beijing Institute of Microbiology and Epidemiology

Jiajie Tu, Yawei Guo, Pengying Zhang, Yilong Fang, Xiaoyun Chen, Shanshan Lu and Wei Wei
Key Laboratory of Anti-Inflammatory and Immune Medicine, Ministry of Education, Anhui Collaborative Innovation Center of Anti-Inflammatory and Immune Medicine, Institute of Clinical Pharmacology, Anhui Medical University, Hefei, China

Wenming Hong and Xinming Wang
Key Laboratory of Anti-Inflammatory and Immune Medicine, Ministry of Education, Anhui Collaborative Innovation Center of Anti-Inflammatory and Immune Medicine, Institute of Clinical Pharmacology, Anhui Medical University, Hefei, China

Department of Neurosurgery, First Affiliated Hospital of Anhui Medical University, Hefei, China

Index

Lightning Source UK Ltd.
Milton Keynes UK
UKHW050200070922
408456UK00002B/17

9 781639 893379